AF357841

Biodiversity and Conservation
of Forests

Biodiversity and Conservation of Forests

Guest Editors

Konstantinos Poirazidis
Panteleimon Xofis

Basel • Beijing • Wuhan • Barcelona • Belgrade • Novi Sad • Cluj • Manchester

Guest Editors

Konstantinos Poirazidis
Department of Environment
Ionian University
Zakynthos
Greece

Panteleimon Xofis
Department of Forestry and
Natural Environment
International Hellenic
University
Drama
Greece

Editorial Office
MDPI AG
Grosspeteranlage 5
4052 Basel, Switzerland

This is a reprint of the Special Issue, published open access by the journal *Forests* (ISSN 1999-4907), freely accessible at: https://www.mdpi.com/journal/forests/special_issues/biodiversity_conservation_forests.

For citation purposes, cite each article independently as indicated on the article page online and as indicated below:

Lastname, A.A.; Lastname, B.B. Article Title. *Journal Name* **Year**, *Volume Number*, Page Range.

ISBN 978-3-7258-3295-8 (Hbk)
ISBN 978-3-7258-3296-5 (PDF)
https://doi.org/10.3390/books978-3-7258-3296-5

Cover image courtesy of Gregory Azorides

Contents

About the Editors

Konstantinos Poirazidis

Dr. Konstantinos Poirazidis is a Professor of Spatial Ecology and Conservation and the Director of the Laboratory of Environmental Management and Sustainable Development at the Department of Environmental Sciences, School of Environment, Ionian University, Greece. He has been active in biodiversity and spatial ecology, both in Greece and abroad. He has published over 57 scientific articles in many international journals. His main research interests include the interpretation and ecological analysis of the natural environment, the development of geospatial models and spatial modeling of ecological data, the analysis of landscape structure, and the analysis and interpretation of trends in fauna population data.

Panteleimon Xofis

Dr. Panteleimon Xofis is associate professor in the Department of Forest and Natural Environment Sciences of the Democritus University of Thrace, Greece and director of the lab of Forest ecology and Environmental Remote Sensing. He holds a PhD in fire ecology from Imperial College London and from the same university he also holds an MSc in Landscape Ecology Design and management. His research interests include forest and fire ecology, applied remote sensing, landscape ecology, biodiversity and nature conservation and urban ecology. Dr. Xofis has been active in research for more than 25 years, having published 47 articles in scientific journals.

Editorial

Biodiversity and Conservation of Forests

Panteleimon Xofis [1,*], Georgios Kefalas [1] and Konstantinos Poirazidis [2]

[1] Department of Forestry and Natural Environment, International Hellenic University, GR-66100 Drama, Greece; gkefalas@emt.ihu.gr

[2] Department of Environmental Sciences, Ionian University, GR-29100 Zakynthos, Greece; kpoiraz@ionio.gr

* Correspondence: pxofis@for.ihu.gr; Tel.: +30-697-3035-416 or +30-25210-60430

Citation: Xofis, P.; Kefalas, G.; Poirazidis, K. Biodiversity and Conservation of Forests. *Forests* **2023**, *14*, 1871. https://doi.org/10.3390/f14091871

Received: 8 September 2023
Accepted: 10 September 2023
Published: 14 September 2023

Forests are extremely valuable ecosystems, associated with a number of ecosystem services that are of significant importance for human wellbeing. Although biodiversity conservation stands at the top of the list of desired ecosystem services, carbon storage, water regulation and supply, wood and non-wood products, recreation, soil protection, and nutrient cycling are also all important. While forests cover only 30% of the Earth's surface, they host 60,000 tree species, 80% of all known amphibians, 75% of all bird species, and 68% of all mammals [1,2]. Tropical forests alone host 60% of the world's vascular plants [2]. The conservation of biological diversity is of crucial importance, not only for ethical reasons, but also for a number of services that are offered by biodiversity to humans, including the provision of food, medicinal plants, and others [3]. Despite the extremely significant contributions of forests to the conservation of world's biological diversity, they currently face different and often contradicting challenges.

On the one hand, the abandonment of marginally productive agricultural land, observed in Europe, North America, and elsewhere, which is associated with socioeconomic changes during the 20th century, provides an opportunity for degraded or even lost forests to recover and reoccupy their pre-human areas. This results in a significant increase in forest cover and a decrease in forest fragmentation [4,5], with positive consequences for biodiversity conservation, as is proved with the recent increase in the distribution and abundance of Europe's emblematic carnivores [6,7]. Furthermore, increased forest cover has been reported to contribute significantly to increased biomass and carbon stock, with positive consequences for mitigating climate change [8]. The increases in forest cover, however, may also lead to an increased degree of landscape homogeneity, with negative impacts on local biodiversity and species that require a mosaic of forested and open areas to cover their needs in terms of food and protection [9,10].

On the other hand, extensive deforestation of the globally important tropical forests, and land conversion to agriculture, continues to occur, threatening the long-term sustainability of these biodiversity hotspots [11]. This deforestation often occurs at large spatial scales without necessarily ensuring significant economic benefits, while the loss of habitats and biodiversity is undoubtedly huge [12]. As a result, an estimated loss of 420 million hectares has been reported since 1990 and, despite the decreasing trend of deforestation, 10 million hectares of forests are still being lost every year [2]. Forest loss, along with other human activities, results in an estimated one million species of the world's plants and animals (out of an estimated total of eight million) facing the threat of extinction [13], and this leads scientists to suggest that the world is currently facing the sixth mass extinction.

All the above issues stress the need for sustainable forest management, and for reconciling land management and socioeconomic development with the need to conserve the global biodiversity at all levels, from genetic variants to species, populations, and ecosystems. In this Special Issue, a collection of articles is published covering a wide range of topics related to forest and biodiversity conservation.

The conservation of genetic diversity and resources is of crucial importance for biodiversity conservation, especially under conditions of climate change, which transforms habitats across the globe. Chertov et al. [14] studied the genetic diversity and population structure of *Larix sibirica* in the Urals, concluding that the currently observed differentiations between the populations need to be preserved by maintaining non-fragmented populations and preventing the geographical isolation of existing populations. In the same region, Sboeva et al. [15] reported a limited genetic diversity in *Pius silvestris*, which is explained by the biogeography of the established populations. A similar study by Gong et al. [16] on *Cinnamomum camphora* in East Asia revealed significant differences in genetic diversity among populations and moderate differentiation between populations. They point out the need of enhancing gene flow to reverse the negative effects of genetic drift and reduce the risk of genetic diversity reduction. Significant differences in the within-population genetic diversity, and high differentiations between populations, were observed by Huang et al. [17] in different populations of *Camellia chekiangoleosa*. They recommend the adoption of appropriate management measures to maintain the diversity in the high-diversity populations and increase the gene exchange in the low-diversity populations to restore genetic diversity. Limited genetic variations among studied populations of *Triadica sebifera* are reported by Zhou et al. [18], along with a relatively low within-population genetic diversity. However, the existence of some rare alleles in some populations may prove extremely important in maintaining the level of genetic diversity. Ex situ conservation is proposed in the above studies as an alternative approach for maintaining diversity in ecologically and economically important species, while Liu et al. [19] provide specific instructions for the preservation of the pollen of *Gleditsia sinensis*. A combination of in situ and ex situ conservation is also proposed by Kinho et al. [20] for the endangered species *Pericopsis mooniana* and its introduction into areas that are expected to become suitable as a result of climate change.

Climate and global changes result in significant land use changes and habitat alterations, with subsequent alterations in species' current and future geographical distributions, as well as in their physiological responses. Zhang et al. [21], studying four alpine *Rhododedron* species, revealed a significant expected shrinkage in their geographical distribution as a result of climate change, as well as an expected movement of species to different biogeographical regions. A significant expected decrease in the currently suitable habitats for *Swietenia macrophylla* (mahogany) was reported under different climate change scenarios by Herrera-Feijoo et al. [22]. At the same time, currently unsuitable areas may become suitable in the future, which demonstrates the dynamic nature of any conservation measures, including the designation of protected areas. Climatic patterns affect a number of physiological responses of species, including flowering and fruiting. Dagnachew et al. [23] point out the need for a better understanding of the relationships between climatic variables and species physiological responses, which vary considerably between species, in order to develop an effective conservation scheme for species of high ecological and economical importance.

The dynamic recovery of European forests was confirmed in the study by Referowska-Chodak and Kornatowska [24], covering a period of 75 years in Poland. Irrespectively of the political regime, the study demonstrates the steady increase in forest cover and biomass, as well as in the areas designated as protected. The latter constitutes an important tool for biodiversity conservation, and the NATURA2000 network of protected areas is one of them. The study by Kermavnar et al. [25], however, points out the need for adopting management and conservation measures not only based on broad habitat type definitions, but also based on the specific characteristics of each habitat and on the need to expand the nature conservation actions beyond the designated protected areas. Uprety et al. [26], for instance, identified ecologically important ecosystems outside of the protected areas in the Chure region of Nepal, and point out the need for active conservation measures in the region. An interesting model for the prioritization of areas where conservation actions need to be applied is presented by Vu et al. [27], which has been tested in Vietnam. The model results in seven criteria and seventeen indicators to determine priority areas for biodiversity conservation. In the same direction is the approach presented by Ette et al. [28], where a novel Biodiversity Composite Index (BCI) is proposed for the assessment of the ecosystem, species, and genetic diversity status. Based on this assessment appropriate conservation priorities and objectives can be adopted. An even simpler index derived by remote sensing data, the Normalized Difference Vegetation Index (NDVI), is proposed as a surrogate for field-based biodiversity assessments by Naunyal et al. [29]. NDVI was found to explain 65% of the variation in plant species diversity, while the same study also demonstrates the great value of remote sensing data and methods for monitoring land use changes caused by anthropogenic activities. The latter, as has already been stated above, constitutes a major threat for the integrity of many ecosystems. Mariscal et al. [30], studying an Ecuadorian Andean cloud forest, point out the need to prevent the further loss of primary forests and the further increase in fragmentation. The same study shows that areas that have been under disturbance regimes of varying intensities do not necessarily proceed towards a composition and structure similar to primary forests in the post-disturbance stage. Instead, alternative stable successional stages are observed. The stability of these successional stages is explained by the fact that plant assemblages at any stage reflect the historical and current environmental conditions. Furthermore, as pointed out by Chen et al. [31], shifts in plant functional traits occur with succession, demonstrating complex and diverse trade-offs that result in variation among the ecological strategy spectra of different successional stages and the respective assemblages.

As already mentioned, forests harbor a wide variety of wildlife species, including mammals, birds, arthropods, and others. The strong relationship between bird communities and forest habitat types is demonstrated in the study by Purevdorj et al. [32] in northern Mongolia. The study points out the need for maintaining diverse forest habitats and restoring forests that have been lost or degraded by anthropogenic activities. Given the complexity of forest habitat types, new forest inhabiting species are constantly detected and described by science, such as the new arthropod species described in the study by Jang et al. [33]. However, it is not only the plant and animal species that benefit from increases in forest cover and decreased fragmentation. Increased biomass leads to increased carbon stock, and Abdul-Hamid et al. [34] presented an interesting approach in estimating biomass in the ecologically vulnerable and extremely important mangrove forests. The approach can also be applied to other forest types and different biogeographical regions. The overall positive effects of increased forest cover and decreased fragmentation in a wide range of ecosystem services, including provisioning, regulating, and cultural, is demonstrated in the study by Kefalas et al. [35] in the eastern Mediterranean region.

Beyond any doubt, forests constitute the main terrestrial biodiversity carrier of the world, compared to any alternative land use/cover type. Their conservation is a global challenge, and has to be the first political and social priority. We believe that the studies presented in the current issue provide important and useful insights towards achieving the target of ensuring a sustainable environment for current and future generations.

Author Contributions: Writing—original draft preparation, P.X., G.K. and K.P. All authors have read and agreed to the published version of the manuscript.

Conflicts of Interest: The authors declare no conflict of interest.

References

1. Hilton-Taylor, C.; Stuart, S.N.; Vie, J.-C. *Wildlife in a Changing World: An Analysis of the 2008 IUCN Red List of Threatened Species*; IUCN: Gland, Switzerland, 2009; p. 184.
2. UNEP; FAO. *The State of the World's Forests 2020: Forests, Biodiversity and People*; FAO: Rome, Italy, 2020; p. 214.
3. Hunter, M.L.J. Biological Diversity. In *Maintaining Biodiversity in Forest Ecosystems*; Hunter, M.L.J., Ed.; Cambridge University Press: Cambridge, UK, 1999.
4. Xofis, P.; Poirazidis, K. Combining different spatio-temporal resolution images to depict landscape dynamics and guide wildlife management. *Biol. Conserv.* **2018**, *218*, 10–17. [CrossRef]
5. Xofis, P.; Spiliotis, A.J.; Chatzigiovanakis, S.; Chrysomalidou, S.A. Long-Term Monitoring of Vegetation Dynamics in the Rhodopi Mountain Range National Park-Greece. *Forests* **2022**, *13*, 377. [CrossRef]

6. Chapron, G.; Kaczensky, P.; Linnell, J.D.C.; Von Arx, M.; Huber, D.; Andrén, H.; López-Bao, J.V.; Adamec, M.; Álvares, F.; Anders, O.; et al. Recovery of large carnivores in Europe's modern human-dominated landscapes. *Science* **2014**, *346*, 1517–1519. [CrossRef] [PubMed]
7. Cimatti, M.; Ranc, N.; Benítez-López, A.; Maiorano, L.; Boitani, L.; Cagnacci, F.; Čengić, M.; Ciucci, P.; Huijbregts, M.A.J.; Krofel, M.; et al. Large carnivore expansion in Europe is associated with human population density and land cover changes. *Divers. Distrib.* **2021**, *27*, 602–617. [CrossRef]
8. Silver, W.L.; Osterag, R.; Lugo, A.E. The potential for carbon sequestration through reforestation of abandoned tropical agricultural and pasture lands. *Restor. Ecol.* **2000**, *8*, 394–407. [CrossRef]
9. Otero, I.; Marull, J.; Tello, E.; Diana, G.L.; Pons, M.; Coll, F.; Martí Boada, M. Land abandonment, landscape, and biodiversity: Questioning the restorative character of the forest transition in the Mediterranean. *Ecol. Soc.* **2015**, *20*, 7. [CrossRef]
10. Zakkak, S.; Radovic, A.; Nikolov, S.C.; Shumka, S.; Kakalis, L.; Kati, V. Assessing the effect of agricultural land abandonment on bird communities in southern-eastern Europe. *J. Environ. Manag.* **2015**, *164*, 171–179. [CrossRef]
11. Hill, S.L.L.; Arnell, A.; Maney, C.; Butchart, S.H.M.; Hilton-Taylor, C.; Ciciarelli, C.; Davis, C.; Dinerstein, E.; Purvis, A.; Burgess, N.D. Measuring forest biodiversity status and changes globally. *Front. For. Glob. Change* **2019**, *2*, 70. [CrossRef]
12. Abram, N.K.; MacMillan, D.C.; Xofis, P.; Ancrenaz, M.; Tzanopoulos, J.; Ong, R.; Goossens, B.; Koh, L.P.; Del Valle, C.; Peter, L.; et al. Identifying Where REDD+ Financially Out-Competes Oil Palm in Floodplain Landscapes Using a Fine- Scale Approach. *PLoS ONE* **2016**, *11*, e0156481. [CrossRef]
13. IPBES. *Summary for Policymakers of the Global Assessment Report on Biodiversity and Ecosystem Services of the Intergovernmental Science-Policy Platform on Biodiversity and Ecosystem Services*; IPBES Secretariat: Bonn, Germany, 2019; p. 56.
14. Chertov, N.; Vasilyeva, Y.; Zhulanov, A.; Nechaeva, Y.; Boronnikova, S.; Kalendar, R. Genetic Structure and Geographical Differentiation of Larix sibirica Ledeb. in the Urals. *Forests* **2021**, *12*, 1401. [CrossRef]
15. Sboeva, Y.; Chertov, N.; Nechaeva, Y.; Valeeva, A.; Boronnikova, S.; Kalendar, R. Genetic Diversity, Structure, and Differentiation of *Pinus sylvestris* L. Populations in the East European Plain and the Middle Urals. *Forests* **2022**, *13*, 1798. [CrossRef]
16. Gong, X.; Yang, A.; Wu, Z.; Chen, C.; Li, H.; Liu, Q.; Yu, F.; Zhong, Y. Employing Genome-Wide SNP Discovery to Characterize the Genetic Diversity in Cinnamomum camphora Using Genotyping by Sequencing. *Forests* **2021**, *12*, 1511. [CrossRef]
17. Huang, B.; Wang, Z.; Huang, J.; Li, X.; Zhu, H.; Wen, Q.; Xu, L.-A. Population Genetic Structure Analysis Reveals Significant Genetic Differentiation of the Endemic Species Camellia chekiangoleosa Hu with a Narrow Geographic Range. *Forests* **2022**, *13*, 234. [CrossRef]
18. Zhou, P.; Zhou, Q.; Dong, F.; Shen, X.; Li, Y. Study on the Genetic Variation of Triadica sebifera (Linnaeus) Small Populations Based on SSR Markers. *Forests* **2022**, *13*, 1330. [CrossRef]
19. Liu, Q.; Yang, J.; Wang, X.; Zhao, Y. Studies on Pollen Morphology, Pollen Vitality and Preservation Methods of *Gleditsia sinensis* Lam. (Fabaceae). *Forests* **2023**, *14*, 243. [CrossRef]
20. Kinho, J.; Suhartati, S.; Husna, H.; Tuheteru, F.D.; Arini, D.I.D.; Lawasi, M.A.; Ura', R.; Prayudyaningsih, R.; Yulianti, Y.; Subarudi, S.; et al. Conserving Potential and Endangered Species of Pericopsis mooniana Thwaites in Indonesia. *Forests* **2023**, *14*, 437. [CrossRef]
21. Zhang, J.-H.; Li, K.-J.; Liu, X.-F.; Yang, L.; Shen, S.-K. Interspecific Variance of Suitable Habitat Changes for Four Alpine Rhododendron Species under Climate Change: Implications for Their Reintroductions. *Forests* **2021**, *12*, 1520. [CrossRef]
22. Herrera-Feijoo, R.J.; Torres, B.; López-Tobar, R.; Tipán-Torres, C.; Toulkeridis, T.; Heredia-R, M.; Mateo, R.G. Modelling Climatically Suitable Areas for Mahogany (Swietenia macrophylla King) and Their Shifts across Neotropics: The Role of Protected Areas. *Forests* **2023**, *14*, 385. [CrossRef]
23. Dagnachew, S.; Teketay, D.; Demissew, S.; Awas, T.; Lemessa, D. The Effects of Monthly Rainfall and Temperature on Flowering and Fruiting Intensities Vary within and among Selected Woody Species in Northwestern Ethiopia. *Forests* **2023**, *14*, 541. [CrossRef]
24. Referowska-Chodak, E.; Kornatowska, B. Effects of Forestry Transformation on the Landscape Level of Biodiversity in Poland's Forests. *Forests* **2021**, *12*, 1682. [CrossRef]
25. Kermavnar, J.; Kozamernik, E.; Kutnar, L. Assessing the Heterogeneity and Conservation Status of the Natura 2000 Priority Forest Habitat Type Tilio–Acerion (9180*) Based on Field Mapping. *Forests* **2023**, *14*, 232. [CrossRef]
26. Uprety, Y.; Tiwari, A.; Karki, S.; Chaudhary, A.; Yadav, R.K.P.; Giri, S.; Shrestha, S.; Paudyal, K.; Dhakal, M. Characterization of Forest Ecosystems in the Chure (Siwalik Hills) Landscape of Nepal Himalaya and Their Conservation Need. *Forests* **2023**, *14*, 100. [CrossRef]
27. Vu, X.D.; Csaplovics, E.; Marrs, C.; Nguyen, T.T. Criteria and Indicators to Define Priority Areas for Biodiversity Conservation in Vietnam. *Forests* **2022**, *13*, 1341. [CrossRef]
28. Ette, J.-S.; Sallmannshofer, M.; Geburek, T. Assessing Forest Biodiversity: A Novel Index to Consider Ecosystem, Species, and Genetic Diversity. *Forests* **2023**, *14*, 709. [CrossRef]
29. Naunyal, M.; Khadka, B.; Anderson, J.T. Effect of Land Use and Land Cover Change on Plant Diversity in the Ghodaghodi Lake Complex, Nepal. *Forests* **2023**, 529. [CrossRef]
30. Mariscal, A.; Thomas, D.C.; Haffenden, A.; Manobanda, R.; Defas, W.; Angel Chinchero, M.; Simba Larco, J.D.; Jaramillo, E.; Roy, B.A.; Peck, M. Evidence for Alternate Stable States in an Ecuadorian Andean Cloud Forest. *Forests* **2022**, *13*, 875. [CrossRef]

31. Chen, C.; Wen, Y.; Ji, T.; Zhao, H.; Zang, R.; Lu, X. Ecological Strategy Spectra for Communities of Different Successional Stages in the Tropical Lowland Rainforest of Hainan Island. *Forests* **2022**, *13*, 973. [CrossRef]
32. Purevdorj, Z.; Munkhbayar, M.; Paek, W.K.; Ganbold, O.; Jargalsaikhan, A.; Purevee, E.; Amartuvshin, T.; Genenjamba, U.; Nyam, B.; Lee, J.W. Relationships between Bird Assemblages and Habitat Variables in a Boreal Forest of the Khentii Mountain, Northern Mongolia. *Forests* **2022**, *13*, 1037. [CrossRef]
33. Jang, C.-M.; Yoo, J.-S.; Kim, S.-T.; Bae, Y.-S. Rocky Area Inhabiting Daddy Long-Legs Spiders, Pholcus Walckenaer, 1805 (Araneae: Pholcidae) in Mountainous Mixed Forests from South Korea. *Forests* **2023**, *14*, 538. [CrossRef]
34. Abdul-Hamid, H.; Mohamad-Ismail, F.-N.; Mohamed, J.; Samdin, Z.; Abiri, R.; Tuan-Ibrahim, T.-M.; Mohammad, L.-S.; Jalil, A.-M.; Naji, H.-R. Allometric Equation for Aboveground Biomass Estimation of Mixed Mature Mangrove Forest. *Forests* **2022**, *13*, 325. [CrossRef]
35. Kefalas, G.; Lorilla, R.S.; Xofis, P.; Poirazidis, K.; Eliades, N.-G.H. Landscape Characteristics in Relation to Ecosystem Services Supply: The Case of a Mediterranean Forest on the Island of Cyprus. *Forests* **2023**, *14*, 1286. [CrossRef]

*forests*

MDPI

Article

Genetic Structure and Geographical Differentiation of *Larix sibirica* Ledeb. in the Urals

Nikita Chertov [1], Yulia Vasilyeva [1], Andrei Zhulanov [1], Yulia Nechaeva [1], Svetlana Boronnikova [1,*] and Ruslan Kalendar [2,3,*]

[1] Faculty of Biology, Perm State University, 614990 Perm, Russia; nikita.chertov22@gmail.com (N.C.); yulianechaeva@mail.ru (Y.V.); aumakua.ru@gmail.com (A.Z.); ulia-2012@mail.ru (Y.N.)
[2] National Laboratory Astana, Nazarbayev University, Nur-Sultan 010000, Kazakhstan
[3] HiLIFE Institute of Biotechnology, University of Helsinki, Biocenter 3, Viikinkaari 1, FI-00014 Helsinki, Finland
* Correspondence: svboronnikova@yandex.ru (S.B.); ruslan.kalendar@helsinki.fi (R.K.); Tel.: +358-294-158-869 (R.K.)

Abstract: The Ural Mountains and the West Eurasian Taiga forests are one of the most important centers of genetic diversity for *Larix sibirica* Ledeb. Forest fragmentation negatively impacts forest ecosystems, especially due to the impact of their intensive use on the effects of climate change. For the preservation and rational use of forest genetic resources, it is necessary to carefully investigate the genetic diversity of the main forest-forming plant species. The *Larix* genus species are among the most widespread woody plants in the world. The Siberian larch (*Larix sibirica*, *Pinaceae*) is found in the forest, forest-tundra, tundra (Southern part), and forest-steppe zones of the North, Northeast, and partly East of the European part of Russia and in Western and Eastern Siberia; in the Urals, the Siberian larch is distributed fragmentarily. In this study, eight pairs of simple sequence repeat (SSR) primers were used to analyse the genetic diversity and population structure of 15 Siberian larch populations in the Urals. Natural populations in the Urals exhibit indicators of genetic diversity comparable to those of Siberia populations (expected heterozygosity, $He = 0.623$; expected number of alleles, $Ne = 4017$; observed heterozygosity, $Ho = 0.461$). Genetic structure analysis revealed that the examined populations are relatively highly differentiated ($Fst = 0.089$). Using various algorithms for determining the spatial genetic structure, the examined populations formed three groups according to geographical location. The data obtained are required for the development of species conservation and restoration programs, which are especially important in the Middle Urals, which is the region with strong forest fragmentation.

Keywords: simple sequence repeats (SSR); multiplex PCR; genetic diversity; population structure; genetic differentiation; *Larix sibirica* Ledeb.

Citation: Chertov, N.; Vasilyeva, Y.; Zhulanov, A.; Nechaeva, Y.; Boronnikova, S.; Kalendar, R. Genetic Structure and Geographical Differentiation of *Larix sibirica* Ledeb. in the Urals. *Forests* **2021**, *12*, 1401. https://doi.org/10.3390/f12101401

Academic Editors: Konstantinos Poirazidis and Panteleimon Xofis

Received: 17 September 2021
Accepted: 13 October 2021
Published: 14 October 2021

1. Introduction

Forest fragmentation negatively impacts forest ecosystems, especially due to the impact of their intensive use on the effects of climate change [1]. The increased anthropogenic load and the intensity of forest exploitation increase forest sensitivity to global climate changes and consequently to a shift in the boundaries of the woody plant area of species. In addition, reduction in forest areas is currently observed worldwide due to adverse natural phenomena (floods, fires), damage to tree species by diseases and pests, and intensive human use of forest resources [2]. In this regard, one of the main tasks is to develop fundamental principles and experimental approaches to identify tree forms that have the potential for climatic adaptation, at which the genetic potential, diversity, and ecological variability of the forest woody plants in different climatic conditions play a key role. In the Urals, the zone of introgression of the genus *Pinaceae* species is located in which unique combinations of genotypes of species and their hybrids are concentrated, and their ecological plasticity is distinguished. The genetic structure and diversity of populations of woody species are formed under the influence of factors such as genetic drift, mutation, migration and selection, and environmental factors [2,3]. In addition, anthropogenic factors also affect the genetic structure of woody plant populations [1]. Consequently, the study of the current state of forest genetic resources and genetic structure of the main forest-forming types of conifer populations is important both in government projects to preserve plant resources and in applied aspects of their rational use in economic interests [4,5]. The primary focus has been on the study of gene pools of forest resource plant species that occupy extensive ranges and have economic importance. One of these types of coniferous plants is the species of *Larix Mill*, which are the primary species of the boreal forests of the Northern hemisphere [6,7]. In the Urals, the *Larix* genus is represented by the West Race of the Siberian Larch or Russian Larch (*Larix sibirica*) [8]. The coniferous wood is characterized by many favorable qualities and is universal in its application. Besides the physicomechanical properties that are valued in the construction industry, larch contains several useful substances that are prospective raw materials for the forestry industry. Larch contains biologically active substances (BAS) such as phenols, polyphenolic compounds, and pectin, among others [9]. The investigation of genetic diversity, structure, differentiation, and identification of *L. sibirica* populations is also important for studying the content of resin acids as promising BAS [10–12]. Molecular genetic research in larch has been actively conducted since the end of 1980, both in Russia and elsewhere, via the application of isozyme and DNA markers. Currently, a large amount of structural data has been accumulated on the genetic diversity and differentiation of populations of various types of Siberian larch [13–22]. Previous studies addressing the genetic diversity of this species revealed lower indices of genetic variability in *L. sibirica*, consistent with its narrow endemic status. Application of only isozyme markers made it possible to study some of the genetic diversity in expressed protein-coding genes, which in the conifer genome amounts to only 1–2%, whereas the main part of the genome has been characterized with a variety of DNA markers and next-generation sequencing (NGS) technologies [23–26]. For example, DNA markers based on microsatellite repeats (simple sequence repeats; SSR) are used to explore genetic polymorphisms of conifer species, including Larix species [27–31]. Microsatellites, or STRs, are repetitive sequences of 2–6 bp tandemly repeated units of DNA that are present throughout the entire genome. Microsatellite markers are codominant and multiallelic and efficient for allelic diversity study [32], which makes them an effective tool in population genetics, to study genetic diversity and structure. A combination of microsatellite markers with capillary electrophoresis significantly increases the accuracy of the analysis. Data from many years of research in the Urals (including in Perm Krai) clearly revealed the fragmentation of larch stands. In addition, in this region, the "larchless tongue" of the western macroslope of the Ural Mountains is located and is marked on all species distribution maps [33–35]. At the same time, it is in this "Permian-Kama Pre-Ural" population that the highest level of species diversity in the Urals is established, as revealed by data from morphological and isozyme analyses [36]. The interpopulation genetic differentiation of Siberian larch in various forest conditions of the Urals is influenced by a combination of factors, such as isolation and natural selection. The spatial genetic structure of populations is also largely determined by the history of the formation of the species' range [37–39]. Thus, currently the genetic diversity and structure of the Ural populations of *L. sibirica* has been relatively well studied only on the basis of polymorphisms of isozyme markers [36] and with inter-simple sequence repeat (ISSR) markers [40,41]. Genetic polymorphisms of other types of molecular markers of species of the genus *Larix* in the Urals, also including microsatellite markers, has been studied fragmentarily and the data obtained are insufficient for comprehensive assessment of the genetic structure and differentiation of the species in this region. Although the genetic diversity and population structure of *L. sibirica* populations have been addressed previously [40,41], here we provide a new approach and obtained more precise genetic diversity characterization by using microsatellites markers. In this regard, study of the genetic structure, population differentiation, and nature of the alleles of *L. sibirica* distribution from the Ural Mountains and West Eurasian Taiga forests based on polymorphism analysis of microsatellite markers. This is an urgent task for the preservation of populations that are productive and resistant to various environmental factors in conditions of fragmentation of the area of the species.

In this study, we used a new collection of samples from 15 populations of the western race of Siberian larch (*Larix sibirica*) from the Northern, Middle, and Southern Urals to systematically analyse the genetic variations and population structure of this collection based on analysis of polymorphisms of microsatellite markers. Explanations are also provided on the role of geographical isolation and environmental heterogeneity in genetic differentiation among regions in the Ural Mountains and West Eurasian Taiga forests.

2. Materials and Methods

2.1. Sample Collection and DNA Extraction

During the 2015 to 2020 growing seasons, we collected young leaves (needles) of individuals from 28 to 32 trees in each of the 15 populations of the western race of the Siberian larch (*Larix sibirica* Ledeb. (*L. sukaczewii* Dyl.)) (Figure S1). The samples selected for the study were from a wide altitudinal-latitudinal gradient; the study region length was more than 1000 km from North to South, and the altitude varied from 190 to 916 meters above sea level (Supplementary Table S1). The studied samples were conditionally divided into mountain (from 500 meters above sea level) and flat (up to 500 meters above sea level). Five populations were from the South Urals in Chelyabinsk district and the Republic of Bashkortostan, five were from the Middle Urals in Sverdlovsk district and in Perm Krai, and the last five populations were from the Northern Urals in the Perm Krai (Supplementary Table S1).

DNA was isolated according to a procedure for complex biological samples [42]. The weighed amount of needles was up to 20 mg. A NanoDrop 2000 spectrophotometer (Thermo Fischer Scientific, Waltham, MA, USA) was used to determine the concentration and quality of DNA. The DNA pellets were dissolved in $1\times$ TE buffer (1 mM EDTA, 10 mM Tris-HCl, pH 8.0). The resultant DNA for each isolate was diluted with $1\times$ TE solution to 10 ng/μL and used as a template for PCR experiments.

2.2. Microsatellite Amplification

In this study, eight polymorphic SSR primer pairs (Table 1) were used for further analysis, which were developed by Isoda and Watanabe (2006) and Wagner (2012) and showed effective differentiation for *L. kaempferi* and *L. decidua* samples [28,43]. These eight SSR primer pairs were organized in multiplexes pairwise according to the expected allele range size for efficient multiplex analysis of the amplification products. An M13 universal adapter sequence ($5'$-TGTAAAACGACGGCCAGT-$3'$) labeled with fluorochromes (FAM and JOE dyes) was used for the amplification of fluorescently labeled PCR products. PCR amplification was conducted in a total volume of 20 μl and contained $1\times$ PCR buffer with 2.5 mM $MgCl_2$, 0.25 μM of reverse primer, 0.2 μM of forward primer with M13 tail, 0.05 μM of fluorescently labeled M13 primer, 0.2 mM each dNTP, 1 unit of SynTaq DNA polymerase (Syntol, Moscow, Russia), and 50 ng DNA template. PCR amplification was performed in a CFX96 (Bio-Rad, Hercules, CA, USA) under the following conditions: initial denaturation step at 94 °C for 10 min followed by 30 cycles at 94 °C for 30 s, 52–58 °C for 45 s, 72 °C for 45 s, 8 cycles at 94 °C for 30 s, 53 °C for 45 s, 72 °C for 45 s, and final extension at 72 °C for 5 min. The amplification products were separated by electrophoresis in 2% agarose gel in $1\times$ TBE buffer, stained with ethidium bromide, and photographed under UV light in a GelDoc XR gel documentation system (Bio-Rad, Hercules, CA, USA) (Figure S2). Molecular weight marker GeneRuler 100 bp Plus DNA Ladder (Thermo Fischer Scientific, Waltham, MA, USA) and Quantity One software (Bio-Rad, USA) were used to determine DNA fragment lengths. The PCR amplicons were separated by capillary electrophoresis on a Genetic Analyzer 3500xl (Applied Biosystems, Waltham, MA, USA) using the POP7 matrix and 600 LIZ (Thermo Fisher Scientific, Waltham, MA, USA) as a dye size standard. GeneMarker v2.6.3 (ABI) software was used to determine the genotypes.

Table 1. Microsatellite primer pairs used to detect polymorphisms of *L. sibirica*.

SSR Locus	Primer Sequence (5′-3′)	Ta (°C)	A	Repeat Motif	Lfr (bp)
bcLK189	F: M13-ACCATACGCATACCCAATAGA R: AGTTTTCCTTTCCCACACAAT	58	17	$(AG)_{17}AT(AG)_6$	155–170
Ld101	F: M13-ACACCAAGGACTCTCTGACTAC R: GGTGATTCCAGAAGCAGGTG	58	16	$(AC)_{12}$	189–225
bcLK228	F: M13-CCCTAACCCTAGAATCCAATAA R: GAGGAAGGCGACAAGTCATT	61	12	$(AG)_{18}$	179–215
Ld56	F: M13-AGCCATCGTGGTTCTTCTTTG R: CTTGTAACTGTGCACCCACC	58	9	$(AC)_{16}$	227–243
Ld50	F: M13-GAAGGCGACTTTACATGCCC R: TCCATCTTTATGTCTCTTCCATGC	58	6	$(CA)_{18}$	161–189
bcLK253	F: M13-AACACCATAGTGCAATGTGC R: TCCTCTTGTTGATGCCACTT	58	4	$(AG)_{17}$	199–204
Ld42	F: M13-TCGTATGCATTGTCCAAATTTCC R: TCCAAGTGAGGTCACACGAG	58	8	$(TG)_{14}$	167–186
bcLK263	F: M13-CGATTGGTATAGTGGTCATTGT R: CCATCATACCTTCTTGAAGAG	58	37	$(TC)_{20}$	191–254

M13 sequence tail (5′-TGTAAAACGACGGCCAGT); F and R, forward and reverse primers, respectively; *Ta*, PCR annealing temperature; A, allele number obtained for all analyzed individuals; Lfr, length of the PCR amplicons in base pairs considering the addition of M13.

Analysis of chromatograms to determine fragment lengths was performed using GeneMarker v2.6.3 software. Computer analysis of DNA polymorphism was performed using Arlequin v3.5.2.2 [44] and a specialized macro GenAlEx v6.5 [45] for MS Excel. Migration test was performed using STRUCTURE v2.3.4 [46] software with the parameters specified in Supplementary Table S2. Calculation of the Goldstein distance matrix [47] and the construction of the minimum spanning tree was performed in EDENetworks v2.18 [48]. Dendrogram construction was performed in MEGA X [49] using UPGMA and least-squares analysis. Analysis using the method of principal coordinates was performed using the PAST v4.06 program [50]. For each sample, 19 basic climatic parameters were obtained from the bioclimatic variables database of the WorldClim service (https://www.worldclim.org/data/worldclim21.html, accessed date August 2021) using the raster v3.4-13 package [51]; on this basis, a distance matrix is formed by calculating Canberra distance. Correlation analysis of genetic and climatic distances (Mantel test) was performed in GenAlEx macro. MavericK v1.0.5 software [52] was used with the parameters specified in Supplementary Table S3 to analyse and visualize the genetic structure and study the correspondence between clusters of genotypes and the studied populations.

3. Results

3.1. Polymorphic SSR Primers

The optimal pairs of primers and fluorescent dyes were determined, which allowed multiplex PCR and efficient separation of amplified fragments of different loci. For FAM dye, such combinations were the bcLK189 and Ld101 and bcLK253 and Ld50 primer pairs. For JOE dye, these were the bcLK228 and Ld56 and bcLK263 and Ld42 primer pairs.

The results of capillary electrophoresis confirmed the success and specificity of the amplification under the selected conditions (Supplementary Figure S3).

3.2. Polymorphism of Microsatellite Markers in 15 Populations of L. sibirica

Eight SSR loci were used to analyse the polymorphisms of *L. sibirica* microsatellite loci. All loci were polymorphic (P95 = 1000) for the studied populations. The total number of alleles per locus for 451 individuals from 15 populations varied from 4 (bcLK253) to 37 (bcLK263). The total number of SSR markers installed was 109. The absolute and effective number of alleles per locus was the largest in the Vsn populations growing in the vicinity of the village Verkhnyaya Sanarka in Chelyabinsk district (Na = 8.500, Ne = 5.276); the smallest was in the Sks populations (Na = 4.250, Ne = 2.283) from the protected areas "Larch Grove" in the Suksunsky district of the Perm Krai. The average values of these indicators were Na = 6.942 and Ne = 4.017 (Table 2). Observed (Ho) and expected heterozygosity (He) for the total populations of *L. sibirica* were 0.461 and 0.623, respectively. These indicators were greatest in the populations Irm (Ho = 0.567) and Gai (He = 0.682) and smallest in the populations Ish (Ho = 0.366) and Sks (He = 0.520). Thus, a high level of genetic diversity of *L. sibirica* was revealed in Ural (observed number of alleles Na = 6.942, expected number of alleles Ne = 4.017, H_O = 0.461, He = 0.623).

Table 2. Genetic diversity of 15 populations of *L. sibirica* based on polymorphisms of microsatellite markers.

Populations	Number of Alleles (*Na*)	Number of Effective Alleles (*Ne*)	Observed Heterozygosity (*Ho*)	Expected Heterozygosity (*He*)
Kar	7.000	3.263	0.522	0.587
Irm	8.250	4.998	0.567	0.676
Vsn	8.500	5.276	0.484	0.643
Kul	7.375	4.034	0.500	0.629
Zil	6.750	3.929	0.460	0.632
Tul	6.375	3.825	0.387	0.634
Ish	4.625	2.762	0.366	0.563
Krv	7.750	4.266	0.433	0.612
Bnd	6.875	4.357	0.463	0.635
Gai	8.125	4.740	0.428	0.682
Kch	8.375	4.584	0.479	0.666
Pol	5.875	3.662	0.562	0.622
Osa	7.000	4.008	0.418	0.626
Sks	4.250	2.283	0.396	0.520
Bil	7.000	4.273	0.458	0.624
For a total sample of 15 populations	6.942	4.017	0.461	0.623

3.3. Analysis of the Genetic Structure and Differentiation of the Studied Populations of Siberian Larch

The genetic structure parameters analysis for the total sample of *L. sibirica* showed that the Ld56 locus made the greatest contribution to the differentiation of the studied populations (Table 3). All studied populations of *L. sibirica* are characterized by a deficiency of heterozygous genotypes. A population structure study using Wright's F-statistics showed that *L. sibirica* in the Urals had on average 26% deficiency of heterozygous genotypes (Fis = 0.259) within the population and 32% deficiency of heterozygotes (Fit = 0.325) between populations. Only 8.9% of all observed variability falls on the interpopulation component. The Ld56 locus made the greatest contribution to the differentiation of the studied populations (Table 3). Analysis of molecular variance (AMOVA) of the genetic diversity distribution in the region revealed that the populations of the Northern, Middle, and Southern Urals are almost the same in the allelic composition of SSR loci (Supplementary Table S4).

Table 3. Genetic structure and differentiation of the studied populations of *L. sibirica*.

Locus	*Fis*	*Fit*	*Fst*
bcLK189	0.138	0.198	0.070
Ld101	0.577	0.609	0.077
bcLK228	0.595	0.628	0.080
Ld56	0.351	0.427	0.117
bcLK253	0.083	0.150	0.073
Ld50	0.357	0.428	0.111
bcLK263	0.014	0.083	0.070
Ld42	−0.045	0.075	0.115
Total	0.259	0.325	0.089

Fis, inbreeding coefficient of individuals in subpopulations; *Fit*, the coefficient of inbreeding of individuals in the population as a whole; *Fst*, coefficient of inbreeding of subpopulations relative to the entire population.

The interregional share of diversity accounts for only 2% and interpopulation 5%, while the main part falls on individual heterozygosity. When comparing the subdivision levels among the populations from different Urals regions, the most differentiated are the populations from the Middle Urals (*Fst* = 7%). The South Ural and North Ural populations have a similar level of differentiation (*PhPT* = 5%). The migration test showed only a small proportion of the admixture of genetic material from other populations (Supplementary Figure S4). The matrix of pairwise F_{ST} values was constructed and visualized (Supplementary Table S5, Figure 1). The Sks sample is clearly distinguished by high values (*Fst* from 0.114 to 0.180).

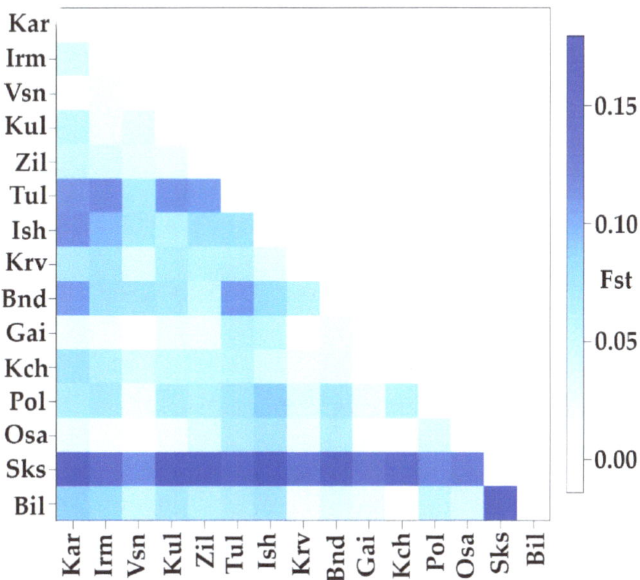

Figure 1. Visualization of the matrix of pairwise F_{ST} values. Color indicates the degree of differentiation (the darker, the stronger the differentiation).

The matrix of the pairwise differences average number between and within populations and Nei's distance were also visualized (Supplementary Figure S5). The largest number of differences was found between the populations Irm and Tul (5.334) and the smallest between the populations Ish and Osa (3.793). The Osa sample is distinguished as the least different from other populations in general. The number of intrapopulation differences is greatest in the Irm sample (5.054) and smallest in the Sks sample (3.181). The distribution between populations Nei's distance coincides with the *Fst* distribution. On the basis of the Goldstein distance matrix using the UPGMA (Unweighted Pair Group Method with Arithmetic mean) method and least-squares analysis, a dendrogram was constructed that reflects the degree of the studied populations' similarity by SSR spectra (Supplementary Figure S6A). Three clusters are clearly distinguished on the dendrogram. The first cluster combines the populations Irm, Kul, Osa, Kar, and Zil (Southern group); the second combines Bnd, Vsn, Bil, and Pol (Middle group); the third combines Ish, Gai, Krv, Kch, and Tul (Northern group). The Sks sample differs markedly from all others and can be considered as a separate group. A minimum spanning tree was also constructed (Figure 2), the structure of which perfectly matches the dendrogram. Principal coordinate analysis (PCoA) also confirmed this subdivision into groups (Supplementary Figure S6B).

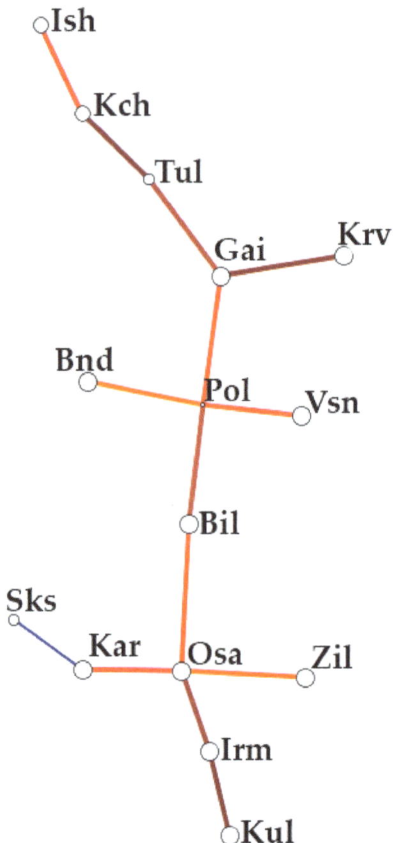

Figure 2. Minimum spanning tree based on Goldstein distances. The rib colors indicate the genetic distance (lighter = shorter distance). The node diameter is proportional to the clonal diversity of the populations.

While correlation analysis did not reveal a close correlation between the geographic and genetic distances among the populations ($r^2 = 0.0006$; $p = 0.060$), a significant (Figure S8) mean positive correlation ($r^2 = 0.570$; $p = 0.040$) in the mountain populations group (Irm, Kul, Kch, Tul, Ish) was revealed. In addition, correlation analysis of climatic and genetic distances was performed, which revealed a weak correlation ($r^2 = 0.0648$, $p = 0.05$). A significant points scatter (Supplementary Figure S7) suggested that for certain groups of populations, the correlation may be strong.

Additional analysis revealed a strong correlation for the southern group of populations (Supplementary Figure S8) ($r^2 = 0.5989$, $p = 0.08$). The correlation was low for the other two groups ($p > 0.25$).

To analyse the population structure, the following three models were compared: no admixture, admixture with fixed alpha = 1, and admixture with variable alpha in the MavericK software [52]. The alpha parameter determines the likelihood of influx of genes from other studied populations; at alpha = 1, the influx of genes from any population is equally probable. The comparison showed that the admixture with the variable alpha model is the most reliable (Supplementary Figure S9-I).

Using the chosen model, the reliability values were calculated for genetic clusters with a possible number of clusters K from 1 to 10. The most reliable selection of four clusters is shown in Supplementary Figure S9-II. For K = 4, the Q-matrix of the distribution of groups of genotypes by sample was calculated and visualized, and a histogram of standard errors of such a distribution was drawn (Figure 3). The error values did not exceed 0.05, which indicates a high reliability of the obtained structure. All genetic clusters are present in all studied populations. There are populations with one dominant cluster (Kar, Tul, Ish, and Sks).

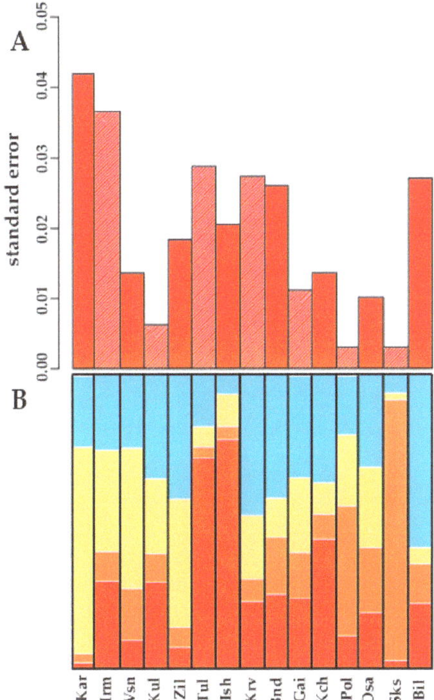

Figure 3. Standard error of calculating genetic structure (**A**) and genetic structure of populations (**B**). Each colour represents genotypes by sample.

4. Discussion

4.1. Polymorphism of Microsatellite Markers in 15 Populations of L. sibirica

In our study of 15 populations of the Siberian larch western race in the Northern, Middle, and Southern Urals, eight SSR loci were used to analyse the polymorphisms of microsatellite loci, which were polymorphic. High values of the genetic diversity of the studied populations were also obtained ($Na = 6.942$; $Ne = 4.017$; $Ho = 0.461$; $He = 0.623$). When compared with the data of other studies of larch and related species from other growing regions, the values of the obtained parameters generally did not show large differences and are at the general level of genetic diversity of larch trees and are slightly inferior only to the populations of Gmelin larch in the Far East [27,31,53].

4.2. Analysis of the Genetic Structure and Differentiation of the Populations of Siberian Larch Studied

The populations studied were divided into three groups according to their geographical location (South, Middle, and North Urals). The Sks sample was an exception and grew in the protected areas "Larch Grove" and separated into an individual group. We assume that this sample may be of artificial origin, which makes it different from other region populations. In addition, analysis of the population structure showed that one genotype is significantly dominant in the Sks sample, which may be a consequence of the sample origin from a group of closely related trees. A similar pattern is observed in the Tul, Ish, and Kar populations. The first two (Tul, Ish) are the most northern of the populations studied and grew at a higher altitude (from 680 to 916 meters). In this regard, the dominance of individual genotypes may be due to the action of long-term natural selection and adaptation to high-altitude conditions. The Kar sample grows in the region that represents an ecological disaster zone due to the emission of pollutants from the production of copper ore [54]. Such anthropogenic load could serve as a key factor for selection and affect the structure of the population's gene pool. Similar results were obtained by us previously using the ISSR method, but a high degree of differentiation was revealed, which is probably associated with the analysis of different structural elements of genomes using SSR and ISSR methods. According to molecular variability (AMOVA) analysis, the main part of genetic diversity falls on the individual heterozygosity of individuals (60%), while interpopulation (5%) and interregional (2%) differences are minimal. This revealed that our study level of subdivision of the western race *L. sibirica* populations for the entire study region (Northern, Middle, and Southern Urals) was mainly higher than that for Siberian larch (and other species of the genus *Larix*) from other growing areas. Thus, 8.9% of all observed variability falls on the interpopulation component, while this indicator in other studies ranges from 6% to 7% and reaches 8% only for *L. gmelinii* in the Far East [27,31,36,53]. Studies of *L. decidua* revealed an almost complete absence of intraspecific differentiation of the studied populations of Romania [31]. Studies of the genetic structure of larch in the Urals using isozyme markers showed a lower degree of differentiation (6.1%) compared to SSR, which can be explained by the fact that different structural elements of the genome were studied [36]. All the populations of *L. sibirica* we studied are characterized by a deficit of heterozygous genotypes at approximately the same level (on average, a 26% deficit within a population and a 32% deficit of heterozygotes between populations were found), as in previous studies of other species of the genus *Larix*. The populations of the Middle Urals have the greatest subdivision (F_{ST} = 7.1%), while in the North and South, they have similar values of about 5%. High differentiation may be a consequence of the fragmented range of *L. sibirica* in the studied region due to the intense anthropogenic load in the last century [36]. This was also confirmed by the migration test, which did not reveal a significant gene flow between the populations. The species range fragmentation reduces the possibilities of gene flow between populations when mixing their gene pools, causes genetic drift, and increases the frequency of closely related crossing [55]. All these processes reduce the adaptive potential of trees [56] due to the influence on the genetic diversity and structure of populations [57]. We revealed the relationship between geographical and genetic distances, but only for mountain populations (Irm, Kul, Kch, Tul, Ish). It is probable that climatic differences, which affect the time of tree dusting in different populations, also contribute to the differentiation. In turn, climatic differences are determined not only by the latitudinal-longitudinal location, but also by the surrounding relief peculiarities, including the proximity of the Ural Mountains. This was partially confirmed by the positive correlation between climatic and genetic distances in the Southern group of populations (r^2 = 0.5989, p = 0.08). Different environmental conditions determine the population structure complexity of any species. At the same time, the complex spatial structure of populations only partially affects their differentiation, while most of the interpopulation differences have a complex multicomponent nature. The uniqueness, diversity, and historical factors of the natural systems development of the Urals also determine the population structure complexity of *L. sibirica* in the region. In addition, the decrease in forest areas and their fragmentation lead to increases in carbon dioxide release into the atmosphere [2], which can accelerate temperature increase on the planet. The data obtained will contribute to the adoption of measures for the conservation of populations of one of the most common species of the genus *Larix*. The revealed genotypes and alleles that are resistant to various environmental factors under conditions of the species range fragmentation will lay the foundation for further marker-mediated selection to create plantations that are resistant to climate change. Thus, our studies have shown that the studied populations of *L. sibirica* in the Urals are relatively highly differentiated. This is due to the high degree of area fragmentation, especially in the Middle Urals. The level of genetic diversity comparable to that in other populations of Siberian larch in Siberia and in other representatives of the genus *Larix* [25,29,51]. Based on knowledge about the variability nature and population structure, it is recommended to select populations of Irm and Pol as having sufficient genetic diversity to preserve the *L. sibirica* gene pool in the region on a population basis. Attention should also be paid to the Kar sample, which grows in a zone of high anthropogenic load, and the Ish and Tul populations, which grow in high-altitude conditions. The data obtained are required for development of conservation and restoration programs for *L. sibirica*, which is especially important in the Middle Urals (a region with strong forest fragmentation). In this region, there is a local spatial structuredness of the gene pools of populations and a general trend toward a decrease in genetic diversity in isolated populations. Further fragmentation of the range and an increase in the number of spatially isolated habitats with extremely small sizes and different genetic pools may pose a threat to the long-term survival of populations and the sustainability of forestry. To preserve the genetic structure of Siberian larch populations, during reforestation, it is necessary to use seed material taken from genetically close populations and to maintain the genetic diversity of populations at a high level to preserve the adaptive potential of the species. The revealed patterns are of interest when planning and conducting forestry activities.

Supplementary Materials: The following are available online at https://www.mdpi.com/article/10.3390/f12101401/s1, Table S1: Studied populations of *L. sibirica*, Table S2: STRUCTURE startup options, Table S3: MavericK startup options, Table S4: Assessment of genetic interregional and interpopulation variability of *L. sibirica* populations based on the results of molecular variance analysis (AMOVA), Table S5: Pairwise F_{ST} matrix for 15 populations of Siberian larch in the Urals, Figure S1: Schematic map of the location of the studied populations of *L. sibirica*, Figure S2: Results of amplification of *L. sibirica* DNA with primer bcLK228 under optimal conditions. Figure S3: Chromatograms of electrophoretic separation of amplification products, Figure S4: STRUCTURE genetic migration test for *L. sibirica*, Figure S5: Average number of pairwise differences, Figure S6: Genetic structure of *L. sibirica* populations, Figure S7: Graph of dependence of genetic and geographical distances, Figure S8: Mantel test for the WorldClim data and Goldstein distance, Figure S9: Probabilistic models for the MavericK program.

Author Contributions: Data curation, S.B. and Y.V.; formal analysis and investigation, N.C., A.Z. and Y.N.; methodology, R.K.; resources, S.B.; supervision, S.B. and R.K.; validation, S.B. and Y.V.; statistical analysis, N.C., A.Z. and Y.N.; writing—original draft, S.B., R.K. and Y.V.; writing—review and editing, S.B. and R.K. All authors have read and agreed to the published version of the manuscript.

Funding: This study was funded within the framework of state assignment no. FSNF-2020-0008 of the Federal State Autonomous Educational Institution for Higher Education "Perm State National Research University" in science and by the Government of Perm Krai, research project No. C-26/174.3 dated 31 January 2019.

Data Availability Statement: Data is contained within the article or Supplementary Material.

Acknowledgments: The authors wish to thank Derek Ho (The University of Helsinki Language Centre) for outstanding editing. Open access funding provided by the University of Helsinki, including the Helsinki University Library.

Conflicts of Interest: The authors declare no conflict of interest.

References

1. Hamrick, J.L.; Godt, M.J.W.; Gonzales, E. Conservation of genetic diversity in old-growth forest communities of the southeastern United States. *Appl. Veg. Sci.* **2006**, *9*, 51–58. [CrossRef]
2. FAO. *Global Forest Resources Assessment: Main Report*; FAOSTAT: Rome, Italy, 2020; p. 184.
3. Balloux, F.; Lugon-Moulin, N. The estimation of population differentiation with microsatellite markers. *Mol. Ecol.* **2002**, *11*, 155–165. [CrossRef] [PubMed]
4. Porth, I.; El-Kassaby, Y. Assessment of the Genetic Diversity in Forest Tree Populations Using Molecular Markers. *Diversity* **2014**, *6*, 283–295. [CrossRef]
5. Yanbaev, Y.; Sultanova, R.; Blonskaya, L.; Bakhtina, S.; Tagirova, A.; Tagirov, V.; Kulagin, A. Gene pool of Scots pine (*Pinus sylvestris* L.) under reforestation in extreme environment. *Wood Res.* **2020**, *65*, 459–470. [CrossRef]
6. Schmidt, W.C. Around the World with Larix: An Introduction. *Ecol. Manag. Larix For. Look Ahead* **1995**, 6–18.
7. Brandt, J.P.; Flannigan, M.D.; Maynard, D.G.; Thompson, I.D.; Volney, W.J.A. An introduction to Canada's boreal zone: Ecosystem processes, health, sustainability, and environmental issues. *Environ. Rev.* **2013**, *21*, 207–226. [CrossRef]
8. Semerikov, V.L. Population structure and molecular systematics of Larix Mill. species. *Ext. Abstr. Dr. (Biol.) Diss.* **2006**.
9. Fedorov, V.S.; Ryazanova, T.V. Bark of Siberian Conifers: Composition, Use, and Processing to Extract Tannin. *Forests* **2021**, *12*, 1043. [CrossRef]
10. Liu, X.; Chen, W.; Liu, Q.; Dai, J. Abietic acid suppresses non-small-cell lung cancer cell growth via blocking IKKbeta/NF-kappaB signaling. *Onco. Targets Ther.* **2019**, *12*, 4825–4837. [CrossRef]
11. Talevi, A.; Cravero, M.S.; Castro, E.A.; Bruno-Blanch, L.E. Discovery of anticonvulsant activity of abietic acid through application of linear discriminant analysis. *Bioorg. Med. Chem. Lett.* **2007**, *17*, 1684–1690. [CrossRef]
12. Ito, Y.; Ito, T.; Yamashiro, K.; Mineshiba, F.; Hirai, K.; Omori, K.; Yamamoto, T.; Takashiba, S. Antimicrobial and antibiofilm effects of abietic acid on cariogenic Streptococcus mutans. *Odontology* **2020**, *108*, 57–65. [CrossRef]
13. Cheliak, W.M.; Pitel, J.A. Inheritance and Linkage of Allozymes in Larix laricina. *Silvae Genet.* **1985**, *34*, 142–147.
14. Fins, L.; Seeb, L.W. Genetic Variation in Allozymes of Western Larch. *Can. J. For. Res.* **1986**, *16*, 1013–1018. [CrossRef]
15. Lewandowski, A.; Berczyk, J.; Mejnartowicz, L. Genetic Structure and the Mating System in an Old Stand of Polish Larch. *Silvae Genet.* **1991**, *40*, 75–79.
16. Semerikov, V.L.; Matveev, A.V. Investigation of Genetic Variation of Allozyme Loci in Siberian Larch Larix sibirica Ledb. *Russ. J. Genet.* **1995**, *34*, 944–949.
17. Shigapov, Z.K.; Putenikhin, V.P.; Shigapova, A.I.; Urazbakhtina, K.A. Genetic Structure of the Ural Populations of Larix sukaczewii. *Russ. J. Genet.* **1998**, *34*, 54–63.
18. Larionova, A.Y.; Yakhneva (Oreshkova), N.V.; Kuz'mina, N.A. Genetic Variation of Siberian Larch in the Lower Angara River Basin. *Lesovedenie* **2003**, *4*, 17–22.
19. Oreshkova, N.V.; Barchenkov, A.P. Genetic Peculiarities and Morphological Variability of Siberian Larch in Altai–Sayany Mountain Region. *Vestn. Krasn. Gos. Agrar. Univ.* **2004**, *40*, 1127–1134.
20. Larionova, A.; Iakhneva, N.V.; Abaimov, A.P. Genetic diversity and differentiation of Gmelin larch Larix gmelinii populations from Evenkia (Central Siberia). *Genetika* **2004**, *40*, 1370–1377. [CrossRef] [PubMed]
21. Babushkina, E.A.; Vaganov, E.A.; Grachev, A.M.; Oreshkova, N.V.; Belokopytova, L.V.; Kostyakova, T.V.; Krutovsky, K.V. The effect of individual genetic heterozygosity on general homeostasis, heterosis and resilience in Siberian larch (Larix sibirica Ledeb.) using dendrochronology and microsatellite loci genotyping. *Dendrochronologia* **2016**, *38*, 26–37. [CrossRef]
22. Oreshkova, N.V.; Putintseva, Y.A.; Sharov, V.V.; Kuzmin, D.A.; Krutovsky, K.V. Development of microsatellite genetic markers in Siberian larch (Larix sibirica Ledeb.) based on the de novo whole genome sequencing. *Russ. J. Genet.* **2017**, *53*, 1194–1199. [CrossRef]
23. Li, S.; Ramakrishnan, M.; Vinod, K.K.; Kalendar, R.; Yrjälä, K.; Zhou, M. Development and Deployment of High-Throughput Retrotransposon-Based Markers Reveal Genetic Diversity and Population Structure of Asian Bamboo. *Forests* **2019**, *11*, 31. [CrossRef]
24. Kalendar, R.; Muterko, A.; Boronnikova, S. Retrotransposable Elements: DNA Fingerprinting and the Assessment of Genetic Diversity. *Methods Mol. Biol.* **2021**, *2222*, 263–286. [CrossRef]
25. Voronova, A.; Rendon-Anaya, M.; Ingvarsson, P.; Kalendar, R.; Rungis, D. Comparative Study of Pine Reference Genomes Reveals Transposable Element Interconnected Gene Networks. *Genes* **2020**, *11*, 1216. [CrossRef]
26. Kalendar, R.; Schulman, A.H. Transposon-based tagging: IRAP, REMAP, and iPBS. *Methods Mol. Biol.* **2014**, *1115*, 233–255. [CrossRef]
27. Oreshkova, N.V.; Belokon, M.M.; Jamiyansuren, S. Genetic diversity, population structure, and differentiation of Siberian larch, Gmelin larch, and Cajander larch on SSR-marker data. *Russ. J. Genet.* **2013**, *49*, 178–186. [CrossRef]
28. Wagner, S.; Gerber, S.; Petit, R.J. Two highly informative dinucleotide SSR multiplexes for the conifer Larix decidua (European larch). *Mol. Ecol. Resour.* **2012**, *12*, 717–725. [CrossRef]
29. Chen, X.B.; Xie, Y.H.; Sun, X.M. Development and characterization of polymorphic genic-SSR markers in Larix kaempferi. *Molecules* **2015**, *20*, 6060–6067. [CrossRef] [PubMed]

30. Dong, M.; Wang, Z.; He, Q.; Zhao, J.; Fan, Z.; Zhang, J. Development of EST-SSR markers in Larix principis-rupprechtii Mayr and evaluation of their polymorphism and cross-species amplification. *Trees* **2018**, *32*, 1559–1571. [CrossRef]
31. Gramazio, P.; Plesa, I.M.; Truta, A.M.; Sestras, A.F.; Vilanova, S.; Plazas, M.; Vicente, O.; Boscaiu, M.; Prohens, J.; Sestras, R.E. Highly informative SSR genotyping reveals large genetic diversity and limited differentiation in European larch (Larix decidua) populations from Romania. *Turk. J. Agric. For.* **2018**, *42*, 165–175. [CrossRef]
32. Liewlaksaneeyanawin, C.; Ritland, C.E.; El-Kassaby, Y.A.; Ritland, K. Single-copy, species-transferable microsatellite markers developed from loblolly pine ESTs. *Theor. Appl. Genet.* **2004**, *109*, 361–369. [CrossRef]
33. Dylis, N.V. Siberian larch. In *Materials for Taxonomy, Geography and History*; Izdatelstvo MOIP: Moscow, Russia, 1947; p. 137.
34. Igoshina, K.N. *Larch in the Urals: Materials on the History of Flora and Vegetation of the USSR*; Nauka: Leningrad, Russia, 1963.
35. Putenikhin, V.P.; Martinsson, O. Present distribution of Larix sukaczewii Dyl. In *Russia*; The Swedish University of Agricultural Sciences, SLU: Umea, Russia, 1995.
36. Putenikhin, V.P.; Farukshina, G.G.; Shigapov, Z.K. Sukachev Larch in the Urals. In *Variability and Population Genetic Structure*; Nauka: Moscow, Russia, 2004.
37. Polezhayeva, M.A.; Semerikov, V.L. *Genetic Diversity of cpSSR Loci in Larix Genus over the Far East Areas*; Bulletin of the North-East Scientific Center, Russian Academy of Sciences, Far East Branch: Moscow, Russia, 2009; Volume 2, pp. 75–83.
38. Zhang, L.; Zhang, H.G.; Li, X.F. Analysis of genetic diversity in Larix gmelinii (Pinaceae) with RAPD and ISSR markers. *Genet. Mol. Res.* **2013**, *12*, 196–207. [CrossRef] [PubMed]
39. Adrianova, I.Y. Genetic variability and differentiation of Kamchatka, Sakhalin and Kuril larch trees. *Sib. For. J.* **2014**, *4*, 110–116.
40. Vasileva, Y.; Prishnivskaya, Y.; Chertov, N.; Zhulanov, A. Analysis of genetic diversity and structure of Urals populations of Western Race of Siberian larch (Larix sibirica Ledeb.) based on intermicrosatellite markers polymorphism. *Bull. Sci. Pract.* **2018**, *4*, 113–124. [CrossRef]
41. Vasilyeva, Y.S.; Sboeva, Y.V.; Boronnikova, S.V.; Chertov, N.V.; Beltyukova, N.N. Genetic diversity, genetic structure and differentiation of Siberian larch populations in the Urals. *Turczaninowia* **2020**, *23*, 67–82. [CrossRef]
42. Kalendar, R.; Boronnikova, S.; Seppanen, M. Isolation and Purification of DNA from Complicated Biological Samples. *Methods Mol. Biol.* **2021**, *2222*, 57–67. [CrossRef]
43. Isoda, K.; Watanabe, A. Isolation and characterization of microsatellite loci from Larix kaempferi. *Mol. Ecol. Notes* **2006**, *6*, 664–666. [CrossRef]
44. Excoffier, L.; Lischer, H.E. Arlequin suite ver 3.5: A new series of programs to perform population genetics analyses under Linux and Windows. *Mol. Ecol. Resour.* **2010**, *10*, 564–567. [CrossRef]
45. Peakall, R.O.D.; Smouse, P.E. genalex 6: Genetic analysis in Excel. Population genetic software for teaching and research. *Mol. Ecol. Notes* **2006**, *6*, 288–295. [CrossRef]
46. Hubisz, M.J.; Falush, D.; Stephens, M.; Pritchard, J.K. Inferring weak population structure with the assistance of sample group information. *Mol. Ecol. Resour.* **2009**, *9*, 1322–1332. [CrossRef]
47. Goldstein, D.B.; Ruiz Linares, A.; Cavalli-Sforza, L.L.; Feldman, M.W. An evaluation of genetic distances for use with microsatellite loci. *Genetics* **1995**, *139*, 463–471. [CrossRef]
48. Kivela, M.; Arnaud-Haond, S.; Saramaki, J. EDENetworks: A user-friendly software to build and analyse networks in biogeography, ecology and population genetics. *Mol. Ecol. Resour.* **2015**, *15*, 117–122. [CrossRef]
49. Kumar, S.; Stecher, G.; Li, M.; Knyaz, C.; Tamura, K. MEGA X: Molecular Evolutionary Genetics Analysis across Computing Platforms. *Mol. Biol. Evol.* **2018**, *35*, 1547–1549. [CrossRef]
50. Hammer, Ø.; Harper, D.A.; Ryan, P.D. PAST: Paleontological statistics software package for education and data analysis. *Palaeontol. Electron.* **2001**, *4*, 9.
51. Hijmans, R.J. Raster: Geographic Data Analysis and Modeling. In 2021. Available online: https://cran.r-project.org/web/packages/raster/index.html (accessed on 22 August 2021).
52. Verity, R.; Nichols, R.A. Estimating the Number of Subpopulations (K) in Structured Populations. *Genetics* **2016**, *203*, 1827–1839. [CrossRef] [PubMed]
53. Amyaga, E.; Nifontov, S. Selection of nuclear microsatellite loci for specific identification of Larix gmélinii Rupr. and comparison of genetic profiles of Larix to solve agricultural problems. *IOP Conf. Ser. Earth Environ. Sci.* **2019**, *316*, 012016. [CrossRef]
54. Decision, On measures for the development of non-ferrous metallurgy region for 1997–2005. In *Decision of the Government of the Chelyabinsk Oblast*; Government of the Chelyabinsk Oblast: Chelyabinsk, Russia, 1997; p. 16.
55. Kremer, A.; Hipp, A.L. Oaks: An evolutionary success story. *New Phytol.* **2020**, *226*, 987–1011. [CrossRef] [PubMed]
56. Cortes, A.J.; Restrepo-Montoya, M.; Bedoya-Canas, L.E. Modern Strategies to Assess and Breed Forest Tree Adaptation to Changing Climate. *Front. Plant. Sci.* **2020**, *11*, 583323. [CrossRef] [PubMed]
57. Vasilyeva, Y.; Chertov, N.; Nechaeva, Y.; Sboeva, Y.; Pystogova, N.; Boronnikova, S.; Kalendar, R. Genetic Structure, Differentiation and Originality of Pinus sylvestris L. Populations in the East of the East European Plain. *Forests* **2021**, *12*, 999. [CrossRef]

Article

Genetic Diversity, Structure, and Differentiation of *Pinus sylvestris* L. Populations in the East European Plain and the Middle Urals

Yana Sboeva [1], Nikita Chertov [1], Yulia Nechaeva [1], Alena Valeeva [1], Svetlana Boronnikova [1,*] and Ruslan Kalendar [2,3,*]

[1] Faculty of Biology, Perm State University, 614990 Perm, Russia
[2] National Laboratory Astana, Nazarbayev University, 53 Kabanbaybatyr Ave., Astana 010000, Kazakhstan
[3] Helsinki Institute of Life Science HiLIFE, University of Helsinki, Biocenter 3, Viikinkaari 1, FI-00014 Helsinki, Finland
* Correspondence: svboronnikova@yandex.ru (S.B.); ruslan.kalendar@helsinki.fi (R.K.); Tel.: +358-294158869 (R.K.)

Abstract: Genetic diversity is important for the long-term survival of species and plays a critical role in their conservation. To manifest the adaptive potential, it is necessary to preserve the allelic diversity of populations, including both typical and region-specific alleles. Molecular genetic analysis of 22 populations of Scotch pine (*Pinus sylvestris* L.; *Pinaceae*) in 10 subjects of the Russian Federation in the East European Plain and the Middle Urals was carried out. Molecular genetic analysis of 22 populations of *P. sylvestris* revealed 182 polymorphic PCR fragments. The studied populations are characterized by a medium level of genetic diversity. A high subdivision coefficient (G_{ST}) of the studied populations was established; the intensity was 0.559. At the same time, the level of subdivision differed for different regions; for the populations from the Middle Urals, it was 15.5% ($G_{ST} = 0.155$), and for the populations from the East European Plain, it was 55.8% ($G_{ST} = 0.558$). The dendrogram of genetic similarity shows five clusters of the studied populations of *P. sylvestris* according to their geographical location. The populations from the East European Plain are mostly characterized by typicality, while the populations from the Middle Urals, on the contrary, are more specific in gene pools. The use of the coefficient of genetic originality to identify populations with typical and specific alleles allows for solving the problem of selecting populations for the conservation of forest genetic resources. The data obtained on genetic diversity, and the structure of populations growing in areas of active logging, are important for determining the geographical origin of plant samples, which is an integral part of the control of illegal logging.

Keywords: inter simple sequence repeats (ISSR); genetic diversity; genetic structure; *Pinus sylvestris* L.

Citation: Sboeva, Y.; Chertov, N.; Nechaeva, Y.; Valeeva, A.; Boronnikova, S.; Kalendar, R. Genetic Diversity, Structure, and Differentiation of *Pinus sylvestris* L. Populations in the East European Plain and the Middle Urals. *Forests* **2022**, *13*, 1798. https://doi.org/10.3390/f13111798

Academic Editors: Panteleimon Xofis and Konstantinos Poirazidis

Received: 5 October 2022
Accepted: 27 October 2022
Published: 29 October 2022

Publisher's Note: MDPI stays neutral with regard to jurisdictional claims in published maps and institutional affiliations.

1. Introduction

Genetic diversity is important for the long-term survival of species and plays a critical role in their conservation [1–3]. To manifest the adaptive potential, it is necessary to preserve the allelic diversity of populations, including both typical and region-specific alleles [4]. Genetic diversity also plays an important role in the response of populations to environmental factors [5,6]. Thus, knowledge about the genetic diversity of populations and their underlying individual and subpopulation components is important for the conservation and rational use of genetic resources, especially in the context of global climate change [7].

Coniferous forests form the basis of boreal ecosystems and are of enormous economic importance. They have a huge local and global impact on ecosystems, playing an important role in the regulation of water flow and soil conservation, being the most important part of the carbon cycle and a tool for cleaning atmospheric air from pollution [8–10]. In addition, the tissues of coniferous plants are rich in biologically active compounds (BAC), such as terpenoids, steroids, alkaloids, flavonoids, a complex of polysaccharides (holocellulose), and others, which are promising raw materials for the pharmaceutical industry [11,12].

Scotch pine (*Pinus sylvestris* L.) is the second most common coniferous species in the world and is of great economic and ecological importance [13]. Pine forests cover 37% of the total land area in the world and about 70% of the land area in the Northern Hemisphere, making Scotch pine one of the most important forest-forming species [13]. The current area of this species is the result of re-colonization events and post-glacial shrinkage of a once-larger distribution area [14,15]. *P. sylvestris* is a species tolerant of a wide range of ecological habitats and plays an important economic and ecological role in the forest ecosystems of Europe [8,9]. Scotch pine has a high genetic diversity that determines quantitative, qualitative, and adaptive traits [16].

Molecular genetic studies of *P. sylvestris* have been carried out both in Europe and Russia [7,12–19]. The results of the analysis of mitochondrial markers indicate that in the entire space from the east of the East European Plain, at least to the Yenisei River, the species is almost genetically homogeneous [18]. However, in the Mediterranean and the southern part of species distribution, there is a significantly greater differentiation of populations [19]. At the same time, analysis of chloroplast DNA revealed a significant genetic heterogeneity of the species throughout its distribution area. This difference can be associated with the use of different types of DNA markers, with different types of inheritance, and is also a consequence of different genetic processes. However, the genetic conservatism of the spacer of internal transcribed ribosomal genes and sequences of chloroplast genes for a particular species makes it unsuitable to study these sequences for determining intraspecific diversity. Molecular markers based on non-coding regions of DNA are usually highly variable and thus provide high-resolution information about genetic diversity within populations and the genetic structure of populations [20–23]. The method of studying the genetic diversity of woody plant species using ISSR (Inter Simple Sequence Repeats) is simple PCR and accessible technique [24]. Due to the large number of copies of microsatellite sequences and their large number in eukaryotic genomes, the use of SSR (Simple Sequence Repeats) sequences as an efficient PCR-based DNA fingerprinting method is convenient [25–27].

Scotch pine is widely used in economic activities, and its timber is actively harvested. To draw up programs for the rational use of forest resources, knowledge about the genetic diversity, and the structure of populations growing in areas of active logging, obtained by identifying polymorphic DNA fragments, is necessary.

In this regard, the study of the molecular genetic diversity and genetic structure of the *P. sylvestris* populations of the Middle Urals and the East European Plain using the ISSR profiling PCR method is promising for the development and optimization of protocols for assessing the state of the gene pools of boreal coniferous species. In addition, this approach is effective for the selection of objects to conserve species of coniferous plants that are productive and resistant to various environmental factors.

Thus, the present work is aimed at a detailed study of the genetic diversity, genetic structure, and differentiation of natural populations of *P. sylvestris* under the conditions of their natural growth over a large area (about 55470 thousand ha.) in two distinct regions, the East European Plain and the Middle Urals.

2. Materials and Methods

2.1. Materials

Twenty-two populations of Scotch pine were chosen as objects of research (*Pinus sylvestris* L.; *Pinaceae*). The studied populations of *P. sylvestris* in Perm Krai were taken from the locations of Kochyovo's (*PS_KOCh*) Gainy's (*PS_SOSN*), Kishert's (*PS_KISH*), Kudymkar's (*PS_LENI*), Cherdyn's (*PS_ChER*) and Berezniki's (*PS_ROMA*) forestries; in the Komi Republic from the locations Lokchim's (*PS_LOKC*) and Sysolsky's (*PS_SYSO*) forestries; in Arkhangelsk Oblast from Krasnoborsk's (*PS_KRAS*) forestry; in Vologda Oblast from Velikoustyugsky's (*PS_VELI*) forestry; in Kostroma Oblast from Pyshchugsky's

(*PS_PYSH*) forestry; in Nizhny Novgorod Oblast from Urensky's (*PS_UREN*) forestry; in Mari El Republic from Korotni's (*PS_KORO*) and Kokshaysk's (*PS_KOKSh*) forestries; in Chuvash Republic from Kirsk's (*PS_KIRS*) forestry; in Ulyanovsk Oblast from Inzensky's (*PS_INZE*) forestry; and in Kirov Oblast from the locations Shabalinsky's (*PS_SHAB*), Yezhikhinsk's (*PS_YEZH*), Darovskoy's (*PS_DARO*), Yuryansky's (*PS_YURY*), Slobodskoy's (*PS_SLOB*) and Belokholunitsky's (*PS_BELO*) forestries (Table S1, Figure 1 and Figure S1).

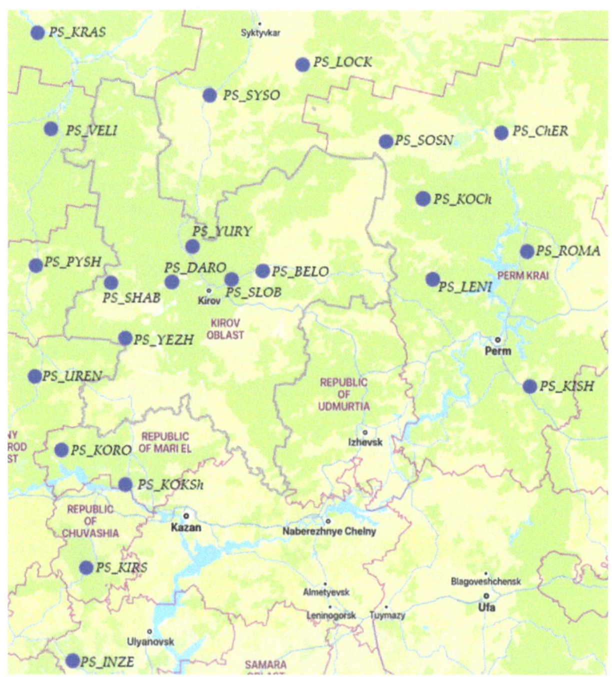

Figure 1. Schematic map of the location of the studied populations of *P. sylvestris*: *PS_KOCh*—Kochyovo; *PS_SOSN*—Gainy; *PS_KISH*—Kishert; *PS_LENI*—Kudymkar; *PS_ChER*—Cherdyn; *PS_ROMA*—Berezniki; *PS_LOKC*—Lokchim; *PS_SYSO*—Sysolsky; *PS_KRAS*—Krasnoborsk; *PS_VELI*—Velikoustyugsky; *PS_PYSH*—Pyshchugsky; *PS_UREN*—Urensky; *PS_KORO*—Korotni; *PS_KOKSh*—Kokshaysk; *PS_KIRS*—Kirsk; *PS_INZE*—Inzensky; *PS_SHAB*—Shabalinsky; *PS_YEZH*—Yezhikhinsk; *PS_DARO*—Darovskoy; *PS_YURY*—Yuryansky; *PS_SLOB*—Slobodskoy; *PS_BELO*—Belokholunitsky.

The collection of plant material was carried out from trees located at a distance of at least 50–150 m from each other [17,28,29]. The geographic distances between the populations varied from a minimum of 49 km (*PS_SHAB* and *PS_DARO* located in Shabalinsky's and Darovskoy's forestries) to a maximum of 941 km between the *PS_ChER* and *PS_INZE* populations. Pairwise geographic distances between all studied populations are presented in Table S2.

2.2. DNA Extraction and PCR

For research, samples of plant tissues were collected individually from 30 to 46 trees in each of the studied populations. DNA isolation was carried out according to the method for complex biological samples [30]. The weight of each sample of dry plant material was approximately 20 mg. To determine the concentration and quality of DNA we used NanoDrop™ 2000 Spectrophotometer (Thermo Fisher Scientific Inc., Waltham, MA, USA). For the Polymerase Chain Reaction (PCR), the DNA concentration of each sample was leveled to 10 ng/μL.

The ISSR profiling analysis method was used to assess the genetic diversity and genetic structure of populations [24,31]. PCR reactions were performed in a 25 µL reaction mixture containing 25 ng of template DNA, 1 × PCR buffer with 2.5 mM MgCl$_2$, 1 µM ISSR primer, 0.25 mM each dNTP, and 2 U Taq DNA polymerase (Sileks M, Russia). PCR amplification was carried out in a SimpliAmp™ Thermal Cycler (Thermo Fisher Scientific Inc., Waltham, MA, USA) under the following conditions: initial denaturation step at 94 °C for 2 min, followed by 32 amplifications at 94 °C for 20 s, at 52–64 °C (depending on primer sequence) [32] for 30 s, and at 72 °C for 60 s, followed by a final extension of 72 °C for 3 min (Table 1) [33].

Table 1. The information of ISSR primers is used to assess the genetic diversity of *P. sylvestris*.

Primer ID	Sequence 5′–3′	Tm (°C)	Ta (°C) *	Total Bands	PIC *
ISSR-1 ((AC)$_8$T)	ACACACACACACACACT	55.0	56	31	0.196
CR-212 ((CT)$_8$TG)	CTCTCTCTCTCTCTCTTG	55.9	56	43	0.260
CR-215 ((CA)$_6$GT)	CACACACACACAGT	52.6	56	33	0.256
M27 ((GA)$_8$C)	GAGAGAGAGAGAGAGAC	54.9	52	33	0.261
X10 ((AGC)$_6$C)	AGCAGCAGCAGCAGCAGCC	72.4	64	42	0.224

* Ta—optimal annealing temperature; Polymorphism Information Content.

All primers were tested to assess the genetic diversity of *P. sylvestris* using PCR amplification for ISSR profiling. PCR products were separated by electrophoresis at 70 V for 5 h in 1.5% agarose gel with 1xTBE buffer, stained with ethidium bromide, and photographed in transmitted ultraviolet light using Gel Doc XR+ Gel Documentation System (Bio-Rad Laboratories, Inc., Hercules, CA, USA) gel documentation system. To determine the length of DNA fragments, a molecular weight marker (100 bp + 1.5 + 3 Kb DNA Ladder, LLC. SibEnzim-M, Moscow, Russia) and the Quantity One 1-D Analysis Software (Bio-Rad Laboratories, Inc., Hercules, CA, USA) were used. In total, polymorphism of ISSR markers with 5 primers was analyzed within 922 individual *P. sylvestris* trees. To check the reliability of the obtained results, a PCR was performed at least three times.

2.3. Data Analysis

To quantify the genetic polymorphism and determine the genetic structure of the twenty-two populations studied, the data were presented in the form of a matrix of binary characters, in which the presence or absence of fragments of the same size in the spectra was considered, respectively, as 1 or 0 state.

Computer processing of the data was performed using the specialized macro GenAlEx for MS Excel to determine the number of alleles (n_a), effective (n_e) number of alleles [34], and expected (*He*) heterozygosity and Shannon's information index (*I*). The following parameters calculated in the POPGENE 1.31 software were used to describe the genetic structure of populations [35]: the expected proportion of heterozygous genotypes in the entire population as a measure of total genetic diversity (H_T); the expected proportion of heterozygous genotypes in a subpopulation, as a measure of intrapopulation diversity (H_S); share of interpopulation genetic diversity in total diversity or the coefficient of gene differentiation (G_{ST}); and AMOVA (Analysis of Molecular Variance) with the calculation of the *PhiPT* index (population subdivision index) using 1000 rounds of permutations [36]. Genetic distances between populations (D_N) were determined using the method of M. Nei [37]. To determine the correlation between pairwise genetic distances (D_N and *PhiPT*) and geographic distances in the general population group, the commonly used Mantel test was used. Regression analysis was carried out in MS Excel.

Based on the binary trait matrix, a genetic distance matrix was calculated, based on which dendrograms reflecting the degree of similarity between the studied populations and trees were generated by the spectrum using the MEGA X program [38].

To identify the structure of intrapopulation diversity, we used the indicator of the proportion of rare alleles (h). In addition, Principal Coordinates Analysis (PCA), implemented in the PAST 4.10 program, was performed to verify the obtained data. In the PAST 4.10 program [39], a detailed dendrogram was constructed for all trees using the Neighbor-joining method, and analysis and visualization were performed using the UMAP (Uniform Manifold Approximation and Projection) method [40]. The specificity of the gene pools of *P. sylvestris* populations was characterized using the genetic originality coefficient (GOC), which makes it possible to characterize populations in terms of the proportion of rare and typical alleles [4].

3. Results

3.1. Genetic Diversity of P. sylvestris

Molecular genetic analysis of twenty-two populations of *P. sylvestris* revealed 182 PCR fragments. The primers used detected between 31 and 43 PCR fragments, and the maximum number of fragments was amplified with the primer CR-212 [(CA)$_6$GT]. On average, one primer identified 36 PCR fragments. The lengths of PCR fragments varied from 150 to 1600 base pairs. The greatest genetic diversity was observed in the populations *PS_DARO* ($I = 0.249$; $H_e = 0.164$; $n_e = 1.268$) from Darovskoy's forestry, *PS_YURY* ($I = 0.263$; $H_e = 0.176$; $n_e = 1.299$) from Yuryansky's forestry and *PS_BELO* ($I = 0.264$; $H_e = 0.177$; $n_e = 1.303$) from Belokholunitsky's forestry. The lowest genetic diversity was observed in the populations *PS_PYSH* ($I = 0.096$; $H_e = 0.063$; $n_e = 1.106$) from Pyshchugsky's forestry, *PS_SHAB* ($I = 0.087$; $H_e = 0.056$; $n_e = 1.092$) from Shabalinsky's forestry, *PS_YEZH* ($I = 0.089$; $H_e = 0.057$; $n_e = 1.092$) from Yezhikhinsk's forestry, *PS_UREN* ($I = 0.092$; $H_e = 0.059$; $n_e = 1.092$) from Urensky's forestry (Table 2). No specific alleles were found in the populations studied (Figure S2).

Table 2. Genetic diversity of the studied populations of *P. sylvestris*.

Populations	H_e	n_e	I	Populations	H_e	n_e	I
PS_KOCh	0.131 (0.012)	1.199 (0.020)	0.213 (0.017)	PS_SHAB	0.056 (0.010)	1.092 (0.017)	0.087 (0.015)
PS_SOSN	0.119 (0.012)	1.182 (0.020)	0.194 (0.017)	PS_YEZH	0.057 (0.010)	1.092 (0.017)	0.089 (0.015)
PS_KISH	0.143 (0.013)	1.228 (0.023)	0.225 (0.019)	PS_UREN	0.059 (0.010)	1.092 (0.016)	0.092 (0.014)
PS_LENI	0.119 (0.011)	1.181 (0.020)	0.195 (0.017)	PS_KORO	0.122 (0.014)	1.207 (0.024)	0.182 (0.020)
PS_ChER	0.142 (0.012)	1.223 (0.022)	0.227 (0.018)	PS_KOKSh	0.117 (0.014)	1.204 (0.025)	0.172 (0.020)
PS_ROMA	0.125 (0.012)	1.194 (0.021)	0.202 (0.017)	PS_KIRS	0.121 (0.014)	1.206 (0.025)	0.180 (0.020)
PS_LOKC	0.112 (0.013)	1.187 (0.023)	0.169 (0.019)	PS_INZE	0.104 (0.013)	1.183 (0.025)	0.152 (0.019)
PS_SYSO	0.106 (0.013)	1.175 (0.023)	0.161 (0.019)	PS_DARO	0.164 (0.014)	1.268 (0.025)	0.249 (0.020)
PS_KRAS	0.129 (0.014)	1.216 (0.024)	0.195 (0.020)	PS_YURY	0.176 (0.015)	1.299 (0.027)	0.263 (0.021)
PS_VELI	0.118 (0.013)	1.193 (0.023)	0.180 (0.019)	PS_SLOB	0.151 (0.014)	1.249 (0.025)	0.230 (0.020)
PS_PYSH	0.063 (0.011)	1.106 (0.019)	0.096 (0.016)	PS_BELO	0.177 (0.015)	1.303 (0.027)	0.264 (0.021)
Total	0.119 (0.003)	1.195 (0.005)	0.183 (0.004)	Total	0.119 (0.003)	1.195 (0.005)	0.183 (0.004)

H_e—expected heterozygosity; n_e—effective number of alleles per locus; I—Shannon's information index; all of the above parameters have standard deviations in standard deviations given in brackets.

3.2. Population Genetic Structure of P. sylvestris

Analysis of the genetic structure of the studied *P. sylvestris* populations revealed that the expected proportion of heterozygous genotypes (H_T) per total sample was 0.270, while the expected proportion of heterozygous genotypes in a subpopulation (H_S) was 0.119. The population subdivision coefficient (G_{ST}) shows that the interpopulation component accounts for 0.559 of the total genetic diversity.

The values of pairwise *PhiPT* genetic distances detected by the AMOVA (Table S3) package ranged from 0.081 (*PS_KOCh/PS_KISH*) to 0.731 (*PS_SHAB/PS_INZE*). For the total sample of *P. sylvestris*, the *PhiPT* index was 0.647, which approximates the G_{ST} = 0.559. Analysis of molecular variability (AMOVA) showed that differences between regions account for 28% of diversity, differences between populations 37%, and intrapopulation differences account for 35% (Table 3). At the same time, the level of subdivision for different regions differed; for the populations from the Middle Urals, it was 15.5% (G_{ST} = 0.155), and for the East European Plain 55.8% (G_{ST} = 0.558).

Table 3. Assessment of genetic intra- and interpopulation variability in *P. sylvestris* populations by AMOVA.

Subdivision Indicator	df	SS	MS	Dispersion	%	*p*
Among groups	1	3442.465	3442.465	10.041	28%	<0.001
Among populations	20	11379.929	568.996	13.168	37%	<0.001
Within populations	900	11373.549	12.637	12.637	35%	<0.001

df—degrees of freedom, SS—the sum of squares, MS—standard deviation, %—the percentage of total genetic diversity, p—significance level when using 1000 rounds of permutation.

The smallest genetic distance was observed between populations *PS_KOCh/PS_KISH* (D_N = 0.012), the largest (D_N = 0.322) between populations *PS_SOSN* and *PS_SHAB* (Table S4). Based on the matrix of pairwise genetic distances (D_N), a cluster analysis was performed using the Neighbor-joining method, and a dendrogram reflecting the degree of similarity in the ISSR profiles of the populations studied was constructed (Figure 2). In the dendrogram, the studied populations were divided into five clusters in accordance with their geographical location: East (I), Center (II), North (III), South (IV), and West (V).

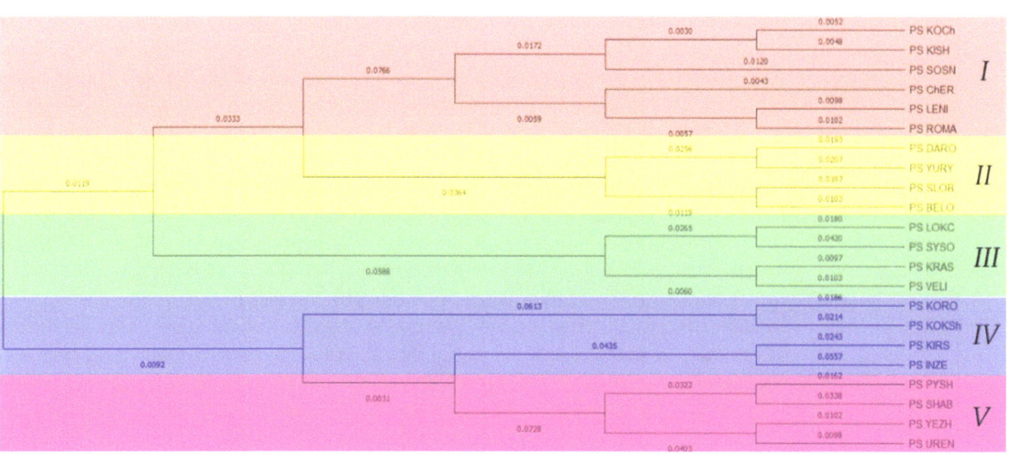

Figure 2. Dendrogram of genetic similarity of twenty-two studied populations of *P. sylvestris*, built based on polymorphism of ISSR profiles by Neighbor-joining method.

The separation of populations into five clusters is supported by the results of the Principal Coordinates Analysis (PCA), based on the *PhiPT* index calculated with the AMOVA package (Figure S3).

STRUCTURE analysis showed the presence of five groups of genotypes in the studied populations. The distribution of genotypes corresponds to the differentiation of populations according to the results of PCA analysis and the Neighbor-joining method (Figure S4).

Analysis of the population structure using UMAP, carried out for populations in the Middle Urals, indirectly confirms their low differentiation, $G_{ST} = 0.155$ (Figure 3). At the same time, the genotypes of individuals from the *PS_KOCh* population are distributed over all groups (Table 3).

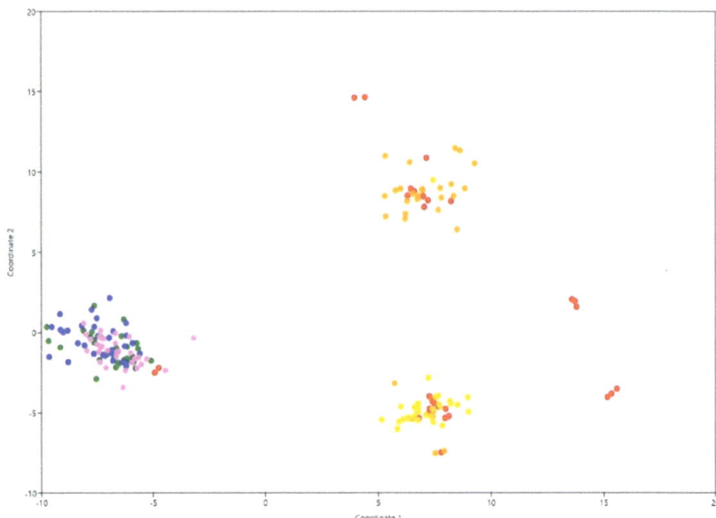

Figure 3. Distribution of individuals in the studied populations of *Pinus sylvestris* L. of the Middle Urals using UMAP.

During the study of *P. sylvesrtis* populations, their spatial and genetic structure was checked for consistency with the "isolation-by-distance" model by the Mantel test. Thus, a pairwise comparison of all twenty-two studied populations revealed the presence of a weak positive correlation ($R^2 = 0.2534$) between geographic and genetic distance (D_N) (Figure S5). A regression analysis was also carried out on these data, and a significant relationship was found, but the R^2 value (0.2534) showed that the correlation was weak (Table S5).

Using a specific and typical allele approach, populations in the studied regions were examined and characterized. The populations from the East European Plain are mostly characterized by the typicality of gene pools (GOC < 1.000); only the *PS_INZE* selection (GOC = 1.041) is characterized by the specificity of the gene pool. The populations from the Middle Urals, on the contrary, are more specific in gene pools (GOC > 1.200), the most specific gene pool belongs to the *PS_KISH* selection (GOC = 1.665).

The proportion of rare alleles indicator (h) assesses the structure of intrapopulation diversity, the lower the value h of the threshold (0.3) level, the more balanced the structure of diversity is characterized by populations. Of the studied populations, h was above the threshold value in *PS_SHAB* ($h = 0.309$) and *PS_YEZ* ($h = 0.304$). Two more populations were close to the threshold, *PS_PYSH* ($h = 0.289$) and *PS_UREN* ($h = 0.290$).

4. Discussion

4.1. Genetic Diversity of P. sylvestris

As a result of the study, a medium level of genetic diversity in *P. sylvestris* populations was revealed ($I = 0.183$; $H_e = 0.119$; $n_e = 1.195$), it is common for populations from the Southern part of the East European Plain, but less than the level of genetic diversity of the South Urals populations [41]. The greatest genetic diversity was revealed in the central populations, *PS_DARO*, *PS_YURY* and *PS_BELO*. The least genetic diversity was observed among the group of Western populations, *PS_PYSH*, *PS_SHAB*, *PS_YEZH* and *PS_UREN*; this may be due to anthropogenic pressure, in particular active logging. In addition, these populations also differ in a less balanced genetic structure in terms of the proportion of rare alleles (*h*), which in turn may also be due to active logging in these regions. The genetic structure and diversity of populations can be greatly affected by random genetic drift, which can lead to the erosion of genetic variation due to the loss of rare alleles [42,43]. No specific alleles were found in the populations studied, which may indicate that these populations are genetically homogeneous.

4.2. Population Genetic Structure of P. sylvestris

The studied populations were divided into five clusters, in accordance with their geographical location. At the same time, the populations from the Middle Urals were distinguished. Differentiation between the populations from the Urals and the East European Plain amounted to about a third (28%) of the observed genetic diversity, another third is due to the interpopulation component (37%), and a third to intrapopulation differences (35%). The data obtained indicate the origin of several genetically differentiated populations and their groups in *P. sylvestris* in the study areas. A similar pattern was observed in previous studies in these regions, on a smaller number of populations [33,44]. The high differentiation may be due to the fragmentation of the area of *P. sylvestris* in the region under study. Significant differentiation between the populations from the East European Plain and the populations from the Middle Urals (28%) may be related to the history of the distribution of the species in the Pleistocene. Populations of the Middle Urals were settled mainly from the South Ural refugium, while this refugium made a smaller contribution to the gene pools of the East European populations [45,46].

In general, there is a genetic homogeneity of *P. sylvestris* populations in the Middle Urals, which is confirmed by the analysis of UMAP and AMOVA. For the *PS_KOCh* population, according to the UMAP analysis, the distribution of genotypes over all identified groups is observed. This may be due to the fact that the population was settled from several different directions. The settlement of the territory of the Urals occurred mainly from the South Ural refugium, but in addition, the settlement came from the Balkan refugium and the refugia of the second order in South Siberia [46]. Correlation analysis between genetic and geographical distances revealed the presence of a medium relationship between them ($R^2 = 0.2534$).

The weak differentiation of the populations in the Middle Urals may be due to the similarity of the habitats of the populations since all the studied populations are located at 180–200 m above sea level. Similar genetic homogeneity of Scotch pine populations was also observed in the study of populations in the territory from the east of the East European Plain at least to the Yenisei River using mitochondrial markers [18].

Within the East European Plain, *P. sylvestris* populations, despite high differentiation, are characterized by typical gene pools. Populations from the Middle Urals are characterized by specific gene pools. Populations with a specific gene pool, such as the selection from the Kishert's forestry (*PS_KISH*), can serve as a source of genetic diversity in reforestation programs. Additionally, populations with a more typical gene pool, having the most common alleles in the region, can be preserved as forest genetic reserves to preserve the genetic resources of the species. Populations from the East European Plain with the most typical gene pools can serve as an example of such populations. Data on the typicality and specificity of gene pools, as well as on the differentiation of populations, can be used in further studies of *P. sylvestris* in the study region.

The obtained data on genetic diversity and the structure of populations growing in areas of active logging are important for drawing up programs for the rational use of forest resources, identifying populations, and determining the geographical origin of plant specimens, including timber, which is an integral part of controlling illegal logging.

The use of the coefficient of genetic originality to identify populations with typical and specific alleles makes it possible to solve the problem of selecting populations for the conservation of forest genetic resources.

Supplementary Materials: The following are available online at https://www.mdpi.com/article/10.3390/f13111798/s1, Supplementary Table S1. The studied populations of *P. sylvestris* were used in ISSR analysis. Supplementary Table S2. Pairwise geographic distances (km) between the studied populations of *P. sylvestris*. Supplementary Table S3. Paired PhiPT genetic distances between the studied populations of *P. sylvestris* by AMOVA. Supplementary Table S4. Pairwise genetic distances (DN) between the populations studied *Pinus sylvestris* L. Supplementary Table S5. Regression analysis of genetic and geographical distances. Supplementary Figure S1. Schematic map of the location of the studied populations of *P. sylvestris*. Supplementary Figure S2. Allele patterns of *P. sylvestris* populations. Supplementary Figure S3. Ordination of the studied populations of *P. sylvestris* using PCA, obtained on the basis of PhiPT matrix of genetic distances. Supplementary Figure S4. The structure of the distribution of genotypes in *P. sylvestris* populations. Supplementary Figure S5. Graph of dependence of genetic (DN) and geographical distances of *P. sylvestris* populations.

Author Contributions: Data curation, S.B. and Y.S.; formal analysis and investigation, N.C., A.V. and Y.N.; methodology, R.K.; resources, S.B.; supervision, S.B. and R.K.; validation, S.B. and Y.S.; statistical analysis, N.C. and Y.N.; writing—original draft, S.B., R.K. and N.C.; writing—review and editing, S.B. and R.K. All authors have read and agreed to the published version of the manuscript.

Funding: This study was funded within the framework of state assignment No. FSNF-2020-0008 of the Federal State Autonomous Educational Institution for Higher Education "Perm State National Research University" in science and by the Government of Perm Krai, research project No. C-26/776 dated 31 March 2022. Open access funding was provided by the University of Helsinki (Finland), including Helsinki University Library, via R.K. All funders provided financial support for the research but did not have any additional role in the study design, data collection, and analysis, decision to publish, or preparation of the manuscript.

Data Availability Statement: Data presented in this article are available on request from the corresponding author.

Conflicts of Interest: The authors declare no conflict of interest.

References

1. Frankham, R. How closely does genetic diversity in finite populations conform to predictions of neutral theory? Large deficits in regions of low recombination. *Heredity* **2012**, *108*, 167–178. [CrossRef] [PubMed]
2. Spielman, D.; Brook, B.W.; Frankham, R. Most species are not driven to extinction before genetic factors impact them. *Proc. Natl. Acad. Sci. USA* **2004**, *101*, 15261–15264. [CrossRef] [PubMed]
3. O'Grady, J.J.; Brook, B.W.; Reed, D.H.; Ballou, J.D.; Tonkyn, D.W.; Frankham, R. Realistic levels of inbreeding depression strongly affect extinction risk in wild populations. *Biol. Conserv.* **2006**, *133*, 42–51. [CrossRef]
4. Potokina, E.K.; Aleksandrova, T.G. Genetic singularity coefficients of common vetch *Vicia sativa* l. Accessions determined with molecular markers. *Russ. J. Genet.* **2009**, *44*, 1309–1316. [CrossRef]
5. Hoelzel, A.R.; Bruford, M.W.; Fleischer, R.C. Conservation of adaptive potential and functional diversity. *Conserv. Genet.* **2019**, *20*, 1–5. [CrossRef]
6. Razgour, O.; Forester, B.; Taggart, J.B.; Bekaert, M.; Juste, J.; Ibanez, C.; Puechmaille, S.J.; Novella-Fernandez, R.; Alberdi, A.; Manel, S. Considering adaptive genetic variation in climate change vulnerability assessment reduces species range loss projections. *Proc. Natl. Acad. Sci. USA* **2019**, *116*, 10418–10423. [CrossRef]
7. Pazouki, L.; Shanjani, P.S.; Fields, P.D.; Martins, K.; Suhhorutšenko, M.; Viinalass, H.; Niinemets, Ü. Large within-population genetic diversity of the widespread conifer *Pinus sylvestris* at its soil fertility limit characterized by nuclear and chloroplast microsatellite markers. *Eur. J. For. Res.* **2015**, *135*, 161–177. [CrossRef]
8. Hogberg, P.; Nordgren, A.; Buchmann, N.; Taylor, A.F.; Ekblad, A.; Hogberg, M.N.; Nyberg, G.; Ottosson-Lofvenius, M.; Read, D.J. Large-scale forest girdling shows that current photosynthesis drives soil respiration. *Nature* **2001**, *411*, 789–792. [CrossRef]
9. Lindén, A.; Heinonsalo, J.; Buchmann, N.; Oinonen, M.; Sonninen, E.; Hilasvuori, E.; Pumpanen, J. Contrasting effects of increased carbon input on boreal som decomposition with and without presence of living root system of *Pinus sylvestris* L. *Plant Soil* **2013**, *377*, 145–158. [CrossRef]
10. Pan, Y.; Birdsey, R.A.; Fang, J.; Houghton, R.; Kauppi, P.E.; Kurz, W.A.; Phillips, O.L.; Shvidenko, A.; Lewis, S.L.; Canadell, J.G.; et al. A large and persistent carbon sink in the world's forests. *Science* **2011**, *333*, 988–993. [CrossRef]
11. Liu, X.; Chen, W.; Liu, Q.; Dai, J. Abietic acid suppresses non-small-cell lung cancer cell growth via blocking ikkbeta/nf-kappab signaling. *Onco. Targets Ther.* **2019**, *12*, 4825–4837. [CrossRef] [PubMed]

12. Dering, M.; Kosiński, P.; Wyka, T.P.; Pers-Kamczyc, E.; Boratyński, A.; Boratyńska, K.; Reich, P.B.; Romo, A.; Zadworny, M.; Żytkowiak, R.; et al. Tertiary remnants and Holocene colonizers: Genetic structure and phylogeography of Scots pine reveal higher genetic diversity in young boreal than in relict Mediterranean populations and a dual colonization of Fennoscandia. *Divers. Distrib.* **2017**, *23*, 540–555. [CrossRef]
13. Floran, V.; Sestras, R.E.; GarcÍA Gil, M.R. Organelle genetic diversity and phylogeography of Scots pine (*Pinus sylvestris* L.). *Not. Bot. Horti Agrobot.* **2011**, *39*, 317–322. [CrossRef]
14. Hebda, A.; Wójkiewicz, B.; Wachowiak, W. Genetic characteristics of scots pine in Poland and reference populations based on nuclear and chloroplast microsatellite markers. *Silva Fenn.* **2017**, *51*, 1721. [CrossRef]
15. Tóth, E.G.; Vendramin, G.G.; Bagnoli, F.; Cseke, K.; Höhn, M. High genetic diversity and distinct origin of recently fragmented scots pine (*Pinus sylvestris* L.) populations along the Carpathians and the Pannonian basin. *Tree Genet. Genomes* **2017**, *13*, 47. [CrossRef]
16. Kavaliauskas, D.; Danusevičius, D.; Baliuckas, V. New insight into genetic structure and diversity of Scots pine (*Pinus sylvestris* L.) populations in Lithuania based on nuclear, chloroplast and mitochondrial DNA markers. *Forests* **2022**, *13*, 1179. [CrossRef]
17. Chertov, N.; Nechaeva, Y.; Zhulanov, A.; Pystogova, N.; Danilova, M.; Boronnikova, S.; Kalendar, R. Genetic structure of Pinus populations in the Urals. *Forests* **2022**, *13*, 1278. [CrossRef]
18. Vidyakin, A.I.; Semerikov, V.L.; Polezhaeva, M.A.; Dymshakova, O.S. Spread of mitochondrial DNA haplotypes in population of Scots pine (*Pinus sylvestris* L.) in northern European Russia. *Russ. J. Genet.* **2012**, *48*, 1267–1271. [CrossRef]
19. Sannikov, S.N.; Petrova, I.V. Phylogenogeography and genotaxonomy of *Pinus sylvestris* L. Populations. *Russ. J. Ecol.* **2012**, *43*, 273–280. [CrossRef]
20. Bouzat, J.L. The population genetic structure of the greater rhea (*Rhea americana*) in an agricultural landscape. *Biol. Conserv.* **2001**, *99*, 277–284. [CrossRef]
21. Grassi, F.; Imazio, S.; Failla, O.; Scienza, A.; Ocete Rubio, R.; Lopez, M.A.; Sala, F.; Labra, M. Genetic isolation and diffusion of wild grapevine Italian and Spanish populations as estimated by nuclear and chloroplast SSR analysis. *Plant Biol.* **2003**, *5*, 608–614. [CrossRef]
22. Pautasso, M. Geographical genetics and the conservation of forest trees. *Perspect. Plant Ecol. Evol. Syst.* **2009**, *11*, 157–189. [CrossRef]
23. Ribeiro, M.M.; Mariette, S.; Vendramin, G.G.; Szmidt, A.E.; Plomion, C.; Kremer, A. Comparison of genetic diversity estimates within and among populations of maritime pine using chloroplast simple-sequence repeat and amplified fragment length polymorphism data. *Mol. Ecol.* **2002**, *11*, 869–877. [CrossRef] [PubMed]
24. Zietkiewicz, E.; Rafalski, A.; Labuda, D. Genome fingerprinting by simple sequence repeat (SSR)-anchored polymerase chain reaction amplification. *Genomics* **1994**, *20*, 176–183. [CrossRef]
25. Kalendar, R.; Muterko, A.; Boronnikova, S. Retrotransposable elements: DNA fingerprinting and the assessment of genetic diversity. *Methods Mol. Biol.* **2021**, *2222*, 263–286. [CrossRef] [PubMed]
26. Kalendar, R.; Shustov, A.V.; Schulman, A.H. Palindromic sequence-targeted (pst) pcr, version 2: An advanced method for high-throughput targeted gene characterization and transposon display. *Front. Plant Sci.* **2021**, *12*, 691940. [CrossRef]
27. Kalendar, R.; Shustov, A.V.; Seppanen, M.M.; Schulman, A.H.; Stoddard, F.L. Palindromic sequence-targeted (PST) PCR: A rapid and efficient method for high-throughput gene characterization and genome walking. *Sci. Rep.* **2019**, *9*, 17707. [CrossRef] [PubMed]
28. Robledo-Arnuncio, J.J.; Gil, L. Patterns of pollen dispersal in a small population of *Pinus sylvestris* L. Revealed by total-exclusion paternity analysis. *Heredity* **2005**, *94*, 13–22. [CrossRef]
29. Ganatsas, P.; Tsakaldimi, M.; Thanos, C. Seed and cone diversity and seed germination of *Pinus pinea* in Strofylia site of the natura 2000 network. *Biodivers. Conserv.* **2008**, *17*, 2427–2439. [CrossRef]
30. Kalendar, R.; Boronnikova, S.; Seppanen, M. Isolation and purification of DNA from complicated biological samples. *Methods Mol. Biol.* **2021**, *2222*, 57–67. [CrossRef]
31. Kalendar, R.; Schulman, A.H. IRAP and REMAP for retrotransposon-based genotyping and fingerprinting. *Nat. Protoc.* **2006**, *1*, 2478–2484. [CrossRef] [PubMed]
32. Kalendar, R. A guide to using FASTPCR software for PCR, in silico PCR, and oligonucleotide analysis. *Methods Mol. Biol.* **2022**, *2392*, 223–243. [CrossRef] [PubMed]
33. Vasilyeva, Y.; Chertov, N.; Nechaeva, Y.; Sboeva, Y.; Pystogova, N.; Boronnikova, S.; Kalendar, R. Genetic structure, differentiation and originality of *Pinus sylvestris* L. Populations in the east of the east European plain. *Forests* **2021**, *12*, 999. [CrossRef]
34. Peakall, R.O.D.; Smouse, P.E. Genalex 6: Genetic analysis in Excel. Population genetic software for teaching and research. *Mol. Ecol. Notes* **2006**, *6*, 288–295. [CrossRef]
35. Francis, Y.; Rongcai, Y.; Timothy, B. Popgene. Available online: https://sites.ualberta.ca/~{}fyeh/popgene_info.html (accessed on 10 October 2022).
36. Nei, M.; Li, W.H. Mathematical model for studying genetic variation in terms of restriction endonucleases. *Proc. Natl. Acad. Sci. USA* **1979**, *76*, 5269–5273. [CrossRef]
37. Nei, M. *Molecular Evolutionary Genetics*; Columbia University Press: Chichester, NY, USA, 1987. [CrossRef]
38. Kumar, S.; Stecher, G.; Li, M.; Knyaz, C.; Tamura, K. Mega x: Molecular evolutionary genetics analysis across computing platforms. *Mol. Biol. Evol.* **2018**, *35*, 1547–1549. [CrossRef]

39. Hammer, Ø.; Harper, D.A.T.; Paul, D.R. Past: Paleontological statistics software package for education and data analysis. *Palaeontol. Electron.* **2001**, *4*, 9.
40. McInnes, L.; Healy, J.; Saul, N.; Großberger, L. UMAP: Uniform manifold approximation and projection. *J. Open Source Softw.* **2018**, *3*, 861. [CrossRef]
41. Khanova, E.; Konovalov, V.; Timeryanov, A.; Isyanyulova, R.; Rafikova, D. Genetic and selection assessment of the Scots pine (*Pinus sylvestris* L.) in forest seed orchards. *Wood Res.* **2020**, *65*, 283–292. [CrossRef]
42. Lande, R. Genetics and demography in biological conservation. *Science* **1988**, *241*, 1455–1460. [CrossRef]
43. Vucetich, J.A.; Waite, T.A. Spatial patterns of demography and genetic processes across the species' range: Null hypotheses for landscape conservation genetics. *Conserv. Genet.* **2003**, *4*, 639–645. [CrossRef]
44. Vidyakin, A.I.; Boronnikova, S.V.; Nechayeva, Y.S.; Pryshnivskaya, Y.V.; Boboshina, I.V. Genetic variation, population structure, and differentiation in Scots pine (*Pinus sylvestris* L.) from the northeast of the Russian plain as inferred from the molecular genetic analysis data. *Russ. J. Genet.* **2015**, *51*, 1213–1220. [CrossRef]
45. Sannikov, S.N.; Petrova, I.V.; Egorov, E.V.; Sannikova, N.S. Searching for and revealing the system of Pleistocene refugia for the species *Pinus sylvestris* L. *Russ. J. Ecol.* **2020**, *51*, 215–223. [CrossRef]
46. Sannikov, S.N.; Petrova, I.V.; Egorov, E.V.; Sannikova, N.S. A system of Pleistocene refugia for *Pinus sylvestris* L. In the southern marginal part of the species range. *Russ. J. Ecol.* **2014**, *45*, 167–173. [CrossRef]

forests

Article

Employing Genome-Wide SNP Discovery to Characterize the Genetic Diversity in *Cinnamomum camphora* Using Genotyping by Sequencing

Xue Gong [1,2], Aihong Yang [1], Zhaoxiang Wu [1], Caihui Chen [1], Huihu Li [1], Qiaoli Liu [1], Faxin Yu [1] and Yongda Zhong [1,*]

[1] Key Laboratory of Horticultural Plant Genetic and Improvement of Jiangxi Province, Institute of Biological Resources, Jiangxi Academy of Sciences, No. 7777, Changdong Road, Gaoxin District, Nanchang 330096, China; gx13864477189@163.com (X.G.); yang1ai2hong3@163.com (A.Y.); wuzhaoxiang@jxas.ac.cn (Z.W.); chenhui_1988@126.com (C.C.); lhh26521@163.com (H.L.); 15270996528@163.com (Q.L.); yufaxin@jxas.ac.cn (F.Y.)

[2] College of Forestry, Jiangxi Agricultural University, Nanchang 330045, China

* Correspondence: zhongyongda@jxas.ac.cn; Tel.: +86-791-8817-5720

Citation: Gong, X.; Yang, A.; Wu, Z.; Chen, C.; Li, H.; Liu, Q.; Yu, F.; Zhong, Y. Employing Genome-Wide SNP Discovery to Characterize the Genetic Diversity in *Cinnamomum camphora* Using Genotyping by Sequencing. *Forests* **2021**, *12*, 1511. https://doi.org/10.3390/f12111511

Academic Editors: Panteleimon Xofis and Konstantinos Poirazidis

Received: 12 October 2021
Accepted: 31 October 2021
Published: 1 November 2021

Abstract: *Cinnamomum camphora* (L.) J.Presl is a representative tree species of evergreen broad-leafed forests in East Asia and has exceptionally high economic, ornamental, and ecological value. However, the excessive exploitation and utilization of *C. camphora* trees have resulted in the shrinking of wild population sizes and rare germplasm resources. In this study, we characterized 171 *C. camphora* trees from 39 natural populations distributed throughout the whole of China and one Japanese population. We investigated genetic diversity and population structure using genome-wide single-nucleotide polymorphism (SNP) identified by genotyping by sequencing (GBS) technology. The results showed the genetic diversity of the *C. camphora* populations from western China > central China > eastern China. Moreover, the Japanese population showed the highest diversity among all populations. The molecular variance analysis showed 92.03% of the genetic variation within populations. The average pairwise F_{ST} was 0.099, and gene flow Nm was 2.718, suggesting a low genetic differentiation among populations. Based on the genetic clustering analysis, the 40 *C. camphora* populations clustered into three major groups: Western China, Central China, and Eastern China + Japan. Eastern China's population had the closest genetic relationship with the Japanese population, suggesting possible gene exchange between the two adjacent areas. This study furthers our understanding of the genetic diversity and genetic structure of *C. camphora* in East Asia and provides genetic tools for developing strategies of *C. camphora* germplasm utilization.

Keywords: *Cinnamomum camphora*; genotyping by sequencing (GBS); genetic diversity; population structure; single nucleotide polymorphism (SNP)

1. Introduction

Cinnamomum camphora is a representative tree species of tropical and subtropical evergreen broad-leafed forests of the genus *Cinnamomum* in the Lauraceae family. It is one of the "bulk" trees with three major uses, including a source of spice, timber, and gardening. *C. camphora* is widely distributed in subtropical China and central and southern Japan [1,2]. Subtropical China is the most abundant wild resource and the main area of production of natural *C. camphora* essential oil globally, accounting for more than 80% of the global production [3]. However, since the 1950s, the destructive harvesting of essential oil by digging the trees and grubbing their roots, and the increasing demand for urban vegetation has led to severe reductions of the resources. At present, the natural distribution of *C. camphora* is scattered, and its mature natural forests are rare, which results in endangered high-quality essential oil containing precious chemical types. Therefore, it is urgent to collect and preserve the germplasm resources of *C. camphora*.

A previous study by Kameyama et al. showed rich genetic diversity in *C. camphora* populations from China and Japan using simple sequence repeat (SSR) markers. The genetic diversity of *C. camphora* populations in China was slightly higher than that in Japan [2]. We previously used transcriptome-based expressed sequence tag–derived SSR (EST-SSR) markers to investigate the genetic diversity and genetic structure of *C. camphora* in southern China. This study showed a moderate genetic diversity, prominent population structure, and genetic differentiation existing mainly within a population [4,5]. The above results revealed the significant genetic differentiation among and within the populations of *C. camphora*. However, due to the large differences in the populations' sources and sizes, previous genetic diversity studies vary significantly. Additionally, the shortcomings of SSR markers, including the unclarity of homozygosity and heterozygosity and invalid alleles, can lead to misidentification of SSR genotypes [6].

Due to their low mutation frequency, fewer alleles, high genetic stability, and distribution throughout the genome, single nucleotide polymorphisms (SNP) have been favoured in plant genetic diversity and genomic analysis [7]. The rapid development of next-generation sequencing (NGS) technology has reduced the cost and improved the efficiency of SNP discovery, which provides great potential for high-throughput SNP genotyping in plant breeding [8]. Wu et al. firstly used the genotyping by sequencing (GBS) technique to identify and genetically analyze plant species [9]. GBS is an NGS-based SNP genotyping technology suitable for species with high diversity and large genomes [10,11]. Due to the lack of need for the reference genome, low cost, simple procedures, and the SNP carrying rich genome-wide information, GBS is widely used in the analysis of plant genetic diversity [12,13], the study of systematic evolution [14], molecular marker localization [15,16], and genetic map construction [17].

The whole genome of *C. camphora* has not been sequenced yet, and the genetic diversity and structure of the *C. camphora* populations across its entire natural distribution area are still unclear. This study characterized 171 *C. camphora* trees from 39 natural populations from the whole distribution region in China and one population in Japan using genome-wide SNP molecular markers obtained by the GBS technique. We aimed to reveal the genetic diversity and structure of *C. camphora* in different regions, understand the sources and patterns of genetic variation, and provide a theoretical basis for protection strategies and genetic improvement of *C. camphora* germplasm resources.

2. Materials and Methods

2.1. Plant Material and DNA Isolation

Leaf samples from 171 *C. camphora* trees were collected in this study from 39 natural populations in the whole range of distribution of *C. camphora* in China and one population in Japan. Sample sources were grouped into four regions based on geographical origin. There were three populations with a total of 14 samples from the western region in China, six populations with a total of 25 samples from the central region, 30 populations with 128 samples from the eastern region, and one population with four samples from Japan (Figure 1 and Table 1). Trees with a diameter at breast height greater than 60 cm were randomly selected, and the distance between them was greater than 50 m. Five healthy leaves were collected from each tree, dried with silica gel. Details on sampling were summarized in Table S1. DNA was extracted using the ionic detergent cetyltrimethylammonium bromide, and the purity and integrity of the DNA were detected by 1% agarose gel electrophoresis. The DNA quality and quantity were determined by a NanoDrop 2000 (Thermo Fisher Scientific Inc., Waltham, MA, USA) and a Qubit® 2.0 Fluorometer (Invitrogen, Carlsbad, CA, USA), respectively.

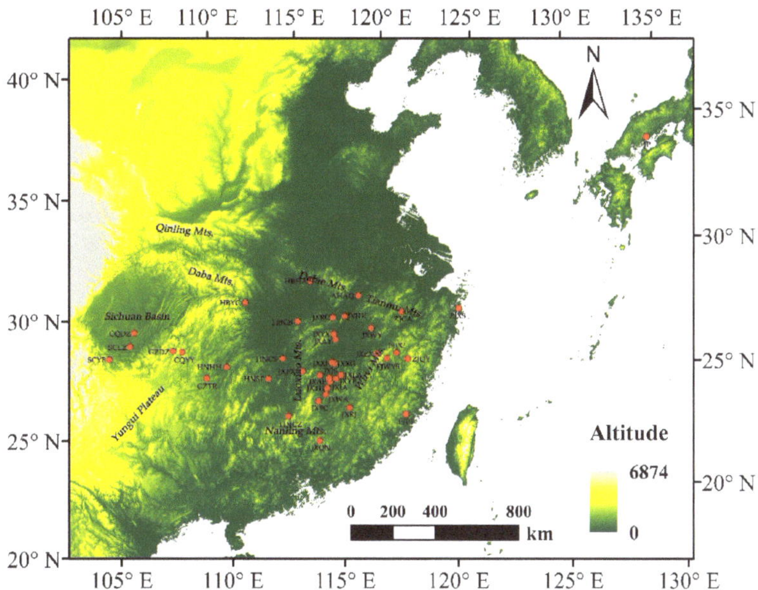

Figure 1. Geographical distribution of the plant materials. Solid dots represent populations collected in natural distribution areas in China and Japan.

2.2. GBS Library Construction and Population SNP Identification

Genomic DNA (0.1–11 µg) was digested with the *Ape*KI restriction endonuclease, and the adaptor and barcode were ligated at both ends by T4 DNA ligase. Then each DNA sample was amplified by PCR. After mixing the samples, the DNA fragments (350–400 bp) were recovered and purified by agarose gel electrophoresis, and the required fragments were selected for the construction of the GBS library. Finally, paired-end sequencing at 150 bp per read was performed on the Illumina HiSeq 2500 sequencing platform (Illumina, San Diego, CA, USA).

Stacks 1.44 [18] was used to cluster the high-quality data of each individual based on sequence similarity. During clustering, low-depth stacks were filtered out, leaving high-depth stacks as the clustering results. The mismatched bases were corrected in each stack to eliminate the impact of sequencing errors and avoid false-positive SNP. Clean data of sample ZJCA-157 and JXAF-115 were selected to be combined and assembled as the reference genome under the parameter (ustacks -t gzfasta -i 1 -m 1 -M 6 -N 6 -p 6). BWA (Burrows-Wheeler-Aligner) was used to compare the high-quality and effective sequencing data against the reference genome (using the assembly results of samples ZJCA-157 and JXAF-115 as the reference genome) [19]. The SNP in the population were detected with SAMTOOLS software. The SNP in the population to be tested were detected using the Bayesian model. The obtained SNP were filtered with the SAMTOOLS [20] settings main parameter -minDP4-max miss0.1-maf0.01. The number of bases support for each SNP was not less than 1. Due to repeat regions in the genome, the number of bases support for each SNP should not be greater than 7000 to reduce the number of false-positive SNP caused by repeat region alignment errors. The genotype quality value was ≥5.

2.3. Statistical Analysis of Data

The distance matrix was calculated using TreeBest-1.9.2 software (http://treesoft.sourceforge.net/treebest.shtml, accessed on 28 October 2021). The phylogenetic tree was constructed through 1000 calculations using the neighbour-joining method [21]. The eigenvectors and eigenvalues were calculated using GCTA (http://cnsgenomics.com/software/gcta/pca.html, accessed on 28 October 2021) software, and the principal component analysis (PCA) distribution map was plotted using R software [22]. The population genetic structure and individual allocation information were constructed using frappe software [23]. The maximum iteration of the expectation-maximization algorithm was set to 10,000 in the Frappe analysis. We predefined the number of genetic clusters from K = 2 to K = 4. Based on the filtered SNP loci, genetic diversity indicators including the observed heterozygosity (Ho), expected heterozygosity (He), interpopulation genetic differentiation coefficient (F_{ST}), and interindividual inbreeding coefficient (F_{IS}) of each population were calculated by Arlequin software [24]. Analysis of molecular variance (AMOVA) was implemented in GenAlEx v6.5 [25]. Analysis of population nucleotide diversity (π) was performed with Stacks 1.44 [18].

Table 1. Geographical locations, sample size and sample numbers of 40 *C. camphora* populations in China and Japan.

Population Code	Sample Size	Locations	Latitude (°N)	Longitude (°E)
Western China	14			
SCLZ	5	Luzhou, Sichuan	28.967	105.433
SCYB	5	Yibing, Sichuan	28.417	104.417
CQDZ	4	Dazu, Chongqing	29.533	105.650
Central China	25			
CQYY	3	Youyang, Chongqing	28.700	108.000
GZDZ	3	Daozhen, Guizhou	28.750	107.567
GZTR	4	Tongren, Guizhou	27.567	109.183
HBYC	5	Yichang, Hubei	30.650	111.283
HNHH	5	Huaihua, Hunan	28.000	110.167
Eastern China	128			
HBCB	4	Chibi, Hubei	29.700	113.817
HBHA	5	Hongan, Hubei	31.333	114.633
AHAQ	5	Anqing, Anhui	30.500	117.000
HNCZ	4	Chenzhou, Hunan	25.783	112.983
HNCS	5	Changsha, Hunan	28.200	112.933
JXYX	3	Yongxiu, Jiangxi	29.033	115.550
JXAY	4	Anyi, Jiangxi	28.800	115.617
JXHK	4	Hukou, Jiangxi	29.717	116.200
JXWY	4	Wuyuan, Jiangxi	29.083	117.450
JXRC	3	Ruichang, Jiangxi	29.717	115.600
JXPX	4	Pingxiang, Jiangxi	27.600	113.833
JXLA	5	Lean, Jiangxi	27.283	115.700
JXZX	4	Zixi, Jiangxi	27.983	117.583
JXQN	3	Quannan, Jiangxi	24.633	114.367
JXRJ	5	Ruijin, Jiangxi	25.883	115.950
JXSC	4	Suichuan, Jiangxi	26.300	114.483
JXTH	3	Taihe, Jiangxi	26.800	114.967
JXWA	5	Wanan, Jiangxi	26.550	114.867
JXAF	4	Anfu, Jiangxi	27.367	114.667
JXXJ	4	Xiajiang, Jiangxi	27.853	115.316
JXXG	5	Xingan, Jiangxi	27.800	115.450
JXYF	5	Yongfeng, Jiangxi	27.150	115.450
JXJA	5	Jian, Jiangxi	27.067	115.133
JXJS	4	Jishui, Jiangxi	27.217	115.117
FJPT	5	Putian, Fujian	25.317	118.567
FJPC	5	Pucheng, Fujian	27.917	118.517
FJWYS	4	Wuyishan, Fujian	27.733	118.017
ZJCA	4	Chunan, Zhejiang	29.600	119.033
ZJQY	4	Qingyuan, Zhejiang	27.600	119.000
ZJXS	5	Xiangshan, Zhejiang	29.367	121.867
Japan	4			
JP	4	Osaka, Japan	34.493	133.246
Total	171			

3. Results

3.1. GBS Analysis and SNP Identification

Total 138.24 Gb bases were generated by GBS sequencing from the 171 *C. camphora* trees (Table S2). After cleaning out low-quality sequences, 138.23 Gb of high-quality sequence data (Q20 $\geq$ 93.34%, Q30 $\geq$ 85.00%) remained, for an average of 0.81 Gb per sample. The average GC content was 38.24%. A total of 960 million clean reads were obtained, ranging from 2,878,540 (HBCB-40) to 13,528,004 (HNCZ-55), with an average of 5,613,668 per sample. An average of 5,389,736 reads (96.18%) was aligned to a reference genome, with the highest (97.18%) alignment rate for GZDZ-18 and the lowest (73.50%) rate for ZJCA-157. The reference genome coverage was in an average of 11.82% 1× genome (between 8.2% and 16.1%), and an average of 5.5% 4× genome (between 4.12% and 7.41%). A total of 130,450 SNP loci were obtained based on the comparison and detection by SAMTOOLS software. After filtering, 14,718 high-quality SNP loci were used for further analysis.

3.2. Population Genetic Structure

To reveal the population genetic structure of *C. camphora*, we divided the 40 populations into four groups, eastern China, central China, western China, and Japan. The neighbour-joining tree showed that, except for the HNCZ-54 trees from the east and HNHH-31 from the west regions, the other 169 *C. camphora* trees were clustered into three clades (Figure 2). Cluster I was composed of 18 *C. camphora* trees, including 16 from the central region and two from the CQDZ population in the western region (11, 12). Cluster II included 15 trees mainly from the west region, in which two trees from the central region (GZDZ-19 and GZDZ-20) and one from the eastern region (HNCZ-54). Cluster III contained the most samples (137), including trees from east China, and a few from the central region (HNSF-35, HNSF-38, HNSF-39, HBYC-26, HBYC-28, and HBYC-29), and all of the trees from Japan.

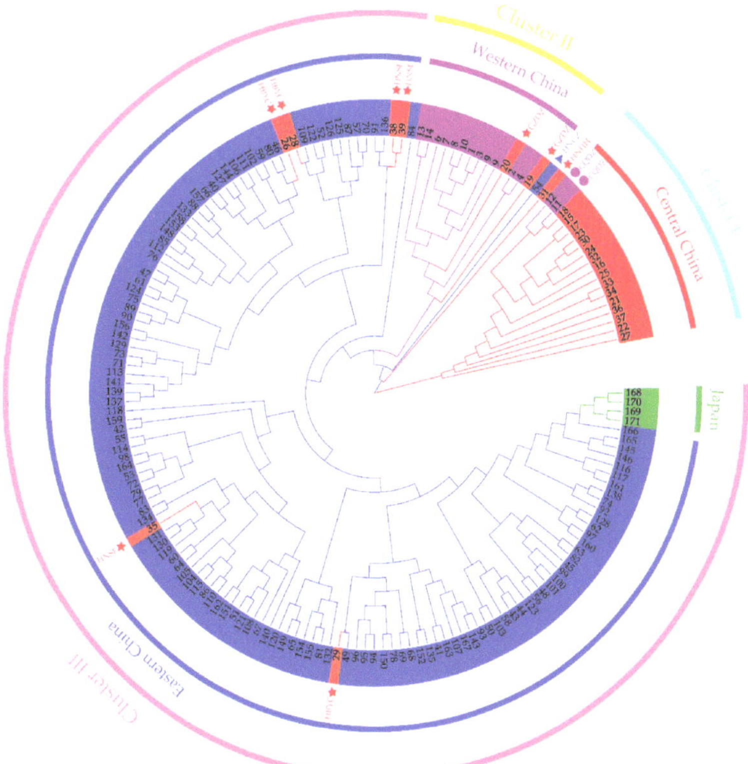

Figure 2. Neighbour-joining phylogenetic tree of *C. camphora* based on all single nucleotide polymorphisms (SNPs), with the evolutionary distances measured by p-distances with TreeBest. The populations from different geographic locations were coloured red (Central China), purple (Western China), blue (Eastern China), and green (Japan). ★, Individuals in Central China; •, Individuals in Western China; ▲, Individuals in Eastern China.

To detect the highest population structure level, we performed the analysis when the parameter *K* was between 2 and 4 (Figure 3). The results showed that when *K* = 2, the trees from central China and the HNCZ-54, CQDZ-11, and CQDZ-12 from the western region were first separated from the others and formed subgroup I. When *K* = 3, trees from the west region were separated from subgroup II and formed subgroup III. When *K* = 4, subgroup IV was further divided into the populations from Japan and eastern China, and the Japanese *C. camphora* population was finally clustered into its own subgroup. These results are consistent with the above phylogenetic tree analysis.

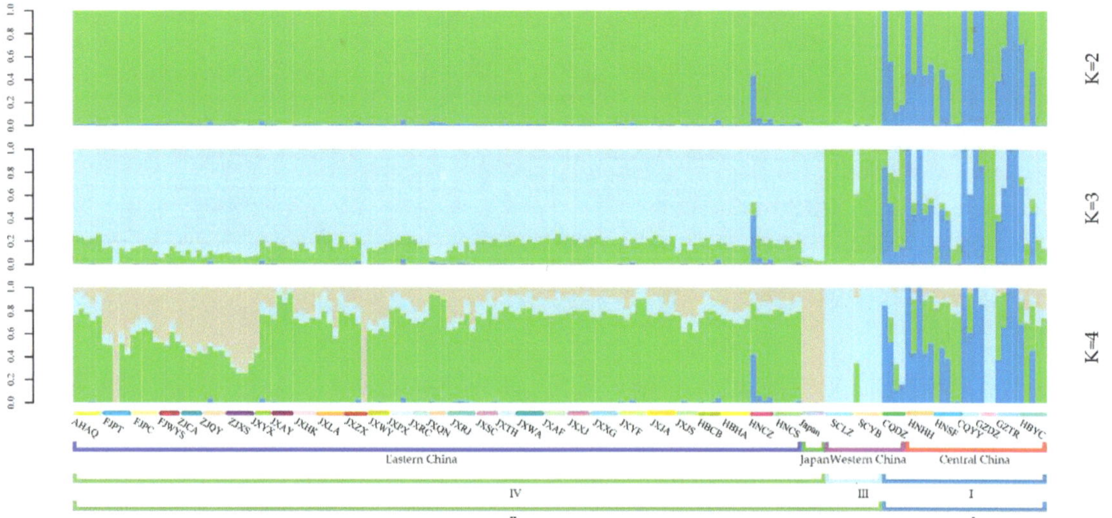

Figure 3. Population genetic structure of 171 *C. camphora* individuals. Each colour represents one cluster. Each individual is represented by a vertical bar, and the length of each coloured segment in each vertical bar represents the proportion contributed by ancestral populations.

PCA showed that the 171 *C. camphora* trees clustered three main groups (Figure 4). All trees can be significantly separated into two groups: central China and eastern China + Japan, according to the first principal component (PC1). In comparison, the western China populations showed a certain level of differentiation from the others two groups on the second principal component (PC2). This result is consistent with the clustering tree (Figure 2) and the population structure analysis (Figure 3). All these data indicated a clear geographical pattern of the genetic structure of the *C. camphora* populations.

3.3. Genetic Diversity in C. camphora

At the species level, the average values of *Ho* and *He* of *C. camphora* were 0.365 and 0.352, respectively (Table 2). The analysis of genetic diversity in each region showed that the population from Japan had the highest level of genetic heterozygosity (*Ho* = 0.504, *He* = 0.442), followed by Western China (*Ho* = 0.333, *He* = 0.333) and Central China (*Ho* = 0.322, *He* = 0.327), and the genetic diversity in eastern China was the lowest (*Ho* = 0.302, *He* = 0.306). The nucleotide diversity (π) of *C. camphora* in the different regions ranged from 0.230 (Japan) to 0.318 (Central China). The inbreeding coefficient F_{IS} ranged from -0.151 (Japan) to 0.020 (Eastern China), with an average of -0.037.

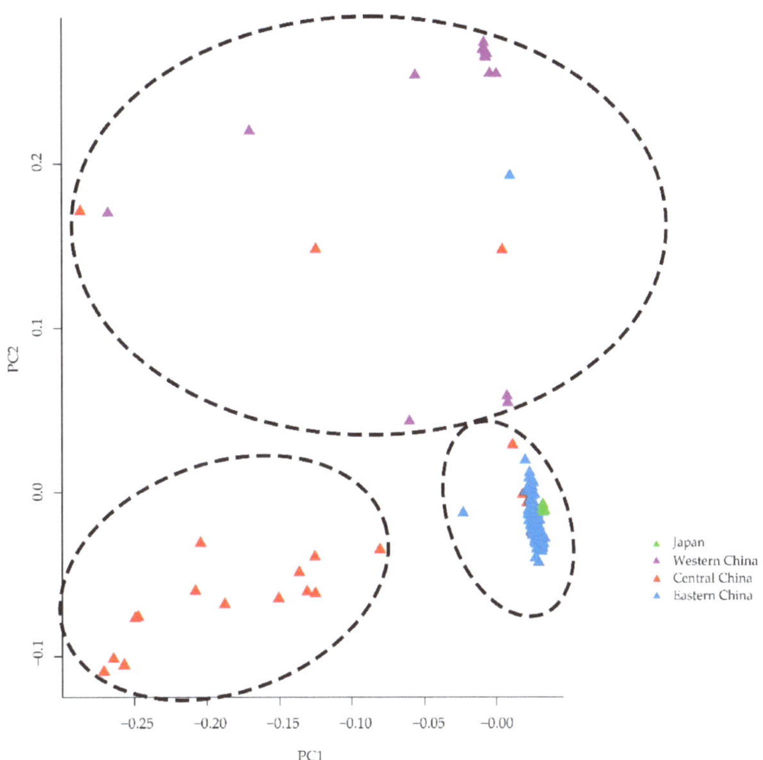

Figure 4. The principal coordinate analysis of *C. camphora* populations using all identified SNPs as markers. The populations from different geographic locations were coloured red (Central China), purple (Western China), blue (Eastern China), and green (Japan).

Table 2. Diversity statistics of four *C. camphora* regions.

Group	*N*	*Ho*	*He*	F_{IS}	π
Western China	14	0.333	0.333	0.011	0.288
Central China	25	0.322	0.327	−0.026	0.318
Eastern China	128	0.302	0.306	0.020	0.305
Japan	4	0.504	0.442	−0.151	0.230
Total	171	0.365	0.352	−0.037	0.285

Note: *N*, Number of individuals; *Ho*, Observed heterozygosity; *He*, Expected heterozygosity; F_{IS}, Fixation index; π, Total nucleotide diversity.

Analysis of molecular variance (AMOVA) of the *C. camphora* populations from the four regions (Table 3) showed that 7.02% of the genetic variation was between the regions. The genetic variation between the populations from the same region was only 0.95%. However, 92.03% of the genetic variation was among the individuals within each population. Among the four regions, the mean genetic differentiation coefficient F_{ST} was 0.099, gene flow (*N*m) was 2.718, indicating a moderate genetic differentiation between populations. The genetic differentiation among eastern, western, and central China was relatively low (F_{ST} = 0.052–0.071, 0.064 on average), less than the differentiation between China and Japan (F_{ST} = 0.096–0.152, 0.133 on average). The genetic differentiation between the eastern China population and the Japanese population was low (0.096). In contrast, a relatively high level of genetic differentiation between western or central China and Japan was observed, 0.152 and 0.151, respectively. Gene exchange between different regions in China was relatively frequent (*N*m = 3.271–4.558), but the gene flow between the Chinese population and the

Japanese population was limited (Nm = 1.395–2.354), indicating that geographical isolation has exacerbated the genetic differentiation between the *C. camphora* populations (Table 4).

Table 3. Analysis of molecular variance (AMOVA) for four regions *C. camphora*.

Source	Degree of Freedom	Sum of Squares	Variance Components	Percentage of Variation %
Among groups	3	27,901.615	154.380	7.02
Within groups among populations	167	345,029.070	20.920	0.95
Within populations	171	346,138.500	2024.202	92.03
Total	341	719,069.184	2199.502	

Table 4. Genetic differentiation of *C. camphora* across four regions (lower diagonal: pairwise F_{ST}; upper diagonal: pairwise Nm).

	Western China	Central China	Eastern China	Japan
Western China	0	3.321	4.558	1.395
Central China	0.070	0	3.271	1.406
Eastern China	0.052	0.071	0	2.354
Japan	0.152	0.151	0.096	0

4. Discussion

4.1. Genotyping by Sequencing

Traditionally, molecular markers such as random amplified polymorphic DNA, amplified fragment length polymorphisms, and SSRs have been widely used to analyze plant genetic diversity. With the rapid development of NGS technology, a new method that is simple but powerful, GBS, has achieved high-efficiency and large-scale genome-wide SNP identification in some species [8]. In 2011, Elshire et al. used GBS to investigate maize (IBM) and barley (Oregon Wolfe Barley) populations. To establish a library with reduced genomic complexity, they identified approximately 200,000 and 25,000 sequence markers, respectively. The analysis showed the potential of this library for marker screening in large-genome species [10]. Pootakham et al. obtained 524,508,111 reads from oil palm (*Elaeis guineensis* Jacg.) using GBS technology, approximately 88% reads were aligned to the reference genome, and 21,471 SNP loci were identified, which was significantly lower than the alignment rate of reads on the reference genome in this study (96.18%) [26]. Gurcan et al. performed SNP detection on apricots by the GBS technique and generated approximately 28 Gb of high-quality data after filtering [27]. Wu et al. applied GBS technology to 75 samples of Lauraceae plants, including 19 samples of *C. kanehirae*, 31 of *C. camphora*, and 25 putative hybrids, for breed identification and genetic analysis. In total, 529,006 SNPs were obtained in the calling analysis, though after filtering, 840 high-quality SNPs were obtained for subsequent analysis. Among them, *C. camphora* groups in eastern and south-western Taiwan had 201 SNP loci, while the western and northern groups (including Vietnam and Japan) had 376 SNP loci [9]. In this study, by sequencing 171 *C. camphora* trees, we obtained 480 million high-quality reads and identified 130,450 SNPs. After filtering, 14,718 high-quality SNPs were used for subsequent genetic diversity and genetic structure analysis. The quality of the obtained reads and SNP loci was as high as the studies from Pootakham et al. and Gurcan et al.

4.2. Genetic Diversity of C. camphora

Genetic diversity refers to the degree of genetic polymorphism between different populations or individuals within a species. Rich genetic diversity is the basis for a species to respond to environmental changes and for people to carry out genetic improvement programs [28,29]. Based on 22 SSR markers, Kameyama evaluated the genetic polymorphism of 104 *C. camphora* samples from three Japanese populations: Meiji Jingu (Shinto Shrine), Kajiya Plantation, and Manazuru Peninsula. They found the average *Ho,* and *He* of each population was 0.53–0.60 and 0.55–0.68, respectively, suggesting a high level of genetic diversity [1]. Kameyama et al. further performed genetic diversity analysis on 817 *C. camphora* individuals from six populations, including three populations (Fujian, Shanghai, and Taiwan) from China and three populations (Eastern and Western Kyushu regions) from Japan. This study showed that a high level of genetic diversity in the *C. camphora* population, in which the Chinse populations (Ho = 0.728, He = 0.878) showed significantly greater diversity than that of the Japanese populations (Ho = 0.614, He = 0.714) [2]. In comparison to the previous study, our previous studies showed a moderate level of genetic diversity (Ho = 0.30–0.61, He = 0.31–0.48) of 41 Chinese *C. camphora* natural populations from the whole range of distribution of *C. camphora* in southern China using EST-SSR markers [4,5]. Interestingly, by using the GBS technique, Wu et al. observed a significantly lower genetic diversity in the population, including 75 individuals from Taiwan, 19 *C. kanehirae*, 31 *C. camphora*, and 25 putative hybrids. The population genetic diversity (Ho = 0.108, He = 0.123) of *C. camphora* in western and northern Taiwan was slightly higher than that in eastern and south-western Taiwan (Ho = 0.085, He = 0.076), and the Ho of *C. camphora* was greater than that of *C. kanehirae* (Ho = 0.076) [9]. In this study, we analyzed the genetic diversity of the Chinese and Japanese *C. camphora* populations through SNP markers generated by the GBS technique. We found that the genetic diversity of the Chinese *C. camphora* populations was relatively high (Ho = 0.319, He = 0.322), which is consistent with our previous findings based on SSR markers [4,5]. However, our populations showed higher genetic diversity than those found by Wu et al. [9]. One possible reason is that the geographical range of populations in Taiwan is small, and the cultivation history of *C. camphora* is long. During the process of the breeding program, a limited number of selected elite trees led to lower diversity [30]. Moreover, the asexual reproduction process using cuttings has further reduced the genetic diversity of *C. camphora*. In this study, a total of 11 Chinese *C. camphora* trees were clustered into different regions (Figure 2). And these trees were obtained from recent cultivations (Table S1). Kameyama et al. showed that human activity had significantly affected gene exchange between *C. camphora* populations [2]. The genetic diversity of a tree species is typically determined by its geographical distribution range and population size [31,32]. Although *C. camphora* is widely distributed in southern China, its origin is limited to East Asia and the Indochina Peninsula. The limited geographical distribution inhibits the gene exchange among populations and enhances genetic diversity within the population [33]. In addition, this study, together with the findings from Wu et al., showed lower genetic diversity of *C. camphora* in China than that of the Japanese population, which was not consistent with the results of Kameyama et al. [2]. This inconsistency may be due to different sampling strategies, such as inconsistencies in the source and size of the population used [32,34]. The difference between the GBS-SNPs and SSRs used in these studies could be another reason for the different findings.

4.3. Genetic Differentiation and Population Genetic Structure of C. camphora

Studying the genetic structure of plant populations is critical to the understanding and maintenance of plant genetic diversity. The genetic structure of populations is affected by the breeding system, genetic drift, population size, seed dispersal, gene flow, evolutionary history, and natural selection [2,35]. As an essential indicator of the population's differentiation, the F_{ST} ranging from 0.05 to 0.15 indicates moderate genetic differentiation between populations [36,37]. In this study, the F_{ST} of *C. camphor* between the four areas was 0.052–0.152 (Table 3), indicating a moderate genetic differentiation of *C. camphor* groups. AMOVA showed 92.03% of the genetic variation was within the population, leaving the genetic variation between regions and populations at only 7.97%. These results are consistent with previous studies on *C. camphora* [2,4] and other tree species in southern China [38,39]. The Nm value also suggested frequent gene exchange between the populations in the region, indicating that the gene exchange offsets the effect of genetic drift to some extent [40]. Considering that *C. camphora* is an insect-pollinated plant, the impact of insects on the long-distance transmission of its pollen may be relatively small. The seeds of *C. camphora* are taken by various birds in the south, so the gene flow between populations is likely to be carried out by long-distance transport of seeds. Hence, the low level of genetic differentiation between populations can be explained by the long-distance transmission and transport of pollen or seeds [41]. In addition, considering the riparian habitat of *C. camphora*, its seed dispersal is easily affected by the regional water system. On the other hand, the insect-pollinated plants with dominant outcrossing provide a new isolation method for the reproductive system, reduce pollen flow between populations, and thus increase genetic differentiation and erosion among populations [4].

The phylogenetic tree showed our 171 *C. camphora* trees from 40 populations clustered into three major groups, indicating an apparent geographical distribution pattern. Southwest China is considered one of the hot spots of species diversity, with rich species diversity, and is also the origin and differentiation centre of many plants and glacial refuges [2,42]. The western regional populations of the Yunnan–Guizhou Plateau constituted subgroup I, with the highest genetic diversity, including YB and LZ in Sichuan and DZ in Chongqing. 137 *C. camphora* trees from the Japanese population and eastern China (excluding only HNCZ-761) constituted the largest subgroup (Figure 2, Cluster III), with the lowest genetic diversity. These regional populations in China are in Jiangxi, Anhui, and Hunan Provinces. Their border areas with neighbouring provinces have the most extensive distribution and the most extant ancient *C. camphora* populations in China. The *C. camphora* population in central China was clustered into a single subgroup II, with two exceptionals of the Hubei YC and Hunan SF populations into subgroup III, indicating a certain degree of gene exchange within eastern China. Similar results were obtained from PCA and population structure analysis. Mountains play a crucial role in the extinction and preservation of subtropical Chinese species and the generation of new species and new population structures. Due to the geographical isolation caused by mountains, gene exchange across mountains is hindered, resulting in significant differentiation of genetic structure and increased genetic diversity [43–46]. It was speculated that the Japanese *C. camphora* originated from China [47]. Moreover, according to our study, the *C. camphora* species formation was relatively late, approximately in the late Tertiary period (unpublished data). Multiple transgressions and regressions in the East China Sea in the fourth century affected the segregation and connection of eastern China, southern Japan, and the Korean Peninsula, which significantly impacted the genetic structure of plants [48–51]. This genetic exchange led to the close genetic relationship between the *C. camphora* populations in eastern China and Japan.

4.4. Conservation Strategies for C. camphora

In situ conservation is an effective measure to protect wild germplasm resources [52]. The native habitat of *C. camphora* is widely distributed throughout East Asia and China in the region with the most abundant wild resources. In recent decades in southern China, many wild *C. camphora* has been cut down to extract essential oil by destructive methods, such as digging trees and grubbing roots. Meanwhile, many ancient *C. camphora* (defined as diameter at breast height exceeding 100 cm) were transplanted into the city during the urban vegetation process. The ageing, physiological function decline, and habitat deterioration have also aggravated the weakening and death of some ancient *C. camphora* trees. Altogether, these factors caused the shrinking of existing wild *C. camphora* resources. Gene flow is a crucial factor affecting the genetic diversity and genetic differentiation of species [53]. The breeders' exchange of seedlings and seeds has significantly affected gene flow between *C. camphora* populations in China and Japan [2]. Transplantation of large and ancient trees should be avoided to maintain the genetic structure of *C. camphora* in each region. At the same time, as an insect-pollinated plant, factors such as genetic erosion and genetic drift may also affect the level of genetic diversity of the offspring of *C. camphora*. There are frequent gene introgressions between *C. camphora* and other species of the same genus [9], so ex-situ conservation is another important way to protect wild species. We should include as many individuals with different haplotypes and evolutionarily significant units as possible for ex-situ introduction. Attention should also be given to the mutual introduction of various populations to increase their gene exchange and diversity, to resist the negative effects of genetic drift, thereby reducing the risk of genetic diversity reduction in wild populations.

5. Conclusions

This is the first study to analyze the genetic diversity and genetic structure of *C. camphora* in East Asia, including 39 populations in its whole distribution range in China and one Japanese population, based on the GBS technique. The results showed that the genetic diversity of the *C. camphora* populations in China were significantly lower than that of the Japanese population. The AMOVA, pairwise F_{ST}, and Nm analysis indicated that the genetic differentiation between populations was quite low, as the genetic variation was mainly within each population. Phylogenetic tree, PCA, and population structure analysis showed that the 40 populations were clustered into three major groups, i.e., western China, central China, and eastern China + Japan, showing a clear geographical distribution pattern. The phylogenetic analysis showed the closest relationship between the populations in eastern China and Japan. There is significant gene exchange between adjacent regions, and due to geographical isolation, the gene exchange between Chinese and Japanese populations is relatively weak. Because the Japanese population sample size in this study was small, further studies will be needed with increased sample size from the Japanese population, and populations from Taiwan and Vietnam to further confirm the findings of existing studies. In any case, this study's results will help further understand the current status of genetic diversity and genetic structure of *C. camphora* in East Asia and provide a theoretical basis for forming appropriate conservation measures and genetic improvement strategies.

Supplementary Materials: The following are available online at https://www.mdpi.com/article/10.3390/f12111511/s1, Table S1: Location information of the 171 sampled *C. camphora* individuals; Table S2: Sequencing quality of the 171 sampled *C. camphora* individuals.

Author Contributions: Conceptualization, Y.Z.; methodology, X.G., Y.Z. and A.Y.; software, A.Y., X.G. and Z.W.; validation, C.C., Q.L. and H.L.; formal analysis, X.G.; investigation, Y.Z., A.Y. and F.Y.; resources, A.Y., Z.W. and Y.Z.; data curation, C.C.; writing—original draft preparation, X.G. and Y.Z.; writing—review and editing, X.G., A.Y. and Y.Z.; supervision, F.Y.; project administration, Y.Z.; funding acquisition, Y.Z. All authors have read and agreed to the published version of the manuscript.

Funding: This research was funded by the National Natural Science Foundation of China (no. 31860079 and 32060354), and Important S & T Specific Projects of Jiangxi Provincial Department of Science and Technology (20203ABC28W016).

Institutional Review Board Statement: Not applicable.

Informed Consent Statement: Not applicable.

Acknowledgments: We are grateful to Enying Liu for her help in revising the manuscript and for her constructive advice.

Conflicts of Interest: The authors declare that they have no conflict of interest.

References

1. Kameyama, Y. Development of microsatellite markers for *Cinnamomum camphora* (Lauraceae). *Am. J. Bot.* **2012**, *99*, e1–e3. [CrossRef]
2. Kameyama, Y.; Furumichi, J.; Li, J.; Tseng, Y.-H. Natural genetic differentiation and human-mediated gene flow: The spatiotemporal tendency observed in a long-lived *Cinnamomum camphora* (Lauraceae) tree. *Tree Genet. Genomes* **2017**, *13*, 38. [CrossRef]
3. Shi, S.; Wu, Q.; Su, J.; Li, C.; Zhao, X.; Xie, J.; Gui, S.; Su, Z.; Zeng, H. Composition analysis of volatile oils from flowers, leaves and branches of *Cinnamomum camphora* chvar. Borneol in china. *J. Essent. Oil Res.* **2013**, *25*, 395–401. [CrossRef]
4. Zhong, Y.; Yang, A.; Li, Z.; Zhang, H.; Liu, L.; Wu, Z.; Li, Y.; Liu, T.; Xu, M.; Yu, F. Genetic Diversity and Population Genetic Structure of *Cinnamomum camphora* in South China Revealed by EST-SSR Markers. *Forests* **2019**, *10*, 1019. [CrossRef]
5. Li, Z.; Zhong, Y.; Yu, F.; Xu, M. Novel SSR marker development and genetic diversity analysis of *Cinnamomum camphora* based on transcriptome sequencing. *Plant Genet. Resour.-Charact. Util.* **2018**, *16*, 568–571. [CrossRef]
6. Wu, X.; Wang, L.; Zhang, D.; Wen, Y. Microsatellite null alleles affected population genetic analyses: A case study of Maire yew (Taxus chinensis var. mairei). *J. For. Res.* **2019**, *24*, 230–234. [CrossRef]

7. Mammadov, J.; Aggarwal, R.; Buyyarapu, R.; Kumpatla, S. SNP markers and their impact on plant breeding. *Int. J. Plant Genom.* **2012**, *2012*, 728398. [CrossRef]
8. Yang, T.Y.; Gao, T.X.; Meng, W.; Jiang, Y.-L. Genome-wide population structure and genetic diversity of Japanese whiting (*Sillago japonica*) inferred from genotyping-by-sequencing (GBS): Implications for fisheries management. *Fish. Res.* **2020**, *225*, 105501. [CrossRef]
9. Wu, C.C.; Chang, S.H.; Tung, C.W.; Ho, C.K.; Gogorcena, Y.; Chu, F.H. Identification of hybridization and introgression between *Cinnamomum kanehirae* Hayata and *C. camphora* (L.) Presl using genotyping-by-sequencing. *Sci. Rep.* **2020**, *10*, 15995. [CrossRef]
10. Elshire, R.J.; Glaubitz, J.C.; Sun, Q.; Poland, J.A. A Robust, Simple Genotyping-by-Sequencing (GBS) Approach for High Diversity Species. *PLoS ONE* **2011**, *6*, e19379. [CrossRef]
11. Poland, J.A.; Rife, T.W. Genotyping-by-Sequencing for Plant Breeding and Genetics. *Plant Genome* **2012**, *5*, 92–102. [CrossRef]
12. Lee, K.J.; Lee, J.R.; Sebastin, R.; Shin, M.J.; Kim, S.H.; Cho, G.T.; Hyun, D.Y. Genetic Diversity Assessed by Genotyping by Sequencing (GBS) in Watermelon Germplasm. *Genes* **2019**, *10*, 822. [CrossRef] [PubMed]
13. Lee, H.Y.; Jang, S.; Yu, C.R.; Kang, B.C.; Chin, J.H.; Song, K. Population Structure and Genetic Diversity of *Cucurbita moschata* Based on Genome-Wide High-Quality SNPs. *Plants* **2021**, *10*, 56. [CrossRef]
14. Solomon, A.M.; Han, K.; Lee, J.H.; Lee, H.Y.; Jang, S.; Kang, B.C. Genetic diversity and population structure of Ethiopian *Capsicum* germplasms. *PLoS ONE* **2019**, *14*, e0216886. [CrossRef] [PubMed]
15. Nuñez-Lillo, G.; Balladares, C.; Pavez, C.; Urra, C.; Sanhueza, D.; Vendramin, E.; Dettori, M.T.; Arús, P.; Verde, I.; Blanco-Herrera, F.; et al. High-density genetic map and QTL analysis of soluble solid content, maturity date, and mealiness in peach using genotyping by sequencing. *Sci. Hortic.* **2019**, *257*, 108734. [CrossRef]
16. Zhang, L.; Guo, D.; Guo, L.; Guo, Q.; Wang, H.; Hou, X. Construction of a high-density genetic map and QTLs mapping with GBS from the interspecific F_1 population of *P. ostii* 'Fengdan Bai' and *P. suffruticosa* 'Xin Riyuejin'. *Sci. Hortic.* **2019**, *246*, 190–200. [CrossRef]
17. Carrasco, B.; Gonzalez, M.; Gebauer, M.; Garcia-Gonzalez, R.; Maldonado, J.; Silva, H. Construction of a highly saturated linkage map in Japanese plum (*Prunus salicina* L.) using GBS for SNP marker calling. *PLoS ONE* **2018**, *13*, e0208032. [CrossRef]
18. Catchen, J.M.; Amores, A.; Hohenlohe, P.; Cresko, W.; Postlethwait, J.H. Stacks: Building and genotyping Loci de novo from short-read sequences. *G3-Genes Genomes Genet.* **2011**, *1*, 171–182. [CrossRef]
19. Jo, H.; Koh, G. Faster single-end alignment generation utilizing multi-thread for BWA. *Bio-Med. Mater. Eng.* **2015**, *26*, S1791–S1796. [CrossRef]
20. Etherington, G.J.; Ramirez-Gonzalez, R.H.; MacLean, D. Bio-samtools 2: A package for analysis and visualization of sequence and alignment data with SAMtools in Ruby. *Bioinformatics* **2015**, *31*, 2565–2567. [CrossRef]
21. Levy, D.; Yoshida, R.; Pachter, L. Beyond pairwise distances: Neighbor-joining with phylogenetic diversity estimates. *Mol. Biol. Evol.* **2006**, *23*, 491–498. [CrossRef]
22. Yang, J.; Lee, S.H.; Goddard, M.E.; Visscher, P.M. GCTA: A tool for genome-wide complex trait analysis. *Am. J. Hum. Genet.* **2011**, *88*, 76–82. [CrossRef] [PubMed]
23. Tang, H.; Peng, J.; Wang, P.; Risch, N.J. Estimation of Individual Admixture: Analytical and Study Design Considerations. *Genet. Epidemiol.* **2005**, *28*, 289–301. [CrossRef]
24. Excoffier, L.; Lischer, H.E. Arlequin suite ver 3.5: A new series of programs to perform population genetics analyses under Linux and Windows. *Mol. Ecol. Resour.* **2010**, *10*, 564–567. [CrossRef]
25. Peakall, R.; Smouse, P.E. GenAlEx 6.5: Genetic analysis in Excel. Population genetic software for teaching and research—An update. *Bioinformatics* **2012**, *28*, 2537–2539. [CrossRef] [PubMed]
26. Pootakham, W.; Jomchai, N.; Ruang-Areerate, P.; Shearman, J.R.; Sonthirod, C.; Sangsrakru, D.; Tragoonrung, S.; Tangphatsornruang, S. Genome-wide SNP discovery and identification of QTL associated with agronomic traits in oil palm using genotyping-by-sequencing (GBS). *Genomics* **2015**, *105*, 288–295. [CrossRef]
27. Gurcan, K.; Teber, S.; Ercisli, S.; Yilmaz, K.U. Genotyping by Sequencing (GBS) in Apricots and Genetic Diversity Assessment with GBS-Derived Single-Nucleotide Polymorphisms (SNPs). *Biochem. Genet.* **2016**, *54*, 854–885. [CrossRef]
28. Li, M.; Chen, S.; Shi, S.; Zhang, Z.; Liao, W.; Wu, W.; Zhou, R.; Fan, Q. High genetic diversity and weak population structure of *Rhododendron jinggangshanicum*, a threatened endemic species in Mount Jinggangshan of China. *Biochem. Syst. Ecol.* **2015**, *58*, 178–186. [CrossRef]
29. Wang, X.; Zhao, W.; Li, L.; You, J.; Ni, B.; Chen, X. Clonal plasticity and diversity facilitates the adaptation of *Rhododendron aureum* Georgi to alpine environment. *PLoS ONE* **2018**, *13*, e0197089. [CrossRef]
30. Quintana, J.; Contreras, A.; Merino, I.; Vinuesa, A.; Orozco, G.; Ovalle, F.; Gomez, L. Genetic characterization of chestnut (*Castanea sativa* Mill.) orchards and traditional nut varieties in El Bierzo, a glacial refuge and major cultivation site in northwestern Spain. *Tree Genet. Genomes* **2014**, *11*, 826. [CrossRef]
31. Li, X.; Li, M.; Hou, L.; Zhang, Z.; Pang, X.; Li, Y. De Novo Transcriptome Assembly and Population Genetic Analyses for an Endangered Chinese Endemic *Acer miaotaiense* (Aceraceae). *Genes* **2018**, *9*, 378. [CrossRef]
32. Ferrer, M.M.; Eguiarte, L.E.; Montana, C. Genetic structure and outcrossing rates in *Flourensia cernua* (Asteraceae) growing at different densities in the South-western Chihuahuan Desert. *Ann. Bot.* **2004**, *94*, 419–426. [CrossRef]
33. Shen, Y.; Cheng, Y.; Li, K.; Li, H. Integrating Phylogeographic Analysis and Geospatial Methods to Infer Historical Dispersal Routes and Glacial Refugia of *Liriodendron chinense*. *Forests* **2019**, *10*, 565. [CrossRef]

34. Liu, F.; Hong, Z.; Xu, D.; Jia, H.; Zhang, N.; Liu, X.; Yang, Z.; Lu, M. Genetic Diversity of the Endangered *Dalbergia odorifera* Revealed by SSR Markers. *Forests* **2019**, *10*, 225. [CrossRef]
35. Kulshreshtha, V.P. Plant population Genetics, Breeding and Genetic Resources. *Sinauer Assoc.* **1990**, *10*, 43–63.
36. Wright, S. The interpretation of population structure by F-statistics with special regard to systems of mating. *Evolution* **1965**, *19*, 395–420. [CrossRef]
37. Slatkin, M. Gene flow and the geographic structure of natural populations. *Science* **1987**, *236*, 787–792. [CrossRef]
38. Tan, J.; Zhao, Z.-G.; Guo, J.-J.; Wang, C.-S.; Zeng, J. Genetic Diversity and Population Genetic Structure of *Erythrophleum fordii* Oliv., an Endangered Rosewood Species in South China. *Forests* **2018**, *9*, 636. [CrossRef]
39. Hartvig, I.; So, T.; Changtragoon, S.; Tran, H.T.; Bouamanivong, S.; Theilade, I.; Kjær, E.D.; Nielsen, L.R. Population genetic structure of the endemic rosewoods *Dalbergia cochinchinensis* and *D. oliveri* at a regional scale reflects the Indochinese landscape and life-history traits. *Ecol. Evol.* **2017**, *8*, 530–545. [CrossRef] [PubMed]
40. Wright, S. Evolution in mendelian populations. *Genetics* **1931**, *16*, 97–159. [CrossRef] [PubMed]
41. Albaladejo, R.G.; Carrillo, L.F.; Aparicio, A.; Fernandez-Manjarres, J.F.; Gonzalez-Varo, J.P. Population genetic structure in *Myrtus communis* L. in a chronically fragmented landscape in the Mediterranean: Can gene flow counteract habitat perturbation? *Plant Biol.* **2009**, *11*, 442–453. [CrossRef]
42. Myers, N.; Mittermeier, R.A.; Mittermeier, C.G.; Fonseca, G.; Kent, J. Biodiversity hotspots for conservation priorities. *Nature* **2000**, *403*, 853–858. [CrossRef]
43. Tian, S.; Kou, Y.; Zhang, Z.; Yuan, L.; Li, D.; Lopez-Pujol, J.; Fan, D.; Zhang, Z. Phylogeography of *Eomecon chionantha* in subtropical China: The dual roles of the Nanling Mountains as a glacial refugium and a dispersal corridor. *BMC Evol. Biol.* **2018**, *18*, 20. [CrossRef]
44. Sun, Y.; Hu, H.; Huang, H.; Vargas-Mendoza, C.F. Chloroplast diversity and population differentiation of *Castanopsis fargesii* (Fagaceae): A dominant tree species in evergreen broad-leaved forest of subtropical China. *Tree Genet. Genomes* **2014**, *10*, 1531–1539. [CrossRef]
45. Qiu, Y.X.; Fu, C.X.; Comes, H.P. Plant molecular phylogeography in China and adjacent regions: Tracing the genetic imprints of Quaternary climate and environmental change in the world's most diverse temperate flora. *Mol. Phylogenetics Evol.* **2011**, *59*, 225–244. [CrossRef] [PubMed]
46. Zhang, Y.H.; Wang, I.J.; Comes, H.P.; Peng, H.; Qiu, Y.X. Contributions of historical and contemporary geographic and environmental factors to phylogeographic structure in a Tertiary relict species, *Emmenopterys henryi* (Rubiaceae). *Sci. Rep.* **2016**, *6*, 24041. [CrossRef] [PubMed]
47. Hammer, M.F.; Karafet, T.M.; Park, H.; Omoto, K.; Harihara, S.; Stoneking, M.; Horai, S. Dual origins of the Japanese: Common ground for hunter-gatherer and farmer Y chromosomes. *J. Hum. Genet.* **2006**, *51*, 47–58. [CrossRef] [PubMed]
48. Kimura, M. Paleography of the Ryukyu Islands. *Tropics* **2000**, *10*, 5–24. [CrossRef]
49. Haq, B.U.; Hardenbol, J.A.N.; Vail, P.R. Chronology of fluctuating sea levels since the triassic. *Science* **1987**, *235*, 1156–1167. [CrossRef]
50. Sakaguchi, S.; Qiu, Y.X.; Liu, Y.H.; Xin-Shuai, Q.I.; Kim, S.H.; Han, J.; Takeuchi, Y.; Worth, J.; Yamasaki, M.; Sakurai, S. Climate oscillation during the Quaternary associated with landscape heterogeneity promoted allopatric lineage divergence of a temperate tree *Kalopanax septemlobus* (Araliaceae) in East Asia. *Mol. Ecol.* **2012**, *21*, 3823–3838. [CrossRef]
51. Lambeck, K.; Chappell, J. Sea level change through the last glacial cycle. *Science* **2001**, *292*, 679–686. [CrossRef] [PubMed]
52. Shen, S.K.; Wang, Y.H.; Wang, B.Y.; Ma, H.Y.; Shen, G.Z.; Han, Z.W. Distribution, stand characteristics and habitat of a critically endangered plant *Euryodendron excelsum* H. T. Chang (Theaceae): Implications for conservation. *Plant Species Biol.* **2009**, *24*, 133–138. [CrossRef]
53. Setoguchi, H.; Mitsui, Y.; Ikeda, H.; Nomura, N.; Tamura, A. Genetic structure of the critically endangered plant *Tricyrtis ishiiana* (Convallariaceae) in relict populations of Japan. *Conserv. Genet.* **2010**, *12*, 491–501. [CrossRef]

Article

Population Genetic Structure Analysis Reveals Significant Genetic Differentiation of the Endemic Species *Camellia chekiangoleosa* Hu. with a Narrow Geographic Range

Bin Huang [1,2], Zhongwei Wang [3], Jianjian Huang [1], Xiaohui Li [4], Heng Zhu [4], Qiang Wen [1,*] and Li-an Xu [2]

[1] Jiangxi Provincial Key Laboratory of Camellia Germplasm Conservation and Utilization, Jiangxi Academy of Forestry, Nanchang 330047, China; huangbin007@yeah.net (B.H.); huang9931@126.com (J.H.)

[2] Co-Innovation Center for Sustainable Forestry in Southern China, Nanjing Forestry University, Nanjing 210037, China; laxu@njfu.edu.cn

[3] Institute of Botany, Jiangsu Province and Chinese Academy of Sciences, Nanjing 210014, China; wangzhongwei@cnbg.net

[4] Forestry Science Institute of Shangrao City, Shangrao 334000, China; lixiaohui666666@163.com (X.L.); zhuheng6666@163.com (H.Z.)

* Correspondence: jxwenqiang2021@163.com; Tel.: +86-139-0708-7913

Abstract: In order to protect and utilize the germplasm resource better, it is highly necessary to carry out a study on the genetic diversity of *Camellia chekiangoleosa* Hu. However, systematic research on population genetics analysis of the species is comparatively rare. Herein, 16 highly variable simple sequence repeat (SSR) markers were used for genetic structure assessment in 12 natural *C. chekiangoleosa* populations. The genetic diversity of *C. chekiangoleosa* was low (h = 0.596), within which, central populations (such as Damaoshan (DMS), Sanqingshan (SQS), and Gutianshan (GTS)) at the junction of four main mountain ranges presented high diversity and represented the center of the *C. chekiangoleosa* diversity distribution; the Hengshan (HS) population in the west showed the lowest diversity, and the diversity of the eastern and coastal populations was intermediate. *C. chekiangoleosa* exhibited a high level of genetic differentiation, and the variation among populations accounted for approximately 24% of the total variation. The major reasons for this situation are the small population scale and bottleneck effects in some populations (HS and Lingshan (LS)), coupled with inbreeding within the population and low gene flow among populations (Nm = 0.796). To scientifically protect the genetic diversity of *C. chekiangoleosa*, in situ conservation measures should be implemented for high-diversity populations, while low-diversity populations should be restored by reintroduction.

Keywords: *Camellia chekiangoleosa*; SSR; genetic diversity; genetic differentiation; population genetic structure

Citation: Huang, B.; Wang, Z.; Huang, J.; Li, X.; Zhu, H.; Wen, Q.; Xu, L.-a. Population Genetic Structure Analysis Reveals Significant Genetic Differentiation of the Endemic Species *Camellia chekiangoleosa* Hu. with a Narrow Geographic Range. *Forests* **2022**, *13*, 234. https://doi.org/10.3390/f13020234

Academic Editors: Konstantinos Poirazidis and Panteleimon Xofis

Received: 20 December 2021
Accepted: 2 February 2022
Published: 3 February 2022

Publisher's Note: MDPI stays neutral with regard to jurisdictional claims in published maps and institutional affiliations.

1. Introduction

The genus *Camellia* L. includes many valuable shrubs, economic trees, and endangered species belonging to the family Theaceae, among which *Camellia japonica* L. is an internationally known ornamental flower [1]. In addition, *C. oleifera* Abel., a species of oil tea camellia, is considered one of the four major woody oil tree species [2]. Oil tea camellia generally refers to a tree species in the genus *Camellia* with a high oil content and high production and cultivation value [3]. In China, oil tea camellia has a cultivation history of more than 2300 years, and it has been used as an oil tree species for more than 1000 years [4]. Currently, there are more than 20 species of oil tea camellias that are mainly planted in China [5]. As the two major oil tea camellia species, *C. oleifera* and *C. chekiangoleosa* Hu. have been usually cultivated in southern China [6]. However, compared with the former, no varieties or superior clones of *C. chekiangoleosa* have been developed, and it is urgent to utilize the germplasm resource of this species [7].

Camellia chekiangoleosa is a typical diploid oil tea camellia species (2 n = 30). The species is a relative of *C. japonica* and *C. oleifera* with excellent ornamental and oil value due to its beautiful flower color and excellent oil quality [8,9]. This species has become an important economic tree in southern China [7]. *C. chekiangoleosa* is still growing in the wild, without sufficient studies on related breeding and horticultural culture [8,10]. Early plant taxonomists believed that *C. chekiangoleosa* was naturally distributed in the mountains at an altitude of 600~1400 m (117° 34′~120°14′ E and 28°08′~29°13′ N) at the junction of five provinces (southern Zhejiang, southern Anhui, northeastern Jiangxi, southern Hunan, and northern Fujian), and its distribution once was narrow but continuous [11]. However, because of low natural fruit yield and great ornamental value, *C. chekiangoleosa* are often felled and transplanted in the natural environment. The large-scale excavation of mountains for farmland has also seriously damaged the natural habitat of *C. chekiangoleosa*, resulting in a sharp reduction in its natural resources. As early as the 1990s, due to the shrinking wild habitat, the concentrated distribution of *C. chekiangoleosa* is very rare. Nowadays, it has been listed as a precious and endangered plant in Zhejiang Province, China, and has been listed as a key protected plant in Fujian Province, China [12]. Today, the survival situation of *C. chekiangoleosa* is even very serious, and it is urgent to protect its natural resources.

The key to protecting biodiversity is to maintain the genetic diversity or evolutionary potential of species. The genetic variation levels and population genetic structures of plant populations result from the comprehensive actions of various factors, such as their evolutionary history, distribution range, life form, breeding mode, and seed dispersion mechanism, which are closely related to adaptability and evolutionary potential [13]. Therefore, the assessment of genetic variation levels and spatial structure of plants is the basis for exploring their adaptability, speciation process, patterns, and evolutionary mechanisms, and is related to the formulation of strategies and measures for species protection and population restoration [14,15]. A large amount of evidence shows that the average level of genetic variation in narrowly distributed species is significantly lower than that in widely distributed species [16,17]. However, even among narrowly distributed species, diversity levels and genetic structures are very different [18]. The population genetic structure of a species is also a product of its long-term evolution, and the population genetic structure observed in the endemic habitats of many species may reflect distinct events in their evolutionary history. For example, species with fragmented habitats are more prone to inbreeding depression and reduced diversity [19]. Therefore, the formulation of species protection strategies, on the basis of a full understanding of species diversity levels and population genetic structures, is a reasonable scientific approach. In addition, understanding the population genetic structures of cultivated species and wild relatives can help us to maintain their genetic integrity and diversity during breeding [20], thus allowing the in-depth discovery and utilization of various genes and genotype resources in the population and the prediction and scientific use of the variation of important economic traits.

The genetic diversity of *C. chekiangoleosa*, based on inter-simple sequence repeat (ISSR) markers, was carried out [12]. The previous research results are worth referencing; however, because of the limitations of the applied research methods and sample numbers, the results are relatively limited. Due to highly polymorphic, reproducible, abundant, inherited, co-dominant, and distributed genomes, simple sequence repeat (SSR) markers are very effective molecular markers in population genetics [17]. In recent years, there have been many research reports on the population genetic structure of species in the genus *Camellia* using SSR markers, such as those of *C. petelotii* (C.W. Chi), *C. sinensis* (L.) (Kuntze), *C. huana* (T.L. Ming and W.J. Zhang), and *C. oleifera* [21–24]. Therefore, under the premise of clarifying the resource distribution, the purpose of this study was to use SSR markers to analyze population genetic structure of *C. chekiangoleosa*. This approach can provide a molecular basis for the protection of *C. chekiangoleosa* natural resources and provide a technical reference for their rational utilization.

2. Materials and Methods

2.1. Population Sample Information

Samples were taken from 12 existing natural populations of *C. chekiangoleosa* within a certain scale in 4 provinces within the Jiangxi, Zhejiang, Fujian, and Hunan Provinces (Supplementary Materials Table S1). An appropriate sample size was selected according to the population size, but at least 30 single plants were sampled in each population; the distance between individuals was more than 20~30 m; the sampled plants showed normal growth without diseases or pests. Young tissue samples (young leaves, buds, or flower buds) were collected and stored via the silica gel drying method. Based on the longitude and latitude of the sampling points indicated by a GarminGPS60 device (WGS84 coordinate system), and the sampling map (Figure 1) was drawn by using Global Mapper [25]. The distribution of the geographic information and sample quantities at each sampling point (correspondence between population name and code) are also shown in Table S1.

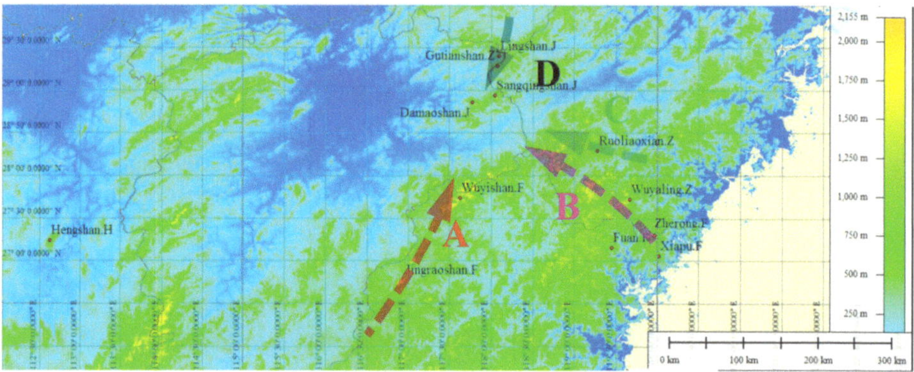

Figure 1. Geographic distribution of the 12 sampled populations of *C. chekiangoleosa*. The dots in the figure are the collection points of each test sample. **A**—Wuyi mountain range population; **B**—Donggong mountain range population; **C**—Xianxia mountain range population; **D**—Huaiyu mountain range population.

2.2. Experimental Methods

Total DNA was extracted via the cetyl trimethylammonium bromide (CTAB) method [26], and a NanoDrop-1000 spectrophotometer (NanoDrop Technologies, Wilmington, DE, USA) was used to detect the concentration of DNA. After electrophoresis in 0.8% agarose gelatin, the integrity of the purified DNA was determined, and qualified DNA samples were stored at $-20\,°C$.

In this study, 16 pairs of highly polymorphic EST-SSR primers (Supplementary Materials Table S2) for *C. chekiangoleosa* were selected out from our previous research [11]. All primers were synthesized by Shanghai Generay Biotech Co., Ltd. The polymerase chain reaction (PCR) system had a volume of 10 µL, which included 1 µL of 10 × PCR buffer, 1.5 mM Mg2+, 100 µM dNTPs, upstream and downstream primers at 0.5 µM, 1.25 U of Taq polymerase (5 U/µL), and 50 ng of DNA. The amplification reaction procedure was as follows: pre denaturation at 94 °C for 5 min; 30 cycles of 94 °C for 30 s, 55 °C for 30 s, and 72 °C for 30 s; a final extension at 72 °C for 1 min; then, storage at 8 °C. During the test, the annealing temperature was increased or decreased according to the melting temperature (Tm) of each primer. The PCR products were uniformly added to 8% polyacrylamide gels, along with a 50 bp labeled fragment size marker (TIANGEN, Beijing, China); standard electrophoresis was performed at a 120 V constant voltage for 2 h; the product was detected by silver staining. Digital photographs were recorded, and Quality One software (Bio–Rad Laboratories Inc., California, USA) was used to quantitatively analyze the fragment size of each detection site according to the standard molecular weight of the 50 bp marker; finally, the data were finally counted.

2.3. Data Analysis

The electrophoretic bands were read according to codominant markers, and the sizes of the allele fragments at each point were recorded. The data were imported into Excel according to the requirements. The polymorphism information content index (PIC) and linkage disequilibrium were calculated by using PowerMarker 3.25 (bioinformatics research center; North Carolina State University, Raleigh, NC, USA) [27], and Bonferroni correction was performed to adjust the *P* value. Popgene 1.32 software [28] was used to calculate the following: number of observed alleles (Na); number of effective alleles (Ne); Nei's gene diversity (h); the Shannon diversity index (I); observed heterozygosity (Ho); expected heterozygosity (He); the inbreeding index (F). The genetic differentiation index (G_{ST}, based on He; F_{ST}, based on an infinite allele model, IAM; R_{ST}, based on a stepwise mutation model—SMM), gene flow (Nm = $(1-F_{ST})/4F_{ST}$), and allele richness (AR) were calculated across all populations at each locus and over all loci using FSTAT 2.9.3 [29].

The genotype data format was transformed by using Genelex 6.4 [30]; an analysis of molecular variance (AMOVA) was carried out for the sources of different levels of genetic variation within and among populations using Arlequin 3.5 [31]. The results of the analysis of the genetic relationships between populations were displayed by factorial correspondence analysis (FCA) with the analysis software Genetix 4.05 (CNRSUMR 5000; Universite Montpellier II, Montpellier, France) [32]. The isolation by distance pattern (IBD) was detected by Mantel tests with 1000 permutations based on matrices of pairwise genetic differentiation (Multilocus estimates of genetic differentiation, Fst, expressed as Fst/(1−Fst)), and geographic distance among populations were performed in NTSYS 2.10e [33–35]. STRUCTURE 2.3.1 software [36] was used to infer the population genetic structure with the allelic model. Moreover, the number of distinguishable groups was initially set between k = 1 and k = 15, and the false positioning points were independent. The length of the burn-in period at the beginning of the Markov chain Monte Carlo simulation algorithm (MCMC) was set to 50,000 times, and the MCMC after non-counting iteration was then set to 50000 times for the estimation of the α value to determine whether a separate population existed ($\alpha \leq 0.2$ was the standard for the classification of the test samples). A suitable value of was selected according to the size of ΔK. ΔK is the difference in LnP(D) based on adjacent K values. Finally, Bottleneck 1.2 software [37] was used to determine whether each group had experienced a bottleneck effect in its recent history. A two-phase mutation model (TPM) and a stepwise mutation model (SMM) were used for Wilcoxon signed rank tests. The parameter settings included a 90% SMM and 10% TPM with 12% variation in TPMs and 1000 repeats.

3. Results

3.1. Genetic Diversity

Using 16 pairs of primers to detect 528 individuals in 12 geographical populations of *C. chekiangoleosa* (Table 1), a total of 74 alleles were obtained. The amplified alleles of a single primer ranged from 3 to 7. The average and effective allele values were 4.625 and 2.682, respectively. The highest amplified effective allele value obtained was for CC_eSSR03 (3.311), and the lowest obtained was for CC_eSSR16 (1.545). The average PIC value of the 16 loci was 0.616, reflecting a high polymorphism level, among which CC_eSSR16 showed the lowest PIC (0.352), and CC_eSSR03 showed the highest (0.698). These results were consistent with the effective equivalent gene comparison result, and the average h at the species level reached 0.596.

Table 1. Information on the polymorphism of 16 microsatellite loci amplified in *C. chekiangoleosa*.

Locus	Na	Ne	PIC	h	G_{ST}	R_{ST}	F_{ST}	Nm
CC_eSSR03	5	3.311	0.698	0.695	0.288	0.532	0.295	0.597
CC_eSSR15	5	1.985	0.496	0.468	0.213	0.086	0.215	0.913
CC_eSSR16	5	1.545	0.352	0.329	0.255	0.340	0.264	0.697
CC_eSSR29	4	2.641	0.621	0.625	0.352	0.486	0.372	0.422
CC_eSSR37	4	3.236	0.691	0.690	0.266	0.229	0.254	0.734
CC_eSSR41	4	2.643	0.622	0.623	0.181	0.316	0.189	1.073
CC_eSSR48	4	1.933	0.483	0.460	0.195	0.143	0.201	0.994
CC_eSSR49	5	2.560	0.609	0.589	0.183	0.260	0.188	1.080
CC_eSSR55	4	2.534	0.605	0.618	0.358	0.358	0.347	0.470
CC_eSSR83	3	2.283	0.562	0.562	0.169	0.140	0.171	1.212
CC_eSSR85	4	2.834	0.647	0.646	0.194	0.148	0.203	0.982
CC_eSSR87	5	3.057	0.673	0.660	0.371	0.447	0.376	0.415
CC_eSSR89	4	2.776	0.640	0.641	0.210	0.122	0.215	0.913
CC_eSSR92	5	2.421	0.595	0.569	0.195	0.157	0.206	0.964
CC_eSSR95	7	3.090	0.806	0.668	0.240	0.169	0.249	0.754
CC_eSSR101	6	4.056	0.761	0.699	0.067	0.113	0.071	3.271
Mean	4.625	2.682	0.616	0.596	0.234	0.253	0.239	0.796

Na—observed number of alleles; Ne—effective number of alleles; h—Nei's diversity index; PIC—polymorphism information content; I—Shannon's information index; G_{ST}—genetic differentiation index based on Nei's diversity; F_{ST}—genetic differentiation index based on IAM; R_{ST}—genetic differentiation index based on SMM.

The results of the genetic diversity analysis of each geographical population of *C. chekiangoleosa* are shown in Table 2. Overall, the genetic diversity levels of different geographical populations of *C. chekiangoleosa* were quite different. AR ranged from 1.625 (HS) to 3.982 (GTS), with an average of 3.098, and the average He was 0.458. The SQS population presented the highest level of genetic diversity and the highest value of He, at 0.575. The lowest He value was found in the HS group, at 0.160. The 12 populations were ranked from high to low, according to the values He, as follows: SQS > DMS > GTS > LS/RLX > WYS > XP > tr > JRS > WYL > FA > HS. The diversity of LS and RLX was similar. The diversity parameter I ranged from 0.290 to 1.029 in the populations, with an average of 0.787. Both He and I show that the diversity of each population was maintained at an intermediate level. The results for the 12 populations sorted according to He and I were different, but the highest values appeared among DMS, SQS, and GTS, and the lowest values were obtained in the HS population, which also showed the lowest percentage of polymorphic loci among the 12 populations (43.75%). The above results indicate the reliability of genetic diversity evaluation.

The fixed index (F), also known as the inbreeding coefficient, can reflect the gene flow and inbreeding of the population. The F value of the natural *C. chekiangoleosa* population in this study was greater than 0, and the F value of the HS population was 0.087, which was the closest value to 0. The fixed indexes (F) of the RLX population and WYL population were relatively large, at 0.345 and 0.360, respectively, but those of the other populations were lower; the average F was 0.205, which generally indicated that the population deviated from the Hardy–Weinberg equilibrium, as reflected in the excess of homozygotes (Table 2).

Table 2. Genotypic variation and heterozygosity statistics of 16 loci of *C. chekiangoleosa*.

Pop	Sample Size	Allelic Richness	Ho [b]	He [b]	I	F	Bottleneck Test r	
							TPM	SMM
GTS	50	3.829	0.439	0.566	1.009	0.216 *	0.39098	0.86026
HS	36	1.625	0.160	0.177	0.290	0.087	0.03906 ▲	0.07813
WYS	50	3.245	0.369	0.509	0.873	0.268 *	0.10458	0.21143
SQS	50	3.694	0.478	0.575	1.005	0.162 *	0.17535	0.40375
LS	48	3.224	0.458	0.539	0.912	0.140 *	0.02139 ▲	0.03864 ▲
JRS	48	2.589	0.309	0.412	0.667	0.242 *	0.07300	0.12054
ZR	30	2.688	0.360	0.418	0.673	0.124	0.34839	0.40375
FA	32	2.793	0.285	0.357	0.606	0.188 *	0.33026	0.15143
WYL	40	2.961	0.258	0.408	0.705	0.360 *	0.85522	0.54163
RLX	48	3.508	0.349	0.539	0.936	0.345 *	0.32251	0.97995
XP	48	3.040	0.372	0.425	0.739	0.114	0.80396	0.59949
DMS	48	3.982	0.444	0.569	1.029	0.211 *	0.93988	0.63217
Mean	44	3.098	0.357	0.458	0.787	0.205		

Na—number of alleles; AR—allelic richness; Ho—observed heterozygosity; He—expected heterozygosity; F—inbreeding coefficient; TPM—a two-phase mutation model; SMM—stepwise mutation model; [b]—heterozygosity at the population level; *—significant deviation from the Hardy–Weinberg equilibrium; ▲—significant deviation from Wilcoxon signed rank tests ($p < 0.05$).

3.2. Genetic Differentiation

The level of population genetic differentiation was analyzed based on Nei's genetic diversity, IAM, and SMM. The statistics showed that the average GST, FST, and RST were 0.234, 0.239, and 0.253, respectively. It indicated that there was a high level of genetic differentiation among the *C. chekiangoleosa* populations (Table 1). Additionally, the AMOVA of the components within and among populations performed with Arlequin 3.5 software also showed that there was significant genetic variation ($p < 0.001$), where 24.15% of the genetic variation existed among populations, while 75.85% of the genetic variation occurred within populations (Table 3). In addition, the gene flow (Nm), estimated based on F_{ST}, was consistent with that estimated based on G_{ST} (Table 1; Supplementary Materials Table S3), with values of 0.796 and 0.731, respectively, representing the numbers of individuals that entered or exited a population in each generation among all populations studied. The measured values showed that the gene flow among populations was low, indicating relatively little exchange of genetic information among populations. The pairwise F_{ST} values between populations are shown in Supplementary Materials Table S4. The genetic differentiation between the SQS and DMS populations was lowest, at only 0.0901. The close geographical distances of the populations and their altitude may be the reasons for this phenomenon.

Table 3. Analysis of molecular variance (AMOVA) within and among populations.

Source of Variation	d.f.	Sum of Squares	Variance Components	Percentage Variation	Probability
Among populations	11	1191.53	1.191	24.15	<0.001
Within populations	1044	3906.02	3.741	75.85	<0.001
Total	1055	5097.55	4.932		

In this study, the TPM model and the SMM model of the Wilcoxon signed rank test [37] were used to analyze the bottleneck effect. According to the data (see Table 2), with the exception of the HS and LS groups, which recently experienced bottleneck effects ($p < 0.05$), no other groups experienced bottleneck effects.

According to 3D graphics of factorial correspondence analysis, the 12 populations were divided into 3 groups: Group I, Group II, and Group III. Group I was the western cluster, and only contained HS groups; Group II was a central cluster with DMS at its core, which was distributed on the ridge connecting the Huaiyu Mountain Range to Wuyi Mountain Range in the south, and included the LS, GTS, SQS, DMS, RLX, WYS, and LRS populations; Group III was an eastern coastal cluster with WYL at its core, including the WYL, ZR, FA, and XP populations (Figure 2). There were certain genetic differences among the groups, especially between the HS population and the other populations.

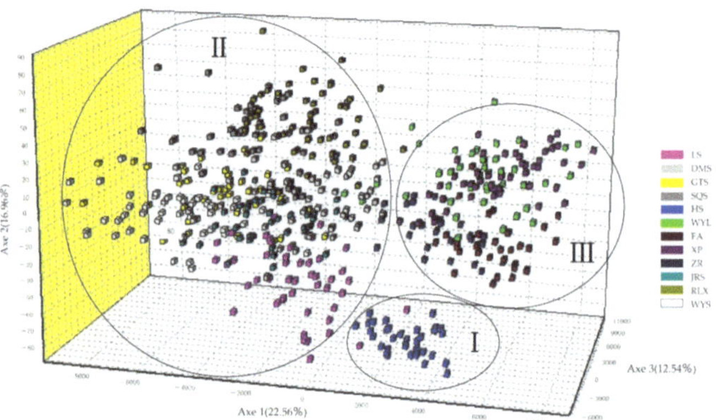

Figure 2. Factorial correspondence analysis of the *C. chekiangoleosa* population. Axes 1, 2, and 3 explain 22.56, 16.96, and 12.54% of the genetic variation, respectively.

In addition, the correlation analysis between genetic differentiation ($F_{ST}/1 - F_{ST}$) and spatial geographical distance was conducted (Supplementary Materials Table S3) to check whether isolation by distance (IBD) existed. The correlation analysis showed that the correlation between the 2 groups was very significant ($r = 0.763$, $p < 0.001$), indicating that geographical isolation is the main cause of genetic differentiation (Figure 3).

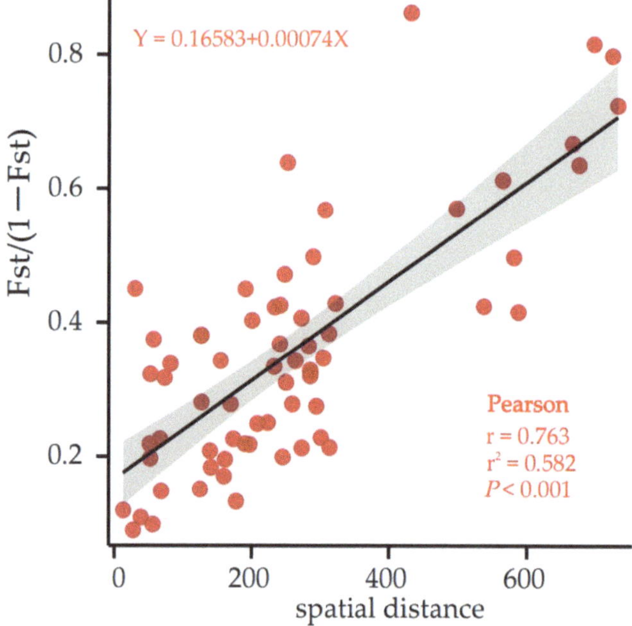

Figure 3. Matrix plot of the Mantel test of genetic differentiation and geographic distance.

3.3. Genetic Structure

The genetic structure of the 12 studied populations was inferred using the STRUC-TURE software. The most probable division with the highest ΔK value was detected at K = 10 (Supplementary Materials, Figure S1a,b). Figure 4 showed the genetic structure of the 12 *C. chekiangoleosa* populations, simulated based on the mathematical model. Each color represents a genetic cluster, and the colored segments show the individual's estimated ancestry proportion to each of the genetic clusters. The 12 populations were clustered into 10 clusters. The DMS population presented the most gene introgression from the GTS, WYS, SQS, LS, and XP populations. The GTS population presented the most gene introgression from the WYS population and LS population. However, the ZR, FA, and WYL populations were mixed, and the degree of differentiation between these populations was extremely weak. Genes from the XP and LS populations mainly infiltrated the mixed group composed of the last three populations. In the SQS, LS, and XP populations, there were also some individuals with similar genetic backgrounds to the GTS population, and the HS population was the most independent.

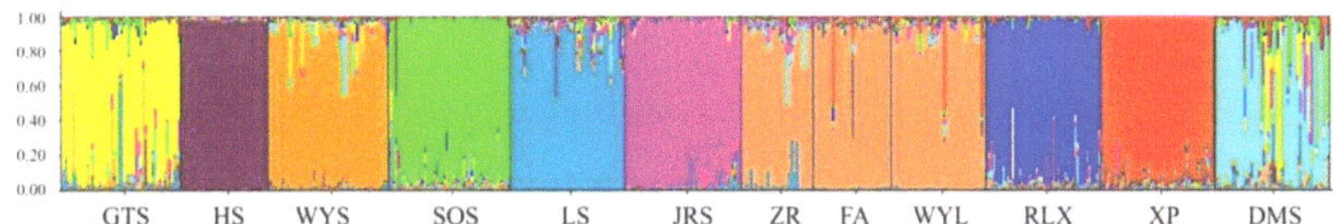

Figure 4. Population genetic structure based on the Bayesian clustering model among 528 samples of *C. chekaingoleosa* at K = 10. Each color represents a genetic cluster.

4. Discussion

4.1. Evaluation of Genetic Diversity in Camellia chekiangoleosa

Previous research has shown that the geographical distribution range of a species often presents a strong correlation with its genetic diversity [23]. Especially, compared with widely distributed plant species, species with narrow geographical ranges or endangered species theoretically show lower genetic diversity [38]. In this study, *C. chekiangoleosa* showed low genetic diversity (h = 0.596) at the species level, that was higher than the average He (when the calculation was based only on polymorphism and double allele loci, this parameter was equivalent to Nei's genetic diversity h) of endangered species (0.420) but lower than that of widely distributed species (0.620) [17]. The results are consistent with previous studies. The diversity level according to similar markers was even lower than that in other widely distributed related species. A total of 96 pairs of EST-SSR primers were used to evaluate the genetic diversity of *C. sinensis* in the main production areas of China, and the value of h reached 0.64 [39]. When 4 pairs of SSR molecular markers were used to evaluate the genetic diversity of the ancient *C. japonica* population, the He value reached 0.84 [40], which was higher than the genetic diversity of the wild populations of *C. chekiangoleosa*. Meanwhile, the assessment of population genetic diversity can reveal the species adaptability to the different environment [41]. Among the *C. chekiangoleosa* populations, the SQS, DMS, LS, and GTS populations, with high genetic diversity, are concentrated in the area linking the Wuyi Mountain Range population and the Huaiyu Mountain Range population in the north, which we believe is suitable establish areas for *C. chekiangoleosa*. In addition, because of the barrier imposed by the Heng Mountains, the diversity of the HS population in the west was lowest under the pressure accumulation of environmental selection.

It is helpful to analyze the reasons for the low genetic diversity of *C. chekiangoleosa* especially when compared with related species. In the genus *Camellia*, some endangered tree species with narrow distributions also show low genetic diversity [42]. When researchers evaluated the diversity of 12 wild populations of the narrowly distributed tree species *C. huana* with chloroplast fragments and 12 pairs of SSR primers, the obtained He, based on SSRs, was 0.466 [22]. Chen analyzed the genetic diversity of 7 populations of 3 endangered

Camellia species (*C. chrysanthoides* (H.T. Chang), *C. micrantha* (S. Yun Liang and Y.C. Zhong), *C. parvipetala* (J.Y. Liang and Z.M. Su)) based on SSRs, and the calculated He values were 0.379~0.543 [43]. Li and Lu carried out genetic diversity studies on the endangered tree species *C. petelotii* and its variant *C. petelotii* var. *microcarpa* (S.L. Mo) (T.L. Ming and W.J. Zhang), and the obtained He values based on SSR markers were 0.546 and 0.533, respectively [20,44]. The genetic diversity of the above related species was thus shown to be similar or slightly higher than that recorded in our study. This is related to the fact that most populations of *C. chekiangoleosa* are not only affected by land loss but are also seriously disturbed by human beings. In addition, certain inbreeding phenomena were identified in each population, and the F values of the WYL and RLX populations reached 0.36 and 0.345, indicating that excess homozygosity is one of the internal reasons for the reduction in genetic diversity [45]. Some individual populations, such as the HS population, have experienced bottleneck effects and genetic drift, resulting in the loss of genetic diversity.

4.2. Population Genetic Structure Analysis Reveals Significant Differentiation of Populations

Compared with the genetic differentiation coefficients of 106 plant species based on SSR molecular markers [17], the genetic differentiation level of *C. chekiangoleosa* populations (F_{ST} = 0.239) was close to the average level of narrowly distributed species (F_{ST} = 0.23), but lower than widely distributed species (F_{ST} = 0.25). Therefore, *C. chekiangoleosa* presents a high differentiation level as a narrowly distributed species. AMOVA showed that 24.15% of the observed variation occurred among the populations of *C. chekiangoleosa*, which was similar to findings in the related species *C. huana* with a narrow distribution [23]. The result also reflected the fact that the population variation of *C. chekiangoleosa* was mainly attributed to the variation between individuals within the population. Moreover, our results (G_{ST} = 0.234) are lower than previous studies on the genetic differentiation of *C. chekiangoleosa* (G_{ST} = 0.3758) [12], which may be caused by different markers, and the population chosen in our study had higher coverage density.

Natural selection is the most important evolutionary driving force leading to population differentiation, while gene flow is an important factor hindering population differentiation [46]. In addition to the distribution characteristics mentioned above, the diffusion mechanism of plant pollen and seeds also impacts the average gene flow between populations [47]. The seeds of *C. chekiangoleosa* exhibit similar large and heavy traits to *C. japonica* and *C. flavida* (H.T. Chang), which are mainly transmitted by gravity [40,48]. *C. chekiangoleosa* is mainly distributed in high mountains at altitudes of 600~1400 m and is pollinated by insects. Therefore, mountains hinder the diffusion of seeds and pollen to a certain extent, thus limiting the spread of seeds and pollen [49]. The F_{ST}-based gene flow value (Nm) of *C. chekiangoleosa* was only 0.796, which is in line with the above statement. Slatkin concluded that, if the Nm of migrating individuals per generation is less than 1, genetic drift will become the dominant factor dividing the genetic structure of a population [50,51]. In this study, there was a significant positive correlation between spatial distance and genetic differentiation that reflected a significant geographical isolation effect among the populations of *C. chekiangoleosa*, which was consistent with the division of population genetic structure caused by lower gene flow [51].

The small-scale distribution of genetic variation within populations, and the large-scale spatial distribution among populations, are two of the important characteristics of population genetic structure [41]. This study showed that 12 geographical populations of *C. chekiangoleosa* were divided into 10 groups, and there was gene infiltration among the 10 groups, which also reflected the continuity of the population historical distribution of *C. chekiangoleosa*. The HS population was independent of several other populations, and the genotypes within individuals of this population were relatively distinct due to genetic drift. In addition, HS was located far from other populations, so it can be expected that further genetic drift will be revealed in this population. From the perspective of gene diversity and the Shannon index, it is found that the DMS population showed a relatively high level of diversity. The reason may be that the frequent activities of insects lead to the intensification of gene interaction in the low altitude environment with high average temperature. In addition, the combination of the FA, ZR, and WYL populations may have been related to the fact that they were all located in the Donggong mountain range, and these individuals may have come from the same or similar ancestors. In summary, we infer that *C. chekiangoleosa* populations can be basically divided into 3 distribution clusters according to the division of mountain ranges: a central cluster, with DMS as the center, which was distributed on the ridge connected with 4 mountain ranges (Figure 1A–D); an eastern coastal cluster with WYL as the center; a western cluster with HS as the center.

4.3. Conservation of Camellia chekiangoleosa Genetic Resources

In summary, this study indicates that the fundamental reason for the endangered status of *C. chekiangoleosa* is the restriction of gene flow among its populations, resulting in a reduction in genetic diversity and an inability to maintain the ability to adapt to the changing environment [52]. First, similar to other *Camellia* species, the pollen transmission mode of *C. chekiangoleosa* depends mainly on insects, while its seeds are mainly transmitted by gravity and animals, which are affected by mountain isolation [48]. Additionally, human activities have exacerbated the fragmentation of the natural habitat of *C. chekiangoleosa*, which is bound to further reduce the pollen flow between groups. In addition, according to the research of Yang [53], the pollen of *C. chekiangoleosa* shows a short survival time and a long pollen tube germination time, which is likely to result in abortion if flowers are not pollinated in a timely manner. According to our field investigation, due to the low average temperature in high-altitude mountainous areas, the flowering period of *C. chekiangoleosa* is long, from November to March of the next year. However, the flowering branches growing on the sunny side of the outer layer of the canopy bloom earlier and often bloom in the inner layer of the canopy, while the flowers in the outer layer are overmature. Therefore, the asynchronous flowering periods within a given stand affect the fruit setting of *C. chekiangoleosa*, imposing high environmental selection pressure on the existing population. If the protection of the species is not strengthened, its genetic differentiation will further increase, and the selfing rate of the remaining population and the probability of genetic drift will also increase, which will continue to reduce genetic diversity.

According to the results of the genetic diversity assessment and the genetic structure analysis, priority should be given to the local protection of populations with high diversity, such as the DMS, SQS, LS, and GTS populations. Protected areas or stations can be established to strengthen management and eliminate logging and damage. Second, the habitats suitable for the growth and population reproduction of *C. chekiangoleosa* should be restored with the aim of increasing the heterogeneity and stability of the habitat, which is particularly important for small populations. For populations with low genetic diversity (HS, FA), seeds should be collected within the population (or the most genetically similar population) whenever possible, and artificial seedling cultivation should be carried out according to the biological characteristics of the plants. These plants can then be reintroduced to the original habitat so that the population can slowly recover and grow—on the basis of preserving the original genotype.

5. Conclusions

In our research, natural populations of *C. chekiangoleosa*, covering the main distributed range, were used to study genetic diversity and differentiation. *C. chekiangoleosa*, as an endemic species with narrow distribution, has low genetic diversity, but its central population still has relatively high genetic diversity. In addition, the level of genetic differentiation among populations is high, and the genetic structure analysis also found that Hengshan population, central populations, and eastern populations can be divided into independent groups, indicating that the isolation of mountains may be the main reason for the formation of genetic differentiation. Therefore, considering the genetic diversity difference and significant differentiation of *C. chekiangoleosa* population, appropriate diversity protection strategies should be adopted according to the genetic diversity level and geographical distribution of different populations.

Supplementary Materials: The following supporting information can be downloaded at: https://www.mdpi.com/article/10.3390/f13020234/s1, Table S1: Geographical distribution of 12 *C. chekiangoleosa* geographic populations, Table S2: Repeat motif, primer sequence, fragment size, Tm and data sources information for 16 EST-SSR loci, Table S3: Genetic differentiation in *C. chekiangoleosa* populations by Ney's, Table S4: Pair-wise F_{ST} value and spatial distance matrix of *C. chekiangoleosa*, Figure S1: (a) The mean log-likelihood value of the data was based on ten repetitions for each K value, (b) The delta K value was changed with each K value.

Author Contributions: This study was carried out with collaboration among all authors. L.-a.X. conceived and designed the experiments; Q.W. and Z.W. performed the experiments; J.H. and Q.W. checked the experimental results; X.L. and H.Z. provided the experimental materials; B.H. and Q.W. wrote the paper. All authors have read and agreed to the published version of the manuscript.

Funding: This work was supported by the National Natural Science Foundation of China (31860179 and 31260184), Key Research and Development Program of Jiangxi Province, China (20201BBF61003 and 20161BBF60122), Science and Technology Innovation Bases Program of Jiangxi province, China (20212BCD46002), Doctor Initial Project of JiangXi Academy of Forestry (2021521001), Oil-tea special research project of Jiangxi Provincial Department of Forestry (YCYJZX20220103).

Acknowledgments: We are very grateful to the Li-an Xu for initiating research ideas.

Conflicts of Interest: The authors declare no conflict of interest.

References

1. Fan, M.L.; Kai, Y.; Rui, Z.; Liu, H.Q.; Xiao, G.; Sun, K.Y. Temporal transcriptome profiling reveals candidate genes involved in cold acclimation of *Camellia japonica* (Naidong). *Plant Physiol. Biochem.* **2021**, *167*, 795–805. [CrossRef]
2. Zhang, J.; Ying, Y.; Yao, X. Effects of turning frequency on the nutrients of *Camellia oleifera* shell co-compost with goat dung and evaluation of co-compost maturity. *PLoS ONE* **2019**, *14*, e0222841. [CrossRef] [PubMed]
3. Zhou, L.Y.; Wang, X.N.; Wang, L.P.; Chen, Y.Z.; Jiang, X.C. Genetic diversity of oil-tea *camellia* germplasms revealed by ISSR analysis. *Int. J. Biomath.* **2015**, *5*, 331–354. [CrossRef]
4. Zhou, J.; Ai, Z.; Wang, H.; Niu, G.; Yuan, J. Phosphorus Alleviates Aluminum Toxicity in *Camellia oleifera* Seedlings. *Int. J. Agric. Biol.* **2019**, *21*, 237–243. [CrossRef]
5. Fan, L.L.; Huang, J.S. Investigation on Species and Verieties pattern of *Camellia* sp. in Suichuan County. *South China For. Sci.* **2001**, *4*, 16–18. [CrossRef]
6. Wang, X.; Zeng, Q.; María, D.; Wang, L.J. Profiling and quantification of phenolic compounds in *Camellia* seed oils: Natural tea polyphenols in vegetable oil. *Food Res. Int.* **2017**, *102*, 184–194. [CrossRef] [PubMed]
7. He, Y.C.; Wu, M.J.; Dong, L.; Wen, Q.; Li, T.; Li, X.H. Analysis of kernel oil content and variation of fatty acid composition of *Camellia chekiangoleosa* in the main producing areas. *Non-Wood For. Res.* **2020**, *3*, 37–45. [CrossRef]
8. Guo, H.; Tan, H.Y.; Zhou, J.P. Proximate composition of *Camellia chekiangoleosa* Hu fruit and fatty acid constituents of its seed oil. *J. Zhejiang Univ.* **2010**, *36*, 662–669. [CrossRef]
9. Xie, Y.; Wang, X. Comparative transcriptomic analysis identifies genes responsible for fruit count and oil yield in the oil tea plant *Camellia chekiangoleosa*. *Sci. Rep.* **2018**, *8*, 6637. [CrossRef]
10. Zhou, W.C.; Wang, Z.W.; Dong, L.; Wen, Q.; Huang, W.Y.; Li, T.; Ye, J.S.; Xu, L.A. Analysis on the character diversity of fruit and seed of *Camellia chekiangoleosa*. *J. Nanjing For. Univ. Nat. Sci. Ed.* **2021**, *45*, 51–59. [CrossRef]
11. Wen, Q.; Xu, L.; Gu, Y.; Huang, M.; Xu, L.A. Development of polymorphic microsatellite markers in *Camellia chekiangoleosa* (Theaceae) using 454-ESTs. *Am. J. Bot.* **2012**, *99*, e203–e205. [CrossRef] [PubMed]
12. Xie, Y.; Li, J.Y.; Zhang, D.L. Assessment of Genetic Diversity and Population Structure of Endangered *Camellia chekiangoleosa* Hu Using Issr Markers. *Pak. J. Bot.* **2018**, *50*, 1965–1970.
13. Hughes, A.R.; Inouye, B.D.; Johnson, M.T.J.; Underwood, N.; Vellend, M. Ecological consequences of genetic diversity. *Ecol. Lett.* **2008**, *11*, 609–623. [CrossRef] [PubMed]
14. Kobori, H. Conservation Education Based on the Principles of Conservation Biology. *Jpn. J. Environ. Educ.* **2009**, *19*, 77–78. [CrossRef]
15. Gentili, R.; Fenu, G.; Mattana, E.; Citterio, S.; Mattia, F.D.; Bacchetta, G.; Peeters, T. Conservation genetics of two island endemic Ribes spp. (Grossulariaceae) of Sardinia: Survival or extinction? *Plant Biol.* **2015**, *17*, 1085–1094. [CrossRef] [PubMed]
16. Fu, Y.B. Understanding crop genetic diversity under modern plant breeding. *Theor. Appl. Genet.* **2015**, *128*, 2131–2142. [CrossRef]
17. Nybom, H. Comparison of different nuclear DNA markers for estimating intraspecific genetic diversity in plants. *Mol. Ecol.* **2004**, *13*, 1143–1155. [CrossRef]

18. Erfanian, M.B.; Sagharyan, M.; Memariani, F.; Ejtehadi, H. Predicting range shifts of three endangered endemic plants of the Khorassan-Kopet Dagh floristic province under global change. *Sci. Rep.* **2021**, *11*, 9159. [CrossRef]

19. Oakley, C.G.; Winn, A.A. Effects of population size and isolation on heterosis, mean fitness, and inbreeding depression in a perennial plant. *New Phytol.* **2012**, *196*, 261–270. [CrossRef]

20. Gentili, R.; Solari, A.; Diekmann, M.; Duprè, C.; Citterio, S. Genetic differentiation, local adaptation and phenotypic plasticity in fragmented populations of a rare forest herb. *Peer J.* **2018**, *6*, e4929. [CrossRef]

21. Lu, X.L.; Chen, H.L.; Liang, X.Y.; Tang, S.Q. Genetic Diversity Analysis of *C. nitidissima* var. *microcarpa* and Peripheral Population of *Camellia nitidissima*. *Mol. Plant Breed.* **2019**, *17*, 301–306. [CrossRef]

22. Liu, S.; Liu, H.; Wu, A.; Hou, Y.; An, Y.; Wei, C.L. Construction of fingerprinting for tea plant (*Camellia sinensis*) accessions using new genomic SSR markers. *Mol. Breed.* **2017**, *37*, 93. [CrossRef]

23. Li, S.; Liu, S.L.; Pei, S.Y.; Ning, M.M.; Tang, S.Q. Genetic diversity and population structure of *Camellia huana* (Theaceae), a limestone species with narrow geographic range, based on chloroplast DNA sequence and microsatellite markers. *Plant Divers.* **2020**, *42*, 343–350. [CrossRef] [PubMed]

24. He, Z.; Liu, C.; Wang, X.; Wang, R.; Tian, Y. Assessment of genetic diversity in Camellia oleifera Abel. accessions using morphological traits and simple sequence repeat (SSR) markers. *Breed. Sci.* **2020**, *70*, 586–593. [CrossRef]

25. Liu, F.L.; Xiao, B. Global Mapper System and Its Application in Marine Geological Survey. *Ocean Technol.* **2011**, *30*, 24–26.

26. Cota-Sánchez, J.H.; Remarchuk, K.; Ubayasena, K. Ready-to-use DNA extracted with a CTAB method adapted for herbarium specimens and mucilaginous plant tissue. *Plant Mol. Biol. Rep.* **2006**, *24*, 161–167. [CrossRef]

27. Liu, K.J.; Muse, S.V. PowerMarker: An integrated analysis environment for genetic marker analysis. *Bioinformatics* **2005**, *21*, 2128–2129. [CrossRef]

28. Yeh, F.; Yang, R.; Boyle, T. *POPGENE Version 1.32 Microsoft Windows-Based Freeware for Populations Genetic Analysis*; University of Alberta: Edmonton, AB, Canada, 1999.

29. Goudet, J. FSTAT, a Program to Estimate and Test Gene Diversities and Fixation Indices, Version 2.9.3. 2001. Available online: http://www.unil.ch/popgen/softwares/fstat.htm (accessed on 20 June 2021).

30. Peakall, R.; Smouse, P.E. Genalex 6: Genetic analysis in Excel. Population genetic software for teaching and research. *Mol. Ecol. Notes* **2006**, *6*, 288–295. [CrossRef]

31. Excoffier, L.; Lischer, H. Arlequin suite ver 3.5: A new series of programs to perform population genetics analyses under Linux and Windows. *Mol. Ecol. Resour.* **2010**, *10*, 564–567. [CrossRef]

32. Belkhir, K.; Borsa, P.; Chikhi, L.; Raufaste, N.; Bonhomme, F. *GENETIX (ver.4.02): Logiciel sous Windows TM pour la Genetique des Populations, Laboratoire Genome, Populations, Interactions CNRSUMR 5000*; Universite Montpellier II: Montpellier, France, 2004.

33. Manel, S.; Schwartz, M.K.; Luikart, G.; Taberlet, P. Landscape genetics combining landscape ecology and population genetics. *Trends Ecol. Evol.* **2003**, *18*, 189–197. [CrossRef]

34. Travadon, R.; Sache, I.; Dutech, C.; Stachowiak, A.; Marquer, B.; Bousset, L. Absence of isolation by distance patterns at the regional scale in the fungal plant pathogen *Leptosphaeria maculans*. *Fungal Biol.* **2011**, *115*, 649–659. [CrossRef] [PubMed]

35. Rohlf, F.J. *NTSYS-pc: Numerical Taxonomy and Multivariate Analysis System, Version 2.2, 2.1*; Department of Ecoloy and Evolution, State University of New York: New York, NY, USA, 2005.

36. Evanno, G.; Regnaut, S.; Goudet, J. Detecting the number of clusters of individuals using the software STRUCTURE: A simulation study. *Mol. Ecol.* **2005**, *14*, 2611–2620. [CrossRef] [PubMed]

37. Piry, S.; Luikart, G.; Cornuet, J.M. BOTTLENECK: A computer program for detecting recent reductions in the effective population size using allele frequency data. *J. Hered.* **1999**, *90*, 502–503. [CrossRef]

38. Solórzano, S.; Arias, S.; Dávila, P. Genetics and Conservation of Plant Species of Extremely Narrow Geographic Range. *Diversity* **2016**, *8*, 31. [CrossRef]

39. Yao, M.Z.; Ma, C.L.; Qiao, T.T.; Jin, J.Q.; Liang, C. Diversity distribution and population structure of tea germplasms in China revealed by EST-SSR markers. *Tree Genet. Genomes* **2012**, *8*, 205–220. [CrossRef]

40. Ueno, S.; Tomaru, N.; Yoshimaru, H.; Manabe, T.; Yamamoto, S. Genetic structure of *Camellia japonica* L. in an old-growth evergreen forest, Tsushima, Japan. *Mol. Ecol.* **2000**, *9*, 647–656. [CrossRef]

41. Zhou, P.Y.; Hui, L.X.; Huang, S.-J.; Ni, Z.X.; Yu, F.-X.; Xu, L.A. Study on the Genetic Structure Based on Geographic Populations of the Endangered Tree Species: Liriodendron chinense. *Forests* **2021**, *12*, 917. [CrossRef]

42. Garg, K.M.; Chattopadhyay, B. Gene Flow in Volant Vertebrates: Species Biology, Ecology and Climate Change. *J. Indian Inst. Sci.* **2021**, *101*, 165–176. [CrossRef]

43. Chen, H.; Xuelin, L.U.; Quanqing, Y.E.; Tang, S.Q. Genetic diversity and structure of three yellow *Camellia* species based on SSR markers. *Guihaia* **2019**, *39*, 318–327. [CrossRef]

44. Li, X.L.; Wang, J.; Fan, Z.Q.; Li, J.Y.; Yin, H.F. Genetic diversity in the endangered *Camellia nitidissima* assessed using transcriptome-based SSR markers. *Trees* **2020**, *34*, 543–552. [CrossRef]

45. Shi, X.; Wen, Q.; Cao, M.; Guo, X.; Xu, L.A. Genetic Diversity and Structure of Natural Quercus variabilis Population in China as Revealed by Microsatellites Markers. *Forests* **2017**, *8*, 495. [CrossRef]

46. Cristian, T.D.; Eduardo, R.; Fidelina, G.; Glenda, F.; Cavieres, L.A. Genetic Diversity in *Nothofagus alessandrii* (Fagaceae), an Endangered Endemic Tree Species of the Coastal Maulino Forest of Central Chile. *Ann. Bot.* **2007**, *100*, 75–82. [CrossRef]

47. Miao, C.Y.; Yang, J.; Mao, R.L.; Li, Y. Phylogeography of Achyranthes bidentata (Amaranthaceae) in China's Warm-Temperate Zone Inferred from Chloroplast and Nuclear DNA: Insights into Population Dynamics in Response to Climate Change During the Pleistocene. *Plant Mol. Biol. Rep.* **2017**, *35*, 166–176. [CrossRef]

48. Peng, G.; Tang, S. Fine-scale spatial genetic structure and gene flow of *Camellia flavida*, a shadetolerant shrub in karst. *Acta Ecol. Sin.* **2017**, *37*, 7313–7323. [CrossRef]

49. Gao, Y.; Ai, B.; Kong, H.H.; Hang, H.W. Geographical pattern of isolation and diversification in karst habitat islands: A case study in the Primulina eburnea complex. *J. Biogeogr.* **2015**, *42*, 2131–2144. [CrossRef]

50. Slatkin, M. A measure of population subdivision based on microsatellite allele frequencies. *Genetics* **1995**, *139*, 457–462. [CrossRef] [PubMed]

51. Slatkin, M. Gene flow and the geographic structure of natural populations. *Science* **1987**, *236*, 787–792. [CrossRef]

52. Bhattacharyya, P.; Kumaria, S. Molecular characterization of *Dendrobium nobile* Lindl., an endangered medicinal orchid, based on randomly amplified polymorphic DNA. *Plant Syst. Evol.* **2015**, *301*, 201–210. [CrossRef]

53. Yang, Z.L.; Ji-Yuan, L.I.; Fan, Z.Q. Effect of Storage Temperature on Pollen Germination of Sect. *Camellia* Species and *C. Japonica* Cultivars. *J. Zhejiang For. Sci. Technol.* **2004**, *24*, 1–3. [CrossRef]

MDPI

Article

Study on the Genetic Variation of *Triadica sebifera* (Linnaeus) Small Populations Based on SSR Markers

Pengyan Zhou [1,†], Qi Zhou [2,†], Fengping Dong [2], Xin Shen [2] and Yingang Li [2,*]

[1] Co-Innovation Center for Sustainable Forestry in Southern China, Nanjing Forestry University, Nanjing 210037, China
[2] Zhejiang Academy of Forestry, 399 Liuhe Road, Hangzhou 310023, China
* Correspondence: hzliyg@126.com; Tel.: +86-0571-87798027
† These authors contributed equally to this work.

Abstract: *Triadica sebifera* (Linnaeus) Small is a tree species native to China. The seeds of *T. sebifera* are rich in oil and are widely used in industrial fields. To explore the genetic diversity and genetic differentiation of *T. sebifera* germplasm resources, 10 pairs of microsatellite markers were applied to 203 samples collected from eight populations. Forty-three alleles were detected. The average expected heterozygosity (He = 0.491) revealed a low level of genetic diversity. The genetic differentiation among *T. sebifera* populations was low (Fst = 0.026), which might be related to high gene flow (average Nm = 11.151). Genetic distance and structure results further confirmed that the genetic compositions of the eight populations were quite similar. One of the possible reasons for this phenomenon is that the early introduction and cultivation of *T. sebifera* were common, so gene exchange was frequent among populations. However, UPGMA clustering results indicated that the eight *T. sebifera* populations could still be divided into three categories. The classification was related to their geographical location: the southwestern group (ZY), central group (HG and XY) and eastern group (LS, HS, LX, XZ and LY). The reason for this differentiation might be severe deforestation following the decline in *T. sebifera* economic status. In addition, the central XY population had the largest number of rare alleles (4). In conclusion, although *T. sebifera* germplasm resources had a low level of genetic diversity, several rare alleles were detected in the central populations, which are valuable for breeding. These resources should be conserved to maintain genetic diversity in the *T. sebifera* populations. Moreover, geographical distances were important reasons for the limited genetic variations among the populations.

Keywords: *Triadica sebifera* (Linnaeus) Small; genetic diversity; rare allele; germplasm resource conservation

Citation: Zhou, P.; Zhou, Q.; Dong, F.; Shen, X.; Li, Y. Study on the Genetic Variation of *Triadica sebifera* (Linnaeus) Small Populations Based on SSR Markers. *Forests* **2022**, *13*, 1330. https://doi.org/10.3390/f13081330

Academic Editors: Konstantinos Poirazidis and Panteleimon Xofis

Received: 18 July 2022
Accepted: 18 August 2022
Published: 20 August 2022

Publisher's Note: MDPI stays neutral with regard to jurisdictional claims in published maps and institutional affiliations.

1. Introduction

Triadica sebifera (Linnaeus) Small is a tree species native to China that belongs to the genus *Sapium* (*Triadica*) in the family Euphorbiaceae. It is mainly distributed in Jiangsu, Zhejiang, Fujian, Henan and other provinces (regions) [1]. The seeds of *T. sebifera* are rich in oil and have been widely used in soap, paint and other industrial products, making it an ideal tree species for oil production [2]. However, the demand for *T. sebifera* gradually decreased due to the impact of imported oil. Therefore, many *T. sebifera* trees have been cut down, which has resulted in a sharp decrease in the germplasm resources of this species [3]. Recently, *T. sebifera* has been planted widely in gardens because of its high ornamental value. The success of *T. sebifera* breeding depends on the available germplasm resources. It is necessary to extensively collect *T. sebifera* germplasm resources and analyse their distribution patterns and genetic differentiation levels to protect and utilize them more scientifically and rationally.

At present, research on *T. sebifera* is mainly focused on phylogenetic analysis, chemical composition, biological activity and other economic traits [4–7]. However, there have been few studies on its germplasm resources and genetic diversity, with the exception of studies on the screening of elite clones, introduction tests, etc. [8–10]. Although there have been studies on the genetic diversity of *T. sebifera*, most of them used intersimple sequence repeat (ISSR) markers [11,12], which are dominant markers and cannot distinguish heterozygotes. This approach is not conducive to the genetic evaluation of *T. sebifera* germplasm resources because of the limited information provided. In contrast, simple sequence repeat (SSR) markers have the advantages of large numbers, a wide distribution, and codominance. They have been widely used to study the genetic aspects of many woody plants, such as poplar, bamboo, and *Eucalyptus robusta* Smith [13–15]. When molecular markers are used for genetic analysis, more widely collected samples can reveal genetic differentiation and genetic diversity more comprehensively and accurately. Dewalt et al. [16] performed research on *T. sebifera* populations using SSR markers, but the number of samples was small, making it difficult to obtain comprehensive genetic characteristics of the *T. sebifera* populations.

Because of these research limitations, many questions remain, such as the following: What are the levels of genetic diversity and genetic structure in existing *T. sebifera* forests? How can the existing *T. sebifera* forest resources be protected? With these perspectives in mind, 10 pairs of SSR markers were used to explore the genetic variation of eight *T. sebifera* populations in the distribution area. The aim was not only to comprehensively explore the existing genetic resources of *T. sebifera* populations from a population genetics perspective, including aspects of genetic diversity and genetic structure, but also to explore the characteristics and origin of the differentiation of *T. sebifera* populations. This study is expected to provide a more reliable basis for formulating protection and utilization measures for *T. sebifera*.

2. Materials and Methods

2.1. Sample Information

Germplasm resources were collected in the main distribution areas of *T. sebifera*, covering seven provinces, namely, Guizhou, Zhejiang, Anhui, Hubei, Henan, Jiangsu and Shandong (Figure 1). Two-year-old branches were collected from different geographical regions, and the sample information for each population was recorded. A total of 203 samples from eight populations were ultimately collected, with 19 to 34 samples from each population. The samples were preserved by grafting in the *T. sebifera* Germplasm Resource Garden, which is located at the Zhejiang Academy of Forestry Sciences (30°13′09″ N, 120°01′44″ E). The information for each population is provided in Figure 1 and Table 1. Young leaves were collected and stored at −80 °C for later use.

Table 1. The origins of the eight *T. sebifera* (Linnaeus) populations.

Population Code	Location	Longitude (E)	Latitude (N)	Altitude (m a.s.l.)	Annual Rainfall (mm)	N
ZY	Zunyi, Guizhou	27°41′57″	106°54′41″	209	1200	19
HG	Huanggang, Hubei	31°26′38″	114°24′06″	124	1400	24
XY	Xinyang, Henan	31°39′35″	115°23′35″	229	1200	28
HS	Huangshan, Anhui	30°01′16″	118°0′7″	224	2395	27
LY	Linyi, Shandong	35°15′40″	117°58′19″	174	840	20
XZ	Xuzhou, Jiangsu	34°16′37″	118°26′41″	198	876	34
LX	Lanxi, Zhejiang	29°19′19″	119°41′56″	98	1158	28
LS	Lishui, Zhejiang	28°41′18″	119°17′08″	326	1350	23
Total						203

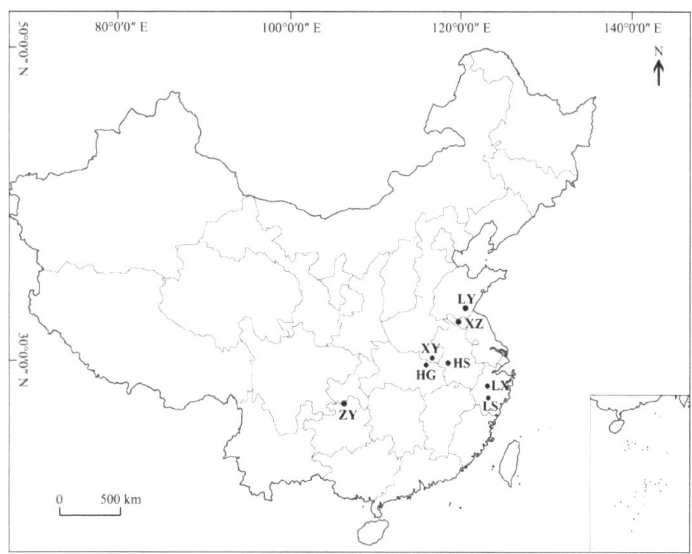

Figure 1. Geographical distribution map of the eight *T. sebifera* (Linnaeus) populations.

2.2. Experimental Method

The cetyl trimethyl ammonium bromide (CTAB) method was used to extract total genomic DNA [17]. A total of 120 pairs of SSR primers were designed by our team based on transcriptome sequences, and 10 of them were selected for this study. The primer information is shown in Table 2.

Table 2. Information on the 10 SSR loci.

Locus Code	Repeat Motif	Primer Sequence (5′~3′)	Fragment Size (bp)	TM (°C)
E-SSR25	(AAG)10	AGGTTGACGACTTCTGTGTT AGTTAGCCTGACCATTTCTC	351	53 °C
E-SSR29	(CT)11	ACCTTGCGAATGTTTATCC GGGGAAAAACAGATGGAAT	339	50 °C
E-SSR52	(AG)12	CTTTTACCTTTGATGTCGG GTTTCGGCAATTTCTCTGT	491	53 °C
E-SSR53	(AG)10	AAACAAGTGAAGTGCCCAT TTAGCCCAGCCCATTATTA	380	51 °C
E-SSR55	(TCT)10	GCGTACCTTCTTCAATGCTC TTCAACTTCTCTTTCCGTCA	428	53 °C
E-SSR58	(AGA)11	TCCACCTAGCGAAGTTTTG TTGATTCCTCCCCTTGTTT	295	52 °C
E-SSR61	(AAG)12	GGTTTCTTTTGCTCTCTTC CCGGTTACTGCATTTCATA	279	50 °C
E-SSR85	(CT)10	TTGCTCTTGGGACCTATTA TTCTTCCCTTGTGAGTTGT	290	50 °C
E-SSR103	(TC)10	CTACCCAATCACCTCTTTC TTCTTCTCTGTTCTGGCTC	287	50 °C
E-SSR106	(AGG)10	TCCCAGTTGACTGACGAACA CGAGGGTGAGGTCAGAGAAG	197	55 °C

Polymerase chain reactions (PCRs) were performed using the GeneAmp PCR System 9700 (Applied Biosystems Inc., Carlsbad, CA, USA) after the upstream sequence of the SSR primer was fluorescently labelled with FAM. The total PCR volume was 20 µL, which included 60 ng of genomic DNA, 10 µL of 2× TSINGKE Master Mix, and 1.5 µL (10 µmol/L) of EST-SSR forward primer and reverse primer. Finally, ddH2O was added to reach a total volume of 20 µL. PCR products were separated on an ABI 3730xl instrument for short tandem repeat (STR) typing. The typing results were read by Peak Scanner version 1.0 software (Applied Biosystems, Foster City, CA, USA) [18].

2.3. Data Analysis

POPGENE version 1.32 software (University of Alberta, Edmonton, Canada) was used to calculate population genetic parameters, such as Shanno's information index (I), the average number of observed alleles (Na), the effective number of alleles (Ne), expected heterozygosity (He), and the genetic differentiation coefficient (Fst) [19].

PowerMarker version 3.25 software (North Carolina State University, Raleigh, NC, USA) [20] was used to calculate the polymorphism information content (PIC) and linkage disequilibrium. Allelic richness (AR) was analysed by FSTAT version 2.9.3 software (University of Lausanne, Lausanne, Switzerland) [21]. GenAlEx version 6 software (Rutgers University, New Brunswick, NJ, USA) was used for analysis of molecular variance (AMOVA) and principal coordinate analysis (PCoA) [22].

Population genetic structure was analysed by Structure version 2.3.4 software (Stanford University, San Francisco, USA) [23]. For clustering from K = 1 to K = 9 (number of populations + 1), the Markov chain Monte Carlo simulation (MCMC) algorithm [24] was applied under admixed ancestry and an allele frequency model. The Evanno method was used to determine the optimal K value as implemented in STRUCTURE HARVESTER [25]. CLUMPP version 1.1.2 (Stanford University, Stanford, CA, USA) and DISTRUCT version 1.1 (Stanford University, San Francisco, USA) [26,27] were used to plot the results. The unweighted pair group method with arithmetic mean (UPGMA) cluster diagram and Mantel test were completed using NTSYS version 2.10 software (State University of New York, NY, USA) [28].

3. Results

3.1. Genetic Diversity

Ten pairs of SSR primers were applied to PCR amplification of 203 samples from eight populations. The STR typing results are shown in Figure 2. A total of 43 alleles were detected, with an average of 4.3 alleles (and range of 3 to 5) per SSR locus (Table 3). Overall, the 203 *T. sebifera* samples showed a moderate level of genetic diversity. The average Ne, He and I of each locus were 2.034, 0.491, and 0.854, respectively. Among the 10 SSR loci, E-SSR58 had the lowest polymorphism level. Its Ne, He, I and PIC were the smallest, at 1.410, 0.292, 0.598, and 0.269, respectively. In contrast, E-SSR85 had the highest polymorphism level. Its Ne, He, I and PIC were the highest, at 2.899, 0.657, 1.214, and 0.614, respectively. The level of polymorphism shown by E-SSR85 was the highest, followed by that of E-SSR25. In addition, except for the E-SSR58 locus, all the loci significantly deviated from Hardy–Weinberg equilibrium ($p < 0.05$).

The genetic diversity levels of the eight *T. sebifera* populations were not significantly different (average He = 0.486) (Table 4). The XZ population had the most genetic diversity, with the largest effective allele number (Ne = 2.079) and the highest degree of genetic diversity (AR = 3.185) and expected heterozygosity (He = 0.503), followed by the XY population. In contrast, the LS population had the lowest level of genetic diversity, with the lowest Na (1.908) and He (0.466). The average inbreeding coefficient (F) was −0.203 among the eight *T. sebifera* populations. All populations showed F < 0, indicating excess heterozygotes and insufficient homozygotes.

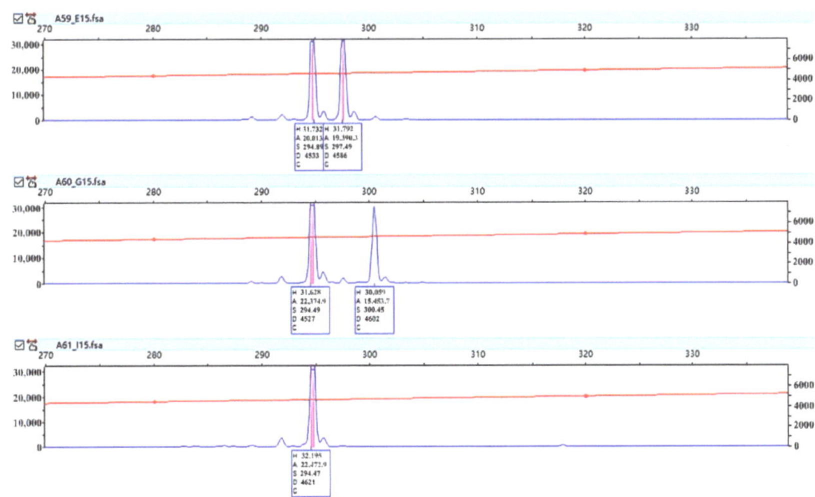

Figure 2. Amplification map of the SSR marker E-SSR58.

Table 3. Polymorphism information of the 10 SSR loci.

Locus	Na	Ne	He	I	PIC	Fst	Nm	HWE
E-SSR25	5	2.411	0.587	1.043	0.520	0.042	5.706	**
E-SSR29	5	1.662	0.399	0.730	0.329	0.037	6.484	*
E-SSR52	3	2.084	0.521	0.797	0.408	0.017	14.091	**
E-SSR53	5	1.929	0.483	0.861	0.427	0.014	18.192	**
E-SSR55	4	2.208	0.5 48	0.891	0.446	0.024	10.312	**
E-SSR58	4	1.410	0.292	0.598	0.269	0.038	6.389	
E-SSR61	4	1.685	0.408	0.762	0.370	0.022	10.910	**
E-SSR85	5	2.899	0.657	1.214	0.614	0.039	6.239	**
E-SSR103	3	1.930	0.483	0.798	0.443	0.019	13.279	**
E-SSR106	5	2.124	0.531	0.846	0.417	0.012	19.911	**
Mean	4.3	2.034	0.491	0.854	0.424	0.026	11.151	

Na—Observed number of alleles, Ne—Effective number of alleles, He—Expected heterozygosity, PIC—Polymorphism information content, Fst—Genetic differentiation index, Nm—Gene flow, I—Shannon's information index, HWE—Hardy–Weinberg Equilibrium, *: Significant ($p < 0.05$), **: Extremely significant ($p < 0.01$).

Table 4. Information on genetic diversity in the eight populations based on 10 SSR loci.

Population Code	Na	Ne	AR	Ho	He	F
ZY	2.9	2.047	2.890	0.563	0.492	−0.146
HG	3.3	1.974	3.293	0.613	0.497	−0.233
XY	3.5	2.021	3.245	0.554	0.498	−0.113
HS	3.3	1.990	3.192	0.556	0.480	−0.157
LY	3.2	1.989	3.415	0.595	0.485	−0.228
XZ	3.4	2.079	3.185	0.641	0.503	−0.276
LX	3.3	1.969	3.125	0.596	0.471	−0.265
LS	3.0	1.908	3.295	0.565	0.466	−0.212
Mean	3.24	1.997	3.205	0.585	0.486	−0.203

AR—Allelic richness, Ho—Observed heterozygosity, F—Inbreeding coefficient.

3.2. Genetic Differentiation and Genetic Structure

The genetic differentiation coefficient (Fst) of the eight *T. sebifera* populations was 0.026, indicating low genetic differentiation among them (Table 3). The results of AMOVA further revealed different levels of genetic variation between regions and populations. However, the variation was mainly concentrated within the populations, reaching an extremely significant level ($p < 0.01$, Table 5). Four percent of the genetic variation occurred among regions (southwest and mideast, $p = 0.001$). One percent of the total variation occurred among populations within regions, and 95% of the genetic variation occurred within populations ($p = 0.001$).

Table 5. AMOVA of 203 samples from eight *T. sebifera* populations.

Source of Variance	Variance Component	Percentage of Total	*p* Value
Among regions	0.155	4%	**
Among populations within regions	0.038	1%	
Within populations	3.823	95%	**
Total	4.016	100%	

**: Extremely significant ($p < 0.01$).

The common ancestor relationship of *T. sebifera* populations was analysed for further research on population genetic structure. According to the maximum ΔK value, the optimal K was 3 (Figure 3A). When K = 3, the genetic information divided the *T. sebifera* populations into three different ancestral populations (Figure 3B). However, the genetic information sources of the eight populations were quite similar when K = 2, 3 or 8. In addition, a cluster diagram (Figure 3C) was constructed based on the genetic distances between the populations (Table 6). There was a certain degree of genetic differentiation among the eight populations. They could be roughly divided into three categories: the ZY population in the southwest was divided into a separate class, the HG and XY populations in the central region were divided into one class, and the LS, HS, LX, XZ and LY populations were grouped into one category. The classification results of the eight populations and their distribution positions were associated. The results of the Mantel test performed on the eight *T. sebifera* populations further proved this result, implying a significant association between the genetic and geographical distances between populations (r = 0.6697, $p = 0.9860$, Figure 4).

Table 6. Pairwise genetic distances between *T. sebifera* populations.

	ZY	LS	LX	HS	HG	XY	XZ	LY
ZY	0							
LS	0.043	0						
LX	0.051	0.025	0					
HS	0.038	0.009	0.019	0				
HG	0.059	0.026	0.026	0.035	0			
XY	0.047	0.031	0.032	0.028	0.024	0		
XZ	0.036	0.021	0.013	0.015	0.030	0.024	0	
LY	0.051	0.017	0.016	0.014	0.027	0.031	0.013	0

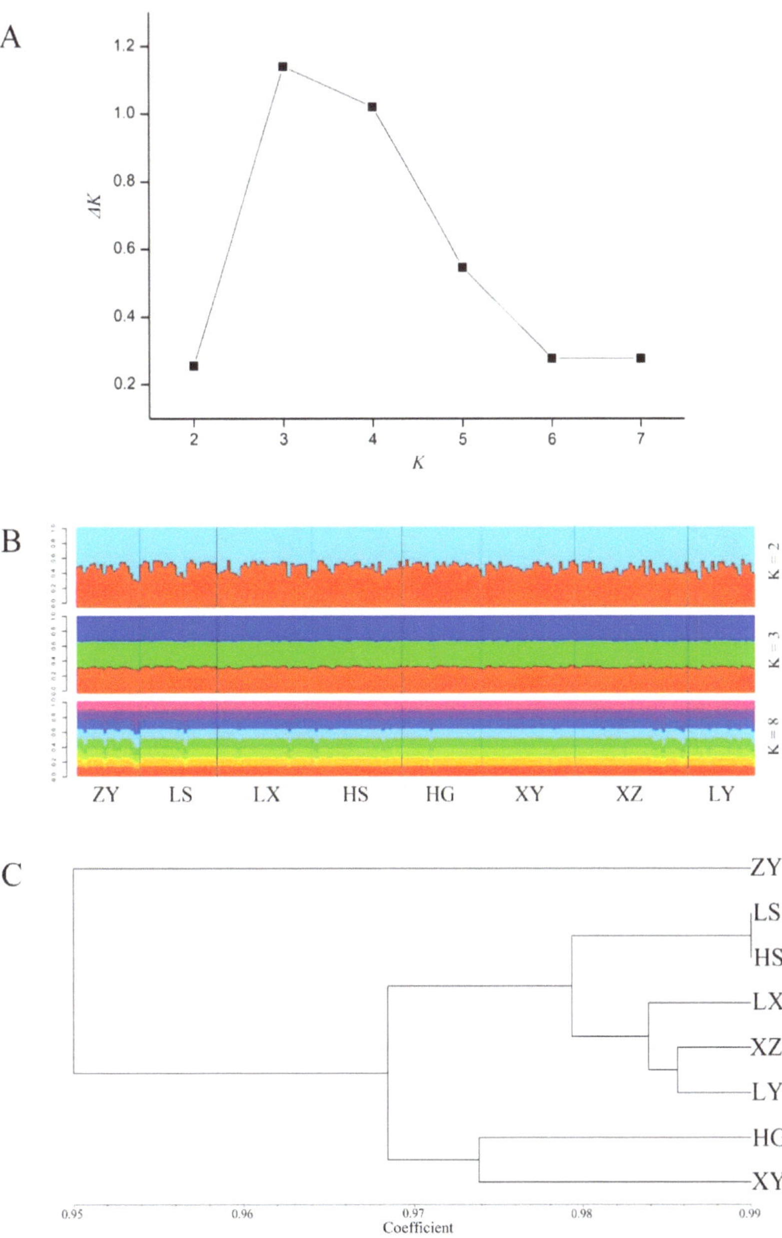

Figure 3. Genetic structure of eight *T. sebifera* populations. (**A**) Relationship between the number of clusters (K) and the corresponding ΔK statistic; (**B**) Results of the structure analysis of *T. sebifera* populations when K = 2, 3, and 8. Different colors (light-blue, red, blue, green, dark purple, purple, light-green, and orange) represent different genetic compositions; (**C**) Cluster diagram of the eight populations.

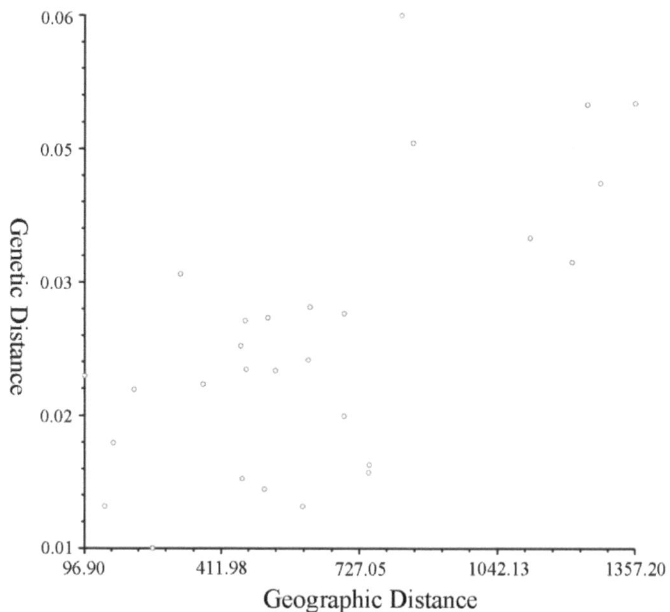

Figure 4. The result of the Mantel test.

3.3. Rare Alleles

The frequencies of the 43 alleles ranged from 0.0049 to 0.8350, with an average of 0.2326. The minimum allele frequencies (MAFs) of the 10 SSR loci were 0.0049 to 0.0591, and the average was 0.0195. Rare alleles with a frequency less than 0.1 were selected, and the frequency of each rare allele in the populations was calculated (Table 7). Six rare alleles were found among the 10 SSR loci, which were distributed at 5 SSR loci. Two rare alleles were detected at the G-SSR106 locus, and one rare allele each was detected at the other four loci. The central XY population and HG population had the largest numbers of rare alleles, with four and three, respectively. In contrast, the western and eastern populations had 0~2 rare alleles.

Table 7. Rare allele frequencies within the *T. sebifera* populations.

	E-SSR29-E	E-SSR55-E	E-SSR61-E	E-SSR85-C	E-SSR106-D	E-SSR106-E
ZY	0.079					
HG			0.021	0.042		0.021
XY	0.018	0.018		0.018	0.018	
HS						
LY					0.075	
XZ						0.044
LX		0.018				
LS		0.022	0.022			

4. Discussion

4.1. Genetic Diversity

Genetic diversity is an important indicator used to measure the ability of a species to resist changes in the external environment. It is affected by selection, genetic drift, and breeding system. Research on the genetic diversity of *T. sebifera* populations could provide a theoretical reference for the protection and utilization of *T. sebifera*. Li [11] used ISSR markers to study 32 *T. sebifera* leaf samples from six populations, and the H and I values were 0.2822 and 0.4321, respectively. Zhang [12] also used ISSR markers to analyse 72 samples in four main production areas and obtained similar results (He = 0.281). In this study, the average He of 203 *T. sebifera* samples from eight populations based on SSR markers was 0.486, which was higher than that in previous studies. There are two possible explanations for this discrepancy. First, the marker types used were different. Compared with ISSR markers, SSR markers can reveal more genetic variation. Second, there were differences in the number of samples. The number of samples selected in this study was relatively large. In addition, the distribution range covered the main distribution areas of *T. sebifera*. DeWalt et al. [16] used six pairs of SSR markers to analyse 129 samples from 12 *T. sebifera* populations in China, and the values were higher than those of this study (average He = 0.70). The reason was inferred to be sample differences. In recent years, many older trees might have been transplanted or traded with the gradual popularization of *T. sebifera* application. This contributed to the loss of *T. sebifera* resources and a decrease in the genetic diversity of the *T. sebifera* populations.

Compared with the results from SSR markers in other angiosperms, the genetic diversity of *T. sebifera* (He = 0.491) was significantly lower than that of *Ginkgo biloba* Linn (He = 0.808) [29], *Quercus variabilis* BI. (He = 0.707) [30], *Cunninghamia lanceolata* (Lamb.) Hook (He = 0.557) [31] and other species. *T. sebifera* has been widely promoted and planted as an important oil tree species. However, with industrial development and changes in the national industrial structure, the economic status of *T. sebifera* has declined, and the natural populations have been reduced excessively. Therefore, its genetic resources are difficult to preserve, resulting in a significant reduction in genetic diversity. Therefore, protecting the genetic diversity of *T. sebifera* is an important goal for preserving its natural resources.

4.2. Genetic Differentiation and Genetic Structure

Natural selection contributes to population differentiation and is the most important evolutionary force. Gene flow plays an important role in selection. For perennial outcrossing woody plants, genetic differentiation among populations is generally low. When Fst is between 0 and 0.05, there is basically no differentiation among populations. There is a moderate degree of differentiation when Fst is 0.05~0.15. When Fst is 0.15~0.25, the degree of differentiation is relatively high. There is strong differentiation when Fst is greater than 0.25 [32]. The differentiation of *T. sebifera* populations detected in this study (Fst = 0.026) was much weaker than that of *Liriodendron chinense* (Hemsl.) Sarg. (Fst = 0.302) [33], *Q. variabilis* (Fst = 0.067) [30], *Pinus massoniana* Lamb (Fst = 0.072) [34], etc. Our results were quite different from those of Li [11] (Gst = 0.439), which might be due to the use of different markers and numbers of samples.

The results of the AMOVA further indicated that the variation among *T. sebifera* populations was quite low, and most of the genetic variation existed within the populations. Cluster analysis also showed that the eight *T. sebifera* populations had similar ancestral origins. The small genetic distances (the highest was only 0.051) also indicated close kinship among the populations. *T. sebifera* was widely distributed in the early stage because of its high economic value. The mutual introduction and cultivation in different regions led to frequent genetic exchange among geographic groups. This also resulted in similar origins among different geographic populations.

However, there was still a certain degree of genetic differentiation among the eight *T. sebifera* populations. The UPGMA clustering results showed that the eight populations could be divided into three categories, which were related to their geographical location, namely, the southwestern group (ZY), the central group (HG and XY) and the eastern group (LS, HS, LX, XZ and LY). The results of the Mantel test further indicated that there was a significant positive correlation between genetic distance and geographic distance. Many *T. sebifera* resources have been removed, contributing to the differentiation among geographical populations, while introduction and cultivation have been greatly reduced due to the decline of the *T. sebifera* industry.

4.3. Genetic Resource Conservation Strategy of T. sebifera

Abundant germplasm resources are the key to successful forest tree breeding, and they guarantee the continuation of tree species [35]. The overall genetic diversity of *T. sebifera* was low, and there has been a downward trend in recent years. Even though quite similar amounts of genetic information were observed between populations, six rare alleles were detected in this study. The populations in the central regions had more rare alleles. These resources with rare alleles are not only important for maintaining species diversity but also valuable breeding materials. Therefore, the protection and utilization of various *T. sebifera* populations should be strengthened in the future, especially for the central populations with more rare alleles.

5. Conclusions

In general, our study included a genetic evaluation of eight populations of *T. sebifera*. The genetic compositions of the eight populations were similar, and the genetic diversity was low. One possible reason was that introduction was common a long time ago, so gene exchange was frequent among populations. However, the UPGMA results still classified these populations into three categories, indicating that they began to diverge geographically. Finally, rare alleles were detected in some populations, which will be of great help in maintaining the level of genetic diversity in *T. sebifera* populations.

Author Contributions: Conceptualization, Y.L. and Q.Z.; methodology, F.D.; software, X.S.; validation, F.D., X.S. and P.Z.; investigation, F.D.; data curation, Q.Z.; writing—original draft preparation, P.Z.; writing—review and editing, P.Z.; supervision, Q.Z.; project administration, Y.L. and Q.Z.; funding acquisition, Y.L. and Q.Z. All authors have read and agreed to the published version of the manuscript.

Funding: This research was funded by Special Support Funds of Zhejiang for Scientific Research Institutes (2021F1065-12) and a Key Scientific and Technological Grant of Zhejiang for Breeding New Agricultural Varieties (2021C02070-7).

Institutional Review Board Statement: Not applicable.

Informed Consent Statement: Not applicable.

Data Availability Statement: Not applicable.

Acknowledgments: We are grateful to Li-an Xu and Ming Qi for their help in the sampling involved in the field and the laboratory work.

Conflicts of Interest: The authors declare no conflict of interest.

References

1. Jin, D.J.; Huang, H.K.; Tang, R.Q.; Tong, Q.Y.; Shi, D.Y.; Hou, Z.S. Investigation and research on the resources of *Triadica sebifera* varieties in China. *Guangxi Plants* **1997**, *4*, 345–362.
2. Li, D.L.; Huang, D.; Wang, J.; Jin, Y.Q. A review of research on *Triadica sebifera*. *Jiangsu For. Technol.* **2009**, *36*, 43–47.
3. Zheng, K.; Zhao, L.; Lang, N.J.; Peng, M.J. Research progress on germplasm resources and cultivation of *Triadica sebifera*, a woody biomass energy tree. *Biomass Chem. Eng.* **2006**, *S1*, 360–366.
4. Wang, S.; Chen, Y.; Yang, Y.; Wu, W.; Liu, Y.; Fan, Q.; Zhou, R. Phylogenetic relationships and natural hybridization in *Triadica* inferred from nuclear and chloroplast DNA analyses. *Biochem. Syst. Ecol.* **2016**, *64*, 142–148. [CrossRef]
5. DeWalt, S.J.; Siemann, E.; Rogers, W.E. Microsatellite markers for an invasive tetraploid tree, Chinese tallow (*Triadica sebifera*). *Mol. Ecol. Notes* **2006**, *6*, 505–507. [CrossRef]
6. Gao, R.; Su, Z.; Yin, Y.; Sun, L.; Li, S. Germplasm, chemical constituents, biological activities, utilization, and control of Chinese tallow (*Triadica sebifera* (L.) Small). *Biol. Invasions* **2016**, *18*, 809–829. [CrossRef]
7. Zhi, Y.; Taylor, M.C.; Campbell, P.M.; Warden, A.C.; Shrestha, P.; El Tahchy, A.; Rolland, V.; Vanhercke, T.; Petrie, J.R.; White, R.G.; et al. Comparative lipidomics and proteomics of lipid droplets in the mesocarp and seed tissues of Chinese tallow (*Triadica sebifera*). *Front. Plant Sci.* **2017**, *8*, 1339. [CrossRef]
8. Feng, Y.; Luo, J.X.; Du, B.; Gu, Y.J.; Wang, H.L.; Xiang, Q. Study on the phenotypic diversity of seed traits in natural populations of *Triadica sebifera* in Sichuan-Chongqing area. *Sichuan For. Technol.* **2011**, *32*, 19–27. [CrossRef]
9. Yang, J. Investigation on the Growth of Introduced *Triadica sebifera* in Rizhao City. *Shandong For. Technol.* **1996**, *2*, 17–18.

10. Wang, H.; Yang, J. Analysis of Ecological Adaptation of *Sapium sabiferum* Introduced to Shandong Province. *Chin. J. Ecol.* **1997**, *16*, 17–19+27.

11. Li, M.; Li, C.Z.; Wang, L.Y.; Jiang, L.J. Growth characteristics of *Triadica sebifera* from different provenances at seedling stage. *Econ. For. Res.* **2012**, *30*, 75–79. [CrossRef]

12. Zhang, S.; Wang, X.G.; Du, K.B.; Zhang, Y.Q.; Luo, Z.J.; Tu, B.K. Correlation analysis among several quantitative characters of *Triadica sebifera*. *Econ. For. Res.* **2010**, *28*, 61–66. [CrossRef]

13. He, X.D.; Zheng, J.W.; Sun, C.; He, K.Y.; Wang, B.S. Construction of fingerprints for 33 varieties in Salicaceae. *J. Nanjing For. Univ. (Nat. Sci. Ed.)* **2021**, *45*, 35–42. [CrossRef]

14. Yuan, J.L.; Ma, J.X.; Zhong, Y.B.; Yue, J.J. SSR-based hybrid identification, genetic analyses and fingerprint development of hybridization progenies from sympodial bamboo (Bambusoideae, Poaceae). *J. Nanjing For. Univ. (Nat. Sci. Ed.)* **2021**, *45*, 10–18. [CrossRef]

15. Li, F.; Gan, S.; Zhang, Z.; Weng, Q.; Xiang, D.; Li, M. Microsatellite-based Genotyping of the Commercial Eucalyptus Clones Cultivated in China. *Silvae Genet.* **2011**, *60*, 216. [CrossRef]

16. DeWalt, S.J.; Siemann, E.; Rogers, W.E. Geographic distribution of genetic variation among native and introduced populations of Chinese tallow tree, *Triadica sebifera* (Euphorbiaceae). *Am. J. Bot.* **2011**, *98*, 1128–1138. [CrossRef]

17. Cota-Sánchez, J.H.; Remarchuk, K.; Ubayasena, K. Ready-to-use DNA extracted with a CTAB method adapted for herbarium specimens and mucilaginous plant tissue. *Plant Mol. Biol. Rep.* **2006**, *24*, 161–167. [CrossRef]

18. Palero, F.; González-Candelas, F.; Pascual, M. MICROSATELIGHT–pipeline to expedite microsatellite analysis. *J. Hered.* **2011**, *102*, 247–249. [CrossRef]

19. Nei, M. Definition and Estimation of Fixation Indices. *Evolution* **1986**, *40*, 643–645. [CrossRef]

20. Liu, K.J.; Muse, S.V. PowerMarker: An integrated analysis environment for genetic marker analysis. *Bioinformatics* **2005**, *21*, 2128–2129. [CrossRef]

21. Goudet, J. FSTAT, A Program to Estimate and Test Gene Diversities and Fixation Indices, version 2.9.3. 2001. Available online: https://www.unil.ch/popgen/softwares/fstat.Htm (accessed on 18 October 2021).

22. Peakall, R.; Smouse, P.E. GENALEX 6: Genetic analysis in Excel. Population genetic software for teaching and research. *Mol. Ecol. Notes* **2006**, *6*, 288–295. [CrossRef]

23. Evanno, G.; Regnaut, S.; Goudet, J. Detecting the number of clusters of individuals using the software STRUCTURE: A simulation study. *Mol. Ecol.* **2005**, *14*, 2611–2620. [CrossRef] [PubMed]

24. Falush, D.; Stephens, M.; Pritchard, J.K. Inference of population structure using multilocus genotype data: Linked loci and correlated allele frequencies. *Genetics* **2003**, *64*, 1567–1587. [CrossRef]

25. Earl, D.A.; vonHoldt, B.M. STRUCTURE HARVESTER: A website and program for visualizing STRUCTURE output and implementing the Evanno method. *Conserv. Genet. Resour.* **2012**, *4*, 359–361. [CrossRef]

26. Jakobsson, M.; Rosenberg, N.A. CLUMPP: A cluster matching and permutation program for dealing with label switching and multimodality in analysis of population structure. *Bioinformatics* **2007**, *23*, 1801–1806. [CrossRef] [PubMed]

27. Rosenberg, N.A. Distruct: A program for the graphical display of population structure. *Mol. Ecol Notes* **2004**, *4*, 137–138. [CrossRef]

28. Rohlf, F.J. *NTSYS-pc: Numerical Taxonomy and Multivariate Analysis System*, version 2.2, 2.1; Department of Ecoloy and Evolution, State University of New York: New York, NY, USA, 2005.

29. Zhou, Q.; Mu, K.; Ni, Z.; Liu, X.; Li, Y.; Xu, L.-A. Analysis of genetic diversity of ancient Ginkgo populations using SSR markers. *Ind. Crops Prod.* **2020**, *145*, 111942. [CrossRef]

30. Shi, X.; Wen, Q.; Cao, M.; Guo, X.; Xu, L.A. Genetic diversity and structure of natural *Quercus variabilis* population in China as revealed by microsatellites markers. *Forests* **2017**, *8*, 495. [CrossRef]

31. Chen, X.; Xu, H.; Xiao, F.; Sun, S.; Lou, Y.; Zou, Y.; Xu, X. Genetic diversity and paternity analyses in a 1.5th generation seed orchard of Chenshan red-heart Chinese fir. *J. Nanjing For. Univ. (Nat. Sci. Ed.)* **2021**, *45*, 87–92. [CrossRef]

32. Li, H.; Chen, L.; Liang, C.; Huang, M. Experimental study on provenance of *Liriodendron* species. *For. Sci. Technol. Dev.* **2005**, *05*, 13–16.

33. Zhou, P.-Y.; Hui, L.-X.; Huang, S.-J.; Ni, Z.-X.; Yu, F.-X.; Xu, L.-A. Study on the Genetic Structure Based on Geographic Populations of the Endangered Tree Species: *Liriodendron chinense*. *Forests* **2021**, *12*, 917. [CrossRef]

34. Feng, Y.H.; Li, H.G.; Yang, Z.Q.; Wu, D.S. Study on the genetic structure of natural populations of *Pinus massoniana* provenance. *Guangxi Plant* **2016**, *36*, 1275–1281, 1395.

35. Qi, M.; Zhou, Q.; Ni, Z.; Wu, Y.; Han, X.; Xu, L. Genetic structure analysis of ancient *Ginkgo biloba* L. populations based on SSR markers. *Chin. J. Ecol.* **2019**, *38*, 2902–2910. [CrossRef]

Article

Studies on Pollen Morphology, Pollen Vitality and Preservation Methods of *Gleditsia sinensis* Lam. (Fabaceae)

Qiao Liu [1,2,3,4], Ju Yang [1,2,3,4], Xiurong Wang [1,*] and Yang Zhao [1,2,3,4]

1   College of Forestry, Guizhou University, Guiyang 550025, China
2   Institute for Forest Resources & Environment of Guizhou, Guizhou University, Guiyang 550025, China
3   Key Laboratory of Forest Cultivation in Plateau Mountain of Guizhou Province, Guiyang 550025, China
4   Key Laboratory of Plant Resource Conservation and Germplasm Innovation in Mountainous Region (Ministry of Education), Guizhou University, Guiyang 550025, China
*   Correspondence: xrwang@gzu.edu.cn

Citation: Liu, Q.; Yang, J.; Wang, X.; Zhao, Y. Studies on Pollen Morphology, Pollen Vitality and Preservation Methods of *Gleditsia sinensis* Lam. (Fabaceae). *Forests* **2023**, *14*, 243. https://doi.org/10.3390/f14020243

Academic Editors: Konstantinos Poirazidis and Panteleimon Xofis

Received: 18 December 2022
Revised: 21 January 2023
Accepted: 22 January 2023
Published: 28 January 2023

Abstract: *Gleditsia sinensis* Lam. (Fabaceae) is an endemic species in China, which has a wide range of ecological functions and high economic value. *G. sinensis* belongs to androdioecy, and the stamens of perfect flowers are aborted, meaning that a perfect flower is a functional female flower. Understanding the dynamic process of flowering and the characteristics of pollen morphology effectively determine the viability of pollen vitality, and the suitable conditions for short-term storage of pollen can provide theoretical basis and technical reference for hybrid breeding and germplasm conservation of *G. sinensis*. In this study, the male plants of *G. sinensis* in Guiyang area were used as research materials. The flowering dynamic process of male flowers was recorded through field observation. The morphology of pollen was observed and analyzed with a scanning electron microscope (SEM). The germination characteristics of pollen were studied with an in vitro germination method, and the pollen vitality was also determined using four staining methods. The effects of different storage temperatures and water contents on pollen germination rate were discussed. The results showed that the male flowers of *G. sinensis* had a short, single flowering period, lasting 2–3 days from the opening to the shedding. The dynamic opening process of a single flower was artificially divided into five stages. Pollen grains of *G. sinensis* are oblate spheroidal, tricolporate with equatorial elongated endoapertures and the sporoderm surface is reticulate. The MTT (Thiazolyl Blue Tetrazolium Bromide) staining method could accurately and quickly determine the pollen vitality of *G. sinensis*. The highest pollen germination rate was 65.89% ± 3.41%, and the length of the pollen tube was 3.96 mm after cultured in 15% sucrose + 100 mg/L boric acid + 20 mg/L calcium chloride for 24 h. It was necessary to collect the pollen at the big bud stage, which was conducive to improving the efficiency of pollen collection because the pollen had been mature with high pollen vitality at this stage. When it came to pollen preservation, the pollen germination rate was significantly affected by storage time, storage temperature and pollen water content. The pollen still had high vitality after being stored at −80 °C for 30 days when the moisture content of the pollen decreased to 9%, and the pollen germination rate only decreased by 28.84% compared with that before storage. In conclusion, this study has comprehensively and systematically studied the morphology, vitality determination and preservation methods of the pollen of *G. sinensis*, providing a theoretical basis for the cross regional breeding and the conservation and utilization of germplasm resources.

Keywords: *Gleditsia*; pollen viability; pollen germination; pollen preservation

1. Introduction

In seed plants, pollen is of great importance for sexual reproduction as a male gametophyte containing genetic information [1]. Pollen viability is a prerequisite for successful pollination, playing an important role in fertilization and fruit setting of flowering plants [2,3], embryo development [4,5], seed quality [6,7] and breeding efficiency. Due to this importance, researchers have conducted many studies on pollen, such as the pollen development process [8], electron microscopy palynology [9], pollen viability detection and pollen preservation methods [10,11], signal transduction and related regulation during pollen tube growth [12–14]. All studies above are indispensable for plant cross breeding from practice to theory [15].

Gleditsia sinensis Lam. (Fabaceae) belongs to genus *Gleditsia* in Fabaceae, which is an endemic species widely distributed in China [16,17]. It is an ideal tree for economic forest, timber forest, shelterbelt and landscaping because of its tall and beautiful shape, strong resistance and barren resistance. The extract of saponin and pod is an excellent Chinese medicinal material [18] and industrial raw material [19], which has high application value and broad prospect. *G. sinensis* belonged to androdioecy, and its bisexual flowers are functionally female, with most stamens aborted [20] and with a breeding system dominated by outcrossing [21]. In this respect, the fruit and seed set are almost entirely dependent on the pollens provided by male plants during synchronous flowering. Dioecious plants have been under great reproductive pressure due to frequent occurrences of missed flowering periods, lack of pollinators, and imbalanced sex ratios in recent years [22,23]. This necessitates the conservation of pollen resources. In the past, studies of *G. sinensis* focused mostly on its medicinal components [18], cultivation techniques [24], germplasm resource investigation [25] and stress physiology [26,27] while only a few studies examined its reproductive process [28] like the pollen morphology observed [29] and observation of flowering characteristics [30]. In contrast, pollen-related research of *G. sinensis* is fragmented and largely unknown, which is not conducive to the popularization and application of this plant.

Therefore, the observation of pollen morphological characteristics by scanning electron microscope, investigation of single male flower flowering dynamic process, optimized in vitro germination and the effects of different storage conditions on pollen preservation have been analyzed in this study, and the results based on the above studies are supposed to provide reference for the research of the reproductive biology of *G. sinensis*, the preservation of germplasm resources and the hybridization and breeding.

2. Materials and Methods

2.1. Study Site

Experiments were conducted at Guiyang's planting base (26.45 N, 106.56 E) from March to May of 2021 and 2022. In the experiment field, the average annual temperature was 15.7 °C, and the average annual precipitation was 1215.7 mm. Our test plants were twelve-year-old male plants that had consistently grown and flowered in the plantation.

2.2. Observation of Male Single Flower Flowering Dynamic Process

A study of the flowering dynamics of a male single flower of *G. sinensis* was conducted in 2021 and 2022. A total of 30 male flowers at the full bud stage were placed on each test plant, and 6 trees at full bloom stage were employed for observation in total. We observed and photographed the flower opening process several times over the course of several days, from 8:00 a.m. to 18:00 p.m., and the viability of pollen was determined by collecting pollen at corresponding stages.

2.3. Pollen Morphology Observation

2.3.1. Light Microscopic Studies

Mature anthers collected were preserved in glass vials containing acetic alcohol (1:3). The material was then centrifuged and washed twice with distilled water to remove traces of acetic alcohol. After this, acetolysis after preservation in glacial acetic acid was used as recommended by Reitsma [31] and Raynal A and Raynal J [32]. The material was given three washings, and residual material was mounted in glycerin jelly that was already stained in 1% safranine as recommended by Kisser [33] and Erdtman [34]. In the Leica research microscope, 5–7 slides were prepared to study. The measurement of pollen characters was made from 50 grains taken at random. Microphotographic work was carried out using Leica DM300 Photomicroscope fixed in Institute for Forest Resources & Environment of Guizhou, Guizhou University. The terminology used was in accordance with the studies by Erdtman [34] and Halbritter [35].

2.3.2. Scanning Electron Microscopy

The anthers of newly opened male flowers on the test plants were randomly selected at the peak of the flowering period in 2022 and then soaked in glutaraldehyde solution (prepared with pH 8.0 phosphate buffer) at 4 °C for 24 h. The fixed pollen grains were then dehydrated and dried with an acetone series (30%, 50%, 70%, 80%, 90%, 95%, 100% and 100%). Afterward, they were mounted on metallic stubs, coated with gold-palladium, and then observed under a scanning electron microscope (SEM). A Hitachi JSM-SU8100 scanning electron microscope was used to take photos of pollen grains.

2.4. Study on Methods for Determining Pollen Viability

2.4.1. In Vitro Germination of *G. sinensis*

The Effects of Different Medium Components on the In Vitro Germination of *G. sinensis*

Using a single factor experimental design (Table 1), we examined the influence of sucrose, boric acid (H_3BO_3) and $CaCl_2$ on pollen germination to determine the concentrations of media components that are conducive to in vitro germination. The specific operation was as follows: In a 2 mL centrifuge tube, fresh anthers were mixed with 1 mL of different liquid medium formulations. After homogenizing shock, anthers were filtered for pollen suspensions, which were incubated for 4 h at a constant temperature of 25 °C. A suspension of cultured pollen was mixed and dropped onto the slide. Once covered, the slide was observed under a microscope and photographed. For each slide, five visual fields were randomly selected, and each treatment was repeated 6 times. The length of pollen tubes at germination exceeded the diameter of pollen grains was used to judge pollen germination. The following formula was used to calculate the in vitro germination rate of *G. sinensis*: pollen germination rate (%) = germination pollen number/observed total pollen number × 100.

Table 1. In vitro pollen germination media.

Media Type	Control (CK)	Sucrose (%)				Boric Acid (mg/L)				Calcium Chloride (mg/mL)			
Sucrose	0	5	10	15	20								
Boric acid	0					50	100	150	200				
Calcium chloride	0									50	100	150	200
Distilled water	100 mL												

Screening of Pollen Liquid Medium

Three factors and three levels (Table 2) were selected based on the concentration of media components tested by a single factor test. In addition, an orthogonal test (L9 (34)) was designed to identify the medium formula best suited for determining pollen in vitro germination rates. The specific operation was the same as described in single factor experimental, except extending the culture time to 12 h.

Table 2. Factor levels of liquid medium screening for in vitro germination of *G. sinensis*.

Levels	Test Factors		
	Sucrose (%)	Boric Acid (mg/L)	Calcium Chloride (mg/L)
	A	B	C
1	5	50	20
2	10	100	40
3	15	150	60

Effect of Incubation Time on The In Vitro Germination Rate

During the experiment, the pollen germination rate of *G. sinensis* increased as culture time increased. A study was conducted to determine the germination rate and pollen tube length after different incubation times (2 h, 4 h, 8 h, 12 h, 24 h) at 25 °C constant temperature based on the selected liquid culture medium. Its function was to select the culture time suitable for in vitro germination of *G. sinensis* and exclude the difference of pollen germination rate caused by insufficient culture time.

2.4.2. Pollen Viability by Dyeing Methods

The pollen vitality was determined using four staining methods: TTC (2.3.5-Triphenyl Tetrazolium Chloride) staining method [36], MTT (Thiazolyl Blue Tetrazolium Bromide) staining method [37], KI-I_2 staining method [38] and Alexander staining method [36]. In vitro germination rate of pollen was used as a control to select the suitable staining methods for *G. sinensis*. The specific operation was as follows: Four kinds of dye were dropped on the slide in advance, and a small amount of pollen was taken with tweezers and mixed with the dye. Following the covering of the glass slide, the following treatments were performed according to different staining methods (Table 3). An optical microscope (Leica DM750) was used to observe and photograph the corresponding slides. A random selection of 5 visual fields with a minimum of 50 pollen grains per visual field was used on each slide, and each method was repeated 6 times. The pollen vitality determined by four staining methods was then calculated based on observation results and the following formula:

Table 3. Treatment methods after dyeing.

Type of Dye Method	Post-Treatment	Judgment Basis of Active Pollen
TTC	2 h of dark culture at 37 °C	Red
MTT	Stand for 5–10 min	Purple
KI-I_2	Observe immediately	Blue
Alexander	Observe immediately	cell wall green, protoplast red

Pollen vitality (%) = number of dyed pollen grains/total number of observed pollen grains $\times$ 100.

2.5. Study on The Pollen Preservation Method of G. sinensis

2.5.1. Material Selection for Preservation

A large number of fresh male flowers at different flowering stages were collected and brought back to the laboratory in 2022. Then pollen grains isolated from different flowering stages were measured by in vitro germination method to screen suitable pollen preservation materials.

2.5.2. Pollen Storage Conditions

For pollen collection, fresh male flowers were collected based on the selected preservation materials above. The collected pollen was divided into small portions and wrapped into pollen bags by the weighing paper, which was put in a sealed silica gel dryer, drying for 3 h, 6 h and 9 h to reduce the pollen water content to different gradients. After drying, pollen bags of different water contents were placed into 2 mL centrifuge tubes and stored at four different temperatures (RT, 4 °C, −20 °C and −80 °C). After storage, the pollen viability was determined by in vitro germination at 0 day, 7 days, 15 days and 30 days.

2.6. Statistical Analysis of Data

The obtained test data were cleared up using Excel 2010. Data were analyzed using one-way analysis of variance, three-factor analysis of variance and orthogonal experimental design in IBM SPSS (version 18.0, IBM, Armonk, NY, USA). R (version 4.1.2) and GraphPad Prism (version8.0.2, Graphpad, San Diego, CA, USA) were used for data visualization.

3. Results

3.1. Flowering Dynamics Observation of Male Flower

Field studies indicated that the male flowers of *G. sinensis* were in racemes with about 50 small flowers per inflorescence, opening upward from the base (Figure 1a). From the opening to the shed, a single male flower of *G. sinensis* had a short flowering period, lasting 2–3 days. A five-stage dynamic process of the opening of single male flower was artificially categorized: big bud stage (stage I), about 2 h before flower opening, at which the corollas were slightly dilated and the anthers did not extend the corolla, indicating the flowers were about to open (Figure 1b). At the early dispersed-powder stage (stage II), anthers just started dehiscing and dispersing powder. At this time, flowers were half open, with several anthers gathering in the crown and slightly extending out of the corolla. In addition, pollen was visible along the longitudinally split abdominal sutures of the anther (Figure 1c). About 2 h after scattering was the mass scattering stage (stage III), at which the corolla diameter reached its maximum and all anthers cracked loose powder with a large amount of yellow pollens visible to the naked eye. In addition, anthers were yellowish brown and almost perpendicular to the filaments (Figure 1d). At 4 h after dispersed powder was the end dispersed-powder stage (stage IV), during which the corolla began to lose water and the anther color changed to reddish-brown or even black from yellow with some pollen grains still in its dehiscent chamber (Figure 1e). The corolla wipeout stage (stage V) was about 24 h after powder dispersion, where the filaments wilt due to water loss and the corolla was about to fall off (Figure 1f). In brief, male flowers usually have a short single flower life (2–3 days), and the dispersing powder peaks within 4 h following anthers.

Figure 1. Male flowers at different stages: (**a**) Male flower inflorescence, (**b**) Big bud stage, (**c**) Early dispersed pollen stage, (**d**) Mass scattering stage, (**e**) End dispersed pollen stage, (**f**) Corolla wipeout stage.

3.2. Pollen Grain Morphological Characteristics

The results showed that the average pole axis of pollen was 24.03 ± 2.67 um and the average equatorial diameter was 25.29 ± 4.21 um. According to the ratio of its polar axis length (P) to equatorial (P/E = 0.95), pollen grains are oblate spheroidal, belonging to small or medium-sized grains (Figure 2a). Pollen is tricolporate with equatorial elongated apertures (Figure 2b–d), and pollen grains were evenly distributed with thick reticular outer wall ornamentation except for the position close to the margo (Figure 2c,d). The meshes were mostly round, nearly round or irregular polygon among reticular outer wall ornamentation, in which size was irregularly distributed.

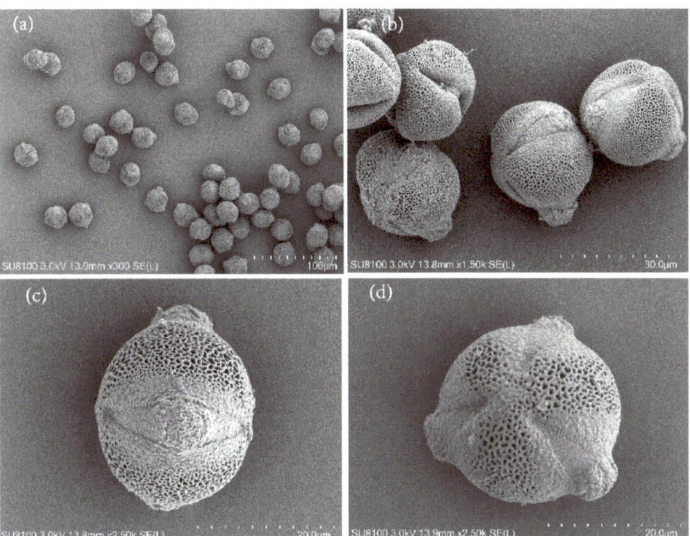

Figure 2. Scanning electron microscope (SEM) micrograph of *G. sinensis* pollen grains. Overall shape of pollen grains (**a**,**b**), pollen grain in equatorial view (**c**) and in polar view (**d**).

3.3. Research on the Method of Pollen Vitality Determination
3.3.1. In Vitro Germination Method to Determine the Pollen Vitality
Effect of Different Medium Components on In Vitro Germination

The addition of sucrose, $CaCl_2$ and H_3BO_3 to the liquid medium significantly affected in vitro germination (Figure 3). The pollen germination increased significantly as the sucrose concentration increased, reaching its highest rate at 15%, but it decreased significantly when sucrose concentrations reached 20%, indicating the appropriate concentration of sucrose promoting pollen germination. The pollen germination rate both decreased with the increase concentration of $CaCl_2$ and H_3BO_3 if the concentration was more than 100 mg/L, which showed that more calcium chloride and boric acid were not conducive to the pollen germination.

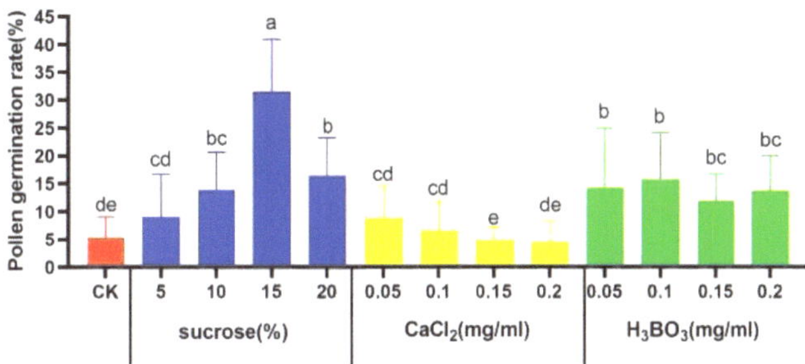

Figure 3. Effect of different concentration of Sucrose, $CaCl_2$ and H_3BO_3 on in vitro pollen germination of *G. sinensis*.

To sum up, the optimum concentrations of the sucrose, $CaCl_2$ and H_3BO_3 were 15% (30.90%), 50 mg/L (9.22%) and 100 mg/L (17.15%), respectively, all significantly higher than CK (5.21%). Too low or too high concentration would reduce the pollen germination rate of *G. sinensis*.

Selection of Liquid Culture Medium Formula

The results of an orthogonal test design and its data variance analysis (Table 4) indicated that sucrose (F = 106.147, $p < 0.001$) was most effective in the germination of pollen, followed by calcium chloride (F = 15.786, $p < 0.001$) and boric acid (F = 42.304, $p < 0.001$). During this experiment, 15% sucrose, 20 mg/L calcium chloride and 100 mg/L boric acid were found to be the best combination of medium. The pollen germination rate increased significantly with the increase of sucrose concentration (Figure 4). In response to boric acid concentration, pollen germination rate varied. B2 was significantly higher than B1 and B3, but B1 and B3 did not differ significantly, indicating that 100 mg/L boric acid was more beneficial to the in vitro germination. Pollen germination was also affected by the concentration of $CaCl_2$. A significant difference did not exist between C1 and C3, but both were significantly higher than C2, indicating that 20 mg/L and 40 mg/L $CaCl_2$ significantly promoted pollen germination.

Table 4. The range analysis of germination rates using an Orthogonal Assay Test Strategy (OATS). L Combinations (Levels Factors) = L9 (34) on different media.

Combinations	Sucrose (%) A	Boric Acid (mg/L) B	Calcium Chloride (mg/L) C	Empty Column	Pollen Germination Rate
1	S1	B1	C1	1	25.50% ± 0.55%
2	S1	B2	C2	2	19.65% ± 0.46%
3	S1	B3	C3	3	19.97% ± 0.22%
4	S2	B1	C2	3	17.68% ± 0.27%
5	S2	B2	C3	1	33.36% ± 0.31%
6	S2	B3	C1	2	23.85% ± 0.36%
7	S3	B1	C3	2	41.20% ± 0.21%
8	S3	B2	C1	3	47.40% ± 0.5%
9	S3	B3	C2	1	29.71% ± 0.27%
K1	65.12%	84.38%	**96.75%**	88.57%	
K2	74.89%	**100.41%**	67.04%	84.70%	
K3	**118.31%**	73.53%	94.53%	85.05%	
$\bar{x}1$	21.71%	28.13%	**32.25%**	29.52%	
$\bar{x}2$	24.96%	**33.47%**	22.35%	28.23%	
$\bar{x}3$	**39.44%**	24.51%	31.51%	28.35%	
R	53.19%	26.88%	29.71%	3.87%	
Order of influencing factors			A > C > B		
Best levels	A3	B2	C1		
Excellent combination			A3 B2 C1		

Ki is the sum of the different levels for the i th factor, xi is the mean of the different levels for the ith factor, and R the range difference between the max and mini values. Three levels of sucrose (S): S1 = 5 g/100 mL, S2 = 10 g/100 mL, S3 = 15 g/100 mL; boric acid (B): B1 = 50 mg/L, B2 = 100 mg/L, B3 = 150 mg/L; Calcium nitrate (C): C1 = 20 mg/L, C2 = 40 mg/L, C3 = 60 mg/L. Also see Table 3. Bold numbers are the highest values and bold letters are the best components.

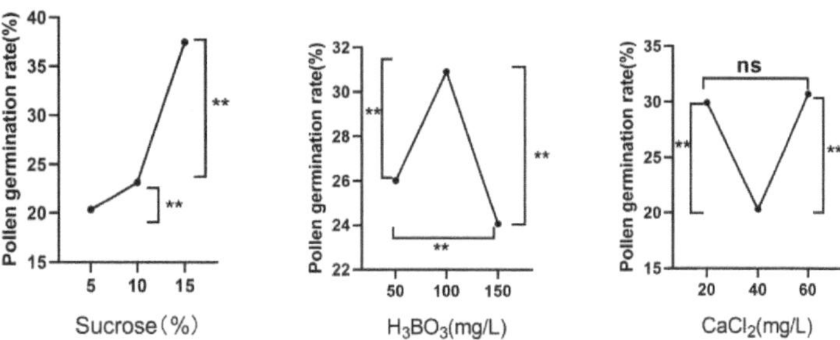

Figure 4. Effect of combination of sucrose, H_3BO_3 and $CaCl_2$ on pollen germination of *G. sinesis* in liquid medium, ns—non-significant; ** significant at $p \leq 0.01$.

In Vitro Germination Rate of Different Culture Durations

A significant increase in pollen germination rate and pollen tube length was observed with prolonged incubation (Figure 5A) when pollen grains were cultured in the liquid medium (15% sucrose concentration + 20 mg/L calcium chloride + 100 mg/L boric acid) at a constant temperature of 25 °C. As a result, the highest germination rate was 65.89% ± 3.41% (Figure 5B), and the pollen tube reached to 3.96 mm ± 0.06 mm after 24 h in culture. The observation results of pollen tube dynamics indicate that the pollen tube along the style path reached the ovule 24 h after pollination (unpublished). Therefore, in this study, the germination rate of pollen cultured in liquid medium for 24 h was taken as the final pollen germination rate of *G. sinensis*. Thus, the vitality of fresh pollen was 65.89% ± 3.41%.

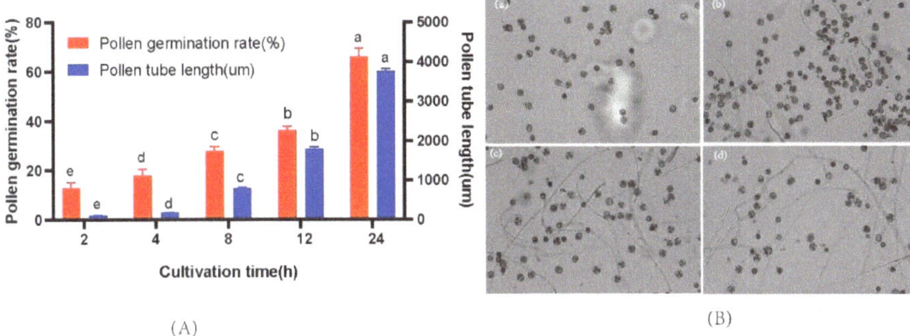

(A)　　　　(B)

Figure 5. (**A**) Effect of incubation time on in vitro pollen germination and tube growth; (**B**) The four different photos show the pollen germination and pollen tube growth status at different time of incubation (a) 1 h, (b) 4 h, (c) 12 h, and (d) 24 h.

3.3.2. Pollen Vitality of *G. sinensis* Measured by Four Different Staining Methods

In three of four dyeing methods, active pollen could be differentiated from inactive pollen except in KI-I$_2$, which was the only dyeing method that could not distinguish between active and inactive pollens (Figure 6d). The MTT staining method, TTC staining method and Alexander staining method revealed pollen vitality of 68.19%, 36.87% and 90.40%, respectively, and they differed significantly from one another (Table 5). Comparing the in vitro germination rate with dyeing methods allowed us to decide that the MTT staining method was suitable for the rapid determination of the pollen vitality of *G. sinensis*. As compared to a control sample, MTT staining method did not significantly alter pollen vitality.

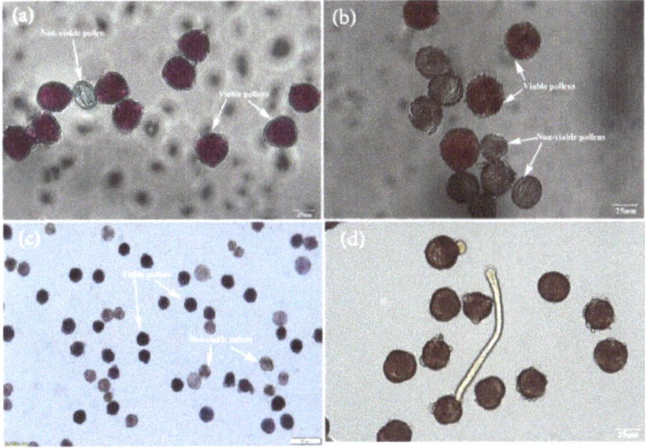

Figure 6. Pollen grains were stained by four staining method, (**a**) Alexander staining method, (**b**) TTC staining method, (**c**) MTT staining method and (**d**) KI-I$_2$ staining method.

Table 5. Comparison of pollen viability results determined with four methods.

Methods		Pollen Viability ± Standard Error
In vitro germination		65.89% ± 3.41% b
staining	TTC	36.87% ± 1.80% c
	Alexander	90.40% ± 1.54% a
	MTT	68.19% ± 2.09% b

A significant difference between methods was indicated by lowercase letters ($p < 0.05$). Values are means ± SE.

3.4. Pollen Preservation Method for G. sinensis

3.4.1. Effect of Pollens at Different Flowering Stages on In Vitro Germination Rate

The results showed that pollen germination rates increased gradually with incubation time for four flowering stages (Figure 7). In addition, the final pollen germination rate among four stages was not significantly different after 24 h of pollen culture. Based on these results, the pollen viability and germination trend were basically the same for the four stages. Despite this, the germination rate of pollen at different flowering stages varied significantly from 2–4 h, with higher germination rate of the fourth stage and lower germination rate of stage 2. This might be explained by the different water content of pollen grains during dispersion due to their brief dormancy. As a result, pollens of stage 2 were best suited to instant pollination since all anthers had cracked and there were a large number of pollen grains exposed. On the basis of the above results, pollen at the large bud stage can be collected to improve pollen collection efficiency without worrying about pollen viability.

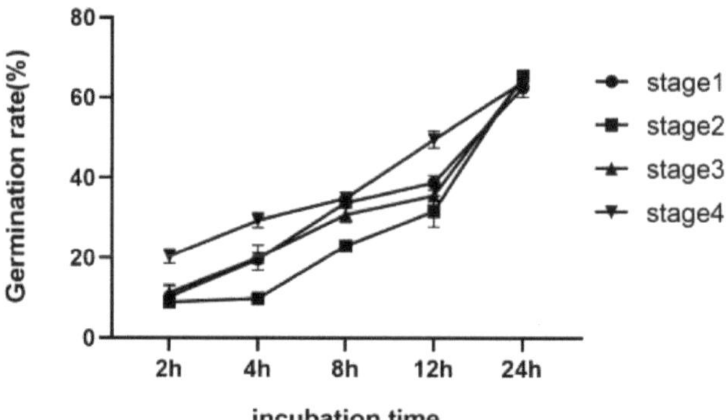

Figure 7. Effects of pollen at different flowering stages on pollen germination rate.

3.4.2. Short-Term Preservation of Pollen

Using the three-factor ANOVA method, the results indicated that storage time (F = 2298.6, $p < 0.0001$), storage temperature (F = 891.01, $p < 0.0001$) and pollen water content (F = 17.45, $p < 0.0001$) were significant factors affecting pollen germination. In addition, there are two interactions among the three influencing factors: storage time * storage temperature (F = 164.49, $p < 0.0001$), storage time * pollen water content (F = 38.19, $p < 0.0001$), storage temperature * pollen water content (F = 13.81, $p < 0.0001$). The interaction of three factors, storage time * storage temperature * pollen water content (F = 11.86, $p < 0.0001$), also existed.

Effect of Pollen Water Content on In Vitro Germination

After 3 h, 6 h and 9 h of silica gel drying, pollen water content was reduced to 14%, 9% and 5%, and the germination rate of pollens were 59.03% ± 3.20%, 57.97% ± 2.02% and 52.13% ± 3.47%, which were lower than that of fresh pollens by 10.41%, 12.02% and 20.88%, respectively. Based on the results above, pollen germination was greatly affected by the pollen water content and decreased significantly with decreasing water content.

The pollen preservation results showed all pollen germination rates under RT treatment were almost zero after 30 days of storage (Figure 8). When stored at 4 °C for 30 days, pollen with a higher water content of 14% had a significantly lower germination rate than that with a lower water content of 9% and 5%. It meant that pollen could be preserved for longer if its water content was reduced when stored under 4 °C. With a reduction in storage temperature to −20 °C or −80 °C, there was no significant difference in pollen germination rates with different water contents after 30 days, but after 30 days, the germination rate of pollen stored at −20 °C was much lower than that of those stored at −80 °C. In spite of the significant effect of pollen water content on the germination rate after storage, the effect gradually weakened with the storage temperature decreased. This indicated that the water content of pollen (5%–14%) was no longer the main factor affecting the germination rate after storage, if the storage temperature reduced to −20 °C or below.

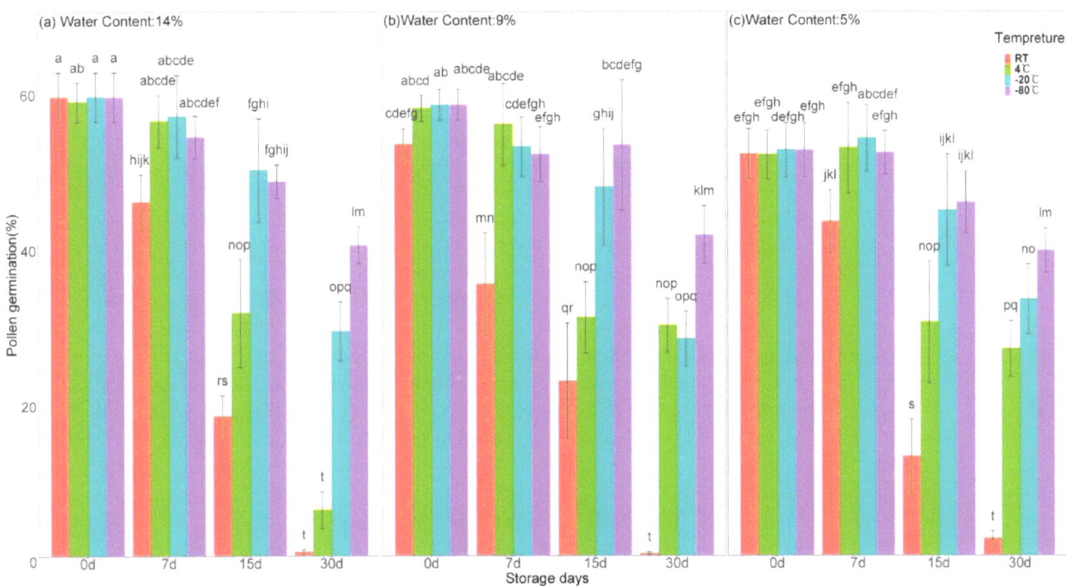

Figure 8. Effects of storage temperature, storage time and pollen water content on the pollen preservation of *G. sinensis*. A significant difference between methods was indicated by lowercase letters ($p < 0.05$) using Multiple Testing.

Effect of Storage Temperature on In Vitro Germination

As the storage time increased, the germination rate of all treatments decreased significantly. The pollen germination rate was almost zero after 30 days of storage under RT treatment. However, pollen stored at 4 °C, −20 °C and −80 °C were treatments that were effective in delaying the decreasing trend of pollen germination, and a temperature of −80 °C was most effective for preserving pollen.

Effect of Storage Time on In Vitro Germination

Despite the aforementioned observations, all treatments showed a gradual decline in pollen germination with the extension of storage time (Figure 8). Pollen germination rates of different treatments decreased at different rates due to the effects of storage temperature, pollen water content, storage time and interactions between them. Pollen germination decreased the fastest under RT treatment. Within 7 days of storing, the germination rate of pollen stored at other low temperatures remained similar to that before storage, except for RT treatment. This indicated that the pollen germination rate remained unchanged at low temperature after short-term storage for 7 days. Although the pollen germination rate decreased with the extension of the storage time to 15 or 30 days, the lower the storage temperature, the higher the pollen germination rate. That is to say that storage temperature gradually became the main factor affecting the germination rate of pollen if the storage time was over 7 days.

The highest germination rate was observed after 30 days of cryopreservation at $-80\,^\circ$C for *G. sinensis* pollen, if the pollen water content was reduced to 9%, which was only 28.84% lower than that before storage. This could be used as a suitable method for preserving pollen.

4. Discussion

4.1. Pollen Morphological Characteristics of G. sinensis

Palynology is often used in plant classification and phylogenetic studies [39–41], the size of pollen grains according to the appearance and shape of pollen [41], pollen outer wall ornamentation [40,42] and number and location of germination pores [43]. Observations of the microscopic morphology of pollen revealed that pollen grains of *G. sinensis* were oblate spheroidal, tricolporate with equatorial elongated endoapertures and a reticulate sporoderm surface. This is consistent with its classified status because pollen of the majority of *caesalpinioid* species is isopolar and tricolporate with perforate or reticulate surface ornamentation, according to Hannah Banks & Paula J. Rudall [44]. The small or medium pollen size is also consistent with the widespread type of pollen in *caesalpinioid*, in which the widespread type of pollen is small to medium in size [45]. The equatorial elongated endoapertures may be a strategy help to preserve the structural integrity of the pollen wall and, therefore, the viability of the pollen grain [44]. Endoapertures have been found to be positively correlated with pollen germination rate in studies [46]. The pollen of *G. sinensis* germinated rapidly after 1 h in culture, consistent with this. In this article, pollen grains have reticulate ornamentation (with sculpturing elements forming an open network or reticulum over the pollen surface), which was related to foldable structures and the natural design of pollen grains during pollen dehydration [47]. In short, the shape and appearance of pollen and pollen apertures are closely related to the classification of this species [44,45].

4.2. Methods for the Determination of Pollen Viability

In vitro germination was considered to accurately predict pollen viability [11,48–50]. Due to the specificity of plant species, the most important step of in vitro germination was selecting the appropriate medium for pollen germination. An addition of 15% sucrose, 100 mg/L boric acid and 20 mg/L calcium chloride enhanced pollen germination in this study. This was consistent with the existing research results because sucrose was the most effective carbon source [51,52] and an important signaling substance during the in vitro pollen germination [53]. As well as promoting pollen germination, boron contributes to pollen tube elongation through signal transduction [54,55]. The mineral calcium (Ca^{2+}) regulates ion balance and plays a critical role in pollen tube growth [14,56].

The concentration of sucrose required for pollen germination varied with plant species [50,57]. Similarly to previous studies in other plants [58,59], 15% sucrose added to pollen germination of *G. sinensis* was most beneficial. According to the single factor experiment, variations in boron concentrations did little to promote pollen germination since the amount of boron required in the pollen medium had the smallest effect [60]. A slight deviation could cause insufficient [54] or excessive toxicity [61], which was consistent with the results of previous Fragallah [62,63]. Moreover, 50 mg/L $CaCl_2$ increased pollen germination, but the pollen germination rate decreased with the increase of $CaCl_2$ concentration. This may be related to the regulation of pollen tube calcium signal [13] and the inhibition of pollen tube growth by high Ca^{2+} concentration [64].

In order to detect pollen viability, the staining method was considered to be the easiest and fastest method [38,65,66]. The pollen viability of *G. sinensis* was assessed using four staining methods, and it varied with different methods. In comparison to in vitro germination rate, TTC's pollen viability was lower while Alexander's was higher. KI-I$_2$ did not distinguish non-viable pollen. Only MTT was a suitable method for accurate determination of pollen viability of *G. sinensis*. A major reason for the difference between four staining methods was the specificity of pollen [36,67]. The results of this study were similar to those of other pollen viability studies [10,38,57]. TTC reacts with dehydrogenase in normal pollen and appears red, which is considered to be the most common technique [68]. However, it is a fat-soluble photosensitive complex, susceptible to environmental influences. In this article, it did not stain pollen grains of *G. sinensis* well, just as the results described in *Sinobeam* [69] and *Rhododendron* [37], which may be caused by the failure to meet the conditions due to operating errors. In spite of the fact that Alexander's staining method was able to distinguish viable pollen grains and its result was much higher than those of the control, the reason of it may be attributed to weak ability to distinguish pollen viability. The KI-I$_2$ staining method relies on starch to stain pollen. Moreover, the staining effect was affected not only by the amount of starch, but also by the ratio of amylopectin to amylose. However, pollen with vitality cannot be determined using this method in this study due to all pollen grains stained with the same color. Viable pollen grains could be stained to purple by MTT staining method based on the activity of dehydrogenase in pollen grains. This method is suitable for the determination of pollen viability of *G. sinensis* in this study, just as *Rhododendron* [37].

4.3. Pollen Preservation of G. sinensis

In certain storage conditions, such as low temperature, low oxygen and low humidity, enzyme activity and related metabolic activities may be reduced or inhibited, thereby extending pollen grain life [70–74]. It was found that the pollen germination rate was easily affected by storage temperature, storage time and pollen water content during pollen storage of *G. sinensis* in this study. A common feature of all pollen preservation treatments was a decrease in germination rate with increasing storage time, such as plum [75], pecan [49], rose [15], litchi [76] and chrysanthemum [77]. It was almost impossible for pollen stored at RT to germinate after 30 days, but it was still possible to maintain pollen vitality at low temperatures. Especially when preserved at $-80\,^{\circ}$C, the germination rate only decreased by 28.84%, which was the same as the results of previous studies on hickory walnut and chrysanthemum [57,76]. The three water contents of pollens set in this study played a certain role in the pollen preservation, but when the temperature was reduced to $-20\,^{\circ}$C or below, pollen viability was no longer largely determined by the pollen water content. It may be caused by an insufficient gap between water content gradients [71,75]. A suitable method for short-term pollen preservation of *G. sinensis* was identified in this study. However, further pollination tests were needed to more accurately evaluate the feasibility of this preservation method in the breeding of *G. sinensis*.

5. Conclusions

The study results showed that male flowers of *G. sinensis* lasted 2–3 days during their single flowering period. The dynamic process of single male flower opening was artificially divided into five stages. Pollen grains of *G. sinensis* are oblate spheroidal, tricolporate with equatorial elongated endoapertures and the sporoderm surface is reticulate. The fresh pollen germination rate of *G. sinensis* reached its maximum value (65.89% $\pm$ 3.41%), and the length of pollen tube was 3.96 mm after 24 h incubation at 25 $^{\circ}$C based on the selected liquid medium: 15% sucro e + 100 mg/L boric acid + 20 mg/L calcium chloride. In comparison with the other three staining methods, the MTT staining method could accurately and rapidly determine pollen viability. A short-term preservation method for *G. sinensis* can be achieved by reducing the water content of pollen to 9% and then storing it at $-80\,^{\circ}$C for 30 days.

Author Contributions: Q.L. and J.Y. designed and carried out the experiments. Q.L. performed the data analyses and drafted the manuscript. X.W. and Y.Z. supported this research and made revisions to the manuscript. All authors have read and agreed to the published version of the manuscript.

Funding: This work was supported by the Characteristic Forestry Industry Research Project of Guizhou Province (GZMC-ZD20202098); Science and Technology Plan Project of Guizhou Province ([2020]1Y056); Guizhou Characteristic Forestry Industry scientific Research Project-Research and Demonstration of Key Technology of Directional Cultivation of Gleditsia sinensis, No. Te LinYan: 2020.9-2023.12; Guizhou Science and Technology Planning Project (Guizhou Science and Technology Cooperation Support [2020]1Y058).

Data Availability Statement: The data presented in this study are available on request from the corresponding author.

Acknowledgments: The author would like to thank College of forestry and the Institute of Forest Resources & Environment of Guizhou to provide Laboratory platform and experiment management teachers for their assistance with microscopy.

Conflicts of Interest: The authors declare no conflict of interest.

References

1. Boavida, L.C.; Vieira, A.M.; Becker, J.D.; Feijo, J.A. Gametophyte Interaction and Sexual Reproduction: How Plants Make a Zygote. *Int. J. Dev. Biol.* **2005**, *49*, 615–632. [CrossRef] [PubMed]
2. Jiang, Y.; Lahlali, R.; Karunakaran, C.; Warkentin, T.D.; Davis, A.R.; Bueckert, R.A. Pollen, Ovules, and Pollination in Pea: Success, Failure, and Resilience in Heat: Heat Stress and Reproduction in Pea. *Plant Cell Environ.* **2019**, *42*, 354–372. [CrossRef]
3. Fan, Y.-F.; Jiang, L.; Gong, H.-Q.; Liu, C.-M. Sexual Reproduction in Higher Plants I: Fertilization and the Initiation of Zygotic Program. *J. Integr. Plant Biol.* **2008**, *50*, 860–867. [CrossRef] [PubMed]
4. Aloisi, I.; Distefano, G.; Antognoni, F.; Potente, G.; Parrotta, L.; Faleri, C.; Gentile, A.; Bennici, S.; Mareri, L.; Cai, G.; et al. Temperature-Dependent Compatible and Incompatible Pollen-Style Interactions in Citrus Clementina Hort. Ex Tan. Show Different Transglutaminase Features and Polyamine Pattern. *Front. Plant Sci.* **2020**, *11*, 1018. [CrossRef] [PubMed]
5. Katano, K.; Oi, T.; Suzuki, N. Failure of Pollen Attachment to the Stigma Triggers Elongation of Stigmatic Papillae in Arabidopsis Thaliana. *Front. Plant Sci.* **2020**, *11*, 989. [CrossRef]
6. Mao, X.; Fu, X.-X.; Huang, P.; Chen, X.-L.; Qu, Y.-Q. Heterodichogamy, Pollen Viability, and Seed Set in a Population of Polyploidy Cyclocarya Paliurus (Batal) Iljinskaja (Juglandaceae). *Forests* **2019**, *10*, 347. [CrossRef]
7. Delph, L.F.; Mutikainen, P. Testing Why the Sex of the Maternal Parent Affects Seeding Survival in a Gynodioecious Species. *Evolution* **2003**, *57*, 231. [CrossRef]
8. Yao, W.; Li, C.; Lin, S.; Wang, J.; Fan, T.; Zhao, W. The Structures of Floral Organs and Reproductive Characteristics of Ornamental Bamboo Species, Pleioblastus Pygmaeus. *Hortic. Plant J.* **2022**, S2468014122000553. [CrossRef]
9. Attique, R.; Zafar, M.; Ahmad, M.; Zafar, S.; Ghufran, M.A.; Mustafa, M.R.U.; Yaseen, G.; Ahmad, L.; Sultana, S.; Nabila; et al. Pollen Morphology of Selected Melliferous Plants and Its Taxonomic Implications Using Microscopy. *Microsc. Res. Tech.* **2022**, *85*, 2361–2380. [CrossRef]
10. Kumari, M.; Prasad, A.; Rahman, L.; Mathur, A.K.; Mathur, A. In Vitro Germination, Storage and Microscopic Studies of Pollen Grains of Four Ocimum Species. *Ind. Crops Prod.* **2022**, *177*, 114445. [CrossRef]
11. Wani, M.S.; Hamid, M.; Tantray, Y.R.; Gupta, R.C.; Munshi, A.H.; Singh, V. In Vitro Pollen Germination of Betula Utilis, a Typical Tree Line Species in Himalayas. *South Afr. J. Bot.* **2020**, *131*, 214–221. [CrossRef]
12. Althiab-Almasaud, R.; Chen, Y.; Maza, E.; Djari, A.; Frasse, P.; Mollet, J.; Mazars, C.; Jamet, E.; Chervin, C. Ethylene Signaling Modulates Tomato Pollen Tube Growth through Modifications of Cell Wall Remodeling and Calcium Gradient. *Plant J.* **2021**, *107*, 893–908. [CrossRef] [PubMed]
13. Liu, J.; Liu, J.; Zhang, X.; Wei, H.; Ren, J.; Peng, C.; Cheng, Y. Pollen Tube in Hazel Grows Intermittently: Role of Ca2+ and Expression of Auto-Inhibited Ca2+ Pump. *Sci. Hortic.* **2021**, *282*, 110032. [CrossRef]
14. Dumais, J. Mechanics and Hydraulics of Pollen Tube Growth. *New Phytol.* **2021**, *232*, 1549–1565. [CrossRef] [PubMed]
15. Jeong, N.-R.; Park, K.-Y. Rose Pollen Management Methods to Improve Productivity. *Agronomy* **2022**, *12*, 1285. [CrossRef]
16. Lan, Y.P.; Gu, W.C. Geographical Variation of Morphologic Characteristics of Gleditsia sinensis Seeds and Legumes in the North Region. *Sci. Silvae Sin.* **2006**, *42*, 47–51.
17. Li, W.; Lin, F.G.; Zheng, Y.Q.; Li, B. Phenotypic diversity of pods and seeds in natural populations of *Gleditsia sinensis* in southern China. *Chin. J. Plant Ecol.* **2013**, *37*, 61–69. [CrossRef]
18. Yu, J.; Xian, Y.; Li, G.; Wang, D.; Zhou, H.; Wang, X. One New Flavanocoumarin from the Thorns of *Gleditsia Sinensis*. *Nat. Prod. Res.* **2017**, *31*, 275–280. [CrossRef]
19. Lan, Y.P.; Zhou, L.D.; Li, S.Y.; Cao, Q.C.; Lan, W.Z. Advances in Research of Gleditsia and Its Prospect of Industrializational Development. *World For. Res.* **2004**, *6*, 17–21. [CrossRef]
20. Cerino, M.C.; Castro, D.C.; Richard, G.A.; Exner, E.L.; Pensiero, J.F. Functional Dioecy in Gleditsia Amorphoides (Fabaceae). *Aust. J. Bot.* **2018**, *66*, 85. [CrossRef]
21. Li, J.J.; Ye, C.L.; Shang, X.C.; Wang, J.; Zhang, B.; Wang, Z.Z.; Zhang, G.T. Study on breeding and pollination characteristics of Gleditsia sinensis. *China J. China Mater. Med.* **2018**, *43*, 4831–4836. [CrossRef]

22. Stehlik, I.; Friedman, J.; Barrett, S.C.H. Environmental Influence on Primary Sex Ratio in a Dioecious Plant. *Proc. Natl. Acad. Sci. USA* **2008**, *105*, 10847–10852. [CrossRef] [PubMed]
23. Hultine, K.R.; Grady, K.C.; Wood, T.E.; Shuster, S.M.; Stella, J.C.; Whitham, T.G. Climate Change Perils for Dioecious Plant Species. *Nat. Plants* **2016**, *2*, 16109. [CrossRef] [PubMed]
24. Li, B.H.; Zhang, W.Q.; Zhao, L.Q.; Yan, F.Y.; Zheng, Q.; Zheng, J.W.; Geng, Y.X. The present situation and industrial development strategy of Gleditsis microphylla resources in Hebei province. *For. Ecol. Sci.* **2020**, *35*, 432–435. [CrossRef]
25. Fan, D.C.; Zhang, S.A.; Liu, Y.; Luo, Y. Genetic Diversity and Fingerprints of Glediysia sinensis Germplasms Based on RSAP. *J. Henan Agric. Sci.* **2017**, *46*, 103–107. [CrossRef]
26. Yu, Z.Y.; Yan, H.B.; Zhang, H.F.; Yang, X.Q. Effect of different salt stresses on seed germination and seedling physiological characteristics of Gleditsia sinensis. *J. Northeast. Agric. Univ.* **2020**, *51*, 28–35. [CrossRef]
27. Ma, B. *Selection for Plant Species during Ecological Rehabilitation in Limestone Mountains of Shanxi*; Shanxi Normal University: Xi'an, China, 2020. [CrossRef]
28. Wang, L. *Study on the Reproductive Characteristics of Gleditsia Officinalis*; Shandong University of Chinese Medicine: Jinan, China, 2020. [CrossRef]
29. Fernandez, R.D.; Ceballos, S.J.; Malizia, A.; Aragón, R. Gleditsia Triacanthos (Fabaceae) in Argentina: A Review of Its Invasion. *Aust. J. Bot.* **2017**, *65*, 203. [CrossRef]
30. Wang, L.; Liu, Z.Y.; Zhou, Y.; Liu, Q.; Zhang, Y.Q. Observation and Analysis on Flowering Characteristics of Zaojia (*Gleditsia* Plant). *J. Shandong Univ. TCM* **2021**, *45*, 388–392. [CrossRef]
31. Reitsma, T. Size modification of recent pollen grains under different treatments. *Rev. Palaeobot. Palynol.* **1969**, *9*, 175–202. [CrossRef]
32. Raynal, A.; Raynal, J. Une techinique de préparation des graines de pollen fragiles. *Adansonia* **1971**, *11*, 77–79.
33. Kisser, J. Bemerkungen zum Einschluß in Glyzerin-gelatine. Ztschr. f. wiss. Mikrosk. *Technology* **1934**, *51*, 372–374.
34. Erdtman, G. *Pollen Morphology and Plant Taxonomy: An Introduction to Palynology (Vol. 4)*; Almqvist & Wiksell: Stockholm, Sweden, 1952.
35. Halbritter, H.; Ulrich, S.; Grímsson, F.; Weber, M.; Zetter, R.; Hesse, M.; Buchner, R.; Svojtka, M.; Frosch-Radivo, A.; Halbritter, H.; et al. Pollen Morphology and Ultrastructure. In *Illustrated Pollen Terminology*; Springer: Berlin/Heidelberg, Germany, 2018.
36. Alexander, M.P. Differential Staining of Aborted and Nonaborted Pollen. *Stain. Technol.* **1969**, *44*, 117–122. [CrossRef] [PubMed]
37. Zhang, C.Y.; Geng, X.M. Comparative study on methods for determining pollen viability of six Rhododendron species. *J. Plant Sci.* **2012**, *30*, 92–99. [CrossRef]
38. Luo, S.; Zhang, K.; Zhong, W.-P.; Chen, P.; Fan, X.-M.; Yuan, D.-Y. Optimization of in Vitro Pollen Germination and Pollen Viability Tests for Castanea Mollissima and Castanea Henryi. *Sci. Hortic.* **2020**, *271*, 109481. [CrossRef]
39. Khan, A.; Ahmad, M.; Zafar, S.; Abbas, Q.; Arfan, M.; Zafar, M.; Sultana, S.; Ullah, S.A.; Khan, S.; Akhtar, A.; et al. Light and Scanning Electron Microscopic Observation of Palynological Characteristics in Spineless *Astragalus* L. (Fabaceae) and Its Taxonomic Significance. *Microsc. Res. Tech.* **2022**, *85*, 2409–2427. [CrossRef]
40. Claxton, F.; Banks, H.; Klitgaard, B.B.; Crane, P.R. Pollen Morphology of Families Quillajaceae and Surianaceae (Fabales). *Rev. Palaeobot. Palynol.* **2005**, *133*, 221–233. [CrossRef]
41. Al-Hakimi, A.S.; Maideen, H.; Saeed, A.A.; Faridah, Q.Z.; Latiff, A. Pollen and Seed Morphology of Justicieae (Ruellioideae, Acanthaceae) of Yemen. *Flora* **2017**, *233*, 31–50. [CrossRef]
42. Lopes, A.C.V.; de Souza, C.N.; Saba, M.D.; Gasparino, E.C. Pollen Morphology of Malvaceae *s.l.* from Cerrado Forest Fragments: Details of Aperture and Ornamentation in the Pollen Types Definition. *Palynology* **2022**, *46*, 1–15. [CrossRef]
43. Zhang, J.; Huang, D.; Zhao, X.; Hou, X.; Di, D.; Wang, S.; Qian, J.; Sun, P. Pollen Morphology of Different Species of *Iris Barbata* and Its Systematic Significance with Scanning Electron Microscopy Methods. *Microsc. Res. Tech.* **2021**, *84*, 1721–1739. [CrossRef]
44. Banks, H.; Rudall, P.J. Pollen Structure and Function in Caesalpinioid Legumes. *Am. J. Bot.* **2016**, *103*, 423–436. [CrossRef]
45. Banks, H.; Klitgaard, B.B. Palynological contribution to the systematics of detarioid legumes (Leguminosae: Caesalpinioideae). *Adv. Legume Syst.* **2000**, *9*, 79–106.
46. Lukšić, K.; Zdunić, G.; Mucalo, A.; Marinov, L.; Ranković-Vasić, Z.; Ivanović, J.; Nikolić, D. Microstructure of Croatian Wild Grapevine (Vitis Vinifera Subsp. Sylvestris Gmel Hegi) Pollen Grains Revealed by Scanning Electron Microscopy. *Plants* **2022**, *11*, 1479. [CrossRef] [PubMed]
47. Katifori, E.; Alben, S.; Cerda, E.; Nelson, D.R.; Dumais, J. Foldable Structures and the Natural Design of Pollen Grains. *Proc. Natl. Acad. Sci. USA* **2010**, *107*, 7635–7639. [CrossRef] [PubMed]
48. Alexander, L.W. Optimizing Pollen Germination and Pollen Viability Estimates for Hydrangea Macrophylla, Dichroa Febrifuga, and Their Hybrids. *Sci. Hortic.* **2019**, *246*, 244–250. [CrossRef]
49. Peng, H.-Z.; Jin, Q.-Y.; Ye, H.-L.; Zhu, T.-J. A Novel in Vitro Germination Method Revealed the Influence of Environmental Variance on the Pecan Pollen Viability. *Sci. Hortic.* **2015**, *181*, 43–51. [CrossRef]
50. Ferreira, M.d.S.; Soares, T.L.; Costa, E.M.R.; da Silva, R.L.; de Jesus, O.N.; Junghans, T.G.; Souza, F.V.D. Optimization of Culture Medium for the in Vitro Germination and Histochemical Analysis of Passiflora Spp. Pollen Grains. *Sci. Hortic.* **2021**, *288*, 110298. [CrossRef]

51. Laggoun, F.; Ali, N.; Tourneur, S.; Prudent, G.; Gügi, B.; Kiefer-Meyer, M.-C.; Mareck, A.; Cruz, F.; Yvin, J.-C.; Nguema-Ona, E.; et al. Two Carbohydrate-Based Natural Extracts Stimulate in Vitro Pollen Germination and Pollen Tube Growth of Tomato Under Cold Temperatures. *Front. Plant Sci.* **2021**, *12*, 552515. [CrossRef]
52. Reinders, A. Fuel for the Road—Sugar Transport and Pollen Tube Growth. *EXBOT J.* **2016**, *67*, 2121–2123. [CrossRef]
53. Hirsche, J.; Garcï¿½a Fernï¿½ndez, J.M.; Stabentheiner, E.; Groï¿½kinsky, D.K.; Roitsch, T. Differential Effects of Carbohydrates on Arabidopsis Pollen Germination. *Plant Cell Physiol.* **2017**, *58*, 691–701. [CrossRef]
54. Fang, K.F.; Du, B.S.; Zhang, Q.; Xing, Y.; Cao, Q.Q.; Qin, L. Boron Deficiency Alters Cytosolic Ca^{2+} Concentration and Affects the Cell Wall Components of Pollen Tubes in *Malus Domestica*. *Plant Biol J* **2019**, *21*, 343–351. [CrossRef]
55. Muengkaew, R.; Chaiprasart, P.; Wongsawad, P. Calcium-Boron Addition Promotes Pollen Germination and Fruit Set of Mango. *Int. J. Fruit Sci.* **2017**, *17*, 147–158. [CrossRef]
56. Yang, H.; You, C.; Yang, S.; Zhang, Y.; Yang, F.; Li, X.; Chen, N.; Luo, Y.; Hu, X. The Role of Calcium/Calcium-Dependent Protein Kinases Signal Pathway in Pollen Tube Growth. *Front. Plant Sci.* **2021**, *12*, 633293. [CrossRef] [PubMed]
57. Wang, X.; Wu, Y.; Lombardini, L. In Vitro Viability and Germination of Carya Illinoinensis Pollen under Different Storage Conditions. *Sci. Hortic.* **2021**, *275*, 109662. [CrossRef]
58. Liu, X.; Yin, Y. Effects of Different solid Medium formulations on pollen germination and analysis of Cross compatibility of Miticola Formosa. *J. Plant Resour. Environ.* **2022**, *31*, 49–56.
59. Baloch, M.J.; Lakho, A.R.; Bhutto, H.; Solangi, M.Y. Impact of Sucrose Concentrations on in Vitro Pollen Germination of Okra, Hibiscus Esculentus. *Pak. J. Biol. Sci.* **2001**, *4*, 402–403. [CrossRef]
60. Tushabe, D.; Rosbakh, S. A Compendium of in Vitro Germination Media for Pollen Research. *Front. Plant Sci.* **2021**, *12*, 709945. [CrossRef]
61. Fang, K.; Zhang, W.; Xing, Y.; Zhang, Q.; Yang, L.; Cao, Q.; Qin, L. Boron Toxicity Causes Multiple Effects on Malus Domestica Pollen Tube Growth. *Front. Plant Sci.* **2016**, *7*, 208. [CrossRef]
62. Fragallah, S.; Lin, S.; Li, N.; Ligate, E.; Chen, Y. Effects of Sucrose, Boric Acid, PH, and Incubation Time on in Vitro Germination of Pollen and Tube Growth of Chinese Fir (*Cunnighamial Lanceolata* L.). *Forests* **2019**, *10*, 102. [CrossRef]
63. Fragallah, S.A.D.A.; Wang, P.; Li, N.; Chen, Y.; Lin, S. Metabolomic Analysis of Pollen Grains with Different Germination Abilities from Two Clones of Chinese Fir (*Cunninghamia Lanceolata* (Lamb) Hook). *Molecules* **2018**, *23*, 3162. [CrossRef]
64. Zheng, R.H.; Su, S.D.; Xiao, H.; Tian, H.Q. Calcium: A Critical Factor in Pollen Germination and Tube Elongation. *Int. J. Mol. Sci.* **2019**, *12*, 420. [CrossRef]
65. Atlagić, J.; Terzić, S.; Marjanović-Jeromela, A. Staining and Fluorescent Microscopy Methods for Pollen Viability Determination in Sunflower and Other Plant Species. *Ind. Crops Prod.* **2012**, *35*, 88–91. [CrossRef]
66. Wang, Z.; Yin, M.; Creech, D.L.; Yu, C. Microsporogenesis, Pollen Ornamentation, Viability of Stored Taxodium Distichum Var. Distichum Pollen and Its Feasibility for Cross Breeding. *Forests* **2022**, *13*, 694. [CrossRef]
67. D'souza, L. Staining Pollen Tubes in the Styles of Cereals with Cotton Blue; Fixation by Ethanol-Lactic Acid for Enhanced Differentiation. *Stain. Technol.* **1972**, *47*, 107–108. [CrossRef] [PubMed]
68. Rodriguez-Riano, T.; Dafni, A. A New Procedure to Asses Pollen Viability. *Sex. Plant Reprod.* **2000**, *12*, 241–244. [CrossRef]
69. Fu, Q.C.; Wang, J.; Zhang, X.Y.; Liu, C.; Liu, F. Study on Detection Methods for Pollen Viability of *Sinojackia sacocarpa* L.Q.Luo. *Mol. Plant Breed.* **2015**, *13*, 1146–1150. [CrossRef]
70. Lora, J.; de Oteyza, M.A.P.; Fuentetaja, P.; Hormaza, J.I. Low Temperature Storage and in Vitro Germination of Cherimoya (*Annona Cherimola* Mill.) Pollen. *Sci. Hortic.* **2006**, *108*, 91–94. [CrossRef]
71. Ćalić, D.; Milojević, J.; Belić, M.; Miletić, R.; Zdravković-Korać, S. Impact of Storage Temperature on Pollen Viability and Germinability of Four Serbian Autochthon Apple Cultivars. *Front. Plant Sci.* **2021**, *12*, 709231. [CrossRef]
72. Iovane, M.; Cirillo, A.; Izzo, L.G.; Di Vaio, C.; Aronne, G. High Temperature and Humidity Affect Pollen Viability and Longevity in *Olea europaea* L. *Agronomy* **2021**, *12*, 1. [CrossRef]
73. Kulus, D. Managing Plant Genetic Resources Using Low and Ultra-Low Temperature Storage: A Case Study of Tomato. *Biodivers Conserv.* **2019**, *28*, 1003–1027. [CrossRef]
74. Oliver, S.N.; Van Dongen, J.T.; Alfred, S.C.; Mamun, E.A.; Zhao, X.; Saini, H.S.; Fernandes, S.F.; Blanchard, C.L.; Sutton, B.G.; Geigenberger, P.; et al. Cold-Induced Repression of the Rice Anther-Specific Cell Wall Invertase Gene OSINV4 Is Correlated with Sucrose Accumulation and Pollen Sterility. *Plant Cell Environ.* **2005**, *28*, 1534–1551. [CrossRef]
75. Đorđević, M.; Vujović, T.; Cerović, R.; Glišić, I.; Milošević, N.; Marić, S.; Radičević, S.; Fotirić Akšić, M.; Meland, M. In Vitro and In Vivo Performance of Plum (*Prunus domestica* L.) Pollen from the Anthers Stored at Distinct Temperatures for Different Periods. *Horticulturae* **2022**, *8*, 616. [CrossRef]
76. Wang, L.; Wu, J.; Chen, J.; Fu, D.; Zhang, C.; Cai, C.; Ou, L. A Simple Pollen Collection, Dehydration, and Long-Term Storage Method for Litchi (*Litchi Chinensis* Sonn.). *Sci. Hortic.* **2015**, *188*, 78–83. [CrossRef]
77. Miler, N.; Wozny, A. Effect of Pollen Genotype, Temperature and Period of Storage on In Vitro Germinability and In Vivo Seed Set in Chrysanthemum—Preliminary Study. *Agronomy* **2021**, *11*, 2395. [CrossRef]

Review

Conserving Potential and Endangered Species of *Pericopsis mooniana* Thwaites in Indonesia

Julianus Kinho [1,*], Suhartati [2], Husna [3], Faisal Danu Tuheteru [3], Diah Irawati Dwi Arini [1], Moh. Andika Lawasi [4], Resti Ura' [1], Retno Prayudyaningsih [5], Yulianti [2], Subarudi [6], Lutfy Abdulah [1], Ruliyana Susanti [1], Totok Kartono Waluyo [7], Sona Suhartana [7], Andianto [8], Marfuah Wardani [2], Titi Kalima [1], Elis Tambaru [9], Wahyudi Isnan [1], Adi Susilo [1], Ngatiman [10], Laode Alhamd [1], Dulsalam [7] and Soenarno [7]

[1] Research Center for Ecology and Ethnobiology, National Research and Innovation Agency, Jalan Raya Jakarta-Bogor Km 46, Cibinong 16911, Indonesia
[2] Research Center for Plant Conservation, Botanic Gardens, and Forestry, National Research and Innovation Agency, Jalan Raya Jakarta-Bogor Km 46, Cibinong 16911, Indonesia
[3] Department of Forestry, Faculty of Forestry and Environment Science, Halu Oleo University, Jl. HEA Mokodompit, Kendari 93121, Indonesia
[4] Research Center for Society and Culture, National Research and Innovation Agency, Jl. Gatot Subroto No. 10, Mampang Prapatan, Jakarta 12710, Indonesia
[5] Research Centre for Applied Microbiology, National Research and Innovation Agency, Jalan Raya Jakarta-Bogor Km 46, Cibinong 16911, Indonesia
[6] Research Center for Population, National Research and Innovation Agency, JL. Gatot Subroto No. 10, Mampang Prapatan, Jakarta 12710, Indonesia
[7] Research Center for Biomass and Bioproducts, National Research and Innovation, Jalan Raya Jakarta-Bogor Km 46, Cibinong 16911, Indonesia
[8] Research Center for Biosystematics and Evolution, National Research and Innovation Agency, Jalan Raya Jakarta-Bogor Km 46, Cibinong 16911, Indonesia
[9] Department of Biology, Faculty of Mathematics and Natural Sciences, Hasanuddin University, Jl. Perintis Kemerdekaan Km 10, Makassar 90245, Indonesia
[10] Research Center for Applied Zoology, National Research and Innovation Agency, Jalan Raya Jakarta-Bogor Km 46, Cibinong 16911, Indonesia
* Correspondence: julianus.kinho@brin.go.id

Citation: Kinho, J.; Suhartati; Husna; Tuheteru, F.D.; Arini, D.I.D.; Lawasi, M.A.; Ura', R.; Prayudyaningsih, R.; Y.; S.; et al. Conserving Potential and Endangered Species of *Pericopsis mooniana* Thwaites in Indonesia. *Forests* **2023**, *14*, 437. https://doi.org/10.3390/f14020437

Academic Editors: Konstantinos Poirazidis and Panteleimon Xofis

Received: 31 December 2022
Revised: 12 February 2023
Accepted: 17 February 2023
Published: 20 February 2023

Abstract: Indonesia has around 4000 wood species, and 10% (400) of species are categorized as commercial wood. One species is kayu kuku (*Pericopsis mooniana* Thwaites), native to Southeast Sulawesi. This species is considered a fancy wood used for sawn timber, veneer, plywood, carving, and furniture. The high demand for wood caused excessive logging and threatened its sustainability. In addition, planting *P. mooniana* has presented several challenges, including seedling production, viability and germination rate, nursery technology, and silviculture techniques. As a result, the genera of *Pericopsis*, including *P. elata* (Europe), *P. mooniana* (Sri Lanka), and *P. angolenses* (Africa), have been listed in the Convention on International Trade in Endangered Species (CITES) Appendix. Based on The International Union for Conservation of Nature (IUCN) Red List of Threatened Species, *P. mooniana* is categorized as Vulnerable (A1cd). This conservation status has raised issues regarding its biodiversity, conservation, and sustainability in the near future. This paper aims to review the conservation of potential and endangered species of *P. mooniana* and highlight some efforts for its species conservation and sustainable use in Indonesia. The method used is a systematic literature review based on *P. mooniana's* publication derived from various reputable journal sources and additional literature sources. The results revealed that the future demand for *P. mooniana* still increases significantly due to its excellent wood characteristics. This high demand should be balanced with both silviculture techniques and conservation efforts. The silviculture of *P. mooniana* has been improved through seed storage technology, improved viability and germination rates, proper micro and macro propagation, applying hormones, in vitro seed storage, improved nursery technology, and harvesting techniques. *P. mooniana* conservation can be conducted with both in situ and ex situ conservation efforts. In situ conservation is carried out by protecting its mother trees in natural conditions (i.e., Lamedae Nature Reserve) for producing good quality seeds and seedlings. Ex situ conservation is realized by planting seeds and seedlings to produce more wood through rehabilitating and restoring critical forests and lands due to its ability to adapt to marginal land and mitigate climate change. Other actions required for supporting ex situ conservation are preventing illegal logging, regeneration, conservation education, reforestation, agroforestry system applied in private and community lands, and industrial forest plantations.

Keywords: fancy wood; economic value; sustainable management; conservation

1. Introduction

Indonesia has around 4000 wood species that can be used for building construction. However, only 10% (400) of wood species have economic value, and 260 species are commercially traded in national and international markets [1]. One of the commercial and tropical wood species is kayu kuku (*Pericopsis mooniana* Thwaites.), a native species in Southeast Sulawesi [2]. This species has good characteristics such as being a luxurious, fancy, and expensive wood with high economic value in the international market [2,3]. Its wood can be used as household tools and veneer and is suitable for heavy construction such as ship decks, bridges, railway wood sleepers, frame wood, and truck tailgate [4].

Another "tropical wood" species is *P. elata* (Harms) Meeuwen., known as African satinwood and gold teak [5]. This species is quite famous for its timber due to its similarity of texture and other wood characteristics to those of teak (*Tectona grandis* L.). This timber is traded internationally for exterior and interior work [5]. Like *P. elata* timber in Africa, *P. mooniana* timber from Indonesia has been traded and utilized increasingly without replanting efforts [3], exported since 1972, and exploited uncontrollably [4]. Consequently, the species' population is decreasing, as Rain Forest Action reported in 2004 [6]. Due to its high demand and excessive harvesting in the international market, trees from the genera *Pericopsis* have been included on the IUCN red list. For example, *P. elata* is listed as Endangered [5,7], *P. mooniana* as Vulnerable [8,9], and *P. laxiflora* (Benth. ex Baker) Meeuwen. as the Least Concern species [10].

The *Pericopsis* trees grow naturally in the Lamedae forest, including the natural reserve forest, Kolaka district, and Southeast Sulawesi [9,11]. Generally, the trees are found in beach forests, near the river bank, and in forests with an elevation of 200–350 m asl [3]. Nkulu et al. [12] found *P. angolensis* (Baker) Meeuwen. dominant in a habitat based on the Mg concentration above 140 μg.g^{-1} and soil Al concentration at a threshold value of 220 μg.g^{-1}. The natural distribution of Afrormosia (*P. elata*) ranges from Ivory Coast, Ghana, Cameroon, DRC, Congo-Brazzaville, CAR, and Nigeria [5]. *P. mooniana* is a native species found in wet zone floodplain forests in Sri Lanka [13].

Indonesia owns 120 million hectares of forest, accounting for 64% of the country's total land area [14], including 14 million ha of critical forests and land. Meanwhile, *P. mooniana* is well suited for rehabilitation, particularly for ex-mining reclamation land [15]. This tree is a pioneer species due to its adaptability to marginal and unproductive lands [16,17]. This has been supported by research findings of Perala and Wulandari [18] that the tree has a high survival rate of 63% in gold-tailing lands, reaching 92% with additional treatment of vermicompost, rhizobium, and mycorrhiza. This indicated that the trees could grow and develop well in all forest areas, implying the potency of this species will become a mainstay and superior tree in the near future in Indonesia.

P. mooniana seed germination is only 68% [19], and its seedlings have been predicted to be well growing (100%) with various land media, tailings, and combinations between tailing and manure [16]; however, it is still questionable regarding its growth after planting in various types of forest lands. Therefore, research on seed production and the silviculture techniques of *P. mooniana* is needed to optimize its growth and species conservation in response to its high use and demand in the future. This paper aims to review the conserving potential and endangered species of *P. mooniana* and highlight some efforts for this species' conservation and sustainable use in Indonesia.

2. Methods

The method used is the systematic literature review. The review was conducted based on publications of *P. mooniana* derived from various reputable sources and additional literature sources to capture the state of conserving this species as potential and endangered species. Some keywords in English and Indonesian were used to find relevant issues by employing a search engine. An intensive search for online publications for 1990–2022 was carried out in August–September 2022. Literature references were collected through search with the Google search engine, Google Scholar, Science Direct, ResearchGate, Crossref, Scopus, PubMed, and other relevant databases. The stages of searching and screening the publications are shown in Figure 1.

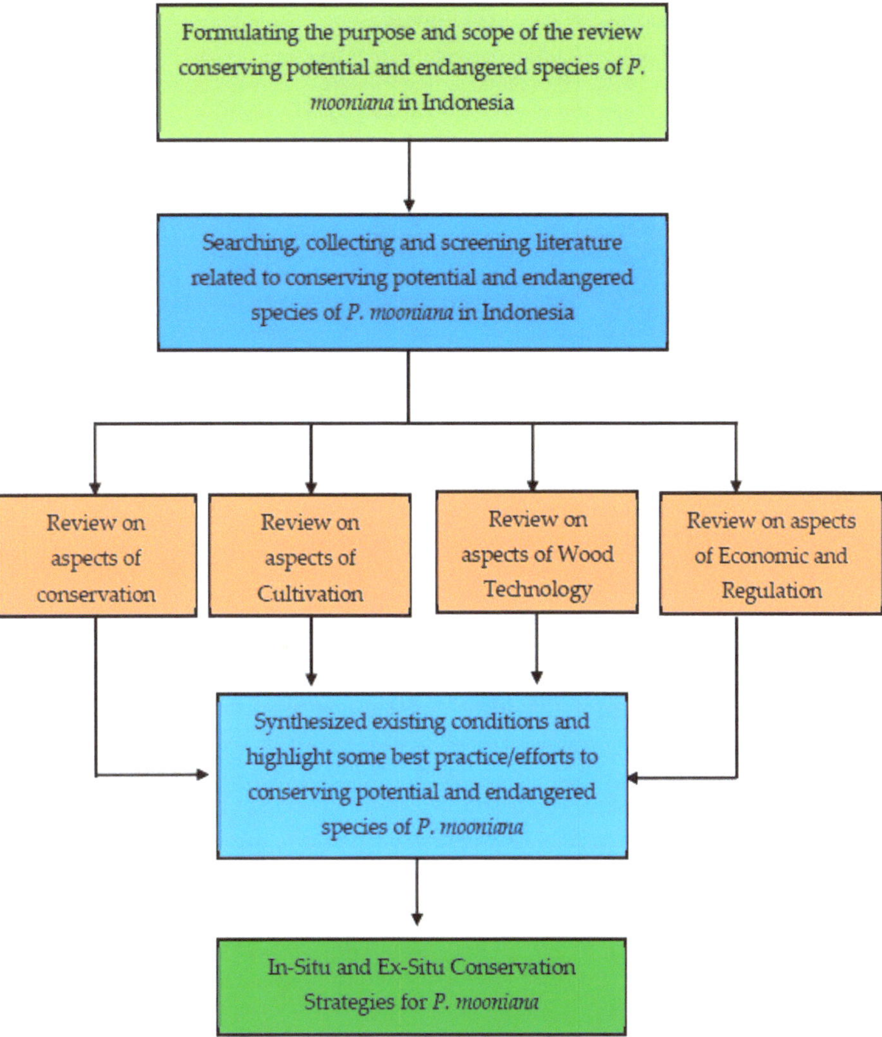

Figure 1. Stages in conducting the review.

3. Results

3.1. Conservation

3.1.1. Botanical Description of Kayu Kuku (*P. mooniana*)

Classification of kayu kuku (*P. mooniana*) [20]:

Kingdom: Plantae
Division: Spermatophytae
Sub division: Angiospermae
Class: Magnoliopsida
Order: Fabales
Family: Fabaceae
Genus: *Pericopsis*
Species: *Pericopsis mooniana* Thwaites.

P. mooniana has a tree habitus and medium size. In natural forests, the trunk height reaches 30–40 m; the branch-free trunk (clear bole) can reach 3/4 part of its total height, and the diameter reaches 35–100 cm [21]. The main stem is straight, shallowly grooved, and notched at the base; the stem is thin and smooth reddish, and the leaves are crossed opposite. The bark is light brown. This species has brighter colored sapwood than its reddish-brown heartwood [6] and belongs to the I-II durable class wood type [22]. The wooden surface is slippery and shiny and has decorative lines [23]. *P. mooniana* has an alternate arrangement of leaves and shoots, ovate to elliptical, rounded at the base of the leaf, and pointed to tapered at the tip of the leaf; the surface of the leaf is glabrous; the veins of the leaves total seven pairs; it has caducous scales; small stipules are not even present as shown in Figure 2 [24]. The oblong-shaped flowers are purple-black, and the petals are greenish [20].

Figure 2. Picture of *P. mooniana*, natural stands (**A**); flowering twigs, flower parts, and pods (**B**); fruit morphology (**C,D**) [4,20].

P. mooniana has a fruiting season in September and October [6]. The fruit is classified as a boxed fruit with seed chambers or boxes of up to four chambers (generally one, two, and three seed chambers). Each chamber contains one seed, hard fruit skin (has thick cork tissue), and a single fruit or single true fruit. Young fruits are light green [20]. When ripe, the skin splits (*dehiscence*), is inseparable from the fruit stem, and is pale white or brownish. The fruit is rounded at the base and tapered at the end [4]. The pod-shaped fruit belongs to the *indehiscent* fruit type. One pod holds 1–6 seeds, measuring 4–16 cm long and 2.5–5.0 cm wide, round at the base, and tapering at the end. The unripe pods are green and turn brownish when ripe [22]. Physiologically mature pods are brown with button-shaped seeds 1 cm in diameter and 4 mm thick [6]. The size dimension of the seed considered medium, around $0.7 \times 1.6 \times 0.5$ cm [20]. Seeds are orthodox, which means they have a hard seed coat and are difficult to germinate because they experience seed dormancy. The structure of seeds consists of *Spermodermis, Testa, Tegmen, Funiculus,* and *Nucleus semminis* [4]. Seeds of this species are classified as dicotyledonous seeds (two pieces) and have a length of 1.0–1.3 cm and a width of 0.7–0.8 cm [22].

3.1.2. Habitat Characteristics and Distribution

P. mooniana is a single species in Southeast Asia, synonymous with *Ormosia villamilii* Merr., *Pericopsis ponapensis* Hosok. The vernacular names of *P. mooniana* are nandu wood, nedun tree (En), kayu kuku (Indonesia), kayu besi papus (Sulawesi), nani laut (Irian Jaya), kayu laut (Sabah Peninsula), merbau laut (peninsula, Malaysia), and makapilite (Bisaya, Philippines) [24,25].

Pericopsis species were categorized as luxury wood consisting of five species, four found in the African Continent. *P. mooniana* has wide distribution covering Sri Lanka, Southeast Asia (Malaysia, Indonesia, the Philippines), and Oceania (Papua New Guinea)., The species distribution of *P. mooniana* in Indonesia includes Sumatera, Kalimantan, Sulawesi, Maluku, Halmahera, and Papua (Figure 3) [4,6,24–26].

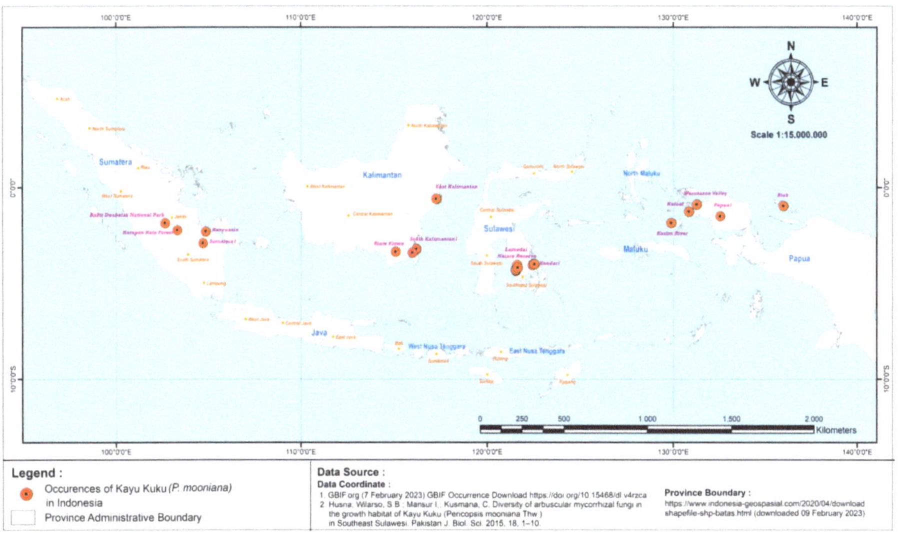

Figure 3. Distribution map of *P. mooniana* in Indonesia.

P. mooniana grows naturally in coastal, riparian, evergreen, and semi-deciduous forests [21] with relatively fertile regosol soil. This species distributed at 200–350 m asl, with annual rainfall around 750–2000 mm. This species is suitable to grow on non-stagnant soils, loamy, hilly topography with gentle slopes at an altitude of <30 m asl [4,6].

3.1.3. Genetic Resources of *P. mooniana*

P. mooniana is a rare and endangered tree species and potentially a priority species requiring further protection for evaluating, conserving, and managing forest genetic resources [8,27]. Deforestation, mining activities, over-exploitation of timber, forest conversion to agriculture, and settlement threaten the habitat of *P. mooniana* [28]. Widyatmoko [29] reported that in 2018 more than 600 plant species in Indonesia were almost threatened with extinction. This pressure caused the population of species *P. mooniana* in their natural habitat to decline, resulting in a potential risk of extinction. Therefore, *P. mooniana* was proposed for inclusion in Annex II of the 1992 CITES convention, which requires all trade in a species to be registered (UNEP-WCMC 2014).

In addition to the exploitation and habitat disturbance, *P. mooniana* is also threatened by genetic issues such as inbreeding, loss of genetic diversity, and accumulation of gene mutations that can lead to extinction. Therefore, genetic conservation plays a vital role because it is associated with significant factors contributing to species and population extinction [30]. Genetic materials from various sources are required to conserve *P. mooniana*. Knowledge of genetic diversity is the first step in conserving species to reduce the risk of extinction by developing conservation strategies and restoration practices [31–33].

Research on the genetics of *P. mooniana* in the context of conservation in Indonesia is limited to the Lamedae Nature Reserve area of Southeast Sulawesi and the Pulau Laut of South Kalimantan [34–36]. The genetic diversity of *P. mooniana* in the four populations studied in the Lamedae Nature Reserve and surrounding populations shows moderate genetic diversity [34]. However, the genetic diversity of *P. mooniana* at the seedling level shows a reasonably high genetic diversity [35]. Meanwhile, the genetic diversity of *P. mooniana* found on the Pulau Laut of South Kalimantan shows low genetic diversity [36]. Nowadays, *P. mooniana* stands quickly disappear due to logging and land clearing. Only a few stands of *P. mooniana* remain in Papua [37], Pulau Laut South Kalimantan [38], Jambi South Sumatera [39], and Southeast Sulawesi (Lamedae Nature Reserve) [40], and this species tends to survive continuously in forest areas.

This genetic conservation activity not only keeps the number of trees and populations of *P. mooniana* but also needs for plants' development and adaptation to their environment [31,41]. Genetic conservation is one of the methods to maintain maximum genetic diversity and give opportunities for the species to adapt and evolve [31]. The results of the genetic diversity analysis of the endangered *P. mooniana* are as follows: (1) the genetic diversity ranges from low to moderate [42], (2) the distribution of genetic diversity is mainly within the population [43], (3) population distribution based on genetic distance is closely related to geographic distance [43,44] and (4) mating occurs randomly if the number of parent trees is large enough [34].

The genetic diversity of *P. mooniana* is obtained using various molecular markers, namely isozymes, RFLP (restriction fragment length polymorphism), DNA markers (deoxyribose nucleic acid), and sequencing [45–47]. In addition, it can also be used for species identification (DNA barcoding) [48], clone identification [49], and population genetic diversity [50–52]. Random amplified polymorphism DNA (RAPD) markers have been used to analyze the genetic diversity of *P. mooniana* from four populations in Southeast Sulawesi (Lamedae Nature Reserve and the villages of Lamedae, Balijaya, and Tangketada). The degree of genetic diversity is nearly identical across all populations with an average He value of =0.361 and is, therefore, classified as moderate. Lamedae Nature Reserve has the highest genetic diversity (He = 0.383 ± 0.031). The population of *P. mooniana* from the three villages has a close genetic relationship [34]. Yuskianti et al. [36] observed that the genetic diversity of *P. mooniana* in four populations on the Pulau Laut of South Kalimantan showed that the diversity was low (He = 0.191 ± 0.013). They are closely genetically related, indicating they come from a common origin.

3.2. Cultivation of P. mooniana

Efforts to preserve and develop *P. mooniana* trees are strongly influenced by advances in cultivation techniques. The extinction of a species is mainly due to over-exploitation that is not balanced with a proper cultivation system, thus causing a decrease in natural germplasm sources and resulting in the existence of a species becoming rare or endangered [53]. This also works for high economic value species due to their luxury wood and being liked by the community [3,9,34]. Efforts to preserve *P. mooniana* in the context of sustainability are strongly influenced by the cultivation level and its technological developments. The scope of cultivation includes seed technology, nursery techniques, generative and vegetative propagation, and identification and control of pests and diseases that attack the *P. mooniana* trees.

3.2.1. Seed Technology

Plant cultivation begins with conceiving seed technology, starting from understanding the phenology of flowers and fruit to the right time to collect fruit [54]. Research on the flowering and fruiting period of *P. mooniana* is limited. Knowledge of the flowering period in forest plant species is essential if breeding activities are carried out by knowing the right time for fertilization or crosses [55]. *P. mooniana* fruits every year, meaning the potential for generative propagation is very high because it is not constrained in the fruiting period. However, until now, there is no information about the potential production of *P. mooniana* seed per tree, but data on the number of seeds in one pod are available [22].

The weight of 1000 seeds is 2546 g, and 1 kg contains ± 4000 seeds. The seed is classified as an orthodox seed, with the seed's initial water content being below 10% [22]. Therefore, *P. mooniana* seeds can be stored for a long time in cold storage conditions. The moisture content of the seeds for storage can be decreased below 8% without significantly reducing its viability. Even seeds of the same genus, namely *P. elata*, are predicted to be able to be stored for up to 243 years [56]. This longevity of storage indicates that the seeds of the genus *Pericopsis* can be stored for a long time while retaining their viability. It is advantageous in supporting conservation efforts. However, until now, there has been no research on the longevity storage of *P. mooniana* seeds.

Anatomically, the seed of *P. mooniana* has a reasonably hard seed coat with a waxy layer covering it; this is one of the obstacles in the germination of the seed [19]. The hard coat causes seed dormancy, which is classified as physical dormancy, and this condition is also found in seeds of the same genus, namely *P. elata* [56]. However, an attractive condition is seen in *P. angolensis*; although it is in the same genus, no dormancy happened [57]. Breaking dormancy is the first step to accelerate the germination of *P. mooniana* seeds. Seeds without breaking dormancy have a germination of about 2%; meanwhile, those breaking dormancy have an average germination of up to 60% more than the control [22]. The most effective breaking dormancy for *P. mooniana* seeds is wounding or sacrificing on the surface of the seed coat or soaking in hot water (80 °C–90 °C) for 10 min, followed by soaking in cold water for 48 h [22]. This technique can increase germination by more than 60%. Seed selection and treatment can speed up the germination rate. By soaking yellow and brown seeds in hot water to "scarify" them, the first germination time can be shortened, and the viability increases to 76% [23]. The scarification technique by broken seed skin leads to significant differences in the percentage of sprouts, germination, average days of germination, vigor index, number of leaves, and seedling height [58].

The average germination rate of *P. mooniana* seeds is 15–37 days, but if the dormancy is broken, the germination time will be faster, which is 6–8 days on average [19,22,59]. The speed and uniformity of germination affect the process of procuring seeds; this is related to the amount and time required to procure seeds for planting. Another method that can be used to estimate the potential viability of *P. mooniana* seeds is a rapid test method; this is possible because direct viability testing takes quite a long time. Rapid seed viability tests can be carried out in various ways, such as radiographic analysis, tetrazolium test, cutting test, and other methods [53].

Another significant activity in maintaining seed viability is the post-harvest seed storage process because of its orthodox seed. It uses an airtight container stored in a low-temperature room. Along with advances in science and technology, techniques for increasing seed vigor or invigoration can be carried out in various ways: through priming, gamma ray irradiation, and ultrafine bubbles [60–62]. However, until now, there has been no research on seed invigoration of *P. mooniana* seeds, so this is an opportunity to increase the viability of *P. mooniana* seeds to preserve them.

3.2.2. Propagation Techniques

Seed procurement can be done by generative methods such as tillers and stumps, vegetative propagation, such as macro propagation (cuttings, air layering grafting, and budding), and micropropagation (tissue culture).

Generative

Cultivation of *P. mooniana* with generative propagation and procurement seeds should be obtained from tree sources over 15 years old, with a percentage of sprouts ranging from 80% [20]. The time required from the imbibition process to the release of the cotyledons is ±48 days, and it is ready to be weaned into polybags [19]. Generative propagation, including viability and germination rate, has been described in Section 3.2.1.

Vegetative Propagation

Several studies on vegetative propagation for this species include shoot-cutting techniques and tissue culture, especially in manufacturing seedlings from genetically identical superior clones [63–65]. The technique can also overcome the scarcity of seedlings, especially for species that do not fruit yearly. The shoot-cutting technique can be applied by adding as much as 60 ppm of the hormone IBA [64].

Material for tissue culture (in vitro techniques) must be well prepared. The material derived from the seeds affects the ability to sprout. Seeds that have been removed pericarp and stored at room temperature showed the highest shoot length compared to seed pericarp removal stored at 10 °C [63].

Seed germination and shoot propagation of *P. mooniana* have been investigated using in vitro techniques. Seeds were sowed in Murashige and Skoog (MS) medium without plant growth regulators and showed 100% germination. The MS + BA 0.75 mg/L treatment showed the best media treatment for shoot propagation, as indicated by the highest number of 2.3 shoots and 6.8 cm height on average at 12 weeks of observation. The tissue culture requires sterilized seeds [59], especially for the explant of *P. mooniana*; it is necessary to choose healthy, brightly colored, and not wrinkled seeds [65].

In multiple shoot cultures, establishment of seedling shoot tips *of P. mooniana* would be improved by using an MS medium with several doses of growth hormone BAP (benzyl amino purin) and NAA (naphthalene acetic acid) [66]. The highest level of micropropagation depends not only on the medium and selection of a suitable explant but also on the combination of growth and hormone [67].

3.2.3. Seedling Improvement Technique

The optimal temperature and humidity for germination of *P. mooniana* are at temperatures between 25 °C−30 °C with humidity between 60%−70%. Meanwhile, the best medium for germinating *P. mooniana* seeds is a mixture of soil and sand (*V/V*: 1/1) that has been sterilized. Normal *P. mooniana* germination is characterized by the appearance of two healthy, sturdy leaves. *P. mooniana* seedlings can be weaned in polybags when the seedlings look sturdy, the roots are well formed, and the seedlings are ready to wean around 3–4 weeks after sowing. The media must contain enough nutrients to support the growth of seedlings until they become ready to plant in the field [17]. *P. mooniana* are pioneer plants that grow well without shade [68]. Seedling of *P. mooniana* is maintained in the nursery for about 3–4 months until the stems are woody and the seedling medium is compact.

3.2.4. Pest and Disease of *P. mooniana*

Observations showed that 1-year-old *P. mooniana* seedlings in the nurseries had three pests: *Zeuzera* larva (coffee), leaf-rolling caterpillars, and grasshoppers (*Valanga* sp.). The symptoms of the attack by larvae of *Z. coffeae* were borer on young stems, causing dry leaves and plant death. Plant death is due to the larvae burrowing the branch and moving in a vertical direction, thus damaging the xylem and phloem tissues of the plant [69]. The pest attacks were found in *P. mooniana* seeds from Tanggetada Natural Forest, with an attack of 1.65% [70]. *Z. coffeae* often attacks annual plants and tiny seedlings in forests, horticultural nurseries, and young plants, including *P. mooniana* [71].

Pests that attack young leaves of *P. mooniana* are larvae *Sylepta* sp., which causes the leaves to curl or fold, containing some larvae that are actively eating the leaves [72], and *Valanga* sp., causing leaf tearing in the leaf edge or leaf with a hole in the middle [70]. The attacks of these pests were found in the Lamedae Nature Reserve and Tanggetada Natural Forest. The leaf-folding caterpillar attack was also found on the *P. mooniana* plant in KHDTK, Malili, East Luwu, and South Sulawesi [40].

3.2.5. Mycorrhizal Fungi Application

Arbuscular mycorrhizal fungi (AMF) have been developed and used as a biofertilizer to support various uses, including for extensive agricultural development, forest reforestation [73], restoration of degraded land and ecosystems [74,75], land phytoremediation polluted [76], and conservation programs for endangered tropical tree species [77,78]. A total of 75 species of AMF from 23 genera and 11 families are found in Indonesia and are in symbiosis with various plants in various land use types [79]. Fifteen species of AMF from five families and nine genera have been found in the rhizosphere of *P. mooniana* in natural habitats and development areas in Southeast Sulawesi. Glomeraceae, including the dominant family and four AMF species, were first reported in Indonesia, namely *Glomus canadense*, *G. halonatum*, *Racocetra gregaria*, and *Ambispora appendicula* [80,81]. AMF isolated from the rhizosphere of *P. mooniana* has been collected, reproduced, and tested on its viability and growth both on a nursery/greenhouse and field scale (Table 1, Figures 4 and 5).

Table 1. Review of AMF symbiotic research with *P. mooniana*.

AMF Species/Treatment	Media	Period (Month)	Effect	Experiment Type	References
Septoglomus constrictum + two days of watering	Gold tailings media	4	Increase shoot P levels and uptake	Greenhouse	[82]
G. claroideum and *G. coronatum*	Gold tailings media	4	Increase the height diameter, number of nodules, dry weight of plants, and uptake of N and P in the roots	Greenhouse	[83]
G. claroideum + field capacity 100 and 75 %	Gold tailings media	4	Increase plant height, diameter, total leaf area, nodulation, and dry weight	Greenhouse	[84]
G. claroideum and *G. coronatum*	Gold tailings media	4	Increase the uptake of N, P, Mn, and Fe leaves	Greenhouse	[85]
5 g AMF inoculum Mycofer (mix. *Glomus manihotis*, *Glomus etunicatum*, *Acaulospora tuberculata*, *Gigaspora margarita*) and addition sago waste 20 g	Soil media of nickel post-mining site	3	Increase height, dry weight, root nodules, and absorption of K and Ca and reduce Ni by 32%	Greenhouse	[86]
AMF native from rhizosfer P. mooniana	Soil media of nickel post-mining site	5	Increase growth and dry weight, absorption of N, P, and K in three parts of the plants, of Ca (in stems and leaves) and of Mg in leaf tissues and root nodules; and reduce Ni content	Greenhouse	[87]
C. etunicatum/Ha	Soil media of nickel post-mining site	5	Increase the height, diameter, number, and area of leaves, plant dry weight, and Seedlimh quality index	Greenhouse	[88]
C. etunicatum/Ha + without treatment of Urea	Soil media of nickel post-mining site	5	Increase the number of leaves, nodulation, and dry weight of shoots and roots	Greenhouse	[89]
AMF local from rhizosfer *P. mooniana*	Overburden coal media		Increase in the growth and dry weight of plants and also tend to increase the C, N, P, K, Ca, and Mg accumulation	Greenhouse	[90]

Table 1. *Cont.*

AMF Species/Treatment	Media	Period (Month)	Effect	Experiment Type	References
Mycofer IPB (*Glomus etunicatum*, *G. manihotis*, *Acaulospora tuberculata*, and *Gigaspora rosea*)	Ultisol	3	Increase growth and dry weight	Greenhouse	[91]
Glomus sp. (HA)	Ultisol	3	Increase height, diameter, number of leaves, dry weight, root nodules, root, and leaf length	Greenhouse	[75]
Acaulospora sp. 1 (HA)	Ultisol	3	Increase growth in height, diameter, number, and area of leaves	Greenhouse	[92]
Local AMF from *P. mooniana* rhizosphere	Inceptisol	5	Increase in height, diameter, number of leaves and root nodules, total dry weight, total chlorophyll, P, K, Ca, and Mg uptake	Greenhouse	[93]
AMF from *P. mooniana* rhizosphere from Lamedai NR (non-serpentine) and PT. Vale Indonesia (serpentine)	Nickel post-mining land	3	Increase survival rate, growth, dry weight, and accumulation of N, P, and K	Field	[94]
		24	Increase height and diameter		[95]
		36	Increase growth in height, diameter, number, length, and width of leaves and dry weight of wet and dry leaves		[96]
G. coronatum	Gold post-mining land	4	Increase the height, diameter, and dry weight of the leaves as well as the uptake of N, P, K, Mn, and Fe	Field	[83]

Figure 4. Visualization of growth performance of mycorrhizal and non-mycorrhizal *P. mooniana* at greenhouse and nursery scales. (**A**) Coal overburden (OB) media [90], (**B**) gold tailings media [83], (**C**) serpentine soil media/post-nickel mining [87], and (**D**) serpentine soil media [2].

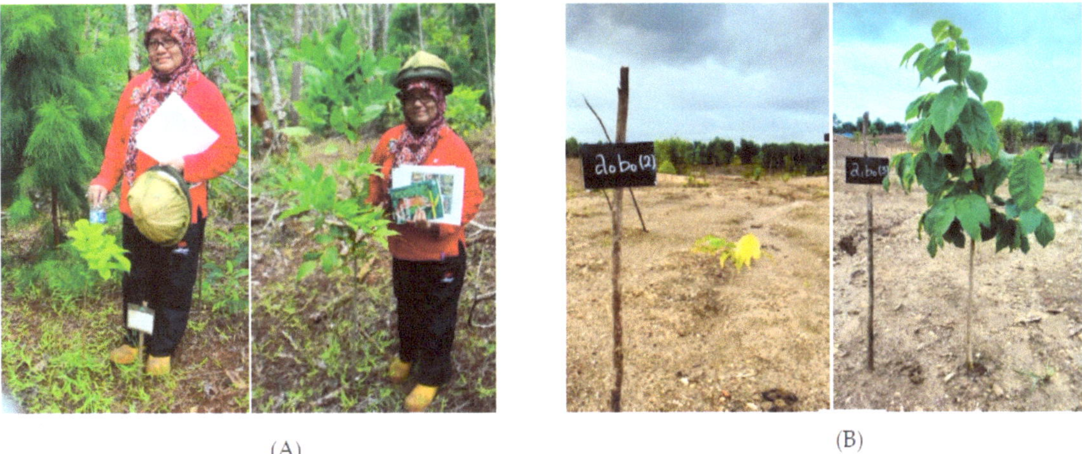

Figure 5. Visualization of growth performance of *P. mooniana* with and without mycorrhizal in a post-nickel field at 12 months (**A**) and a post-gold mine at four months after planting (**B**) (Photo: F.D. Tuheteru and Husna, 2022).

Table 1 shows that the application of AMF increased the growth, dry weight, nodulation, and nutrient uptake of *P. mooniana* at the nursery scale in various growing media. Based on these different research results, AMF has the potential to be developed as a biological fertilizer to support conservation programs for endangered tropical tree species and, at the same time, accelerate the success of *P. mooniana* planting in disturbed and damaged land restoration programs.

Besides AMF, *P. mooniana* also has symbiosis with rhizobium. There were 18 species of rhizobia associated with *P. mooniana* in Sri Lanka [97]. Rhizobia isolated from the roots of *P. mooniana* was effective in improving nodulation and nitrogen fixation and increasing 50% of plant dry matter under conditions of low N fertilization levels for 12 months of observation. In addition to single inoculation, double inoculation of AMF and rhizobium can increase plant growth. The synergy of AMF and rhizobium increased *P. mooniana* 2–4 fold growth compared to the control [98]. The increasing number of mycorrhizal *P. mooniana* root nodules (Figure 6) is related to the improvement in phosphorus uptake by AMF required by rhizobium [99].

3.2.6. Silviculture Techniques

Silvicultural practices such as determining plant spacing and basic and advanced fertilization greatly determine plant growth. The planting hole is made wider on post-mining land than on mineral soil. In addition, it uses the intensive application of basic and advanced fertilizers that increase the incremental growth in plant height and diameter. Wide spacing increases plant diameter. The following is a summary of publications, interviews, and observations on-site related to silvicultural practices and their effects on plant growth, namely:

1. In the nickel post-mining area, planting holes were $60 \times 60 \times 60$ cm. Plants were planted with a distance between planting holes of 4 m x 4 m. Before planting, the soil is mixed with compost and urea (5 kg and 100 g, respectively), one-fifth of the standard carried out by PT. Vale Indonesia Tbk and mixing sago pulp with soil media was carried out, with a height and diameter increment of 0.44 m and 0.8 cm [2,96]. *P. mooniana* was planted between reclaimed trees at a spacing of 4×5 m and filled with 25 kg of active compost and 0.4 kg of urea + TSP + KCl + dolomite each per planting hole. In the first and second 6 months of maintenance, 0.4 kg each of urea + TSP + KCl was added, and in the third period, pruning was conducted and

mulch was placed at the base of the stem circle. This treatment increases height and diameter increment of 0.93 m and 1.06 cm, respectively (personal communication with Mr. Guntur Sambernyowo, Retired PT Vale Indonesia).

2. In gold post-mining land, the spacing is 2 × 2 m with a planting hole of 40 × 40 × 40 cm. *P. mooniana* seedlings with a height of ±30 cm were transferred and planted in the holes with 4 kg of compost. A total of 50 g of NPK fertilizer per plant was added after one day of planting and then treated with watering and weed control. It increased height and diameter averages of 0.2 m and 0.36 cm, respectively [83].

3. In the land occupied by *Imperata cylindrica*, the planting distance was 2.5 × 2.5 m with a planting hole of 30 × 30 × 30 cm and basic fertilizer of 1 kg of compost. The mean increase in height and diameter is 0.89 m and 1.20 cm at 17 years old (observation on site).

4. In the green open space of Halu Oleo University, the planting spacing was 3 × 3 m with 30 × 30 × 30 cm planting holes and no further fertilization. The mean increase in height and diameter at 13 years old is 0.84 m and 1.75 cm (observation on site).

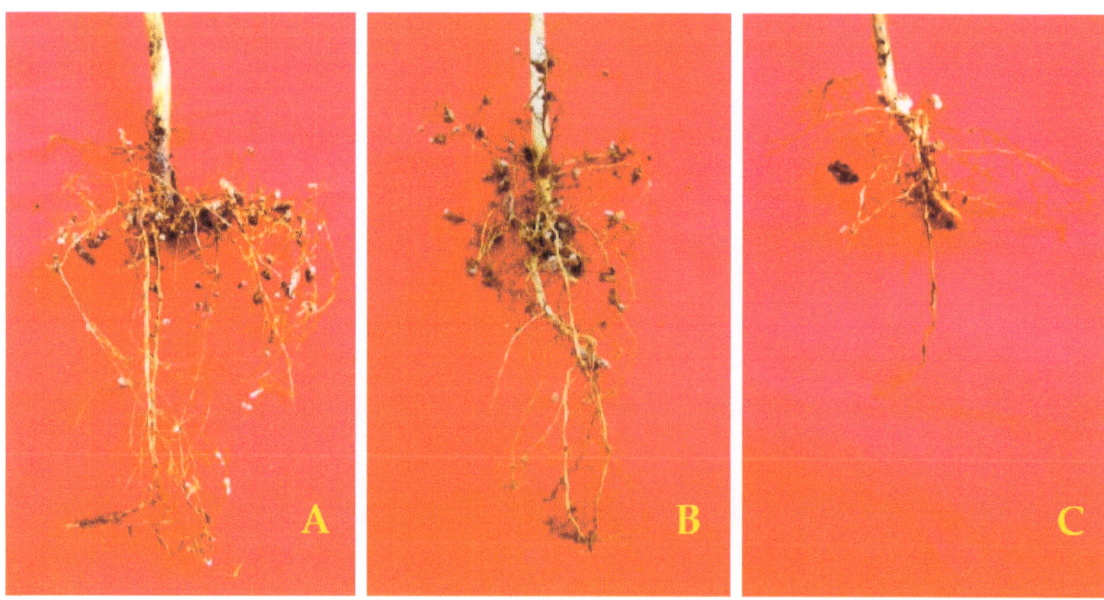

Figure 6. Visualization of root nodules of *P. mooniana* plant mycorrhizal (**A,B**) and non-mycorrhizal (**C**) [2].

3.2.7. Future Opportunities in Improving Seed and Seedling Quality

There is still a technological gap in *P. mooniana* cultivation in Indonesia, while natural conditions indicate that *P. mooniana* stands to decline. Therefore, it is necessary to dig more profound the technology that can fill the gap. Regarding conservation, it is essential to use quality seeds and seedlings. Suppose we identify several aspects that need to be encouraged to be developed to support conserving *P. mooniana* species. We begin with the seed source of *P. mooniana*, which until now is limited. The existing seed source is the identified seed stands and is the lowest rank in the classification of seed sources. Classification of seed sources determines the genetic quality of the plants. The genetic diversity of *P. mooniana* from several populations in Indonesia shows that originating from populations in Papua has a high level of diversity compared to other populations from Sulawesi, Kalimantan, and Java [100]. This information is essential if we develop a seed source because it will be related to the quality of the seed produced from that source.

Seed quality is also determined by post-harvest handling because it is related to the physical and physiological quality of the seed. The physiological quality of seeds can be improved by applying invigoration technology. Until now, research on the invigoration of *P. mooniana* seeds has never been done; this condition is a challenge for forest tree seed researchers to research improving the physiological quality of *P. mooniana* seeds. Seed invigoration techniques on various forestry plant species have increased seed germination [101]. If the invigoration technique is applied to *P. mooniana* seeds, it is very promising to increase their viability when their vigor has decreased.

Regarding seed technology for conserving *P. mooniana* species, a seed coating technique can also be used. Seed coating is wrapping seeds by mixing various microbes such as plant growth-promoting bacteria, rhizobia, arbuscular mycorrhizal fungi, and Trichoderma [102]. All of them are beneficial for the growth at the seedlings level and the plants in the field. This effort can help improve the quality of seeds and seedlings and support successful planting to preserve *P. mooniana*.

3.2.8. Harvesting Technique

To find out the general picture of the impact of harvesting *P. mooniana* on sustainability and the environment, knowing the activities of harvesting natural forest wood is a learning process for better management. Sustainable forest management practices also implicate timber harvesting techniques that can improve productivity and wood utilization efficiency and minimize wood waste and residual stand damages. Therefore, reduced impact logging (RIL) must be implemented in *P. mooniana* harvesting, especially in Indonesia's production forests [14].

RIL is one of the harvesting techniques that could cause minimum damage to forest harvesting activity [103,104]. RIL is a low-impact timber harvesting technique through a comprehensive approach to planning, performing, monitoring, and evaluating [105]. The RIL implementation aimed to reduce the negative impact on the environment and improve harvesting efficiency by decreasing the wood harvesting waste volume and harvesting cost. Additionally, it can enhance the production cost; create a conducive growing space; improve the company income, occupational health, and safety; and become a prerequisite for sustainable forest management. By implementing RIL, it is expected that the forest damage in the timber harvesting area could be reduced.

An exemplary implementation of RIL could improve timber production and promote sustainable forests, including biodiversity [106]. Sustainable forest management is an effort of emission reduction through a sustainable innovation to provide a low impact on the environment, including RIL technique implementation and implementation of other practices such as reducing the destructive impact of harvesting activity and the implementation of Sustainable Production Forest Management (SPFM) certification [107]. Learning from various studies in natural forests shows that the application of RIL techniques can increase the efficiency of wood utilization by an average of 12% [108] and reduce residual damage by 13.62% [109,110]. From a conservation perspective, silvicultural techniques of selective logging guarantee more forest sustainability [111] because it can increase environmental stability through the presence of remaining stands resulting in new regeneration [112]. Subtropical forests revealed that new regeneration is significant in establishing vegetation, thereby contributing to biodiversity and continued timber production [113]. Therefore, we should do better to overcome the scarcity of *P.mooniana* in Indonesia, and it is necessary to carry out sustainable forest management by implementing reduced impact logging (RIL). Reduced impact logging activities include improving felling and bucking techniques to make them more efficient and adjusting felling intensity and allowable diameter limits. In addition, it is necessary to carry out reforestation and revegetation activities outside the forest area [3].

Therefore, regarding climate change and global warming issues, the effort to maintain the ecological function of the forest is to treat and protect the forest vegetation from possible damage (deforestation and degradation). After using the RIL technique, the potential for carbon storage was reduced by 2.57 tons C/ha or 9.43 tons CO_2-equivalent [114]. The application of RIL can reduce residual stand damage by 50%, so it can maintain forest carbon of 12.5%–25% greater than conventional logging (CL) and produce a high-quality residual stand in the next cycle. Furthermore, implementing RIL could increase the absorption capacity of CO_2 from the air by 1.7 times that of CL [115], and can potentially reduce emissions by approximately 1–7 tons of CO_2/ha/year [116]. This may act as the basis for future planned forest management activities involving RIL, carbon, and forest certification through the hierarchy of production forest management. The implementation of both RIL and forest certification can be facilitated through the binding of carbon financial incentives [116]. The implementation of RIL and CL has not guaranteed the species' existence. This is because species arrangement in the harvesting is limited for the species group, so scarcity of specific species becomes uncontrollable.

The research showed that the increment in the growth of the stand is greater than the wood consumption rate. If the yield regulation system only considers incremental growth, there will be an excess supply of wood, which threatens to reduce the economic value of the wood itself. Forests will be converted to plantations, settlements, fields, and others. To overcome this problem, yield regulation must consider the wood consumption rate. This term has not yet been developed, but databases related to monitoring consumption rates already exist, one of which is TEINIT (Timber Product Inventory on Information Technology) [117–120]. The simulation results show that the consumption rate of wood products for house and furniture construction is lower than that of wood harvested from natural forests using the incremental approach [121]. For this reason, a harvesting system that considers the consumption rate of certain tree species will help maintain the existence of these species and increase their economic value.

3.3. Importance of P. mooniana for the Wood Industry
3.3.1. Wood Properties

It is necessary to know the nature and characteristics of the wood, which are related to its quality the designation [122]. *P. mooniana* wood has a beautiful pattern resembling teak, as shown in Figure 7 [123,124]. The genus Pericopsis consists of four species, three from Africa and one from tropical Asia. *P. mooniana* is heavily exploited for wood trade in Southeast Asia [125]. It is categorized as the fancy wood class II in the Decree of the Minister of Forestry No. 163/Kpts-II/2003 [126]. This wood species has a beautiful pattern, so it is often used as a substitute for teak (*Tectona grandis*) to make furniture, cabinets, and other construction purposes [127,128].

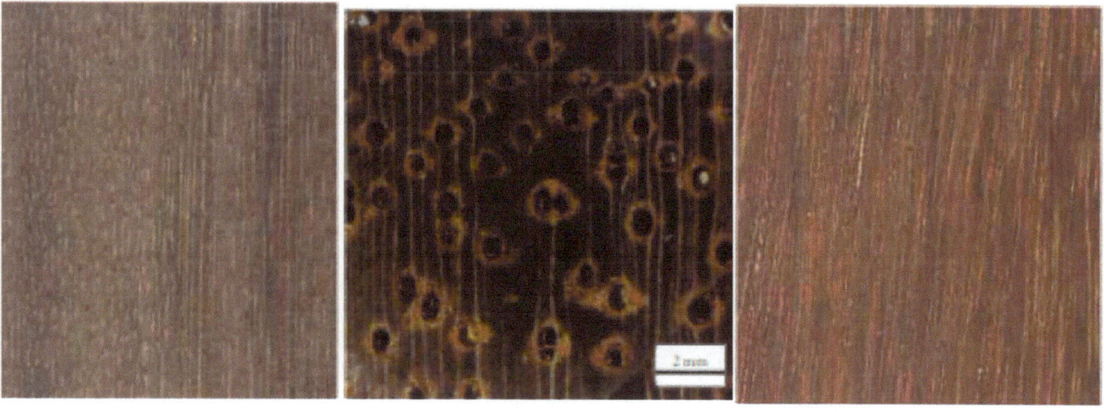

Radial section   Cross section   Tangential section

Figure 7. *P. mooniana* wood in macroscopic section, reproduced with permission from Krisdianto and Dewi [124].

The wood has physical properties, including specific gravity of 0.87, strong class II, radial shrinkage of 2%, and tangential shrinkage of 2%. Thus, it is classified as a low-shrinkage wood and relatively stable. Wood durability is indicated as resistant toward destructive organisms and grouped into class II. The wood's mechanical properties include a bending strength of 954 kg/cm^2 (wet) and 1473 kg/cm^2 (dry), hardness of 752 kg, maximum compressive strength of 485 kg/cm^2 (wet) and 699 kg/cm^2 (dry), and stiffness of 95,000 kg/cm^2 (wet) and 110,000 kg/cm^2 (dry). Regarding machining and workmanship, this wood has poor quality in drilling, engraving, molding, trimming, turning, and polishing [124,129]. The wood properties and characteristics are related to the wood quality. Anoop et al. [128] stated that the higher the specific gravity of the wood, the higher the fiber volume and the lower the vessel volume. The color difference between sapwood and heartwood is very conspicuous. The heartwood is brown-purple-black with light brown sapwood [130]. It differs from *P. elata* heartwood, which is brown to yellowish with streaks or without streaks, and the sapwood is distinct from the heartwood color [131].

3.3.2. Wood Anatomy Structure

The tangential vessel diameter of this wood is 116.7 μm and belongs to a relatively small diameter class which causes a fine texture [130]. The wood anatomy structures included vessels in a round to oval form that are diffuse-porous, arranged solitarily, and joined radially by 2–3 vessels, the number of vessels: 7–8/mm^2, ray height: 234–237.9 micrometers (μm), ray width: 29.6–30.4 μm, fiber length: 1142.6–1255.2 μm, fiber lumen diameter: 7.4–8.6 μm, and fiber wall thickness: 4.9–5.1 μm. The rays' cells of *P. mooniana* are arranged in an orderly manner so that they have a distinctive and attractive pattern known as the fancy wood group with a ripple mark. It has wing-shaped (aliform) to confluent parenchyma cells. Its rays' cells are uniseriate, biseriate, and multiseriate with a heterocellular composition that consists primarily of procumbent, and there are cells that are upright like a cube. The rays on the tangential section appear in a regular and distinctive arrangement [132]. This wood fiber has medium fiber lengths and thick cell walls [124,130]. Ishiguri et al. [127] stated that the wood structure is a vessel diameter of 160–170 μm, vessel frequency of 8–10 vessels/mm^2, fiber length of 1.0–1.7 mm, fiber cell diameter of 15–20 μm, and fiber cell wall thickness of 3.5 μm.

3.3.3. Wood Uses

P. mooniana is known as a luxury wood with a specific gravity of 0.87, strong class II, and durable class II [126], so it is suitable for indoor and outdoor furniture, window and door panels, veneer, luxury plywood, decorative, parquet flooring, bridges, and construction materials [1,124]. It has a beautiful pattern and is often used as a substitute for teak [126–128]. It is a pride and prestige for the upper middle class to use luxurious and expensive wood furniture. It offers a sense of safety and environmental friendliness and is picturesque and easy to shape. In recent years, most people have preferred wooden furniture due to its stylish interior design, accentuating aesthetic appearance, and long lifetime [133]. It can be used for a beautiful veneer with a beautiful wood pattern that is made using a slicing method to maintain its beautiful pattern and minimize incision waste by about 5% [4,134]. It is used for surface coating furniture, interior pillars, decorative walls, and a face veneer for composite wood products, including plywood, particle board, and fiberboard [135].

3.4. Economics and Regulation

As mentioned, *P. mooniana* wood is luxurious, fancy, and expensive wood with high economic value in the international market [2,136]. The estimated price is similar to teak wood products [136]. However, Suhartati et al. [4] pointed out that the price of exported *P. mooniana* wood is 2–3 times the price of teak wood due to its excellent wood characteristics. Since then, the wood has been commercially categorized as fancy and luxury wood similar to bongin (*Irvingia malayana* Oliv), bungur (*Lagerstroemia speciosa*), cempaka (*Michelia* spp, and cendana (*Santalum album*) as stated in Forestry Minister No. 163/Kpts-II/2003.

3.4.1. Trading and Its Regulations

The *Pericopsis* wood products (Figure 8) have been traded locally, so it is difficult to get quantitative data on its wood supply and demand. The wood is utilized increasingly without replanting efforts [3] and exported since 1972 and exploited uncontrollably [4], so a decrease in the population of *P. mooniana was* reported by Rain Forest Action in the year 2004 [6]. This condition has put the trees as vulnerable species (VU A1cd ver 2,3) in the IUCN Red List of Threatened Species. These species are also included as tree species in the Southeast Asia region and are immediately saved for protection from vulnerable threats by the UNEP World Conservation Monitoring Centre [9].

Figure 8. Wood product of *P. mooniana* from Southeast Sulawesi, Indonesia (Photo: F.D. Tuheteru, 2023).

The Natural Resources Conservation Institute Sulawesi Tenggara in 2012 reported that the potency of *Pericopsis* trees at the Nature Reserve (NR) of Lamedae significantly declined, was no longer dominant [3], and became vulnerable tree species [6]. In order to prevent the population decline, the Government of Indonesia, through the Ministry of Forestry, has issued its regulation number: 209/kpts- II/1994. This regulation has appointed the NR of Lamedae as a location for sustaining the *Pericopsis* stand population [6].

The history of the appointment of Lamedae NR as a location for sustaining the *Pericopsis* stands population has been well described [9,137]. It began with the letter of the Director of Nature Protection and Preserving, Ministry of Agriculture No: 1058/IV-i/IV/6/1972, dated 22 August 1972, proposing to Governor to appoint the Lamedae-Tangketada forest compartment that grows abundant *P. mooniana* trees as NR areas. Thus, the Governor agreed by issuing a recommendation letter No: PTA/4/1/11. On February 16, 1974, the Ministry of Agriculture determined the Lamedae-Tangketada forest as NR with a total area of 500 Ha. In 1987, the total area of Lamedae NR became 635,15 Ha after completing its border areas. Based on its final border areas, the Ministry of Forestry issued a Forestry Minister No. 209/KPTS-II/1994, dated April 30, 1994, for determining Lamedae forest as NR [9] and that the Lamedae NR is only one of the in situ conservation for *Pericopsis* trees in Indonesia, particularly in Southeast Sulawesi.

3.4.2. Sustainable Wood Supply

The vulnerability of *P. mooniana* will threaten its wood supply to wood processing industries. In order to increase the population of *P. mooniana*, the species can be chosen as the main species for rehabilitating and restoring the critical forest and land areas in its near-native habitat. There are many ways to rehabilitate critical areas, one of which is to implement an appropriate and effective agroforestry system to restore critical land by using the agri-horti-silviculture model [138]. This model is a technique for cultivating local forest trees with certain food plant species and conserving local species more than producing commercial food [138]. This method allows local species with vulnerable categories to be conserved while combining them with appropriate annual crops.

In conservation, local tree species such as *P. mooniana* have a relatively high probability of being productively conserved with agroforestry techniques using the agri-horti-silviculture model [138]. Due to its pioneering nature and being able to grow on degraded soils, its combination with other plants in an agroforestry cropping pattern will produce at least two main outputs, namely: (1) preservation of the studied species in a productive conservation model; (2) restoration of land quality through revegetation activities.

Using *P. mooniana* as shade stands in agroforestry systems with the aim of revegetation of critical lands has an excellent opportunity to be actualized. However, few reports are available regarding *P. mooniana* agroforestry practices by local farmers [136]. They have experimented with agroforestry patterns using *P. mooniana* as shade trees and patchouli (*Pogestemon cablin* Benth.) as its intercrop. The results showed that the shade treatment significantly affected growth in stem height, the number of leaves, and the concentration of biomass of patchouli plants [136].

3.5. Discussion and Conservation Strategies for P. mooniana

The main obstacles to *P. mooniana* cultivation in Indonesia are (i) a decline in its natural stands and (ii) a technology gap for its cultivation. Hence, it requires several aspects to support its species conservation. The priority for species conservation needs appropriate planning in determining its conservation strategy. It begins with genetic conservation efforts that are carried out by establishing in situ and ex situ conservation areas with proper methods, planning, monitoring, and evaluation. The preparation of conservation plans fully requires information on species taxonomy, species biology, species distribution, and current population numbers in nature [139]. In this case, the genetic diversity of *P. mooniana* is a fundamental element for increasing the productivity of forest plantations and their availability in the future. In addition, genetic-based ecosystem restoration is a new strategy for genetically *P. mooniana*.

There are two genetic resource conservations: in situ and ex situ. The in situ conservation for genetic conservation can be done by (1) selecting a representative population of each natural distribution and (2) selecting populations with variations in morphological characters and high genetic diversity. It is more effective and realistic than ex situ, as well as protects the reservoir of genes for potential use in the future at conservation areas. A major problem with in situ conservation is the conflict between reserves and local people. Hence, involving local communities in conservation efforts becomes essential. Although the interests of local people and the conservation managing authorities differ, efficient management of the buffer-zone areas has become a possible solution. The next step is maintaining the number of individuals and periodically managing genetic analysis. The in situ conservation of *P. moniana* has been supported legally in Southeast Sulawesi through the Forestry Minister degree No. 209/kpts- II/1994 appointing the NR of Lamedae as a location for sustaining *Pericopsis* stand population [6].

Ex situ conservation is also an integral part of tree improvement activities. These include gene banks for seed and pollen, clone banks, breeding populations, and cryopreservation. *P. mooniana* seeds and seedlings were collected from remaining natural populations in Southeast Sulawesi and South Kalimantan and were utilized to build an ex situ conservation plot on the island of Java and provide material for the *P. mooniana* tree breeding program.

Based on the condition of *P. mooniana* in Indonesia, in situ and ex situ conservation of *P. mooniana* is a top priority to prevent genetic erosion [140]. Furthermore, coordination between the two conservation stakeholders is critical for long-term conservation sustainability through effective conservation planning [141]. The importance of in situ genetic conservation is implemented to overcome the weakness of ex situ conservation, which cannot imitate evolutionary processes [142,143]. In situ genetic conservation is carried out by selective and adaptive processes that give rise to new genetic traits in the face of environmental challenges [144]. Furthermore, it must be based on an efficient network of protected areas and have the legal powers established by law [145]. Meanwhile, ex situ conservation is implemented as an alternative strategy when in situ conservation cannot be implemented adequately [146]. The ex situ conservation strategy of *P. mooniana* is carried out by taking and preserving samples of species, subspecies, or varieties as living plant collections in field gene banks, botanical gardens, and arboretums, or as samples of seeds, ovules, and pollen or DNA under controlled conditions [140].

Other ex situ conservation actions can be conducted through several real actions in field sites, such as preventing illegal logging, regeneration, conservation education, reforestation, agroforestry actions, and industrial forest plantation. Illegal logging can be reduced by good cooperation between the local government and law enforcers from the Ministry of Environment and Forestry [9]. Regeneration and increasing wood production for *P. mooniana* can be done by assessing the knowledge of its potency, tree distribution, wood characteristics, end-use, and type of fruits and seeds [4]. Conservation education can raise the local community's awareness of conserving *P. mooniana* by clearly explaining why this tree should be conserved. For example, it needs conservation because this tree is categorized as a vulnerable species. Its wood demand is still high due to its luxury, price, beautiful decorative pattern, and multi-used timber. Reforestation can be real action for the wide distribution of *P. mooniana* through planting in the community lands and critical forest land in Indonesia. This reforestation would assist the ex situ conservation of *Pericopsis* [9]. An agroforestry system can also be used to conserve this species. Farmers can plant seasonal crops with this species as a shading function. The research found that *P. mooniana* is reasonably correlated with other trees, so this tree is suitable as the main tree in the agroforestry system. The development of industrial plantation with *P. mooniana* as the main trees will increase the wood supply for wood processing industries.

4. Conclusions

The sustainability of *P. mooniana* is threatened by internal (low germination rate, limited genetic resources, and limited silviculture technique) and external (land conversion, illegal logging, and excessive logging due to its high demand) interferences. The solution for its sustainability is a strategic combination of its high demand, silviculture, and conservation effort. Its wood future demand would increase significantly due to its excellent wood characteristics. This demand should be balanced by silviculture and conservation efforts. The silviculture of *P. mooniana* has been improved through seed storage technology, improved viability and germination rate, proper micro and macro propagation, applying hormones, in vitro seed storage, and improved nursery technology and harvesting techniques. *P. mooniana* conservation can be conducted using both in situ and ex situ conservation efforts. In situ conservation is carried out by protecting its mother trees in natural conditions (i.e., Lamedae Nature Reserve) for producing good quality seeds and seedlings. Ex situ conservation is realized by planting seeds and seedlings to produce more wood through rehabilitating and restoring critical forests and lands because it can adapt to marginal land. On the other hand, it also has uses for mitigating climate change. Other actions required for supporting ex situ conservation are preventing illegal logging, regeneration, conservation education, reforestation, agroforestry systems applied in private and community lands, and industrial forest plantation.

Author Contributions: Each author (J.K., S. (Suhartati), H., F.D.T., D.I.D.A., M.A.L., R.U., R.P., Y., S. (Subarudi), L.A. (Lutfy Abdulah), R.S., T.K.W., S.S., A., M.W., T.K., E.T., W.I., A.S., N., L.A. (Laode Alhamd), D., and S. (Soenarno)) equally contributed as main contributors to the design and conceptualization of the manuscript, conducted the literature reviews, performed the analysis, prepared the initial draft, and revised and finalized the manuscript. All authors have read and agreed to the published version of the manuscript.

Funding: This research received no external funding.

Data Availability Statement: Not applicable.

Conflicts of Interest: The authors declare no conflict of interest.

References

1. Soerianegara, I.; Lemmens, R. *Sumber Daya Nabati Asia Tenggara: Pohon Penghasil Kayu Perdagangan Yang Utama*, 1st ed.; Prosea-Balai Pustaka: Jakarta, Indonesia, 2002; Volume 5.
2. Husna. *Budi Daya Dan Konservasi Kayu Kuku*, 1st ed.; Tuheteru, F.D., Elviandri, Y., Eds.; IPB Press: Bogor, Indonesia, 2015.
3. Alfaizin, D. Potensi Kayu Kuku (*Pericopsis mooniana* THW) untuk revegetasi lahan kritis. In *Prosiding Seminar Biologi from Basic Science to Comprehensive Education*; Universitas Islam Negeri Alauddin: Makassar, Indonesia, 2016; pp. 219–225.
4. Suhartati; Nursyamsi; Alfaizin, D. Mengenal morfologi, tipe buah dan biji pada pohon Kayu Kuku (*Pericopsis mooniana* THW). *Info Tek. Eboni* **2015**, *12*, 87–96.
5. Affre, A.; Kathe, W.; Raymakers, C. Looking Under the Veneer. In *Implementation Manual on EU Timber Trade Control: Focus on CITES-Listed Trees*; TRAFFIC Europe: Brussels, Belgium, 2004.
6. Akbar, A.; Rusmana. Membangkitkan Primadona yang Mulai Langka: Kayu Kuku (*Pericopsis mooniana* Thw). *Bekantan* **2013**, *1*, 4–6.
7. Hills, R. Pericopsis elata. Available online: https://www.gbif.org/species/176802781 (accessed on 16 August 2022).
8. Asian Regional Workshop. Pericopsis mooniana. Available online: https://www.gbif.org/species/176802779 (accessed on 20 August 2022).
9. Yuskianti, V.; Salihin. Sebaran Kayu Kuku (*Pericopsis mooniana*) di tegakan benih teridentifikasi (TBT) L. Gani di Anaiwoi, Sulawesi Tenggara. *Wana Benih* **2016**, *17*, 9–15.
10. Botanic Gardens Conservation International. Pericopsis laxiflora. Available online: http://www.worldfloraonline.org/taxon/wfo-0000212085 (accessed on 21 August 2022).
11. Micheneau, C.; Dauby, G.; Bourland, N.; Doucet, J.L.; Hardy, O.J. Development and Characterization of Microsatellite Loci in *Pericopsis elata* (Fabaceae) Using A Cost-efficient Approach. *J. Bot* **2011**, *98*, 268–270. [CrossRef]
12. Nkulu, S.N.; Meerts, P.; Wa Ilunga, E.I.; Shutcha, M.N.; Bauman, D. Medicinal *Vitex species* (Lamiaceae) occupy different niches in Haut-Katanga Tropical Dry Woodlands. *Plant Ecol. Evol.* **2022**, *155*, 236–247. [CrossRef]
13. Gunatilleke, N.; Pethiyagoda, R.; Gunatilleke, S. Biodiversity of Sri Lanka. *J. Natl. Sci. Found. Sri Lanka* **2008**, *36*, 25–61. [CrossRef]
14. Ministry of Environment and Forestry(MoEF). *The State of Indonesia's Forests*, 1st ed.; Efransjah, S., Murniningtyas, Erwinsyah, Damayanti, E., Eds.; Ministry of Environment and Forestry: Jakarta, Indonesia, 2022.
15. Setyowati, R.D.N.; Amala, N.A.; Aini, N.N.U. Studi pemilihan tanaman revegetasi untuk keberhasilan reklamasi lahan bekas tambang. *Al-Ard J. Tek. Lingkung* **2017**, *3*, 14–20. [CrossRef]
16. Saputra, R.A.; Riniarti, M. Pertumbuhan dan mutu fisik bibit Kayu Kuku (*Pericopsis mooniana*) pada media tanam tailing pertambangan emas rakyat dengan pemberian amelioran. *ULIN J. Hutan Trop.* **2020**, *4*, 58–69. [CrossRef]
17. Suhartati; Alfaizin, D. Teknik pembibitan spesies Kayu Kuku (*Pericopsis mooniana*) untuk reklamasi lahan bekas tambang tanah liat. *J. Faloak* **2018**, *2*, 103–114. [CrossRef]
18. Perala, I.; Wulandari, A.S. Kayu Kuku (*Pericopsis mooniana* Thw.) seedlings growth response to tailing media added with vermicompost, rhizobium, and arbuscular mycorrhizalf. *IOP Conf. Ser. Earth Environ. Sci.* **2019**, *394*. [CrossRef]
19. Alfaizin, D.; Suhartati; Kurniawan, E. Benih dan perkecambahan kayu kuku(*Pericopsis mooniana* THW). *Info Tek. Eboni* **2016**, *13*, 1–11.
20. Suhartati; Alfaizin, D.; Kurniawan, E. Budidaya tanaman Kayu Kuku (*Pericopsis mooniana* Thw). In *Budidaya dan pemanfaatan spesies pohon kayu lokal Sulawesi*; Nas Media Pustaka: Makassar, Indonesia, 2020; pp. 61–97.
21. Heyne, K. *Tumbuhan Berguna Indonesia*; Yayasan Sarana Wana Jaya: Jakarta, Indonesia, 1987.
22. Wulandari, A.S.; Farzana, A.R. Mutu fisik dan teknik pematahan dormansi benih Kayu Kuku (*Pericopsis mooniana* (Thw.) Thw.). *J. Trop. Silvic.* **2020**, *11*, 199–205. [CrossRef]
23. Suhartati; Alfaizin, D. Seed germination of *Pericopsis mooniana* Thw. based on color and scarification techniques. *J. Perbenihan Tanam. Hutan* **2017**, *5*, 115–124. [CrossRef]
24. Lemmens, R.H.M.; Soerianegara, I.; Wong, W. *Plant Resources of South-East Asia. Timber Tress: Minor Commercial Timbers*; Backhuys Publishers: Leiden, The Netherlands, 1994; Volume 5.
25. Sidiyasa, K.; Soerianegara, I.; Martawijaya, A.; Sudo, S. *Plant Resources of South-East Asia. Timber Tree: Major Commercial Timbers*; PROSEA: Bogor, Indonesia, 1994; Volume 5.
26. Yuniarti, N.; Syamsuwida, D. Kayu kuku (*Pericopsis mooniana* Thw). In *Atlas Benih Tanaman Hutan Indonesia*, 2nd ed.; Balai Penelitian Teknologi Perbenihan Tanaman Hutan: Bogor, Indonesia, 2011; Volume 5, pp. 32–34.
27. Hardiyanto, E.B.; Na'iem, M. Present Status of Conservation, Utilization, and Management of Forest Genetic Resources in Indonesia. Available online: https://www.fao.org/3/ac648e/ac648e05.htm#TopOfPage (accessed on 17 July 2022).
28. Lestari, D.; Santoso, W. Inventory and habitat study of orchids species in Lamedai Nature Reserve, Kolaka, Southeast Sulawesi. *J. Biol. Divers.* **2011**, *12*, 28–33. [CrossRef]

29. Widyatmoko, D. Strategi dan inovasi konservasi tumbuhan Indonesia untuk pemanfaatan secara berkelanjutan. In Proceedings of the Seminar Nasional Pendidikan Biologi dan Saintek (SNPBS) ke-IV, Surakarta, Indonesia, 27 April 2019; pp. 1–22.
30. Frankham, R. 2019 Conservation Genetics. In *Encyclopedia of Ecology*, 2nd ed; Elsevier: Oxford, UK, 2019; Volume 1, pp. 382–390.
31. Coker, O.M. Importance of genetics in conservation of biodiversity. *Niger. J. Wildl. Manag.* **2017**, *1*, 11–18.
32. Maia, R.T. Introductory Chapter: Genetic Variation-The Source of Biological Diversity. Available online: https://www.intechopen.com/chapters/75504 (accessed on 23 August 2022).
33. Li, Y.; Liu, C.; Wang, R.; Luo, S.; Nong, S.; Wang, J.; Chen, X. Applications of molecular markers in conserving endangered. *J. Species Biodivers. Sci.* **2020**, *28*, 367–375.
34. Nurtjahjaningsih, I.; Widyatmoko, A.; Rimbawanto, A. Genetic diversity of kayu kuku (*Pericopsis mooniana* (Thwaites) Thwaites) population at Lamedai forest Southeast Sulawesi based on RAPD markers. *J. Pemuliaan Tanam. Hutan* **2019**, *13*, 25–32. [CrossRef]
35. Yudohartono, T.P.; Fiani, A.; Yuliah. Genetic diversity of kayu kuku (Pericopsis mooniana Thw) from Southeast Sulawesi provenance observed at seedling stage. *J. Pemuliaan Tanam. Hutan* **2019**, *13*, 11–17. [CrossRef]
36. Yuskianti, V.; Sulistyawati, P. Genetic diversity of Pericopsis mooniana from South Kalimantan based on Random Amplified Polymorphism DNA (RAPD) Markers. *IOP Conf. Ser. Earth Environ. Sci.* **2021**, *914*, 012029. [CrossRef]
37. KPHP Model Sorong. Rencana Pengelolaan Hutan Jangka Panjang KPHP Model Sorong. Available online: http://kph.menlhk.go.id/sinpasdok/public/RPHJP/RPHJP_SORONG_SELATAN.pdf (accessed on 25 August 2022).
38. Yuskianti, V. Genetic conservation of a threatened species, Kayu kuku (Pericopsis mooniana) in Indonesia. In *IUFRO-INAFOR Joint International Conference 2017 Promoting Sustainable Resources from Plantation for Economic Growth and Community Benefits*; FORDIA Ministry of Environment and Forestry Republic of Indonesia: Bogor, Indonesia, 2018; pp. 403–410.
39. Amandita, F.Y.; Rembold, K.; Vornam, B.; Rahayu, S.; Siregar, I.Z.; Kreft, H.; Finkeldey, R. DNA barcoding of flowering plants in Sumatra, Indonesia. *Ecol. Evol.* **2019**, *9*, 1858–1868. [CrossRef]
40. Suhartati; Alfaizin, D. Cultivation of *Pericopsis mooniana* Thw Case Study: KHDTK Malili, Luwu Timur Regency. *IOP Conf. Ser. Earth Environ. Sci.* **2020**, *486*, 012106. [CrossRef]
41. Heywood, V.H. An overview of in situ conservation of plant species in the Mediterranean Flora. *J. Mediterr* **2014**, *24*, 5–24.
42. Sulistyawati, P.; Widyatmoko, A. Keragaman genetik populasi Kayu Merah (*Pterocarpus indicus* Willd) menggunakan penanda Random Amplified Polymorphism DNA. *J. Pemuliaan Tanam. Hutan* **2017**, *11*, 67–76. [CrossRef]
43. Widyatmoko, A.Y.P.B.C.; Aprianto, A. Keragaman genetik *Gonystylus bancanus* (miq.) kurz berdasarkan penanda RAPD (Random Amplified polymorphic DNA). *J. Pemuliaan Tanam. Hutan* **2013**, *7*, 53–71. [CrossRef]
44. Hakim, L.; Widyatmoko, A.Y.P.B.C. Strategi konservasi ex-situ jenis ulin (*Eusideroxylon zwageri*), Eboni (*Diospyros celebica*), dan Cempaka (*Michelia* spp.). In *Status Konservasi Dan Formulasi Strategi Konservasi Jenis-Jenis Pohon Terancam Punah (Ulin, Eboni Dan Michelia)*, 1st ed.; Bismark, M., Murniati, Eds.; Pusat Penelitian dan Pengembangan Konservasi dan Rehabilitasi Badan Litbang Kehutanan dan ITTO: Bogor, Indonesia, 2011; pp. 163–177.
45. Byrne, M. A molecular journey in conservation genetics. *Pacific. Conserv. Biol.* **2018**, *24*, 235–243. [CrossRef]
46. Singh, P.; Sharma, H.; Nag, A.; Bhau, B.S.; Sharma, R.K. Development, and characterization of polymorphic microsatellites markers in endangered *Aquilaria malaccensis*. *Conserv. Genet. Resour.* **2015**, *7*, 61–73. [CrossRef]
47. Lee, C.T.; Lee, S.L.; Tnah, L.H.; Ng, K.K.S.; Ng, C.H.; Cheng, S.; Tani, N. Isolation and characterization of 16 microsatellite markers in Intsia palembanica, a high-value tropical hardwood species. *Conserv. Genet. Resour.* **2014**, *6*, 389–391. [CrossRef]
48. Rimbawanto, A.; Widyatmoko, A. Identifikasi *Aquilaria malaccensis* dan *A. microcarpa* menggunakan Penanda RAPD. *J. Pemuliaan Tanam. Hutan* **2011**, *5*, 23–30. [CrossRef]
49. Widyatmoko, A.Y.P.B.C.; Shiraishi, S. Development of SCAR markers for identification of teak clones. *AIP Conf. Proc.* **2019**, *2120*, 030001. [CrossRef]
50. Widyatmoko, A.Y.P.B.C.; Rimbawanto, A.; Suharyanto. Keragaman genetik *Araucaria cuminghamii* menggunakan penanda RAPD. In *Proceedings Seminar Nasional Peningkatan Produktivitas Hutan: Peran Konservasi Sumberdaya Genetik, Pemuliaan dan Silvikultur dalam Mendukung Rehabilitasi Hutan*; UGM dan ITTO: Yogyakarta, Indonesia, 2005; pp. 397–408.
51. Widyatmoko, A.Y.P.B.C.; Afritanti, R.; Taryono; Rimbawanto, A. Keragaman genetik lima populasi *Gyrinops verstegii* di Lombok menggunakan penanda RAPD. *J. Pemuliaan Tanam. Hutan* **2009**, *3*, 1–10. [CrossRef]
52. Hartati, D.; Rimbawanto, A.; Taryono; Sulistyaningsih, E.; Widyatmoko, A.Y.P.B.C. Pendugaan keragaman genetik di dalam dan antar provenan pulai (Alstonia scholaris (L.) R. Br.) menggunakan penanda RAPD. *J. Pemuliaan Tanam. Hutan* **2007**, *1*, 89–98.
53. Liu, U.; Cossu, T.A.; Davies, R.M.; Forest, F.; Dickie, J.B.; Breman, E. Conserving orthodox seeds of globally threatened plants ex situ in the Millennium Seed Bank Royal Botanic Gardens, Kew, UK: The status of seed collections. *Biodivers. Conserv.* **2020**, *29*, 2901–2949. [CrossRef]
54. Laure, N.M.; Samuel, G.L.; Francois, M.; Emmanuel, E.; Din, N. Phenological Behavior of *Pericopsis elata* in a Production Forest at Libongo (Southeast Cameroon). *Int. J. Sci. Res. Methodol.* **2019**, *12*, 9–26.
55. Angbonda, D.M.A.; Monthe, F.K.; Bourland, N.; Boyemba, F.; Hardy, O.J. Seed and pollen dispersal and fine-scale spatial genetic structure of a threatened tree species: *Pericopsis elata* (HARMS) Meeuwen (Fabaceae). *Tree Genet. Genomes* **2021**, *17*, 1–14. [CrossRef]
56. Tandoh, P.; Banful, B.; Gaveh, E.; Amponsah, J. Effects of packaging materials and storage periods on seed quality and longevity dynamics of *Pericopsis elata* seeds. *Environ. Earth Ecol.* **2017**, *1*, 27–38. [CrossRef]

57. Nkengurutse, J.; Khalid, A.; Mzabri, I.; Kakunze, A.C.; Masharabu, T.; Berrichi, A. Germination optimization study of five indigenous fabaceae tree species from Burundi miombo woodlands. *J. Mater. Environ. Sci.* **2016**, *7*, 4391–4403.
58. Romdyah, N.L.; Riniarti, M.; Asmarahman, C.; Yuwono, S.B. Skarifikasi Awal Dan Penambahan Beberapa Jenis Zat Pengatur Tumbuh Untuk Percepatan Perkecambahan Benih Kayu Kuku (*Pericopsis moonianna* Thw). *EnviroScienteae* **2020**, *16*, 296–308. [CrossRef]
59. Nursyamsi. Teknik skarifikasi benih Kayu Kuku (*Pericopsis Mooniana* Thw) untuk mematahkan dormansi melalui kultur jaringan. In Proceedings of the Seminar Nasional from Basic Science to Comprehensive Education, Makassar, Indonesia, 26 August 2016.
60. Araújo, S.d.S.; Paparella, S.; Dondi, D.; Bentivoglio, A.; Carbonera, D.; Balestrazzi, A. Physical methods for seed invigoration: Advantages and challenges in seed technology. *Front. Plant Sci.* **2016**, *7*, 1–12. [CrossRef]
61. Adetunji, A.E.; Sershen; Varghese, B.; Pammenter, N. Effects of exogenous application of five antioxidants on vigour, viability, oxidative metabolism and germination enzymes in aged cabbage and lettuce seeds. *S. Afr. J. Bot.* **2021**, *137*, 85–97. [CrossRef]
62. Siregar, I.Z.; Muharam, K.F.; Purwanto, Y.A.; Sudrajat, D.J. Seed germination characteristics in different storage time of gmelina arborea treated with ultrafine bubbles priming. *Biodiversitas* **2020**, *21*, 4558–4564. [CrossRef]
63. Prasetyawati, C.A.; Nursyamsi; Alfaizin, D. Effect of seed storage methods on germination growth of Pericopsis mooniana thw. through in-vitro technique. *IOP Conf. Ser. Earth Environ. Sci.* **2020**, *473*, 6–11. [CrossRef]
64. Sukendro, A.; Ayuning, D. Study of vegetative propagation on *Pericopsis Mooniana* Thw with cutting. *J. Silvikultur Trop.* **2016**, *7*, 53–57.
65. Nursyamsi; Toaha, A.Q. Tahapan sterilisasi dan skarifikasi benih kayu kuku (*Pericopsis mooniana* THW) untuk mempercepat perkecambahan secara in vitro. *J. Info Tek. Eboni* **2017**, *14*, 11–22.
66. Abeyratne, W.; Bandara, D.; Senanayake, Y.D. In vitro Propagation of Nadun (*Pericopsis mooniana*). *Trop. Agric. Res.* **1990**, *2*, 20–35.
67. Indravathi, G. In vitro propagation studies of *Albizia amara* (Roxb.) Boiv. *African J. Plant Sci.* **2013**, *7*, 1–8. [CrossRef]
68. Pereira, S.R.; Laura, V.A.; Souza, A.L.T. Establishment of Fabaceae tree species in a tropical pasture: Influence of seed size and weeding methods. *Restor. Ecol.* **2013**, *21*, 67–74. [CrossRef]
69. Aini, F.N.; Sulistyowati, E. Mewaspadai hama minor pada Kakao: *Zeuzera coffae* Nietner. *War. Pus. Penelit. Kopi dan Kakao Indones.* **2016**, *28*, 24–29.
70. Fiani, A.; Yuliah; Pamungkas, T. Inventarisasi jenis hama yang menyerang bibit Kayu Kuku (*Pericopsis mooniana*) umur satu tahun di persemaian. In Proceedings of the Seminar Nasional Pendidikan Biologi Dan Saintek (Snpbs), Surakarta, Indonesia, 27 April 2019; pp. 59–65.
71. Roychoudhury, N.; Mishra, R.K. Coffee borer, Zeuzera coffeae: Its pest profile and use as tribal food. *Van Sangyan* **2022**, *9*, 24–26.
72. Wiratno; Deciyanto, S. Ciri-ciri dan siklus hidup serangga penggulung daun nilam, Sylepta sp. (Lepidoptera:Pyralidae). *Bul. Littro* **1991**, *1*, 15–19.
73. Tawaraya, K.; Turjaman, M. Use of arbuscular mycorrhizal fungi for reforestation of Degraded Tropical Forests. In *Mycorrhizal Fungi: Use in Sustainable Agriculture and Land Restoration*; Springer: Berlin/Heidelberg, Germany, 2014; pp. 357–373.
74. Asmelash, F.; Bekele, T.; Birhane, E. The potential role of arbuscular mycorrhizal fungi in the restoration of degraded lands. *Front. Microbiol.* **2016**, *7*, 1–15. [CrossRef]
75. Husna; Tuheteru, F.D.; Arif, A. Arbuscular mycorrhizal fungi and growth on serpentine solis. In *Arbuscular Mycorrhizas and Stress Tolerance of Plants*; Springer Nature: Singapore, 2017; pp. 293–303.
76. Riaz, M.; Kamran, M.; Fang, Y.; Wang, Q.; Cao, H.; Yang, G.; Deng, L.; Wang, Y.; Zhou, Y.; Anastopoulos, I.; et al. Arbuscular mycorrhizal fungi-induced mitigation of heavy metal phytotoxicity in metal contaminated soils: A critical review. *J. Hazard. Mater.* **2021**, *402*, 123919.
77. Faad, H.; Tuheteru, F.D.; Arif, A. Arbuscular mycorrhizal fungi symbiosis and conservation of endangered tropical Legume Trees. In *Root Biology, Soil Biology 52*; Giri, B., Prasad, R., Varma, A., Eds.; Springer International Publishing: Singapore, 2018; pp. 465–486.
78. Husna; Tuheteru, F.D.; Arif, A. The potential of arbuscular mycorrhizal fungi to conserve Kalappia celebica, an endangered endemic legume on gold mine tailings in Sulawesi, Indonesia. *J. For. Res.* **2021**, *32*, 675–682. [CrossRef]
79. Husna; Tuheteru, F.D.; Arif, A.; Basrudin; Albasari. Biodiversity of arbuscular mycorrhizal fungi in tropical Indonesia. In *An Introduction to Microorganisms*, 1st ed.; Wu, Q., Zou, Y., Zhang, F., Shu, B., Eds.; Nova Science Publisher Inc: Hauppauge, NY, USA, 2021; pp. 201–247.
80. Husna; Wilarso, S.B.; Mansur, I.; Kusmana, C.; Kramadibrata, K. Fungi mikoriza arbuskula pada rizosfer *Pericopsis mooniana* (thw.) thw. di Sulawesi Tenggara. *Ber. Biol.* **2014**, *13*, 263–273.
81. Husna; Wilarso, S.B.; Mansur, I.; Kusmana, C. Diversity of arbuscular mycorrhizal fungi in the growth habitat of Kayu Kuku (*Pericopsis mooniana* Thw.) in Southeast Sulawesi. *Pakistan. J. Biol. Sci.* **2015**, *18*, 1–10.
82. Husna; Tuheteru, F.D.; Basri, A.; Arif, A.; Basruddin; Umar, Y. Serapan hara tanaman kayu kuku (*Pericopsis moonia* Thw) bermikoriza pada interval penyiraman berbeda. *J. Hutan Trop.* **2020**, *15*, 88–101.
83. Husna; Tuheteru, F.D.; Arif, A. Arbuscular mycorrhizal fungi to enhance the growth of tropical endangered species *Pterocarpus indicus* and *Pericopsis mooniana* in post gold mine field in Southeast Sulawesi, Indonesia. *Biodiversitas* **2021**, *22*, 3844–3853.
84. Kusuma, G.A.; Husna; Mamma, S. The Response of kayu kuku (*Pericopsis mooniana* (Thw).Thw.) to mycorrhizae and the level of water availability in gold tailing media. *J. Berk. Penelit. Agron.* **2022**, *10*, 68–84.

85. Tuheteru, F.D.; Husna; Nuranisa; Basrudin; Arif, A.; Albasri; Safitri, I.; Karepesina, S.; Nurdin, W. Effects of arbuscular mycorrhizal fungi on nutrient uptake by Pericopsis mooniana in media post gold mining land. *J. Trop. Mycorrhiza* **2022**, *1*, 9–16.
86. Husna. *Pertumbuhan Bibit Kayu Kuku (Pericopsis Mooniana Thw) Melalui Aplikasi Fungi Mikoriza Arbuskula (FMA) Dan Ampas Sagu Pada Media Bekas Tambang Nikel*; Halu Oleo University: Kendari, Indonesia, 2010.
87. Husna; Budi, S.W.; Mansur, I.; Kusmana, C. Growth and nutrient status of kayu kuku [*Pericopsis mooniana* (Thw.) Thw] with mycorrhiza in soil media of nickel post mining site. *Pakistan. J. Biol. Sci.* **2016**, *19*, 158–170. [CrossRef]
88. Husna; Tuheteru, F.D.; Salim, M.I.; Adam, L.A.; Nirwan; Arif, A.; Basrudin; Albasri. Pemanfaatan pupuk hayati mikoriza lokal Sulawesi Tenggara untuk memacu pertumbuhan Kayu Kuku pada Media tanah pascatambang nikel. In *Proceedings Seminar Nasional Mikoriza: Mikoriza Untuk Pembangunan Pertanian dan Kehutanan Berkelanjutan*; UHO Edu Press: Kendari, Indonesia, 2021; pp. 101–114.
89. Husna; Tuheteru, F.D.; Wulandari, J.; Arif, A.; Basrudin; Albasari; Nurdin, W. Respons pertumbuhan semai Kayu Kuku (*Pericopsis Mooniana* (Thw .) Thw .) dengan inokulasi fungi mikoriza arbuskula (FMA) lokal dan urea pada media tanah serpentine. In *Proceedings Seminar Nasional: Mikoriza Untuk Pembangunan Pertanian Dan Kehutanan Berkelanjutan*; UHO Edu Press: Kendari, Indonesia, 2021; pp. 129–148.
90. Husna; Mansur, I.; Budi, S.W.; Tuheteru, F.D.; Arif, A.; Tuheteru, E.J.; Albasri. Effects of arbuscular mycorrhizal fungi and organic material on growth and nutrient uptake by pericopsis mooniana in coal mine. *Asian J. Plant Sci.* **2019**, *18*, 101–109. [CrossRef]
91. Iskandar, F. *Peningkatan Kualitas Bibit Kayu Kuku (Pericopsis Mooniana Thwaites) Yang Diberi Fungi Mikoriza Arbuskula Dan Tepung Tulang*; Skripsi, Haluoleo University: Kendari, Indonesia, 2010.
92. Husna; Raharjo, W.P.; Tuheteru, F.D.; Arif, A.; Basrudin; Albasri; Nurdin, W. Perbaikan kualitas bibit kayu kuku (*Pericopsis Mooniana* (Thw .) Thw. dengan aplikasi fungi mikoriza arbuskula (Fma) lokal Sulawesi Tenggara. In *Proceedings Seminar Nasional: Mikoriza Untuk Pembangunan Pertanian Dan Kehutanan Berkelanjutan*; UHO Edu Press: Kendari, Indonesia, 2021; pp. 251–264.
93. Husna; Wilarso, S.; Mansur, I.; Kusmana, C. Respon pertumbuhan bibit kayu kuku (*Pericopsis mooniana* (Thw.)Thw) terhadap inokulasi fungi mikoriza arbuskula lokal. *J. Pemuliaan Tanam. Hutan* **2015**, *9*, 131–148. [CrossRef]
94. Husna. Potensi Fungi Mikoriza Arbuskula (FMA) Lokal Dalam Konservasi Ex-Situ Jenis Terancam Punah Kayu Kuku [Pericopsis mooniana (Thw.) Thw.]. Disertasi, Bogor Agricultural University, Bogor, Indonesia, 2015.
95. Husna; Tuheteru, F.D.; Wigati, E. Short communication: Growth response and dependency of endangered nedun tree species (*Pericopsis mooniana*) affected by indigenous arbuscular mycorrhizal fungi inoculation. *Nusant. Biosci.* **2017**, *9*, 57–61. [CrossRef]
96. Husna; Saputra, A.; Tuheteru, F.D.; Arif, A. Keberhasilan hidup tanaman kayu kuku bermikoriza umur tiga tahun pada lahan pascatambang nikel PT Vale Indonesia. In *Proceedings Seminar Nasional: Mikoriza untuk Pembangunan Pertanian Dan Kehutanan Berkelanjutan*; UHO Edu Press: Kendari, Indonesia, 2021; pp. 237–250.
97. Wijesundara, T.I.L.; Van Holm, L.H.J.; Kulasooriya, S.A. Rhizobiology and nitrogen fixation of some tree legumes native to Sri Lanka. *Biol. Fertil. Soils* **2000**, *30*, 535–543. [CrossRef]
98. Husna; Hamirul, H.; Arief, L. *Uji Efektivitas VA Mikoriza Dan Rhizobium Terhadap Pertumbuhan Semai Kayu Kuku (Pericopsis mooniana Thw.)*; Departemen Pendidikan dan Kebudayaan: Jakarta, Indonesia, 1994.
99. Javaid, A. Microbes for legume improvement microbes legum. In *Microbes for Legume Improvement*; Khan, M.S., Zaidi, A., Musarrat, J., Eds.; Springer: Vienna, Austria, 2010; pp. 409–426.
100. Prihatini, I.; Widyatmoko, A.Y.P.B.C.; Nurtjahjaningsih, I.L.G.; Yuskianti, V.; Danarto, S.A. DNA barcoding of *Pericopsis mooniana* from two different populations in Indonesia based on rDNA ITS (Internal Transcribed Spacer). *IOP Conf. Ser. Mater. Sci. Eng.* **2020**, *935*, 012024. [CrossRef]
101. Zanzibar, M.; Nugraheni, Y.M.M.A.; Putri, K.P.; Yuniarti, N.; Sudrajat, D.J.; Yulianti; Syamsuwida, D.; Aminah, A.; Rustam, E.; Sukendro, A. Toona sureni (blume) merr. seeds invigoration using 60Co Gamma-Rays irradiation. *Forest. Sci. Technol.* **2021**, *17*, 80–87.
102. Rocha, M.; Singh, N.; Ahsan, K.; Beiriger, A.; Prince, V.E. Neural crest development: Insights from the zebrafish. *Dev. Dyn.* **2020**, *249*, 88–111. [CrossRef]
103. Elias. Dampak penerapan RIL terhadap simpanan karbon di hutan pegunungan tropis. In *Proceedings of the Seminar Hasil Penelitian Teknologi Dan Inovasi Keteknikan Kehutanan Dan Pengolahan Hasil Hutan Dalam Menunjang Industri Pengolahan Hasil Hutan*; Badan Penelitian dan Pengembangan dan Inovasi, Kementerian Lingkungan Hidup dan Kehutanan: Bogor, Indonesia, 2015; pp. 3–14.
104. Ellis, P.W.; Gopalakrishna, T.; Goodman, R.C.; Putz, F.E.; Roopsind, A.; Umunay, P.M.; Zalman, J.; Ellis, E.A.; Mo, K.; Gregoire, T.G.; et al. Reduced-impact logging for climate change mitigation (RIL-C) can halve selective logging emissions from tropical forests. *For. Ecol. Manag.* **2019**, *438*, 255–266. [CrossRef]
105. Suhartana, S.; Yuniawati; Rahmat. Pemanenan kayu ramah lingkungan di hutan tanaman lahan kering di Sumatera, Kalimantan dan Jawa Barat. In *Proceedings of the Seminar Nasional MAPEKI 16: Pemanfaatan sumberdaya terbarukan untuk Kesejahteraan Manusia dan Kelestarian Lingkungan*; Masyarakat Peneliti Kayu Indonesia: Balikpapan, Indonesia, 2014; pp. 402–406.
106. Bicknell, J.E.; Struebig, M.J.; Edwards, D.P.; Davies, Z.G. Improved timber harvest techniques maintain biodiversity in tropical forests. *Curr. Biol.* **2014**, *24*, 119–120. [CrossRef]
107. Direktorat Jenderal Pengelolaan Hutan Produksi *Standar dan pedoman pelaksanaan penilaian kinerja pengelolaan hutan produksi lestari (PHPL) dan verifikasi legalitas kayu (VLK)*. Available online: https://silk.menlhk.go.id/app/Upload/hukum/20160801/24fbdaeef1ae1cd12488cb4051740dce.pdf (accessed on 26 October 2022).

108. Soenarno; Yuniawati; Dulsalam; Suhartana, S. Determination of standardized natural forest exploitation factor and logging waste of Sub Region Central Kalimantan Province. *J. Penelit. Has. Hutan* **2021**, *39*, 155–169.

109. Soenarno; Endom, W.; Bustomi, S. Residual stand damage due to timber harvesting on hilly tropical forest inCentral Kalimantan. *J. Penelit. Has. Hutan* **2017**, *35*, 273–288.

110. Mousavi, M.R. Comparison of damage to residual stand due to applying two different harvesting methodsin the Hyrcanian forest of Iran: Cut-to-length vs. tree length. *Casp. J. Environ. Sci.* **2017**, *15*, 13–27.

111. Sist, P.; Nolan, T.; Bertault, J.G.; Dykstra, D. Harvesting intensity versus sustainability in Indonesia. *For. Ecol.Manage.* **1998**, *108*, 251–260. [CrossRef]

112. Rodríguez-Calcerrada, J.; P´erez-Ramos, I.M.; Ourcival, J.-M.; Limousin, J.-M.; Joffre, R.; Rambal, S. Is selective thinning an adequate practice for adapting Quercus ilex coppices to climate change? *Ann. For. Sci.* **2011**, *68*, 575–585. [CrossRef]

113. Pakhriazad, H.; Shinohara, T.; Nakama, Y.; Ykutake, K. A Selective Management System (SMS): A case study in the implementation of SMS in managing the dipterocarp forests of Peninsular Malaysia. *Kyushu J.For.Res* **2004**, *57*, 39–44.

114. Yuniawati; Dulsalam; Andini, S. Potential carbon storage and CO2 emissions due to felling in Papua natural forest. *J. Penelit. Has. Hutan* **2022**, *40*, 61–73.

115. Elias. *Pengertian dan perkembangan IPTEKS pemanenan kayu*; PT. Media Anugrah Perkasa: Jakarta, Indonesia, 2015.

116. Galante, M.V.; Dutschke, M.; Patenaude, G.; Vickers, B. Climate change mitigation through reduced-impact logging and the hierarchy of production forest management. *Forests* **2012**, *3*, 59–74. [CrossRef]

117. Abdulah, L. *Metode Pendugaan Konsumsi Kayu Untuk Penggunaan Rumah Tangga*; Bogor Agricultural University: Bogor, Indonesia, 2020.

118. Abdulah, L.; Suhendang, E.; Purnomo, H.; Matangaran, J.R. The role of urban household wood product consumption on forest management and its impact: A system modelling simulation approach in Bogor City. *IOP Conf. Ser. Earth Environ. Sci.* **2021**, *800*, 1–12. [CrossRef]

119. Abdulah, L.; Suhendang, E.; Purnomo, H.; Matangaran, J.R. Measuring the sustainability of wood consumption at the household level in indonesia: Case study in Bogor, Indonesia. *Biodiversitas* **2020**, *21*, 457–464. [CrossRef]

120. Abdulah, L.; Irawan, A.; Arini, D.I.D.; Suryaningsih, R. The Use of Marketing Monitoring Web System Application to Estimate Cempaka Wood Product Consumption Level in North Sulawesi, Indonesia. *IOP Conf. Ser. Earth Environ. Sci.* **2021**, *891*, 1–13. [CrossRef]

121. Abdulah, L.; Yulianti, M.; Mindawati. How the growth of natural forests meets Indonesia's demand for wood for construction and household furniture? *IOP Conf. Ser. Earth Environ. Sci.* **2022**, *959*, 012070. [CrossRef]

122. Seng, O. *Specific Gravity of Indonesian Woods and Its Significance for Parctical Use*; Pusat Penelitian dan Pengembangan Hasil Hutan, Departemen Kehutanan: Jakarta, Indonesia, 1990.

123. Munandar Budidaya Tanaman Kehutanan Jenis Kuku (Pericopsis mooniana Thwaites). Available online: https://www.mounandar.blogspot.com/2010/06/budidaya-tanaman-kehutananjenis-kuku.html (accessed on 14 July 2015).

124. Krisdianto; Dewi, L.M. *Jenis Kayu Untuk Mebel*, 1st ed.; Pusat Penelitian dan Pengembangan Keteknikan Kehutanan dan Pengolahan Hasil Hutan, Badan Penelitian dan Pengembangan Kehutanan-Kementerian Kehutanan: Bogor, Indonesia, 2012.

125. Pericopsis elata (Harms) Meeuwen. Available online: https://www.prota4u.org/database/protav8.asp?g=psk&p=Pericopsis%20elata (accessed on 18 September 2022).

126. Djarwanto; Damayanti, R.; Balfas, J.; Basri, E.; Jasni, I.M.S.; Andianto; Martono, D.; Pari, G.; Sopandi, A.; Mardiansyah. *Pengelompokan Jenis Kayu Perdagangan*, 1st ed.; Forda Press: Bogor, Indonesia, 2017; p. 26.

127. Ishiguri, F.; Wahyudi, I.; Takeuchi, M.; Takashima, Y.; Iizuka, K.; Yokota, S.; Yoshizawa, N. Wood properties of Pericopsis mooniana grown in a plantation in Indonesia. *J. Wood Sci.* **2011**, *57*, 241–246. [CrossRef]

128. Anoop, E.V.; Sindhumathi, C.R.; Jijeesh, C.M.; Jayasree, C.E. Radial variation in wood properties of nedun (*Pericopsis mooniana*), an introduced species to South India. *J. Trop. Agric.* **2016**, *54*, 27–34.

129. Martawijaya, A.; Kartasujana, I. *Ciri Umum, Sifat Dan Kegunaan Jenis-Jenis Kayu Indonesia*, 1st ed.; Lembaga Penelitian Hasil Hutan: Bogor, Indonesia, 1977.

130. Pandit, I.K.N. Karakteristik struktur anatomi Kayu Kuku (*Pericopsis mooniana* Thwaiters). *Ilmu Teknol. Kayu Trop.* **2005**, *3*, 1–5.

131. Pericopsis elata (Harms) van Meeuwen (Afrormosia, Kokrodua)-CITES II. Available online: https://www.delta-intkey.com/citeswood/en/www/fabpeel.htm (accessed on 18 September 2022).

132. Soerianegara, I.; Lemmens, R. *Plant Resources of Southeast Asia. No 5(1): Timber Trees: Major Commercial Timbers*; Pudoc Scientific Publishers: Wageningen, The Netherlands, 1993.

133. Murodovna, R.; Sergeevna, K. Wood-As a Decorative Material of the Interior. *Int. J. Integr. Educ.* **2022**, *5*, 310–315.

134. Burley, J.; Evans, J.; Youngquist, J. *Encyclopedia of Forest Sciences*, 1st ed.; Elsevier Academic Press: Oxford, UK, 2004.

135. Zanuttini, R.; Negro, F. Wood-based composites: Innovation towards a sustainable future. *Forests* **2021**, *12*, 10–14. [CrossRef]

136. Sabaruddin, B. Nilai Ekonomi Kayu Kuku. Available online: https://www.forestdigest.com/detail/933/nilai-ekonomi-kayu-kuku (accessed on 17 October 2022).

137. Heropolo. Lamedae Nature Reserve. Available online: https://www.indonesianforest.or.id/lamedae-nature-reserve/ (accessed on 18 October 2022).

138. Hermawan, B. Sustainable agroforestry models for proposed food production in post-mined land sites of South Sumatera. *Int. J. Adv. Sci. Eng. Inf. Technol.* **2016**, *6*, 245–251. [CrossRef]

139. Guidelines for Species Conservation Planning. Available online: https://portals.iucn.org/library/sites/library/files/documents/2017-065.pdf (accessed on 5 September 2022).
140. Maxted, N. In Situ, Ex Situ Conservation. *Encycl. Biodivers. Second Ed.* **2013**, *4*, 313–332.
141. Rahman, W.; Brehm, J.M.; Maxted, N.; Phillips, J.; Cotreras-Toledo, A.R.; Faraji, M.; Quijano, M.P. Gap analyses of priority wild relatives of food crop in current ex situ and in situ conservation in Indonesia. *Biodivers. Conserv.* **2021**, *30*, 2827–2855. [CrossRef]
142. Malhotra, N.; Panatu, S.; Singh, B.; Negi, N.; Singh, D.; Singh, M.; Chandora, R. Genetic Resources: Collection, Conservation, Characterization and Maintenance. In *Lentils*, 1st ed.; Singh, M., Ed.; Academic Press: Cambridge, MA, USA, 2019; pp. 21–41.
143. Rajpurohit, D.; Jhang, T. In Situ and Ex Situ conservation of plant genetic resources and traditional knowledge. In *Plant Genetic Resources and Traditional Knowledge for Food Security*, 1st ed.; Salgotra, R.K., Gupta, B.B., Eds.; Springer Science Business Media: Singapore, 2015; pp. 137–162.
144. National Research Council and Board on Agriculture. *In Situ Conservation of Genetic Resources Managing Global Genetic Resources: Agricultural Crop Issues and Policies*; National Academic Press: Washington, DC, USA, 1993; pp. 1–17.
145. Meyer, S.; Birnbaum, P.; Bruy, D.; Cazé, H.; Garnier, D.; Gâteblé, G.; Lannuzel, G.; McCoy, S.; Tanguy, V.; Veillon, J.M. The New Caledonia Plants RLA: Bringing Botanists Together for the Conservation of the archipelago's Crown Jewel. In *Reference Module in Earth Systems and Environmental Sciences*; Elsevier: Amsterdam, The Netherlands, 2022; pp. 859–874, in press.
146. Ajayi, S.S.S. Principles for the management of protected areas. In *Wildlife Conservation in Africa*; Elsevier Inc.: Amsterdam, The Netherlands, 2019; pp. 85–93.

Article

Interspecific Variance of Suitable Habitat Changes for Four Alpine *Rhododendron* Species under Climate Change: Implications for Their Reintroductions

Jin-Hong Zhang [1,2], Kun-Ji Li [1,2], Xiao-Fei Liu [3], Liu Yang [1,2] and Shi-Kang Shen [1,2,*]

[1] Yunnan Key Laboratory of Plant Reproductive Adaptation and Evolutionary Ecology, School of Ecology and Environmental Sciences, Yunnan University, Kunming 650091, China; zhangjinhong0503@163.com (J.-H.Z.); lkkkky2021@163.com (K.-J.L.); yangliu@ynu.edu.cn (L.Y.)

[2] School of Life Sciences, Yunnan University, Kunming 650091, China

[3] Institute of International Rivers and Eco-Security, Yunnan University, Kunming 650091, China; xfliu@ynu.edu.cn

* Correspondence: ssk168@ynu.edu.cn

Abstract: Rapid temperature changes in mountain ecosystems pose a great threat to alpine plant species and communities. *Rhododendron* species, as the major component of alpine and sub-alpine vegetation, have been demonstrated to be sensitive to climate changes. Therefore, understanding how alpine *Rhododendron* species spread to new habitats and how their geographical distribution range shifts is crucial for predicting their response to global climate change and for facilitating species conservation and reintroduction. In this study, we applied MaxEnt modeling and integrated climate, topography, and soil variables in three periods under three climate change scenarios to predict the suitable habitat for four *Rhododendron* species in China. We measured the potential distribution change in each species using the change ratio and the direction of centroid shifts. The predicted results showed that (1) the threatened species *R. protistum* would have a maximum decrease of 85.84% in its distribution range in the 2070s under RCP 8.5, and *R. rex* subsp. *rex* as a threatened species would experience a distribution range expansion (6.62–43.10%) under all of the three climate change scenarios in the 2070s. (2) *R. praestans* would experience a reduction in its distribution range (7.82–28.34%) under all of the three climate change scenarios in the 2070s. (3) The four *Rhododendron* species would be moved to high latitudes in the north-westward direction as a whole in the future, especially the two threatened species *R. protistum* and *R. rex* subsp. *rex*. (4) Aside from climate variables, soil factors also exert an important influence on the distribution of *Rhododendron* species. This study revealed the species-specific response of *Rhododendron* species to climate change. The results can not only provide novel insights into conservation strategies of *Rhododendron* species, but also propose a valuable method for the habitat selection during the reintroduction of endangered species.

Keywords: plant conservation; geographical range shift; *Rhododendron*; alpine species; reintroduction

Citation: Zhang, J.-H.; Li, K.-J.; Liu, X.-F.; Yang, L.; Shen, S.-K. Interspecific Variance of Suitable Habitat Changes for Four Alpine *Rhododendron* Species under Climate Change: Implications for Their Reintroductions. *Forests* **2021**, *12*, 1520. https://doi.org/10.3390/f12111520

Academic Editors: Konstantinos Poirazidis and Panteleimon Xofis

Received: 9 October 2021
Accepted: 1 November 2021
Published: 3 November 2021

Publisher's Note: MDPI stays neutral with regard to jurisdictional claims in published maps and institutional affiliations.

1. Introduction

Human-induced global changes are the most severe phenomena since the Last Glacial Maximum. Rapid global climate warming has led to distribution pattern changes for numerous species since the original habitats of these species may no longer be appropriate for their growth and survival [1]. Climate change plays a driving role in biodiversity loss, changes in the spatial patterns of species, and threatened species' survival, and it can increase the risk of extinction for endangered plants. Numerous studies have shown that the habitat range of most species is decreasing [2,3], and the geographic distributions of these species are expected to move towards high altitudes and high elevations to expand to new favorable areas under climate warming [1]. Moreover, growing evidence shows that the speed of warming is amplified by elevation under the background of global warming, which means that temperatures change more rapidly at high elevations than at low elevations [4]. This rapid change in temperature poses a great threat to alpine plant species and communities [5,6]. Therefore, understanding how alpine species spread to new habitats and how the geographical distribution range changes is crucial for predicting their response to global climate change and facilitating species conservation and reintroduction [7].

Rhododendron L., which comprises about 1025 species, is one of the largest woody plant genera. This genus is widely distributed from the northern temperate region throughout southeast Asia to northeastern Australia [8,9]. About 571 *Rhododendron* species exist in China, and 70% of them are endemic in the southwestern region of China. Therefore, the mountain land of Southwest China is considered the center of origin or evolution of modern *Rhododendron* [8]. However, with the ever-increasing impact of human activities and global climate change, the survival of *Rhododendron* resources in China is facing a serious threat [10]. In addition, as the major component of alpine and sub-alpine vegetation, the *Rhododendron* species have been demonstrated to be sensitive to climate change [11–13]. Yu et al. (2019) [12] applied the presence-only ecological niche model MaxEnt and predicted the distribution of 10 narrow-ranging and 10 wide-ranging *Rhododendron* species. The authors discovered the negative effect of climatic and land use changes on the distribution of species. Lu et al. (2021) [13] assessed the effect of protected areas and tourist attractions on the current and future suitable habitat distributions of seven species of the genus *Rhododendron* in Northeast China and provided conservation suggestions.

Species distribution models (SDMs), as an effective tool in predicting species' suitable habitats under climate change, play important roles in studying the processes of species' ecological evolution and in conservation planning [13,14]. On the basis of species occurrence data and environmental variables, SDMs can predict the potential distribution of species at different spatial and temporal scales, especially under scenarios of global climate change [15]. At present, SDMs based on different theoretical and analysis methods, such as surface range envelope (BIOCLIM), artificial neural network (ANN), maximum entropy (Maxent), random forest (RF), generalized linear model (GLM), and generalized additive model (GAM), are widely used in ecology and biogeography. The Maxent model is widely recognized as one of the most effective tools for predicting species' suitable habitats and changes in future distributions due to its advantages. In particular, it requires only species presence data, can deal effectively with limited occurrence data and small sample sizes, and can use continuous and categorical environmental data as input variables. Numerous studies have predicted changes in animal and plant distributions in response to contemporary climate changes using the Maxent model [7,16–21].

Reintroduction is one of the rescue methods of plant conservation, especially for endangered species threatened by habitat destruction and loss under global changes. Predicting habitat suitability and the potential habitat distribution range is critical to the reintroduction of endangered plants [22,23]. Here, we integrate climate and soil variables to predict the suitable habitat for four *Rhododendron* species in China. We hypothesize that the suitable range for the four *Rhododendron* species would contract and shift to high altitudes in the future since the original distribution areas may become climatically unsuitable under climate change omission scenarios. Furthermore, we predict that inter-specific variance would occur in the future. The results of this study can provide novel insights into the creation of conservation strategies and habitat selection during the reintroduction of *Rhododendron* species.

2. Materials and Methods

2.1. Species Occurrence Data

Four alpine *Rhododendron* species that belong to *Rhododendron* Sect. *Ponticum* G. Don., namely, *Rhododendron protistum* Balf.f. & Forrest (including its variant), *Rhododendron rex* H.Lév. subsp. *rex*, *Rhododendron praestans* Balf.f. & W.W.Sm. and *Rhododendron sinogrande* Balf.f. & W.W.Sm., were selected in the present study. *Rhododendron rex* subsp. *rex* is an endangered plant endemic to China [8]. *Rhododendron protistum* is listed as a critically endangered species in the China Plant Red Data Book [24]. Four *Rhododendron* species are evergreen trees that are distributed in forest ecosystems in the southwest plateau mountains in China.

Occurrence data were obtained from published studies, the China Digital Herbarium (http://www.cvh.ac.cn/, accessed on 1 October 2021), and GBIF (https://www.gbif.org/, accessed on 1 October 2021). In accordance with the biological and ecological characteristics of these species, we removed unreasonable sites to ensure geographic accuracy. Each sample point was at least 2 km apart to reduce the effect of spatial autocorrelation. Then, we obtained 24, 22, 15, and 39 effective localities of *Rhododendron protistum*, *Rhododendron rex* subsp. *rex*, *Rhododendron praestans*, and *Rhododendron sinogrande* in Southwest China for the subsequent model construction and analysis (Figure 1).

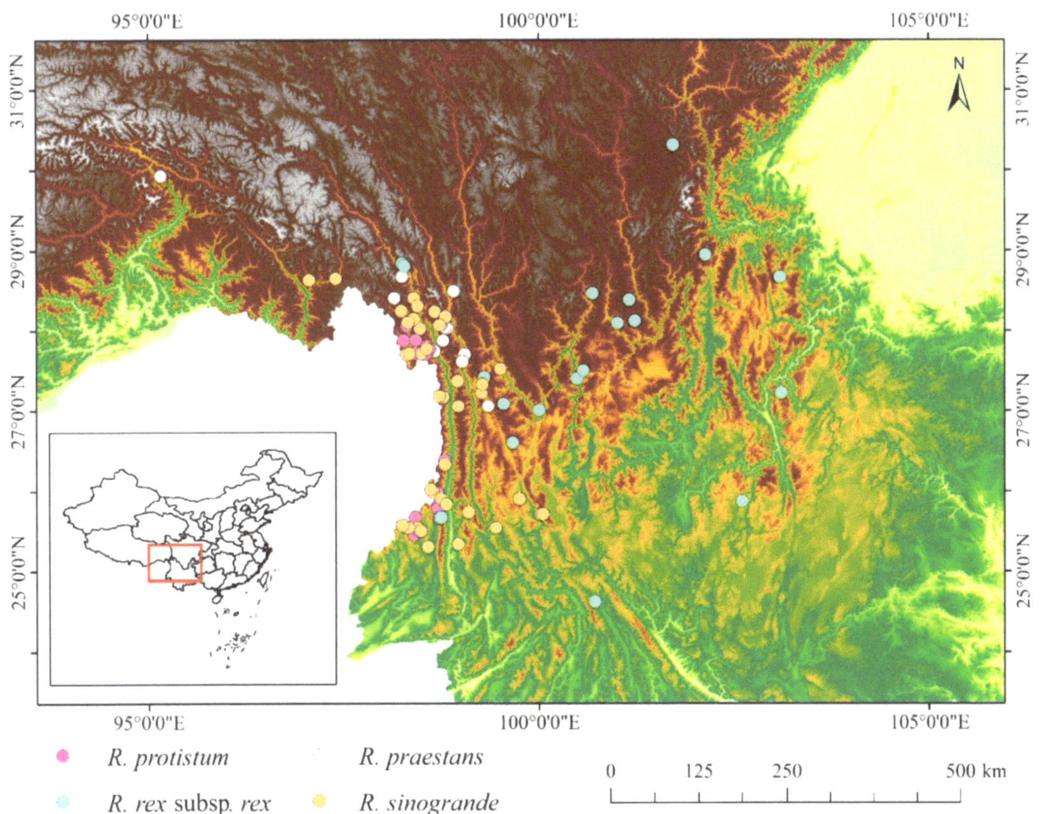

Figure 1. Occurrence points of four *Rhododendron* species.

2.2. Environment Data

Climate variables are the major predictors in the MaxEnt model since they are the primary factors regulating species' geographic distributions. However, recent studies have found that other factors, including soil types and characteristics, land use changes, and vegetation types, have a significant influence on distribution predictions. These factors are often correlated with climate change [15,16]. Therefore, combining climate and other factors that affect species' biological characteristics in MaxEnt model building can reveal a suitable habitat under the scenarios of global climate change. This suitable habitat can be used for species reintroduction and rehabilitation.

The environment data included 19 climate variables, 3 topography variables, and 16 soil variables. Climate data (Bio1–Bio19) on current climate conditions (1960–1990), the 2050s (average for 2041–2060), and the 2070s (average for 2061–2080) were downloaded from the Worldclim Database (https://www.worldclim.org/, accessed on 1 October 2021) at a resolution of 30 arc-seconds (~1 km^2). For future climatic projection, we used the Beijing Climate Center Climate System Model (BCC-CSM1-1), which is a widely utilized global climate model (GCM) in China provided by IPCC5 [25]. The representative concentration pathways (RCPs) form a set of greenhouse gas concentration and emission pathways which are widely used to research the impacts and potential policy of the responses to climate change. For this GCM, we selected three RCPs of emission scenarios from low to high, namely, RCP 2.6, RCP 4.5, and RCP 8.5. Furthermore, we used topsoil variables at depth intervals of 0–30 cm in Maxent model building for precise suitable habitat prediction. The topsoil variable data were derived from the Harmonized World Soil Database (http://www.iiasa.ac.at/web/home/research/researchPrograms/water/HWSD.html, accessed on 1 October 2021) at the same resolution (1 km^2). Digital elevation model (DEM) data were obtained from the SRTM Database (http://srtm.csi.cgiar.org/, accessed on 1 October 2021) with a 90 m resolution. Meanwhile, the slope and aspect were extracted from DEM data in ArcGIS 10.2 and resampled into a 1 km spatial resolution.

2.3. Study Area and Environment Variables Selection

To approach a more realistic prediction, we delimited the study area of each *Rhododendron* species by drawing a minimum convex polygon (MCP) with a buffer distance of 1 degree (~111 km) around the localities and obtained the environment factors within this region by the Wallace package in R. All of the analyses include variables and the selections were accomplished within the buffer areas. In addition, we explored the potential distribution of four *Rhododendron* species in Southwest China.

The collinearity among environmental factors causes overfitting and uncertainty in the predicted results [26]. To establish a good performance model with few variables for four *Rhododendron* species, first, we used 38 variables to prebuild the MaxEnt model three times in succession and eliminated the environment factors with no contribution or importance to the models. Moreover, we eliminated four variables that combined temperature and precipitation since they have artificial discontinuities [27,28]. Therefore, we performed a correlation analysis by the Band Collection Statistics tool of ArcGIS for the remaining environment variables, except for the categorical factor. In addition, we retained one factor which has a high contribution from each set of highly cross-correlated variables ($R^2 > 0.8$) for further modeling (Tables 1, 2 and A1, Tables A2–A4).

Table 1. Environmental predictors used for modeling the habitat suitability distribution of four *Rhododendron* species.

Species	Variables Abbreviation
R. protistum	Bio3, Bio4, Bio13, Bio15, Bio17, Slo, Asp, T_CaCO$_3$, T_CLAY, T_GRAVEL, T_OC
R. rex subsp. *rex*	Bio4, Bio10, Bio12, Bio14, Slo, Asp, T_USDA_TEX_CLASS
R. praestans	Bio2, Bio7, Bio15, Slo, T_ECE, T_OC, T_USDA_TEX_CLASS
R. sinogrande	Bio2, Bio3, Bio4, Bio13, Bio17, Slo, Asp, T_CaCO$_3$, T_CLAY, T_USDA_TEX_CLASS

Table 2. List of environmental predictors used for modeling the habitat suitability distribution of four *Rhododendron* species.

Data Sources	Abbreviation	Description	Units
Worldclim Database (https://www.worldclim.org/, accessed on 1 October 2021) original resolution 30 arc-s (~1 km^2)	Bio2	Mean Diurnal Range (Mean of monthly (max temp–min temp))	°C
	Bio3	Isothermality (Bio2/Bio7×100)	-
	Bio4	Temperature Seasonality (standard deviation ×100)	C of V
	Bio7	Temperature Annual Range	°C
	Bio10	Mean Temperature of Warmest Quarter	°C
	Bio12	Annual Precipitation	mm
	Bio13	Precipitation of Wettest Month	mm
	Bio14	Precipitation of Driest Month	mm
	Bio15	Precipitation Seasonality (Coefficient of Variation)	C of V
	Bio17	Precipitation of Driest Quarter	mm
Harmonized World Soil Database (http://www.iiasa.ac.at/web/home/research/researchPrograms/water/HWSD.html, accessed on 1 October 2021) original resolution 30 arc-s (~1 km^2)	T_CaCO3	Topsoil calcium carbon	% weight
	T_CLAY	Topsoil clay fraction	% weight
	T_GRAVEL	Topsoil gravel content	% vol
	T_OC	Topsoil organic carbon	% weight
	T_ECE	Topsoil salinity	dS/m
	T_USDA_TEX	Topsoil USDA texture classification	name
SRTM Database (http://srtm.csi.cgiar.org/, accessed on 1 October 2021) original resolution 90 m	Slo	Slope	°
	Asp	Aspect	°C

2.4. Model Evaluation and Model Selection

When selecting the model parameters, we used models with regularization multiplier (RM) values ranging from 0.5 to 4.5 (increments of 0.5) and with six feature class (FC) combinations (L, LQ, H, LQH, and LQHP, with L = linear, Q = quadratic, H = hinge, and p = product). In accordance with the lowest delta, the AICc score and the predictor variables (RM and FC) were used in the final model. All of the analyses were performed using the Wallace package in R [29–31].

We inputted the species and environment data into MaxEnt 3.4.1 [32]. The feature combination and regularization multiplier were set based on Table 3. For each run, we set the output format as "Cloglog" and the output file type as "asc." The model was estimated using 75% of the occurrence data for model calibration, and 25% was used for model testing. The replicate was set as "10" with a replicated run type as "Crossvalidate" to reduce uncertainty. The remaining parameters adopted default settings. The average of 10 predictions was taken as the result. The area under the ROC curve (AUC) and the true skill statistic (TSS) are the two metrics that we used to measure model performance. The range of the AUC is from 0.5 to 1 and TSS ranges between −1 and +1. The closer the assessment value is to 1, the better the model performance [33,34].

Table 3. The models calculate the results of Wallace package, as well as the values of AUC and TSS for four *Rhododendron* species. FC: Feature classes (H = Hinge, L = Linear, Q = Quadratic, p = Hinge); RM: Regularization multiplier; avg.test.AUC: The average result of n iterations of predicted values for the test localities; avg.test.or10 pct: The proportion of test localities with suitability values that is lower than what is excluded in the 10% of training localities with the lowest predicted suitability; delta. AICc: Akaike Information Criterion corrected.

Species	FC	RM	Avg.Test. AUC	Avg.Test. or 10 pct	Delta.AICc	AUC	TSS
R. protistum	LQH	2.5	0.99	0.33	0	0.996 ± 0.002	0.94 ± 0.16
R. rex subsp. *rex*	LQHP	2.5	0.92	0.15	0	0.918 ± 0.050	0.72 ± 0.23
R. praestans	LQHP	1.5	0.94	0.13	0	0.922 ± 0.105	0.66 ± 0.41
R. sinogrande	LQH	1.5	0.96	0.28	0	0.975 ± 0.012	0.79 ± 0.18

2.5. Geospatial Analyses

To estimate the similarity of environment conditions between the buffered background area and projection area and to know where the climates are novel, we calculated the multivariate environmental similarity surfaces (MESS) following the guideline from Elith et al. [35]. The novelty of environment (i.e., extrapolation, negative values in MESS result) indicated where the prediction areas are questionable and extrapolated in ecologically unrealistic ways [35–37]. Therefore, we masked out these extrapolation areas (negative values in the MESS map output). As a result, a predictive map with four classes of habitat suitability for the four *Rhododendron* species was defined using the reclass tool in ArcGIS 10.2. The four classes were unsuitable (<0.10), low suitability (0.10–0.33), moderate suitability (0.33–0.67), and high suitability (>0.67). To measure the predicted distribution changes and centroid shifts for each species, we projected the binary species distribution models (SDMs) onto WGS_1984_UTM_Zone_47N projection in ArcGIS 10.2. Then, we calculated the change ratio between the current and future distribution and centroid change direction vectors of the current and future binary SDMs using the SDM Toolbox [38].

3. Results

3.1. Model Selection and Evaluation

In accordance with the lowest delta.AICc, we obtained optimal parameters for each *Rhododendron* species (Table 3). On the basis of these parameters, current suitable habitat distributions of *R. protistum*, *R. rex* subsp. *rex*, *R. praestans*, and *R. sinogrande* in the current period were constructed. The AUC values for the models of the four *Rhododendron* species were between 0.920 and 0.988, and the TSS ranged from 0.66 to 0.94 (Table 3).

3.2. Current Habitat Distribution and Dominant Environment Variables

After masking out the outlier areas which have negative values in MESS (Figure A1, Table A5), about 10% to 20% of projection grid cells were predicted as a suitable potential area under current climatic conditions. The current potential distribution prediction of the four *Rhododendron* species (Figure 2) indicated that *R. protistum* is mainly distributed in the Gaoligong Mountain of northwestern Yunnan Province in China. In addition, Tibet bordering northern Myanmar has a small suitable area. *R. rex* subsp. *rex* has a wide, highly suitable distribution range and is mainly distributed in northeastern Yunnan Province and southwestern Sichuan Province. *R. praestans* has a major current habitat suitability in northwestern Yunnan Province and southeastern Tibet. Moreover, the Hengduan Mountainous Region is a potential current habitat. The current habitat distribution range of *R. sinogrande* is similar to *R. protistum*. In other words, it is mainly distributed in western Yunnan Province and southeastern Tibet (Figure 2). The current distribution areas of *R. protistum*, *R. rex* subsp. *rex*, *R. praestans*, and *R. sinogrande* are 10,440.58, 178,394.88, 132,745.11, and 38,819.5 km^2, respectively.

The jackknife test on the variables' contributions to Maxent revealed the influence of the environmental factors on the spatial distribution of the species. High values of the regularized training gain indicate a highly significant contribution of variables [21]. The results of the jackknife test on the regularized training gain for the four *Rhododendron* species are shown in Figure A2. For *R. protistum*, Bio13 provided very high gains (>1.0) when utilized independently, indicating that the precipitation of the wettest month is the dominant factor in habitat conditions. Bio15, Bio17, Bio3, Bio4, T_CLAY, T_GRAVEL, and T_OC provided a moderate gain when used independently, suggesting that they can affect the habitat suitability distribution of *R. protistum*. For *R. rex* subsp. *rex*, only Bio4 and soil type (T_USDA_TEX_CLASS) provided a moderate gain when used independently. Meanwhile, Bio15, Bio2, Bio7, and the slope provided moderate gains for *R. praestans* when used independently. Bio2 and Bio4 provided high gains (>1.0) for *R. sinogrande*, and Bio13, Bio17, Bio3, Slope, T_CLAY, and T_USDA_TEX_CLASS provided moderate gains when used independently.

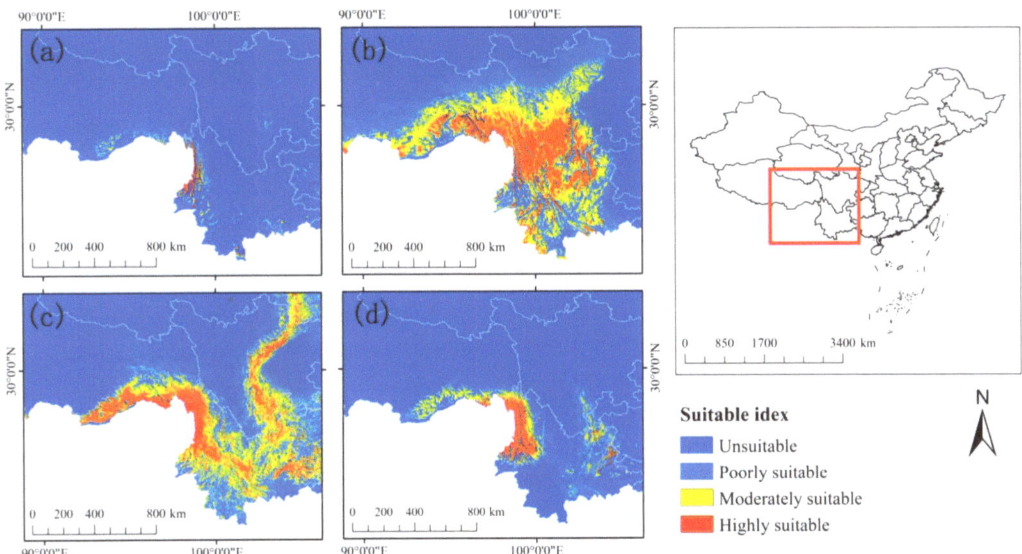

Figure 2. The potential current distribution of four *Rhododendron* species: (**a**) *R. protistum*, (**b**) *R. rex* subsp. *rex*, (**c**) *R. praestans*, (**d**) *R. sinogrande*.

3.3. Projected Changes in Habitat Suitability for Future Periods

Prediction of the potential distribution of four *Rhododendron* species in China under different future climatic scenarios (RCP 2.6, RCP 4.5, and RCP 8.5) in the 2050s and 2070s are shown in Figure 3. Compared with the current high-suitability habitat distribution range (Figure 4) in the 2070s, *R. protistum* will show a decreasing–increasing–decreasing trend with intensified emission scenarios, and it will have a maximum decrease of 85.84% in the 2070s under RCP 8.5. *R. rex* subsp. *rex* would experience a distribution range expansion under all of the three scenarios in the 2070s. However, *R. praestans* will show an opposite performance as *R. rex* subsp. *rex*. In addition, the distribution range of *R. sinogrande* will decrease under RCP 2.6 and expand under RCP 4.5 and RCP 8.5.

After a comparison of the high-suitability habitat change ratio of each species between two periods (current–2050s and 2050s–2070s) under RCP 2.6, RCP 4.5, and RCP 8.5 (Figure 5), we found that the high-suitability habitat of *R. protistum* and *R. rex* subsp. *rex* will increase in the two periods under RCP 4.5. However, *R. protistum* will exhibit a trend of shrinkage–expansion and expansion–shrinkage under RCP 2.6 and RCP 8.5, respectively. The high-suitability habitat will decrease sharply in the 2050s–2070s under RCP 8.5. Meanwhile, the high-suitability habitat of *R. rex* subsp. *rex* will increase by at most 44.38% within the current–2050s under RCP 2.6. The high-suitability habitat of *R. praestans* will increase first, then decrease under RCP 2.6 and RCP 4.5, and it will show a persistent shrinkage trend under RCP 8.5. On the contrary, the high-suitability habitat of *R. sinogrande* will decrease first, then increase under RCP 2.6 and RCP 4.5.

3.4. Shifts of the Centroids of the High-Suitability Habitat in the Future

The overall trend of the centroid change of *R. protistum* will continue to be at high latitudes in the period current–2050s and 2050s–2070s under the three scenarios (Figure 6a), indicating that the climate sensitivity of the species will survive. *R. rex* subsp. *rex* will migrate to the northwest of the high latitude under RCP 4.5 and RCP 8.5, but under RCP 2.6, its centroid will move to low and then to high latitude (Figure 6b). For the centroid shift of *R. praestans*, it will move to high latitude, except for the period current–2050s under RCP 8.5 (Figure 6c). However, *R. sinogrande* is different from the former species, its centroid will change to high latitude first and then to low latitude under RCP 2.6 and RCP 4.5 (Figure 6d).

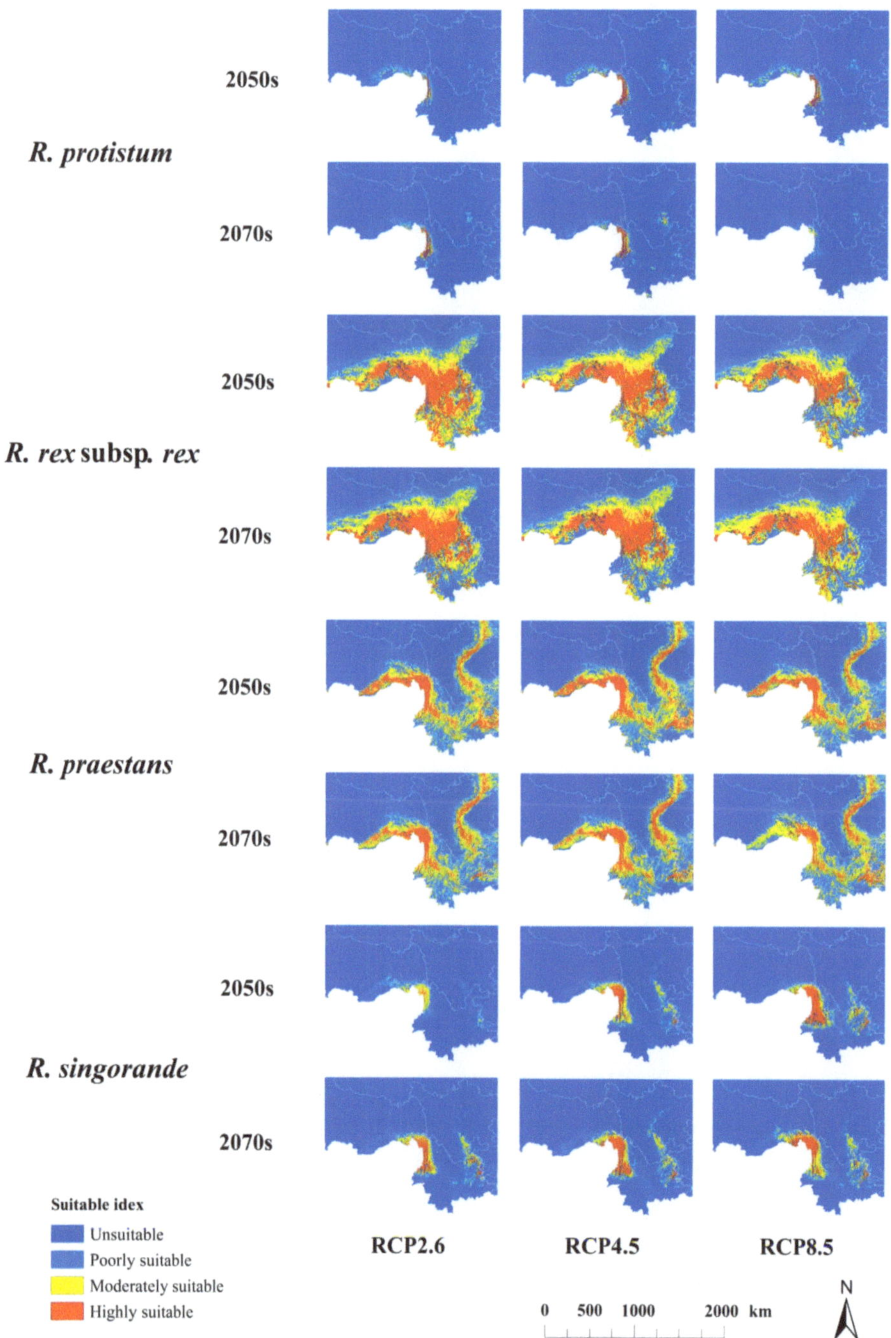

Figure 3. Prediction of the potential distribution of four *Rhododendron* species in China under different future climatic scenarios (RCP 2.6, RCP 4.5, and RCP 8.5) in the 2050s and 2070s.

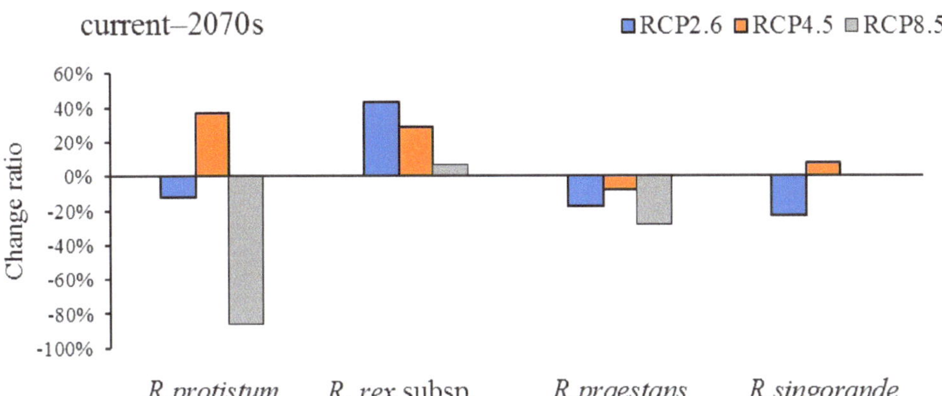

Figure 4. Change ratio of high-suitability habitat distribution range for four *Rhododendron* species in the 2070s under RCP 2.6, RCP 4.5, and RCP 8.5.

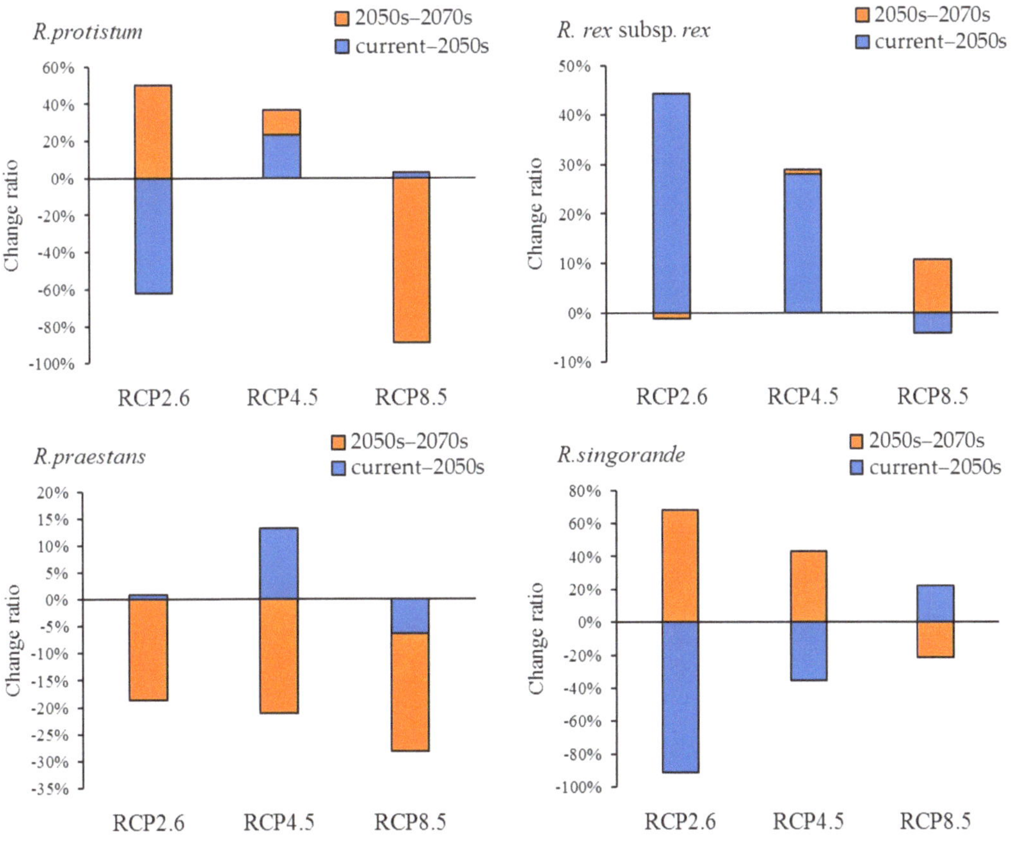

Figure 5. Change ratio of high-suitability habitat distribution range for four *Rhododendron* species in two periods (current-2050s and 2050s–2070s) under RCP 2.6, RCP 4.5, and RCP 8.5.

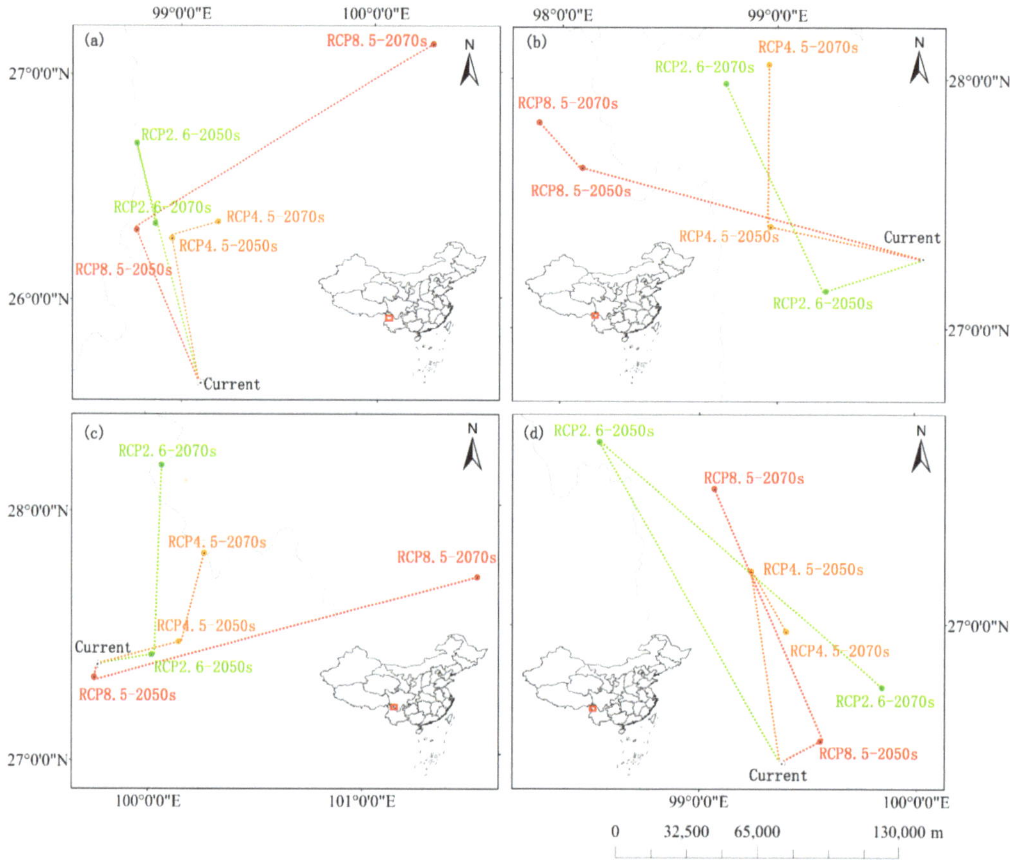

Figure 6. Map of centroid shifts of high-suitability habitats for four *Rhododendron* species from current–2050s and 2070s under RCP 2.6, RCP 4.5, and RCP 8.5. The green line depicts the shifting path of high-suitability habitats under RCP 2.6. The orange line depicts the shifting path of high-suitability habitats under RCP 4.5. The red line depicts the shifting path of high-suitability habitats under RCP 8.5. (**a**) *R. protistum*, (**b**) *R. rex* subsp. *rex*, (**c**) *R. praestans*, (**d**) *R. sinogrande*.

4. Discussion

MAXENT is reliable in predicting the potential geographic distributions of species [19–21,23]. Ardestani reported that [22] the species with a wide distribution range tend to have low AUC values. Our results on the four *Rhododendron* species' AUC values support this viewpoint. From small to large, the current habitat distribution ranges of the four focused species are in the order of *R. protistum*, *R. sinogrande*, *R. praestans*, and *R. rex* subsp. *rex*, and the AUC values are 0.996, 0.975, 0.922, and 0.918, respectively. In other words, *R. protistum* has the smallest distribution range with the largest AUC value (AUC = 0.996), and *R. rex* subsp. *rex* has the widest distribution range and the smallest AUC value.

The niche theory suggests that the species with a limited distribution range size often has a narrow niche breadth and low tolerance and adaptability to changeable environments [39]. The species with a wide distribution range tend to have a broad niche breadth, which allows them to adapt to climate change [13]. In the present study, the geographical ranges of *R. protistum* and *R. sinogrande* were locally distributed. These species have strict requirements for habitat environments. In other words, the predicted probability of presence will be greater than 0.632 if precipitation or other environment variables reach a certain value [40]. Moreover, the distribution sizes of *R. rex* subsp. *rex* and *R. praestans* were larger than the former. From the jackknife test on regularized training gain, we found that *R. rex* subsp. *rex* and *R. praestans* have a broad range of environment requirements. However, the suitable habitat proportion of limited-distribution plants *R. protistum* and *R. sinogrande* did not decrease under different future climate conditions according to the niche theory. On the contrary, their suitable habitat area will increase in the future. This prediction is consistent with the results of Yu et al. (2019) [12] and Lu et al. (2021) [13], in which the narrow-ranging species may experience range expansion under climate change. It is possible that the species occupying a warm climatological niche will benefit from climate change [41]. In addition, the distribution range of *R. praestans* was wider than the former, and its habitat size will decrease in the future. This result is consistent with the result of Yu et al. (2019) [12], in which the wide-ranging *Rhododendron* species will be adversely affected by environment changes. However, the distribution shift trend of *R. rex* subsp. *rex* was contrary to *R. praestans*. These species-specific responses of *Rhododendron* species imply specific conservation and reintroduction strategies for different species.

In general, the species would tend to distribute to high latitudes or elevations in the future to achieve a suitable habitat under continuous climate warming. Previous studies have demonstrated that numerous species change their geographic distribution range towards high latitudes and altitudes [1,16,17]. However, an increasing number of reports have found other types of range changes, such as east–west directions across longitudes or towards low elevations and tropical latitudes [42]. In this study, the four *Rhododendron* species were predicted to move towards high latitudes in the north-westward direction as a whole, especially the two threatened species *R. protistum* and *R. rex* subsp. *rex*. These results are in line with the universal migration rule of alpine plant species. However, the different species showed different movement trends (towards low latitudes or east–west direction across longitudes) under the different climate scenarios. For instance, the centroid shifts of the highly suitable habitat of *R. sinogrande* were predicted to move to high latitudes and to low latitudes in RCP 2.6 and RCP 4.5, respectively. This result implies that the species *R. sinogrande* has stronger adaptability to climate warming than the other species.

An organism's habitat is the combination of the space it occupies and all of the ecological factors in that space, including the abiotic environment and the biotic environment, which are necessary for the survival of the other organisms [21]. Therefore, before conducting the predictions, we constructed a buffer area around the species occurrence localities to delimit the study region. This can greatly improve the accuracy and truthfulness of the predictions since the area includes environments that are accessible to the species, given its limitations and the configuration of barriers [43]. If possible, we could base this on a biological justification of the factor, physiological information from laboratory experiments, and natural history information when selecting the predictor variables to make the predictions realistic [44]. Previous studies have reported that aside from climate variables, other environmental factors, such as soil, topography and land use, vegetation dynamics, and inter-specific competition, affect the response of *Rhododendron* species to climate change and their future distribution [11,45,46]. Feng et al. (2020) [16] built ecological niche models using climate and soil variables when predicting the habitat distribution of *Camptotheca acuminate*. The authors found that soil factors can accurately limit the distribution range of species on the basis of climate factors, and the accuracy of climate and soil models for *C. acuminata* is high. Abdelaal [23] suggested that combining climate variables with soil variables can predict the distribution of species accurately. In the present study, climate, topographic, and soil variables were combined to explore the potential habitat distribution shift in current and future periods of four alpine *Rhododendron* species under rapid climate change. The results of the regularized training gain of each species showed that soil factors affect the distribution of *Rhododendron* species. T_CLAY, T_GRAVEL, and T_OC are important factors in predicting the distribution of *R. protistum*. T_USDA_TEX_CLASS has a moderate influence on *R. rex* subsp. *rex*. T_CLAY and T_USDA_TEX_CLASS can affect the habitat distribution of *R. sinogrande*. The clay content (T_CLAY) provides a high soil organic carbon, and the gravel content in the topsoil (T_GRAVEL) affects the respiration and water absorption of the roots. The content of topsoil organic carbon (T_OC) is a significant indicator in evaluating soil fertility. Moreover, the soil type (T_USDA_TEX_CLASS) can reflect the soil physical properties. Overall, we reasonably speculate that the nutrient-holding and water-holding capacities of soil significantly affect the distribution of *Rhododendron* species.

Rhododendron species have high ecological and ornamental values, but 70% of the plants in this group are classified as vulnerable, threatened, endangered or critically endangered [9,10]. Therefore, conservation recommendations must be made in accordance with the characteristics of the species, especially under future climate change scenarios. SDMs are easy-to-use tools for selecting the probable distribution and areas suitable for the reintroduction of endangered species [20,21,23]. Moreover, our results indicate that aside from climate factors, the contents of clay, gravel, and organic carbon in the topsoil have a significant influence on the endangered plant *R. protistum*. This result implies that the conservation and future reintroduction of *R. protistum* should consider not only the climate factors, but also the characteristics of topsoil in the micro-habitat. In addition, we found that *R. protistum* will face a severe contraction across its highly suitable distribution area in the 2070s under the RCP 8.5 climate change scenario. Notably, the remaining habitat area is still concentrated in the national natural conservation zone in the Gaoligong Mountain. Therefore, we reasonably speculate that the survival of *R. protistum* depends on the specific environment or local microhabitats, and *R. protistum* may have nearly no adaptive capacity to climate change. We recommend protecting *R. protistum* in situ where its preference is. In addition, we recommend that the habitat selection of future reintroduction and population recovery of *R. protistum* should focus on the Gaoligong Mountain. *R. rex* subsp. *rex* is an endangered plant endemic to China. However, the present study showed that this species will experience distribution range expansion under all of the three scenarios in the 2070s. This finding indicates that *R. rex* subsp. *rex* may have a certain adaptive capacity to future global warming. Meanwhile, a previous study has reported that the main reason for the endangerment of *R. rex* subsp. *rex* is exotic anthropogenic disturbance [8,47]. Furthermore, the present study found that the distribution of *R. rex* subsp. *rex* is mainly affected by the temperature seasonality standard deviation and soil type. Therefore, we suggest that *R. rex* subsp. *rex* should be given priority by building a conservation area to protect it from high-frequency human disturbances and implement reintroduction strategies for its future suitable habitat. Although *R. praestans* and *R. sinogrande* are not listed as threatened plants in China, the distribution size of *R. praestans* will decrease by 28.34% in the future. Therefore, we suggest that comprehensive studies and additional other alpine *Rhododendron* species should be given urgent attention under the scenario of climate change.

5. Conclusions

Global climate change causes distribution changes in organisms, which significantly affect biodiversity conservation and ecosystem safety. Therefore, predicting the trends of distribution changes of plants can provide a basis for plant conservation and habitat selection in species' reintroduction strategies. The present study applied MaxEnt modeling and integrated climate, topography, and soil variables in three periods under three climate change scenarios to predict the suitable habitat for four *Rhododendron* species in China. We revealed species-specific responses to climate change and the influence of topsoil factors on the distribution of *Rhododendron* species. Moreover, we presented corresponding conservation strategies for different species.

Author Contributions: Conceptualization, S.-K.S., L.Y. and X.-F.L.; methodology, J.-H.Z., X.-F.L. and K.-J.L.; formal analysis, J.-H.Z. and K.-J.L.; writing—original draft preparation, J.-H.Z.; writing—review and editing, S.-K.S., L.Y. and X.-F.L.; supervision, S.-K.S.; project administration, S.-K.S.; funding acquisition, S.-K.S. All authors have read and agreed to the published version of the manuscript.

Funding: This research was funded by the National Natural Science Foundation of China (31870529, 31560224), the Science and Technology Basic Resources Investigation Program of China (2017FY100100), the Young Academic and Technical Leader Raising Foundation of Yunnan Province (2018HB035), the Program for Excellent Young Talents, Yunnan University, and the Scientific research fund project of Education Department of Yunnan (2021Y045).

Conflicts of Interest: The authors declare no conflict of interest.

Appendix A

Table A1. Multi-collinearity test by utilizing correlations among climate variables for *R. protistum*.

Factor	Bio2	Bio3	Bio4	Bio6	Bio9	Bio13	Bio15	Bio17	Asp	Ele	Slo	T_CaCO$_3$	T_CLAY	T_GRAVEL	T_OC
Bio2	1.00														
Bio3	0.59	1.00													
Bio4	0.27	−0.59	1.00												
Bio6	−0.85	−0.19	−0.61	1.00											
Bio9	−0.77	−0.07	−0.68	0.99	1.00										
Bio13	−0.67	0.03	−0.72	0.84	0.85	1.00									
Bio15	0.75	0.45	0.20	−0.68	−0.64	−0.50	1.00								
Bio17	−0.78	−0.33	−0.39	0.77	0.75	0.70	−0.73	1.00							
Asp	−0.02	−0.01	−0.02	0.01	0.01	0.02	−0.02	0.02	1.00						
Ele	0.87	0.44	0.33	−0.93	−0.89	−0.74	0.73	−0.73	−0.02	1.00					
Slo	−0.07	0.19	−0.32	0.15	0.16	0.23	−0.13	0.04	0.03	−0.07	1.00				
T_CaCO3	−0.23	−0.31	0.15	0.19	0.16	0.05	−0.17	0.03	−0.01	−0.30	−0.11	1.00			
T_CLAY	−0.53	−0.09	−0.42	0.68	0.68	0.59	−0.48	0.57	0.01	−0.64	0.09	0.08	1.00		
T_GRAVEL	0.14	0.10	0.02	−0.14	−0.12	−0.11	0.18	0.00	0.00	0.18	−0.10	−0.09	−0.22	1.00	
T_OC	0.14	0.11	−0.02	−0.10	−0.08	−0.04	0.07	−0.04	0.00	0.11	−0.05	−0.10	0.07	0.14	1.00

Table A2. Multi-collinearity test by utilizing correlations among climate variables for *R. rex subsp. rex*.

Factor	Bio4	Bio10	Bio12	Bio14	Bio19	Asp	Slo
Bio4	1.00						
Bio10	−0.51	1.00					
Bio12	−0.42	0.49	1.00				
Bio14	−0.56	0.63	0.72	1.00			
Bio19	−0.53	0.55	0.74	0.95	1.00		
Asp	−0.01	−0.05	0.01	0.00	0.00	1.00	
Slo	0.21	−0.20	−0.13	−0.21	−0.15	0.00	1.00

Table A3. Multi-collinearity test by utilizing correlations among climate variables for *R. praestans*.

Factor	Bio2	Bio5	Bio7	Bio15	Bio16	Slo	T_ECE	T_OC
Bio2	1.00							
Bio5	−0.83	1.00						
Bio7	0.67	−0.59	1.00					
Bio15	0.75	−0.70	0.48	1.00				
Bio16	−0.68	0.70	−0.84	−0.53	1.00			
Slo	−0.07	0.05	−0.25	−0.13	0.24	1.00		
T_ECE	0.04	0.02	0.09	0.01	−0.07	−0.08	1.00	
T_OC	0.14	−0.12	0.05	0.07	−0.04	−0.05	−0.05	−0.10

Table A4. Multi-collinearity test by utilizing correlations among climate variables for *R. singorand*.

Factor	Bio2	Bio3	Bio4	Bio5	Bio7	Bio11	Bio13	Bio16	Bio17	Asp	Ele	Slo	T_CaCO$_3$	T_Clay
Bio2	1.00													
Bio3	0.59	1.00												
Bio4	0.27	−0.59	1.00											
Bio5	−0.83	−0.43	−0.29	1.00										
Bio7	0.67	−0.18	0.89	−0.59	1.00									
Bio11	−0.78	−0.08	−0.68	0.90	−0.87	1.00								
Bio13	−0.67	0.03	−0.72	0.69	−0.84	0.86	1.00							
Bio16	−0.68	0.01	−0.71	0.70	−0.84	0.87	1.00	1.00						
Bio17	−0.78	−0.33	−0.39	0.70	−0.67	0.73	0.70	0.71	1.00					
Asp	−0.02	−0.01	−0.02	0.00	−0.03	0.01	0.02	0.02	0.02	1.00				
Ele	0.87	0.44	0.33	−0.98	0.63	−0.91	−0.74	−0.76	−0.73	−0.02	1.00			
Slo	−0.07	0.19	−0.32	0.05	−0.25	0.18	0.23	0.24	0.04	0.03	−0.07	1.00		
T_CaCO$_3$	−0.23	−0.31	0.15	0.31	0.02	0.17	0.05	0.05	0.03	−0.01	−0.30	−0.11	1.00	
T_CLAY	−0.53	−0.09	−0.42	0.64	−0.56	0.68	0.59	0.60	0.57	0.01	−0.64	0.09	0.08	1.00

Table A5. The property of the extrapolation area to the projection area for four *Rhododendron* species.

Species	Current	RCP 2.6		RCP 4.5		RCP 8.5	
		2050s	2070s	2050s	2070s	2050s	2070s
R. protistum	90.74%	91.59%	90.56%	90.25%	91.27%	91.40%	91.89%
R. rex subsp. *rex*	71.17%	68.31%	67.77%	68.67%	68.18%	69.85%	69.51%
R. praestans	77.41%	74.93%	76.74%	76.61%	77.39%	77.10%	75.06%
R. sinogrande	87.70%	86.78%	85.54%	86.24%	86.42%	86.17%	86.29%

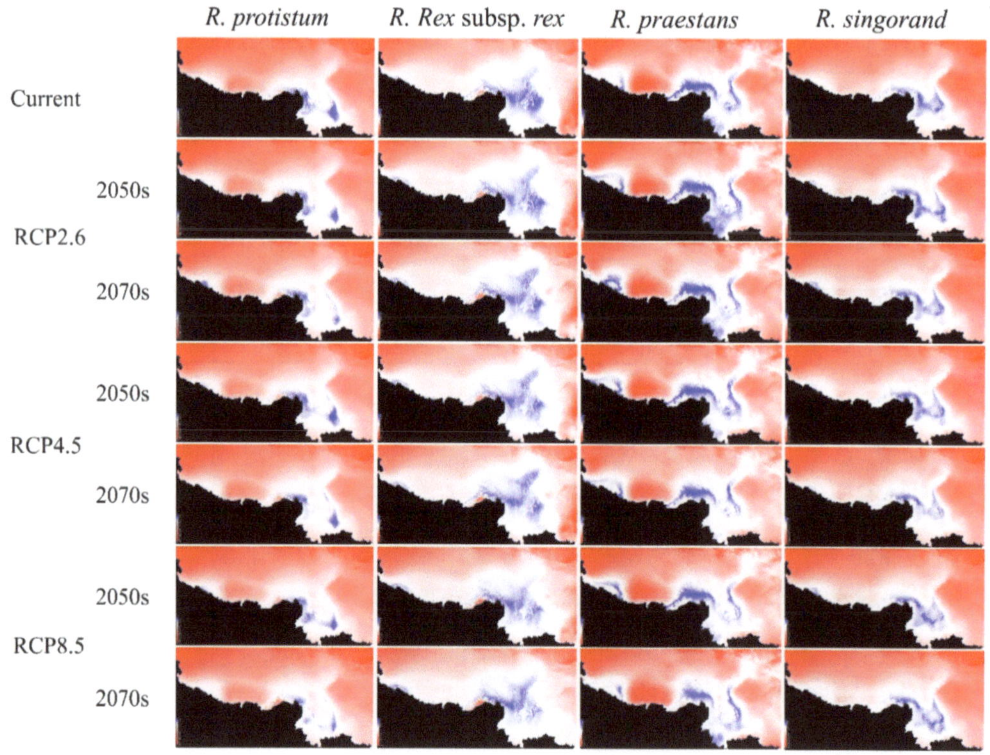

Figure A1. The multivariate environmental similarity surface (MESS) map output for four *Rhododendron* species in different periods under RCP 2.6, RCP 4.5, and RCP 8.5. Areas with different degrees of similarity by the white-blue colors (comprised between 0 and 1) and extrapolation areas are displayed in red.

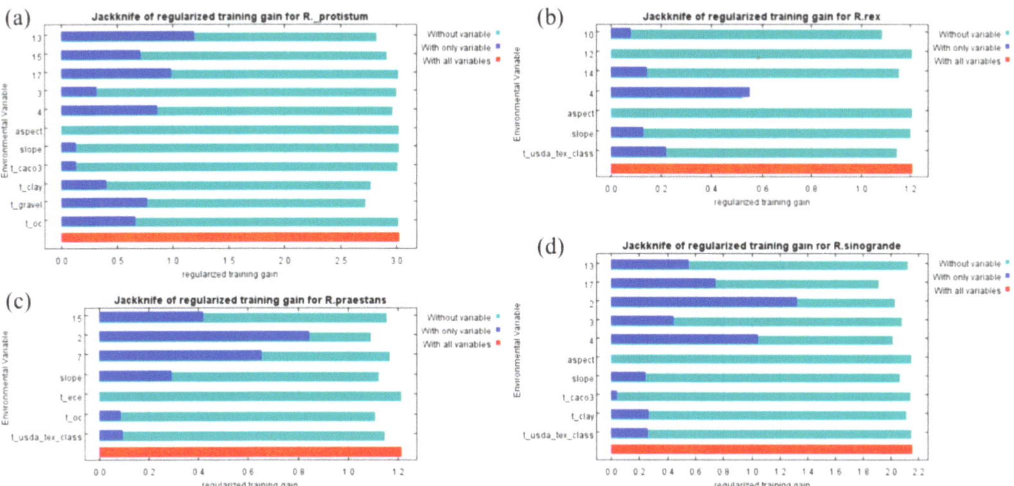

Figure A2. The Jackknife test of variables' contribution for four *Rhododendron* species: (**a**) *R. protistum*, (**b**) *R. rex* subsp. *rex*, (**c**) *R. praestans*, (**d**) *R. sinogrande*.

References

1. Chen, I.C.; Hill, J.K.; Ohlemüller, R.; Roy, D.B.; Thomas, C.D. Rapid range shifts of species associated with high levels of climate warming. *Science* **2011**, *333*, 1024–1026. [CrossRef] [PubMed]
2. Bertrand, R.; Lenoir, J.; Piedallu, C.; Dillon, G.R.; De Ruffray, P.; Vidal, C.; Pierrat, J.C.; Gégout, J.C. Changes in plant community composition lag behind climate warming in lowland forests. *Nature* **2011**, *479*, 517–520. [CrossRef] [PubMed]
3. Moritz, C.; Agudo, R. The future of species under climate change: Resilience or decline? *Science* **2013**, *341*, 504–508. [CrossRef] [PubMed]
4. Pepin, N.; Bradley, R.S.; Diaz, H.F.; Baraër, M.; Caceres, E.B.; Forsythe, N.; Fowler, H.; Greenwood, G.; Hashmi, M.Z.; Liu, X.D. Mountain Research Initiative Elevation-dependent warming in mountain regions of the world. *Nat. Clim. Chang.* **2015**, *5*, 424–430. [CrossRef]
5. Dirnböck, T.; Essl, F.; Rabitsch, W. Disproportional risk for habitat loss of high-altitude endemic species under climate change. *Glob. Chang. Biol.* **2011**, *17*, 990–996. [CrossRef]
6. Hülber, K.; Wessely, J.; Gattringer, A.; Moser, D.; Kuttner, M.; Essl, F.; Leitner, M.; Winkler, M.; Ertl, S.; Willner, W.; et al. Uncertainty in predicting range dynamics of endemic alpine plants under climate warming. *Glob. Chang. Biol.* **2016**, *22*, 2608–2619. [CrossRef] [PubMed]
7. Wyatt, G.E.; Hamrick, J.L.; Trapnell, D.W. Ecological niche modelling and phylogeography reveal range shifts of pawpaw, a North American understorey tree. *J. Biogeogr.* **2021**, *48*, 974–989. [CrossRef]
8. Zhang, Y.; Zhang, X.; Wang, Y.H.; Shen, S.K. De novo assembly of transcriptome and development of novel EST-SSR markers in *Rhododendron rex* Lévl. Through illumina sequencing. *Front. Plant Sci.* **2017**, *8*, 1–12. [CrossRef]
9. Shrestha, N.; Wang, Z.H.; Su, X.Y.; Xu, X.T.; Lyu, L.; Liu, Y.P.; Dimitrov, D.; Kennedy, J.D.; Wang, Q.G.; Tang, Z.Y.; et al. Global patterns of *Rhododendron* diversity: The role of evolutionary time and diversification rates. *Glob. Ecol. Biogeogr.* **2018**, *27*, 913–924. [CrossRef]
10. Ma, Y.P.; Nielsen, J.; Chamberlain, D.F.; Li, X.Y.; Sun, W.B. The conservation of *Rhododendrons* is of greater urgency than has been previously acknowledged in China. *Biodivers. Conserv.* **2014**, *23*, 3149–3154. [CrossRef]
11. Kumar, P. Assessment of impact of climate change on *Rhododendrons* in Sikkim Himalayas using Maxent modelling: Limitations and challenges. *Biodivers. Conserv.* **2012**, *21*, 1251–1266. [CrossRef]
12. Yu, F.Y.; Wang, T.J.; Groen, T.A.; Skidmore, A.K.; Yang, X.F.; Ma, K.P.; Wu, Z.F. Climate and land use changes will degrade the distribution of *Rhododendrons* in China. *Sci. Total Environ.* **2019**, *659*, 515–528. [CrossRef]
13. Lu, Y.P.; Liu, H.C.; Chen, W.; Yao, J.; Huang, Y.P.; Zhang, Y.; He, X.Y. Conservation planning of the genus *Rhododendron* in Northeast China based on current and future suitable habitat distributions. *Biodivers. Conserv.* **2021**, *30*, 673–697. [CrossRef]
14. Kozak, K.H.; Graham, C.H.; Wiens, J.J. Integrating GIS-based environmental data into evolutionary biology. *Trends Ecol. Evol.* **2008**, *23*, 141–148. [CrossRef] [PubMed]
15. Guo, Y.L.; Li, X.; Zhao, Z.F.; Nawaz, Z. Predicting the impacts of climate change, soils and vegetation types on the geographic distribution of *Polyporus umbellatus* in China. *Sci. Total Environ.* **2019**, *648*, 1–11. [CrossRef]
16. Feng, L.; Sun, J.J.; Shi, Y.B.; Wang, G.B.; Wang, T.L. Predicting suitable habitats of *Camptotheca acuminata* considering both climatic and soil variables. *Forests* **2020**, *11*, 891. [CrossRef]

17. Fragnière, Y.; Pittet, L.; Clément, B.; Bétrisey, S.; Gerber, E.; Ronikier, M.; Parisod, C.; Kozlowski, G. Climate Change and Alpine Screes: No Future for Glacial Relict *Papaver occidentale* (Papaveraceae) in Western Prealps. *Diversity* **2020**, *12*, 346. [CrossRef]

18. Hunter-ayad, J.; Seddon, P.J.; Hunter-ayad, J. Reintroduction modelling: A guide to choosing and combining models for species reintroductions. *J. Appl. Ecol.* **2020**, *57*, 1233–1243. [CrossRef]

19. Tariq, M.; Bhatt, S.K.N.I.D.; Bhavsar, D.; Roy, A.; Pande, V. Phytosociological and niche distribution study of *Paris polyphylla* smith, an important medicinal herb of Indian Himalayan region. *Trop. Ecol.* **2021**, *62*, 163–173. [CrossRef]

20. Wani, I.A.; Verma, S.; Mushtaq, S.; Alsahli, A.A.; Alyemeni, M.N.; Tariq, M.; Pant, S. Ecological analysis and environmental niche modelling of *Dactylorhiza hatagirea* (D. Don) Soo: A conservation approach for critically endangered medicinal orchid. *Saudi J. Biol. Sci.* **2021**, *28*, 2109–2122. [CrossRef] [PubMed]

21. Yi, Y.J.; Cheng, X.; Yang, Z.F.; Zhang, S.H. Maxent modeling for predicting the potential distribution of endangered medicinal plant (*H. riparia* Lour) in Yunnan, China. *Ecol. Eng.* **2016**, *92*, 260–269. [CrossRef]

22. Ardestani, E.G.; Tarkesh, M.; Bassiri, M.; Vahabi, M.R. Potential habitat modeling for reintroduction of three native plant species in central Iran. *J. Arid Land* **2015**, *7*, 381–390. [CrossRef]

23. Abdelaal, M.; Fois, M.; Dakhil, M.A.; Bacchetta, G.; El-Sherbeny, G.A. Predicting the Potential Current and Future Distribution of the Endangered Endemic Vascular Plant *Primula boveana* Decne. ex Duby in Egypt. *Plants* **2020**, *9*, 957. [CrossRef] [PubMed]

24. Fu, L.G.; Jin, J.M. *Red List of Endangered Plants in China*; Science Press: Beijing, China, 1992.

25. *Climate Change 2013: The Physical Science Basis. Contribution of Working Group I to the Fifth Assessment Report of the Intergovernmental Panel on Climate Change*; Stocker, T.F.; Qin, D.; Plattner, G.-K.; Tignor, M.; Allen, S.K.; Boschung, J.; Nauels, A.; Xia, Y.; Bex, V. (Eds.) Cambridge University Press: Cambridge, UK; New York, NY, USA, 2013; Volume AR5, p. 1535.

26. Graham, M.H. Confronting multicollinearity in ecological multiple regression. *Ecology* **2003**, *84*, 2809–2815. [CrossRef]

27. Escobar, L.E.; Lira-Noriega, A.; Medina-Vogel, G.; Townsend Peterson, A. Potential for spread of the white-nose fungus (*Pseudogymnoascus destructans*) in the Americas: Use of Maxent and NicheA to assure strict model transference. *Geospat. Health* **2014**, *9*, 221–229. [CrossRef] [PubMed]

28. Zhu, G.P.; Peterson, A.T. Do consensus models outperform individual models? Transferability evaluations of diverse modeling approaches for an invasive moth. *Biol. Invasions* **2017**, *19*, 2519–2532. [CrossRef]

29. Team, R.C. R: A Language and Environment for Statistical Computing. 2018. Available online: https://www/R-project.org (accessed on 1 October 2021).

30. Warren, D.L.; Seifert, S.N. Ecological niche modeling in Maxent: The importance of model complexity and the performance of model selection criteria. *Ecol. Appl.* **2011**, *21*, 335–342. [CrossRef]

31. Kass, J.M.; Vilela, B.; Aiello-Lammens, M.E.; Muscarella, R.; Merow, C.; Anderson, R.P. Wallace: A flexible platform for reproducible modeling of species niches and distributions built for community expansion. *Methods Ecol. Evol.* **2018**, *9*, 1151–1156. [CrossRef]

32. Phillips, S.J.; Dudík, M.; Schapire, R.E. Maxent Software for Modeling Species Niches and Distributions (Version 3.4.1). Available online: http://biodiversityinformatics.amnh.org/open_sourse/maxent/ (accessed on 1 October 2021).

33. Swets, J. Measuring the accuracy of diagnostic systems. *Science* **1988**, *240*, 1285–1293. [CrossRef] [PubMed]

34. Allouche, O.; Tsoar, A.; Kadmon, R. Assessing the accuracy of species distribution models: Prevalence, kappa and the true skill statistic (TSS). *J. Appl. Ecol.* **2006**, *43*, 1223–1232. [CrossRef]

35. Elith, J.; Kearney, M.; Phillips, S. The art of modelling range-shifting species. *Methods Ecol. Evol.* **2010**, *1*, 330–342. [CrossRef]

36. Feng, X.; Lin, C.; Qiao, H.; Ji, L. Assessment of climatically suitable area for Syrmaticus reevesii under climate change. *Endanger. Species Res.* **2015**, *28*, 19–31. [CrossRef]

37. Guillaumot, C.; Moreau, C.; Danis, B.; Saucède, T. Extrapolation in species distribution modelling. Application to Southern Ocean marine species. *Prog. Oceanogr.* **2020**, *188*, 102438. [CrossRef]

38. Brown, J.L. SDMtoolbox: A python-based GIS toolkit for landscape genetic, biogeographic and species distribution model analyses. *Methods Ecol. Evol.* **2014**, *5*, 694–700. [CrossRef]

39. Slatyer, R.A.; Hirst, M.; Sexton, J.P. Niche breadth predicts geographical range size: A general ecological pattern. *Ecol. Lett.* **2013**, *16*, 1104–1114. [CrossRef] [PubMed]

40. Phillips, S.J.; Anderson, R.P.; Dudík, M.; Schapire, R.E.; Blair, M.E. Opening the black box: An open-source release of Maxent. *Ecography* **2017**, *40*, 887–893. [CrossRef]

41. Frishkoff, L.O.; Karp, D.S.; Flanders, J.R.; Zook, J.; Hadly, E.A.; Daily, G.C.; M'Gonigle, L.K. Climate change and habitat conversion favour the same species. *Ecol. Lett.* **2016**, *19*, 1081–1090. [CrossRef] [PubMed]

42. Lenoir, J.; Svenning, J. Climate-related range shifts—A global multidimensional synthesis and new research directions. *Ecography* **2015**, *38*, 15–28. [CrossRef]

43. Anderson, R.P.; Raza, A. The effect of the extent of the study region on GIS models of species geographic distributions and estimates of niche evolution: Preliminary tests with montane rodents (genus *Nephelomys*) in Venezuela. *J. Biogeogr.* **2010**, *37*, 1378–1393. [CrossRef]

44. Guevara, L.; Gerstner, B.E.; Kass, J.M.; Anderson, R.P. Toward ecologically realistic predictions of species distributions: A cross-time example from tropical montane cloud forests. *Glob. Chang. Biol.* **2018**, *24*, 1511–1522. [CrossRef] [PubMed]

45. Eşen, D.; Zedaker, S.M.; Kirwan, J.L.; Mou, P. Soil and site factors influencing purple-flowered rhododendron (*Rhododendron ponticum* L.) and eastern beech forests (*Fagus orientalis* Lipsky) in Turkey. *For. Ecol. Manag.* **2004**, *203*, 229–240. [CrossRef]

46. Tokuoka, Y.; Hayakawa, H.; Hashigoe, K. Spatial distribution and environmental preferences of a threatened species (*Rhododendron uwaense*) and two common species *(R. dilatatum* var decandrum and R. weyrichii) in southwestern Japan. *J. For. Res.* **2020**, *25*, 113–119. [CrossRef]
47. Zhang, X.; Liu, Y.H.; Wang, Y.H.; Shen, S.K. Genetic diversity and population structure of *Rhododendron rex* Subsp. *rex* inferred from microsatellite markers and chloroplast DNA sequences. *Plants* **2020**, *9*, 338. [CrossRef]

Article

Modelling Climatically Suitable Areas for Mahogany (*Swietenia macrophylla* King) and Their Shifts across Neotropics: The Role of Protected Areas

Robinson J. Herrera-Feijoo [1,2,3], Bolier Torres [4,5,*], Rolando López-Tobar [1,6], Cristhian Tipán-Torres [7], Theofilos Toulkeridis [8], Marco Heredia-R [9] and Rubén G. Mateo [3,10,*]

[1] Facultad de Ciencias Agrarias y Forestales, Universidad Técnica Estatal de Quevedo (UTEQ), Quevedo Av. Quito km, 1 1/2 Vía a Santo Domingo de los Tsáchilas, Quevedo 120550, Ecuador
[2] Escuela de Doctorado, Centro de Estudios de Posgrado, Universidad Autónoma de Madrid, C/Francisco Tomás y Valiente, n° 2, 28049 Madrid, Spain
[3] Departamento de Biología, Facultad de Ciencias, Universidad Autónoma de Madrid, Ciudad Universitaria de Cantoblanco, 28049 Madrid, Spain
[4] Facultad de Ciencia de la Vida, Universidad Estatal Amazónica (UEA), Pastaza 160101, Ecuador
[5] Unidad de Posgrado, Universidad Técnica Estatal de Quevedo (UTEQ), Quevedo Av. Quito km, 1 1/2 Vía a Santo Domingo de los Tsáchilas, Quevedo 120550, Ecuador
[6] Escuela Técnica Superior de Ingeniería de Montes, Forestal y del Medio Natural, Universidad Politécnica de Madrid, 28040 Madrid, Spain
[7] Ochroma Consulting & Services, Tena 150150, Ecuador
[8] Department of Earth Sciences and Construction, University of the Armed Forces ESPE, Av. General Rumiñahui S/N, Sangolquí 171103, Ecuador
[9] Facultad de Ciencias Pecuarias y Biológicas, Universidad Técnica Estatal de Quevedo (UTEQ), Quevedo Av. Quito km, 1 1/2 Vía a Santo Domingo de los Tsáchilas, Quevedo 120550, Ecuador
[10] Centro de Investigación en Biodiversidad y Cambio Global (CIBC-UAM), Universidad Autónoma de Madrid, 28049 Madrid, Spain
[*] Correspondence: btorres@uea.edu.ec (B.T.); rubeng.mateo@uam.es (R.G.M.)

Citation: Herrera-Feijoo, R.J.; Torres, B.; López-Tobar, R.; Tipán-Torres, C.; Toulkeridis, T.; Heredia-R, M.; Mateo, R.G. Modelling Climatically Suitable Areas for Mahogany (*Swietenia macrophylla* King) and Their Shifts across Neotropics: The Role of Protected Areas. *Forests* **2023**, *14*, 385. https://doi.org/10.3390/f14020385

Academic Editors: Konstantinos Poirazidis and Panteleimon Xofis

Received: 9 December 2022
Revised: 7 February 2023
Accepted: 10 February 2023
Published: 14 February 2023

Abstract: Mahogany (*Swietenia macrophylla* King) is a species with great economic interest worldwide and is classified as vulnerable to extinction by the IUCN. Deforestation and climate change are the main hazards to this species. Therefore, it is vital to describe possible changes in distribution patterns under current and future climatic conditions, as they are important for their monitoring, conservation, and use. In the current study, we predict, for the very first time, the potential distribution of Mahogany based on data that reflect the total distribution of the species, climatic and edaphic variables, and a consensus model that combines the results of three statistical techniques. The obtained model was projected to future climatic conditions considering two general circulation models (GCM), under two shared socioeconomic pathways (SSP245 and SSP585) for 2070. Predictions under current climatic conditions indicated wide adequate areas in Central American countries such as Mexico and demonstrated a coverage of up to 28.5% within the limits of the protected areas. Under future scenarios, drastic reductions were observed in different regions, particularly in Venezuela, Perú, and Ecuador, with losses of up to 56.0%. On the other hand, an increase in suitable areas for the species within protected areas was also detected. The results of this study are certainly useful for identifying currently unrecorded populations of Mahogany, as well as for identifying locations that are likely to be suitable both now and in the future for conservation management planning. The methodology proposed in this work is able to be used for other forest species in tropical zones as a tool for conducting dynamic conservation and restoration strategies that consider the effects of climate change.

Keywords: mahogany; *Swietenia macrophylla*; neotropic; protected areas; forest monitoring

1. Introduction

Neotropical forests are characterized by hosting multiple biodiversity hotspots [1], particularly in plant terms [2,3]. Current findings indicate that there could be approximately 73,000 species of trees on the planet; of this percentage, it is suggested that some 43% are encountered in the neotropics [4]. These forests are of fundamental relevance in the conservation of biodiversity and mitigation of the effects of climate change, since it is estimated that they allow the capture of 13% of the total annual emissions of anthropogenic CO_2 worldwide [5]. Therefore, various studies have demonstrated that forests in tropical areas provide numerous alternatives focused on the planning and implementation of sustainable forest conservation strategies [4,5]. However, this potential has been limited due to different natural and anthropic disturbances, among the most relevant being the effect of climate change. Recent studies have found that the global mean surface temperature from 1961 to 2019 increased by 0.66 °C [6]. There is scientific evidence that these changes cause shifts and contractions in the altitude range of tropical species [7,8]. For example, a recent investigation, performed on Chimborazo vegetation in Ecuador, indicated a strong displacement of seed plants towards an altitude of 5185 m, which is >500 m to the 4600 m reported by Humboldt in 1802 [9]. On the other hand, global deforestation for agricultural purposes has been very intense in recent decades [10,11], where some 178 million hectares (ha) of forest have been lost from 1990 to 2020, at a rate of 7.8 million ha/year in the period 1990–2000, 5.2 million ha/year in 2000–2010, and 4.7 million between 2010–2020 [12]. Different scientific works have yielded that these tropical deforestation processes cause a reduction in evapotranspiration levels and precipitation at the local level, hence indirectly affecting a wide variety of plant species [13,14]. In many areas, this change in land use begins with the selective exploitation of timber species of high commercial value, an activity that also affects species of global commercial interest, such as Mahogany (*Swietenia macrophylla* King) [15,16], a timber tree whose native range covers much of the Neotropics.

Botanical records indicate that Mahogany is encountered natively in Belize, Brazil, Colombia, Costa Rica, Dominica, Ecuador, El Salvador, French Guiana, Guatemala, Guyana, Honduras, Mexico, Nicaragua, Panama, Peru, Venezuela, and Bolivia [17,18]. Mahogany wood is desired in international markets in the United States, Japan, and Europe due to its high demand for home construction, plywood, boards, boats, and fine cabinetry [19]. In addition, it has chemical properties that are of great importance for multiple uses, such as the production of cosmetics, control of hypertension, and production of antidiabetic, anticancer, and anti-inflammatory substances [20–22]. Due to these multiple uses, it is considered one of the most commercially important plant species worldwide [23], reaching prices between USD 1700 and USD 11,000 per m^3 of wood [24,25]. Its great commercial interest has increased its overexploitation [15]. Mass extraction has generated significant negative impacts on the habitat of this species, with less than one individual per hectare found in small and dispersed populations [26,27]. This species presents a slow growth, and therefore an ineffective natural regeneration, a period of 30 years being necessary for it to be able to develop completely [28]. In addition, it presents a great vulnerability to climatic conditions that affect its germination or pests that limit its growth [29,30]. For all these reasons, it is considered vulnerable to extinction according to the IUCN evaluation categories and is one of the most threatened forest species in the world [31,32]. In addition, due to its conservation status, it was included in Appendix II of the Convention on International Trade in Endangered Species of Wild Fauna and Flora (CITES) [33]. Therefore, there is a strong need to implement forest restoration strategies that allow for the correct management and conservation of Mahogany in the neotropics.

On a global scale, various forest restoration initiatives focused on achieving multiple objectives are being conducted [34,35], including anticipation of the possible effects of climate change [4,36], recovery of biodiversity [5,37], and the enhancement of rural livelihoods [38,39]. We could highlight the Bonn Challenge [40] and the United Nations Decade for Restoration [41], which aim to reforest 350 million hectares by 2030 [42,43].

However, effective planning is necessary to select suitable species for the future, since most countries, especially tropical ones, do not perform climate vulnerability studies in order to select the species and sites to restore. If they would perform these studies, they would achieve effective development and adaptation to a changing world in a context of climate change [44].

Climate change represents a challenge for ecosystems and biodiversity in areas with potential for forest restoration [4,45]. Climate variability, intensity, and increased frequency of extreme weather events could increase tree mortality [46], change phenology [47] and physiology [48], or cause migration to higher and cooler elevations [49] and even an accelerated increase in extinction rates [50]. Different recent studies estimate that between 39 and 41% of plant species are at risk of extinction globally [51,52]. Biological extinctions are advancing at an exceptional rate; this being an important point of interest in the scientific community, some authors even mention the sixth great extinction of species [53]. It is estimated that a large portion of species modify their geographic distribution as a response to changes in climatic conditions [54]. However, the most influential threat facing trees in reforested locations is related to their dispersal capacity [55,56]. The dispersal ability of most trees is limited to external factors such as wind and living organisms [57]. In this sense, a greater incidence of climatic changes and a decrease in dispersal capacity could increase the vulnerability of forest species [58]. This vulnerability also depends on their ability to resist and adapt to the existing environmental conditions [59,60]. For example, generalist species have a higher resistance to various climatic conditions compared to specialist species [61]. Under this premise, it is known that native forest species are able to be both generalists and specialists, which is why they have a high potential to be considered in forest restoration programs [62].

However, restoration programs usually consider exotic forest species, generating large extensions of forest monoculture [63,64]. These practices are not recommended because a forest monoculture presents low carbon sequestration rates compared to those seen in native forests [65]. In addition, they present low biodiversity and potentially high vulnerability to climate change [66]. It is important to point out that exotic species can often become invasive, which is a worrying phenomenon since it is estimated that they rank second in reasons for the current global biodiversity crisis [67,68]. For example, these species tend to compete with native species, reducing local biodiversity, and in some cases their eradication is unfeasible because it is considered impracticable in economic terms [69]. Therefore, international standards recommended by the Society for Ecological Restoration (SER), utilized by researchers, professionals, land managers, community leaders, and decision makers to restore ecosystems in seventy countries, suggest giving greater emphasis to the use of native forest species that are often vulnerable to extinction in tropical areas [70,71]. In addition, it is important to note that these species are usually characterized by their dense wood, which is related to greater carbon sequestration, as is the case of Mahogany [72].

Therefore, the identification of areas where the environmental conditions are suitable for the implementation of these reforestation strategies is of vital importance [44]. Therefore, as an optimal tool for this objective, we propose here the use of ecological modelling techniques, commonly known as species distribution models (SDM [73]). The theoretical principle of these models lies in the fact that they correlate data on the presence of species (georeferenced locations) with environmental variables (climate, soil, etc.) to predict the suitable space for the development of an organism for different periods of time (past, present, future) [73–75]. Among the main applications of these models are projections to future climatic conditions to assess how climate change affects plant species [76,77]. They are undoubtedly the most widely used tools to predict the potential distribution of species in the future, but despite their potential to be used in restoration plans, their use has been very limited in this regard. For example, they have made it possible to identify the distribution of timber forest species facilitating their adequate management and conservation in certain territories [78,79], and recently have facilitated ecological restoration strategies principally in the European continent [79–81].

Previous studies have attempted to model the potential geographic distribution of Mahogany in Mexico [29,82,83] and Brazil [84,85]. However, in these countries, only a partial fraction of the climatic niche of the species was considered (only the presence records for the species in one country). This practice has been habitual, but it is not the most appropriate option to generate reliable SDMs [86], especially to project the potential distribution of the species into the future, since the models created under these considerations generate biased predictions [87–89]. Therefore, in the present work, for the first time the entire known native distribution range of Mahogany is considered, allowing the capture of all the environmental variability available for the species and generating more reliable climate change predictions [90].

In this context, due to the current state of conservation of Mahogany and its commercial importance, this study proposes the identification of suitable areas under present conditions and different climate change scenarios with the potential to be used in restoration and/or reforestation with Mahogany in its native distribution area. Furthermore, the areas predicted by the models were compared with the network of protected areas (PA) with the aim to propose forests management strategies to stakeholders in tropical countries that allow for the conservation of the species in the short and long term.

2. Materials and Methods

2.1. Study Area

The extent of the calibration area has a key impact on the results of the SDMs [91, 92]. We designed our study area based on prior knowledge of the actual distribution of the species in the region from 33° N to 35° S and from 119° W to 35° W (Figure 1). Specifically, we bounded the study area to the south to include Brazil, to the north to include Mexico, to the west by the Pacific Ocean, and to the east by the Atlantic Ocean. The study region is considered a global biodiversity hotspot [93,94]. In this zone the environmental conditions stand out for presenting an annual mean temperature that oscillates between 1 to 29.3 °C [95], annual precipitation between 0 to 7150 mm [95], and altitude in an approximate range of 0 to 6650 m above sea level (m.a.s.l.) [96].

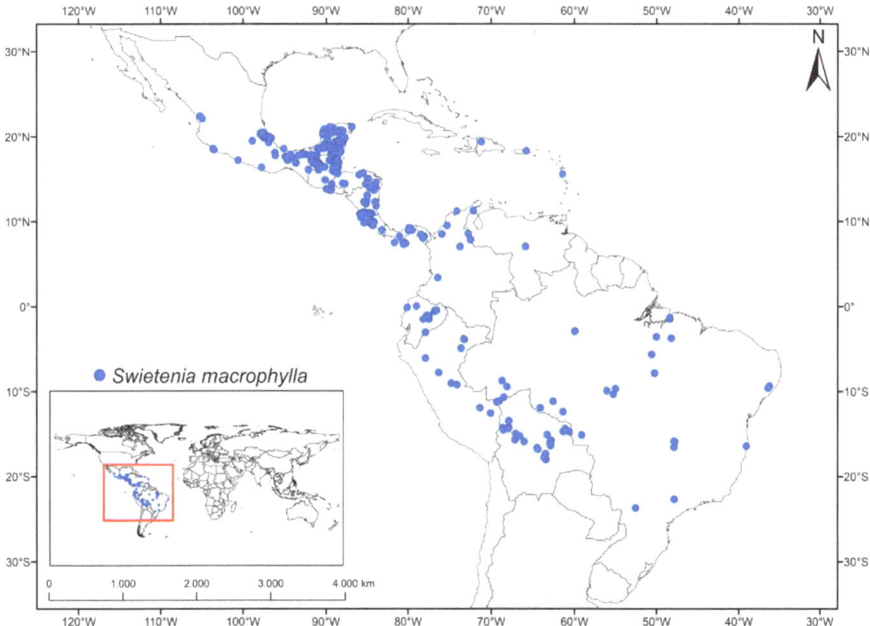

Figure 1. Study area. Presence records of Mahogany (blue dots) in its native distribution area.

2.2. Presence Records

Some 4569 presence records for Mahogany were collected from two types of information sources; online biodiversity databases, such as the Global Biodiversity Information Facility (GBIF; www.gbif.org accessed on 16 January 2023), the Integrated Digitized Biocollections (iDigBio, https://www.idigbio.org accessed on 16 January 2023), sistema distribuído de informação que integra dados primários de coleções científicas (SpeciesLink, https://splink.cria.org.br accessed on 16 January 2023), the Botanical Information and Ecology Network (BIEN, https://bien.nceas.ucsb.edu/bien accessed on 16 January 2023), BIOWEB (https://bioweb.bio/portal accessed on 16 January 2023), and the Programa Nacional de Monitoreo de la Biodiversidad de Ecuador (SINMBIO, https://bndb.sisbioecuador.bio/bndb/index.php accessed on 16 January 2023), as well as scientific publications [17,97–101]. The data download was performed using the following packages: OccCite [102], PlantR [103], BIEN [104], and Ridigbio [105]. In the process of obtaining the data, only records reported in botanical collections were considered, as they are of vital importance to clarify the taxonomy of plants and identify their main habitats. Additionally, these present great potential for the creation of ecological models focused on the conservation of threatened species [106–108].

To ensure the quality of the presence data [109], we relied on the existing literature that provides methodological suggestions with the aim of generating a reliable database that can be used in ecological modelling [86,110,111]. Consequently, a protocol based on three steps was applied, the first step being to initially mitigate the effect of seasonality collection [112] through preserving records collected in years after 1950. Secondly, to take into account that presences collect duplicate samples of the same individuals that are stored in different institutions, records that presented the same geographic coordinates were eliminated. Finally, the possible sampling bias associated with collection patterns was attenuated [113] by applying a distance of 1 km between each record. All the steps conducted in this process were executed through the use of packages developed in the R environment (spThin [114], SpeciesGeoCoder [115]). The final database included 407 presence records.

2.3. Environmental Data

Two databases were used as independent variables. On the one hand, the 19 bioclimatic variables available in the Worldclim 2.1 climate data repository (https://www.worldclim.org/ accessed on 16 January 2023) were used at a resolution of 30 s (approximately 1 km^2) [95]. In order to predict future changes in the distribution areas of *S. macrophylla*, the projections of the general circulation models (GCM) until the year 2070 from the Coupled Model Intercomparison Project Phase 6 (CMIP6, [116]) were considered. Two GCMs (MRI-ESM2.0 [117], MIROC6 [118]) were chosen to consider the variability and uncertainty associated with the creation of future climate models and their influence on spatial predictions [119,120]. Additionally, two shared socioeconomic pathways (SSPs) were used; specifically, SSP2-4.5 and SSP5-8.5 for each GCM. The selected SSPs represent conservative (SSP2-4.5 predicts an increase of up to 2.7 °C by 2100) and pessimistic (SSP5-8.5 predicts an increase of up to 5 °C by 2100) conditions.

On the other hand, layers of different soil properties at a depth of 0 to 5 cm were obtained from the SoilGrids (https://www.soilgrids.org accessed on 16 January 2023) database at a resolution of 1 km^2 [121]. In relation to the selection of variables used in the creation of models, the initial set was reduced to mitigate the effects related to multicollinearity [122], possible pseudo-estimates of predicted areas [89], and interpretative complexity of the models [123]. A Pearson correlation coefficient (r > 0.8) was applied between pairs of variables [124]. Highly correlated variables were removed based on the threshold mentioned above. Finally, the models generated for Mahogany included eleven environmental variables (see Table 1).

Table 1. Environmental variables used in the creation of SDMs for Mahogany.

Category	Variable	Description	Unit
Bioclimatic variables	bio3	Isothermality	$^{\circ}$C
	bio5	Max temperature of warmest month	$^{\circ}$C
	bio6	Min temperature of coldest month	$^{\circ}$C
	bio17	Precipitation of driest quarter	mm
	bio18	Precipitation of warmest quarter	mm
	bio19	Precipitation of coldest quarter	mm
Soil variables	bdod	Bulk density of the fine earth fraction	kg dm^{-3}
	nitrogen	Total nitrogen (N)	g kg^{-1}
	phh2o	pH (H$_2$O)	-
	sand	Sand (>0.05 mm) in fine earth	%
	soc	Soil organic carbon in fine earth	g kg^{-1}

2.4. Species Distribution Models

Consensus models [125] were generated using the biomod2 package [126] by combining predictions generated by three statistical techniques; generalized additive models (GAM, [127]), boosted regression trees (BTR, [128]), and random forests (RF, [129]). Following Brun et al. [123], three different parametrization complexity options were compared. For GAMs we established the degree of freedom in the smooth terms to 1.5, 3, and 10 for simple, intermediate, and complex parameterization options. Simple, intermediate, and complex RF models were trained with the minimum number of observations in the terminal nodes to 40, 20 and 1, respectively. Finally, the complexity of BRTs was varied by training 100, 300, and 10,000 trees for simple, intermediate, and complex options. The predictive accuracy of the models was evaluated by selecting a repeated random sample (ten times) where 80% of the data was used to train the models and the remaining 20% to evaluate them. For each replicate, the accuracy of the model was assessed using the Area Under the ROC Curve (AUC) statistic [130]. The evaluation of the predictive performance of the models aims to assess the accuracy of the predictions, thereby guaranteeing confidence in the results obtained. Those models that presented an AUC value < 0.8 [131] were eliminated. The relative importance of each independent variable was calculated by a repeated random permutation test [126]. Finally, for each parametrization option a consensus model was generated through the weighted average of their AUC values. The three parametrization options were compared by means of different statistics: AUC [130], partial AUC [132], TSS [133], and omission value [134]. The complex option obtained better results (see Table 2). Subsequently, the complex consensus model was projected to each GCM (MRI-ESM2.0, MIROC6) and their respective SPPs (SSP2-4.5 and SSP5-8.5). Next, the median of the replicates was calculated and applied for each future scenario, following the procedure used by Simoes et al. [135]. To calculate the area predicted by the models and conduct the respective comparisons between the current and future predictions, all the models were binarized using a commission error of 5% [136].

Table 2. Comparison of the three parametrization options (complex, intermediate, and simple) employed to generated ensemble models thought different statistics.

Model	AUC Media	TSS	Partial AUC	Omision Rate 5%
Complex	0.93	0.74	1.85	0.050
Intermediate	0.93	0.74	1.71	0.049
Simple	0.93	0.74	1.69	0.051

2.5. Changes in the Potential Geographic Distribution of Mahogany

The differences in the predicted areas between the current model and the two future scenarios (SSP2-4.5 and SSP5-8.5) were calculated. Specifically, areas with potential range loss (i.e., potential areas where the species is currently present but is likely to be absent in the future), gain (i.e., potential areas where the species is currently absent, but is likely to be present in the future), and stability (i.e., potential areas where the species is potentially present now and is likely to be present in the future) were considered. On the other hand, the potential changes in the altitudinal ranges in the potential areas predicted by the future models were also estimated. In this sense, the pixels predicted by the models were used as a mask to extract their corresponding altitude values. In order to perform this process, a 250 m resolution Digital Terrain Model (DTM) derived from the Shuttle Radar Topographic Mission (SRTM) was downloaded [96]. The DTM was resampled to a resolution of 1 km².

2.6. Representativeness of Distribution Areas of Mahogany in the Network of Protected Areas

An intersection was realized between current and future models and a map of the global network of PA was obtained from https://www.protectedplanet.net/en (accessed on 16 January 2023). This occurred based on the premise that PAs provide a greater survival advantage in terms of conservation to endangered species, in addition to being a more effective in situ conservation strategy in the short and long term [137–139].

3. Results
3.1. Statistical Performance of the Models

Three types of approaches used in the generation of the consensus model for Mahogany were evaluated, the complex approach being the one that was statistically reliable, indicating a robust predictive performance since it obtained a mean AUC value = 0.93, TSS = 0.74, partial AUC = 1.85 and an omission value of 0.050 (Table 2). On the other hand, the environmental variables with major relative importance to the SDMs were min temperature of the coldest month (Bio 6), pH, and nitrogen (see Supplementary Figure S1).

3.2. Current Potential Distribution of Mahogany

Under current climatic conditions, the predictions of the binary model yielded climatically suitable areas for Mahogany of approximately 1,250,321 km² (Figure 2), which are found in an altitudinal pattern between 0 and 2663 m.a.s.l. A large part of the predicted areas was distributed in the eastern, southern and western zones of Mexico, Nicaragua, Honduras, Guatemala, Costa Rica, Panama, Bolivia, Belize, southern Colombia, and the coastal region and centre of the Ecuadorian Amazon. Particularly, Mexico stands out for being the country with the highest proportion of suitable areas with 28.5% of the total potential distribution of the species.

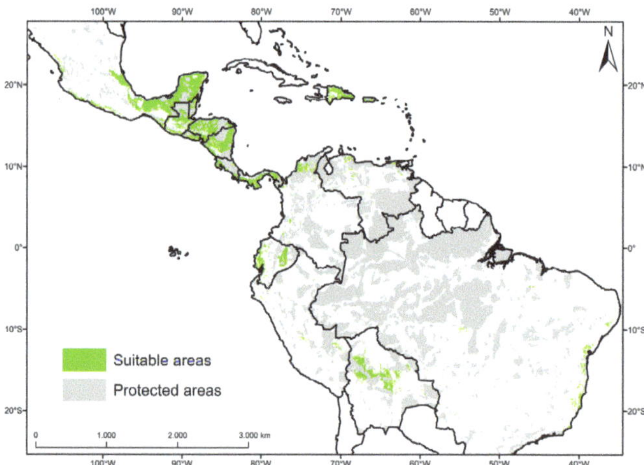

Figure 2. Potentially suitable areas for Mahogany under current conditions predicted in the binary model.

3.3. Possible Changes in the Environmentally Suitable Areas Predicted by Future Mahogany Models

Considering that the resulting models predict fractional and distant potential areas between each country included in the study area, the areas predicted by the models in Ecuador were selected as an example because they predicted significant potential changes (stability, loss, gain) in the future distribution of Mahogany for this country (Figure 3). However, all areas predicted by countries are presented in Table 3.

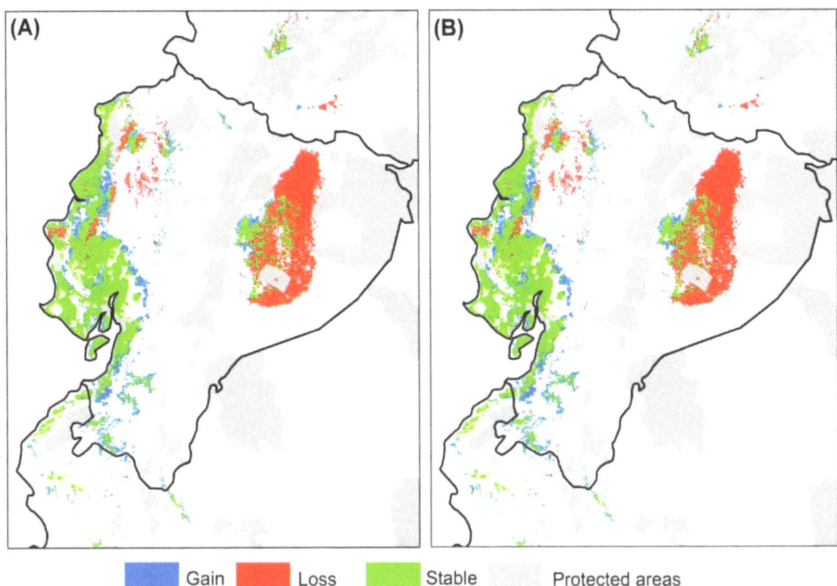

Figure 3. Changes in the potential distribution of Mahogany in the future: (**A**) Potential changes predicted in future climate scenario SSP2-4.5; (**B**) Potential changes predicted in future climate scenario SSP5-8.5.

Table 3. Changes in the potential future distribution of Mahogany in its native range.

Country	Current	SSP2-4.5						SSP5-8.5					
		Stable	%	Loss	%	Gain	%	Stable	%	Loss	%	Gain	%
Belize	26,251	26,196	99.8	55	0.2	0.0	0.0	26,181	99.7	70	0.3	1	0.004
Bolivia	15,2876	114,041	74.6	38,835	25.4	47,652	31.2	118,038	77.2	34,838	22.8	41,876	27.4
Brazil	64,690	54,111	83.6	10,579	16.4	127,285	196.8	47,707	73.7	16,983	26.3	102,627	158.6
Colombia	65,862	54,472	82.7	11,390	17.3	23,827	36.2	47,849	72.7	18,013	27.3	33,427	50.8
Costa Rica	38,259	35,473	92.7	2786	7.3	1606	4.2	33,641	87.9	4618	12.1	2542	6.6
Dominica	834	771	92.4	63	7.6	0.0	0.0	706	84.7	128	15.3	1	0.1
Ecuador	56,900	33,772	59.4	23,128	40.6	11,050	19.4	32,817	57.7	24,083	42.3	20,670	36.3
El Salvador	19,146	19,091	99.7	55	0	3474	18.1	18,945	99.0	201	1.0	3452	18.0
Guatemala	89,556	86,531	96.6	3025	3.4	11,110	12.4	85,836	95.8	3720	4.2	15,023	16.8
Honduras	116,168	115,535	99.5	633	0.5	10,008	8.6	113,661	97.8	2507	2.2	12,299	10.6
México	355,755	342,038	96.1	13,717	3.9	58,797	16.5	324,593	91.2	31,162	8.8	59,811	16.8
Nicaragua	132,988	130,865	98.4	2123	1.6	2469	1.9	129,975	97.7	3013	2.3	3149	2.4
Perú	18,544	10,896	58.8	7648	41.2	17,550	94.6	11,631	62.7	6913	37.3	25,814	139.2
Panamá	61,645	54,667	88.7	6978	11.3	1693	2.7	56,104	91.0	5541	9.0	3672	6.0
Puerto Rico	6928	5517	79.6	1411	20.4	424	6.1	4269	61.6	2659	38.4	219	3.2
Venezuela	43,919	27,086	61.7	16,833	38.3	18,111	41.2	19,326	44.0	24,593	56.0	34,255	78.0
Total	1,250,321	1,111,062	88.9	139,259	11.1	335,056	26.8	1,071,279	85.7	179,042	14.3	358,838	28.7

Based on future climatic conditions with a moderate emissions scenario (SSP2-4.5), the models suggest that around 88,9% of suitable habitat for Mahogany would remain stable (i.e., suitable in the current period and in the future). It is important to point out that the largest proportion of stable areas would be found in Mexico with approximately

342,038 km^2. However, we encountered that in Belize it is expected that 99.8% of the suitable areas will remain stable in the future. On the other hand, Peru is the country with the lowest number of stable future areas with 58.8%. Similarly, in a severe emissions scenario (SSP5-8.5), a 3.1% reduction in suitable stable areas is anticipated compared to that predicted under the conservative scenario (SSP2-4.5), reflecting a percentage of habitat stability for 88,8% of species. Specifically, patterns similar to those described above were identified in the SSP2-4.5 scenario, with Belize once again being the country with the highest proportion of possibly stable areas in the future with 99.7%. Similarly, in El Salvador the species is also expected to have large areas of suitable areas in the future with 99.0%.

On the other hand, in relation to the loss of suitable areas, there was a drastic loss of 41.2% of suitable areas in Peru and 40.6% in Ecuador in a conservative scenario (SSP2-4.5). Particularly in Ecuador, these reductions are expected to be more evident in the Amazon region of Ecuador; specifically, in the provinces of Pastaza and Napo. Similarly, in a gloomy emissions scenario (SSP5-8.5), an increase of 3.2% is expected compared to the expected losses reported by the model for the scenario (SSP2-4.5). Overall, the models foresee possible reductions equivalent to 14.3% in the species' range. Particularly, in this scenario, Venezuela is expected to lose 56.0% of the areas suitable for the species. On the other hand, in second place would again be Ecuador with a predicted loss rate of 42.3% of suitable areas for Mahogany.

The models also forecast possible expansion (gain) of suitable areas. Under an optimistic emissions scenario (SSP2-4.5), potentially suitable habitat gains for the species are anticipated to be 26.8% across its entire native range. The largest proportion of forecast gains are again centered in Brazil, Peru, Venezuela, and Colombia. Furthermore, it should be emphasized that the models did not predict suitable new areas in Belize and Dominica, both of which had a success rate of 0%. The potential increases in suitable areas under a pessimistic emissions scenario (SSP5-8.5) are expected to be greater than those evidenced in comparison to the conservative scenario (SSP2-4.5) with an increase of 1.9%. In general, the models predict gains of potentially suitable areas for the species of 28.7%. On the other hand, the models forecast possible slight area gains in Belize and Dominica with 0.004% and 0.1, respectively. In this context, it is expected that the species will have new areas for adaptation and survival through a progressive ascent of 282 m in an optimistic scenario (SSP2-4.5) and 792 m in a pessimistic scenario (SSP5-8.5).

3.4. Representativeness of Distribution Areas of Mahogany in the Network of Protected Areas

From the intersection of the potentially suitable areas predicted by the models under current and future conditions with the PA network, it was demonstrated that the proportion of suitable habitat for Mahogany under current conditions that would be available within the PA limits would be some 28.5% (Table 4).

Specifically, this coverage of suitable areas was found in greater proportion in Venezuela, Colombia, and Costa Rica. These findings suggest that the species may currently be under conservation in the areas predicted by the models. However, a slight coverage of suitable areas under conservation for the species was also observed in Puerto Rico and Ecuador, with only 4.7% and 5.9%, respectively.

In future climate conditions with an optimistic scenario (SSP2-4.5), the models demonstrated a broad coverage of stable areas within PAs of 89.24%. Thus, in Belize 99.7% of the suitable areas previously predicted by the current model would remain stable. On the other hand, in a dim scenario (SSP5-8.5) a slight reduction (3.51%) in suitable stable areas is anticipated compared to the forecasts provided by the scenario (SSP2-4.5). In general, a stability of suitable habitat within PAs of 85.73% is expected.

Table 4. Changes in the potential future distribution of Mahogany within the network of protected areas.

Country	Current	Protected Areas	%	SSP2-4.5						SSP5-8.5					
				Stable	%	Loss	%	Gain	%	Stable	%	Loss	%	Gain	%
Belice	26,251	9712	37.0	9681	99.7	31	0.3	0	0.0	9676	99.6	36	0.4	0	0.0
Bolivia	152,876	55,563	36.3	44,194	79.5	11,369	20.5	15,142	27.3	44,577	80.2	10,986	19.8	16,745	30.1
Brasil	64,690	13,714	21.2	10,687	77.9	3027	22.1	53,598	390.8	10,251	74.7	3463	25.3	43,045	313.9
Colombia	65,862	30,973	47.0	26,192	84.6	4781	15.4	11,095	35.8	24,142	77.9	6831	22.1	14,073	45.4
Costa Rica	38,259	17,295	45.2	15,985	92.4	1310	7.6	1330	7.7	15,177	87.8	2118	12.2	2171	12.6
Dominica	834	189	22.7	183	96.8	6	3.2	0	0.0	185	97.9	4	2.1	1	0.5
Ecuador	56,900	3384	5.9	1873	55.3	1511	44.7	582	17.2	1958	57.9	1426	42.1	911	26.9
El Salvador	19,146	2909	15.2	2877	98.9	32	1.1	877	30.1	2837	97.5	72	2.5	1008	34.7
Guatemala	89,556	37,037	41.4	36,836	99.5	201	0.5	906	2.4	36,645	98.9	392	1.1	1429	3.9
Honduras	116,168	31,004	26.7	30,846	99.5	158	0.5	2907	9.4	30,397	98.0	607	2.0	3900	12.6
México	355,755	61,281	17.2	58,697	95.8	2584	4.2	6537	10.7	54,174	88.4	7107	11.6	5830	9.5
Nicaragua	132,988	50,831	38.2	49,997	98.4	834	1.6	739	1.5	49,127	96.6	1704	3.4	631	1.2
Perú	18,544	6255	33.7	3809	60.9	2446	39.1	5738	91.7	3869	61.9	2386	38.1	7278	116.4
Panamá	61,645	12,590	20.4	9828	78.1	2762	21.9	541	4.3	10,276	81.6	2314	18.4	1248	9.9
Puerto Rico	6928	324	4.7	232	71.6	92	28.4	57	17.6	192	59.3	132	40.7	44	13.6
Venezuela	43,919	23,190	52.8	16,016	69.1	7174	30.9	11,670	50.3	11,931	51.4	11,259	48.6	24,905	107.4
Total	1,250,321	356,251	28.5	317,933	89.24	38,318	10.76	111,719	31.36	305,414	85.73	50,837	14.27	123,219	34.59

Based on future scenarios, the models indicated small reductions (loss) of habitat for the species within PAs. In the moderate emissions scenario (SSP2-4.5), a coverage loss of 10.76% was reported. Of this percentage, the largest proportion of reductions is expected to be visible in Ecuador with a loss rate of 44.7% of suitable areas. Furthermore, with a scenario of not very encouraging emissions (SSP5-8.5), a double of losses is forecasted compared to what is indicated in the SSP2-4.5 scenario. Hence, the models suggest losses of suitable habitat of 14.27% within PAs for the species. In particular, it was shown that Venezuela would have significant reductions in suitable areas under conservation with predicted losses of 48.6%.

Finally, the results predict that the species will have new areas potentially suitable for its conservation within the PA network currently in force in the future. A coverage of 31.36% was evidenced considering the models under the emissions scenario SSP2-4.5. Particularly, these area increases are expected to be most notable in Brazil, Peru, and Venezuela. On the contrary, it was also possible to notice null rates (0%) of gain of areas under conservation for the species in Belize and Dominica. Nonetheless, considering the emissions scenario SSP5-8.5, the models foresee an increase in suitable areas possibly under conservation in the future within PAs of 34.59%. It is expected that these increases will be observed in greater proportion in Brazil, Peru, Venezuela, and Colombia. However, in Ecuador a radical increase in suitable habitat was evidenced with a coverage of 313.9% within PAs for the species.

4. Discussion

4.1. Predictive Performance and Strengths of the Models

In the last decade, multiple recommendations have been formulated to improve the quality of species distribution models, thereby improving their reliability, replicability, and use in different forest restoration initiatives [80]. In the present study, ecological modelling techniques that take into consideration global recommendations for the reliable application of species distribution models were used [110,140]. All this facilitated the mitigation of possible problems, such as (1) errors associated with sampling bias and spatial aggregation [141]; (2) effects related to multicollinearity [122]; possible pseudo-estimates of predicted areas [91]; and interpretive complexity of the models [123].

In addition, it has been demonstrated that there is a wide variability in the predictions resulting from the models when using various algorithms individually. This is due to the fact that they are susceptible to the configurations and parameterizations used [142]. Previous studies conducted about Mahogany use the maximum entropy algorithm (Maxent) as the only algorithm to carry out the modelling of the species [29,82–85]. We imply that this selection may be based mainly on the fact that Maxent is one of the algorithms most explored and used in the existing literature [143]. However, recent research also suggests that Maxent's predictive power could be unreliable and possibly biased when making projections to novel future climates, leading to overestimates of predicted areas beyond previously known ranges for the species and under unfavorable climatic conditions for its adaptation [144,145]. Therefore, here, a consensus model approach was used to model the current and future distribution of Mahogany [125]. This approach has been evaluated and the findings suggest that combining predictions from multiple algorithms could improve predictions compared to models generated by separate algorithms [146,147]. Furthermore, our models considered the inclusion of soil variables, since recent evidence suggests that the use of soil variables helps improve current predictions and reduces overestimations in future predictions [148–150]. All of the above could be verified, since the selected modelling approach, after the evaluation process of its predictions, was statistically reliable (see Table 2). These qualities give high confidence in the predicted adequate area for this species in its native distribution range, and they could thus be considered for use in different approaches to conservation and forest restoration. However, it will be essential to confirm the actual presence of Mahogany through the implementation of forest inventories to evaluate the present findings.

One of the main achievements of this study was to include in the creation of the models the entire known native distribution range of Mahogany for the first time. Previous research has been done to model the potential geographic distribution of Mahogany in Mexico [17,29,83,99,100] and Brazil [84,85]. However, locally calibrated SDMs are limited by partial representation of the ecological niche of the species and therefore future projections may be biased [87–89]. Nonetheless, the models generated in this study fully represent the climatic niche of Mahogany, which allows for the capture of all the environmental variability available to the species [54] and therefore allows for more reliable models for the present and projections with less future over-prediction or under-prediction errors [90]. In this way, we consider that the information presented by our models is extremely important in terms of forest restoration and adaptation to the possible effects of climate change, since it allows for the characterization of potentially suitable areas in current and future conditions for the species, which facilitates the maximization of the potential success of conservation strategies and forest landscape restoration programs [65].

4.2. Model Limitations

It is necessary to point out that there could be certain limitations related to the presence set and the non-consideration of biotic interactions in the generation of our model. Initially, we highlight that the set of occurrences compiled for Mahogany for its native distribution area considered all possible open access databases of biodiversity and records published in scientific articles. However, the uncertainty associated with the spatial precision associated with the data must be considered [151,152]. Hence, the possible sampling biases associated with the greater number of botanical inventories in regions, mainly in Central America, must be acknowledged. These patterns are fundamentally associated with accessibility (roads, rivers, etc.) to the sampling sites [153]. All these points were considered in our study in the data cleaning process based on the methodological suggestions recommended for the development of species distribution models [110,111,140], with the aim of achieving the most reliable model possible. However, we ascertain that our model did not consider the possible biotic interactions that could exist for the species in its native area. For example, a forest inventory conducted in native populations of Mahogany in the province of Pastaza yielded that 93% of the sampled seedlings indicated signs of herbivory in 50% of their leaves [154]. This demonstrates that the species presents a great vulnerability to herbivores that limits its growth and the survival of individuals in its populations [29,30]. Research focused on the phytosanitary control of Mahogany indicated that the most relevant biological threat to the species would be the insect *Hypsipyla grandella* (Lepidoptera: Pyralidae) [155,156]. The larvae of this species attack the terminal buds, particularly of young trees, generating a defective growth of the species which decreases its commercial value. Even if the attacks are not frequent, the intensity of the attacks can cause the death of the plant [157,158].

In addition, it is known that a species can be influenced by different biotic interactions that are able to prevent it from occupying the entire area that corresponds to its ecological niche [159]. Therefore, considering all the possible biotic interactions known for the species to be modelled will provide additional information that could be key to anticipating whether these interactions could be positive or negative for the survival of the species [160,161]. Furthermore, recent research indicates that the predictive power of species distribution models could be improved when biotic interactions between species are considered [162,163]. However, at present, integrating these interactions is still a challenge, as more research is needed to determine whether the incorporation of these variables should be continuous or binary in the development of models [164]. We suggest that it would be necessary to explore in greater depth the influence of this pest on the distribution of Mahogany through the participation of experts in different forest disciplines [165], with a view to precisely characterizing the possible negative effects of this interaction in the future.

4.3. Current Potential Distribution of Mahogany

Based on current conditions, our model suggests that the species could have climatically suitable areas in 16 countries from Central America to South America, with suitable areas being found in greater proportion in Mexico and Bolivia. These findings are consistent with the known distribution ranges previously described for the species [17,99,100]. Mexico has been one of the countries with the largest number of investigations focused on understanding the distribution patterns of Mahogany [82,83,166–168]. This is mainly related to the more significant number of occurrence records for the species in this country. Regarding the patterns described in this study, the current model demonstrates adequate areas in 18 states of Mexico, representing 28.5% of the total distribution of the species. The states with the highest proportion of suitable areas were Campeche, Quintana Roo, and Yucatán. These areas reported by our models are similar to those predicted by models developed in previous research in Mexico and the province of Yucatán [82,83,167].

In relation to the predicted areas in Brazil, it was possible to demonstrate that these only represent 5.2% of the total distribution of the species. In addition, it is important to point out that the areas suitable for the species indicate a widely discontinuous and fragmented pattern, which is similar to that reported in previous studies where it is established that the species would be almost in danger of extinction due to the fact that its populations are very restricted and fragmented [169,170]. Furthermore, considering the spatial distribution of suitable areas in Brazil, previous studies reported that the state of Acre had the most suitable climatic conditions for the species [84,85]. These areas do not coincide with the results obtained in this study, as our models suggest that the largest proportion of areas reported for Brazil would be encountered in the states of Bahia and Espírito Santo. However, both studies indicated a concordance in the prediction of the absence of suitable areas for the species in the north-western region of Amazonas. This pattern could be associated with the presence of large extensions of tributaries and rivers, as is the case of the Amazon [171,172]. On the other hand, the model also reported similar patterns of absence in the north-central region of Colombia. These findings seem to be induced by the presence of large foothills with altitudes which the species had not been able to access for biogeographical reasons and by the lack of suitable environments under current conditions [17,168,173].

4.4. Changes in the Potential Geographic Distribution of Mahogany in the Future

Overall, under future climatic conditions, our models predicted that 88.9% of the areas suitable for Mahogany will remain stable throughout most of its range. For example, the stable areas reported in Mexico were similar to those shown in recent studies where they mention that areas such as Quintana Roo, Yucatán, and Calakmul would be potentially suitable for the habitat of the species in the face of climate change [26,29,83]. On the contrary, our models allowed us to identify that Peru and Ecuador were the countries with the lowest proportion of potentially stable areas with 58.8% and 59.4%, respectively. These reductions are expected to be detected mainly in the Amazon region of Ecuador, specifically in the provinces of Pastaza and Napo. The reductions reported by our models should be of great interest in decision making by local managers since, in Ecuador, it is estimated that under current conditions the suitable areas available in its extent of occurrence (EOO) would have been reduced by a percentage equal to or greater than 80% over the last thirty years. The species would thus be currently considered critically endangered (CR A2cd) in this country [154]. In this context, our findings suggest that Mahogany populations currently reported in the provinces of Pastaza, Napo, Orellana, and Morona Santiago [154] would be potentially prone to extinction.

Nevertheless, our models forecast possible gains from suitable areas going forward, reporting gains of 26.8% to 28.7% for the conservative (SSP2-4.5) and pessimistic (SSP5-8.5) emissions scenarios, respectively. These gains were forecast in greater proportion in Brazil, Peru, Colombia, Venezuela, and Bolivia. However, an encouraging scenario was also reported for Ecuador with small gains in areas in the coastal region, specifically in the provinces of Guayas, Manabí, Esmeraldas, and El Oro. It was demonstrated that the gains in areas occurred in an altitude range higher than that previously reported by our model under current conditions, which was from 0 to 2663 m.a.s.l. Thus, it is expected that the species will have new areas for its adaptation and survival through a progressive ascent of 792 m over time. It is estimated that the increase in global temperature would induce a displacement of forest species towards areas that were previously colder, where environmental circumstances may be more conducive to the adaptation of populations of the species in the future [174,175]. However, this process may be slow considering that the transition of the flora towards higher altitude ranges depends directly on its dispersal capacity [176–178].

This consideration could not be beneficial for the species as the pressure imposed by human activities advances at an accelerated rate compared to biological adaptation processes. For example, in the Amazonian region of Ecuador (ARE), activities such as deforestation [179], land use change [180], mining [181], and urban expansion [182] could have an even greater impact on the reduction of suitable accessible habitat for Mahogany. Therefore, it is interesting to mention that evidence of the influence of these activities has recently been discovered, since recent population studies conducted in the province of Pastaza found only seven individuals in three plots of 1.7 hectares [154], which indicates that the populations of the species could be under great anthropic pressure. However, these figures are even more worrying in inventories performed in Brazil and Mexico, where less than one individual per hectare was found in small and scattered populations [26,27].

4.5. Representativeness of the Potential Distribution of Mahogany in the Network of Protected Areas

Among all the investigations previously conducted for the species considering the use of SDMs in Mexico and Brazil [29,82–85], the current study stands out for being the first to perform an intersection analysis between the predicted areas under current and future conditions and the PA network. This analysis was performed in order to identify areas where the species could be protected under immediate short-term and long-term conservation statuses, as this could be one of the best in situ conservation strategies [139]. In this context, with the current climatic conditions, our models suggest that 28.5% of the areas predicted for Mahogany would be under conservation within the limits of the PA network, with Venezuela, Colombia, and Costa Rica being the countries where the species would find greater opportunity for immediate conservation. These results are extremely important, given that PAs have been considered one of the best strategies for biodiversity conservation, particularly for threatened forest species [78,183,184]. The effectiveness of PAs lies in the fact that they provide adequate conditions for the establishment and survival of many species from different taxonomic groups, in addition to mitigating the effects of overexploitation and selective felling of forest species associated with deforestation processes [185,186]. This is demonstrated by recent studies which have shown greater diversity within PAs than outside them [187]. Therefore, the results evidenced by the model for Mahogany suggest that the species would have optimal conditions for its conservation in these areas.

Subsequently, a slight proportion of suitable areas under conservation was also identified in Ecuador, where our models suggest that 5.9% of suitable areas would be found within the existing PAs for this country. However, this percentage disagrees with that suggested by Iglesias et al. [154], who indicate that the species has not yet been detected within the national system of PAs of Ecuador (SNAP, according to its acronym in Spanish). The models detected habitat reductions (losses) for the species within PAs throughout its native range. Such reductions are expected to be 44.7% to 48.6% for the SSP2-4.5 and SSP5-8.5 scenarios, respectively. Interestingly, the models indicate that possibly part of these area reductions would occur in Ecuador.

In Ecuador, the SNAP covers about 20% of Ecuador's surface with 69 PAs distributed in its four regions (coastal lowland, highlands, Amazonian lowland, and Galápagos). In addition, Ecuador in 2008 implemented the "Programa Socio Bosque", aimed at the conservation of forest resources inside and outside PA and buffer zones [188]. Despite all these initiatives, Ecuador is one of the countries in South America with the greatest expansion of road infrastructure inside and outside the SNAP [182,189]. Globally, an estimated 14.8% of tree species would fall within areas under high human pressure, even within existing PAs [190]. In Ecuador, it is estimated that 72% of the 4437 threatened native vascular plants would be found outside the areas delimited in the current SNAP [191,192]. Therefore, it is known that the least protected species are usually the most threatened [193], as is the case of Mahogany. We suggest that this could be even more serious for Ecuador, considering the increase in mining activities [181]. Additionally, it is necessary to emphasize that Mahogany has a very slow growth rate, fully developing in an average of 30 years [28]. For this reason, PAs play a key role in its in-situ conservation, since they should be areas under continuous conservation, which could provide the species with a safe habitat for its survival and development in the short and long term.

Within such future climatic conditions, our models evidenced a possible increase of 390.8% for the SSP2-4.5 scenario and 313.9% for the SSP5-8.5 scenario. In both scenarios, the models anticipate that profits will be higher in Brazil, Peru, and Venezuela. Particularly, in Ecuador a radical increase in suitable habitat was evidenced with a coverage of 26.9% within PAs for the species. These results could be very encouraging considering the forecasts predicted by our models for the species in the future. Ecuador is one of the countries with the highest global conservation priority that currently requires more connectivity between existing PA areas in the SNAP [192,194]. Recent studies suggest that improving connectivity between PAs could be one of the best initiatives to safeguard current biodiversity and enhance adaptive capacity in the context of future climate change [79,195,196]. However, it is also necessary to improve monitoring within PAs through new current initiatives based on the use of remote sensing for the detection of deforestation hotspots and selective logging [197,198], such as drones applied to monitoring the current state of forests [199,200]. Nonetheless, it is also necessary to encourage environmental education to raise awareness and promote the conservation of forest resources, knowledge of the impacts of human activities on forests, alternative uses of non-timber forest resources, and the possible effects induced by global climate change [201].

4.6. Considerations for Current and Future Monitoring of Mahogany Populations

Herbarium specimens provide essential information that can be used to increase knowledge of Mahogany distribution patterns. However, the limited availability of digitized herbarium collections available for plants is one of the current problems [106,202]. It is estimated that ~36% of plants have little information on their distribution in global herbaria, and that between 11.2 and 36.5% of these species have less than five reported observations [203]. Therefore, it would be essential to increase botanical sampling efforts for the species in tropical areas. In addition, in Ecuador, for example, it is necessary to increase efforts to digitize existing botanical collections in national herbaria. In turn, these data can be shared with the scientific community through open access databases in Ecuador (BIOWEB, SISBIO) and worldwide (GBIF, IDIGBIO, etc.). On the other hand, the continuous monitoring of all forest species through the National Forestry Inventory of Ecuador (ENFI, according to acronym in Spanish) should be promoted [204].

Nonetheless, currently many of the efforts focused on the survey and inventory of specimens are based on geographical decisions that consider distance, accessibility, and available economic resources [205]. As a result, in most inventoried species there is evidence of different types of sampling bias that limit biological knowledge of them outside of inaccessible areas which could be climatically suitable [153,206,207]. In this way, the use of methodologies that allow for the identification of optimal areas under geographic and environmental considerations that facilitate the documentation of the greatest number of specimens with fewer economic resources and less human effort [208] is key to obtaining a completer and more enriched database.

Given this urgent need, new methodologies implemented in packages developed for the R programming environment have now emerged which could help address current necessities. One of these packages is WhereNext [209], which is based on the use of a generalized model of dissimilarity. The principle of this model is the understanding of the compositional dissimilarity of two sites through the differences between their conditions and the geographic distance between them. This allows it to identify areas that decrease the average distance between the previously sampled sites and those not yet sampled [210]. On the other hand, another current alternative is the Biosurvey package, which uses a methodology that considers the environmental conditions available in the study areas and records of the presence of the species [211]. In addition, Biosurvey provides the opportunity to evaluate previously selected post-analysis sites, with climatically suitable areas identified through species distribution models [211]. However, the authors suggest that the areas described by the SDM to be used should be previously verified by a group of experts in the species of interest [165]. Based on this context, in this study we suggest considering the areas predicted by our models under current conditions to be used in future research that could be implemented through the previously described packages, thus achieving the detection of new sites that complement the collections available for the species.

Supplementary Materials: The following supporting information can be downloaded at: https://www.mdpi.com/article/10.3390/f14020385/s1, Figure S1: Relative importance of each independent variable inside the species distribution models. We calculated the mean value (10 replicates models) for each sampling strategy, variable, modelling technique (GLM, BRT, and RF).

Author Contributions: Conceptualization, R.J.H.-F., R.G.M. and B.T.; methodology, R.J.H.-F. and R.G.M.; software, R.J.H.-F. and R.G.M.; validation, R.J.H.-F., B.T. and R.G.M.; formal analysis, R.J.H.-F. and R.G.M.; investigation, R.J.H.-F., B.T., C.T.-T., T.T., M.H.-R., R.L.-T. and R.G.M.; data curation R.J.H.-F.; writing—original draft preparation, R.J.H.-F., C.T.-T., T.T., M.H.-R. and R.L.-T.; writing—review and editing, R.J.H.-F., B.T., C.T.-T., T.T., M.H.-R., R.L.-T. and R.G.M.; supervision R.J.H.-F., B.T. and R.G.M. All authors have been involved in developing, writing, commenting, editing, and reviewing the manuscript. All authors have read and agreed to the published version of the manuscript.

Funding: This research received no external funding.

Institutional Review Board Statement: Not applicable.

Informed Consent Statement: Not applicable.

Data Availability Statement: This is not applicable as the data are not in any data repository of public access; however, if editorial committees needs access, we will happily provide them with it. Please use this email: rherreraf2@uteq.edu.ec.

Acknowledgments: The authors would like to thank the Faculty of Agricultural and Forestry Sciences, Quevedo State Technical University (UTEQ), and Department of Biology, Universidad Autónoma de Madrid, for supporting all the production stages of this document, as well as to all botanists for their efforts in botanical collections in the neotropics.

Conflicts of Interest: The authors declare no conflict of interest.

References

1. Raven, P.H.; Gereau, R.E.; Phillipson, P.B.; Chatelain, C.; Jenkins, C.N.; Ulloa Ulloa, C. The distribution of biodiversity richness in the tropics. *Sci. Adv.* **2020**, *6*, eabc6228. [CrossRef]
2. Ulloa Ulloa, C.; Acevedo-Rodríguez, P.; Beck, S.; Belgrano, M.J.; Bernal, R.; Berry, P.E.; Brako, L.; Celis, M.; Davidse, G.; Forzza, R.C. An integrated assessment of the vascular plant species of the Americas. *Science* **2017**, *358*, 1614–1617. [CrossRef] [PubMed]
3. Ter Steege, H.; Prado, P.I.; de Lima, R.A.F.; Pos, E.; de Souza Coelho, L.; de Andrade Lima Filho, D.; Salomão, R.P.; Amaral, I.L.; de Almeida Matos, F.D.; Castilho, C.V. Biased-corrected richness estimates for the Amazonian tree flora. *Sci. Rep.* **2020**, *10*, 10130. [CrossRef]
4. Koch, A.; Kaplan, J.O. Tropical forest restoration under future climate change. *Nat. Clim. Chang.* **2022**, *12*, 279–283. [CrossRef]
5. Da Rosa, C.M.; Marques, M.C.M. How are biodiversity and carbon stock recovered during tropical forest restoration? Supporting the ecological paradigms and political context involved. *J. Nat. Conserv.* **2021**, *65*, 126115. [CrossRef]
6. Valipour, M.; Bateni, S.M.; Jun, C. Global surface temperature: A new insight. *Climate* **2021**, *9*, 81. [CrossRef]
7. Zu, K.; Wang, Z.; Zhu, X.; Lenoir, J.; Shrestha, N.; Lyu, T.; Luo, A.; Li, Y.; Ji, C.; Peng, S. Upward shift and elevational range contractions of subtropical mountain plants in response to climate change. *Sci. Total Environ.* **2021**, *783*, 146896. [CrossRef]
8. Ramírez-Barahona, S.; Cuervo-Robayo, Á.P.; Feeley, K.; Ortiz-Rodríguez, A.; Vásquez-Aguilar, A.; Ornelas, J.F.; Rodríquez-Correa, H. Climate change and deforestation drive the displacement and contraction of tropical montane cloud forests. *Biol. Sci.* **2021**. [CrossRef]
9. Morueta-Holme, N.; Engemann, K.; Sandoval-Acuña, P.; Jonas, J.D.; Segnitz, R.M.; Svenning, J.-C. Strong upslope shifts in Chimborazo's vegetation over two centuries since Humboldt. *Proc. Natl. Acad. Sci. USA* **2015**, *112*, 12741–12745. [CrossRef]
10. Habel, J.C.; Rasche, L.; Schneider, U.A.; Engler, J.O.; Schmid, E.; Rödder, D.; Meyer, S.T.; Trapp, N.; Sos del Diego, R.; Eggermont, H. Final countdown for biodiversity hotspots. *Conserv. Lett.* **2019**, *12*, e12668. [CrossRef]
11. Trew, B.T.; Maclean, I.M.D. Vulnerability of global biodiversity hotspots to climate change. *Glob. Ecol. Biogeogr.* **2021**, *30*, 768–783. [CrossRef]
12. FAO. UN Global Forest Resources Assessment 2020: Key Findings. 2020. Available online: https://www.fao.org/forest-resources-assessment/en/ (accessed on 28 January 2023).
13. Khaine, I.; Woo, S.Y. An overview of interrelationship between climate change and forests. *Forest Sci. Technol.* **2015**, *11*, 11–18. [CrossRef]
14. Allen, K.; Dupuy, J.M.; Gei, M.G.; Hulshof, C.; Medvigy, D.; Pizano, C.; Salgado-Negret, B.; Smith, C.M.; Trierweiler, A.; Van Bloem, S.J.; et al. Will seasonally dry tropical forests be sensitive or resistant to future changes in rainfall regimes? *Environ. Res. Lett.* **2017**, *12*, 23001. [CrossRef]
15. Myster, R.W. Effects of selective-logging, litter and tree species on forests in the Peruvian Amazon: Seed predation, seed pathogens, germination. *New Zeal. J. For. Sci.* **2021**, *51*. [CrossRef]
16. Whitman, A.A.; Brokaw, N.V.L.; Hagan, J.M. Forest damage caused by selection logging of mahogany (*Swietenia macrophylla*) in northern Belize. *For. Ecol. Manage.* **1997**, *92*, 87–96. [CrossRef]
17. Gillies, A.C.M.; Navarro, C.; Lowe, A.J.; Newton, A.C.; Hernandez, M.; Wilson, J.; Cornelius, J.P. Genetic diversity in Mesoamerican populations of mahogany (*Swietenia macrophylla*), assessed using RAPDs. *Heredity* **1999**, *83*, 722–732. [CrossRef]
18. Krisnawati, H.; Kallio, M.H.; Kanninen, M. *Swietenia macrophylla King: Ecology, Silviculture and Productivity*; CIFOR: Bogor, Indonesia, 2011; ISBN 6028693391.
19. Telrandhe, U.B.; Kosalge, S.B.; Parihar, S.; Sharma, D.; Hemalatha, S. Collection and Cultivation of *Swietenia macrophylla* King. *Sch. Acad. J. Pharm.* **2022**, *1*, 13–19. [CrossRef]
20. Mahendra, C.K.; Goh, K.W.; Ming, L.C.; Zengin, G.; Low, L.E.; Ser, H.-L.; Goh, B.H. The Prospects of *Swietenia macrophylla* King in Skin Care. *Antioxidants* **2022**, *11*, 913. [CrossRef]
21. Wang, G.-K.; Sun, Y.-P.; Jin, W.-F.; Yu, Y.; Zhu, J.-Y.; Liu, J.-S. Limonoids from *Swietenia macrophylla* and their antitumor activities in A375 human malignant melanoma cells. *Bioorg. Chem.* **2022**, *123*, 105780. [CrossRef]
22. Yudhani, R.D.; Nugrahaningsih, D.A.A.; Sholikhah, E.N.; Mustofa, M. The molecular mechanisms of hypoglycemic properties and safety profiles of *Swietenia macrophylla* seeds extract: A review. *Open Access Maced. J. Med. Sci.* **2021**, *9*, 370–388. [CrossRef]
23. Grogan, J.; Landis, R.M.; Free, C.M.; Schulze, M.D.; Lentini, M.; Ashton, M.S. Big-leaf mahogany *Swietenia macrophylla* population dynamics and implications for sustainable management. *J. Appl. Ecol.* **2014**, *51*, 664–674. [CrossRef]
24. Jhou, H.C.; Wang, Y.N.; Wu, C.S.; Yu, J.C.; Chen, C.I. Photosynthetic gas exchange responses of *Swietenia macrophylla* King and *Melia azedarach* L. plantations under drought conditions. *Bot. Stud.* **2017**, *58*, 57. [CrossRef]

25. Urrunaga, A.; Orbegozo, I.; Mulligan, F.J.J. *La Máquina Lavadora: Cómo el Fraude y la Corrupción en el Sistema de Concesiones Están Destruyendo el Futuro de los Bosques del Perú*; Environmental Investigation Agency (EIA): London, UK, 2012; Volume 1.
26. Garza-López, M.; Ortega-Rodríguez, J.M.; Zamudio-Sánchez, F.J.; López-Toledo, J.F.; Domínguez-Álvarez, F.A.; Sáenz-Romero, C. Calakmul como refugio de *Swietenia macrophylla* King ante el cambio climático. *Bot. Sci.* **2016**, *94*, 43–50. [CrossRef]
27. De Oliveira, S.S.; Campos, T.; Sebbenn, A.M.; d'Oliveira, M.V.N. Using spatial genetic structure of a population of *Swietenia macrophylla* King to integrate genetic diversity into management strategies in Southwestern Amazon. *For. Ecol. Manage.* **2020**, *464*, 118040. [CrossRef]
28. Chuquizuta, P.D.; Rodríguez, O.A.V. Crecimiento de plántulas de caoba (*Swietenia macrophylla* King) en respuesta a extractos vegetales. *Agrociencia* **2020**, *54*, 673–681. [CrossRef]
29. Sampayo-Maldonado, S.; Ordoñez-Salanueva, C.A.; Mattana, E.; Way, M.; Castillo-Lorenzo, E.; Dávila-Aranda, P.D.; Lira-Saade, R.; Téllez-Valdés, O.; Rodriguez-Arevalo, N.I.; Ulian, T. Thermal Niche for Seed Germination and Species Distribution Modelling of *Swietenia macrophylla* King (Mahogany) under Climate Change Scenarios. *Plants* **2021**, *10*, 2377. [CrossRef]
30. Darko, C.B.; Opuni-Frimpong, E.; Owusu, S.A.; Kyere, B.; Storer, A.J. Sustainability of Mahogany Production in Plantations: Does Resource Availability Influence Susceptibility of Young Mahogany Plantation Stands to Hypsipyla robusta Infestation? *Int. J. For. Res.* **2022**, *2022*, 5588184. [CrossRef]
31. Donald, L.G.; Pete, B.; Jacek, P.S.; Kevin, B. Chapter 2—Forest regions of the world. In *Introduction to Forestry and Natural Resources*; Academic Press: Cambridge, MA, USA, 2022; pp. 21–79. [CrossRef]
32. Grogan, J.; Loveless, M.D. Flowering phenology and its implications for management of big-leaf mahogany *Swietenia macrophylla* in Brazilian Amazonia. *Am. J. Bot.* **2013**, *100*, 2293–2305. [CrossRef] [PubMed]
33. Grogan, J.; Barreto, P. Big-leaf mahogany on CITES Appendix II: Big challenge, big opportunity. *Conserv. Biol.* **2005**, *19*, 973–976.
34. Urzedo, D.; Westerlaken, M.; Gabrys, J. Digitalizing forest landscape restoration: A social and political analysis of emerging technological practices. *Environ. Polit.* **2022**, 1–26. [CrossRef]
35. Indrajaya, Y.; Yuwati, T.W.; Lestari, S.; Winarno, B.; Narendra, B.H.; Nugroho, H.Y.S.H.; Rachmanadi, D.; Turjaman, M.; Adi, R.N.; Savitri, E. Tropical Forest Landscape Restoration in Indonesia: A Review. *Land* **2022**, *11*, 328. [CrossRef]
36. Jones, G.M.; Keyser, A.R.; Westerling, A.L.; Baldwin, W.J.; Keane, J.J.; Sawyer, S.C.; Clare, J.D.J.; Gutiérrez, R.J.; Peery, M.Z. Forest restoration limits megafires and supports species conservation under climate change. *Front. Ecol. Environ.* **2022**, *20*, 210–216. [CrossRef]
37. Prieto, P.V.; Bukoski, J.J.; Barros, F.S.M.; Beyer, H.L.; Iribarrem, A.; Brancalion, P.H.S.; Chazdon, R.L.; Lindenmayer, D.B.; Strassburg, B.B.N.; Guariguata, M.R. Predicting landscape-scale biodiversity recovery by natural tropical forest regrowth. *Conserv. Biol.* **2022**, *36*, e13842. [CrossRef]
38. Adiwinata, A.; Wicaksono, S.A.; Ichsan, A.C.; Yumn, A.; Goib, B.K.; Muslimah, S.; Susanti, F.N.; Purwanto, E. *A Policy Framework to Facilitate Integrated Forest Landscape Restoration (FLR) to Enhance Local Livelihoods in Indonesia*; CIFOR: Bogor, Indonesia, 2022.
39. Erbaugh, J.T.; Oldekop, J.A. Forest landscape restoration for livelihoods and well-being. *Curr. Opin. Environ. Sustain.* **2018**, *32*, 76–83. [CrossRef]
40. Verdone, M.; Seidl, A. Time, space, place, and the Bonn Challenge global forest restoration target. *Restor. Ecol.* **2017**, *25*, 903–911. [CrossRef]
41. ONU Decenio de las Naciones Unidas Sobre la Restauración de los Ecosistemas. Available online: https://www.decadeonrestoration.org/es (accessed on 28 January 2023).
42. Stanturf, J.A. Forest landscape restoration: Building on the past for future success. *Restor. Ecol.* **2021**, *29*, e13349. [CrossRef]
43. Romijn, E.; Coppus, R.; De Sy, V.; Herold, M.; Roman-Cuesta, R.M.; Verchot, L. Land restoration in Latin America and the Caribbean: An overview of recent, ongoing and planned restoration initiatives and their potential for climate change mitigation. *Forests* **2019**, *10*, 510. [CrossRef]
44. Di Sacco, A.; Hardwick, K.A.; Blakesley, D.; Brancalion, P.H.S.; Breman, E.; Cecilio Rebola, L.; Chomba, S.; Dixon, K.; Elliott, S.; Ruyonga, G. Ten golden rules for reforestation to optimize carbon sequestration, biodiversity recovery and livelihood benefits. *Glob. Chang. Biol.* **2021**, *27*, 1328–1348. [CrossRef]
45. Mori, A.S.; Dee, L.E.; Gonzalez, A.; Ohashi, H.; Cowles, J.; Wright, A.J.; Loreau, M.; Hautier, Y.; Newbold, T.; Reich, P.B.; et al. Biodiversity–productivity relationships are key to nature-based climate solutions. *Nat. Clim. Chang.* **2021**, *11*, 543–550. [CrossRef]
46. Bauman, D.; Fortunel, C.; Delhaye, G.; Malhi, Y.; Cernusak, L.A.; Bentley, L.P.; Rifai, S.W.; Aguirre-Gutiérrez, J.; Menor, I.O.; Phillips, O.L. Tropical tree mortality has increased with rising atmospheric water stress. *Nature* **2022**, *608*, 528–533. [CrossRef]
47. Zheng, W.; Liu, Y.; Yang, X.; Fan, W. Spatiotemporal Variations of Forest Vegetation Phenology and Its Response to Climate Change in Northeast China. *Remote Sens.* **2022**, *14*, 2909. [CrossRef]
48. Shahid, M.; Pinelli, E.; Dumat, C. Tracing trends in plant physiology and biochemistry: Need of databases from genetic to kingdom level. *Plant Physiol. Biochem.* **2018**, *127*, 630–635. [CrossRef]
49. Boisvert-Marsh, L.; de Blois, S. Unravelling potential northward migration pathways for tree species under climate change. *J. Biogeogr.* **2021**, *48*, 1088–1100. [CrossRef]
50. Nic Lughadha, E.; Bachman, S.P.; Leão, T.C.C.; Forest, F.; Halley, J.M.; Moat, J.; Acedo, C.; Bacon, K.L.; Brewer, R.F.A.; Gâteblé, G. Extinction risk and threats to plants and fungi. *Plants People Planet* **2020**, *2*, 389–408. [CrossRef]
51. Haevermans, T.; Tressou, J.; Kwon, J.; Pellens, R.; Dubéarnès, A.; Veron, S.; Bel, L.; Dervaux, S.; Dibie-Barthelemy, J.; Gaudeul, M. Global Plant Extinction Risk Assessment Informs Novel Biodiversity Hotspots. *bioRxiv* **2021**. [CrossRef]

52. Antonelli, A.; Smith, R.J.; Fry, C.; Simmonds, M.S.J.; Kersey, P.J.; Pritchard, H.W.; Abbo, M.S.; Acedo, C.; Adams, J.; Ainsworth, A.M. State of the World's Plants and Fungi. 2020. Available online: https://www.kew.org/science (accessed on 28 January 2023).

53. Cowie, R.H.; Bouchet, P.; Fontaine, B. The Sixth Mass Extinction: Fact, fiction or speculation? *Biol. Rev.* **2022**, *97*, 640–663. [CrossRef]

54. Littlefield, C.E.; Krosby, M.; Michalak, J.L.; Lawler, J.J. Connectivity for species on the move: Supporting climate-driven range shifts. *Front. Ecol. Environ.* **2019**, *17*, 270–278. [CrossRef]

55. Aavik, T.; Helm, A. Restoration of plant species and genetic diversity depends on landscape-scale dispersal. *Restor. Ecol.* **2018**, *26*, S92–S102. [CrossRef]

56. Camargo, P.H.S.A.; Pizo, M.A.; Brancalion, P.H.S.; Carlo, T.A. Fruit traits of pioneer trees structure seed dispersal across distances on tropical deforested landscapes: Implications for restoration. *J. Appl. Ecol.* **2020**, *57*, 2329–2339. [CrossRef]

57. Liang, Y.; Duveneck, M.J.; Gustafson, E.J.; Serra-Diaz, J.M.; Thompson, J.R. How disturbance, competition, and dispersal interact to prevent tree range boundaries from keeping pace with climate change. *Glob. Chang. Biol.* **2018**, *24*, e335–e351. [CrossRef]

58. Sajjad, H.; Kumar, P.; Masroor, M.; Rahaman, M.H.; Rehman, S.; Ahmed, R.; Sahana, M. Forest Vulnerability to Climate Change: A Review for Future Research Framework. *Forests* **2022**, *13*, 917.

59. Barragán, G.; Wang, T.; Rhemtulla, J.M. Trees planted under a global restoration pledge have mixed futures under climate change. *Restor. Ecol.* **2022**, e13764. [CrossRef]

60. Pecl, G.T.; Araújo, M.B.; Bell, J.D.; Blanchard, J.; Bonebrake, T.C.; Chen, I.-C.; Clark, T.D.; Colwell, R.K.; Danielsen, F.; Evengård, B. Biodiversity redistribution under climate change: Impacts on ecosystems and human well-being. *Science* **2017**, *355*, eaai9214. [CrossRef]

61. Sverdrup-Thygeson, A.; Skarpaas, O.; Blumentrath, S.; Birkemoe, T.; Evju, M. Habitat connectivity affects specialist species richness more than generalists in veteran trees. *For. Ecol. Manag.* **2017**, *403*, 96–102. [CrossRef]

62. Davis, M.A.; Chew, M.K.; Hobbs, R.J.; Lugo, A.E.; Ewel, J.J.; Vermeij, G.J.; Brown, J.H.; Rosenzweig, M.L.; Gardener, M.R.; Carroll, S.P. Don't judge species on their origins. *Nature* **2011**, *474*, 153–154. [CrossRef] [PubMed]

63. Brancalion, P.H.S.; Amazonas, N.T.; Chazdon, R.L.; van Melis, J.; Rodrigues, R.R.; Silva, C.C.; Sorrini, T.B.; Holl, K.D. Exotic eucalypts: From demonized trees to allies of tropical forest restoration? *J. Appl. Ecol.* **2020**, *57*, 55–66. [CrossRef]

64. Weidlich, E.W.A.; Flórido, F.G.; Sorrini, T.B.; Brancalion, P.H.S. Controlling invasive plant species in ecological restoration: A global review. *J. Appl. Ecol.* **2020**, *57*, 1806–1817. [CrossRef]

65. Lewis, S.L.; Wheeler, C.E.; Mitchard, E.T.A.; Koch, A. Restoring natural forests is the best way to remove atmospheric carbon. *Nature* **2019**, *568*, 25–28. [CrossRef]

66. Clark, J.S.; Bell, D.M.; Hersh, M.H.; Nichols, L. Climate change vulnerability of forest biodiversity: Climate and competition tracking of demographic rates. *Glob. Chang. Biol.* **2011**, *17*, 1834–1849. [CrossRef]

67. Pyšek, P.; Hulme, P.E.; Simberloff, D.; Bacher, S.; Blackburn, T.M.; Carlton, J.T.; Dawson, W.; Essl, F.; Foxcroft, L.C.; Genovesi, P. Scientists' warning on invasive alien species. *Biol. Rev.* **2020**, *95*, 1511–1534. [CrossRef]

68. Seebens, H.; Bacher, S.; Blackburn, T.M.; Capinha, C.; Dawson, W.; Dullinger, S.; Genovesi, P.; Hulme, P.E.; van Kleunen, M.; Kühn, I. Projecting the continental accumulation of alien species through to 2050. *Glob. Chang. Biol.* **2021**, *27*, 970–982. [CrossRef]

69. Cuthbert, R.N.; Diagne, C.; Hudgins, E.J.; Turbelin, A.; Ahmed, D.A.; Albert, C.; Bodey, T.W.; Briski, E.; Essl, F.; Haubrock, P.J. Biological invasion costs reveal insufficient proactive management worldwide. *Sci. Total Environ.* **2022**, *819*, 153404. [CrossRef] [PubMed]

70. Gann, G.D.; McDonald, T.; Walder, B.; Aronson, J.; Nelson, C.R.; Jonson, J.; Hallett, J.G.; Eisenberg, C.; Guariguata, M.R.; Liu, J. International principles and standards for the practice of ecological restoration. *Restor. Ecol.* **2019**, *27*, S1–S46. [CrossRef]

71. Horák, J.; Brestovanská, T.; Mladenović, S.; Kout, J.; Bogusch, P.; Halda, J.P.; Zasadil, P. Green desert?: Biodiversity patterns in forest plantations. *For. Ecol. Manag.* **2019**, *433*, 343–348. [CrossRef]

72. Brancalion, P.H.S.; Bello, C.; Chazdon, R.L.; Galetti, M.; Jordano, P.; Lima, R.A.F.; Medina, A.; Pizo, M.A.; Reid, J.L. Maximizing biodiversity conservation and carbon stocking in restored tropical forests. *Conserv. Lett.* **2018**, *11*, e12454. [CrossRef]

73. Guisan, A.; Thuiller, W.; Zimmermann, N.E. *Habitat Suitability and Distribution Models: With Applications in R*; Cambridge University Press: Cambridge, UK, 2017; ISBN 0521765137.

74. Peterson, A.T.; Soberón, J.; Pearson, R.G.; Anderson, R.P.; Martínez-Meyer, E.; Nakamura, M.; Araújo, M.B. *Ecological Niches and Geographic Distributions (MPB-49)*; Princeton University Press: Princeton, NJ, USA, 2011; ISBN 1400840678.

75. Mateo, R.G.; Felicisimo, A.M.; Munoz, J. Species distributions models: A synthetic revision. *Rev. Chil. Hist. Nat.* **2011**, *84*, 217–240. [CrossRef]

76. Sanczuk, P.; De Lombaerde, E.; Haesen, S.; Van Meerbeek, K.; Van der Veken, B.; Hermy, M.; Verheyen, K.; Vangansbeke, P.; De Frenne, P. Species distribution models and a 60-year-old transplant experiment reveal inhibited forest plant range shifts under climate change. *J. Biogeogr.* **2022**, *49*, 537–550. [CrossRef]

77. Lima, V.P.; de Lima, R.A.F.; Joner, F.; Siddique, I.; Raes, N.; Ter Steege, H. Climate change threatens native potential agroforestry plant species in Brazil. *Sci. Rep.* **2022**, *12*, 2267. [CrossRef]

78. Cotrina Sánchez, A.; Rojas Briceño, N.B.; Bandopadhyay, S.; Ghosh, S.; Torres Guzmán, C.; Oliva, M.; Guzman, B.K.; Salas López, R. Biogeographic Distribution of Cedrela spp. Genus in Peru Using MaxEnt Modeling: A Conservation and Restoration Approach. *Diversity* **2021**, *13*, 261. [CrossRef]

79. Goicolea, T.; Mateo, R.G.; Aroca-Fernández, M.J.; Gastón, A.; García-Viñas, J.I.; Mateo-Sánchez, M.C. Considering plant functional connectivity in landscape conservation and restoration management. *Biodivers. Conserv.* **2022**, *31*, 1591–1608. [CrossRef]

80. Gastón, A.; García-Viñas, J.I.; Bravo-Fernández, A.J.; López-Leiva, C.; Oliet, J.A.; Roig, S.; Serrada, R. Species distribution models applied to plant species selection in forest restoration: Are model predictions comparable to expert opinion? *New For.* **2014**, *45*, 641–653. [CrossRef]

81. Carrillo-García, C.; Girola-Iglesias, L.; Guijarro, M.; Hernando, C.; Madrigal, J.; Mateo, R.G. Ecological niche models applied to post-megafire vegetation restoration in the context of climate change. *Sci. Total Environ.* **2023**, *855*, 158858. [CrossRef]

82. Ramírez-Magil, G.; Botello, F.; Navarro-Martínez, M.A. Idoneidad de hábitat para *Swietenia macrophylla* en escenarios de cambio climático en México. *Madera Bosques* **2020**, *26*, e2631954. [CrossRef]

83. Navarro-Martínez, A.; Ellis, E.A.; Hernández-Gómez, I.; Romero-Montero, J.A.; Sánchez-Sánchez, O. Distribution and abundance of big-leaf mahogany (*Swietenia macrophylla*) on the Yucatan Peninsula, Mexico. *Trop. Conserv. Sci.* **2018**, *11*, 1940082918766875. [CrossRef]

84. Silva, M.C. Da Modelo de Distribuição de Nicho Ecológico Para *Swietenia macrophylla* King na Amazônia Brasileira. 2020. Available online: http://repositorio.ufra.edu.br/jspui/handle/123456789/1112 (accessed on 28 January 2023).

85. Milagres, V.A.C.; Machado, E.L.M. Potential distribution modeling of useful Brazilian trees with economic importance. *J. Agric. Sci. Technol.* **2016**, *6*, 400–410.

86. Sillero, N.; Barbosa, A.M. Common mistakes in ecological niche models. *Int. J. Geogr. Inf. Sci.* **2021**, *35*, 213–226. [CrossRef]

87. Carretero, M.A.; Sillero, N. Evaluating how species niche modelling is affected by partial distributions with an empirical case. *Acta Oecologica* **2016**, *77*, 207–216. [CrossRef]

88. Petitpierre, B.; McDougall, K.; Seipel, T.; Broennimann, O.; Guisan, A.; Kueffer, C. Will climate change increase the risk of plant invasions into mountains? *Ecol. Appl.* **2016**, *26*, 530–544. [CrossRef]

89. Mateo, R.G.; Gaston, A.; Aroca-Fernández, M.J.; Broennimann, O.; Guisan, A.; Saura, S.; García-Viñas, J.I. Hierarchical species distribution models in support of vegetation conservation at the landscape scale. *J. Veg. Sci.* **2019**, *30*, 386–396. [CrossRef]

90. Chevalier, M.; Zarzo-Arias, A.; Guélat, J.; Mateo, R.G.; Guisan, A. Accounting for niche truncation to improve spatial and temporal predictions of species distributions. *Front. Ecol. Evol.* **2022**, *10*, 760. [CrossRef]

91. Mendes, P.; Velazco, S.J.E.; de Andrade, A.F.A.; Júnior, P.D.M. Dealing with overprediction in species distribution models: How adding distance constraints can improve model accuracy. *Ecol. Modell.* **2020**, *431*, 109180. [CrossRef]

92. Velazco, S.J.E.; Ribeiro, B.R.; Laureto, L.M.O.; Júnior, P.D.M. Overprediction of species distribution models in conservation planning: A still neglected issue with strong effects. *Biol. Conserv.* **2020**, *252*, 108822. [CrossRef]

93. Noroozi, J.; Talebi, A.; Doostmohammadi, M.; Rumpf, S.B.; Linder, H.P.; Schneeweiss, G.M. Hotspots within a global biodiversity hotspot-areas of endemism are associated with high mountain ranges. *Sci. Rep.* **2018**, *8*, 10345. [CrossRef]

94. Myers, N.; Mittermeier, R.A.; Mittermeier, C.G.; da Fonseca, G.A.B.; Kent, J. Biodiversity hotspots for conservation priorities. *Nature* **2000**, *403*, 853–858. [CrossRef]

95. Fick, S.E.; Hijmans, R.J. WorldClim 2: New 1-km spatial resolution climate surfaces for global land areas. *Int. J. Climatol.* **2017**, *37*, 4302–4315. [CrossRef]

96. Jarvis, A. Hole-Filed Seamless SRTM Data. 2008. Available online: http//srtm.csi.cgiar.org (accessed on 28 January 2023).

97. Lemes, M.R.; Dick, C.W.; Navarro, C.; Lowe, A.J.; Cavers, S.; Gribel, R. Chloroplast DNA microsatellites reveal contrasting phylogeographic structure in mahogany (*Swietenia macrophylla* King, Meliaceae) from Amazonia and Central America. *Trop. Plant Biol.* **2010**, *3*, 40–49. [CrossRef]

98. Wightman, K.E.; Ward, S.E.; Haggar, J.P.; Santiago, B.R.; Cornelius, J.P. Performance and genetic variation of big-leaf mahogany (*Swietenia macrophylla* King) in provenance and progeny trials in the Yucatan Peninsula of Mexico. *For. Ecol. Manag.* **2008**, *255*, 346–355. [CrossRef]

99. Degen, B.; Ward, S.E.; Lemes, M.R.; Navarro, C.; Cavers, S.; Sebbenn, A.M. Verifying the geographic origin of mahogany (*Swietenia macrophylla* King) with DNA-fingerprints. *Forensic Sci. Int. Genet.* **2013**, *7*, 55–62. [CrossRef] [PubMed]

100. Navarro, C.; Hernández, G. Progeny test analysis and population differentiation of mesoamerican mahogany (*Swietenia macrophylla*). *Agron. Costaric.* **2004**, *28*, 37–51.

101. Andino, J.E.G.; Pitman, N.C.A.; Ulloa, C.U.; Romoleroux, K.; Fernández-Fernández, D.; Ceron, C.; Palacios, W.; Neill, D.A.; Oleas, N.; Altamirano, P.; et al. Trees of Amazonian Ecuador: A taxonomically verified species list with data on abundance and distribution. *Ecology* **2019**, *100*, e02894. [CrossRef]

102. Owens, H.L.; Merow, C.; Maitner, B.S.; Kass, J.M.; Barve, V.; Guralnick, R.P. occCite: Tools for querying and managing large biodiversity occurrence datasets. *Ecography* **2021**, *44*, 1228–1235. [CrossRef]

103. De Lima, R.A.F.; Sánchez-Tapia, A.; Mortara, S.R.; ter Steege, H.; de Siqueira, M.F. plantR: An R package and workflow for managing species records from biological collections. *Methods Ecol. Evol.* **2021**, *14*, 332–339. [CrossRef]

104. Maitner, B.S.; Boyle, B.; Casler, N.; Condit, R.; Donoghue, J.; Durán, S.M.; Guaderrama, D.; Hinchliff, C.E.; Jørgensen, P.M.; Kraft, N.J.B. The bien r package: A tool to access the Botanical Information and Ecology Network (BIEN) database. *Methods Ecol. Evol.* **2018**, *9*, 373–379. [CrossRef]

105. Michonneau, F.; Collins, M.; Chamberlain, S. Ridigbio: An Interface to iDigBio's Search API That Allows Downloading Specimen Records. R Package Version 0.3.2. 2016. Available online: https://github.com/iDigBio/ridigbio (accessed on 28 January 2023).

106. Albani Rocchetti, G.; Armstrong, C.G.; Abeli, T.; Orsenigo, S.; Jasper, C.; Joly, S.; Bruneau, A.; Zytaruk, M.; Vamosi, J.C. Reversing extinction trends: New uses of (old) herbarium specimens to accelerate conservation action on threatened species. *New Phytol.* **2021**, *230*, 433–450. [CrossRef]

107. Lang, P.L.M.; Willems, F.M.; Scheepens, J.F.; Burbano, H.A.; Bossdorf, O. Using herbaria to study global environmental change. *New Phytol.* **2019**, *221*, 110–122. [CrossRef]

108. Nic Lughadha, E.; Walker, B.E.; Canteiro, C.; Chadburn, H.; Davis, A.P.; Hargreaves, S.; Lucas, E.J.; Schuiteman, A.; Williams, E.; Bachman, S.P. The use and misuse of herbarium specimens in evaluating plant extinction risks. *Philos. Trans. R. Soc. B* **2019**, *374*, 20170402. [CrossRef]

109. Fei, S.; Yu, F. Quality of presence data determines species distribution model performance: A novel index to evaluate data quality. *Landsc. Ecol.* **2016**, *31*, 31–42. [CrossRef]

110. Zurell, D.; Franklin, J.; König, C.; Bouchet, P.J.; Dormann, C.F.; Elith, J.; Fandos, G.; Feng, X.; Guillera-Arroita, G.; Guisan, A. A standard protocol for reporting species distribution models. *Ecography* **2020**, *43*, 1261–1277. [CrossRef]

111. Sillero, N.; Arenas-Castro, S.; Enriquez-Urzelai, U.; Vale, C.G.; Sousa-Guedes, D.; Martínez-Freiría, F.; Real, R.; Barbosa, A.M. Want to model a species niche? A step-by-step guideline on correlative ecological niche modelling. *Ecol. Modell.* **2021**, *456*, 109671. [CrossRef]

112. Boakes, E.H.; McGowan, P.J.K.; Fuller, R.A.; Chang-qing, D.; Clark, N.E.; O'Connor, K.; Mace, G.M. Distorted views of biodiversity: Spatial and temporal bias in species occurrence data. *PLoS Biol.* **2010**, *8*, e1000385. [CrossRef]

113. Baker, D.J.; Maclean, I.M.D.; Goodall, M.; Gaston, K.J. Correlations between spatial sampling biases and environmental niches affect species distribution models. *Glob. Ecol. Biogeogr.* **2022**, *31*, 1038–1050. [CrossRef]

114. Aiello-Lammens, M.E.; Boria, R.A.; Radosavljevic, A.; Vilela, B.; Anderson, R.P. spThin: An R package for spatial thinning of species occurrence records for use in ecological niche models. *Ecography* **2015**, *38*, 541–545. [CrossRef]

115. Töpel, M.; Zizka, A.; Calió, M.F.; Scharn, R.; Silvestro, D.; Antonelli, A. SpeciesGeoCoder: Fast categorization of species occurrences for analyses of biodiversity, biogeography, ecology, and evolution. *Syst. Biol.* **2017**, *66*, 145–151. [CrossRef]

116. Eyring, V.; Bony, S.; Meehl, G.A.; Senior, C.A.; Stevens, B.; Stouffer, R.J.; Taylor, K.E. Overview of the Coupled Model Intercomparison Project Phase 6 (CMIP6) experimental design and organization. *Geosci. Model Dev.* **2016**, *9*, 1937–1958. [CrossRef]

117. Yukimoto, S.; Kawai, H.; Koshiro, T.; Oshima, N.; Yoshida, K.; Urakawa, S.; Tsujino, H.; Deushi, M.; Tanaka, T.; Hosaka, M. The Meteorological Research Institute Earth System Model version 2.0, MRI-ESM2. 0: Description and basic evaluation of the physical component. *J. Meteorol. Soc. Japan. Ser. II* **2019**, *97*, 931–965. [CrossRef]

118. Tatebe, H.; Ogura, T.; Nitta, T.; Komuro, Y.; Ogochi, K.; Takemura, T.; Sudo, K.; Sekiguchi, M.; Abe, M.; Saito, F. Description and basic evaluation of simulated mean state, internal variability, and climate sensitivity in MIROC6. *Geosci. Model Dev.* **2019**, *12*, 2727–2765. [CrossRef]

119. Thuiller, W.; Guéguen, M.; Renaud, J.; Karger, D.N.; Zimmermann, N.E. Uncertainty in ensembles of global biodiversity scenarios. *Nat. Commun.* **2019**, *10*, 1446. [CrossRef]

120. Neupane, N.; Zipkin, E.F.; Saunders, S.P.; Ries, L. Grappling with uncertainty in ecological projections: A case study using the migratory monarch butterfly. *Ecosphere* **2022**, *13*, e03874. [CrossRef]

121. Poggio, L.; De Sousa, L.M.; Batjes, N.H.; Heuvelink, G.; Kempen, B.; Ribeiro, E.; Rossiter, D. SoilGrids 2.0: Producing soil information for the globe with quantified spatial uncertainty. *Soil* **2021**, *7*, 217–240. [CrossRef]

122. Feng, X.; Park, D.S.; Liang, Y.; Pandey, R.; Papeş, M. Collinearity in ecological niche modeling: Confusions and challenges. *Ecol. Evol.* **2019**, *9*, 10365–10376. [CrossRef]

123. Brun, P.; Thuiller, W.; Chauvier, Y.; Pellissier, L.; Wüest, R.O.; Wang, Z.; Zimmermann, N.E. Model complexity affects species distribution projections under climate change. *J. Biogeogr.* **2020**, *47*, 130–142. [CrossRef]

124. Dormann, C.F.; Elith, J.; Bacher, S.; Buchmann, C.; Carl, G.; Carré, G.; Marquéz, J.R.G.; Gruber, B.; Lafourcade, B.; Leitão, P.J. Collinearity: A review of methods to deal with it and a simulation study evaluating their performance. *Ecography* **2013**, *36*, 27–46. [CrossRef]

125. Araújo, M.B.; New, M. Ensemble forecasting of species distributions. *Trends Ecol. Evol.* **2007**, *22*, 42–47. [CrossRef]

126. Thuiller, W.; Lafourcade, B.; Engler, R.; Araújo, M.B. BIOMOD—A platform for ensemble forecasting of species distributions. *Ecography* **2009**, *32*, 369–373. [CrossRef]

127. Hastie, T.J.; Tibshirani, R.J. *Generalized Additive Models*; Routledge: London, UK, 2017; ISBN 020375378X.

128. Friedman, J.H. Greedy function approximation: A gradient boosting machine. *Ann. Stat.* **2001**, *29*, 1189–1232. [CrossRef]

129. Breiman, L. Random forests. *Mach. Learn.* **2001**, *45*, 5–32. [CrossRef]

130. Shabani, F.; Kumar, L.; Ahmadi, M. Assessing accuracy methods of species distribution models: AUC, specificity, sensitivity and the true skill statistic. *Glob. J. Hum. Soc. Sci.* **2018**, *18*, 6–18.

131. Marmion, M.; Parviainen, M.; Luoto, M.; Heikkinen, R.K.; Thuiller, W. Evaluation of consensus methods in predictive species distribution modelling. *Divers. Distrib.* **2009**, *15*, 59–69. [CrossRef]

132. Peterson, A.T.; Papeş, M.; Soberón, J. Rethinking receiver operating characteristic analysis applications in ecological niche modeling. *Ecol. Modell.* **2008**, *213*, 63–72. [CrossRef]

133. Allouche, O.; Tsoar, A.; Kadmon, R. Assessing the accuracy of species distribution models: Prevalence, kappa and the true skill statistic (TSS). *J. Appl. Ecol.* **2006**, *43*, 1223–1232. [CrossRef]

134. Anderson, R.P.; Lew, D.; Peterson, A.T. Evaluating predictive models of species' distributions: Criteria for selecting optimal models. *Ecol. Modell.* **2003**, *162*, 211–232. [CrossRef]
135. Simoes, M.; Romero-Alvarez, D.; Nuñez-Penichet, C.; Jiménez, L.; Cobos, M.E. General theory and good practices in ecological niche modeling: A basic guide. *Biodivers. Inform.* **2020**, *15*, 67–68. [CrossRef]
136. Scherrer, D.; D'Amen, M.; Fernandes, R.F.; Mateo, R.G.; Guisan, A. How to best threshold and validate stacked species assemblages? Community optimisation might hold the answer. *Methods Ecol. Evol.* **2018**, *9*, 2155–2166. [CrossRef]
137. Watson, J.E.M.; Dudley, N.; Segan, D.B.; Hockings, M. The performance and potential of protected areas. *Nature* **2014**, *515*, 67–73. [CrossRef]
138. Kearney, S.G.; Adams, V.M.; Fuller, R.A.; Possingham, H.P.; Watson, J.E.M. Estimating the benefit of well-managed protected areas for threatened species conservation. *Oryx* **2020**, *54*, 276–284. [CrossRef]
139. Mestanza-Ramón, C.; Henkanaththegedara, S.M.; Vásconez Duchicela, P.; Vargas Tierras, Y.; Sánchez Capa, M.; Constante Mejía, D.; Jimenez Gutierrez, M.; Charco Guamán, M.; Mestanza Ramón, P. In-Situ and Ex-Situ Biodiversity Conservation in Ecuador: A Review of Policies, Actions and Challenges. *Diversy* **2020**, *12*, 315. [CrossRef]
140. Araújo, M.B.; Anderson, R.P.; Barbosa, A.M.; Beale, C.M.; Dormann, C.F.; Early, R.; Garcia, R.A.; Guisan, A.; Maiorano, L.; Naimi, B. Standards for distribution models in biodiversity assessments. *Sci. Adv.* **2019**, *5*, eaat4858. [CrossRef] [PubMed]
141. Dubos, N.; Préau, C.; Lenormand, M.; Papuga, G.; Montsarrat, S.; Denelle, P.; Le Louarn, M.; Heremans, S.; Roel, M.; Roche, P. Assessing the effect of sample bias correction in species distribution models. *arXiv* **2021**, arXiv:2103.07107. [CrossRef]
142. Hallgren, W.; Santana, F.; Low-Choy, S.; Zhao, Y.; Mackey, B. Species distribution models can be highly sensitive to algorithm configuration. *Ecol. Modell.* **2019**, *408*, 108719. [CrossRef]
143. Urbina-Cardona, N.; Blair, M.E.; Londoño, M.C.; Loyola, R.; Velásquez-Tibatá, J.; Morales-Devia, H. Species distribution modeling in Latin America: A 25-year retrospective review. *Trop. Conserv. Sci.* **2019**, *12*, 1940082919854058. [CrossRef]
144. Qiao, H.; Feng, X.; Escobar, L.E.; Peterson, A.T.; Soberón, J.; Zhu, G.; Papeş, M. An evaluation of transferability of ecological niche models. *Ecography* **2019**, *42*, 521–534. [CrossRef]
145. Escobar, L.E.; Qiao, H.; Cabello, J.; Peterson, A.T. Ecological niche modeling re-examined: A case study with the Darwin's fox. *Ecol. Evol.* **2018**, *8*, 4757–4770. [CrossRef] [PubMed]
146. Hao, T.; Elith, J.; Guillera-Arroita, G.; Lahoz-Monfort, J.J. A review of evidence about use and performance of species distribution modelling ensembles like BIOMOD. *Divers. Distrib.* **2019**, *25*, 839–852. [CrossRef]
147. Hao, T.; Elith, J.; Lahoz-Monfort, J.J.; Guillera-Arroita, G. Testing whether ensemble modelling is advantageous for maximising predictive performance of species distribution models. *Ecography* **2020**, *43*, 549–558. [CrossRef]
148. Roe, N.A.; Ducey, M.J.; Lee, T.D.; Fraser, O.L.; Colter, R.A.; Hallett, R.A. Soil chemical variables improve models of understorey plant species distributions. *J. Biogeogr.* **2022**, *49*, 753–766. [CrossRef]
149. Zuquim, G.; Costa, F.R.C.; Tuomisto, H.; Moulatlet, G.M.; Figueiredo, F.O.G. The importance of soils in predicting the future of plant habitat suitability in a tropical forest. *Plant Soil* **2020**, *450*, 151–170. [CrossRef]
150. De Castro Oliveira, G.; Arruda, D.M.; Fernandes Filho, E.I.; Veloso, G.V.; Francelino, M.R.; Schaefer, C.E.G.R. Soil predictors are crucial for modelling vegetation distribution and its responses to climate change. *Sci. Total Environ.* **2021**, *780*, 146680. [CrossRef]
151. Marcer, A.; Chapman, A.D.; Wieczorek, J.R.; Xavier Picó, F.; Uribe, F.; Waller, J.; Ariño, A.H. Uncertainty matters: Ascertaining where specimens in natural history collections come from and its implications for predicting species distributions. *Ecography* **2022**, *2022*, e06025. [CrossRef]
152. Smith, A.B.; Murphy, S.J.; Henderson, D.; Erickson, K.D. Imprecisely georeferenced specimen data provide unique information on species' distributions and environmental tolerances: Don't let the perfect be the enemy of the good. *bioRxiv* **2021**. [CrossRef]
153. Hughes, A.C.; Orr, M.C.; Ma, K.; Costello, M.J.; Waller, J.; Provoost, P.; Yang, Q.; Zhu, C.; Qiao, H. Sampling biases shape our view of the natural world. *Ecography* **2021**, *44*, 1259–1269. [CrossRef]
154. Iglesias, J.; Muñoz, L.; Santiana, J.; Chinchero, M.; Jiménez, D.; Palacios, W.; Jadán, A. Estudio Poblacional de *Swietenia macrophylla* King (Caoba/Ahuano) en la Provincia de Pastaza, Ecuador. In Proceedings of the XII Congreso Latinoamericano de Botánica, Quito, Ecuador, 21 October 2018; Volume 1.
155. Pérez-Salicrup, D.R.; Esquivel, R. Tree infection by Hypsipyla grandella in *Swietenia macrophylla* and *Cedrela odorata* (Meliaceae) in Mexico's southern Yucatan Peninsula. *For. Ecol. Manag.* **2008**, *255*, 324–327. [CrossRef]
156. Pinto, R.C.; Pinheiro, C.; Vidal, E.; Schwartz, G. Technical and financial evaluation of enrichment planting in logging gaps with the high-value species *Swietenia macrophylla* and *Handroanthus serratifolius* in the Eastern Amazon. *For. Ecol. Manag.* **2021**, *495*, 119380. [CrossRef]
157. De Castro, M.T.; Montalvão, S.C.L.; Monnerat, R.G. Breeding and biology of *Hypsipyla grandella* Zeller (Lepidoptera: Pyralidae) fed with mahogany seeds (*Swietenia macrophylla* King). *J. Asia. Pac. Entomol.* **2016**, *19*, 217–221. [CrossRef]
158. Lunz, A.M.; Thomazini, M.J.T.; Moraes, M.C.B.; Neves, E.J.M.; Batista, T.F.C.; Degenhardt, J.; de Sousa, L.A.; Ohashi, O.S. Hypsipyla grandella em mogno (*Swietenia macrophylla*): Situação atual e perspectivas. *Pesqui. Florest. Bras.* **2009**, 45. [CrossRef]
159. Bebber, D.P.; Gurr, S.J. Biotic interactions and climate in species distribution modelling. *BioRxiv* **2019**, 520320. [CrossRef]
160. Dormann, C.F.; Bobrowski, M.; Dehling, D.M.; Harris, D.J.; Hartig, F.; Lischke, H.; Moretti, M.D.; Pagel, J.; Pinkert, S.; Schleuning, M. Biotic interactions in species distribution modelling: 10 questions to guide interpretation and avoid false conclusions. *Glob. Ecol. Biogeogr.* **2018**, *27*, 1004–1016. [CrossRef]

161. Oliveira, M.R.; Tomas, W.M.; Guedes, N.M.R.; Peterson, A.T.; Szabo, J.K.; Júnior, A.S.; Camilo, A.R.; Padovani, C.R.; Garcia, L.C. The relationship between scale and predictor variables in species distribution models applied to conservation. *Biodivers. Conserv.* **2021**, *30*, 1971–1990. [CrossRef]

162. Ashraf, U.; Chaudhry, M.N.; Peterson, A.T. Ecological niche models of biotic interactions predict increasing pest risk to olive cultivars with changing climate. *Ecosphere* **2021**, *12*, e03714. [CrossRef]

163. Anderson, R.P. When and how should biotic interactions be considered in models of species niches and distributions? *J. Biogeogr.* **2017**, *44*, 8–17. [CrossRef]

164. Gábor, L.; Šímová, P.; Keil, P.; Zarzo-Arias, A.; Marsh, C.J.; Rocchini, D.; Malavasi, M.; Barták, V.; Moudrý, V. Habitats as predictors in species distribution models: Shall we use continuous or binary data? *Ecography* **2022**, *2022*, e06022. [CrossRef]

165. Merow, C.; Galante, P.J.; Kass, J.M.; Aiello-Lammens, M.E.; Babich Morrow, C.; Gerstner, B.E.; Grisales Betancur, V.; Moore, A.C.; Noguera-Urbano, E.A.; Pinilla-Buitrago, G.E. Operationalizing expert knowledge in species' range estimates using diverse data types. *Front. Biogeogr.* **2022**, *14*, 2. [CrossRef]

166. Navarro-Martínez, A.; Ramírez-Magil, G. Geographic Information Systems for Forest Species Distribution and Habitat Suitability. In *GIS LATAM Conference*; Springer: Berlin/Heidelberg, Germany, 2020; pp. 125–135.

167. Garza López, M. El Centro de la Peninsula de Yucatán, México, Como Refugio de dos Especies Forestales Ante los Efectos del Cambio Climático. 2015. Available online: https://repositorio.chapingo.edu.mx/items/12260729-5098-46fd-ab05-0afe8954ac02 (accessed on 28 January 2023).

168. Caballero, R.I.A. Nicho Ecológico y Variables Que Intervienen en el Desempeño de Swietenia en México. 2019. Available online: http://riaa.uaem.mx/xmlui/handle/20.500.12055/501 (accessed on 28 January 2023).

169. Grogan, J.; Ashton, M.S.; Galvao, J. Big-leaf mahogany (*Swietenia macrophylla*) seedling survival and growth across a topographic gradient in southeast Pará, Brazil. *For. Ecol. Manag.* **2003**, *186*, 311–326. [CrossRef]

170. Grogan, J.; Matthew Landis, R. Growth history and crown vine coverage are principal factors influencing growth and mortality rates of big-leaf mahogany *Swietenia macrophylla* in Brazil. *J. Appl. Ecol.* **2009**, *46*, 1283–1291. [CrossRef]

171. Ruokolainen, K.; Moulatlet, G.M.; Zuquim, G.; Hoorn, C.; Tuomisto, H. Geologically recent rearrangements in central Amazonian river network and their importance for the riverine barrier hypothesis. *Front. Biogeogr.* **2019**, *11*, e45046. [CrossRef]

172. Nazareno, A.G.; Dick, C.W.; Lohmann, L.G. A biogeographic barrier test reveals a strong genetic structure for a canopy-emergent Amazon tree species. *Sci. Rep.* **2019**, *9*, 18602. [CrossRef] [PubMed]

173. Gullison, R.E.; Panfil, S.N.; Strouse, J.J.; Hubbell, S.P. Ecology and management of mahogany (*Swietenia macrophylla* King) in the Chimanes Forest, Beni, Bolivia. *Bot. J. Linn. Soc.* **1996**, *122*, 9–34. [CrossRef]

174. Boisvert-Marsh, L.; Périé, C.; de Blois, S. Divergent responses to climate change and disturbance drive recruitment patterns underlying latitudinal shifts of tree species. *J. Ecol.* **2019**, *107*, 1956–1969. [CrossRef]

175. Steinbauer, M.J.; Grytnes, J.-A.; Jurasinski, G.; Kulonen, A.; Lenoir, J.; Pauli, H.; Rixen, C.; Winkler, M.; Bardy-Durchhalter, M.; Barni, E. Accelerated increase in plant species richness on mountain summits is linked to warming. *Nature* **2018**, *556*, 231–234. [CrossRef]

176. Román-Palacios, C.; Wiens, J.J. Recent responses to climate change reveal the drivers of species extinction and survival. *Proc. Natl. Acad. Sci. USA* **2020**, *117*, 4211–4217. [CrossRef] [PubMed]

177. Fahad, S.; Sonmez, O.; Saud, S.; Wang, D.; Wu, C.; Adnan, M.; Turan, V. *Climate Change and Plants: Biodiversity, Growth and Interactions*; CRC Press: Boca Raton, FL, USA, 2021; ISBN 1000379787.

178. Sánchez-Fernández, D.; Rizzo, V.; Cieslak, A.; Faille, A.; Fresneda, J.; Ribera, I. Thermal niche estimators and the capability of poor dispersal species to cope with climate change. *Sci. Rep.* **2016**, *6*, 23381. [CrossRef]

179. Kleemann, J.; Zamora, C.; Villacis-Chiluisa, A.B.; Cuenca, P.; Koo, H.; Noh, J.K.; Fürst, C.; Thiel, M. Deforestation in Continental Ecuador with a Focus on Protected Areas. *Land* **2022**, *11*, 268. [CrossRef]

180. Torres, B.; Fischer, R.; JC, V.; Günter, S. Deforestación en Paisajes Forestales tropicales del Ecuador: Bases científicas para perspectivas políticas. *Univ. Estatal Amaz. Inst. Johan Heinrich Thunen. Puyo. Ecuador. Ser. Publ. Misceláneas Ina.* **2020**, *15*, 172.

181. Mestanza-Ramón, C.; Cuenca-Cumbicus, J.; D'Orio, G.; Flores-Toala, J.; Segovia-Cáceres, S.; Bonilla-Bonilla, A.; Straface, S. Gold mining in the Amazon Region of ecuador: History and a review of its socio-environmental impacts. *Land* **2022**, *11*, 221. [CrossRef]

182. Huera-Lucero, T.; Salas-Ruiz, A.; Changoluisa, D.; Bravo-Medina, C. Towards sustainable urban planning for Puyo (Ecuador): Amazon forest landscape as potential green infrastructure. *Sustainability* **2020**, *12*, 4768. [CrossRef]

183. Haight, J.; Hammill, E. Protected areas as potential refugia for biodiversity under climatic change. *Biol. Conserv.* **2020**, *241*, 108258. [CrossRef]

184. De Souza, A.C.; Prevedello, J.A. The importance of protected areas for overexploited plants: Evidence from a biodiversity hotspot. *Biol. Conserv.* **2020**, *243*, 108482. [CrossRef]

185. Kuempel, C.D.; Adams, V.M.; Possingham, H.P.; Bode, M. Bigger or better: The relative benefits of protected area network expansion and enforcement for the conservation of an exploited species. *Conserv. Lett.* **2018**, *11*, e12433. [CrossRef]

186. Geldmann, J.; Barnes, M.; Coad, L.; Craigie, I.D.; Hockings, M.; Burgess, N.D. Effectiveness of terrestrial protected areas in reducing habitat loss and population declines. *Biol. Conserv.* **2013**, *161*, 230–238. [CrossRef]

187. Gray, C.L.; Hill, S.L.L.; Newbold, T.; Hudson, L.N.; Börger, L.; Contu, S.; Hoskins, A.J.; Ferrier, S.; Purvis, A.; Scharlemann, J.P.W. Local biodiversity is higher inside than outside terrestrial protected areas worldwide. *Nat. Commun.* **2016**, *7*, 12306. [CrossRef]

188. De Koning, F.; Aguiñaga, M.; Bravo, M.; Chiu, M.; Lascano, M.; Lozada, T.; Suarez, L. Bridging the gap between forest conservation and poverty alleviation: The Ecuadorian Socio Bosque program. *Environ. Sci. Policy* **2011**, *14*, 531–542. [CrossRef]
189. Andrade-Núñez, M.J.; Aide, T.M. Using nighttime lights to assess infrastructure expansion within and around protected areas in South America. *Environ. Res. Commun.* **2020**, *2*, 21002. [CrossRef]
190. Guo, W.-Y.; Serra-Diaz, J.M.; Schrodt, F.; Eiserhardt, W.L.; Maitner, B.S.; Merow, C.; Violle, C.; Anand, M.; Belluau, M.; Bruun, H.H. High exposure of global tree diversity to human pressure. *Proc. Natl. Acad. Sci. USA* **2022**, *119*, e2026733119. [CrossRef]
191. Kleemann, J.; Koo, H.; Hensen, I.; Mendieta-Leiva, G.; Kahnt, B.; Kurze, C.; Inclan, D.J.; Cuenca, P.; Noh, J.K.; Hoffmann, M.H. Priorities of action and research for the protection of biodiversity and ecosystem services in continental Ecuador. *Biol. Conserv.* **2022**, *265*, 109404. [CrossRef]
192. Cuesta, F.; Peralvo, M.; Merino-Viteri, A.; Bustamante, M.; Baquero, F.; Freile, J.F.; Muriel, P.; Torres-Carvajal, O. Priority areas for biodiversity conservation in mainland Ecuador. *Neotrop. Biodivers.* **2017**, *3*, 93–106. [CrossRef]
193. Fajardo, J.; Lessmann, J.; Bonaccorso, E.; Devenish, C.; Munoz, J. Combined use of systematic conservation planning, species distribution modelling, and connectivity analysis reveals severe conservation gaps in a megadiverse country (Peru). *PLoS ONE* **2014**, *9*, e114367. [CrossRef] [PubMed]
194. Saura, S.; Bertzky, B.; Bastin, L.; Battistella, L.; Mandrici, A.; Dubois, G. Protected area connectivity: Shortfalls in global targets and country-level priorities. *Biol. Conserv.* **2018**, *219*, 53–67. [CrossRef]
195. Brennan, A.; Naidoo, R.; Greenstreet, L.; Mehrabi, Z.; Ramankutty, N.; Kremen, C. Functional connectivity of the world's protected areas. *Science* **2022**, *376*, 1101–1104. [CrossRef]
196. McGuire, J.L.; Shipley, B.R. Dynamic priorities for conserving species. *Science* **2022**, *376*, 1048–1049. [CrossRef]
197. Gorelick, N.; Hancher, M.; Dixon, M.; Ilyushchenko, S.; Thau, D.; Moore, R. Google Earth Engine: Planetary-scale geospatial analysis for everyone. *Remote Sens. Environ.* **2017**, *202*, 18–27. [CrossRef]
198. Zhao, Q.; Yu, L.; Li, X.; Peng, D.; Zhang, Y.; Gong, P. Progress and trends in the application of Google Earth and Google Earth Engine. *Remote Sens.* **2021**, *13*, 3778. [CrossRef]
199. Birk, A. Seeing through the forest and the trees with drones. *Sci. Robot.* **2021**, *6*, eabj3947. [CrossRef] [PubMed]
200. Kocer, B.B.; Ho, B.; Zhu, X.; Zheng, P.; Farinha, A.; Xiao, F.; Stephens, B.; Wiesemüller, F.; Orr, L.; Kovac, M. Forest drones for environmental sensing and nature conservation. In Proceedings of the 2021 Aerial Robotic Systems Physically Interacting with the Environment (AIRPHARO), Biograd na Moru, Croatia, 4–5 October 2021; IEEE: Piscataway, NJ, USA, 2021; pp. 1–8.
201. Veloz Mino, S.P.; del Carmen Villavicencio Narváez, L.; Serrano Avalos, K.V.; Avalos Perez, M.C.; Veloz Mino, M.F.; Lopez Rodriguez, M.A. Impact of educational workshops for the conservation and protection of forests in the environmental education of children. *Dilemas Contemp. Polit. Valores* **2018**, *5*, 2.
202. Jackowiak, B.; Lawenda, M.; Nowak, M.M.; Wolniewicz, P.; Błoszyk, J.; Urbaniak, M.; Szkudlarz, P.; Jędrasiak, D.; Wiland-Szymańska, J.; Bajaczyk, R. Open Access to the Digital Biodiversity Database: A Comprehensive Functional Model of the Natural History Collections. *Diversity* **2022**, *14*, 596. [CrossRef]
203. Enquist, B.J.; Feng, X.; Boyle, B.; Maitner, B.; Newman, E.A.; Jørgensen, P.M.; Roehrdanz, P.R.; Thiers, B.M.; Burger, J.R.; Corlett, R.T. The commonness of rarity: Global and future distribution of rarity across land plants. *Sci. Adv.* **2019**, *5*, eaaz0414. [CrossRef]
204. ENF Inventario Nacional Forestal. Available online: http://enf.ambiente.gob.ec/web_enf/ (accessed on 28 January 2023).
205. Zizka, A.; Antonelli, A.; Silvestro, D. Sampbias, a method for quantifying geographic sampling biases in species distribution data. *Ecography* **2021**, *44*, 25–32. [CrossRef]
206. Daru, B.H.; Park, D.S.; Primack, R.B.; Willis, C.G.; Barrington, D.S.; Whitfeld, T.J.S.; Seidler, T.G.; Sweeney, P.W.; Foster, D.R.; Ellison, A.M. Widespread sampling biases in herbaria revealed from large-scale digitization. *New Phytol.* **2018**, *217*, 939–955. [CrossRef] [PubMed]
207. Engemann, K.; Enquist, B.J.; Sandel, B.; Boyle, B.; Jørgensen, P.M.; Morueta-Holme, N.; Peet, R.K.; Violle, C.; Svenning, J. Limited sampling hampers "big data" estimation of species richness in a tropical biodiversity hotspot. *Ecol. Evol.* **2015**, *5*, 807–820. [CrossRef]
208. SoberónM, J.; LlorenteB, J. The use of species accumulation functions for the prediction of species richness. *Conserv. Biol.* **1993**, *7*, 480–488. [CrossRef]
209. Velásquez-Tibatá, J. Package WhereNext. Available online: https://github.com/jivelasquezt/WhereNext-Pkg (accessed on 28 January 2023).
210. Ferrier, S.; Manion, G.; Elith, J.; Richardson, K. Using generalized dissimilarity modelling to analyse and predict patterns of beta diversity in regional biodiversity assessment. *Divers. Distrib.* **2007**, *13*, 252–264. [CrossRef]
211. Nuñez-Penichet, C.; Cobos, M.E.; Soberón, J.; Gueta, T.; Barve, N.; Barve, V.; Navarro-Sigüenza, A.G.; Peterson, A.T. Selection of sampling sites for biodiversity inventory: Effects of environmental and geographical considerations. *Methods Ecol. Evol.* **2022**, *13*, 1595–1607. [CrossRef]

forests

Article

The Effects of Monthly Rainfall and Temperature on Flowering and Fruiting Intensities Vary within and among Selected Woody Species in Northwestern Ethiopia

Sinework Dagnachew [1,*], Demel Teketay [2], Sebsebe Demissew [3], Tesfaye Awas [1] and Debissa Lemessa [3]

[1] Ethiopian Biodiversity Institute, Addis Ababa P.O. Box 30726, Ethiopia; tesfayeawas@gmail.com
[2] Department of Range and Forest Resources, Botswana University of Agriculture and Natural Resources, Private Bag 0027, Gaborone P.O. Box 47204, Botswana; dteketay@buan.ac.bw
[3] Department of Plant Biology and Biodiversity Management, Addis Ababa University, Addis Ababa P.O. Box 3434, Ethiopia; sebseb.demissew@gmail.com (S.D.); lemdeb@yahoo.com (D.L.)
* Correspondence: sineworkseed@gmail.com; Tel.: +251-911135089

Abstract: The phenological responses of plants to climatic variables are critical for conservation planning; however, it is less understood in an Afrotropical context. Here, we observed how flowering and fruiting phenophases of seven indigenous plant species are related to monthly rainfall and temperature for 24 months in Ethiopia. We employed linear and non-linear models to test the effects on flowering and fruiting intensity. The results of the linear model showed that flowering intensity decreased with increasing monthly temperature for *Maytenus arbutifolia*, *Prunus africana*, and *Solanecio gigas*, but increased for *Bersama abyssinica*, and decreased with increasing monthly rainfall for *Maytenus arbutifolia*. The results of the non-linear model indicated that the flowering intensity of *Brucea antidysenterica*, *Dombeya torrida* and *Rosa abyssinica* decreased, leveled off and increased with increasing monthly temperature. Moreover, the fruiting intensity of *Brucea antidysenterica* and *Rosa abyssinica* decreased with increasing monthly rainfall, but increased for *Bersama abyssinica*; The fruiting intensity increased with increasing monthly temperature for *Brucea antidysenterica* and *Rosa abyssinica*. Altogether, the effects of climatic variables not only vary among the species, but also among the phenophases of a plant species. Hence, considering these varied effects in forest conservation schemes is critical, especially during the epoch of this continuing climate change.

Keywords: flowering patterns; seasonality; tropical trees; climatic variables

Citation: Dagnachew, S.; Teketay, D.; Demissew, S.; Awas, T.; Lemessa, D. The Effects of Monthly Rainfall and Temperature on Flowering and Fruiting Intensities Vary within and among Selected Woody Species in Northwestern Ethiopia. *Forests* **2023**, *14*, 541. https://doi.org/10.3390/f14030541

Academic Editors: Konstantinos Poirazidis and Panteleimon Xofis

Received: 17 January 2023
Revised: 27 February 2023
Accepted: 6 March 2023
Published: 9 March 2023

1. Introduction

Phenology is the study of the periodicity or timing of biological events such as the timing of plant flowering or bird migration, triggered by environmental cues taking place during the year [1,2]. The causes of their timing are affected by biotic and abiotic forces and the interrelation among phases of the same or different species [3]. The climate is the main factor controlling and regulating phenological events in plants, and global warming has affected species distributions and the timing of leaf change and reproduction [4]. How plants respond to changing patterns of temperature and precipitation has major implications for agriculture, forestry, growing season dynamics and ecosystem processes [5].

Plant phenology variations have long been used as indicators of ongoing climate change conditions [6]. Phenology has become important in the context of understanding climate change impacts on ecosystems [7]. Phenology closely tracks climate and also drives many ecological interactions, making the study of phenology essential for predicting how species will respond to climate change [8–10].

Plant phenology is typically quantified by observing the date of onset and the duration of particular phenophases [4]. In general, the study of phenological aspects of plants involves the observation, recording and interpretation of the timing of biological events, such as seed dispersal, germination, vegetative growth, flowering, fruiting and leaf flushing all affected by seasonal changes [8]. Tree phenology observations involve bud growth, leaf emergence, flowering, fruiting and leaf fall activities in seasons with variations in environmental and climatic factors, such as temperature and precipitation [11]. Plant phenology could offer much information on the conditional changes of global climate and the consequent shifts in plant life events, as their recorded dates provide a high-temporal resolution of ongoing changes [4]. Understanding the genetic and physiological mechanisms that plants use for the timing of seasonal responses allows for predicting phenological responses to anthropogenic climate change [12].

A better understanding of plant phenology will assist land managers, seed collectors and other stakeholders to make more informed decisions relating to the management of these species. Plant species phenology is used to realize the vegetative and reproductive potential of the species and in developing viable local conservation strategies [13]. Understanding the phenology of invasive plant species can also help to design control mechanisms [14].

Phenological patterns in plants are influenced by a combination of biotic and abiotic factors that determine the occurrence and inhibition of physiological events [15]. The abiotic and biotic factors are not mutually exclusive, and several are likely to interact to regulate the expression of each phenological phase [3]. Phenology is used to assess the consequences of global warming on plant distributions [16]. However, plant phenology responses to invariant cues, such as photoperiods, may be important in defining the timing, periodicity and particularly the synchrony of plant reproduction, especially in tropical environments where climatic seasonality is low [17].

In tropical ecosystems, phenology might be less sensitive to temperature and photoperiod and more tuned to seasonal shifts in precipitation, as seasons are often marked by differences in rainfall and life-history events occurring in response to water availability [3]. Periodic changes in rainfall, which are caused by movements of the Inter-Tropical Convergence Zone often, play an important role [18]. In forests with a marked dry season, flowering may be more sensitive to seasonal rainfall, changes in water availability and soil moisture, and their fruits ripen towards the end of the dry season or at the beginning of the rainy season [19].

Tropical phenology is characterized by its high diversity due to small annual changes in day length and temperature and the absence of a cool season that hampers growth [20]. Within a tropical forest, species differ in phenological patterns and community-wide phenological patterns differ among regions that differ in climate patterns [21]. Even within a community, phenological cycles vary from species that reproduce multiple times per year to those that reproduce only once in several years [22]. At higher latitudes, since these habitats may be under snow or ice, almost all phenology is tied to a single highly variable event, the timing of snowmelt [23].

Scientific knowledge of the phenology of plant species is basic for the understanding of biological processes and the functioning of the forest ecosystem [24]. Serious effects on biodiversity can result from altered phenologies due to enhanced asynchrony in ecological interactions such as plant–insect and predator–prey interactions, leading to reduced fitness and ultimately extinctions [25]. Moreover, phenology is the major determinant of the plant species range, so it can be used to assess the consequences of global warming on plant distributions [16].

Regarding the phenology of Ethiopian plant species, few studies have been carried out in different floristic regions of Ethiopia. To name a few, the phenology of seven indigenous trees in the dry Afromontane of Munessa-Shashemene Forest, South Ethiopia [24], phenological events in *Commiphora myrrha* and *Boswellia neglecta* in Southeastern Ethiopia [13], phenology of the alien invasive plant species *Prosopis juliflora* in arid and semi-arid [14] and flowering and fruiting phenology of some forest plant species in Western Ethiopia Combretum-Terminalia Woodlands [26]. However, no research findings have been found in the Gojjam floristic region. The present study is the first detailed study on the phenology of flowering and fruiting of seven indigenous tree species categorized under six families of the dry tropical montane forest of this region in Ethiopia for two years from December 2018 to November 2020.

The specific questions addressed in this study were (1) what are the frequency, timing and duration of flowering and fruiting of selected tree species at the study site and (2) which environmental factors (temperature and rainfall) have the greatest influence on triggering these phenological events?

2. Materials and Methods

2.1. Study Area

The study area is located in Machakel woreda, East-Gojjam Zone, Amahara National regional State (ANRS) of Ethiopia (Figure 1) between 10°19′75″–10°41′00″ N and 37°16′46″–37°45′42″ E l. The vegetation of the study area belongs to the dry evergreen Afromontane forest [27]. The maximum monthly temperature is 28.5 °C and the minimum monthly temperature is 9.8 °C. The average monthly temperature is 18 °C and the area receives the mean annual rainfall of 1397 mm with a unimodal pattern that peaks between June to October.

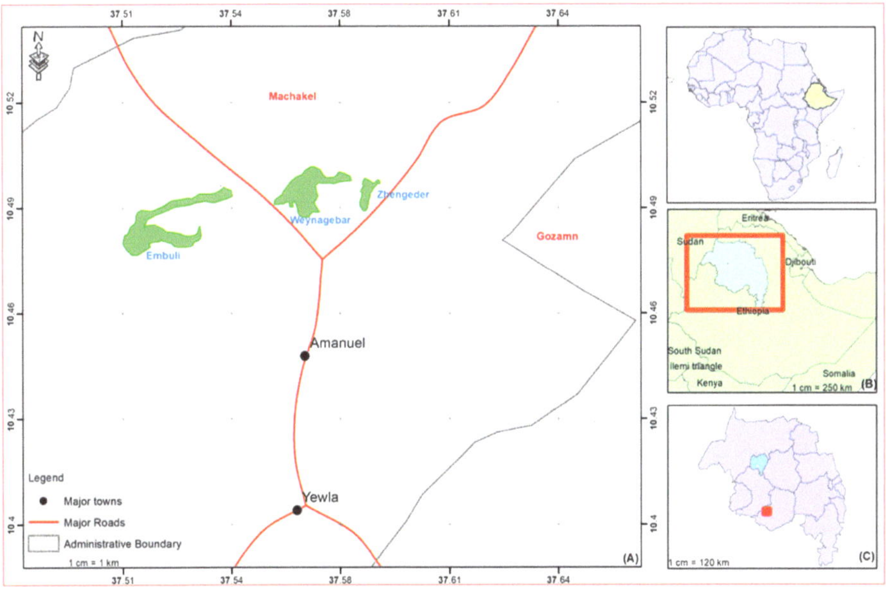

Figure 1. The locations of the three forest study patches in East Gojjam zone, Amhara National Regional State (ANRS) in relation to the map of Ethiopia, (**A**) map of three forest patches, (**B**) map of Ethiopia, (**C**) map of ANRS.

2.2. Study Species

A reconnaissance survey was conducted to select the forest and the study species. Three vegetation patches with similar physiological characteristics and agroecological zones were selected during the reconnaissance survey. To select the study species, from each patch, we used the criteria of (1) dominant woody species in the patches and assumed to play key roles in shaping the ecological system of the patches; (2) species which provide substantial provisional and cultural ecosystem services of wood products, food, medicinal and fodder values; and (3) species whose population are currently declining due to overexploitation for the mentioned services [28–33]. Consequently, seven indigenous tree species, namely *Bersama abyssinca* Fresen., *Brucea antidysenterica* J.F. Mill., *Solanecio gigas* (Vatke) C. Jeffrey, from the Embuli forest patch *Dombeya torrida* (J. F. Gmel.) P. Bamps, *Maytenus arbutifolia*

(A. Rich.) Wilczek, *Rosa abyssinica* Lindley from Zhingder and *Prunus africana* (Hook. f.) Kalkm from Woynagbar were selected for the study. Plant spcies such as *Acacia abyssinica*, *Buddleja polystachya*, *Embelia schimperi*, *Phytolacca dodecandra*, *Vernonia amygdalina*, *Carissa spinarum*, *Rubus* spp. and *Maesa lanceolata* were also observed during the survey.

2.3. Data Collection

A total of 112 reproductive, healthy individual trees/plant species having similar crown sizes (visual estimation) were selected for the study. For each tree species, 16 individuals with their diameter at breast height (DBH) ranging from 10 to 60 cm, were randomly selected 100 m far apart from each other, tagged with plastic ribbon and the locations also recorded using GPS. The data of flowering and fruiting phenophases were collected every month for two years (December 2018–November 2020). During each visit, flowering and fruiting phenophases were recorded through careful observation of the canopy.

The whole crown of each tree was examined and binocular was also used to observe the crowns of tall trees. The intensity of flowering (the number of flower buds and open flowers) and fruiting (the number of unripe and ripe fruits) in each tree crown was assigned to four different classes: 0 (0%), 1 (1%–25%), 2 (26%–50%), 3 (51%–75%) and 4 (>75%) with the percentage values referring to the estimated proportions of each phenophase in the crown by adopting the method used by Tesfaye et al. [24]. The data on monthly rainfall and average monthly temperature of the area were obtained from the National Meteorological Agency of Ethiopia.

2.4. Data Analyses

The data on flowering and fruiting intensity and the effects of monthly rainfall and temperature were analyzed using descriptive, linear and non-linear models. The rainfall data were standardized or scaled to bring the values to a comparable level with the temperature data before running the models and in data visualizations. Before we ran the actual test, we first checked for the multicollinearity between the mean monthly rainfall and temperature explanatory variables. From the scatter plots, we visualized how the relationships between either flowering or fruiting and rainfall/temperature of most species varied not only between species but also among phenophases of the same species. Accordingly, we used the linear model when the relationship between explanatory and response variables is linear, but for non-linear relationships, we used a quadratic polynomial or non-linear regression model to test the effects of monthly temperature and rainfall on flowering and fruiting intensity of *Bersama abyssinica*, *Brucea antidysenterica*, *Dombeya torrida*, *Prunus africana*, *Maytenus arbutifolia*, *Rosa abyssinica* and *Solanecio gigas*. Moreover, we employed the non-linear regression model to test the effects of temperature and rainfall on the flowering intensity of *Brucea antidysenterica*, *Dombeya torrida* and *Rosa abyssinica*, and on the fruiting intensity of *Dombeya torrida*, *Maytenus arbutifolia*, *Prunus africana* and *Solanecio gigas*. For all of the analyses, we used the R program [34].

3. Results

3.1. Flowering Phenology

All species flowered annually during the two-year study period where flowers were observed every year with a unimodal pattern. Observation of flowering periodicity showed both strongly seasonal and continuous flowering patterns throughout the months of the years, ranging from three to twelve months (Figures 2–8).

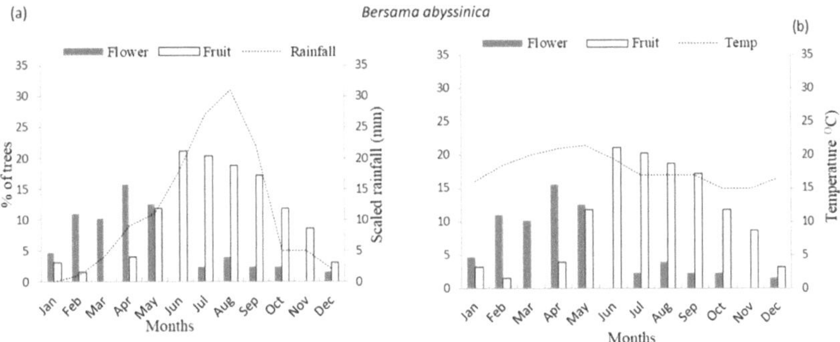

Figure 2. Percentage of flowering and fruiting of *Bersama abyssinica* in relation to: (**a**) monthly precipitation (**b**) average temperature. In which Jan = January, Feb = February, Mar = March, Apr = April, May = May, Jun = June, Jul = July, Aug = August, Sep = September, Oct = October, Nov = November, Dec = December.

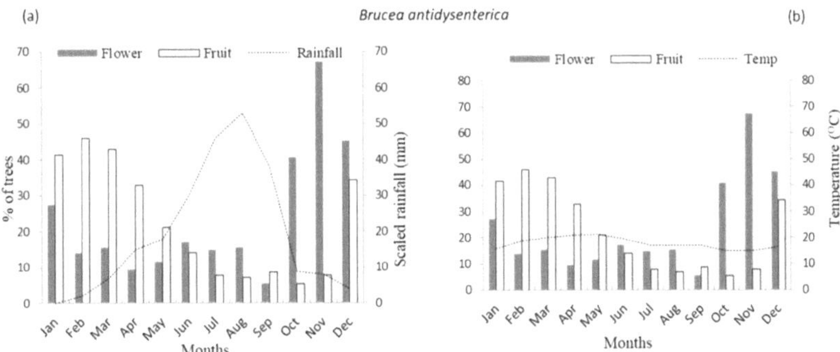

Figure 3. Percentage of flowering and fruiting of *Brucea antidysenterica* in relation to: (**a**) monthly precipitation (**b**) average temperature. In which Jan = January, Feb = February, Mar = March, Apr = April, May = May, Jun = June, Jul = July, Aug = August, Sep = September, Oct = October, Nov = November, Dec = December.

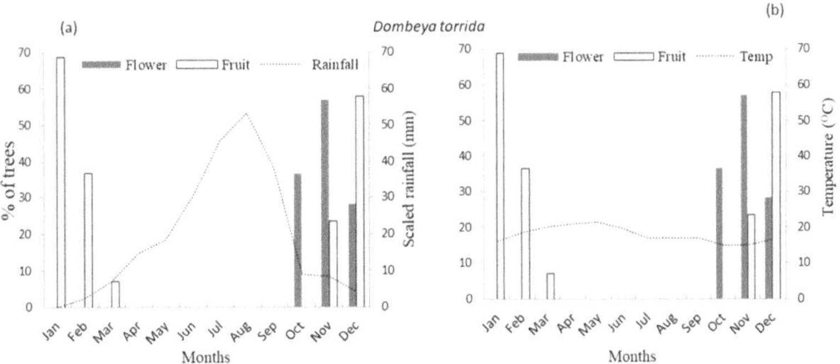

Figure 4. Percentage of flowering and fruiting of *Dombeya torrida* in relation to: (**a**) monthly precipitation and (**b**) average temperature. In which Jan = January, Feb = February, Mar = March, Apr = April, May = May, Jun = June, Jul = July, Aug = August, Sep = September, Oct = October, Nov = November, Dec = December.

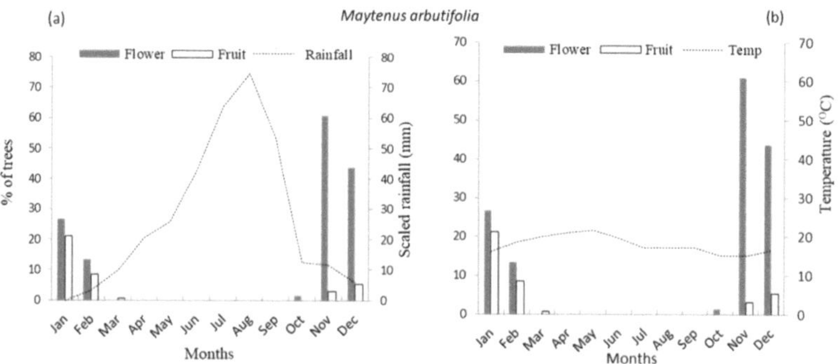

Figure 5. Percentage of flowering and fruiting of *Maytenus arbutifolia*, in relation to: (**a**) monthly precipitation and (**b**) average temperature. In which Jan = January, Feb = February, Mar = March, Apr = April, May = May, Jun = June, Jul = July, Aug = August, Sep = September, Oct = October, Nov = November, Dec = December.

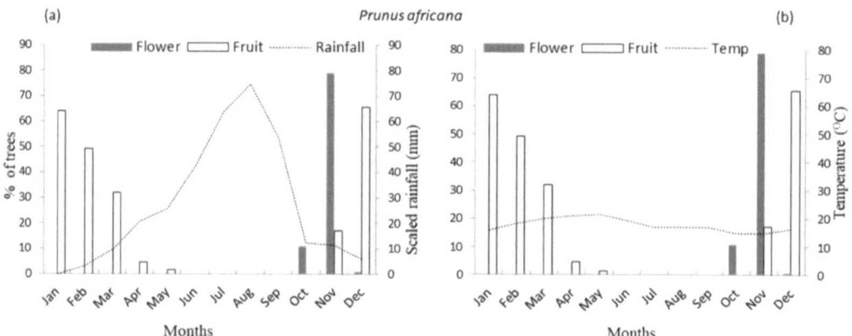

Figure 6. Percentage of flowering and fruiting of *Prunus africana* in relation: (**a**) to monthly precipitation and (**b**) temperature. In which Jan = January, Feb = February, Mar = March, Apr = April, May = May, Jun = June, Jul = July, Aug = August, Sep = September, Oct = October, Nov = November, Dec = December.

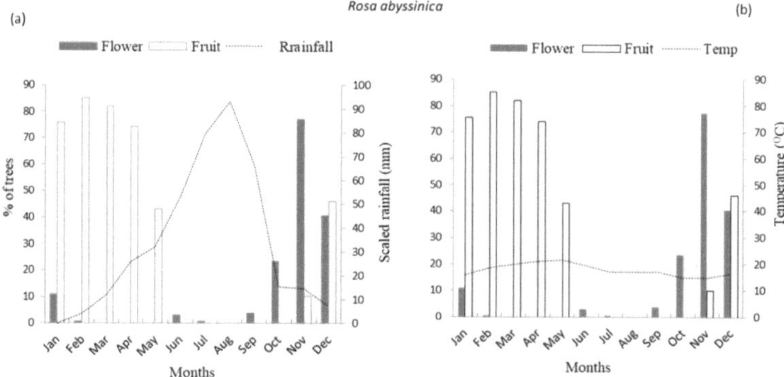

Figure 7. Percentage of flowering and fruiting of *Rosa abyssinica* in relation to: (**a**) monthly precipitation and (**b**) average temperature. In which Jan = January, Feb = February, Mar = March, Apr = April, May = May, Jun = June, Jul = July, Aug = August, Sep = September, Oct = October, Nov = November, Dec = December.

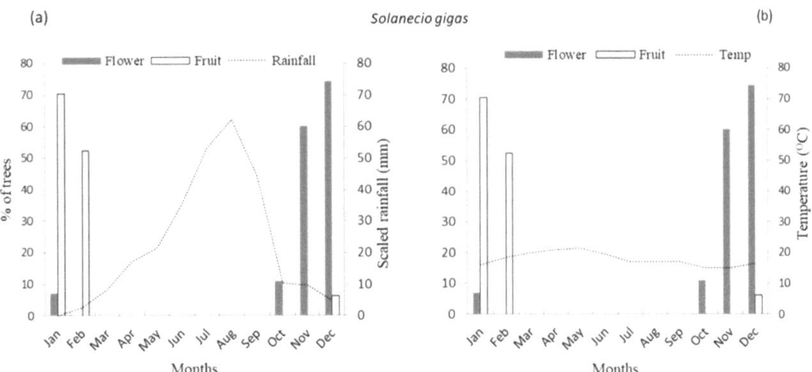

Figure 8. Percentage of flowering of *Solanecio gigas* and fruiting in relation to: (**a**) monthly precipitation and (**b**) average temperature. In which Jan = January, Feb = February, Mar = March, Apr = April, May = May, Jun = June, Jul = July, Aug = August, Sep = September, Oct = October, Nov = November, Dec = December.

Seasonality in the flowering pattern was exhibited by four species: *Dombeya torrida*, *Maytenus arbutifolia*, *Prunus africana* and *Solanecio gigas* where flowering lasted for three to five months with a unimodal pattern (Figures 4, 5 and 8). A relatively shorter flowering period was observed for *Dombeya torrida* and *Prunus africana* as flowering lasted for only three months and, generally, began in early October and continued into late December (Figures 4 and 6). The duration of flowering months for *Maytenus arbutifolia* and *Solanecio gigas* were four and five months, respectively (Figures 2, 5 and 8). The results showed that *Bersama abyssinica*, *Brucea antidysenterica*, and *Rosa abyssinica* exhibited continuous flowering (Figures 2, 3 and 7).

The onset and peak flowering periods of most study species coincided with the end of the rainy season or the beginning of the long dry season (Figures 2–8). The onset of flowering of most seasonal flowering species was October and peak flowering was November (Figures 2–8). However, for *Bersama abyssinica*, even though the flowering is not seasonal, the highest number of individuals flowering was observed at the beginning of the rainy season (Figure 2a). On the contrary, *Brucea antidysenterica* flowering was observed throughout the seasons, but significantly peaked at the end of the rainy season or the beginning of the dry season (Figure 3a).

A linear regression model test showed that the flowering intensity of *Bersama abyssinica* increases with increasing temperature ($p = 0.01$, Table 1, Figure 9a). However, the flowering intensity decreases with increasing temperature for *Maytenus arbutifolia* ($p < 0.01$), *Prunus africana* ($p = 0.04$), and *Solanecio gigas* ($p = 0.036$, Table 1, Figure 9b–d). On the other hand, flowering intensity decreases with increasing rainfall for *Maytenus arbutifolia* ($p = 0.011$, Table 1, Figure 9e).

Table 1. The linear regression model showing the effect of mean monthly temperature (°C) and rainfall (mm) on the flowering percent of species.

No	Species	Variable	Estimate	Standard Error	F-Value	Adjusted R^2	*p*-Value
1	*Bersama abyssinica*	Temp	1.551	0.588	6.958	0.21	0.015
		RF	−0.0127	0.010	4.368	0.23	0.22
2	*Maytenus arbutifolia*	Temp	−4.942	1.518	9.414	0.423	0.004
		RF	−0.073	0.026	9.414	0.423	0.011
3	*Prunus africana*	Temp	−4.330	1.983	4.768	0.14	0.04
		RF	−0.031	0.034	2.763	0.13	0.38
4	*Solaneciogigas*	Temp	−4.71	2.12	3.413	0.17	0.036
		RF	−0.065	0.034	4.554	0.236	0.07

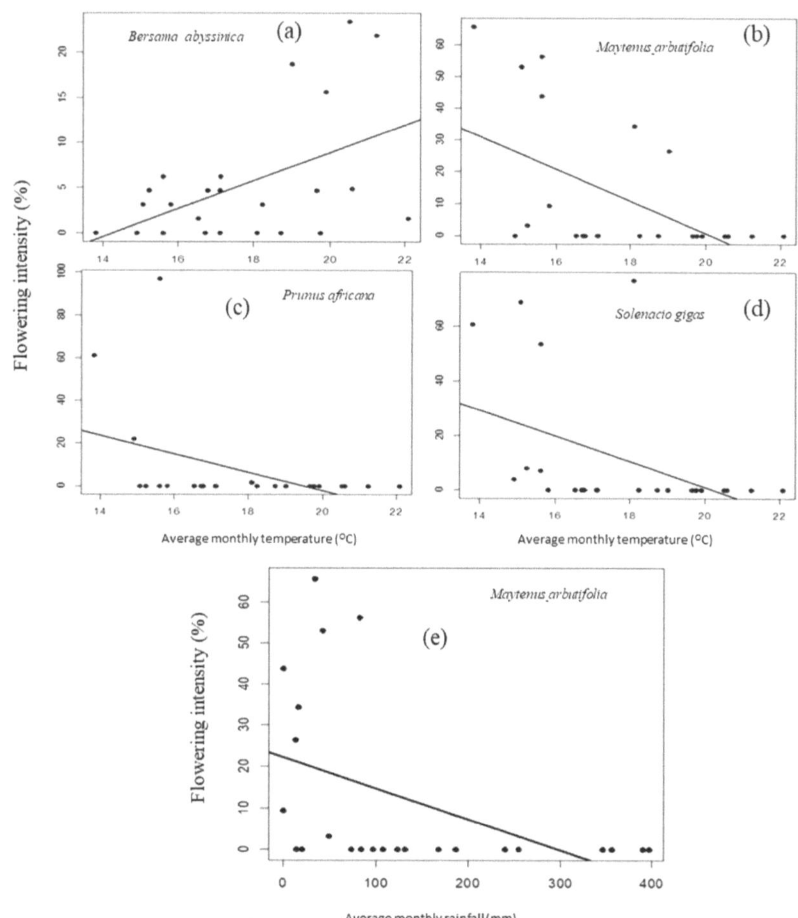

Figure 9. The patterns of the flowering intensity in relation to mean monthly temperature: (**a**) *Bersama abyssina*, (**b**) *Maytenus arbutifolia*, (**c**) *Prunus africana* and (**d**) *Solanecio gigas*, and in relation to monthly rainfall: (**e**) *Maytenus arbutifolia*.

Similarly, the result of the non-linear quadratic polynomial regression model analysis indicated that temperature has both significant decreasing and increasing effects on the flowering percent of the study species (Figure 10). The results of the non-linear models showed that the flowering intensity was decreased, flattened and dropped, and indicated a slightly increasing trend with increasing monthly rainfall for *Dombeya torrida* and *Rosa abyssinica* ($p < 0.01$, Table 2, Figure 10). There was no flowering recorded between the range of temperature around 16 °C and 20 °C (Figures 4 and 7, Figure 10b,c). The temperature shows a significant effect on the flowering of *Brucea antidysentrica* where flowering intensity decreases with increasing temperature ($p = 0.01$, Table 2, Figure 10a). However, the flowering of *Brucea antidysentrica* was observed all over the months of the year (Figure 3).

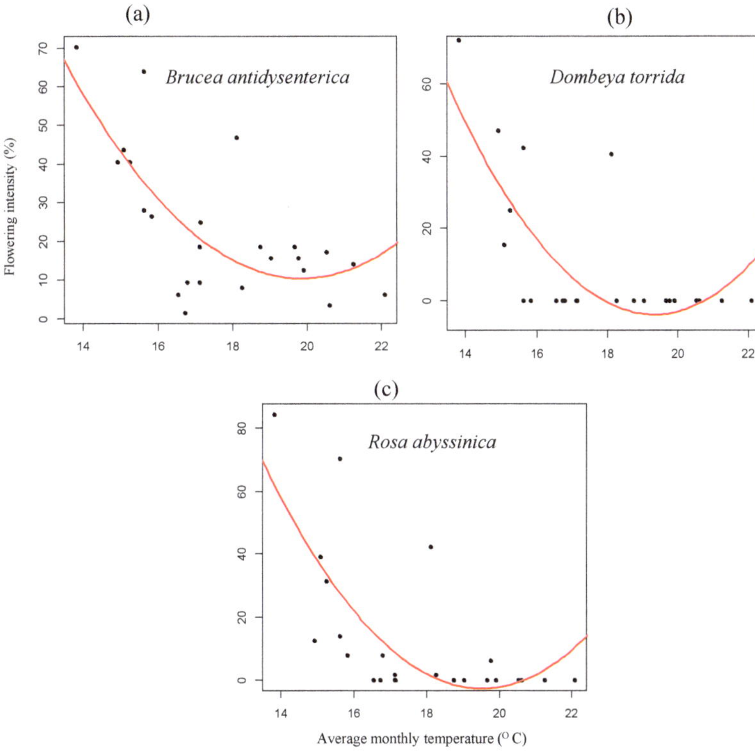

Figure 10. The patterns of flowering intensity in relation to the monthly temperature: (**a**) *Brucea antidysenterica*, (**b**) *Dombeya torrida*, (**c**) *Rosa abyssinica*.

Table 2. The non-linear quadratic polynomial regression model fits showing the effect of average monthly temperature (T) and rainfall (RF) on the flowering percent of tree species.

Species	Factor	Estimates	Standard Error	F-Values	Adjusted R^2	*p*-Values
	T	−55.57	20.45	12.09	0.49	0.01
	T^2	1.4	0.57			0.02
Brucea antidysentrica	RF	−0.1177	0.11	2.41	0.11	0.278
	RF^2	0.00015	0.0002			0.576
	T	−72.75	22.03	12.03	0.48	0.003
	T^2	1.88	0.61			0.006
Dombeya torrida	RF	−0.008	0.12	1.461	0.03	0.498
	RF^2	0.000074	0.0003			0.807
	T	−77.97	26.1	11.51	0.47	0.007
	T^2	1.998	0.73			0.01
Rosa abyssinica	RF	0.131	0.1372	1.649	0.05	0.350
	RF^2	0.0001	0.00035			0.617

3.2. Fruiting Phenology

In most species, fruiting observed extended over several months of the year compared to flowering (Figures 2–8). For most species, the fruiting intensity has been observed to decline towards the beginning of the rainy season (Figures 4–8). Fruits were seen almost year-round in the case of *Bersama abyssinica* and *Brucea antidysenterica*. However, peak fruiting was observed in the middle of the rainy season and at the beginning of the rainy season, respectively (Figures 2 and 3).

The general linear model analysis showed that fruiting was significantly decreased with increasing monthly rainfall for *Brucea antidysenterica* ($p < 0.01$) and *Rosa abyssinica* ($p < 0.01$) (Table 3, Figure 11b,c); whereas, it increased with increasing average monthly temperature ($p < 0.01$, Table 3, Figure 11d,e). The fruiting intensity of *Bersama abyssinica* increased with increasing monthly rainfall ($p < 0.01$, Table 3, Figure 11a).

Table 3. The linear regression model showing the effect of mean monthly temperature (°C) and rainfall on the fruiting percent of species.

No	Species	Variable	Estimate	Standard Error	F-Values	Adjusted R^2	*p*-Values
1	*Bersama abyssinica*	Temp	−0.809	0.858	5.534	0.28	0.356
		RF	0.047	0.015	10.23	0.29	0.004
2	*Brucea antidysenterica*	Temp	3.019	1.07	12.78	0.51	0.010
		RF	−0.079	0.019	12.78	0.51	<0.001
3	*Rosa abyssinica*	Temp	6.97	2.25	15.85	0.56	0.005
		RF	−0.185	0.039	15.85	0.56	<0.001

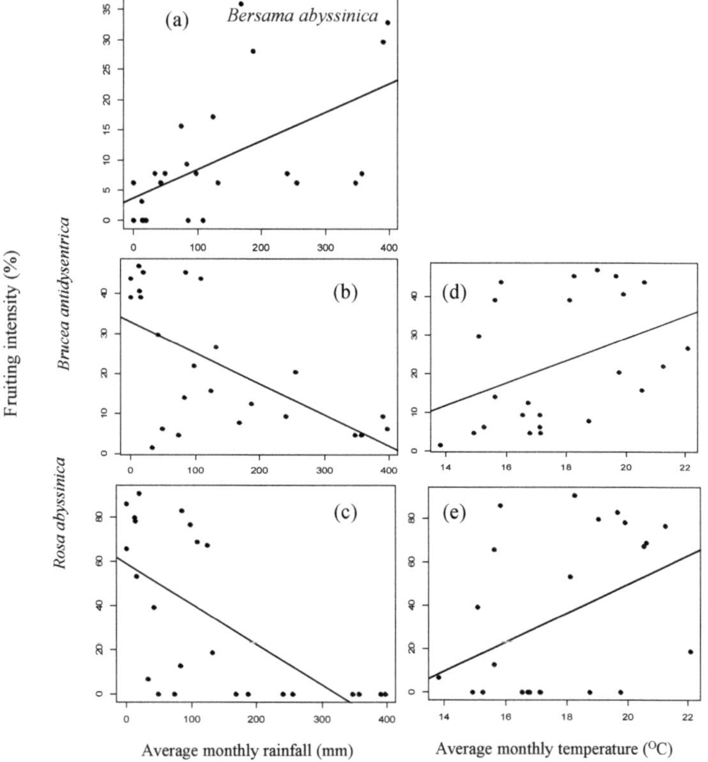

Figure 11. The line graph showing the fruiting intensity in relation to mean monthly temperature and monthly rainfall: (**a**) *Bersama abyssinica*, (**b,d**) *Brucea antidysenterica*, (**c,e**) *Rosa abyssinica*.

The results of the non-linear quadratic polynomial regression model analyses indicated that the average monthly rainfall has significant positive and negative effects on the fruiting percent of *Dombeya torrida, Maytenus arbutifolia, Prunus africana* and *Solanecio gigas* ($p < 0.01$, Table 4, Figure 12a–d). The analyses showed that a high percentage of fruiting was observed at the beginning of the rain, and fruiting percent decreased, leveled off below zero, and slowly increased with increasing monthly rainfall (Figure 12).

Table 4. The non-linear quadratic polynomial regression model fits showing the effect of monthly rainfall (RF) and average monthly temperature (T) on the fruiting percentage of tree species.

Species	Factors	Estimates	Standard Error	F-Values	Adjusted R^2	*p*-Values
Dombeya torrida	RF	−0.44	0.11	11.37	0.47	<0.001
	RF^2	0.0008	0.0003			0.007
	T	8.1143	39.3815	1.45	0.04	0.839
	T^2	−0.3365	1.0942			0.761
Maytenus arbutifolia	RF	−0.1444	0.03285	8.0	0.38	0.002
	RF^2	0.00024	0.000083			0.01
	T	−4.42234	10.57080	1.20	0.02	0.680
	T^2	0.09619	0.29371			0.747
Prunus africana	RF	−0.52	0.088	26.64	0.69	<0.001
	RF^2	0.001	0.0002			<0.001
	T	7.0066	41.4899	0.45	−0.05	0.868
	T^2	−0.2589	1.1528			0.824
Solanecio gigas	RF	−0.42	0.11	8.73	0.40	0.001
	RF^2	0.0009	0.0003			0.006
	T	31.068	38.17	0.74	−0.02	0.425
	T^2	−0.9198	1.0605			0.396

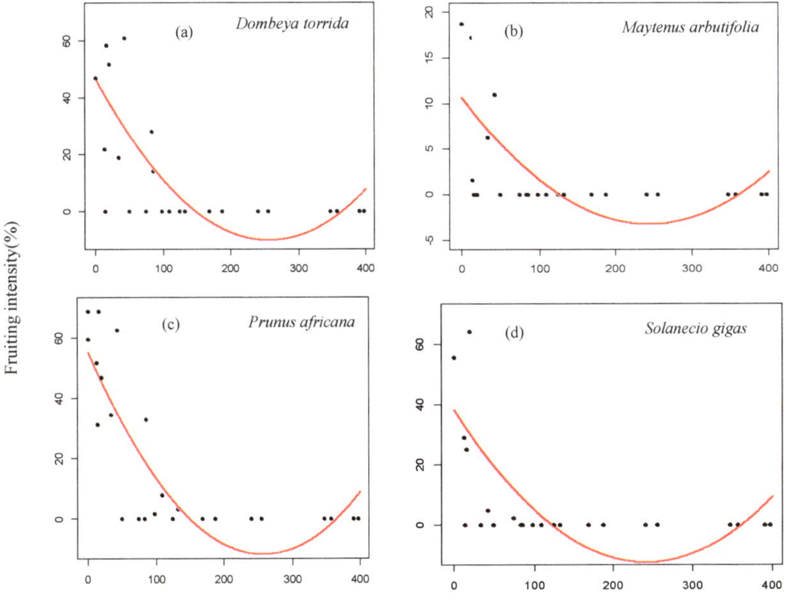

Figure 12. Patterns of the fruiting intensity in relation to monthly rainfall: (**a**) *Dombeya torrida*, (**b**) *Maytenus arbutifolai*, (**c**) *Prunus africana* and (**d**) *Solanecio gigas*.

4. Discussion

4.1. Patterns of Monthly Rainfall and Temperature on Flowering Phenology of Woody Species

Flowering patterns are vital to understanding the dynamics of plant reproduction. The timing of flowering is the result of natural selection mediated by a combination of biotic and abiotic factors, such as temperature and the availability of water, nutrients and light [35]. In tropical ecosystems, seasonal changes in rainfall, temperature and day length are the primary flowering constraints [36,37]. Additionally, flowering patterns are also shaped by biotic factors, mainly plant-pollinator interactions [38]. Therefore, documenting the different flowering responses to plant traits and environmental factors will be important for understanding the strategies of different species to survive and coexist [24,38].

In the present study, the flowering patterns of most study species were predominately annual, unimodal and seasonal in most examined species. However, few species showed continuous flowering patterns throughout the months of the year. The finding is in consistent with the report in our herbarium-based study of the flowering and fruiting phenology of twelve indigenous plant species from Ethiopia [33]. Four species, namely, *Dombeya torrida*, *Maytenus arbutifolia*, *Prunus africana* and *Solanecio gigas* showed seasonal patterns while *Bersama abyssinica*, *Brucea antidysenterica* and *Rosa abyssinica* exhibited continuous flowering. In all cases, the peak flowering periods coincided with the end of the rainy season or the long dry season except *Bersama abyssinica*, which showed peak flowering at the beginning of the rainy season. Several studies have also reported strong seasonality and annual flowering patterns for trees in other dry tropical forests [39–42]. It is also indicated that flower peaks were concentrated in the dry season [24,41,43].

Although both rainfall and temperature affect flowering, the present results showed that the average monthly temperature has more significant effects on the flowering percentage of most study species than rainfall. Tesfaye et al. [24] also reported that flowering significantly correlated to mean monthly temperature. However, the initiation of flowering was observed at the end of the rainy season, which indicates that moisture triggers flowering. Other authors also explained that the flowering phenology of tropical tree species was triggered by moisture [19,21].

For all seasonal flowering, and, even, in most continuous flowering species, flowering peaks at the beginning of the dry season (Figure 3). This may be because the dry season flowering in tropical forests may be enhanced by the higher temperatures and solar irradiance [24]. Berlin [43] also reported that, flowering peaks for most of their study species corresponded to the period of greatest solar irradiance at higher temperatures. According to Janzen [44], tree species in the dry tropical forest flower in the dry season because the wet season is the major period for vegetative growth for species. Therefore, reproduction in the dry season can provide temporal separation of reproductive activity and vegetative growth [44,45]. Since the dry season is characterized by a lack of vegetative competition, this should cause the flowering and fruiting period to shift toward the dry season. The timing of flowering may also be influenced by interactions with other organisms, such as pollinators, seed predators, and herbivores [35]. During the dry season, dry season weather conditions and leaflessness of plants probably favored pollinators (selective force in keeping dry season flowering) and dispersal agents in their pollination activities [44].

Moreover, for seasonal flowering species, the results provided evidence that no flowering individuals of tree species were recorded in the rainy season. This finding is also supported by the fact that flowering during the dry season can benefit a plant by avoiding pollen damage by rain [44]. According to Justiniano and Fredericksen [41], sub-canopy species are less seasonal in their fruiting and flowering. This is possible because of reduced variability in solar radiation, soil moisture and relative humidity in the forest understory [19]. With their lower stature and exposure to lower wind velocities, wind dispersal is not likely to be an effective strategy for sub-canopy species. Hence, seeds of all sub-canopy species are dispersed by animals or gravity [41]. In this study, *Brucea antidysenterica*, which contributes to the sub-canopy stratum in our study site, showed continuous flowering and fruiting.

4.2. Patterns of Monthly Rainfall and Temperature on Fruiting Phenology of Woody Species

Most species exhibited extended fruiting over several months of the year. According to Janzen [44], fruiting extends for several more months because the fruits develop slowly to mature at the possible maximum rates through physiological processes. It has also ecosystem functioning implications in that extended fruiting can ensure the availability of fruit resources for frugivore animals for most parts of the year [41,43]. Extended fruiting was observed, even for species that showed a short period of flowering such as *Dombya torrida* and *Prunus africana*. However, for *Maytenus arbutifolia*, fruiting tends to occur at the same time as flowering. The exact overlapping of flowering and fruiting of specimens has been shown from a herbarium-based study of flowering and fruiting for *Maytenus arbutifolia* [33].

In the seasonal flowering and fruiting species, the peak fruiting was observed through 2–3 months following the peak flowering towards the end of the dry season and/or the beginning of the long rainy season (Figure 3). Such marked dry season fruiting peaks have been reported from dry tropical forests such as Ethiopia [24], Bolivia [41] and Cote d'Ivoire [46].

Fruiting phenology is closely correlated with the seed dispersal mechanism. Most canopy trees had small, wind-dispersed seeds or fruits that matured during the latter part of the dry season. Canopy tree species and species with small seeds may be able to more effectively disperse seeds via wind because they are exposed to stronger dry-season winds at the canopy level [40]. In the current study, *Dobmeya torrida* and *Solanecio gigas* fruiting peaked clearly during the dry season of the study area (January and February). In a study of soil seed banks in dry Afromontane forests, Teketay [47] reported that in the litter layer samples collected at the end of the dry season (March) and in the middle of the rainy period (July) within the same year at Gara Ades, southeastern Ethiopia, there was a marginal difference in the quantity of seeds, but there was a marked difference in species composition. This was attributed to the difference of species in their timing of flowering, fruit maturation and dispersal. Several climax species disperse their seeds on the onset of or during the long rainy period and germinate to form seedling banks on the forest floor while many species disperse their seeds during the dry period [47,48].

It has been shown that rainfall has a significant impact on the fruiting intensity of *Bersama abyssinica*, *Brucea antidysenterica*, *Prunus africana* and *Rosa abyssinica* where the percentage of fruiting intensity dropped towards the beginning of the rainy season. This indicated seeds mature and are dispersed during or at the beginning of the rainy season. These observations of the current study strongly support the idea that fruiting towards the end of the dry season or during the rainy season in tropical forests may ensure the dispersal of seeds when soil moisture conditions are favorable for seed germination, seedling growth, and survival [19,44]. According to Teketay [47] and Tesfaye et al. [49], seedling recruitment in *Prunus africana* was higher in the major rainy season than in the dry season.

5. Conclusions

The results of this study suggest that there is a strong seasonality in the phenological pattern of tree species. The results showed environmental factors, such as rainfall and temperature could influence the reproductive phenology of plants. For *Dombeya torrida*, *Maytenus arbutifolia*, *Prunus africana and Solanecio gigas*, flowering was seasonal and observed for a few months. The timing of flowering was concentrated at the beginning of the dry season or at the end of the rainy season while most species produce fruits during the long dry season. However, *Bersama abyssinica*, *Brucea antidysenterica* and *Rosa abyssinica* flowered or fruited continuously all year round. In most species, flowering was significantly correlated with temperature whereas fruiting was significantly correlated with rainfall. Observations on the phenology of a tree provide basic information since phenological events have many practical implications, such as planning seed collections. It is understood that seed shading/dispersal will take place immediately after peak fruiting; therefore, seed collection before dispersal can be planned based on the results of this study, which could be between January and March for most of the study species.

Author Contributions: S.D. (Sinework Dagnachew) designed the study method, collected data and prepared the manuscript. D.L.: analysed data and contributed to the write-up of the manuscript, S.D. (Sebsebe Demissew) and T.A. contributed to the write-up of the manuscript, and D.T. contributed to the designing of the study method as well as the write up and proof-reading of the manuscript. All authors have read and agreed to the published version of the manuscript.

Funding: The study was supported by "Ethiopian Endemic Plants Project" of the Ethiopian Biodiversity Institute and Addis Ababa University. The research received no external funding. The APC is funded by Demel Teketay.

Data Availability Statement: The datasets generated during the fieldwork and/or analyzed data are available from the corresponding author on reasonable request.

Acknowledgments: The authors are thankful to Ethiopian Biodiversity Institute and Addis Ababa University for financing the project. The authors would like to thank Machakel Wereda Agricultural Development office for the support during field data collection. The first author also thanks National Meteorological Agency of Ethiopia for the provision of all required climate data of the study area.

Conflicts of Interest: The authors declare no conflict of interest.

References

1. Schwartz, M.D. Introduction. In *Phenology: An Integrative Environmental Science*; Schwartz, M.D., Ed.; Kluwer Academic Publishers: Alphen aan den Rijn, The Netherlands, 2003; pp. 3–7.
2. Denny, E.G.; Gers, K.L.; Miller-Rushing, A.J.; Tierney, G.L.; Crimmins, T.M.; Enquist, C.A.F.; Guertin, P.; Rosemartin, A.H.; Schwartz, M.D.; Thomas, K.A.; et al. Standardized phenology monitoring methods to track plant and animal activity for science and resource management applications. *Int. J. Biometeorol.* **2014**, *58*, 591–601. [CrossRef]
3. Sanchez-Azofeifa, A.; Kalacska, M.E.; Quesada, M.; Stoner, K.E.; Jorge, A.; Lob, J.A.; Arroyo-Mora, P. Tropical Dry Climates. In *Phenology: An Integrative Environmental Science*; Schwartz, M.D., Ed.; Kluwer Academic Publishers: Alphen aan den Rijn, The Netherlands, 2003; pp. 121–138.
4. Menzel, A.; Sparks, T.H.; Estrella, N.; Koch, E.; Aasa, A.; Ahas, R.; Alm-Kuebler, K.; Bissolli, P.; Braslavska, O.G.; Briede, A.; et al. European phenological response to climate change matches the warming pattern. *Glob. Chang. Biol.* **2006**, *12*, 1969–1976. [CrossRef]
5. Tang, J.; Körner, C.; Muraoka, H.; Piao, S.; Shen, M.; Thackeray, S.J.; Yang, X. Emerging opportunities and challenges in phenology: A review. *Ecosphere* **2016**, *7*, e01436. [CrossRef]
6. Zhao, M.; Peng, C.; Xiang, W.; Deng, X.; Tian, D.; Zhou, X.; Yu, G.; He, H.; Zhao, Z. Plant phenological modeling and its application in global climate change research: Overview and future challenges. *Environ. Rev.* **2012**, *21*, 1–14. [CrossRef]
7. IPCC. *Climate Change: Impacts, Adaptation, and Vulnerability. Part A: Global and Sectoral Aspects. Contribution of Working Group II to the Fifth Assessment Report of the Intergovernmental Panel on Climate Change*; Cambridge University Press: Cambridge, UK; New York, NY, USA, 2014.
8. Fenner, M. The phenology of growth and reproduction in plants. *Perspect. Plant Ecol. Evol. Syst.* **1998**, *1*, 78–91. [CrossRef]
9. Willis, C.G.; Ellwood, E.R.; Primack, R.B.; Davis, C.C.; Pearson, K.D.; Gallinat, A.S.; Yost, J.M.; Nelson, G.; Mazer, S.J.; Rossington, N.L.; et al. Old Plants, New Tricks: Phenological Research Using Herbarium Specimens. *Trends Ecol. Evol.* **2017**, *32*, 531–546. [CrossRef]
10. Walther, G.R.; Post, E.; Convey, P.; Menzel, A.; Parmesan, C.; Beebee, T.J.C.; Fromentin, J.M.; Hoegh-Guldberg, O.; Bairlein, F. Ecological responses to recent climate change. *Nature* **2002**, *416*, 389–395. [CrossRef]
11. Moza, M.K.; Bhatnagar, A.K. Phenology and climate change. *Curr. Sci.* **2005**, *9*, 243–244.
12. Williams, J.W.; Jackson, S.T.; Kutzbacht, J.E. Projected distributions of novel and disappearing climates by 2100 AD. *Proc. Natl. Acad. Sci. USA* **2007**, *104*, 5738–5742. [CrossRef] [PubMed]
13. Dejene, T.; Mohamed, O.; Yilma, Z.; Eshete, A. Phenology of leaf, flower and fruits of Boswellia neglecta and Commiphora myrrha in Borena Zone, Southeastern Ethiopia. *Hort Flora Res. Spectr.* **2016**, *5*, 269–274.
14. Shiferaw, W.; Bekele, T.; Demissew, S.; Aynekulu, E. Phenology of the alien invasive plant species Prosopis juliflora in arid and semi-arid areas in response to climate variability and some perspectives for its control in Ethiopia. *Pol. J. Ecol.* **2020**, *68*, 37–46. [CrossRef]
15. Lobo, J.; Quesada, M.; Stoner, K.E.; Fuchs, E.J.; Herrerias-Diego, Y.; Rojas, J.; Saborio, G. Factors affecting phenological patterns of Bombaceaous trees in seasonal forests in Costa Rica and Mexico. *Am. J. Bot.* **2003**, *90*, 1054–1063. [CrossRef]
16. Chuine, I.; Beaubien, E.G. Phenology is a major determinant of tree species range. *Ecol. Lett.* **2001**, *4*, 500–510. [CrossRef]

17. Borchert, R.; Robertson, K.; Schwartz, M.D.; Williams-Linera, G. Phenology of temperate trees in tropical climates. *Int. J. Biometeorol.* **2005**, *50*, 57–65. [CrossRef] [PubMed]
18. Igboabuchi, N.A.; Echereme, C.B.; Ekwealor, K.U. Phenology in plants: Concepts and uses. *Int. J. Sci. Res. Methodol.* **2018**, *11*, 8–24.
19. Van Schaik, C.; Terborgh, W.J.; Wright, J.S. The phenology of tropical forests: Adaptive significance and consequences for primary consumers. *Annu. Rev. Ecol. Syst.* **1993**, *24*, 353–377. [CrossRef]
20. Newstrom, L.E.; Frankie, G.W.; Baker, H.G. A new classification for plant phenology based on flowering patterns in lowland tropical rainforest trees at La Selva, Costa Rica. *Biotropica* **1994**, *26*, 141–159. [CrossRef]
21. Sakai, S. Phenological diversity in tropical forests. *Popul. Ecol.* **2001**, *43*, 77–86. [CrossRef]
22. Sakai, S.; Kitajima, K. Tropical phenology: Recent advances and perspectives. *Ecol. Res.* **2019**, *34*, 50–54. [CrossRef]
23. Inouye, D.W.; Wielgolaski, F.E. High altitude climates. In *Phenology: An Integrative Environmental Science*; Schwartz, M.D., Ed.; Kluwer Academic Publishers: Alphen aan den Rijn, The Netherlands, 2003; pp. 75–92.
24. Tesfaye, G.; Teketay, D.; Fetene, M.; Beck, E. Phenology of seven indigenous tree species in a dry Afromontane forest, southern Ethiopia. *Trop. Ecol.* **2011**, *52*, 229–241.
25. Parmesan, C. Influences of species, latitudes and methodologies on estimates of phenological response to global warming. *Glob. Chang. Biol.* **2007**, *13*, 1860–1872. [CrossRef]
26. Mosissa, D. Flowering and Fruiting Phenology of Some Forest Plant Species in the Remnants of Combretum—Terminalia Woodlands of Western Ethiopia. *Acta Sci. Agric.* **2019**, *3*, 126–132. [CrossRef]
27. Friis, I.; Demissew, S.; Breugel, P.V. Atlas of potential vegetation of Ethiopia. In *Biologiske Skrifter*; The Royal Danish Academy of Sciences and Letters: Copenhagen, Denmark, 2010; Volume 58, p. 315.
28. Wube, A.A.; Franz Bucar, F.; Asres, K.; Gibbons, S.; Rattray, L.; Croft, S.L. Antimalarial Compounds from *Kniphofia foliosa* roots. *Phytother. Res.* **2005**, *19*, 472–476. [CrossRef] [PubMed]
29. Teketay, D.; Senbeta, F.; Maclachlan, M.; Bekele, M.; Barklund, P. *Edible Wild Plants in Ethiopia*; Teketay, D., Ed.; Addis Ababa University Press: Addis Ababa, Ethiopia, 2010.
30. Lulekal, E.; Asfaw, A.; Kelbessa, K.; Van Damme, P. Linking Ethnobotany, Herbaria and Flora to Conservation: The Case of Four Angiosperm Families at the National Herbarium of Ethiopia. *J. East Afr. Nat. Hist.* **2012**, *101*, 99–125. [CrossRef]
31. Anza, M.; Worku, F.; Libsu, S.; Mamo, F.; Endale, M. Phytochemical Screening and Antibacterial Activity of Leaves Extract of Bersama abyssinica. *J. Adv. Bot. Zool.* **2015**, *3*, 2348–7313.
32. Meragiaw, M.; Asfaw, Z.; Argaw, M. The Status of Ethnobotanical Knowledge of Medicinal Plants and the Impacts of Resettlement in Delanta, Northwestern Wello, Northern Ethiopia. *Hindawi* **2016**, *2016*, 5060247. [CrossRef] [PubMed]
33. Dagnachew, S.; Teketay, D.; Demissew, S.; Hawas, T.; Kindu, M. Herbarium-based study of phenology of twelve indigenous 563 and endemic species from Ethiopia. *S. Afr. J. Bot.* **2022**, *150*, 260–274. [CrossRef]
34. R Core Team. R: A Language and Environment for Statistical Computing. R Foundation for Statistical Computing. Available online: http://www.R-project.org/ (accessed on 14 July 2022).
35. Ehrlén, J. Selection on flowering time in a life-cycle context. *Oikos* **2015**, *124*, 92–101. [CrossRef]
36. Morellato, L.P.C.; Talora, D.C.; Takahasi, A.; Bencke, C.C.; Romera, E.C.; Zipparro, V.B. Phenology of Atlantic rain forest trees: A comparative study. *Biotropica* **2000**, *32*, 811–823. [CrossRef]
37. Abrahamczyk, S.; Kluge, J.; Gareca, Y.; Reichle, S.; Kessler, M. The Influence of Climatic Seasonality on the Diversity of Different Tropical Pollinator Groups. *PLoS ONE* **2011**, *6*, e27115. [CrossRef]
38. Cortés-Flores, J.; Hernández-Esquivel, K.; González-Rodríguez, A.; Ibarra-Manríquez, G. Flowering phenology, growth forms, and pollination syndromes in tropical dry forest species: Influence of phylogeny and abiotic factors. *Am. J. Bot.* **2017**, *104*, 39–49. [CrossRef] [PubMed]
39. Bendix, J.; Homeier, J.; Ortiz, E.C.; Emck, P.; Breckle, S.W.; Beck, E. Seasonality of weather and tree phenology in a tropical evergreen mountain rainforest. *Int. J. Biometeorol.* **2006**, *50*, 370–384. [CrossRef] [PubMed]
40. Frankie, W.G.; Baker, G.H.; Opler, A.P. Comparative phenological studies of trees in tropical wet and dry forests in the lowlands of Costa Rica. *J. Ecol.* **1974**, *62*, 881–919. [CrossRef]
41. Justiniano, M.J.; Fredericksen, T.S. Phenology of tree species in Bolivian dry forests. *Biotropica* **2000**, *32*, 276–281. [CrossRef]
42. Malizia, L.R. Seasonal fluctuations of birds, fruits, and flowers in a subtropical forest of Argentina. *Condor* **2001**, *103*, 45–61. [CrossRef]
43. Berlin, E.K.; Pratt, K.T.; Simon, C.J.; Kowalsky, R.J. Plant phenology in a Cloud Forest on the Island of Maui, Hawaii. *Biotropica* **2000**, *32*, 90–99. [CrossRef]
44. Janzen, D.H. Synchronization of sexual reproduction of trees within the dry season in Central America. *Evolution* **1967**, *21*, 620–637. [CrossRef]
45. Bawa, K.S.; Kang, H.; Grayum, M.H. Relationship among time, frequency and duration of flowering in tropical rain forest trees. *Am. J. Bot.* **2003**, *90*, 877–887. [CrossRef]
46. Anderson, D.P.; Nordheim, V.E.; Moermond, C.T.; Bi, Z.B.G.; Boesch, C. Factors influencing tree phenology in Tai National Park, Cote d'Ivoire. *Biotropica* **2005**, *37*, 631–641. [CrossRef]
47. Teketay, D. Seed Ecology and Regeneration in Dry Afromontane Forests of Ethiopia. Ph.D. Thesis, Swedish University of Agricultural Sciences, Umeå, Sweden, 1996.

48. Burger, W.C. Flowering periodicity at four levels in eastern Ethiopia. *Biotropica* **1974**, *6*, 38–42. [CrossRef]
49. Tesfaye, G.; Teketay, D.; Fetene, M.; Beck, E. Regeneration of seven indigenous tree species in a dry Afromontane forest, southern Ethiopia. *Flora* **2010**, *205*, 135–143. [CrossRef]

*forests*

Article

Effects of Forestry Transformation on the Landscape Level of Biodiversity in Poland's Forests

Ewa Referowska-Chodak [1,*] and Bożena Kornatowska [2]

[1] Institute of Forest Sciences, Warsaw University of Life Sciences, Nowoursynowska 159, 02-776 Warszawa, Poland
[2] Institute of Environmental Protection—National Research Institute, Krucza 5/11D, 00-548 Warszawa, Poland; bozena.kornatowska@ios.edu.pl
* Correspondence: ewa_referowska_chodak@sggw.edu.pl

Abstract: At all times, historical, political, economic, and social factors have affected the management of forests, with direct and indirect effects on the landscape. This study aimed to trace the impact of Poland's forestry evolution over the last 75 years (1945–2020) on forest biodiversity at the landscape level. Five indicators were selected (forest area, forest fragmentation, protected forests, protective forests, harvesting intensity) to identify directions and dynamics of changes of the forest landscape and their determinants and repercussions. In addition, there were determined forest landscapes threats and recommendations for further action and intervention were formulated. The study period embraced two eras of widely divergent political-economic conditions in Poland (socialism and democracy). In the socialism era (1945–1989), there promptly increased total forest cover, wood resources (total growing stock) and the total area of protective forests (essential for safeguarding biodiversity, including the landscape level). In the era of democracy (1990–2020), average growing stock density increased intensely, and at the same time, a greater emphasis was put on reducing forest fragmentation and clear-cut logging. The results obtained showed equal average increase in the area of protected forests in both eras under the study (most intense at their crossing point). In view of the protection of biodiversity at the forest landscape level, the changes throughout the study period were considered positive, although not without problems and challenging consequences for foresters. The determined pressures to the forest landscapes, requiring legal, political, or financial solutions, include a risk of alteration of the ownership structure of Poland's forests or possibility of operational changes in the State Forests National Forest Holding; outdated forest policies; organizational difficulties in the forest landscape protection; insufficient conservation funding; uneven distribution and further fragmentation of forests; and—last but not least—climate change impacts, including extreme weather events and droughts.

Keywords: SFM indicator; forest area; forest fragmentation; protected forest; protective forest; harvesting intensity

Citation: Referowska-Chodak, E.; Kornatowska, B. Effects of Forestry Transformation on the Landscape Level of Biodiversity in Poland's Forests. *Forests* **2021**, *12*, 1682. https://doi.org/10.3390/f12121682

Academic Editors: Konstantinos Poirazidis and Panteleimon Xofis

Received: 13 November 2021
Accepted: 27 November 2021
Published: 1 December 2021

1. Introduction

Forests play an exceptional role in maintaining biodiversity [1], including that at the landscape level. Fulfillment of this function has always been affected by the way of forests management (e.g., [2]), reliant upon inevitably changing historical, political, economic, and social conditions. As a consequence, forest management has been reflected in functioning and diversity of forests and their role in shaping the landscape. A comprehensive analysis of these relationships, carried out in consideration of conceivable conflicts, the long-term perspective and large spatial scale, can be a source of insights and inspirations—useful for forest management improvement, not only within boundaries of a given country, but also at a level of, e.g., the continent. The present study attempted to depict these issues by the example of Poland (Central Europe—Figure 1).

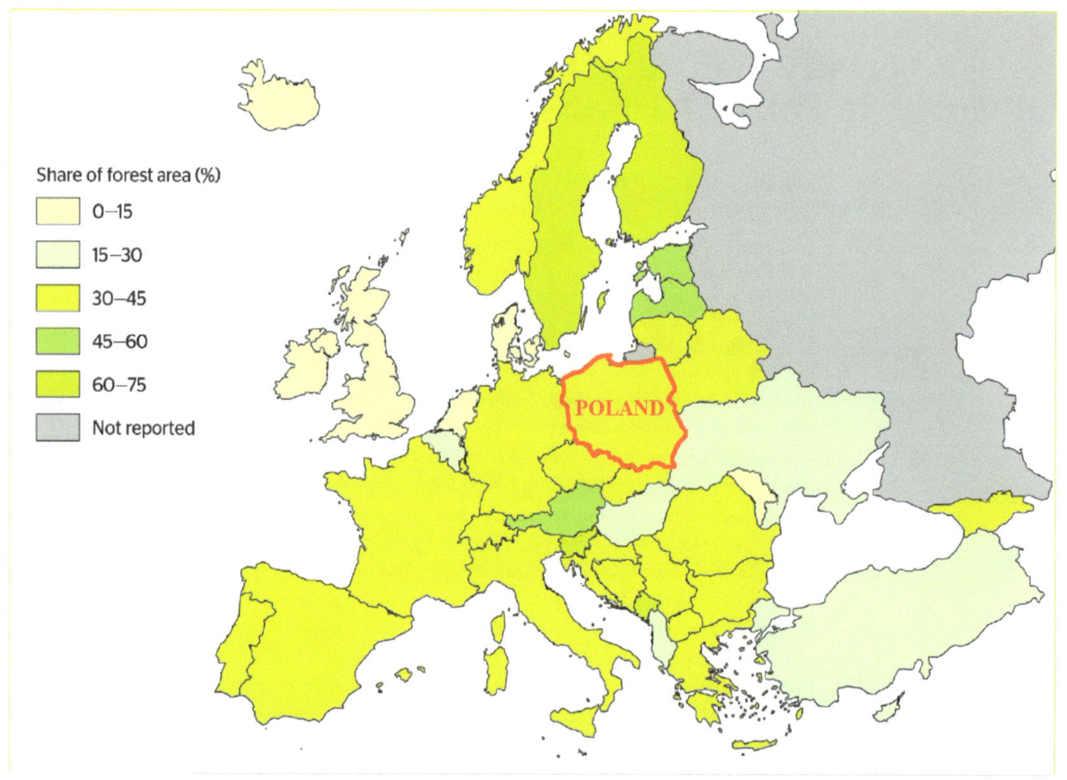

Figure 1. Poland's forest area as compared to other countries of Europe (source: [1], modified).

In the long ago past, almost entire area of then Poland was covered by primeval forests, except for high altitude mountainous region [3]. On account of the country development, already at the end of the 1700s, the forest cover within its then borders was 40% [4]. At that time, Poland was partitioned between the bordering empires (Russia, Prussia and Austria), who intensively exploited local environment for 123 years (1772–1918) [5–7]. It resulted in, inter alia, the simplification and standardization of Poland's forest landscape [8–11].

The foundations for the Polish model of multifunctional forestry were laid between 1918 and 1938. In 1924, there was established the Polish State Forests Enterprise who is still functioning as the State Forests National Forest Holding (State Forests, SF, also in the sense of forests managed by this enterprise). The development of Polish forestry was interrupted by the World War II (WW2). In 1939–1945, Poland's forests were very demolished by warfare and military battles fought on the Eastern Front, as well as were indiscriminately exploited by the German occupying forces. As compared to that at the end of the 1700s, Poland's forest cover as of 1945 was reduced almost by two (from approximately 40% to 20.8%) [12–14].

In the years 1945–1989, there was implemented a socialist model of economy in Poland, centrally planned and regulated, characterized by strong nationalization and industrialization of the country [12,15,16]. Forest management at that time followed a resource-based economic model [8,17]. Sustainability and ecological principles were of secondary importance [17], as only the achievement of production goals mattered [18]. At the same time, increasing environmental pollution entailed a strong deterioration of forest health [19,20], in extreme situations leading to loss of entire stands. Consequently, forest biodiversity at all levels was declining, regardless of efforts undertaken by forest managers to alleviate the problem.

At the end of the 1980s, only the historic change of Poland's political system, i.e., the transition from socialism into democracy, allowed for intensification of efforts towards sustainable management of national forests. This coincided with active participation of Poland in the works of the Convention on Biological Diversity (CBD) and the Ministerial Conference on the Protection of Forests in Europe (MCPFE—now FOREST EUROPE), associated with taking responsibility for the commitments imposed under the framework of these processes. Starting from the beginning of the 1990s, the importance of forest biodiversity and ecosystem services was recognized through the adoption of relevant legislation: the Forest Act [21], which stressed the necessity for permanent preservation of forests and included the concept of sustainable forest management (SFM) as well as emphasized the need of its implementation in Poland's forests; National Forest Policy [22], which gave high priority to efforts towards improving forest biodiversity [23], and the orders by the Director General with regard to the best ecological practices in forest management [24,25].

The beginning of 21st century brought further political changes in Poland—the accession to the European Union (EU) in 2004, preceded, among others, by comprehensive works concerning the implementation of the Habitats and the Birds Directives. Both directives have played a fundamental role in the enforcement of nature conservation in national forests [26]. In the context of the present study, ratification (2003) of the Aarhus Convention [27] was of great importance, as then Poland's society was provided for tools to impose demands for forest management with the use of ecological solutions.

In view of the above, the main objective of the present study was to evaluate effects of evolution of forestry under Poland's conditions on forest biodiversity at the landscape level, with the use selected indicators. In the perspective of 75 years, the specific objectives were distinguished: (1) to determine the direction and dynamics of changes in the forest landscape, (2) to identify the determinants of the observed changes and their impacts (3) to identify threats to the forest landscape and the direction of further action for its benefit.

2. Materials and Methods

2.1. Indicators

At a European level, assessing and reporting progress on sustainable forest management at regional and national levels has been carried out with the use of a set of criteria and indicators (C&I), among which those concerning the status of forest biodiversity reflect the state of more than one of its levels [28,29]. For instance, C&I Criterion C4: *Maintenance, conservation, and appropriate enhancement of biological diversity in forest ecosystems* describes a variety of existing life forms as well as their ecological roles and genetic diversity. In this context, e.g., forest stand species composition (Indicator 4.1 *Diversity of tree species*) can be considered in view of both species diversity and ecosystem diversity, taking into consideration the effect of individual species on the spatial structure of the entire ecosystem. For the purpose of the present study, the influence of forest management evolution in Poland, especially in forests managed by the State Forests National Forest Holding, on forest biodiversity at the landscape level was based on selected indicators that relate to different levels of forest biodiversity, and especially to that at the landscape level.

The effects of Polish forestry evolution on forest biodiversity at the landscape level were evaluated with reference to the indicators in the set of C&I for SFM [28] and those proposed by Mederski et al. [17]. Bearing in mind the influence of the ecological course of action on the forest landscapes, the following indicators were chosen:

1. Forest area—indicator for SFM (C&I CRITERION 1 *Maintenance and Appropriate Enhancement of Forest Resources and their Contribution to Global Carbon Cycles*, Indicator 1.1 *Forest area* [28]). For the purpose of this study the term "forest area" refers to the area (ha) physically covered by forests (or temporarily deprived of them), exclusive of lands associated with forest management, defined as "Land occupied for use for forest management purposes: buildings and structures, forest zoning lines, forest roads, forest nurseries, timber storage areas, water reclamation facilities, land under power lines, forest parking lots, and tourist facilities." (Forest Act [21]). Being the

most complex plant formation, forests constitute an indispensable component of the landscape. Their area represents their share in the landscape, in other words forest cover is understood as the percentage ratio (%) of forest area to the total geodetic area of the country [30];

2. Forest fragmentation—indicator for SFM (C&I, Indicator 4.7 *Forest fragmentation* [28]. There were examined: the number of forest patches (items), average forest patch size (ha), the share of separated forest patches and of continuous forest (%), as well as by the share of forests patches of different size in total forest area (%). The way the forest is shaped (continuous stands vs. isolated patches, smaller vs. larger patches) is not without influence on landscape mosaicity. Forest fragmentation translates into conditions for maintenance and enhancement of ecosystems, species and gene pool;

3. Protected forests—indicator for SFM (C&I, Indicator 4.9 *Protected forests* [28]). This indicator was considered in view of a broader approach assumed by Mederski et al. [17], i.e., "*Forest functions—protection* vs. *economic role*". In this study, the emphasis was put on nature conservation forms that are crucial for the protection of forest biodiversity as well as changes of their area (ha). At the landscape level, protected forests can be distinguished from managed forests, by giving the impression of "more natural". At the same time they provide favorable conditions to support natural processes and safeguard valuable habitats and species;

4. Protective forests—indicator for SFM (C&I, Indicator 5.1 *Protective forests—soil, water, and other ecosystem functions—infrastructure and managed natural resources* [28]). Likewise in the case of C&I Indicator 4.9, this one was considered in view of a broader context proposed by Mederski et al. [17] "*Forest functions—protection* vs. *economic role*". The present paper focused on the protective forests established with the aim to enhance biodiversity conservation, and they were characterized with reference to changes in their surface area (ha). The protective forests analyzed under this study are to a big extent analogous to protected forests;

5. Harvesting intensity—this indicator covers a complex issue, for which the most important indicator is the intensity of the use of annual wood increment (fellings as percent of net annual increment). It refers to: C&I for SFM [28], i.e., CRITERION 3 *Maintenance and Encouragement of Productive Functions of Forests (Wood and Non-Wood)*, Indicator 3.1 *Increment and fellings*. Based on the latter and taking into account total growing stock in m^3 (C&I Indicator 1.2 *Growing stock* [28]), harvested timber volume (net, without bark, in m^3 —as in the indicator *Wood production* proposed by Mederski et al. [17]), as well as average growing stock density (standing timber volume m^3 per ha of forest area), was assessed in view of relationships between harvesting intensity and sustainability and quality of the forest landscape.

2.2. Scope of Analyses

The present study focused on analyzing the effects of forest management evolution in state-owned forests (SF), mainly due to easier access to reliable information and data, as well as the homogeneity of forest management objectives pursued at a large spatial scale. The State Forests is the largest specialized public forest management entity in the EU [31]. Today it manages almost 77% of the total forest area in Poland [30]. The references to the country as a whole were made only when relevant information on state-owned forests (SF) was unavailable or in the cases pertinent for the results presented.

The study period covered the period 1945–2020 of Poland's history. There was assumed that the effects of forestry evolution on forest biodiversity should be presented based on data compiled in 10-year intervals, marked by the following years: 1950 (state reconstruction after the World War II; socialist economy model forced; reforestation/afforestation activities undertaken), 1960 (socialist economy model fully implemented; increased forest cover), 1970 (socialist economy model; deterioration of forest health), 1980 (initial political and economic changes; starting point for environmental protection), 1990 (transformation towards free market economy; instigation of key changes in forest management),

2000 (preparations for Poland's accession to the EU and adoption of EU legislation), 2010 (further greening of forest management and counteracting climate change; continuation of establishing Natura 2000 sites; implementing the Aarhus Convention) and 2020 (continuation of forest sustainable management; increased pressure of the society on the protection of national forests). The analysis of the results obtained focused on the two periods in the history of Polish forestry—the era of socialism (1945–1989) and the era of democracy (1989–2020). The year 1990 was assumed as the milestone in the adopted timeline.

In the case of lacking data for the selected study years, information was supplemented based on available data for the years as close as possible to those studied. The tables show data referring to the end of a given year. In some cases, this did not apply to the year 2020 due to so far absence of statistical summaries, hence, information for the beginning of 2020 was presented. Due to specifics of national statistics system operating in the late 1950s and early 1960s (the marketing year covered the last quarter of the previous year and the first three quarters of the following year), available data on wood resources and harvesting in the year 1960 were calculated as a sum of $\frac{3}{4}$ values of the parameter for the marketing year 1959/1960 and $\frac{1}{4}$ values for the year 1960/1961.

The substantive scope of the work included: evaluating information/data compiled in terms of characteristics of the feature analyzed, taking into account variation over time as well as the direction and dynamics of changes; providing a comprehensive analytical commentary; identifying threats to the forest landscape and indicating directions of beneficial actions.

2.3. Sources of Information

The presented numerical data come from statistical Yearbooks on forestry and environmental protection published mainly by Statistics Poland (GUS), the reports published the State Forests National Forest Holding including those financial/economic, the Forest Data Bank [32], information regarding forest status update [33–36], monographs and articles prepared for the needs of the reports State of Europe's Forests (SoEF) prepared by MCPFE/FOREST EUROPE [1]. Due the lack of relevant studies/reports for the years at the beginning of the study period, some data were not available, nevertheless, the trend and dynamics of changes could still be revealed. For the purpose of the discussion of the results, there were used the results of several articles from Scopus database (keyword: Polish forestry), as well as those found using a snowballing approach.

3. Results and Discussion

3.1. Forest Area

The change in forest area is one of the key elements affecting the increase, maintenance or decline of the number, quality, and intensity of ecosystem services provided by forests, including those associated with the landscape. Ever since until 1945, Poland's forest area had gradually declined, and this general trend could not be reversed even by local forest succession on abandoned agricultural lands [37–39]. In 1945, the country's forest area reached its lowest ever value. At that time, 5408 thousand ha of forests remained under SF administration, which was 83.6% of existing forests in Poland [40,41]. In the following years the area of state-owned forests (SF) as well as of those privately owned progressively increased (Table 1).

Table 1. The area of forests administered by SF as compared to all Poland's forests *.

Year	1950	1960	1970	1980	1990	2000	2010	2020
Forest area [thousand ha]	6915	7684	8432	8622	8694	8865	9121	9260
Forest cover [%]	22.2	24.6	27.0	27.6	27.8	28.4	29.2	29.6
Forests managed by SF [thousand ha]	5740	6136	6503	6716	6805	6953	7072	7121
Share of forests managed by SF in total forest area [%]	83.0	79.9	77.1	78.0	78.3	78.4	77.5	76.9

* sources: [19,41–46].

The area of forests. Taking into consideration the whole study period, in 2020, the area of Polish forests managed by SF (Table 1) was almost 132% of that in 1945 (average annual expansion: 0.42%). However, once evaluated against the background of transition from socialism/state-directed economy to democracy/free market economy, the forest area managed by SF in 1990 was almost 126% of that in 1945 (average annual expansion: 0.57%). In the subsequent 30 years (1990–2020), forest area increased just by 5% (average annual expansion: 0.15%).

Over the study period, the significant increase of the area of Poland's forests (Figure 2) was primarily associated with intensive afforestation (mainly by planting young trees) on state-owned (including nationalized) non-forest lands, carried out in the period from 1945 to the late 1970s [7,12]. In the 1980s, Polish economy began to collapse, followed by the crisis in the country's development, intensified in the second half of the 1980s, which among others, caused a downfall in the timber market and, consequently, an evident decrease in the number of tasks carried out by SF [12]. Another increase in the extent of afforestation (mainly by planting) took place between 1995 and 2005 (free-market economy). This was due to the adoption and implementation of the National Program of the Augmentation of Forest Cover [47]. The main objective of the Program was to increase forest cover to 30% by 2020 and then to 33–34% by 2050 [7,48,49].

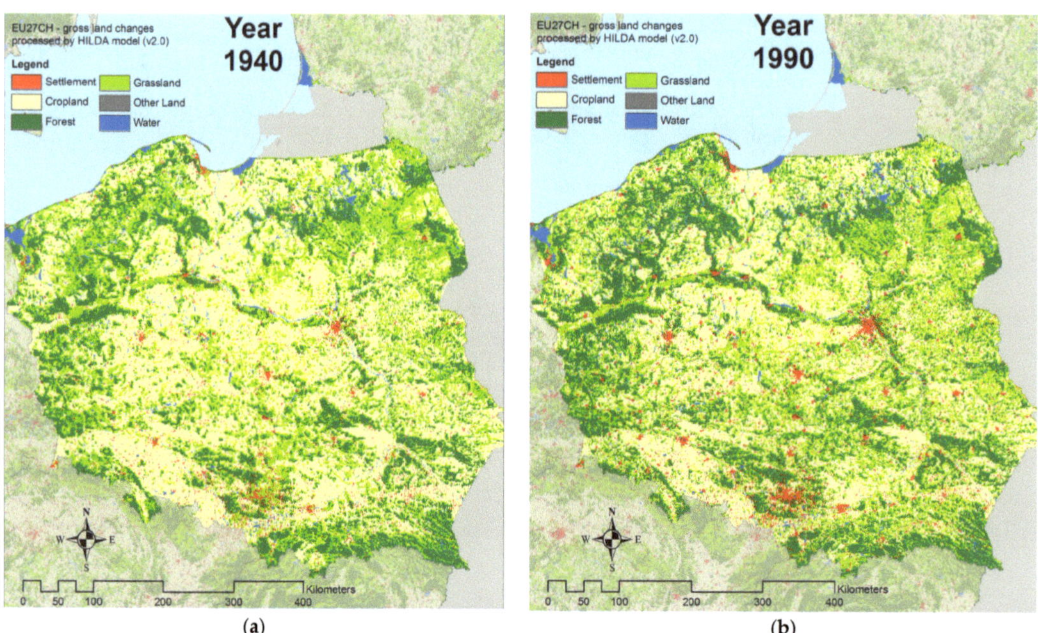

(a) (b)

Figure 2. *Cont.*

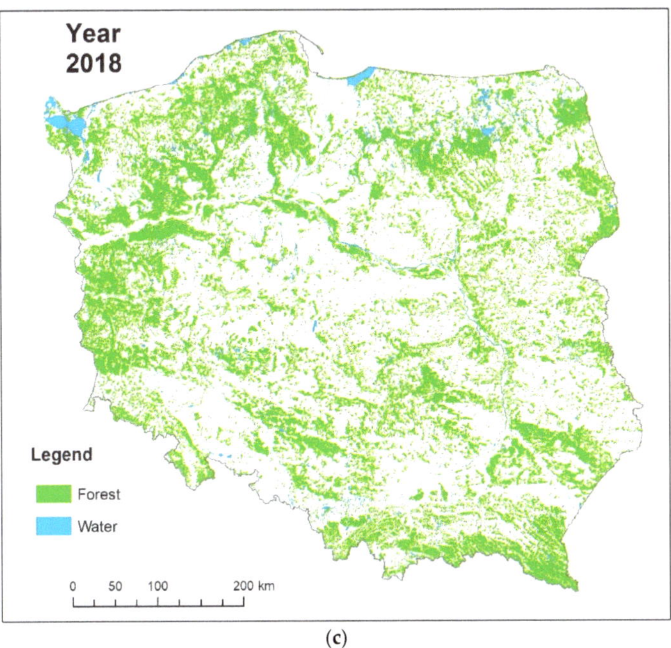

(c)

Figure 2. (**a**) Reconstruction of historic land cover of Poland in 1940 (based on [50]); (**b**) Reconstruction of historic land cover of Poland in 1990 (based on [50]); (**c**) Forest cover of Poland in 2018 (based on [51]).

Over the past years, forest expansion in Poland has slowed down as it has across Europe [1]. In Poland, this has been due to: higher EU subsidies for keeping land in agricultural production; an unfavorable change in the minimum area for afforestation subsidies; reduction of the maximum area designated to reforestation/afforestation; elimination of financial support for afforestation of permanent grassland; restrictions on agricultural land trade; insufficient financial support for farmers; changeability of the financing system, as well as lower and lower supply of state land that could still be afforested [7,41,49,52,53]. The latter is of the greatest importance for the State Forests. The afforestation potential of Poland is estimated to be up to 2 million ha of poor soils which do not guarantee the profitability of agricultural production [48], however, these sites often support habitats of valuable non-forest ecosystems and species.

Forest cover. Efforts undertaken towards augmentation of forest cover in Poland (Figure 2) have resulted in achieving forest cover of 29.6% (Table 1), against the average of 34.8% in Europe, which ranks Poland only 27 on the continent, even with ninth largest forest area (excluding Russia) [1]. It should be noted, however, that official Polish statistics on forest land often do not include abandoned agricultural lands, with noticeable forest vegetation [7]. When these were included in statistics, in 2014, Poland's forest cover was 29.4%, while the real forest cover was 32.0% [7,54]. In view of the above, in Poland, there is a need to classify all lands covered by forests into the category: forest land and to update statistics as regards forest cover [49,54].

Share of forests managed by the State Forests. Although the share of forest area administered by SF declined somewhat over the study period, it is still high, which is characteristic of the former socialist "Eastern Block" countries [1,7,55]. In view of Poland's biodiversity conservation (all levels, including the landscape), state ownership of forests/lands constitutes a pillar of nature conservation [56].

3.2. Forest Fragmentation

The size and location of the forest have an influence on its role in shaping the landscape and determine forest functions associated with supporting ecosystems and providing habitats for species [57,58]. The minimum forest size to ensure the protection of forest interior species is 30–40 ha [59]. However, the best conditions are provided by the core forest area of at least 10,000 ha, covered by continuous forest with a density exceeding 90%. The conditions are also very good, if adjacent forest patches are connected with core forest areas [57].

As estimated, deforestation for agricultural purposes carried out in Poland over many centuries has led not only to the reduction in forest area, but also to its considerable fragmentation [60,61]. Such adverse effects were observed in the socialism era [22]. Due to the lack of SF-only data, Table 2 presents available statistics for all Poland's forests (the vast majority of which is managed by SF—almost 77% in 2020 [46]). From the perspective of SF, the percentage values presented may be partial as they concern also private forests (just over 19% [46]), where fragmentation is higher [39]. The data presented below (Table 2), reflecting the situation in 21st century (democracy era), were compiled consistent with a uniform methodology.

Table 2. Fragmentation of Polish forests *.

Year	1950	1960	1970	1980	1990	2000	2010	2020 [2]
Number of forest patches [items]	n.a.	n.a.	23,020 [1]	n.a.	n.a.	8666	n.a.	8647
Average forest patch size [ha]	n.a.	n.a.	n.a.	n.a.	n.a.	1324	n.a.	1325
Separated forest patches [%]	n.a.	n.a.	n.a.	n.a.	n.a.	4.4	n.a.	4.5
Continuous forest [%]	n.a.	n.a.	n.a.	n.a.	n.a.	95.6	n.a.	95.5
Share of forest patches up to 10,000 ha in total forest area [%]	n.a.	n.a.	n.a.	n.a.	n.a.	16.3	n.a.	16.3
Share of forest patches larger than 10,001 ha [%]	n.a.	n.a.	n.a.	n.a.	n.a.	83.7	n.a.	83.7

* sources: [1,62]; [1] data for 1969 and the SF only, [2] data for 2018; n.a.—information not available.

Number of forest patches. In the process of economy transition, as said by some available information sources, in the forests exclusively under SF management, there occurred "tens of thousands of forest patches" [22]. In contrast, data for the year 2000 presented in Table 2 (for Poland as a whole) indicate the existence of about 9000 forest patches. This considerable discrepancy stems from not the same methods of classification. The efforts to decrease the number of forest patches undertaken in the course of transition are reflected here only to a small extend, even though the reduction of forest fragmentation through coherent forest stands connected with ecological corridors is one of the most important objectives of the National Program for the Augmentation of Forest Cover [47,49,63]. It is also worth noting, that although the forest area increased by about 400 thousand hectares in 21st century (Table 1), the number of forest patches has somewhat decreased (Table 2), which proves a positive direction of change. Nowadays, however, the implementation of the Program faces many difficulties (Section 3.1), which may impede the process of reducing forest fragmentation in Poland, also in forests managed by SF.

Average forest patch size. A negligible difference between the values of average forest patch size for 2000 and 2018 (Table 2) is somewhat perplexing, especially in view of available data, as it indicates that the total forest area in Poland increased (as said above: by almost 400 thousand ha) and the number of patches slightly decreased (by 19 patches). Therefore, for the purpose of this study, own calculations were carried out, with the use of data on the forest area for 2000 (Table 1) and 2018 [30] and data on the number of patches shown in Table 2. The obtained results showed that the average forest patch size was 1023 ha and 1070 ha, for the years 2000 and 2018, respectively. This result better than in

Table 2 reflects the changes that have occurred during the 18 years of the 21st century. When compared with other European countries (the average forest patch size: 763 ha), forest patches in Poland's forests are considerably larger, although much smaller than, e.g., those in Sweden (2687 ha), the Netherlands (2880 ha) and Finland (3371 ha) [1]. A slightly increasing trend in Polish forests can be considered positive, contrary to the average negative trend recorded in European forests [1].

Continuous forest vs. separated patches. The establishment of new forests as a result of reforestation/afforestation or natural succession, in the form of isolated patches or irregular branches of already existing forests, enhances forest fragmentation. In particular, this applies to the areas with low forest cover and fragmented forest-ownership structure [39]. In 21st century, an increasing share of smaller, fragmented forest patches has been observed not only in Poland (Table 2—0.1%), but also in other European regions, especially in central-west Europe and south-west Europe [1]. When compared to other countries, forest fragmentation in Poland (Table 2—4.5%) is still relatively low, which is encouraging—lower values are recorded only in a few countries, e.g., Netherlands (1.3%), Sweden (1.2%) and Finland (1.1%) [1].

Patches of various size. As reported in the SoEF Report [1], the proportions of different size forest patches in total Poland's forest area in 2000 and 2018 (Table 2) are identical, regardless of changes in total forest area as well as in the number of patches. Taking the latter values as correct, the current share of Poland's forests creating the best conditions for forest species preservation [57] is quite high (at a country level—almost 84%). In the case of SF, this value may be even higher, as privately owned forests show greater fragmentation. The value for Poland is higher than the overall average for Europe (76.6%) and lower than that reported by other European countries, e.g., the Netherlands and Sweden (95.5% each) (based on [1]).

3.3. Protected Forests

Focused on safeguarding biological diversity and natural ecological processes, protected areas are one of the oldest tools for the protection of the world's nature and natural resources [1]. They considerably contribute to the landscape values, and in the case of forests can be visually distinguished—in protected forests, there are relatively fewer (or not at all) noticeable traces of cutting/logging which usually result in a decrease of landscape aesthetic and recreational values [64]. Some protected areas had been established within the area of today's Poland before the country regained its independence (before 1918, e.g., nature reserves established in Nawojowa, southern Poland or near Lesko, south-eastern Poland) [65,66], nonetheless, the system of nature protection (including establishment of protected areas) was gradually reorganized in the decades after the year 1945 [65,67]. Since then, nature (forest) protection has undergone qualitative and quantitative development (Table 3).

Table 3. Area of protected forests (PL—Poland, SF—State Forests) *.

Year	PL/SF	1950	1960	1970	1980	1990	2000	2010	2020
National parks	PL	10.5	55.9	66.9	82.9	118.8	190.9	194.7	195.2
[thousand ha]	SF	-	-	-	-	-	-	-	-
Nature reserves	PL	0.02	16.3	>10.4 [1]	>16.7 [1]	>35.9 [1]	84.2	99.2	117.8
[thousand ha]	SF	0.02	n.a.	28.3 [2]	25.4 [3]	42.6	66.0	88.9	104.5
Landscape parks	PL	-	-	-	109.8	687.7	1345.9	>1421.1 [4]	1446.9
[thousand ha]	SF	-	-	-	n.a.	n.a.	n.a.	1137.0 [5]	1256.7 [6]
Landscape protection	PL	-	-	-	283.4	2113.8	2856.5	>2227.9	2942.3
areas [thousand ha]	SF	-	-	-	n.a.	n.a.	n.a.	2244.6 [5]	2467.6 [6]
Natura 2000 network	PL	-	-	-	-	-	-	2767.8 [4]	3243.8
[thousand ha]	SF	-	-	-	-	-	-	2780 [7]	2888 [7]

* sources: [4,7,32–34,36,43,45,48,62,67–74]; [1] data for forest reserves only; [2] data for 1969; [3] data for 1982; [4] data for 2011; [5] data for 2008; [6] data for 2018; [7] non-forest areas included; "-"—form of protection does not exist, n.a.—information not available.

Dynamics of change of protected forest area. During the whole study period, the area of protected forests in Poland (excluding Natura 2000) increased from 0 ha (start of implementation of a new approach towards nature conservation, after 1945) to 4702.2 thousand ha (calculated on the basis of Table 3). Currently it accounts for 50.8% of the total forest area. In the years 1945–1990 (the socialism era), an increase of protected forest area amounted to more than 2956.2 thousand ha (on average more than 65.7 thousand ha/year), whereas in the years 1990–2020 (the democracy era)—the increase was less than 1746 thousand ha (on average less than 58.2 thousand ha/year). Not counting Natura 2000 sites, the average increase in the protected area in the two studied time intervals (socialism and democracy eras) was most likely equal. It should be stressed, however, that in the socialism era, the expansion of nature protection areas in forests was running rather irregularly, as until the early 1980s, it was relatively slow (Table 3). By then, approximately 5.7% of the area of Polish forests was protected. A significant increase in the area of protected forests was observed in the last decade of the socialism era (1980–1989/90) and the first decade of the democracy era (1990s). This was a period of dynamic political, economic, and social changes that resulted in acknowledgement of the importance of environmental protection. The second stimulus for the development of forest protection—already in the democracy era—was Poland's accession to the EU and the need to implement the Natura 2000 network. The established Natura 2000 sites partly overlap with other forms of nature protection in Poland (Figure 3), thus, in the final analysis, they did not significantly affect the expansion of protected forest area as much as actions undertaken already in the 1980s and 1990s.

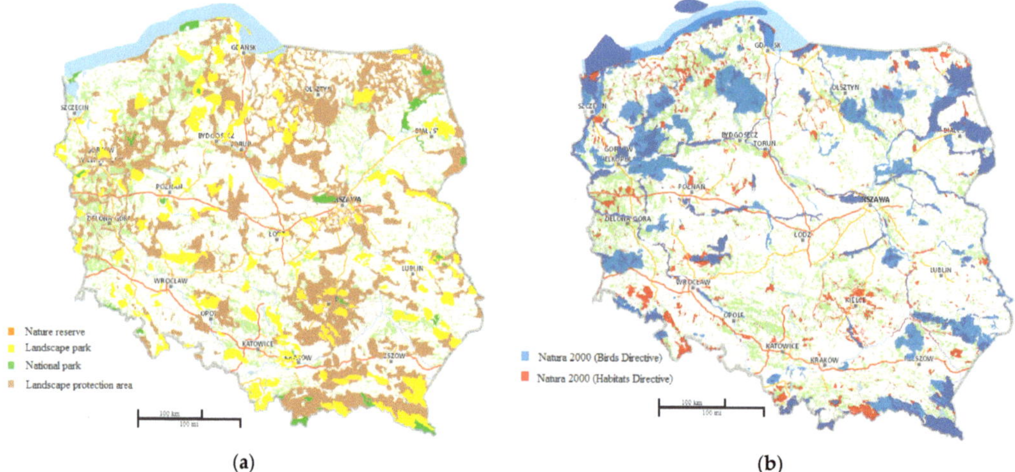

(a) (b)

Figure 3. (**a**) Distribution of protected areas (national parks, nature reserves, landscape parks and landscape protection areas) in Poland (2021); (**b**) Distribution of Natura 2000 areas in Poland (2021) (based on [75]).

Contribution of SF to nature conservation. The majority of Poland's protected areas (excluding national parks) and objects are situated in forests under management of the State Forests (Table 3). At the beginning of 2019, in total 70.6% of the forests managed by SF were under different types of protection [56]. These were predominantly forested areas; however, there were also protected non-forest areas in spatial arrangements with forest areas (e.g., peatlands, water courses and reservoirs, heathlands). The protected areas and objects of protection were gradually designated in SF managed forests throughout the whole period under the study (Table 3), however, only in the democracy era, the spectrum of foresters' activities in this respect was considerably broadened. Since 1998, the nature conservation program comprising information about the protected areas and objects has been a mandatory part of the forest management plan for each forest district [21,26]. Additionally, in many SF units, there were carried out trainings on nature conservation (with special emphasis put on Natura 2000 [76]). On the one hand, this new approach resulted from the increased capacity of foresters and recognition of the necessity to protect forest biodiversity (at all levels, including landscape), and on the other—from public pressure. As in other European countries, in Poland, there has been observed the so-called paradigm shift in nature protection [77]. Consequently, there prevails the preference of Polish society for socio-cultural and ecological functions of public forests as well as assurance of favorable conditions for well-being of forest ecosystems and species [26,78].

FOREST EUROPE classes of protected forests. FOREST EUROPE distinguishes several classes of protected forests: Class 1—protection of biodiversity (Class 1.1—no active intervention, Class 1.2—human intervention limited to minimum, Class 1.3—conservation through active management) and Class 2—protection of landscapes and specific natural elements [1]. Table 4 summarizes data on Polish forests in the democracy era [1], supplemented by own calculations for the socialism era (according to the methodology adopted in Poland for SoEF [1]). The latter—due to scarce information—are approximate.

Table 4. Protected forests (FOREST EUROPE Classes) in Poland *.

Year	1950	1960	1970	1980	1990	2000	2010	2020
Class 1.1 [thousand ha]	6.8	<29.6	<25.8	22.7	30.4	51.3	55.6	72.4
Class 1.2 [thousand ha]	-	-	-	-	-	-	-	-
Class 1.3 [thousand ha]	3.6	<58.9	>61.8	76.9	150.8	226.3	243.8	3019.7
Class 2 [thousand ha]	-	-	-	109.8	687.7	1346	1308	457.2

* Sources: [1,19,68,69,79]; "-"—form of protection does not exist.

A considerable discrepancy between data for Class 1.3 and Class 2 in the year 2020 compared to 2010 is a result of counting forests within Natura 2000 sites in Class 1.3, and excluding from Class 2 forests in landscape parks, where Natura 2000 sites are also present. Currently, on the scale of Europe as a whole, about 24% of reporting countries' forests are protected areas designated for the protection of biodiversity (15%) or landscape and specific natural elements (9%) [1]. In the case of Poland, these areas constitute 33% and 5%, respectively (calculated on the basis of Tables 1 and 4). The high share of Class 1 forests, which ranks Poland 3 in Europe (next to the Republic of Moldova and Italy), is almost entirely due to the dominance of Class 1.3 forests. The area of Class 1.1 forests is very small—countries with similar total forest area to Poland have much more forests with no active intervention, in Italy these grow on 270 thousand ha, and in Ukraine—on 293 thousand ha. The share of Class 2 forests in Poland is lower than Europe's average, and in relation to the countries with similar forest area it is higher than in Ukraine, and lower than in Italy [1]. From the point of view of the protection of forest landscapes, Class 1.1 and Class 2 forests are particularly important.

3.4. Protective Forests

Maintaining forests is a condition for, among others, preserving and regulating water relations as well as protecting nature and landscape [6,7,48]. In this respect some forests play a vital role, therefore, in their case, the protective functions should have higher priority than those productive.

After the World War II, starting from 1957, protective forests were designated in Poland, under six categories, including landscape forests. Over the years, the latter category was step by step neglected [80]. Currently, 10 categories of protective forests are designated [21,81]. Of these, forests designated for the protection of water or those safeguarding natural values, as well as animal refuges are of particular importance for biodiversity conservation (Table 5). The higher priority of the conservation function than the production function has in this case a positive influence on the protection of landscape values.

Table 5. The area of selected categories of protective forests in SF *.

Year	1950	1960	1970	1980	1990	2000	2010	2020
Water conservation forests [thousand ha]	-	n.a.	205.7 [1]	240.2	559.6	1231.8	1490.5	1552.3
Landscape forests [thousand ha]	-	n.a.	637.1 [1]	705.2	654.3	66.6	-	-
Valuable natural forests [thousand ha]	-	-	-	-	-	44.7	139.7	577.1
Animal refuges [thousand ha]	-	-	-	-	-	67.2	73.8	64.1
Total area of the analyzed protective forests [thousand ha]	-	n.a.	842.8	945.4	1213.9	1410.3	1704.0	2193.5

* sources: [19,33–36,43,82]; [1] data for 1975; "-"—form of protection does not exist, n.a.—information not available.

Dynamics of change of protective forest area. During the studied period, the area of the analyzed protective forests increased from 0 ha to 2193.5 thousand ha (Table 5). Currently it amounts to about 30.8% of forest area under SF management. In the socialism era, starting from 1957, protective forest area increased to 1213.9 thousand ha (on average 36.8 thousand ha/year), whereas in the period of democracy, the increase was 979.6 thousand ha (on average 32.7 thousand ha/year). The lower increase rate in the democracy era can be explained by reduced possibilities for designation of additional protective forests, considering absence of a rationale for special protection of some other forests, and a need for keeping the balance between ecological and economic functions of forests (in view of SF self-financing).

Water conservation forests. Forest areas designated to protect surface- and groundwater resources and to regulate hydrological relations in catchment and watershed areas are established in Poland compliant with the Forest Act [21]. Their specific landscape is vulnerable to environmental changes, and especially those connected with water relations. In Poland, numerous valuable forests in terms of water conservation have been degraded as a result of land reclamation (drainage), which was realized until the end of the socialism era [83], even though the protection of exceptional biodiversity of wet forests was already advocated in the 1970s [84]. The protection of wet forest habitats in the form of water-conservation forests began in the socialism era (with greater intensity at the end of this period), however, appropriately intense activities in this regard were undertaken only in the democracy era (Table 5). This resulted from: changes in the forest management model towards ecological dimension; the objectives of biodiversity conservation at a global level [85]; factual implementation of the Ramsar Convention [86]; active participation of Poland in the processes of the Ministerial Conference on the Protection of Forests in Europe (FOREST EUROPE) [55] as well as a recognized need to take into account climate change effects (e.g., periodic droughts [87]).

Animal refuges. This protection form was introduced after 1990 (the democracy era) to safeguard habitats of particular importance for animals under species protection [81], as well as those associated with species/gene pool protection [88]. The trend of changes of animal refuge areas over time does not seem explicit, and has been somewhat decreasing in the last decade (Table 5). This may be due to the fact that concurrently there also exists an analogous (more restrictive) form of nature conservation (selected species protection zones, compliant with Nature Conservation Act [89]) existing since the 1983 and gradually covering more and more taxa. At the end of 2020, in state-owned forests managed by SF, there were designated in total 3990 animal protection zones (an area of 163,000 ha) [46]. These probably fulfilled to some extent the need to designate more animal refuges.

Valuable natural forests. This category of forests was acknowledged after 1990, and defined as valuable fragments of native wildlife [81]. They initially safeguarded valuable/protected plant species and rare/rich/endangered forest ecosystems. Ecosystem services provided relate to the protection of habitats and species/gene pool [88]. Over time, forest of this kind were included into the system of forest management certification by the Forest Stewardship Council (selected categories of High Conservation Values Forests [90]). The area of valuable natural forests significantly increased along with the progress of FSC certification in Poland's forests (by now, FSC certificate has been granted to approx. to 97.7% forests under SF management [91]).

Landscape forests. In the era of socialism, selected forests were designated to protect forest aesthetic and landscape values. Starting from the 1970s, gradual withdrawal of this form of protection coincided with creation of landscape parks (Section 3.3), which by definition protect natural landscapes [89], especially those associated with forests [30]. Thus, it can be considered that this form of nature conservation has, in a way, taken over the function of former protective forests. At the same time, since the beginning of the democracy era, the protection of forests particularly valuable due to their landscape values has been declared in the objectives of Polish forestry [21], which, however, does not entail a separate category of protective forests.

Forest management in protective forests. From the beginning of its functioning in the socialism era, the protective forests were subject to a special management method with positive effects on their landscape values, i.e., clear-cutting was limited, and compound felling with a longer recovery period was recommended; by-product harvesting was limited; water reclamation was restricted; attention was paid to the landscape-forming function of forests [80]. After the political system change, management in protective forests still remained specific, focused on silviculture and protection rather than production [81,92]. This meant, for example, limiting clear-cutting, increasing the age of stands for felling, limiting harvesting of timber and non-timber products [7,48,81] and led to the perception of protective forests as a more restrictive form of nature conservation than, for example, landscape parks [80]. The abovementioned specific management rules in protective forests (applied from the beginning of their functioning) have shaped them somewhat differently as compared to those managed mainly for economic purposes. At the beginning of 2020, the average growing stock density in water conservation forests was 278 m^3/ha, in valuable natural forests—285 m^3/ha, and in animal refuges—302 m^3/ha, whereas in non-protective forests it amounted to 259 m^3/ha. At the same time, the proportion of the areas covered by over 80-year-old stands in the above specified protective forests were 23.8%, 28.2%, and 33.9%, respectively, and in non-protective forests—20.8% (calculated on the basis of [32]). These exemplary parameters and their values show that the structure of protective forests is richer as compared to non-protective forests, and their value for biodiversity conservation is higher. In view of the above, an increase in the protective properties of the protective forests analyzed is clearly visible.

Protective forests in Poland and in Europe. The current share of protective forests in Poland, protecting soil, water, and other forest ecosystem functions (Indicator 5.1 [28]), is quite high (34.6%). Among countries with available data, a higher proportion of such forests occurs in Romania (39.0%), Turkey (39.6%), Moldova (57.4%), Italy (87.5%), and Georgia (100.0%) (based on SoEF [1]). Yet, from the point of view of biodiversity protection, not all protective forests have a similarly high value. It is also worth noting that protective functions are often integrated into multifunctional forestry [1] and the function such as biodiversity conservation can be realized even without designating protective forests.

3.5. Harvesting Intensity

Timber harvesting directly affects forest landscape, on the one hand—through its form (in the extreme case: clear-cutting vs. selection cutting, resulting in two spatially different forest ecosystems [93]), on the other—through harvesting intensity. The intensity/volume of harvesting is planned in relation to growing stock. The change in growing stock in terms of its total amount and per ha of area can be used in the assessment of forest management quality with regard to its impact on forest resources and indirect effects on forest landscape values (e.g., a forest with rich spatial structure vs. forest heavily thinned or composed mainly of young trees). In 1945, in state-owned forests managed by SF, the volume of gross timber amounted to 695 million m^3 (129 m^3/ha) [19,30]. The following subsequent changes in these parameters—under forest use conditions—are shown in Table 6.

Table 6. Characteristics of wood resources of SF and their use *.

Year	1950	1960	1970	1980	1990	2000	2010	2020
Total growing stock [million m^3]	735	819	941	1087	1280	1480	1886	2067
Harvesting [thousand m^3]	14,531	18,040	19,814	20,738	16,947	25,718	33,769	38,232
Relation of timber harvest to annual increment [%]	51.6	57.7	74.6	50.5	49.9	58.7	61.7	73.6 [1]
Average growing stock density [m^3/ha]	128	133	145	162	188	213	267	290

* sources: [4,19,44–46,68,82]. [1] data for 2019.

Harvest volume. Since the beginning of the period under the study, an increase in the volume of timber harvesting has been observed, with one exception in 1990. The intensified crisis in the development of the country at the end of the socialism era caused a collapse in the timber market at that time [12]. Not only in Poland, but also in the entire region of Eastern Europe, with socialism collapse, the wood industry had to be restructured, which resulted in less logging [12,94]. At the same time, the model of forest management and logging was elaborated, which could ensure economic efficiency [18] and guarantee sustainable management of wood resources [18,95,96], so as to maintain forest landscape quality. It should be emphasized that even during the socialism era in Poland, there was harvested no more wood than its annual increment [19,30], in contrast to e.g., the Carpathian part of Ukraine [97]. Nevertheless, in 1947–1956, an increase in harvesting was very rapid, based on not fully justified top-down indicators [12,98].

Dynamics of change in timber resources. In 2020, timber resources (total growing stock) in SF were three times larger than in 1945. Their state at the end of the socialism era was about 180% of the initial state (average annual growth of about 1.8%), while in the democracy era—so far—it has been about 160% (average annual growth of about 1.5%). The decrease of the average growth of the resources with time is caused by the increase of the share of older stands with lower growth rate (based on [36]). Thus, as can be seen, timber resources in SF have grown regardless of increasing timber harvesting, which is due to large-scale post-war afforestation (Section 3.1) and the aforementioned rational harvesting [18,26]. This results from the fact that in the socialism era an average of 62.8% of annual increment was harvested (based on [19]), while in the democracy era an average of 60.3% was harvested (based on [30]). This parameter in the last two decades was lower than the average in Europe—in the years 1999–2019, harvesting in the State Forests amounted to 63% (according to [7]), while in Europe—73% [1]. Thus, despite ranking ninth in Europe (excluding Russia) in terms of forest area, Poland ranks fourth in terms of growing stock [1]. This may indicate the richness and complexity of forest landscapes.

Dynamics of change in average growing stock density. In 2020, the volume of forest stands resources per 1 ha under SF management was more than twice greater (225%) than that recorded in 1945. Stand volume at the end of socialism era was 146% of that in 1945 (average annual increase by 1.0%) and currently, it is 154% of that at the beginning of the democracy era (average annual increase by 1.8%). This was mainly influenced by the increase in the proportion of older age classes of stands (based on [36]), with dominant specimens of mature, thick trees, significantly increasing the landscape value of the forests [99]. The current average growing stock density under SF management (290 m^3/ha—Table 6) is much higher than that average of European forests (169 m^3/ha [1]).

Harvesting mode. In 2020, only 20% of all harvested wood came from clear-cutting [46]. In comparison, in the marketing year 1957/1958 (socialism era), this share was 62% (based on [68]). In view of the landscape values, clear-cuts are negatively perceived by the society [99], hence, their decreasing share in the democracy era can be considered a positive phenomenon. After the change of Poland's political system, as well as the transition of "old" forest management model into pro-environmental, there started to be implemented environmentally friendly technologies, e.g., the use of organic oils in machinery [76]. However, such technologies are in use not always and not everywhere, due to very high costs of their application [100].

3.6. Issues and Directions of the Protection of Forest Landscape in Poland

Form of forest ownership. One of the threats to the forest landscape (and forest, as well) is a proposal, recurring from time to time, to use a part of the State Forests for the restitution of private property (forest and non-forest)—nationalized after the World War II. Forests are a significant component of state property, and in a way they represent a natural candidate for restitution purposes [18]. At the same time, private forests are subject to less demanding regulations as regards biodiversity conservation than those imposed on state-owned forests [21]. Even though since 2001, the state forests have been acknowledged as a national resource that should be preserved [101], legislation can be changed any time. For this reason, it has been postulated that provisions as regard national character of Poland's forests should be included in Poland's Constitution, which has not yet been realized. It would be worth returning to this issue.

Functioning of the State Forests. The State Forests National Forest Holding manages the vast majority of the country's timber resources [30], which arouses the interest of Polish politicians. Proposals are being formulated to include SF finances in the public finance sector, as well as to change the legal formula of SF functioning [78]. Meanwhile, the change of, for example, financial management in SF may result in the change of forest management priorities (e.g., towards boosting a production function) and, in turn, in the deterioration of biodiversity, including landscape diversity. Therefore, the current formula of SF functioning should be maintained.

Forest Policy. The National Forest Policy, adopted in 1997, has no reference to several recent commitments of Poland under the FOREST EUROPE process (e.g., the resolution "Conserving and Enhancing Forest Biological Diversity in Europe") [23], as well as those associated with the accession to the EU (e.g., the Natura 2000 Network, which protects specific habitats/landscapes of particular biogeographical regions, including forests). Although these contents are reflected in the documents organizing the functioning of the State Forests, the lack of a nationwide, up-to-date policy should be considered as negative factor in view of, among others, the protection of forest landscapes. Attempts made in 20th century to create a National Forest Program have not resulted in its legal establishment. This problem should be solved as soon as possible.

Extreme weather events. The presence of forest landscapes can be threatened by extreme events due to climate change, such as catastrophic storms [17,87,102], which haunted Sweden and Central Europe in the past decade [1,31]. A large storm that passed over Poland on 11 August 2017 toppled or broke about 25 million trees in the northwestern part of the country [31]. In 2019, wind damage occurred on 42,300 hectares of forests (in 2020—on 10,700 hectares), being one of the two most important damages from abiotic factors [7,46]. Poland's main forest-forming species, Scots pine (*Pinus sylvestris*), is particularly vulnerable to strong winds [103]. In addition to the destruction of the forest landscape, the extreme weather events cause high financial losses, including lower income from the sale of poorer quality timber. Therefore, the selection of planted tree species in consideration of their resistance to extreme weather events should be of particular importance.

Droughts. Under Poland's conditions, periods of rainfall shortage due to climate change occur more and more frequently [7,87]. Drought threatens forests directly, as water deficiency can lead to death of trees, and indirectly—as it increases a risk of wildfires. Both factors have detrimental effects on the forest landscape. For example, in the very dry year 2015, there broke out 3897 fires in state-owned forests managed by the SF, more than twice many when compared to the year with high precipitation, i.e., 2010 (1777 forest fires recorded) [7,30]. Due to the high risk of fires, affecting over 80% of forest area in Poland, the fire protection system has been developed since 1945 [7,87,104]. As a result, the number of fires and the average area of a single fire have been gradually decreasing over the last 20–30 years, which is a good prognosis in the context of climate change [104]. This trend is worth maintaining. On the contrary, damage and death of tree stands due to drought shows an increasing trend. In 2020, of all the abiotic negative factors, the impact of extreme drought was the most detrimental, and tree stand damages were reported on nearly 62,500 hectares [46]. In response to this problem, since the 1990s, in forests managed by the SF, there have been carried out activities to restore small-scale retention [7,87], disrupted by land melioration implemented almost until the end of the socialism era [83]. In view of the forest landscape conservation needs (also those of habitats and species), it is important to continue retention work in all forest areas where it is possible and needed.

Organizational problems of forest landscape protection. The increase of the legally protected area in forests (Section 3.3) is important in view of benefits for biodiversity [1]. Vegetation in unmanaged and protected forests is more resilient to environmental changes and characteristic of large spatial diversity [105], which translates into the landscape values. On the other hand, however, if protected areas are accumulated within lands administered by a single entity, such as the State Forests, which is the case of Poland, they may pose a number of organizational challenges. Setting out conservation rules and allocation of responsibilities for protected areas designated in state-owned forests managed by SF have been a part of a complex process carried out under the conditions of duties dispersed among various external entities [56]. Another challenge which SF administration has to face concerns approaches taken and communication, for example during development of management plans for Natura 2000 sites The approach assumed by foresters is hardly ever accepted by environmental NGOs, who as a rule demand robust protection of discussed Natura 2000 sites [77], regardless of the fact that, by definition, contemporary nature protection must take into consideration not only ecological dimension, but also those economic and social as well as regional and local circumstances [26].

Financial problems of forest landscape protection. The increase of the protected area in forests (protected forests and protective forests) is also important from the financial perspective. In protected/protective forests, the ability to use their productive function is limited [78,81,89]. At the beginning of 2020, more than 1.25 million hectares of SF managed forests were partially excluded from timber harvesting (17.6% of SF area), and at least 530,000 hectares (7.5%) of this area was fully excluded from harvesting [31]. An increased area of protected sites translates into a decreased area of commercial forests [78]—the main source of income for the State Forests (e.g., [31,106]). The reduction in forest production efficiency, and consequently, lower income can lead to a decline of biodiversity conservation activities carried out by the State Forest [18]. Indeed, it should be emphasized that the direct costs of nature conservation incurred by SF have been still increasing, e.g., in 2009 they amounted to 4.8 million PLN [107], and in 2019—15.3 million PLN [31]. Additionally, since 2012, the State Forests have financially supported national parks (annually 20–50 million PLN [31,106,108–112]). Meanwhile, since 2009, the State Forests have received no targeted subsidies from the state budget for nature protection, even though the law allows such support [56,74]. In the case of protective forests, until 2015, forest management restrictions were compensated by half the forest tax rate, and now they are not [80]. Currently, the area of protective forests constitutes 53.6% of forests under SF management [7], and many of these areas have an analogous rank and function to protected areas designated under the Nature Conservation Act [89]. The above described circumstances pose a financial challenge for the State Forests, given that SF is a self-financing organization [21,26]. In order to resolve these problems, different solutions have been advocated, such as looking for additional financial sources for nature conservation [56], including restoration of subsidies from the state budget to the State Forests [21]. Another solution proposed is to reduce areas under strict protection (in fact, the state of nature is good even in economically used forests) or else to decrease the costs of forest nature conservation [41]. The most important issue is to find a proper balance between conservation of nature perceived as the national heritage and the rules of economy [26]. In addition, it is important to recognize that public expectations on forest nature conservation are high [74]. This is mainly related to much increased public awareness on environmental threats, which has been influenced by increased access to information (the effect of implementation of the Aarhus Convention [27]) and enhanced public knowledge on environmental issues.

Distribution and fragmentation of forests. An increase in the total area of forests in Poland (state- and privately owned forests) has not translated into an increase in uniformity of forest spatial distribution. The forest cover in administration units (a voivodeship level) varies from 21.4% to 49.3% (Figure 4): the highest is in the western, northern, and southeastern parts of Poland, the lowest—in the central and eastern parts. In the latter regions, the share of state-owned forests (under SF management) in total forest area is relatively smaller [46].

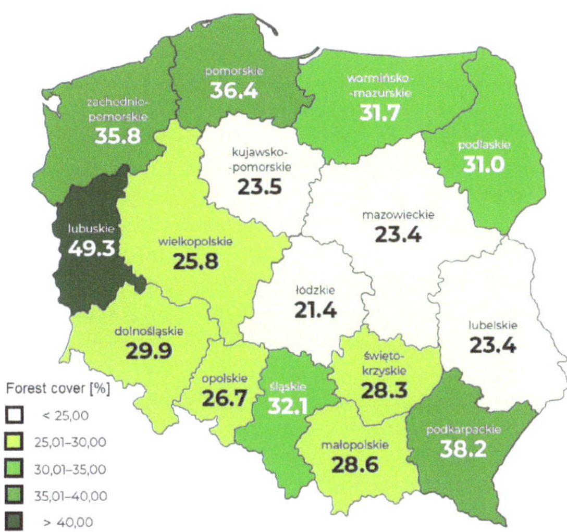

Figure 4. Forest cover of voivodships in Poland (2020) (source: [46], modified).

Low forest cover should therefore be a priority indication for further afforestation. This is because it has been assessed that the uneven distribution of forests, as well as their fragmentation, are one of the reasons for deterioration of ecological coherence in the spatial management of Poland [113]. Even though Raši and Schwarz [114] state that the interpretation of forest fragmentation impact on biodiversity remains limited, the authors of several studies consider fragmentation as one of the most radical disturbances for forests and show explicitly its negative effects [58]. At the landscape level, fragmentation can reduce carbon storage [58,115] and aggravate negative effects of other disturbances and hazards [116,117]—in extreme situations leading to degradation/destruction of forest patch/ecosystem. Forest fragmentation threatens biodiversity also at a species level (especially interior forest species) and can considerably hinder dispersal of specimens [57,58,118,119]. At the same time, it affects genetic biodiversity by causing genetic isolation, weakening genetic diversity, and limiting responses to multiple stresses [57,58,120]. For the above reasons, it is advisable to decrease forest fragmentation. Another factor that worsens ecological coherence is the legislation that, since 2003, has made it easier to cut down state-owned forests to build more roads [26], which is a simpler and cheaper solution than negotiations with private landowners. Improving the situation with regard to forest distribution and fragmentation would require changes in the law and extensive and practical consideration of these issues in land use plans and plans for further afforestation.

Landscape conservation in forest management. Forest landscapes are protected at a national and SF level not only through maintenance of appropriate forest cover, designation of protected areas and protective forests, but also through preservation of specific forest diversity. Since 1952, there has been functioning periodically improved natural-forest regionalization, which since 1990 (after the change of political system) has also taken into account ecological conditions of vegetation development. Currently, 183 mesoregions (which can be assumed as types of forest landscapes) are distinguished all through Poland (including SF) based on geological criteria; landscape types; landforms; climatic conditions; ranges of economically important tree species; and ranges of plant communities [121]. The natural/landscape distinctiveness of landscapes—units higher than mesoregions—has been taken into account in silvicultural principles, especially since 1990, when defining species compositions of stands on particular habitats (this refers to the selection of species and proportions between them [122]).

4. Conclusions

On the one hand, the period of 75 years refers to less than the life span of an average forest stand in Poland, and on the other—it is the period of evolution of Polish forestry, which was marked by two very different political and economic eras: socialism and democracy. The transformation influenced the assessed direction and intensity of changes in the forest landscape, although to different extent, depending on the analyzed indicator. In the socialism era, the total forest area, timber resources (total growing stock), as well as the area of protected forests important for biodiversity safeguarding increased more intensively, whereas in the democracy era, there was observed an intensive increase in average growing stock density, as well as there was put a greater emphasis on reducing forest fragmentation and the use of clear-cutting (disadvantageous for the landscape). The average increase in the area of the protected forests was equal in both epochs, although it was most intensive at their junction. The direction of changes throughout the studied period should be assessed as positive for biodiversity conservation, although not without problems and difficult consequences for the State Forests National Forest Holding, both in organizational and financial terms. Maintaining the quantity and quality of forest landscapes in Poland will require further efforts of foresters, and also legal, political and financial solutions at a national level, as well as actions of communities around the world to reduce the causes of global warming and the accompanying severe weather events and drought.

Author Contributions: Conceptualization, E.R.-C. and B.K.; methodology, E.R.-C.; formal analysis, E.R.-C.; investigation, E.R.-C. and B.K.; writing—original draft preparation, E.R.-C.; writing—review and editing, B.K.; visualization, E.R.-C.; project administration, E.R.-C. All authors have read and agreed to the published version of the manuscript.

Funding: This research received no external funding.

Data Availability Statement: Publicly available datasets were analyzed in this study. This data can be found here: [https://stat.gov.pl/obszary-tematyczne/rolnictwo-lesnictwo/lesnictwo] (accessed on 10 November 2021), [https://stat.gov.pl/obszary-tematyczne/srodowisko-energia/srodowisko] (accessed on 30 October 2021), [https://www.bdl.lasy.gov.pl/portal/publikacje] (accessed on 5 November 2021, [https://www.bdl.lasy.gov.pl/portal/tworzenie-zestawienia-ru] (accessed on 5 November 2021).

Acknowledgments: We would like to thank very much Małgorzata Bidłasik (MSc) from the Institute of Environmental Protection—National Research Institute, for her advice and assistance in the development of Figure 2c. Special thanks should be given to the Reviewers of our work for their efforts and valuable comments that greatly helped to improve the manuscript.

Conflicts of Interest: The authors declare no conflict of interest. The funders had no role in the design of the study; in the collection, analyses, or interpretation of data; in the writing of the manuscript, or in the decision to publish the results.

References

1. Forest Europe. *State of Europe's Forests 2020*; Ministerial Conference on the Protection of Forests in Europe FOREST EUROPE; Liaison Unit Bratislava: Bratislava, Slovakia, 2020.
2. Van Calster, H.; Baeten, L.; De Schrijver, A.; De Keersmaeker, L.; Rogister, J.E.; Verheyen, K.; Hermy, M. Management Driven Changes (1967–2005) in Soil Acidity and the Understorey Plant Community Following Conversion of a Coppice-with-Standards Forest. *For. Ecol. Manag.* **2007**, *241*, 258–271. [CrossRef]
3. Solon, J. Różnorodność Ponadgatunkowa—Krajobrazy [Over-Species Diversity—Landscapes]. In *Różnorodność Biologiczna Polski [Biodiversity of Poland]*; Andrzejewski, R., Weigle, A., Eds.; NFOŚ: Warsaw, Poland, 2003; pp. 155–159, ISBN 83-85908-75-7. (In Polish)

4. Broda, J. *Dzieje Lasów, Leśnictwa i Drzewnictwa w Polsce [History of Forests, Forestry and Timber Industry in Poland]*, 1st ed.; Żabko-Potopowicz, A., Ed.; PWRiL: Warsaw, Poland, 1965. (In Polish)
5. Jażdżewski, K.; Modrzejewski, A. Polityka leśna państwa pruskiego na obecnych ziemiach polskich w latach 1772−1914 [Forest policy of Prussian state in the contemporary Polish lands in years 1772−1914]. *Sylwan* **2013**, *157*, 63–70. [CrossRef]
6. Nienartowicz, A.; Lewandowska-Czarnecka, A.; Ortega, E.; Deptuła, M.; Filbrandt-Czaja, A.; Kownacka, M. Afforestation of Heathlands and Its Influence on the Land Cover, Accumulation of Plant Biomass and Energy Flow in the Landscape: An Example from Zaborski Landscape Park. *EQ* **2015**, *21*, 91–99. [CrossRef]
7. Zajączkowski, G.; Jabłoński, M.; Jabłoński, T.; Szmidla, H.; Kowalska, A.; Małachowska, J.; Piwnicki, J. *Raport o Stanie Lasów w Polsce 2019 [Report on the Condition of Forests in Poland in 2019]*; CILP: Warsaw, Poland, 2020; p. 105. (In Polish)
8. Rykowski, K. Forest Policy Evolution in Poland. *J. Sustain. For.* **1997**, *4*, 119–126. [CrossRef]
9. Węgiel, A.; Jaszczak, R.; Rączka, G.; Strzeliński, P.; Sugiero, D.; Wierzbicka, A. An Economic Aspect of Conversion of Scots Pine (*Pinus sylvestris* L.) Stands in the Polish Lowland. *J. For. Sci.* **2009**, *55*, 293–298. [CrossRef]
10. Keeton, W.S.; Crow, S.M. Sustainable Forest Management Alternatives for the Carpathian Mountain Region: Providing a Broad Array of Ecosystem Service. In *Ecological Economics and Sustainable Forest Management: Developing a Trans-Disciplinary Approach for the Carpathian Mountains*; Soloviy, I., Keeton, W.S., Eds.; Ukrainian National Forestry University Press: Lviv, Ukraine, 2009; pp. 109–127, ISBN 978-966-397-109-0.
11. Dietze, E.; Brykała, D.; Schreuder, L.T.; Jażdżewski, K.; Blarquez, O.; Brauer, A.; Dietze, M.; Obremska, M.; Ott, F.; Pieńczewska, A.; et al. Human-Induced Fire Regime Shifts during 19th Century Industrialization: A Robust Fire Regime Reconstruction Using Northern Polish Lake Sediments. *PLoS ONE* **2019**, *14*, e0222011. [CrossRef]
12. Łuczak, M.; Paschalis-Jakubowicz, P. Uwarunkowania rynku drzewnego w Lasach Państwowych w latach 1918−2008 [Determinants of the wood market of the State Forests in the years 1918−2008]. *Sylwan* **2013**, *157*, 506–515. [CrossRef]
13. Banach, J.; Skrzyszewska, K.; Skrzyszewski, J. Reforestation in Poland: History, Current Practice and Future Perspectives. *Reforesta* **2017**, *3*, 185–195. [CrossRef]
14. Szujecki, A. *Lata Wojny i Okupacji [Years of War and Occupation]*, 1st ed.; Karlikowski, T., Ed.; Z dziejów Lasów Państwowych i leśnictwa polskiego 1924–2004 [The history of the State Forests and the Polish Forestry 1924–2004]; CILP: Warsaw, Poland, 2006; Volume 2, ISBN 83-88478-93-1. (In Polish)
15. Bałtowski, M. *Gospodarka Socjalistyczna w Polsce: Geneza—Rozwój—Upadek [The Socialist Economy in Poland: Genesis—Development—Decline]*, 1st ed.; Wydawnictwo Naukowe PWN: Warsaw, Poland, 2009; ISBN 978-83-01-16044-9. (In Polish)
16. Swadźba, S. System Gospodarczy Polski w Latach 1918–2018 [Poland's Economic System from 1918 to 2018]. *Optimum* **2019**, *1*, 19–31. [CrossRef]
17. Mederski, P.; Jakubowski, M.; Karaszewski, Z. The Polish Landscape Changing Due to Forest Policy and Forest Management. *iForest* **2009**, *2*, 140–142. [CrossRef]
18. Siry, J.P.; Newman, D.H. A Stochastic Production Frontier Analysis of Polish State Forests. *For. Sci.* **2001**, *47*, 526–533. [CrossRef]
19. Rozwałka, Z. *Las w Liczbach [Forest in Numbers]*; Rozwałka, Z., Ed.; ARW A. Grzegorczyk: Warsaw, Poland, 1997; ISBN 83-86902-11-6. (In Polish)
20. *Effects of Air Pollution on Forest Health and Biodiversity in Forests of the Carpathian Mountains*; Szaro, R.C.; Bytnerowicz, A.; Oszlányi, J.; Godzik, B. (Eds.) NATO Science Series; IOS; Ohmsha: Amsterdam, The Netherlands; Washington, DC, USA; Tokyo, Japan, 2002; ISBN 978-1-58603-258-6. Available online: https://books.google.co.jp/books?id=FRz1t6G16I4C&printsec=frontcover&dq=Effects+of+Air+Pollution+on+Forest+Health+and+Biodiversity+in+Forests+of+the+Carpathian+Mountains&hl=zh-TW&sa=X&redir_esc=y#v=onepage&q=Effects%20of%20Air%20Pollution%20on%20Forest%20Health%20and%20Biodiversity%20in%20Forests%20of%20the%20Carpathian%20Mountains&f=false (accessed on 11 November 2021).
21. The Forest Act, 1991 [Dz. U. 1991.101.444 as Amended]. Available online: http://isap.sejm.gov.pl/isap.nsf/download.xsp/WDU19911010444/U/D19910444Lj.pdf (accessed on 4 March 2020). (In Polish)
22. Ministerstwo Ochrony Środowiska, Zasobów Naturalnych i Leśnictwa. In *Polityka Leśna Państwa [National Forest Policy]. Document Approved by the [Polish] Council of Ministers on 22 April 1997*, 1st ed.; MOŚZNiL: Warsaw, Poland, 1997.
23. Kaliszewski, A.; Gil, W. Cele i priorytety "Polityki leśnej państwa" w świetle porozumień procesu Forest Europe (dawniej MCPFE) [Goals and priorities of the 'National Forest Policy' in the light of the Forest Europe (formerly MCPFE) commitments]. *Sylwan* **2017**, *161*, 648–658. [CrossRef]
24. Director-General of the SF Zarządzenie Nr 11 Dyrektora Generalnego Lasów Państwowych z Dnia 14 Lutego 1995 r. w Sprawie Doskonalenia Gospodarki Leśnej Na Podstawach Ekologicznych [Order No. 11 of the Director General of the State Forests of 14 February 1995 on Improving Forest Management on an Ecological Basis]. Signature: ZZ-710-13/95 1995. Available online: https://www.wroclaw.lasy.gov.pl/documents/21578137/0/Zarzadzenie+Nr+11.pdf/3b10b05b-1803-46b9-9461-d76d2bb20c5f (accessed on 10 June 2021). (In Polish)
25. Director-General of the SF Zarządzenie [Order] Nr 11a Dyrektora Generalnego Lasów Państwowych z Dnia 11 Maja 1999 r., Zmieniające Zarządzenie Nr 11 Dyrektora Generalnego Lasów Państwowych z Dnia 14 Lutego 1995 Roku w Sprawie Doskonalenia Gospodarki Leśnej Na Podstawach Ekologicznych. Signature: ZG -7120-2/99. 1999. Available online: https://www.wroclaw.lasy.gov.pl/documents/21578137/0/Zarz_11A_DGLP_11.05.1999_r.pdf/4edb8ccd-1e91-4166-ae9d-07adc7e1cbf0 (accessed on 25 May 2021). (In Polish)

26. Czerepko, J.; Geszprych, M.; Gołos, P. Basic Assumptions for Forest Management and Nature Conservation from Axiological, Legal, and Economic Perspective. *Folia For. Pol.* **2017**, *59*, 68–78. [CrossRef]
27. Aarhus Convention: Convention on Access to Information, Public Participation in Decision-Making, and Access to Justice in Environmental Matters, Adopted on 25 June 1998 in Aarhus. Polish Version: 2003 [Dz. U. 2003.78.706]. Available online: http://isap.sejm.gov.pl/DetailsServlet?id=WDU20030780706 (accessed on 10 November 2021).
28. MCPFE Updated Pan-European Indicators for Sustainable Forest Management. Annex 1 to Madrid Ministerial Declaration. In Proceedings of the 7th Ministerial Conference, Madrid, Spain, 20–21 October 2015.
29. Forest Europe. SFM Criteria Indicators. Available online: https://foresteurope.org/sfm-criteria-indicators (accessed on 11 November 2021).
30. Leśnictwo [Forestry]. Statistical Yearbook of 2019, 2020. Available online: https://stat.gov.pl/obszary-tematyczne/roczniki-statystyczne/roczniki-statystyczne/rocznik-statystyczny-lesnictwa-2020,13,3.html (accessed on 15 March 2020).
31. Directorate-General of the State Forests. *Sprawozdanie Finansowo-Gospodarcze za 2019 rok*; Report on Financial and Economic State of State Forests in 2019; DGLP: Warsaw, Poland, 2020; p. 56. (In Polish)
32. BULiGL Bank Danych o Lasach [Forest Data Bank]—A Database on Resources and Condition of Polish Forests. Available online: https://www.bdl.lasy.gov.pl (accessed on 11 November 2021).
33. BULiGL Wyniki Aktualizacji Stanu Powierzchni Leśnej i Zasobów Drzewnych w Lasach Państwowych Na Dzień 1 Stycznia 1991 r. [Results of the Update of Forest Area and Timber Resources in the State Forests as of 1 January 1991]. 1991. Available online: https://www.bdl.lasy.gov.pl (accessed on 11 November 2021).
34. BULiGL Wyniki Aktualizacji Stanu Powierzchni Leśnej i Zasobów Drzewnych w Lasach Państwowych Na Dzień 1 Stycznia 2001 r. [Results of the Update of Forest Area and Timber Resources in the State Forests as of 1 January 2001]. 2001. Available online: https://www.bdl.lasy.gov.pl (accessed on 11 November 2021).
35. BULiGL Wyniki Aktualizacji Stanu Powierzchni Leśnej i Zasobów Drzewnych w Lasach Państwowych Na Dzień 1 Stycznia 2011 r. [Results of the Update of Forest Area and Timber Resources in the State Forests as of 1 January 2011]. 2011. Available online: https://www.bdl.lasy.gov.pl (accessed on 11 November 2021).
36. BULiGL Wyniki Aktualizacji Stanu Powierzchni Leśnej i Zasobów Drzewnych w Lasach Państwowych Na Dzień 1 Stycznia 2020 r. [Results of the Update of Forest Area and Timber Resources in the State Forests as of 1 January 2020]. 2021. Available online: https://www.bdl.lasy.gov.pl (accessed on 11 November 2021).
37. Kozak, J. Forest Cover Changes and Their Drivers in the Polish Carpathian Mountains since 1800. In *Reforesting Landscapes: Linking Pattern and Process*; Nagendra, H., Southworth, J., Eds.; Landscape Series; Springer: Dordrecht, The Netherlands; London, UK; New York, NY, USA, 2010; pp. 253–273, ISBN 978-1-4020-9655-6.
38. Sewerniak, P. Survey of Some Attributes of Post-Agricultural Lands in Polish State Forests. *EQ* **2016**, *22*, 9–16. [CrossRef]
39. Kozak, J.; Ziółkowska, E.; Vogt, P.; Dobosz, M.; Kaim, D.; Kolecka, N.; Ostafin, K. Forest-Cover Increase Does Not Trigger Forest-Fragmentation Decrease: Case Study from the Polish Carpathians. *Sustainability* **2018**, *10*, 1472. [CrossRef]
40. Lawrence, A. Forestry in Transition: Imperial Legacy and Negotiated Expertise in Romania and Poland. *For. Policy Econ.* **2009**, *11*, 429–436. [CrossRef]
41. Rozwałka, Z. *Leśna Statystyka 1997–2007 [Forestry Statistics 1997–2007]*; CILP: Warsaw, Poland, 2010; ISBN 978-83-61633-24-2. (In Polish)
42. *Rocznik Statystyczny Leśnictwa i Przemysłu Drzewnego 1964 [Statistical Yearbook of Forestry and Wood Industry 1964]*; PWRiL: Warsaw, Poland, 1966. Available online: https://stat.gov.pl/ (accessed on 11 November 2021). (In Polish)
43. *Rocznik Statystyczny Leśnictwa i Gospodarki Drewnem 1979 [Statistical Yearbook of Forestry and Wood Economy]*; GUS: Warsaw, Poland, 1979. Available online: https://stat.gov.pl/ (accessed on 11 November 2021). (In Polish)
44. Leśnictwo [Forestry]. *Statistical Yearbook of 2000*; GUS: Warsaw, Poland, 2001. Available online: https://stat.gov.pl/ (accessed on 11 November 2021).
45. Leśnictwo [Forestry]. *Statistical Yearbook of 2010*; GUS: Warsaw, Poland, 2011. Available online: https://stat.gov.pl/ (accessed on 11 November 2021).
46. Zajączkowski, G.; Jabłoński, M.; Jabłoński, T.; Szmidla, H.; Kowalska, A.; Małachowska, J.; Piwnicki, J. *Raport o Stanie Lasów w Polsce 2020*; Report on the Condition of Forests in Poland in 2020; CILP: Warsaw, Poland, 2021; p. 162. (In Polish)
47. Forest Research Institute (Poland) Krajowy Program Zwiększania Lesistości [The National Program of the Augmentation of Forest Cover]. 1995. Available online: https://archiwum.mos.gov.pl/fileadmin/user_upload/mos/srodowisko/lesnictwo/Krajowy_Program_Zwiekszania_Lesistosci.pdf (accessed on 28 November 2021). (In Polish)
48. Forest Research Institute. *Raport o Stanie Lasów w Polsce 2010*; Report on the Condition of Forests in Poland in 2010; CILP: Warsaw, Poland, 2011.
49. Kaliszewski, A. Krajowy Program Zwiększania Lesistości—Stan i Trudności Realizacji z Perspektywy Lokalnej [National Program for Expanding of Forest Cover—Implementation and Its Difficulties from a Local View]. *Studia I Mater. Cent. Edukac. Przyr. Leśnej* **2016**, *49B*, 7–19.
50. Wageningen University and Research The HIstoric Land Dynamics Assessment (HILDA). Available online: https://www.wur.nl/en/Research-Results/Chair-groups/Environmental-Sciences/Laboratory-of-Geo-information-Science-and-Remote-Sensing/Models/Hilda.htm (accessed on 24 November 2021).

51. IGiK CORINE Land Cover 2018—A Database on Types of Land Cover in Poland. Project Financed by the European Union. Available online: https://clc.gios.gov.pl/index.php/geoportal (accessed on 11 November 2021).

52. Kaliszewski, A.; Młynarski, W.; Gołos, P. Prospects for Agricultural Lands Afforestation in Poland until 2020. *Folia For. Pol.* **2016**, *58*, 163–169. [CrossRef]

53. Kurowska, K.; Kryszk, H. Opłacalność zalesień gruntów rolnych w ramach Programu Rozwoju Obszarów Wiejskich [Profitability of the farmland afforestation within the Rural Development Programme]. *Sylwan* **2017**, *161*, 1035–1045. [CrossRef]

54. Hościło, A.; Mirończuk, A.; Lewandowska, A. Określenie rzeczywistej powierzchni lasów w Polsce na podstawie dostępnych danych przestrzennych [Determination of the actual forest area in Poland based on the available spatial datasets]. *Sylwan* **2016**, *160*, 627–634. [CrossRef]

55. Zając, S.; Kaliszewski, A.; Młynarski, W. Forests and Forestry in Poland and Other EU Countries. *Folia For. Pol.* **2014**, *56*, 185–193. [CrossRef]

56. Referowska-Chodak, E. The Organization of Nature Conservation in State-Owned Forests in Poland and Expectations of Polish Stakeholders. *Forests* **2020**, *11*, 796. [CrossRef]

57. Kapos, V.; Lysenko, I.; Lesslie, R. *Assessing Forest Integrity and Naturalness in Relation to Biodiversity. Forest Resources Assessment Programme, Working Paper 54*; Forestry Department FAO: Rome, Italy, 2002.

58. Thompson, I.D.; Guariguata, M.R.; Okabe, K.; Bahamondez, C.; Nasi, R.; Heymell, V.; Sabogal, C. An Operational Framework for Defining and Monitoring Forest Degradation. *Ecol. Soc.* **2013**, *18*, 20. [CrossRef]

59. Hladnik, D.; Pirnat, J. Urban Forestry—Linking Naturalness and Amenity: The Case of Ljubljana, Slovenia. *Urban. For. Urban. Green.* **2011**, *10*, 105–112. [CrossRef]

60. *Ochrona Środowiska w Gospodarce Przestrzennej*, 1st ed.; Ryszkowski, L.; Kędziora, A. (Eds.) ZBŚRiL PAN: Poznań, Poland, 2005. Available online: https://zbkiks.ug.edu.pl/downloads/2012/jc/gp_pugp-materialy-4-system_i_formy_ochrony_srodowiska. pdf (accessed on 11 November 2021)ISBN 83-89887-26-6. (In Polish)

61. Edman, T.; Angelstam, P.; Mikusiński, G.; Roberge, J.-M.; Sikora, A. Spatial Planning for Biodiversity Conservation: Assessment of Forest Landscapes' Conservation Value Using Umbrella Species Requirements in Poland. *Landsc. Urban. Plan.* **2011**, *102*, 16–23. [CrossRef]

62. *Rocznik Statystyczny Leśnictwa 1971 [Forestry Statistical Yearbook 1971]*; GUS: Warsaw, Poland, 1971. Available online: https://stat.gov.pl/ (accessed on 11 November 2021). (In Polish)

63. Skolud, P. *Zalesianie Gruntów Rolnych i Opuszczonych Terenów Rolniczych—Poradnik Właściciela [Afforestation of Agricultural Land and Abandoned Agricultural Sites—An Owner's Guide]*, 2nd ed.; CILP: Warsaw, Poland, 2008; ISBN 978-83-89744-82-1. (In Polish)

64. Tahvanainen, L.; Tyrväinen, L.; Ihalainen, M.; Vuorela, N.; Kolehmainen, O. Forest Management and Public Perceptions—Visual versus Verbal Information. *Landsc. Urban. Plan.* **2001**, *53*, 53–70. [CrossRef]

65. Szafer, W. Zarys historii ochrony przyrody w Polsce [Outline of the history of nature conservation in Poland]. In *Ochrona Przyrody i jej Zasobów—Problemy i Metody [Conservation of Nature and Its Resources—Problems and Methods]*; Szafer, W., Ed.; ZOP PAN: Kraków, Poland, 1965; Volume 1, pp. 53–123. (In Polish)

66. Szafer, W. Rezerwaty w Polsce [Reserves in Poland]. In *Skarby Przyrody i ich Ochrona [Natural Treasures and Their Protection]*; Szafer, W., Ed.; PROP: Warsaw, Poland, 1932; pp. 294–317. (In Polish)

67. Denisiuk, Z.; Pilipowicz, W.; Przybylski, J. Ocena aktualnego stanu rezerwatów [Assessment of the current status of nature reserves]. In *Ochrona Rezerwatowa w Polsce—Stan Aktualny i Kierunki Rozwoju [Reserve Protection in Poland—Current Status and Development Directions]*; Studia Naturae; ZN im. Ossolińskich: Wrocław, Poland, 1990; Volume 35, pp. 17–29, ISBN 83-04-03699-1. (In Polish)

68. *Rocznik Statystyczny Leśnictwa i Przemysłu Drzewnego 1961 [Statistical Yearbook of Forestry and Wood Industry 1961]*; PWRiL: Warsaw, Poland, 1963. Available online: https://stat.gov.pl/ (accessed on 11 November 2021). (In Polish)

69. *Ochrona Środowiska [Environment]. Statistical Yearbook of 1989*; GUS: Warsaw, Poland, 1989. Available online: https://stat.gov.pl/ (accessed on 11 November 2021).

70. *Ochrona Środowiska [Environment]. Statistical Yearbook of 2000*; GUS: Warsaw, Poland, 2001. Available online: https://stat.gov.pl/ (accessed on 11 November 2021).

71. *Leśnictwo [Forestry]. Statistical Yearbook of 2001*; GUS: Warsaw, Poland, 2002. Available online: https://stat.gov.pl/ (accessed on 11 November 2021).

72. Referowska-Chodak, E. Skala i intensywność ochrony przyrody w lasach [Scale and intensity of nature conservation in forests]. In *Problemy Ochrony Przyrody w Lasach [Problems of Nature Conservation in Forests]*; IBL: Sękocin Stary, Poland, 2010; pp. 291–313, ISBN 978-83-87647-91-9.

73. *Leśnictwo [Forestry]. Statistical Yearbook of 2011*; GUS: Warsaw, Poland, 2012. Available online: https://stat.gov.pl/cps/rde/xbcr/ gus/rl_lesnictwo_2012.pdf (accessed on 15 June 2021).

74. Referowska-Chodak, E. *Ochrona Przyrody w Lasach Państwowych—Potrzeby i Oczekiwania Różnych Grup Społecznych Oraz ich Konsekwencje [Nature Protection in the State Forests—Needs and Expectations of Various Social Groups and Their Consequences]*, 1st ed.; SGGW: Warsaw, Poland, 2020; ISBN 978-83-7583-976-0.

75. GDOŚ Geoserwis Maps—Maps of Poland with Nature Protection Areas. Available online: http://geoserwis.gdos.gov.pl/mapy/ (accessed on 24 November 2021).

76. Blicharska, M.; Angelstam, P.; Elbakidze, M.; Axelsson, R.; Skorupski, M.; Węgiel, A. The Polish Promotional Forest Complexes: Objectives, Implementation and Outcomes towards Sustainable Forest Management? *For. Policy Econ.* **2012**, *23*, 28–39. [CrossRef]

77. Warchalska-Troll, A. Natura 2000 Sites in the Polish Carpathians vs Local Development: Inevitable Conflict? *Ecomont* **2018**, *10*, 50–58. [CrossRef]

78. Gołos, P.; Kaliszewski, A. Społeczne i ekonomiczne uwarunkowania realizacji publicznych funkcji lasu w Państwowym Gospodarstwie Leśnym Lasy Państwowe [Social and economic conditions for providing public forest services in the State Forests National Forest Holding]. *Sylwan* **2016**, *160*, 91–99. [CrossRef]

79. *Ochrona Środowiska [Environment]. Statistical Yearbook of 1990*; GUS: Warsaw, Poland, 1991. Available online: https://www.worldcat.org/title/ochrona-srodowiska/oclc/1112389282 (accessed on 15 June 2021).

80. Nowakowska, J.; Orzechowski, M. Lasy ochronne w Polsce—zarys historii na tle Europy [Protective forests in Poland—an outline of history in comparison to Europe]. *Sylwan* **2018**, *162*, 598–609. [CrossRef]

81. Regulation on Detailed Rules and Procedures for Recognizing Forests as Protective and Detailed Rules for Forest Management in Them, 1992 [Dz.U. 1992.67.337, in Polish]. Available online: http://isap.sejm.gov.pl/isap.nsf/download.xsp/WDU19920670337/O/D19920337.pdf (accessed on 6 May 2018).

82. *Rocznik Statystyczny Leśnictwa i Gospodarki Drewnem 1981 [Statistical Yearbook of Forestry and Wood Economy]*; GUS: Warsaw, Poland, 1981. Available online: http://agro.icm.edu.pl/agro/element/bwmeta1.element.agro-f4ba2e50-20e1-487c-98a0-11abbef6d14e/c/1-10.pdf (accessed on 15 June 2021). (In Polish)

83. Babiński, S.; Białkiewicz, F.; Krajewski, T. Stan melioracji wodnych w lasach [State of water reclamation in forests]. *Zesz. Probl. Postępu Nauk Rol.* **1989**, *375*, 127–138.

84. Jasnowski, M. Aktualny stan i program ochrony torfowisk w Polsce [Current status and conservation programme for peatlands in Poland]. *Chrońmy Przyr. Ojczystą* **1977**, *33*, 18–29.

85. CBD Convention: The Convention on Biological Diversity, Adopted on 5 June 1992 in Rio de Janeiro. Polish Version: 1995 [Dz. U. 1995.118.565, 2002.184.1532]. Available online: http://isap.sejm.gov.pl/isap.nsf/DocDetails.xsp?id=WDU19951180565 (accessed on 10 November 2021).

86. Ramsar Convention: Convention on Wetlands, Adopted on 2 February 1971 in Ramsar. Polish Version: 1978 [Dz. U. 1978.7.24-25]. Available online: http://isap.sejm.gov.pl/isap.nsf/DocDetails.xsp?id=WDU19780070024 (accessed on 10 November 2021).

87. Czerniak, A.; Grajewski, S.; Krysztofiak-Kaniewska, A.; Kurowska, E.E.; Okoński, B.; Górna, M.; Borkowski, R. Engineering Methods of Forest Environment Protection against Meteorological Drought in Poland. *Forests* **2020**, *11*, 614. [CrossRef]

88. Stępniewska, M.; Zwierzchowska, I.; Mizgajski, A. Capability of the Polish Legal System to Introduce the Ecosystem Services Approach into Environmental Management. *Ecosyst. Serv.* **2018**, *29*, 271–281. [CrossRef]

89. The Nature Conservation Act, 2004 [Dz. U. 2004.92.880 as amended, in Polish]. Available online: http://isap.sejm.gov.pl/isap.nsf/download.xsp/WDU20040920880/U/D20040880Lj.pdf (accessed on 20 March 2020).

90. Grupa Robocza FSC-Polska Kryteria Wyznaczania Lasów o Szczególnych Walorach Przyrodniczych (High Conservation Value Forests) w Polsce 2006. Available online: https://pl.fsc.org/sites/default/files/2021-04/Kryteria%20wyznaczania%20Las%C3%B3w%20o%20szczeg%C3%B3lnych%20walorach%20przyrodniczych%20%28HCVF%29%20w%20Polsce.pdf#viewer.action=download (accessed on 15 June 2021).

91. FSC Poland The Website of the FSC Certification Scheme in Poland. Available online: http://pl.fsc.org (accessed on 17 December 2020).

92. *Instrukcja Urządzania Lasu [Forest Management Planning Instruction]*; Święcicki, Z. (Ed.) CILP: Warsaw, Poland, 2012; Volume 1. Available online: https://www.lasy.gov.pl/pl/publikacje/copy_of_gospodarka-lesna/urzadzanie/iul/zarzadzanie-nr-83-2012/view (accessed on 11 November 2021)ISBN 978-83-61633-69-3. (In Polish)

93. Jaworski, A. *Sposoby Zagospodarowania, Odnawianie Lasu, Przebudowa i Przemiana Drzewostanów [Management Practices, Regeneration, Conversion and Transformation of Forest Stands]*, 2nd ed.; Hodowla lasu [Silviculture]; PWRiL: Warsaw, Poland, 2018; Volume 1, ISBN 978-83-09-01113-2. (In Polish)

94. Csóka, P. Capital Management—The Forests in Countries in Transition-Welfare Impacts. In *Forestry and Environmental Change: Socioeconomic and Political Dimensions*; Report No.5 of the IUFRO Task Force on Environmental Change; Innes, J.L., Hickey, G.M., Hoen, H.F., Eds.; CABI: Wallingford, CT, USA, 2005; pp. 125–141, ISBN 978-0-85199-002-6.

95. Kuemmerle, T.; Hostert, P.; Radeloff, V.C.; Perzanowski, K.; Kruhlov, I. Post-Socialist Forest Disturbance in the Carpathian Border Region of Poland, Slovakia, and Ukraine. *Ecol. Appl.* **2007**, *17*, 1279–1295. [CrossRef]

96. Durak, T.; Holeksa, J. Biotic Homogenisation and Differentiation along a Habitat Gradient Resulting from the Ageing of Managed Beech Stands. *For. Ecol. Manag.* **2015**, *351*, 47–56. [CrossRef]

97. Griffiths, P.; Kuemmerle, T.; Baumann, M.; Radeloff, V.C.; Abrudan, I.V.; Lieskovsky, J.; Munteanu, C.; Ostapowicz, K.; Hostert, P. Forest Disturbances, Forest Recovery, and Changes in Forest Types across the Carpathian Ecoregion from 1985 to 2010 Based on Landsat Image Composites. *Remote Sens. Environ.* **2014**, *151*, 72–88. [CrossRef]

98. Broda, J. *Lasy Państwowe w Polsce w Latach 1944–1990 [State Forests in Poland 1944–1990]*, 1st ed.; Broda, J., Ed.; PWN: Warsaw-Poznań, Poland, 1997; ISBN 83-01-12359-1.

99. Referowska-Chodak Management Elbnt and Social Problems Linked to the Human Use of European Urban and Suburban Forests. *Forests* **2019**, *10*, 964. [CrossRef]

100. Dudek, T.; Zięba, W. Wybrane aspekty zrównoważonego użytkowania lasu w nawiązaniu do programu zrównoważonego rozwoju—przykład Polski [Selected aspects of sustainable forest utilization with regard to the sustainable development program—Polish example]. *Sylwan* **2018**, *162*, 469–478. [CrossRef]
101. The Strategic Resources Act, 2001 [Dz. U. 2001.97.1051 as Amended]. Available online: http://isap.sejm.gov.pl/isap.nsf/DocDetails.xsp?id=WDU20010971051 (accessed on 4 March 2020). (In Polish)
102. Lindner, M.; Maroschek, M.; Netherer, S.; Kremer, A.; Barbati, A.; Garcia-Gonzalo, J.; Seidl, R.; Delzon, S.; Corona, P.; Kolström, M.; et al. Climate Change Impacts, Adaptive Capacity, and Vulnerability of European Forest Ecosystems. *For. Ecol. Manag.* **2010**, *259*, 698–709. [CrossRef]
103. Szwagrzyk, J.; Gazda, A.; Dobrowolska, D.; Chećko, E.; Zaremba, J.; Tomski, A. Tree Mortality after Wind Disturbance Differs among Tree Species More than among Habitat Types in a Lowland Forest in Northeastern Poland. *For. Ecol. Manag.* **2017**, *398*, 174–184. [CrossRef]
104. Grajewski, S. Effectiveness of Forest Fire Security Systems in Poland. *Infrastruktura i Ekologia Terenów Wiejskich* **2017**, *IV/2*, 1563–1576. [CrossRef]
105. Durak, T. Long-Term Trends in Vegetation Changes of Managed versus Unmanaged Eastern Carpathian Beech Forests. *For. Ecol. Manag.* **2010**, *260*, 1333–1344. [CrossRef]
106. Directorate-General of the State Forests. *Sprawozdanie Finansowo-Gospodarcze za 2016 rok [Report on Financial and Economic State of State Forests in 2016]*; DGLP: Warsaw, Poland, 2017; p. 43.
107. Directorate-General of the State Forests. *Sprawozdanie Finansowo-Gospodarcze za 2009 rok [Report on Financial and Economic State of State Forests in 2009]*; DGLP: Warsaw, Poland, 2010; p. 34.
108. Directorate-General of the State Forests. *Sprawozdanie Finansowo-Gospodarcze za 2013 rok [Report on Financial and Economic State of State Forests in 2013]*; DGLP: Warsaw, Poland, 2014; p. 46.
109. Directorate-General of the State Forests. *Sprawozdanie Finansowo-Gospodarcze za 2014 rok [Report on Financial and Economic State of State Forests in 2014]*; DGLP: Warsaw, Poland, 2015; p. 46.
110. Directorate-General of the State Forests. *Sprawozdanie Finansowo-Gospodarcze za 2015 rok [Report on Financial and Economic State of State Forests in 2015]*; DGLP: Warsaw, Poland, 2016; p. 40.
111. Directorate-General of the State Forests. *Sprawozdanie Finansowo-Gospodarcze za 2017 rok [Report on Financial and Economic State of State Forests in 2017]*; DGLP: Warsaw, Poland, 2018; p. 49.
112. Directorate-General of the State Forests. *Sprawozdanie Finansowo-Gospodarcze za 2018 rok [Report on Financial and Economic State of State Forests in 2018]*; DGLP: Warsaw, Poland, 2019; p. 45.
113. Resolution No. 239 of the Council of Ministers of 13 December 2011 "Concept of Spatial Planning of the Country 2030", 2011 [M. P. 2012.0.252, in Polish]. Available online: http://isap.sejm.gov.pl/isap.nsf/DocDetails.xsp?id=WMP20120000252 (accessed on 10 October 2021).
114. Raši, R.; Schwarz, M. *Pilot Study: Forest Fragmentation Indicator*, 1st ed.; FOREST EUROPE Liaison Unit Bratislava: Zvolen, Slovakia, 2019; ISBN 978-80-8093-300-5.
115. Pan, Y.; Birdsey, R.A.; Phillips, O.L.; Jackson, R.B. The Structure, Distribution, and Biomass of the World's Forests. *Annu. Rev. Ecol. Evol. Syst.* **2013**, *44*, 593–622. [CrossRef]
116. Lehvävirta, S. Non-Anthropogenic Dynamic Factors and Regeneration of (Hemi)Boreal Urban Woodlands—Synthesising Urban and Rural Ecological Knowledge. *Urban. For. Urban. Green.* **2007**, *6*, 119–134. [CrossRef]
117. Golivets, M. Ecological and Biological Determination of Invasion Success of Non-Native Plant Species in Urban Woodlands with Special Regard to Short-Lived Monocarps. *Urban. Ecosyst.* **2014**, *17*, 291–303. [CrossRef]
118. Fischer, J.; Lindenmayer, D.B. Landscape Modification and Habitat Fragmentation: A Synthesis. *Glob. Ecol. Biogeogr.* **2007**, *16*, 265–280. [CrossRef]
119. Keinath, D.A.; Doak, D.F.; Hodges, K.E.; Prugh, L.R.; Fagan, W.; Sekercioglu, C.H.; Buchart, S.H.M.; Kauffman, M. A Global Analysis of Traits Predicting Species Sensitivity to Habitat Fragmentation: Species Sensitivity. *Glob. Ecol. Biogeogr.* **2017**, *26*, 115–127. [CrossRef]
120. Ledig, F.T. Human Impacts on Genetic Diversity in Forest Ecosystems. *Oikos* **1992**, *63*, 87–108. [CrossRef]
121. Zielony, R.; Kliczkowska, A. *Regionalizacja Przyrodniczo-Leśna Polski 2010 [Natural-Forest Regionalization of Poland 2010]*; CILP: Warsaw, Poland, 2012; ISBN 978-83-61633-62-4. (In Polish)
122. *Zasady Hodowli Lasu [Silvicultural Principles]*, 1st ed.; Haze, M. (Ed.) CILP: Warsaw, Poland, 2012; ISBN 978-83-61633-65-5. Available online: https://www.lasy.gov.pl/pl/pro/publikacje/copy_of_gospodarka-lesna/hodowla/zasady-hodowli-lasu-dokument-w-opracowaniu, (accessed on 11 November 2021). (In Polish)

Article

Assessing the Heterogeneity and Conservation Status of the Natura 2000 Priority Forest Habitat Type *Tilio–Acerion* (9180*) Based on Field Mapping

Janez Kermavnar [1], Erika Kozamernik [1,2,*] and Lado Kutnar [1]

[1] Department of Forest Ecology, Slovenian Forestry Institute, 1000 Ljubljana, Slovenia
[2] Department of Forestry and Renewable Forest Resources, Biotechnical Faculty, University of Ljubljana, 1000 Ljubljana, Slovenia
* Correspondence: erika.kozamernik@gozdis.si

Abstract: Priority habitat types (HTs) within the Natura 2000 network are of the highest importance for conservation in Europe. However, they often occur in smaller areas and their conservation status is not well understood. One such HT is that of the *Tilio–Acerion* forests of slopes, screes and ravines (9180*). The Natura 2000 study site, Boč–Haloze–Donačka gora, in the Sub-Pannonian region of eastern Slovenia is characterized by a matrix of European beech forests and includes rather small, fragmented areas covered by *Tilio–Acerion* forests. The goal of this research was to examine the heterogeneity and conservation status of the selected HT through field mapping, which was performed in the summer of 2020. As the conservation of HT calls for a more detailed approach, we distinguished between the following four pre-defined habitat subtypes: (i) *Acer pseudoplatanus-Ulmus glabra* stands growing mostly in concave terrain, (ii) *Fraxinus excelsior* stands growing on slopes, (iii) *Tilia* sp. stands with thermophilous broadleaves occurring on ridges and slopes, (iv) *Acer pseudoplatanus* stands occurring on more acidic soils with an admixture of *Castanea sativa*. Field mapping information was complemented with the assessment of habitat subtype characteristics using remote sensing data. The results showed that habitat subtypes differed significantly in terms of area, tree species composition, forest stand characteristics, relief features and the various threats they experienced (e.g., fragmentation, tree mortality, ungulate browsing pressure). The differences between subtypes were also evident for LiDAR-derived environmental factors related to topography (i.e., terrain steepness and Topographic Position Index). This study provides a baseline for setting more realistic objectives for the conservation management of priority forest HTs. Due to the specificities of each individual habitat subtype, conservation activities should be targeted to the Natura 2000 habitat subtype level.

Keywords: forest habitat subtype; monitoring; biodiversity conservation; LiDAR; Slovenia

Citation: Kermavnar, J.; Kozamernik, E.; Kutnar, L. Assessing the Heterogeneity and Conservation Status of the Natura 2000 Priority Forest Habitat Type *Tilio–Acerion* (9180*) Based on Field Mapping. *Forests* **2023**, *14*, 232. https:// doi.org/10.3390/f14020232

Academic Editors: Konstantinos Poirazidis and Panteleimon Xofis

Received: 20 December 2022
Revised: 20 January 2023
Accepted: 20 January 2023
Published: 26 January 2023

1. Introduction

The Natura 2000 framework is designed to identify, maintain and protect sites of high biodiversity value [1,2]. Globally, Natura 2000 serves as one of the main strategies to mitigate the decline of biodiversity [3]. Slovenia has the highest percentage of Natura 2000 sites among all EU Member States, with such sites covering almost 38% of the country's terrestrial area [4], with a strong prevalence of forest ecosystems within this area [5]. The Habitats Directive [6] requires the assessment of the conservation status of habitat types [2], which results in an overall estimation of the habitat quality in terms of nature conservation [7] and is usually done by implementing a specific monitoring scheme. Preserving habitats in a favourable conservation status is one of the most effective ways of maintaining biodiversity [8], and a great deal of monitoring effort is required to achieve this important, but often challenging, goal in nature conservation.

The initial step in the monitoring process is to obtain information on where a targeted habitat type occurs. Such data serves as a baseline for all further evaluations and actions related to the conservation status of the targeted habitat, but other key factors beyond geographical location need to be considered as well. According to the Habitats Directive [6], the conservation status of a natural habitat is determined by the sum of the influences acting on a habitat and its typical species, that may affect the habitat's long-term natural distribution, structure and functions, as well as the long-term survival of the typical species. The following defines when the conservation status of a natural habitat is "favourable": (i) its natural range and the areas it covers within that range are stable or increasing, (ii) the specific structure and functions necessary for its long-term maintenance exist and are likely to continue to exist in the foreseeable future, and (iii) the conservation status of its typical species is favourable.

As the more extensive forest habitat types, but sometimes also small-scale habitat types, are quite heterogeneous, Kovač et al. [9] proposed a hierarchical approach in the current concept of forest habitat types with the inclusion of the subtypes of the forest habitat type, which can be distinguished based on ecological and vegetation characteristics. Although the assessment of the conservation status should refer to the forest habitat type (by implementing appropriate indicators), this evaluation can be based on the level of habitat subtypes. In this novel conceptual framework, the habitat subtype is assumed to be much more homogeneous than the habitat type, restricted to a less broad range of ecological factors, and much more clearly defined in terms of its distribution, structure, and functions, as well as the composition of its typical species. Therefore, the assessment of the conservation status of the habitat subtype is much more accurate and reliable, allowing conservation measures to be more targeted and efficient.

A forest habitat type (hereafter HT) is a vegetation system composed of certain species that provides a living environment for various organisms. It may or may not be composed of two or more interacting forest habitat subtypes (hereafter HsTs). All forest sites that are subject to monitoring and assessment must first be identified in the field and afterwards evaluated to produce valid estimates for HT and HsT [9].

In the mountainous regions of temperate Europe, erosion gullies, ravines, gorges, cliffs, steep rocky slopes and rocky outcrops are covered by a particular type of azonal forest vegetation, which is included in the *Tilio platyphylli–Acerion pseudoplatani* Klika 1955 alliance [10]. *Tilio–Acerion* forests of slopes, screes and ravines (Natura 2000 habitat code 9180*) are a priority forest HT. From an ecological perspective, forests belonging to the 9180* HT can be divided into two groups: dry and warm sites with lime trees (*Tilia platyphyllos* Scop., *T. cordata* Mill.) and humid and cool sites with a dominance of Sycamore maple (*Acer pseudoplatanus* L.) [11]. On the European scale, *Tilio–Acerion* forests represent an understudied and data-scarce HT and have been studied mostly using conventional phytosociological methods [10,12,13]. According to preliminary estimates [14], it was concluded that trustworthy information on the spatial distribution and characteristics of this habitat type was lacking. Owing to the limited availability of detailed maps and accurate spatial data, the assessment of the conservation status is itself rather unreliable [5].

Evaluating the conservation status of the HT of EU interest is essential for preserving these areas in favourable conditions [1]. Similar to more frequent forest HTs found in Slovenia, such as 91K0 Illyrian *Fagus sylvatica* forests, 9110 *Luzulo–Fagetum* beech forests and 91L0 Illyrian oak-hornbeam forests, *Tilio–Acerion* forests are facing unprecedented challenges in biodiversity conservation. Due to their small areas and the various threats, they experience, the conservation status of this HT is either unfavourable or simply unknown [14]. Climate change, fragmentation and overexploitation have been identified as the main potential threats to the existence of this HT [14]. According to National Forest Inventory data from 2006, only 0.04% of Slovenian forest area was covered by *Tilio–Acerion* forests. Such minor HTs are more endangered than those covering larger areas.

Natura 2000 habitat types need to be mapped and monitored in order to assess their conditions [15]. Assessing the quality of forest habitats over larger areas is a complex monitoring task [16], particularly in the case of prominent natural heterogeneity within the habitat type (i.e., subtype variation of species composition and associated specificities).

While field-based mapping of HTs represents the main pillar of conservation efforts, remote sensing techniques have been increasingly contributing to monitoring and assessing forest biodiversity-related characteristics and functions [17,18]. Remote sensing has a wide application in nature conservation and has a high potential for reinforced monitoring of forest habitats to support forest biodiversity assessment and sustainable forest management [19]. Remote sensing-derived data are not only an alternative to traditional methods (in-field assessments) but can also offer complementary information when both approaches are combined [20]. LiDAR (Light Detection and Ranging) has been frequently used for mapping natural forest habitats and their conservation status [15,16]. LiDAR is a data-rich source and has the potential to address some of the drawbacks associated with traditional vegetation monitoring methods, especially in remote areas [21,22].

In order to illustrate the characteristics of the 9180* HT and the potential for distinguishing its HsTs, a case-study was conducted in the Natura 2000 site of Boč–Haloze–Donačka gora in Slovenia. This site is representative of the targeted HT, but the degree of conservation of the natural structure and functions of the HT is lower due to a lack of adequate data and the presence of small, isolated patches in remote areas with diverse topography. According to the latest formal report for Natura 2000 [23], the conservation status of the 9180* HT in the Boč–Haloze–Donačka gora site is unfavourable and exhibiting a declining trend. Moreover, the information on the locations and quality of the current Natura 2000 zones is inadequate, as they are defined only by the area where the HT may be located. Improving knowledge through field mapping and reliable surveys of this HT is necessary.

The primary objective of this study was to assess the conservation status of the HT and HsTs in the selected Natura 2000 site using field mapping. Additionally, the characteristics of HsTs were compared using LiDAR-derived data of the stand structure and environmental factors related to topography. The subtype approach employed in this research was intended to assist nature conservationists and forest managers in decision-making and to contribute to the sustainable management of HT 9180*. The general conservation goal is to maintain the current extent of these habitats, which requires the implementation of field mapping to determine their actual areas and the implementation of appropriate measures to maintain or improve their conservation status.

2. Materials and Methods

2.1. Study Area

The study area was a forested landscape located in eastern Slovenia (46.294° N, 15.734° E; Figure 1). As part of the Natura 2000 network, this site is referred to as Boč–Haloze–Donačka gora (site code SI3000118; hereafter referred to as the BHD site). The BHD site is home to 15 Natura 2000 species, including invertebrates (e.g., *Rosalia alpina* L.), mammals, amphibians and flowering plants (e.g., *Pulsatilla vulgaris* subsp. *grandis* (Wender.) Zāmelis). Additionally, it serves to protect seven Habitats Directive HTs [6].

The total surface area of the BHD site is 10,882 ha, and the area of forest zones (nature conservation units) assigned to the priority HT *Tilio–Acerion* is 853 ha. The altitude ranges from 240 m (Dravinja valley) to 978 m (summit of Boč mountain). The BHD site is diverse in terms of relief and geology, resulting in a variety of forest vegetation types. The forests mainly belong to Illyrian beech forests (91K0) and central European acidophilous beech forests (9110) HT. The proportion of forest stands (forest management units) belonging to HT 9180* is 1.3% of the entire BHD site [24]. The structural characteristics are generally favourable, such as a sufficient amount of standing or lying deadwood. However, there is a lack of younger developmental stages, although noble broadleaves often occur as pioneers on primarily beech sites [24].

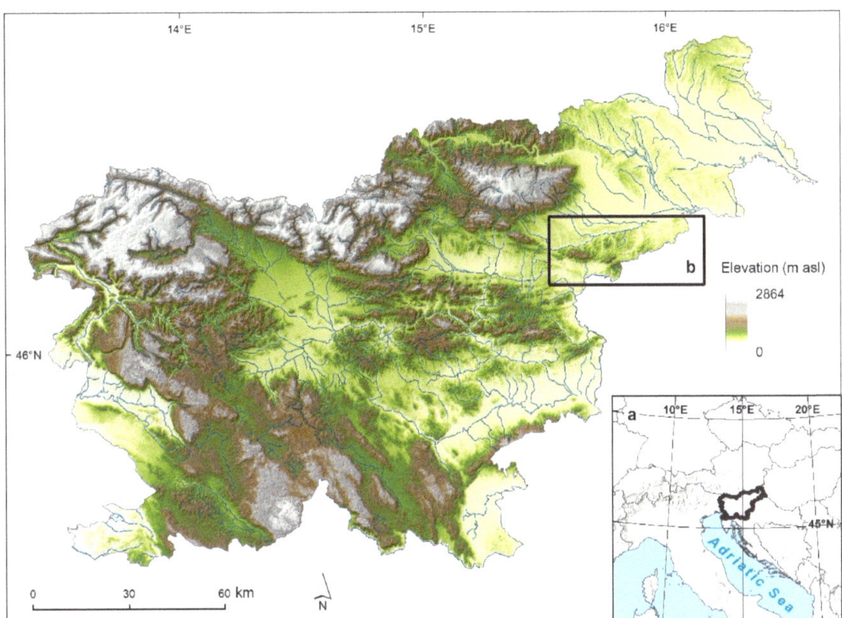

Figure 1. (**a**) The location of Slovenia. (**b**) The Boč–Haloze–Donačka gora (Natura 2000 site) study area in eastern (Sub-Pannonian region) Slovenia.

Boč mountain, which is located in the western part of the BHD site, is an isolated mountain complex with extremely diverse lithological structure and high biodiversity. Geologically and geomorphologically, it represents the easternmost part of the Karavanke mountain range. The mountain is composed of Triassic limestones and dolomites in the western part and Miocene limestones and quartz sandstones in the eastern part. The northern slopes are steeper than the southern slopes, with more than half of the surfaces having slopes of above 20°. The predominant relief forms are peaks, ridges, slopes and small plains. There is no developed river network in the area, as water quickly sinks into the karst underground on the predominantly carbonate parent material. The diversified geological and topographical conditions are manifested in a mosaic of different soil types. Brown calcareous soils, rendzina and rankers alternate depending on the parent material and slope aspect. Pseudogleys predominate on the slopes in the Haloze hills and Donačka gora (both areas located in eastern part of the BHD site), while deep soils are found on clayey and silty deposits in the valleys [25].

The BHD site has a Sub-Pannonian climate, with the majority of rainfall occurring in spring and summer and there being a deficit in winter. Due to the higher altitudes and rugged terrain, the climate on Boč mountain is harsher compared to that of the surrounding hills and valleys. The average annual precipitation amount on Boč is around 1200 mm. Average temperatures in the growing season are around 15 °C in the Haloze hills, and only 8 °C at the top of Boč. Due to the diversity of soils and climatic conditions in the area, floristic elements of the Mediterranean, Central European, Alpine, Illyrian and steppe regions are present [25].

2.2. Description of the Forest Habitat Type and Its Subtypes

Tilio–Acerion (9180*) forests of slopes, screes and ravines are a typical example of azonal forest vegetation. In Slovenia, they commonly occur in small areas, mainly in stony or rocky gullies, in dolines, other depressions and ravines, on torrential fans, and on the gravelly bases of slopes, but also on moist rocks and more sun-exposed ridges at an altitude from the colline to the altimontane vegetation belts [26]. The soil types in these forests are colluvial–deluvial soils, rare rendzinas and brown calcareous soils, and sometimes also dystric brown soils and ranker or eutric brown soils. They are usually biologically very active, with relatively fast decomposition of litter resulting in high nutrient availability in the soil. These forest stands occur on sites that can be quite extreme for the growth of forest vegetation (i.e., very rocky and steep; [26]). Forest stands belonging to HT 9180* have important nature conservation features and are particularly relevant as habitats for rare and protected plant species and other organisms [5]. Within the EUNIS habitat classification [27,28], these forests are classified as Illyrian ravine forests (code T1F63), but the current classification scheme does not include more detailed information at the subtype level.

The tree layer of these forests is composed of noble broadleaves, such as sycamore maple (*Acer pseudoplatanus*), Norway maple (*Acer platanoides* L.), wych elm (*Ulmus glabra* Huds.), European ash (*Fraxinus excelsior* L.), large-leaved lime (*Tilia platyphyllos*) and small-leaved lime (*Tilia cordata*). Floristically, these forests are relatively similar to beech communities, but they have a greater abundance of hygrophilous and nitrophilous plant taxa. The diagnostic species of the understory layer are mainly mesophilous tall forbs and ferns that have high requirements for nutrients, soil moisture and air humidity, such as *Lunaria rediviva* L., *Asplenium scolopendrium* L., *Polystichum setiferum* (Forssk.) Woyn. and *Urtica dioica* L. Forests dominated by *A. pseudoplatanus* occur on colder and wetter sites, while slightly more thermophilous *Tilia* forests exhibit a higher proportion of plant species indicating a warmer and drier microclimate [26]. In this part of Europe, their phytogeographic differentiation from similar forest plant communities in Central Europe is further promoted by the presence of some relict and endemic Illyrian species that survived the Quaternary glaciations in southern European refugia, including typical forest herbs, such as *Lamium orvala* L., *Stellaria nemorum* subsp. *montana* (Pierrat) Berher, *Cardamine waldsteinii* Dyer. and *Scopolia carniolica* Jacq. [10].

Forests of the *Tilio–Acerion* HT encompass diverse site and stand conditions, covering a broad ecological amplitude and an array of forest associations. Whenever an HT displays such heterogeneity, dividing it into HsTs is meaningful and can substantially increase the identification and conservation of the entire HT [9]. We divided the studied HT 9180* into four pre-defined, relatively homogeneous HsTs. The subtypes were defined based on established forest communities (associations), adopting the classification described in up-to-date phytosociological literature [26,29]. Following the classification from the Typology of Forest Sites in Slovenia [30], the four subtypes were *Acer pseudoplatanus-Ulmus glabra* forest stands, growing mostly in concave terrain (hereafter referred to as subtype A; Figure 2a), *Fraxinus excelsior* stands, growing on slopes (hereafter referred to as subtype B; Figure 2b), *Tilia* sp. Stands, with thermophilous broadleaves occurring on exposed ridges and slopes (hereafter referred to as subtype C; Figure 2c), and *Acer pseudoplatanus* stands, occurring on more acidic soils with frequent admixture of *Castanea sativa* Mill. (hereafter referred to as subtype D; Figure 2d).

2.3. Data Collection

In the summer of 2020, we conducted field mapping of the entire study area. All activities were coordinated within the framework of the LIFE Integrated Project for Enhanced Management of Natura 2000 in Slovenia. Our fieldwork primarily focused on the existing (official) Natura 2000 zones of HT 9180*.

Forest stand data from the Slovenian Forest Service database [31] were used as the basis for field mapping. Managed forest stands, with a proportion of noble broadleaves of more than 30% in the total growing stock, were checked, as well as forest reserves (old-growth forests). In the field, a group of trained and calibrated field mappers checked the already known or existing HT 9180* zones and then searched for, and mapped, stands classified under this HT. We used a 0.5 m resolution digital orthophoto (DOF) as the basis for drawing the boundaries of each polygon (i.e., homogenous mapping entity) in the field.

The field mapping was carried out at a scale of 1:5000. Each mapped polygon was assigned to one of the four HsTs described in the previous subsection.

Figure 2. The four forest habitat subtypes of the *Tilio–Acerion* forest habitat type: (**a**) *Acer pseudoplatanus-Ulmus glabra* stands, (**b**) *Fraxinus excelsior* stands, (**c**) *Tilia* sp. dominated stands and (**d**) *Acer pseudoplatanus* stands on more acidic soils with an admixture of *Castanea sativa*. In all subtypes, except *Tilia* sp. dominated stands (characterized by a warmer and drier microclimate), abundant and tall-statured ferns and forbs create the typical physiognomy of the understory vegetation.

For each of the mapped polygons, we estimated important site, stand and nature conservation characteristics, which are explained in more detail in the following section. Tree species composition was evaluated by recognition of each individual tree species and by visually estimating the proportion (%) of noble broadleaves in the total tree layer cover. Additionally, the tree regeneration layer was inspected by identifying the most frequent tree species in the herb (height below 0.5 m) and shrub (height above 0.5 m and below 5 m) layers. The dominant relief form was determined for each polygon, distinguishing between several categories, as follow: flat relief, ridge or top of a hill, slope, bottom of a concave landform, such as a doline, and ditch. Surface rockiness/stoniness was estimated (% of surface covered by rocks/stones), a feature generally associated with bedrock, soil type and topography. The risk levels of various pressures and threats for HsT conservation status were estimated, distinguishing between tree mortality of key tree species, game browsing of juvenile trees, soil disturbance and tree damage due to forest management operations, presence of forest roads, and occurrence of invasive alien species. Based on these threats, and the previously described forest stand and site parameters, the overall estimation of the conservation status of each mapped polygon was determined (favourable, unfavourable or poor status). To summarize, the most important criteria used during the field mapping for assigning polygons to a specific subtype were the following: (i) dominant tree species, (ii) tree species composition, (iii) understory vegetation (vascular plants present in the shrub and herb layer) composition and cover, (iv) tree regeneration layer, (v) dominant relief type, (vi) bedrock type, rockiness/stoniness, and (vii) soil type. For the field mapping of habitat (sub)types, we used a simplified, faster method based only partly on a phytosociological approach, focusing on tree species composition and some characteristic plant species. Screening of vegetation composition, site conditions and conservation status was conducted in a few sampling points randomly distributed across the area of the mapped polygon. The number of sampling points was proportional to the size of the polygon, i.e., more sampling points were selected in larger polygons.

With respect to remote sensing data, LiDAR (ALS – airborne laser scanning) data [32] were collected during the period between 12 March, 2014, and 20 October, 2014. The flight took place at an altitude of 1200 to 1400 m above the ground.

2.4. Data Preparation and Analysis

For data obtained during field mapping, we constructed three binary (0/1) matrices with polygons in rows and descriptive variables in columns. The first matrix was for the tree species composition, where the following species, or species groups, were distinguished: *Acer pseudoplatanus, A. platanoides, Fraxinus excelsior, Ulmus glabra, Tilia platyphyllos, T. cordata, Fagus sylvatica* L., *Picea abies* (L.) H. Karst., *Castanea sativa*, thermophilous broadleaved trees (e.g., *Ostrya carpinifolia* Scop., *Sorbus aria* (L.) Crantz, *Fraxinus ornus* L.) and a group of other trees species (species such as *Abies alba* Mill., *Prunus avium* (L.) L. and *Carpinus betulus* being the most frequent).

The second matrix included data on relief, rockiness/stoniness and soil type. For relief, five categories were distinguished: flat surface, ridge or top of a hill, slope, bottom of concave landform, such as doline, and ditch. Rockiness/stoniness was classified into one of the following five classes: 0%–5%, 5%–10%, 10%–30%, 30%–50% and higher than 50%. Soil data were derived from the Slovenian pedological map 1:250,000 [33] and four soil types were distinguished: brown soils on limestone and dolomite, rendzina, ranker and eutric brown soil. As the soil types often changed over small distances, and some polygons were spread over larger areas with different soil types, the classification of soil types was done by fuzzy logic, with row sums in the matrix equal to 1.

The third matrix was associated with the threats and conservation status of the HsTs (polygon). For threats, we identified seven main categories: game browsing (e.g., visible damage on shoots, leaves and stems of young trees by wild ungulates, most often in the form of deer herbivory and bark stripping), tree mortality of key tree species (e.g., the presence of standing and lying dead trees, the presence of trees with severe crown damage with more dry branches without leaves), small or fragmented area (e.g., smaller patches of HsTs, polygons with a diameter of less than two forest stand heights (ca. 40–60 m) of the surveyed stand), beech competition (e.g., proportion of *Fagus sylvatica* in the mapped polygon greater than 50% of the stand growing stock), impact of forest roads (e.g., distance to the nearest forest road less than one forest stand height (up to ca. 30 m)), presence of invasive alien species (e.g., one or more individuals of invasive alien plant species present either in the tree or understory layer) and other threats, e.g., altered tree species composition (proportion of tree species unsuitable for local site conditions being more than 50% of the stand growing stock; in most cases, this was *Picea abies*), and competitive pressure for tree seedlings and saplings from dense herbaceous vegetation.

The conservation status of the polygon was evaluated as either favourable, unfavourable or poor. The conservation status of the stand in the studied polygon was favourable when the tree species composition was preserved, where regeneration of the key tree species occurred and where there were no obvious pressures or threats to the habitat type. Unfavourable conservation status was determined when only one or two less intense pressures or threats to the habitat type were evident (e.g., partially altered tree species composition, minor crown damage from disease, lack of juvenile stages of key tree species, minor damage to young trees from deer, and insignificantly eroded soil from human activities). The conservation status of the stand was poor when multiple negative factors that posed a serious threat to the long-term existence of habitat type were clearly evident. In addition, two transitional categories were also used: favourable/unfavourable (or unfavourable/favourable) and unfavourable/poor (or poor/unfavourable), resulting in a total of five different categories for the overall estimation of conservation status.

All statistical analyses were performed in the R programming environment version 3.5.2 [34]. A Non-metric Multidimensional Scaling (NMDS) ordination was performed on each matrix, and the results were illustrated in an ordination diagram. The NMDS ordinations were performed in the *vegan* R package [35]. The distance matrix was first calculated with the Jaccard dissimilarity index for binary data using "vegdist" function in the *vegan* package. Then, the distance matrix served as an input for the NMDS ordination. We considered the first two dimensions, since they resulted in a good fit according to the stress values. To quantify the influence of different explanatory variables, the ordination graphs were overlaid with explanatory variables. Significant variables ($p < 0.05$) were projected onto the NMDS biplots by passive fitting. To test whether subtypes differed significantly, in terms of tree species composition and other explanatory variables (relief, soil, threats and conservation status), we employed a permutational multivariate analysis of variance (PERMANOVA; [36]) with 9999 permutations (*vegan* package, function "adonis2"). This statistical method uses a permutation test with pseudo-F ratios. In the case of significant PERMANOVA results, pairwise multilevel comparisons (testing which groups differed from each other) were performed by using a wrapper function, "pairwise.adonis2" [37].

Canonical correspondence analysis (CCA), as a constrained ordination [38], was performed to explore the relationships between tree community and environmental variables. The tree species data were used as a dependent matrix and the other two datasets were used as predictors. We opted for a stepwise selection of significant explanatory variables, starting with an intercept-only model and, then, sequentially adding significant terms at each step, based on an ANOVA-like permutation test [39]. This procedure of model selection was based on Akaike's information criterion. Additionally, we calculated the indicator value of tree species in each subtype by using a statistical method proposed by [40], implemented in the *labdsv* R package [41]. Since the conservation status formed a core concept in our study, additional descriptive statistics of categories describing the conservation status of the studied subtypes were performed.

ALS-derived data were analysed in the post-mapping phase. We first intersected the field-mapped polygons (shapefile creation by digitalization in the ArcGIS 10.6.1. software [42] with two layers: terrain steepness (SLOPE, in degrees) and Topographic Position Index (TPI, unitless variable). The 1×1 m resolution TPI and SLOPE were both derived from ALS raw data [32] with a point density of 2 to 10 per m^2 and a relative horizontal and vertical accuracy of 0.30 and 0.15 m, respectively, and calculated using SAGA GIS and ArcGIS software (tools used: Slope and Terrain Analysis: Topographic Position Index). In the case of TPI, higher values denoted convex topography (ridges, summits), whereas lower TPI values denoted concave terrain (dolines, erosion gullies). The window within which the grid cell values were calculated had a radius of 20 m.

For each pixel (grid cell) in 1×1 m resolution, we extracted a value for SLOPE and TPI (each cell containing one value). This was done by implementing the function "extract" in the *raster* R package [43]. Boxplots were constructed for each parameter and a one-way Analysis of Variance (ANOVA) with Tukey post-hoc procedure was implemented to test the differences between the four habitat subtypes.

3. Results

3.1. Field Mapping

A total of 174 polygons were mapped during our field work. The total area mapped amounted to 314.16 ha (Table 1). More than half of this forest area was identified as subtype A (*Acer pseudoplatanus-Ulmus glabra* stands, 57.8% of the total mapped area), followed by subtype D (*Acer pseudoplatanus* stands on more acidic soils with an admixture of *Castanea sativa*, 32.3%). Polygons belonging to subtype B (*Fraxinus excelsior* stands, 7.9%) and subtype C (*Tilia* sp. stands, 1.9% of the total mapped area) represented a much lower proportion of the total mapped area. Detailed descriptive statistics for each HsT are given in Table 1.

Table 1. Descriptive statistics (number of polygons, total, mean, minimum and maximum area in hectares) for the mapped polygons for each forest habitat subtype: A—*Acer pseudoplatanus-Ulmus glabra* stands, B—*Fraxinus excelsior* stands, C—*Tilia* sp. stands and D—*Acer pseudoplatanus* stands on more acidic soils with an admixture of *Castanea sativa*.

Habitat Subtype	No. of Polygons	Area_sum (ha)	Area_mean (ha)	Area_min (ha)	Area_max (ha)
Subtype A	72	181.74	2.52	0.16	27.05
Subtype B	17	24.95	1.47	0.31	3.16
Subtype C	7	5.92	0.85	0.24	2.36
Subtype D	78	101.55	1.30	0.08	9.35
Total	174	314.16	1.81	0.08	27.05

The four subtypes significantly differed in terms of tree species composition (Figure 3; PERMANOVA test: F = 14.04, R^2 = 20%, $p < 0.001$). Pairwise comparisons revealed that all subtypes differed significantly at $p < 0.001$, except for subtype A and subtype B, which did not differ from each other ($p > 0.05$). Inferred from the F values of the pairwise PERMANOVA tests, the largest significant difference was detected between subtype B and subtype C, while the smallest significant difference was between subtype B and subtype D. The NMDS ordination had a final stress of 0.223. All variables (trees species), except *Fagus sylvatica*, proved significant ($p < 0.05$). The highest explanatory power (corresponding to the length of arrows in the ordination diagram; right panel in Figure 3) was detected for *Fraxinus excelsior* (R^2 = 0.60), followed by the group of the other tree species (0.47), *Ulmus glabra* (0.26), *Picea abies* (0.24), *Tilia* sp. (0.19), thermophilous broadleaves (0.16), *Castanea sativa* (0.15), *Acer platanoides* (0.10) and *Acer pseudoplatanus* (0.08). The low explanatory power of *Acer pseudoplatanus* was related to its presence in almost all of the mapped HsTs (polygons).

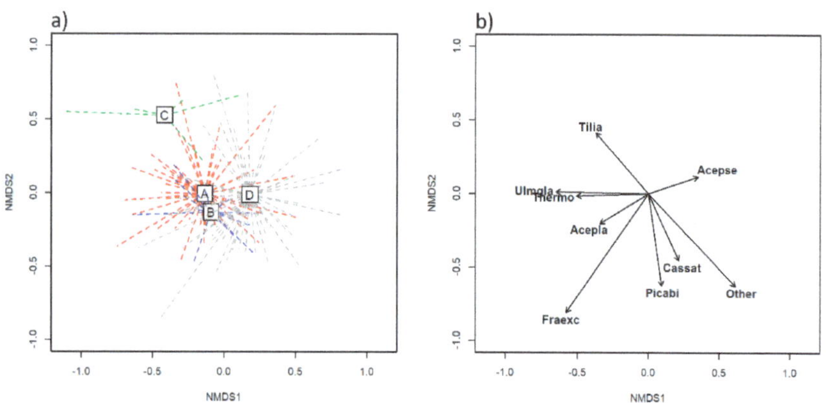

Figure 3. Two-dimensional NMDS ordination based on the composition of tree species. (**a**) The distribution of mapped polygons in the ordination space, distinguishing among four forest habitat subtypes: A (red) – *Acer pseudoplatanus-Ulmus glabra* stands, B (blue) – *Fraxinus excelsior* stands, C (green) – *Tilia* sp. stands and D (grey) – *Acer pseudoplatanus* stands on more acidic soils with an admixture of *Castanea sativa*. Squares with letters inside correspond to the centroids of the habitat subtypes. (**b**) Significant ($p < 0.05$) variables (tree species), which are coded as follows: Acepla—*Acer platanoides*, Acepse—*Acer pseudoplatanus*, Cassat—*Castanea sativa*, Fraexc—*Fraxinus excelsior*, Other—a group of other tree species (e.g., *Abies alba*, *Prunus avium*, *Carpinus betulus*), Picabi—*Picea abies*, Thermo—a group of thermophilous tree species (e.g., *Ostrya carpinifolia*, *Fraxinus ornus*), Tilia—*Tilia platyphyllos*/*T. cordata*, Ulmgla—*Ulmus glabra*.

The two-dimensional NMDS ordination, based on relief, rockiness/stoniness and soil type data (Figure 4), had a final stress of 0.189. According to the results of the PERMANOVA test (F = 17.84, R^2 = 24%, $p < 0.001$), subtypes differed significantly regarding topographic and soil conditions. Pairwise comparisons revealed that all subtypes differed significantly at $p < 0.001$, except for subtype A and subtype B, which did not differ from each other ($p > 0.05$). Inferring from the F values of the pairwise PERMANOVA tests, the largest significant difference was detected between subtype A and subtype D, while the smallest significant difference was between subtype A and subtype C. Only two variables (plane as a relief form and rockiness/stoniness class 5%–10%) did not prove significant. All other predictors were significant and their explanatory power was as follows (listed in descending order according to R^2): brown calcareous soils (R^2 = 0.65): rockiness/stoniness 0%–5% (0.63), rockiness/stoniness 10%–30% (0.57), ditch (0.56), slope (0.41), rendzina (0.29), concave relief (0.28), rockiness/stoniness 30%–50% (0.17), rockiness/stoniness above 50% (0.12), ridge (0.11), eutric brown soil (0.07) and ranker (0.06) (Figure 4).

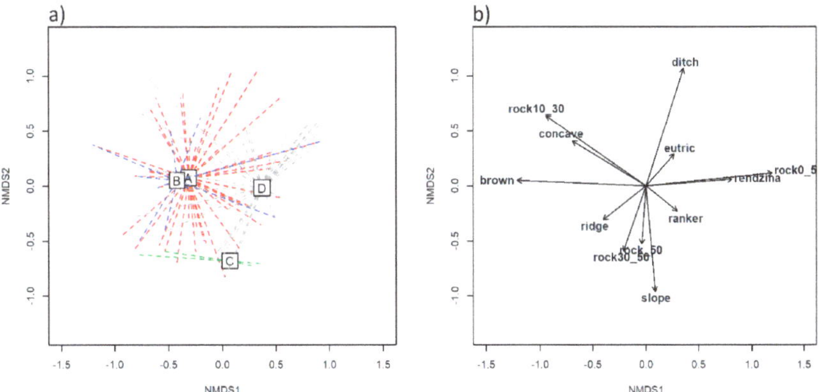

Figure 4. Two-dimensional NMDS ordination based on relief, rockiness/stoniness and soil type data. (**a**) The distribution of mapped polygons in the ordination space, distinguishing four forest habitat subtypes: A (red)—*Acer pseudoplatanus-Ulmus glabra* stands, B (blue)—*Fraxinus excelsior* stands, C (green)—*Tilia* sp. stands and D (grey)—*Acer pseudoplatanus* stands on more acidic soils with an admixture of *Castanea sativa*. Squares with letters inside correspond to the centroids of the habitat subtypes. (**b**) Significant ($p < 0.05$) variables, abbreviated as follows: rock0_5 = rockiness/stoniness 0%–5%, rock10_30 = rockiness/stoniness 10%–30%, rock 30_50 = rockiness/stoniness 30%–50%, rock_50 = rockiness/stoniness above 50%, brown–brown calcareous soils, eutric–eutric brown soils. Labels for other variables are considered to be self-explanatory.

The two-dimensional NMDS ordination (final stress: 0.167), containing information regarding the threats and conservation status, showed a high overlap of centroids for subtypes B, D and A (Figure 5). Nevertheless, the PERMANOVA test suggested statistically significant differences (F = 4.34, $p < 0.001$) between subtypes, although the explained variance was evidently lower (R^2 = 7%) compared to the first two NMDS ordinations. Pairwise comparisons revealed that all four subtypes differed significantly from each other at $p < 0.05$. These comparisons could be ordered as follows (based on the F values of the pairwise PERMANOVA tests): subtype B vs. subtype D ($p < 0.001$), subtype B vs. subtype C ($p < 0.001$), subtype A vs. subtype C ($p < 0.01$), subtype C and subtype D ($p < 0.01$), subtype A vs. subtype D ($p < 0.05$) and subtype A vs. subtype B ($p < 0.05$). The two explanatory variables with the highest r^2 values were unfavourable conservation status (R^2 = 0.68) and favourable conservation status (0.59). The conservation status category, favourable/unfavourable (unfavourable/favourable), also contributed to the explained variation (0.43), as did the category of other threats (0.26) and browsing pressure (0.23). Other significant variables were beech competition (0.11), habitat fragmentation (0.07), invasive alien species (0.05) and the conservation status category unfavourable/poor (poor/unfavourable) (0.04).

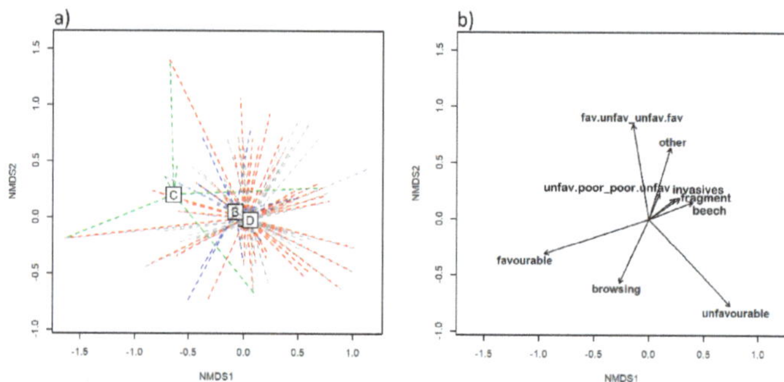

Figure 5. Two-dimensional NMDS ordination based on identified pressures and threats and over-all conservation status. (**a**) The distribution of mapped polygons in the ordination space, distinguishing among four forest habitat subtypes: A (red) – *Acer pseudoplatanus-Ulmus glabra* stands, B (blue) – *Fraxinus excelsior* stands, C (green) – *Tilia* sp. stands and D (grey) – *Acer pseudoplatanus* stands on more acidic soils with an admixture of *Castanea sativa*. Squares with letters inside correspond to the centroids of the habitat subtypes. Note that the centroid of subtype A is hidden behind the centroids of B and D. (**b**) Significant ($p < 0.05$) variables: fav.unfav_unfav.fav = favourable/unfavourable (or unfavourable/favourable) status, unfav.poor_poor.unfav = unfavourable/poor (or poor/unfavourable) status, fragment = habitat fragmentation, beech = competition from *Fagus sylvatica*, invasives = invasive alien plant species, other = other pressures and threats (e.g., altered tree species composition, competitive pressure for tree seedlings and saplings induced by dense herbaceous vegetation).

The final CCA model included eight explanatory variables, which collectively explained 16.2% of the variation in tree community composition. The constraining variables ($p < 0.05$, listed in descending order of importance) were the following: category of rockiness 0%–5%, tree mortality, brown soils on limestone and dolomite, beech competition, concave relief, presence of forest roads, presence of invasive plant species and favourable conservation status (Figure 6). The distribution of samples (polygons) in the CCA diagram revealed that subtype C was clearly positioned away from the other three subtypes and that subtype A and subtype B highly overlapped. Additional patterns inferred from the CCA analysis were a negative association between threats and favourable conservation status (gradient along the CCA axis 2) and that subtype D differed from subtypes A and B mostly in terms of rockiness (correlated with soil type) and the presence of invasive plants (gradient along the CCA axis 1; Figure 6).

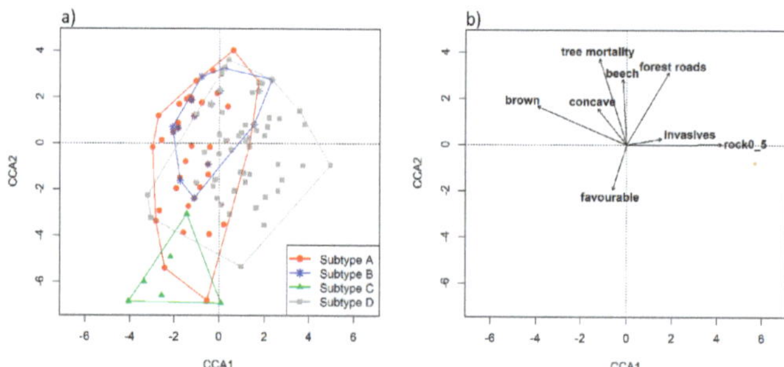

Figure 6. Canonical correspondence analysis (CCA) diagrams showing the main dimensions of tree species compositional variability between four habitat subtypes. (**a**) The distribution of 174 samples in the CCA ordination space with convex hulls constructed around each habitat subtype: A—*Acer pseudoplatanus-Ulmus glabra* stands, B—*Fraxinus excelsior* stands, C—*Tilia* sp. stands and D—*Acer pseudoplatanus* stands on more acidic soils with an admixture of *Castanea sativa*. (**b**) The black arrows denote significant explanatory variables (refer to previous figures for their abbreviations).

We identified at least one indicator tree species significantly associated with subtype B, C or D, whereas subtype A did not have any indicator species (Table 2). *Fraxinus excelsior* was significantly associated with subtype B and *Tilia* sp. and the group of thermophilous broadleaves with subtype C. *Castanea sativa* and *Acer pseudoplatanus* were indicators for subtype D (Table 2).

Table 2. List of indicator tree species significantly associated with different habitat subtypes: A – *Acer pseudoplatanus-Ulmus glabra* stands, B – *Fraxinus excelsior* stands, C – *Tilia* sp. stands and D – *Acer pseudoplatanus* stands on more acidic soils with an admixture of *Castanea sativa*. Significance levels were coded as follows: * $p < 0.05$, ** $p < 0.01$, *** $p < 0.001$.

Habitat Subtype	No. of Indicator Species	Tree Species	Indicator Value (IndVal)	Sig.
Subtype A	0	/	/	/
Subtype B	1	*Fraxinus excelsior*	0.68	***
Subtype C	2	*Tilia* sp.	0.76	***
		thermophilous broadleaves	0.69	**
Subtype D	2	*Castanea sativa*	0.61	**
		Acer pseudoplatanus	0.52	*

The studied subtypes exhibited prominent differences regarding their conservation status (Figure 7). Overall, subtypes could be ranked from those having the most favourable status to those having the least favourable status in the following order: subtype C > subtype A > subtype D > subtype B. More than half of all polygons in subtype C had favourable conservation status and no polygon was assigned to the poor status. In contrast, subtype B had almost 25% of polygons with either poor or unfavourable/poor status. Across all subtypes, the class of unfavourable conservation status had the highest mean proportion (45.9%) (Figure 7).

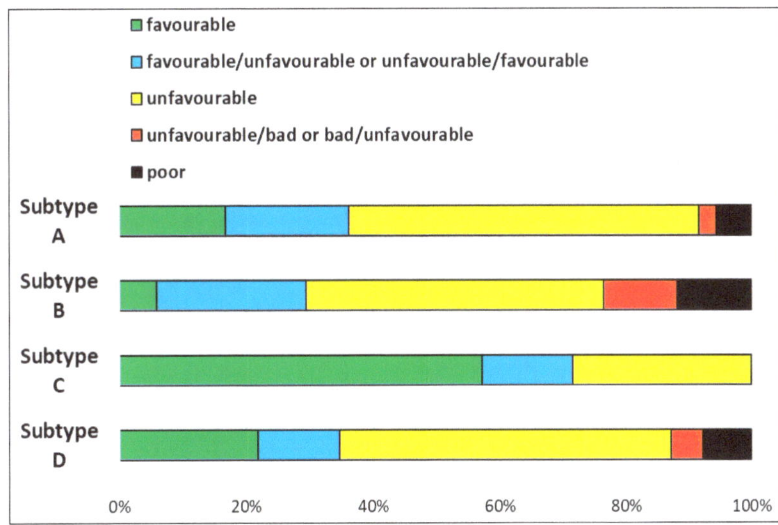

Figure 7. Proportions of the five categories describing conservation status in each habitat subtype: A—*Acer pseudoplatanus-Ulmus glabra* stands, B—*Fraxinus excelsior* stands, C—*Tilia* sp. stands and D—*Acer pseudoplatanus* stands on more acidic soils with an admixture of *Castanea sativa*.

3.2. ALS Data

The ALS-derived datasets were characterized by a large number of datapoints. In the case of SLOPE and TPI, the total number of 1×1 m raster cells was 3,141,648 (subtype A: 1,817,396 cells, subtype B: 249,467, subtype C: 59,200 and subtype D: 1,015,585 cells).

A summarized depiction of each ALS-derived parameter is given in Figure 8. An area in the western and central part of the BHD site, including the Boč mountain summit, was chosen as all four HsT were mapped there, creating a spatially diverse mosaic of four subtypes belonging to the same HT (Figure 8).

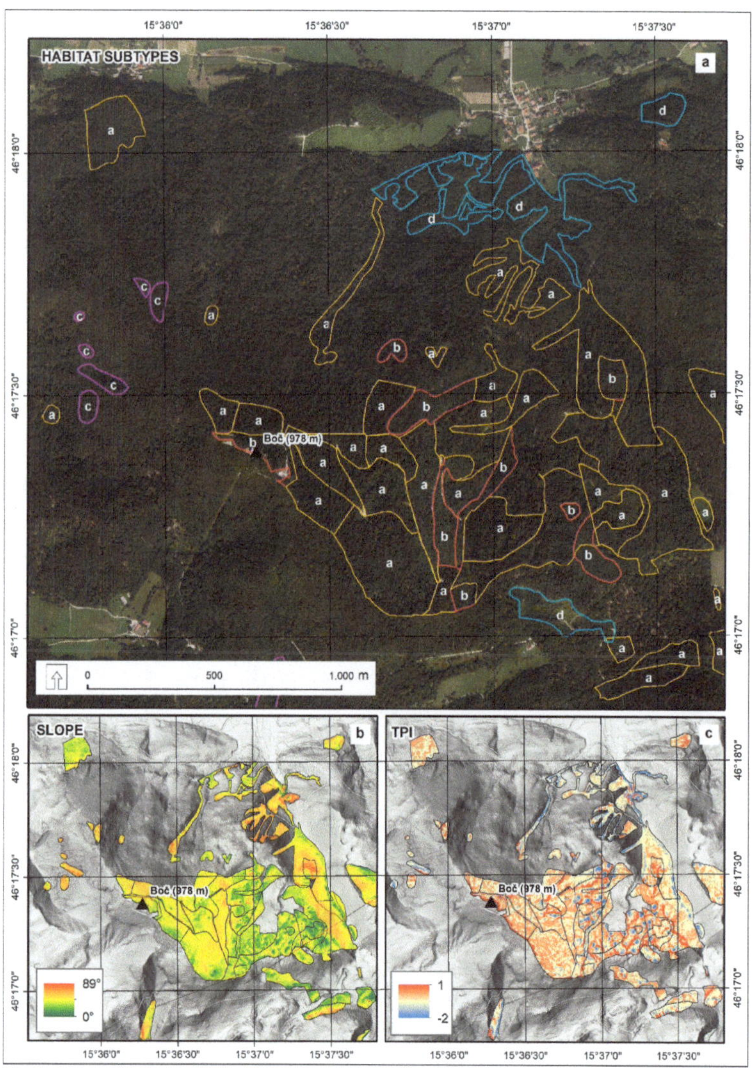

Figure 8. Spatial maps illustrating the digital orthophoto (**a**), terrain steepness (SLOPE, **b**) and Topographic Position Index (TPI, **c**). Higher TPI values denote convex topography (e.g., ridges), whereas lower TPI values were typical for concave terrain (e.g., dolines). The letters within the polygons represent forest habitat subtypes as follows: a—*Acer pseudoplatanus-Ulmus glabra* stands, b—*Fraxinus excelsior* stands, c—*Tilia* sp. stands and d—*Acer pseudoplatanus* stands on more acidic soils with admixture of *Castanea sativa*.

We found significant differences between habitat subtypes for SLOPE and TPI values (Figure 9). Regarding terrain steepness, the most noticeable were the steep slopes for subtype C. Almost 75% of raster cells belonging to subtype C had slope steepness values above 30° (Table 3). These *Tilia*-dominated stands mostly thrived on steep rocky slopes, sometimes even on cliffs. Subtype D was ranked second in terms of slope steepness, with the majority of cells having slopes above 30°. Subtypes A and B exhibited lower slope steepness values (Table 3). Smaller differences between subtypes were found for TPI values. However, Tukey post-hoc tests revealed significant differences for all pairwise comparisons, and, thus, mainly confirming the topographically-induced occurrence of subtypes. The median TPI value was similar for all four subtypes (around 0), but the range distribution of values showed some notable differences (Figure 9). For instance, subtype C had a pronounced portion of cells with a TPI above 1 and even outliers with exceptionally high TPI values, indicating more convex terrain. The other three subtypes had a higher proportion of cells with TPI values below 0. The polygons belonging to subtypes A and D occurred on more concave topographic setups, such as dolines and ditches.

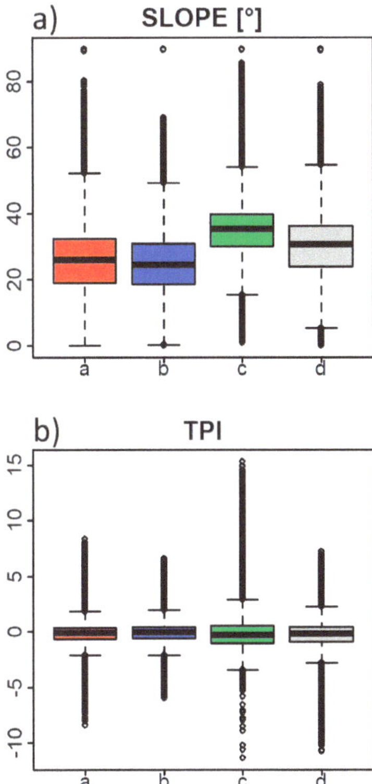

Figure 9. Boxplots showing the distribution of values derived from the ALS data (**a**) SLOPE–terrain steepness and (**b**) TPI – Topographic Position Index, for each of the studied forest habitat subtypes: a (red)—*Acer pseudoplatanus-Ulmus glabra* stands, b (blue)—*Fraxinus excelsior* stands, c (green)—*Tilia* sp. stands and d (grey)—*Acer pseudoplatanus* stands on more acidic soils with an admixture of *Castanea sativa*. Subtypes not sharing the same letter (below boxplot) significantly differ at $p < 0.001$ according to the Tukey post-hoc tests with ANOVA.

Table 3. The distribution of terrain steepness (SLOPE, in degrees) and Topographic Position Index (TPI, unitless) values, expressed in percentages (%) for different categories. The subtypes are A—*Acer pseudoplatanus-Ulmus glabra* stands, B—*Fraxinus excelsior* stands, C—*Tilia* sp. stands and D—*Acer pseudoplatanus* stands on more acidic soils with an admixture of *Castanea sativa*.

SLOPE (in %)				
Category	Subtype A	Subtype B	Subtype C	Subtype D
Below 5°	1.6	1.9	0.1	1.0
5–9.9°	4.7	4.1	0.4	2.5
10–19.9°	21.6	24.5	5.4	11.6
20–29.9°	38.6	41.6	19.3	30.3
Above 30°	33.5	27.9	74.8	54.6
TPI (in %)				
Category	Subtype A	Subtype B	Subtype C	Subtype D
Below −2	5.2	3.7	9.1	11.0
From −2 to −1	11.3	10.8	16.9	12.7
From −1 to 0	39.4	38.0	34.2	33.6
From 0 to 1	36.0	36.6	22.4	34.2
Above 1	8.1	10.9	17.4	8.5

4. Discussion

Proper management of Natura 2000 HTs requires improved knowledge on their ecological characteristics, spatial distribution and conservation status [5]. This study addressed the potential of field mapping of forest stands, based on vegetation typology and ALS-based characterization, to reveal the multi-faceted differences between the habitat subtypes (HsT) of the Natura 2000 priority habitat type *Tilio–Acerion* (9180*). Although these stands covered rather small areas, we discovered substantial variation between HsTs. Since the *Tilio–Acerion* forests are relatively heterogeneous in terms of ecological conditions, the approach based on HsTs could contribute to more reliable conservation assessment and management of an entire HT.

In the BHD Natura 2000 site, we found that the four pre-defined HsTs differ in the majority of the investigated features. The most abundant HsT, covering the largest area, was subtype A (stands dominated by *Acer pseudoplatanus* and *Ulmus glabra*), followed by subtype D (occurring on more acidic soils compared to the other three subtypes). We mapped only seven (out of 174) stands assigned to subtype C (*Tilia platyphyllos* and *T. cordata* as the dominant overstory tree species). These results indicated that the HsTs of the 9180* HT were unevenly represented in the studied site. Furthermore, the average area of mapped polygons varied significantly. It is, therefore, important to use a spatial resolution which can detect subtypes covering smaller areas, such as subtypes C and B in our case.

The four HsTs differed most significantly in terms of tree species composition, as demonstrated by the results derived from the NMDS and CCA. Tree species composition has been established as a cost-effective and easy-to-obtain indicator, with high predictive power in forest ecosystems [8,14] and can serve as an important starting point for assessing the conservation status of a forest HT [5,44]. Despite significant differences in tree species composition between the four HsTs, it is important to note that many noble broadleaves, which are diagnostic species of *Tilio–Acerion* forests, occur in different subtypes and are not strictly confined to a particular subtype. Some of the tree species were identified as good indicators of certain HsTs, according to the IndVal analysis. For example, the tree layer of even-aged forest stands assigned to subtype B was usually strongly dominated by *Fraxinus excelsior*. Similarly, *Tilia platyphyllos* and *T. cordata* were rarely found in subtypes other than subtype C, which frequently exhibited an admixture of thermophilous broadleaves. On the other hand, subtype A did not have any significant tree indicator species, which could be attributed to the normally diverse tree species composition of forest stands in this HsT (frequent admixture of *Ulmus glabra* and *Fagus sylvatica*). *Acer pseudoplatanus* was present in all subtypes, except subtype C. However, it was identified as an indicator species for subtype D, as shown by the IndVal analysis. Compared to subtype A, stands of subtype D were often composed of a pure *A. pseudoplatanus* overstory layer with a smaller admixture of codominant tree species linked to more acidic soil types. Among the subordinate tree species, some could be even used as differentiating species between subtypes. In our study area, this was true for *Castanea sativa*, which was never present on pure calcareous soil and was only present in subtype D (clearly supported by the IndVal analysis). Likewise, *Prunus avium* was also frequently present in stands belonging to subtype D, as these were mainly mapped at lower elevations in the eastern part of the BHD site (i.e., Haloze hills).

With respect to the tree species composition, we noticed that a substantial proportion of mapped stands (with the exception of subtype C) exhibited deviation from the natural or desired species composition in the tree layer. The degree of representativeness of tree species composition was used as one of the core factors for evaluating conservation status during field mapping. Two distinct patterns were identified. Firstly, *Tilio–Acerion* forests in the BHD site were embedded within a large matrix of prevailing *Fagus sylvatica* forests. The threat of beech competition was identified as one of the variables explaining the variation in tree communities between different HsTs (CCA results). In our study area, the 9180* HT was preserved to a greater extent where beech was not competitive. Beech is especially competitive in the regeneration layer, where it can outperform noble broadleaves, due to its high shade tolerance. Secondly, altered tree species composition, in the form of a substantial proportion of *Picea abies*, was a consequence of past forest management. Norway spruce was often planted on forest sites with high soil productivity, where noble broadleaves would naturally dominate. Such human-induced alterations of tree species composition decreased forest naturalness. Altered tree species composition significantly affected the diversity and composition of understory strata and other trophic levels (animals, fungi), and posed a threat to the conservation of the studied HT and its HsTs. A relatively large proportion of spruce in the growing stock was among the main factors indicating the poorer conservation status of this HT. This seems to be a problem for the majority of *Tilio–Acerion* forests in Slovenia. Based on analyses of data from the Slovenia Forest Service, Kutnar et al. [14] found that *Picea abies* accounted for as much as 40% of the total growing stock and *Fagus sylvatica* 25% of the total growing stock in Slovenian *Tilio–Acerion* forest stands. In addition, low vitality, or even complete dieback of edificator tree species, was another contributing factor to the higher discrepancy between the naturally-preserved state of forest stands and observed situations in the field, and, consequently, to poorer conservation status.

Data acquired during field mapping showed that the four subtypes differed in terms of relief features, rockiness/stoniness and soil conditions. These three parameters were intercorrelated. For instance, subtype C, occurring on ridges and steep slopes, also had a higher proportion of rocks/stones on the surface and more shallow soil (e.g., rendzinas). In concave landforms (e.g., dolines), where subtype A was most often found, deeper soil types were formed, such as brown calcareous soils. For subtype D, a general pattern was that it occurred on more eutric brown soils with lower soil pH compared to the other three soil types. The main difference between carbonate (limestone) parent material and mixed or silicate parent material could be quite easily recognized in the field with respect to how much of the surface was covered with rocks and/or stones. All these topographic and soil parameters could serve as surrogates for identification, and differentiation, of HsTs during field mapping. However, the same subtype could vary significantly, and different subtypes, distinguished on stand characteristics (tree species composition), could share similar topographic and soil properties, as these usually change at fine spatial scales.

Even greater similarity between HsTs was found for pressures and threats and the overall evaluation of conservation status. Subtype C somewhat exhibited the most favourable conditions, despite its low frequency of occurrence. This was related to the fact that *Tilia* sp. forest stands were mainly found in more remote areas where anthropogenic pressures were less likely to influence the natural development of forests. We confirmed that among various threats to the conservation status of HsT, habitat fragmentation was common to all subtypes, and stands covering larger areas were rarely found. Small HsT were additionally under pressure because of other threats. Game browsing could be seen as an omnipresent pressure that prevents the ingrowth of key tree species into shrub and tree layers. The Boč mountain is a central area of the mouflon (*Ovis musimon* Pallas) population, a non-native animal species introduced decades ago but preserved for hunting. This species, along with indigenous deer species, represents an important negative factor in the natural development of forests, due to their browsing and worsening of the ecological conditions of native species in the entire BHD site and, particularly, on Boč mountain [25]. Complementary inputs are required from all stakeholders (i.e., foresters, hunters, forest owners, conservationists) to solve the problems of wildlife management [45]. Natural regeneration of site-adapted tree species is key for preserving the favourable conservation status of the HT. A deficit of natural regeneration is a major long-term threat to the studied HT because it defines the tree species composition of mature stands, a feature for which we demonstrated a high importance in the overall identification of the 9180* HT and its subtypes.

Among the most significant threats that contributed to the differentiation between HsT were also invasive alien plant species and pathogens causing tree mortality. *Robinia pseudoacacia* L., known for its invasive potential in a wide range of forest communities, was present as an overstory subordinate species in a few surveyed forest stands, followed by non-native conifers *Pseudotsuga menziesii* (Mirb.) Franco and *Pinus strobus* L. (both artificially promoted by planting). The most frequently recorded invasive plants in the understory layer were fast-growing ruderals *Erigeron annuus* (L.) Desf. and *Impatiens parviflora* DC., but their abundance was rather low in the mapped polygons. The occasional occurrence of these invasive plants was mainly restricted to forest roads, skidding trails and disturbed stands with less canopy closure, as these plants are successful colonizers in canopy gaps with modified ecological conditions. The degree of invasiveness of neophytes should be monitored in the future, as they are often one of the main management concerns.

A major problem for European ash (*Fraxinus excelsior*) is the invasive fungal disease known as ash dieback (*Hymenoscyphus fraxineus*). The threat of tree mortality was among the leading factors explaining tree community composition (see CCA results). It was noted for almost every forest stand belonging to subtype B and often in other subtypes with the presence of individual European ash trees. This pathogen is a pressing conservation biological challenge throughout Europe [46], as it lethally affects ash trees of all age classes and tree mortality levels are high. Other tree species have also been threatened by widespread pest infestations. Dutch elm disease (*Ophiostoma novo-ulmi*) has long been a threat to *Ulmus glabra* trees in the temperate zone.

The relatively unfavourable conservation status of the *Tilio–Acerion* forests in the BHD Natura 2000 site results from a complexity of many different negative factors and their interactions. Pressures from fragmentation, hampered natural regeneration, mortality of key tree species and an unsuitable ratio between developmental (successional) phases [14] pose a serious risk to the studied 9180* HT and, ultimately, undermine the integrity of the existing habitat type to such a degree that the development of a new trajectory is inevitable. As discussed for the 91K0 HT and its HsT [2], it is possible to predict that shade-tolerant beech is likely to prevail over time if regeneration patterns do not change and deer herbivory does not diminish. Forest stands currently composed of noble broadleaves could be replaced by competitively superior beech in the future. Therefore, effective conservation strategies starting with local site-adapted management measures are urgently needed. To ensure the favourable conservation status of the HT, appropriate management and restoration actions should be implemented regularly [47], and, in the case of *Tilio–Acerion* forests, preferably at the level of HsTs.

Forest stands assigned to the *Tilio–Acerion* HT are interesting for both timber production (valuable timber of noble broadleaves) and biodiversity conservation. It is, thus, imperative that the goal-setting process becomes more subtype-specific, aiming at finding trade-offs between economic and biodiversity forest functions. Differentiating between HsTs can contribute to the more sustainable management of these forest stands. Similar to the conclusions made by Kovač et al. [2], we believe that unified management approaches simply do not work because they provide different outcomes in habitat subtypes with different ecological backgrounds, stand characteristics (tree species composition), topographic factors and other specificities (threats in the form of biotic and abiotic stressors) and, thus, in management applications. If the stands belonging to HsTs and forest associations alone are considered as conservation units, and their biodiversity portrayals need to be sustained, only a free-style silvicultural system can be implemented [2].

We showed that information obtained with remote sensing techniques can be used in terms of valuable explanatory and discriminating variables. ALS-derived variables served to characterize individual HsT with respect to topographic variables (slope steepness and terrain concavity and convexity). The application of remote sensing techniques in conservation ecology and forest management is still scarce [48]. ALS data and its derivatives have a strong, yet underexploited, potential to assist in the monitoring of Natura 2000 habitat types and other ecologically significant areas [49]. Overall, the results from Bässler et al. [3] demonstrated that airborne laser scanning is a rapid method of predicting Natura 2000 habitat types with an accuracy similar to time-consuming ground surveys. Multiple arguments were outlined in favour of the application of ALS for conservation efforts. The main advantage of ALS and related methods is that they allow sampling of habitat characteristics at a high spatial resolution over a large spatial extent. However, the authors do not suggest that ALS alone should be used to map Natura 2000 habitat types. In our case, both SLOPE and TPI could be used as input variables to model the probability of occurrence of different HsTs, but any successful application of remotely sensed indicators of habitat characteristics or conservation status requires careful ground truthing [7].

If we are to maintain the favourable condition of forest HT, mapping, monitoring and management should be based on the subtype level [9,50]. The integration of such principles into conservation schemes would help the estimation of habitat type status. Verified estimations based on HsT level can help to identify more detailed threats and subtype-specific stressors, which are usually not identified at the level of habitat types. Such approaches allow for both practical results useful for management as well as more scientific outputs. Assessments made at the subtype level can certainly improve the overall assessment of the conservation status of HTs, because the evaluation is more detailed. Whenever relevant, we recommend transferring the subtype-level approach to other (heterogenous) forest habitat types and regions.

5. Conclusions

In the Boč–Haloze–Donačka gora Natura 2000 site (Slovenia), *Tilio–Acerion* forests of slopes, screes and ravines cover small, fragmented areas within prevailing beech forest communities. The main novelty of this study was the implementation of an HsT approach for *Tilio–Acerion* forests. The results showed that the four studied HsTs differed significantly in terms of the areas they cover and could be distinguished in terms of tree species composition, relief features (concave vs. convex terrain) and the various threats they experienced (habitat fragmentation, mortality of the key tree species, and game browsing of the tree species' regeneration layers). Such pressures put the entire HT into a rather unenviable situation, and appropriate conservation measures are, thus, necessary. We conclude that owing to the specificities of individual HsTs, conservation goals and management activities should not only be targeted at the Natura 2000 HT level. Specific management and conservation measures should address each site-specific HsT of *Tilio–Acerion* forests, which each face different pressures and threats to their stability and long-term persistence.

Author Contributions: Conceptualization, J.K., E.K. and L.K.; methodology, J.K., E.K. and L.K.; statistical analysis, J.K.; spatial analysis, E.K.; field mapping and data curation, J.K., E.K. and L.K.; writing—original draft preparation, J.K., E.K. and L.K.; writing—review and editing, J.K., E.K. and L.K.; visualization, E.K., J.K. All authors have read and agreed to the published version of the manuscript.

Funding: This study was funded by the EU Project LIFE Integrated Project for Enhanced Management of Natura 2000 in Slovenia (LIFE17 IPE/SI/000011) and by the Slovenian Research Agency (ARRS), Research Program P4-0107.

Data Availability Statement: Not applicable.

Acknowledgments: We thank Valerija Babij, Ruben Šprah (Slovenia Forest Service), Aleksander Marinšek, Anica Simčič, Ajša Alagić and David Štefanič (Slovenian Forestry Institute) for their extensive help during the field mapping in the Boč–Haloze–Donačka gora Natura 2000 site and for data processing. We very much appreciate the valuable suggestions for improvement from all reviewers. We are grateful to Philip J. Nagel for proofreading and language editing.

Conflicts of Interest: The authors declare that they have no conflict of interest.

References

1. Velázquez, J.; Tejera, R.; Hernando, A.; Núñez, M.V. Environmental diagnosis: Integrating biodiversity conservation in management of Natura 2000 forest spaces. *J. Nat. Conserv.* **2010**, *18*, 309–317. [CrossRef]
2. Kovač, M.; Hladnik, D.; Kutnar, L. Biodiversity in (the Natura 2000) forest habitats is not static: Its conservation calls for an active management approach. *J. Nat. Conserv.* **2018**, *43*, 250–260. [CrossRef]
3. Bässler, C.; Stadler, J.; Müller, J.; Förster, B.; Göttlein, A.; Brandl, R. LiDAR as a rapid tool to predict forest habitat types in Natura 2000 networks. *Biodivers. Conserv.* **2011**, *20*, 465–481. [CrossRef]
4. EEA. Natura 2000 Barometer Statistics. European Environment Agency, European Environment European Environment Information and Observation Network (Eionet). 2020. Available online: https://www.eea.europa.eu/themes/biodiversity/document-library/natura-2000/natura-2000-network-statistics/natura-2000-barometer-statistics/statistics/barometer-statistics (accessed on 15 June 2020).
5. Kutnar, L.; Dakskobler, I. Ocena stanja ohranjenosti gozdnih habitatnih tipov (Natura 2000) in gospodarjenje z njimi (Evaluation of the conservation status of forest habitat types (Natura 2000) and their forest management). *Gozdarski Vestn.* **2014**, *72*, 419–439.
6. Habitats Directive. *Council Directive 92/43/EEC of 21 May 1992 on the Conservation of Natural Habitats and of Wild Fauna and Flora*; European Commission: Brussels, Belgium, 1992.
7. Schmidt, J.; Fassnacht, F.E.; Neff, C.; Lausch, A.; Kleinschmit, B.; Förster, M.; Schmidtlein, S. Adapting a Natura 2000 field guideline for a remote sensing-based assessment of heathland conservation status. *Int. J. Appl. Earth Obs. Geoinf.* **2017**, *60*, 61–71. [CrossRef]
8. Cantarello, E.; Newton, A.C. Identifying cost-effective indicators to assess the conservation status of forested habitats in Natura 2000 sites. *For. Ecol. Manag.* **2008**, *256*, 815–826. [CrossRef]
9. Kovač, M.; Kutnar, L.; Hladnik, D. Assessing biodiversity and conservation status of the Natura 2000 forest habitat types: Tools for designated forestlands stewardship. *For. Ecol. Manag.* **2016**, *359*, 256–267. [CrossRef]
10. Košir, P.; Čarni, A.; Di Pietro, R. Classification and phytosociological differentiation of broad-leaved ravine forests in southeastern Europe. *J. Veg. Sci.* **2008**, *19*, 331–342. [CrossRef]
11. Steinacker, C.; Beierkuhnlein, C.; Jaeschke, A. Assessing the exposure of forest habitat types to projected climate change—Implications for Bavarian protected areas. *Ecol. Evol.* **2019**, *9*, 14417–14429. [CrossRef]
12. Hrivnák, R.; Slezák, M.; Ujházy, K.; Máliš, F.; Blanár, D.; Ujházyová, M.; Kliment, J. Phytosociological approach to scree and ravine forest vegetation in Slovakia. *Ann. For. Res.* **2019**, *62*, 183–200. [CrossRef]
13. Zukal, D.; Novák, P.; Duchoň, M.; Blanár, D.; Chytrý, M. Calcicolous rock-outcrop lime forests of east-central Europe. *Preslia* **2020**, *92*, 191–211. [CrossRef]
14. Kutnar, L.; Matijašić, D.; Pisek, R. Conservation status and potential threats to Natura 2000 forest habitats in Slovenia. *Šumarski List.* **2011**, *5*, 215–231.
15. Corbane, C.; Lang, S.; Pipkins, K.; Alleaume, S.; Deshayes, M.; Millán, V.E.G.; Strasser, T.; Borre, J.V.; Toon, S.; Michael, F. Remote sensing for mapping natural habitats and their conservation status—New opportunities and challenges. *Int. J. Appl. Earth Obs. Geoinf.* **2015**, *37*, 7–16. [CrossRef]
16. Zlinszky, A.; Deák, B.; Kania, A.; Schroiff, A.; Pfeifer, N. Mapping Natura 2000 Habitat Conservation Status in a Pannonic Salt Steppe with Airborne Laser Scanning. *Remote Sens.* **2015**, *7*, 2991–3019. [CrossRef]
17. Innes, J.L.; Koch, B. Forest biodiversity and its assessment by remote sensing. *Glob. Ecol. Biogeogr. Lett.* **1998**, *7*, 397–419. [CrossRef]
18. Petrou, Z.I.; Manakos, I.; Stathaki, T. Remote sensing for biodiversity monitoring: A review of methods for biodiversity indicator extraction and assessment of progress towards international targets. *Biodivers. Conserv.* **2015**, *24*, 2333–2363. [CrossRef]
19. Vanden Borre, J.; Paelinckx, D.; Mücher, C.A.; Kooistra, L.; Haest, B.; De Blust, G.; Schmidt, A.M. Integrating remote sensing in Natura 2000 habitat monitoring: Prospects on the way forward. *J. Nat. Conserv.* **2011**, *19*, 116–125. [CrossRef]
20. Feilhauer, H.; Dahlke, C.; Doktor, D.; Lausch, A.; Schmidtlein, S.; Schulz, G.; Stenzel, S. Mapping the local variability of Natura 2000 habitats with remote sensing. *Appl. Veg. Sci.* **2014**, *17*, 765–779. [CrossRef]
21. Rechsteiner, C.; Zellweger, F.; Gerber, A.; Breiner, F.T.; Bollmann, K. Remotely sensed forest habitat structures improve regional species conservation. *Remote Sens. Ecol. Conserv.* **2017**, *3*, 247–258. [CrossRef]

22. Anderson, C.T.; Dietz, S.L.; Pokswinski, S.M.; Jenkins, A.M.; Kaeser, M.J.; Hiers, J.K.; Pelc, B.D. Traditional field metrics and terrestrial LiDAR predict plant richness in southern pine forests. *For. Ecol. Manag.* **2021**, *491*, 119118. [CrossRef]

23. EIONET, 2013–2018. European Environment Information and Observation Network. Annex D—Report Format on the 'Main Results of the Surveillance under Article 11' for Annex I Habitat Types. Available online: http://cdr.eionet.europa.eu/Converters/run_conversion?file=si/eu/art17/envxuwnma/SI_habitats_reports-20190819-145416.xml&conv=589&source=remote#9180 (accessed on 20 October 2021).

24. Babij, V.; Danev, G.; Kutnar, L.; Pisek, R. *Splošna Analiza Stanja Gozdov v Gozdnatih IP Območjih, Akcija A.1.1.*; Slovenia Forest Service and Slovenian Forestry Institute: Ljubljana, Slovenia, 2020; 29p.

25. Karlo, T.; Senegačnik, A. *Analiza in Ocena Stanja Projektnega Območja Boč-Haloze-Donačka Gora (Report of detailed analysis of Situation on Nature Conservation Measures for Natura 2000 Sites Boč-Haloze-Donačka Gora)*; Institute of the Republic of Slovenia for Nature Conservation: Ljubljana, Slovenia, 2020; 55p.

26. Dakskobler, I.; Košir, P.; Kutnar, L. *Gozdovi Plemenitih Listavcev v Sloveniji (Broad-Leaved Ravine Forests in Slovenia)*; Silva Slovenica, Gozdarski inštitut Slovenije: Ljubljana, Slovenia; Zveza gozdarskih društev Slovenije—Gozdarska založba Ljubljana: Ljubljana, Slovenia, 2013; 75p.

27. EEA, 2022. EUNIS Habitat Classification. Available online: https://www.eea.europa.eu/data-and-maps/data/eunis-habitat-classification-1 (accessed on 15 June 2022).

28. Chytrý, M.; Tichý, L.; Hennekens, S.M.; Knollová, I.; Janssen, J.A.M.; Rodwell, J.S.; Peterka, T.; Marcenò, C.; Landucci, F.; Danihelka, J.; et al. EUNIS Habitat Classification: Expert system, characteristic species combinations and distribution maps of European habitats. *Appl. Veg. Sci.* **2020**, *23*, 648–675. [CrossRef]

29. Bončina, A.; Rozman, A.; Dakskobler, I.; Klopčič, M.; Babij, V.; Poljanec, A. *Gozdni Rastiščni tipi Slovenije—Vegetacijske, Sestojne in Upravljavske Značilnosti*; Oddelek za gozdarstvo in obnovljive gozdne vire, Biotehniška fakulteta Univerze v Ljubljani in Zavod za gozdove Slovenije: Ljubljana, Slovenia, 2021; 575p.

30. Kutnar, L.; Veselič, Ž.; Dakskobler, I.; Robič, D. Tipologija gozdnih rastišč Slovenije na podlagi ekoloških in vegetacijskih razmer za potrebe usmerjanja razvoja gozdov (Typology of Slovenian forest sites according to ecological and vegetation conditions for the purposes of forest management). *Gozdarski Vestn.* **2012**, *70*, 195–214.

31. ZGS. *Forest Stands Map*; Slovenia Forest Service: Ljubljana, Slovenia, 2019.

32. ARSO LiDAR. Slovenian Environment Agency LiDAR. 2016. Available online: http://gis.arso.gov.si/evode/profile.aspx?id\protect$\relax\protect{\begingroup1\endgroup\@@over4}$atlas_voda_Lidar@Arso (accessed on 6 October 2021).

33. Zupan, M. Digitalna Pedološka Karta Slovenije 1:250 000. Univerza v Ljubljani, Infrastrukturni Center za Pedologijo in Varstvo Okolja. 2011. Available online: https://rkg.gov.si/vstop/ (accessed on 19 October 2021).

34. R Core Team. *R: A Language and Environment for Statistical Computing*; R Foundation for Statistical Computing: Vienna, Austria, 2019.

35. Oksanen, J.; Simpson, G.L.; Blanchet, F.G.; Kindt, R.; Legendre, P.; Minchin, P.R.; O'Hara, R.B.; Solymos, P.; Stevens, M.H.H.; Szoecs, E.; et al. *"Vegan"—Community Ecology Package*. 2019. Available online: https://cran.r-project.org/web/packages/vegan/index.html (accessed on 19 December 2022).

36. Anderson, M.J. Permutational multivariate analysis of variance (PERMANOVA). In *Wiley StatsRef: Statistics Reference Online*; John Wiley & Sons, Ltd.: Hoboken, NJ, USA, 2017; pp. 1–15.

37. Martinez Arbizu, P. *Pairwiseadonis: Pairwise Multilevel Comparison Using Adonis*, R Package Version 0.4; 2020. Available online: https://github.com/pmartinezarbizu/pairwiseAdonis (accessed on 19 December 2022).

38. Ter Braak, C.J.F. The analysis of vegetation-environment relationships by canonical correspondence analysis. *Vegetatio* **1987**, *69*, 69–77. [CrossRef]

39. Oksanen, J. Constrained Ordination: Tutorial with R and Vegan. 2012. Available online: https://www.mooreecology.com/uploads/2/4/2/1/24213970/constrained_ordination.pdf (accessed on 19 December 2022).

40. Dufrêne, M.; Legendre, P. Species assemblages and indicator species: The need for a flexible asymmetrical approach. *Ecol. Monogr.* **1997**, *67*, 345–366. [CrossRef]

41. Roberts, D.W. Package »labdsv«—Ordination and Multivariate Analysis for Ecology. 2022. Available online: https://cran.r-project.org/web/packages/labdsv/labdsv.pdf (accessed on 19 December 2022).

42. ArcGIS 10.6.1. 2018. ESRI. Available online: https://support.esri.com/en/products/desktop/arcgis-desktop/arcmap/10-6-1 (accessed on 23 December 2023).

43. Hijmans, R.J.; van Etten, J.; Sumner, M.; Chen, J.; Baston, D.; Bevan, A.; Bivand, R.; Busetto, L.; Canty, M.; Fasoli, B.; et al. *"Raster" Package—Geographic Data Analysis and Modelling*. 2020. Available online: https://cran.r-project.org/web/packages/raster/index.html (accessed on 19 December 2022).

44. Kovač, M.; Gasparini, P.; Notarangelo, M.; Rizzo, M.; Cañellas, I.; Fernández-de-Uña, L.; Alberdi, I. Towards a set of national forest inventory indicators to be used for assessing the conservation status of the habitats directive forest habitat types. *J. Nat. Conserv.* **2020**, *53*, 125747. [CrossRef]

45. Reimoser, F. Steering the impacts of ungulates on temperate forests. *J. Nat. Conserv.* **2003**, *10*, 243–252. [CrossRef]

46. Pautasso, M.; Aas, G.; Queloz, V.; Holdenrieder, O. European ash (*Fraxinus excelsior*) dieback—A conservation biology challenge. *Biol. Conserv.* **2013**, *158*, 37–49. [CrossRef]

47. Simonson, W.D.; Allen, H.D.; Coomes, D.A. Remotely sensed indicators of forest conservation status: Case study from a Natura 2000 site in southern Portugal. *Ecol. Indic.* **2013**, *24*, 636–647. [CrossRef]
48. Lechner, A.M.; Foody, G.M.; Boyd, D.S. Applications in Remote Sensing to Forest Ecology and Management. *One Earth* **2020**, *2*, 405–412. [CrossRef]
49. Nagendra, H.; Lucas, R.; Honrado, J.P.; Jongman, R.H.G.; Tarantino, C.; Adamo, M.; Mairota, P. Remote sensing for conservation monitoring: Assessing protected areas, habitat extent, habitat condition, species diversity, and threats. *Ecol. Indic.* **2013**, *33*, 45–59. [CrossRef]
50. Tsiripidis, I.; Xystrakis, F.; Kallimanis, A.; Panitsa, M.; Dimopoulos, P. A bottom–up approach for the conservation status assessment of structure and functions of habitat types. Rendiconti Lincei. *Sci. Fis. E Nat.* **2018**, *29*, 267–282. [CrossRef]

forests

Article

Characterization of Forest Ecosystems in the Chure (Siwalik Hills) Landscape of Nepal Himalaya and Their Conservation Need

Yadav Uprety [1,*], Achyut Tiwari [1], Sangram Karki [2], Anil Chaudhary [1], Ram Kailash Prasad Yadav [1], Sushma Giri [3], Srijana Shrestha [3], Kiran Paudyal [3] and Maheshwar Dhakal [3,*]

[1] Central Department of Botany, Tribhuvan University, Kathmandu 44600, Nepal
[2] Forest Research and Training Center, Ministry of Forests and Environment, Babarmahal, Kathmandu 44600, Nepal
[3] President Chure Terai-Madhesh Conservation Development Board, Khumaltar, Lalitpur 44700, Nepal
* Correspondence: yadavuprety@gmail.com (Y.U.); maheshwar.dhakal@gmail.com (M.D.)

Abstract: As a basic component of the forest ecosystem, the forest structure refers to the general distribution of plant species of different life forms and sizes. The characterization of forest structure is the key to understanding the vegetation history, present status, and future development trajectory of the forest ecosystems. The Chure region of Nepal covers about 12.78% of the country's land area and extends east to west along the southern foothills. This biologically rich but geologically fragile region is home to many species and provides many ecosystem services to millions of people. The Chure landscape is severely suffered from anthropogenic disturbances including logging, grazing, fuelwood collection, solid waste disposal, encroachment, forest fire, and excavation of sand, gravel, and boulders. In this study, we aim to characterize the forest ecosystem types outside the protected areas in the Chure region of Nepal and analyze the threat and vulnerability of the landscape from the biodiversity point of view. We sampled 62 sites to study the dominant vegetation type, regeneration status, and major threats to the forest ecosystems. A distribution map of the forest ecosystem types in Chure was prepared. We identified 14 forest ecosystem types in Chure including seven new ones. The newly reported forest ecosystems are *Hymenodictyon excelsum* Forest, *Syzygium cumini* Forest, *Terminalia anogeissiana* Forest, *Schima wallichii–Shorea robusta* Forest, *Pinus roxburghii–Shorea robusta* Forest, *Pinus roxburghii* Forest, and Bamboo thickets. We conclude that intensified human activities including forest encroachment and deforestation are mainly responsible for the ecological imbalance in the Chure region. We emphasize an in-depth analysis of biophysical linkage and immediate conservation efforts for the restoration of the Chure landscape in Nepal.

Keywords: biodiversity; disturbances; ecosystem; forest types; threats

Citation: Uprety, Y.; Tiwari, A.; Karki, S.; Chaudhary, A.; Yadav, R.K.P.; Giri, S.; Shrestha, S.; Paudyal, K.; Dhakal, M. Characterization of Forest Ecosystems in the Chure (Siwalik Hills) Landscape of Nepal Himalaya and Their Conservation Need. *Forests* **2023**, *14*, 100. https://doi.org/ 10.3390/f14010100

Academic Editors: Konstantinos Poirazidis and Panteleimon Xofis

Received: 3 December 2022
Revised: 30 December 2022
Accepted: 31 December 2022
Published: 5 January 2023

1. Introduction

Forest structure is both a product and a key driver of biological diversity and ecosystem processes over a long temporal and spatial scale [1–3]. Forest structure affects forest productivity, tree species diversity, and biological habitat that eventually determines the quality of forest ecosystem goods and services [1,2,4–6]. A particular forest stand structure is a basic unit of a forest community and also a proxy for a specific biological community. Such a community with a dominance of particular species and homogenous environmental parameters with relatively stable conditions forms a unique habitat and represents an ecological facet that evolves as a unique forest ecosystem [7,8]. A local-scale forest ecosystem is characterized by spatially co-occurring vegetation assemblages that share a common ecological gradient, substrate, or process [9]. The interconnections among local-scale ecosystems in different spatial and temporal scales form a hierarchical structure and shape the nature of future ecosystems [10]. The precise spatiotemporal information on forest types and areas at the regional scale is required for their better management, understanding of the carbon cycle, and modeling of biophysical attributes, hydrology, and climate [6,11]. Understanding threats and vulnerabilities to forest biodiversity is needed for actions to slow the current risks and secure ecosystem services for future generation [12].

The Chure mountain range (also called Siwalik hills) forms a more than 2000 km stripe along the outer Himalayas through India to Nepal and into northern Pakistan [13]. Situated between the plains in the south and the Mahabharat hills in the north, the Chure is one of the youngest mountains in the world [14]. The Nepal part of the Chure spreads over 37 districts and covers 12.78% of the country's total land area, forming a stretch of 800 km east to west [15–19]. Besides the lower plains of the Indian subcontinent, Chure also constitutes the larger parts of inner valleys (also called Dun valleys) in Udayapur, Sindhuli, Makawanpur, Chitwan, Surkhet, and Dang districts in Nepal [18]. The average peak is high in west Nepal at about 1800 m and low in east Nepal with a maximum of 700 m altitude [20]. It is the most fragile and vulnerable ecosystem because of natural and anthropogenic factors [14,16,17,21]. Due to the weak and fragile geography, the Chure region is highly vulnerable to erosion and other hazards. For these reasons, coupled with poor water availability, Chure was not inhabited in the past. However, with increasing population pressure from migration after the 1980s the areas of cultivation had begun to appear in these hills which was regarded as a very undesirable trend in view of the extremely fragile nature of the soil in this area [20].

The Chure region is severely affected by various anthropogenic activities. The people living in core Chure forest areas are heavily dependent on forest resources [22]. Along with population growth and migration, the region is threatened by deforestation, grazing, fuelwood collection, encroachment, forest fire, and excavation of sand, gravel, and boulders. Illegal logging and excavation of sand, gravel, and boulders and haphazard development activities without consideration of environmental impacts are mostly responsible for the degradation of the Chure landscape. The environmental integrity of the region has been severely altered because of these drivers of change. Realizing the ecological importance and conservation sensitivity and needs, the Government of Nepal formulated the 20-year Master Plan for sustainable conservation of the region [17]. Chure conservation is one of the national priority programs of the Government of Nepal for the last two decades [23].

A number of studies provide important information about conservation and management issues [24–26], biodiversity, and ecosystem services [16,19,22,27–29], landscape processes [14,30], and agroforestry systems [31] in Chure. The Master Plan has mentioned 11 forest ecosystems outside protected areas (eight in Chure and three in Terai regions) referring to Biodiversity Profile Project (BPP), but the location and conservation status of these ecosystems are unknown [17]. The BPP conducted during 1994–1996 was largely based on the ecological maps produced during 1971–1985 and the reports have highlighted the potential inconsistencies in these ecological maps [15,32]. Considering the basic forest structure as a proxy of specific forest ecosystem type, we aim to identify the forest ecosystems outside protected areas, locate these ecosystems in the map, assess their status, and identify important biodiversity areas in Chure landscape. Moreover, we also present the important knowledge synthesis about the major biological features of the Chure. Characterizing these ecosystems would help deploy proper conservation and management strategies by providing tools for communicating the relevance of ecosystems to the public, and support decision-makers to spatially identify priority areas.

2. Materials and Methods

2.1. Study Area

The study covers the Chure landscape of the Nepal Himalaya (Figure 1). Chure represents the parts of Nepal's tropical and sub-tropical bioclimatic regions and forms the largest and longest landscape. Of the total forest area of Nepal, 23.04% lies in Churia [33]. Chure covers parts of all the seven provinces of Nepal and also the parts of the Terai Arc Landscape (TAL), Chitwan-Annapurna Landscape (CHAL), and Kangchenjunga Landscape

(KL) [34–36]. About 7.7 million (26% of the national population) people live in 37 districts of Chure [37].

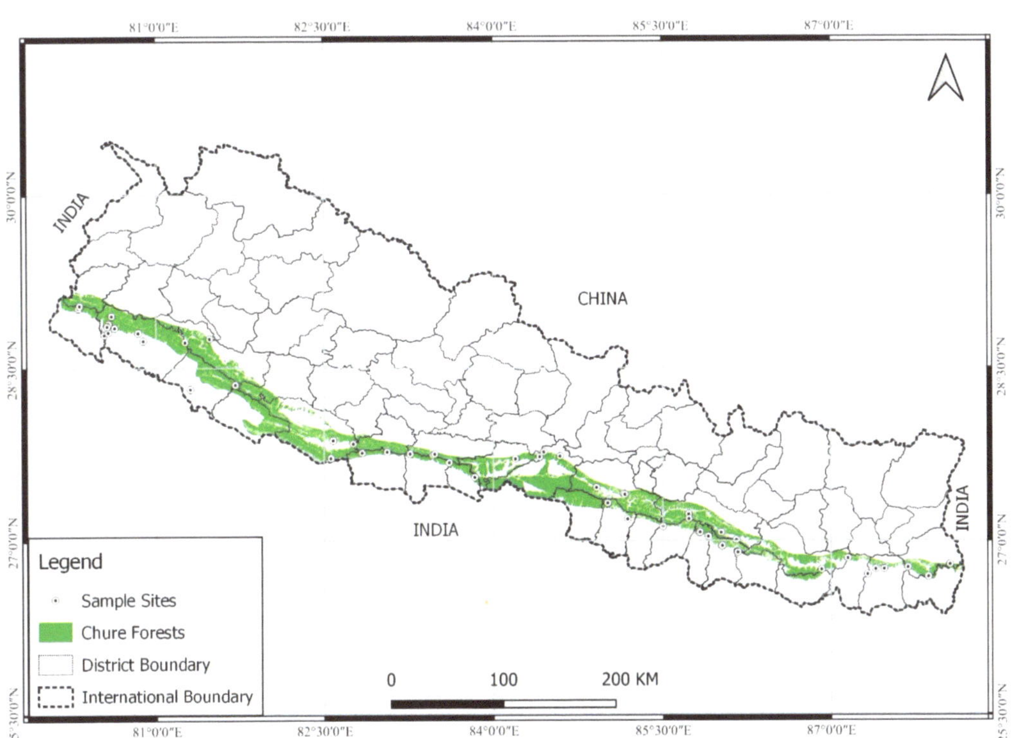

Figure 1. Map of Nepal showing Chure range and forests in green. The black dots represent the sampling sites.

The Chure are formed from soft, very erodible sediments so that gullies and areas of bad-land are frequent. It is believed that Chure hills have been formed from sediments produced by the rising Himalaya during the last 40 million years or so. Chure is geologically young and composed of unconsolidated loose materials originating from soft rocks (e.g., mudstone, sandstone, siltstone, and shale) [13,18].

2.2. Methods

Primary data were collected from the field and secondary information was generated from the various studies including that of Dobremez [38], BPP reports published during 1995–1996 [15,32], and others [16,17,24,25,27,29,39]. Moreover, participatory and consultative approaches were also adopted to identify the sampling sites and threats to the forest ecosystems.

2.2.1. Sampling Procedures

We located forest ecosystems outside the protected areas in the Chure landscape based on the literature [15], satellite images, and expert knowledge. Forests were identified both by the presence of trees (5 m height) and the absence of other predominant land uses following FAO (2000) (land with a tree canopy cover of more than 10 percent and an area of more than 0.5 ha) [8]. Based on this working definition, a forest spreading over 0.5 ha was considered as a distinct forest ecosystem. The forest with more than 60% dominance was named after that species and the forest with the dominance of more than one species was named after those species, for example *Schima wallichii–Shorea robusta* Forest, following the approach used by FRTC [40].

We adopted random-purposive sampling for gathering primary information on forest ecosystems, species composition, and conservation status. A total of 62 sampling sites covering 24 districts were studied (Figure 1). At each sampling site (within the 0.5 ha forest), a plot size of 20 m × 20 m was established. Dominant and co-dominant tree species and their coverage, and other associated species including shrubs and herbs were noted. The sampled forest site located at the lowest elevation was at 134 m asl (Baba Tal, Sarlahi) while the plot at 1355 m asl (Khanidanda Chure Rural Municipality at Kailali district) represents the highest elevation site. Coordinates of each plot were noted using the Global Positioning System and later the distribution map of the forest ecosystem types was prepared. Field visits were conducted from November 2021 to February 2022.

2.2.2. Assessment of Regeneration Status and Disturbances

The regeneration of the forest species and disturbances to the forest ecosystems were studied inside the 20 m × 20 m plots. Each plot was subdivided into 10 m × 10 m and the regeneration status (number of seedlings and saplings) was noted in the clockwise direction within the plots. The regeneration status of the forest is expressed with categorical variables, i.e., low = 1 (less than 20 individuals of seedlings and saplings), moderate = 2 (more than 20 and less than 60 individuals), and high = 3 (more than 60 individuals). Then, the average regeneration score of each forest type was calculated. Such a mixed approach of qualitative and quantitative methods is common in forest status assessments [16,41]. Likewise, disturbances such as forest encroachment and deforestation (based on the evidence of grazing, cut stumps/logging, fuelwood collection, fodder collection), forest fire, and the presence of invasive species were noted and assigned scores from 0 to 4 (0 = no disturbance; no sign of disturbance; 1 = low disturbance; with one evidence of disturbance; 2 = moderate disturbance; with two evidences; 3 = high disturbance; with three evidences, and 4 = very high disturbance; with more than three evidences). Based on the score obtained, the threat level was identified. The disturbance indicators used for this assessment are the common factors responsible for deforestation and forest degradation in Nepal [16,42], and the scale is comparable to Miehe et al. [39].

2.2.3. Stakeholder Consultation

Participatory and consultative (stakeholder involvement) approaches were also used to locate the sampling sites and validate the information collected from the field. For this purpose, we organized seven consultative workshops one in each province. Eight to twelve participants from the communities and forest authorities participated in each consultation workshop and the participants were asked to locate the different forest types on the map. Further, they were asked to highlight the status and observed threats to these forests. The disturbance score and the threat assessment (asterisks) were validated in these consultative workshops.

2.2.4. Plant Identification

In cases where field identification was certain, for example, *Shorea robusta*, *Pinus roxburghii*, and *Schima wallichii*, plant species were identified on the spot. In other cases, field notes, local names, and photographs were taken and herbarium specimens were collected. The specimens were identified with the help of reference collections [43–48] and expert consultation.

3. Results and Discussion

3.1. Forests and Ecosystems Types

Based on the dominant and co-dominant tree species type as one of the major characteristics of forest ecosystem, 14 forest ecosystems outside protected areas in Chure were identified, including the Bamboo thickets in the Siraha district of east Nepal, and located in the map (Tables 1 and S1, Figures 2 and 3). The ecosystem types presented in BPP report [15] were further validated and additional ecosystem types were identified from this study (Table 2). The most common forest ecosystem types in Chure were *Shorea robusta* (39%), followed by Tropical mixed broadleaved forest (25%) whereas *Hymenodictylon excelsum*, *Senegalia catechu*, and *Albizia* forests were poorly represented.

As our study took reference of BPP report [15] (referred also in Master Plan of Chure [17]) to locate the forest ecosystems in Chure (Table S2), we attempted to locate all forest ecosystems including *Alnus nitida* forest in the Chure region. However, our study did not find the *Alnus nitida* forest. The occurrence of this forest species was further verified with the flora of far-west Nepal [47] and experts having knowledge about the local flora of western Nepal. It was confirmed that the species is not present in Chure. This species is a component of Mugu Karnali vegetation in the Humla region at about 2100 m [49]. Hara et al. [44] have also mentioned *Alnus nitida* as a component of western flora. One of the possible reasons why this ecosystem type does not exist could be some of the inconsistencies in the ecological maps produced by Dobremez and others during 1971–1985 as reported by BPP [32].

A total of 118 ecosystems have been identified in Nepal, including 112 forest ecosystems, four cultivation ecosystems, one water body ecosystem, and one glacier/snow/rock ecosystem [15]. Among the five physiographic zones found in Nepal, the Middle Mountains have the highest number (53) of ecosystems. The High Himal and High Mountains combined have 38 ecosystems. The Terai and Siwalik have 14 and 12 ecosystems respectively [15,32]. TISC [50] reduced the 118 types to 36, excluding the Nival zone and the water bodies [51]. This information needs to be updated as natural ecosystems are dynamic in nature, and their characteristics can vary over time [52,53].

3.2. Threats and Vulnerabilities

Of the total 62 sampling sites, forest encroachment and deforestation were found in 56 sites (90% plots) followed by forest fire and invasive species in 16 sites each (26% plots). Further, based on the detailed assessment of disturbance variables, we identified one ecosystem as very highly threatened (with four asterisk marks—*Senegalia catechu* Forest), four ecosystems as highly threatened (with three asterisk marks—*Terminalia* Forest, *Dalbergia sissoo–Senegalia catechu* Forest, Tropical mixed broadleaved forest, *Pinus roxburghii–Shorea robusta* Forest), six as moderately threatened (with two asterisk marks—*Shorea robusta* Forest, *Hymenodictyon excelsum* Forest, *Albizia* Forest, Tropical deciduous riverine Forest, *Schima wallichii –Shorea robusta* Forest, and *Pinus roxburghii* Forest), and three as relatively less disturbed (with one asterisk mark—*Syzygium cumini* Forest, *Terminalia anogeissiana* Forest, and Bamboo thickets) (see asterisk marks in Table 1). Forest disturbances (grazing, logging, fire, flood cutting, encroachment, fuelwood collection) were high in *Terminalia* Forest, *Senegalia catechu* Forest, *Dalbergia sissoo–Senegalia catechu* Forest, Tropical mixed broadleaved Forest, and *Pinus roxburghii–Shorea robusta* Forest, moderate in *Shorea robusta* Forests, *Hymenodictyon excelsum* Forest, *Albizia* Forest, Tropical deciduous riverine Forest, *Schima wallichii–Shorea robusta* Forest, and *Pinus roxburghii* Forest, whereas low in *Syzygium cumini* Forest, *Terminalia anogeissiana* Forest, and Bamboo thickets (Figure 3). It was found that the existence of the *Senegalia catechu* forest is limited in very few localities of Chure as fragmented patches. These forests are one of the highly threatened forest ecosystems in the Chure region mainly due to forest encroachment, deforestation, road construction, and invasion of other species.

Forest ecosystems plagued with the high level of threat caused by disturbances such as logging (mostly illegal) and poor regeneration and survival (in the case of *Senegalia catechu*, *Dalbergia sissoo–Senegalia catechu*) are more vulnerable. The mature individuals of *Senegalia catechu* in the Chure forest are extremely rare, mostly due to the illegal felling. The conservation priorities should be focused on the protection of this threatened species along with other associated species such as *Adina cordifolia* and *Shorea robusta*.

Table 1. Forest ecosystem types with their characteristic vegetation and general description in Chure (see Table S1 for location of the ecosystem types and GPS Coordinates).

SN	Ecosystem Type and Threat Level [#]	Representative Location(s) in Chure	Characteristic Vegetation	Description
1	*Shorea robusta* forest (**Sal ban**) (128–1110 m asl) **	Kanchanpur (Daiji, Bedkot-1), Kailali (Syaule), Dang (Rapti, Bhalubang), Arghakhachi (Pirapani), Palpa (Bhutkhola, Dovan), Rupandehi (Debdaha), Nawalparasi (Daunee, Debchuli, Maulakalika), Tanahun (Pipaltar), Makwanpur (Jayasingh Manohari), Bara (Niggad), Sindhuli (Fulbari Marine Gaupalika, Maddovan, Marin, Ranibas, Kalapani), Mahottari (Tuteshor), Dhanusa (Bhatighari), Udayapur (Sundarpur), Sunsari (Latijoda), Ilam (Shikharkateri).	**Trees**: *Shorea robusta, Terminalia alata, Aegle marmelos, Trewia nudiflora, Lagestroemia parviflora, Ziziphus mauritiana, Engelhardia spicata, Syzygium cumini, Mallotus philippensis* **Shrubs**: *Carissa carandas, Woodfordia fruticosa, Justicia adhatoda, Clerodendron viscosum, Cycas pectinata*	*Shorea robusta* forms the pure forest at lower elevations and predominates the flat places from east to west, at 900 m to 1100 m elevations. Sal forest associates with *Schima wallichii,* another semi-deciduous species in Central and Eastern Nepal at around 1100 m.
2	*Hymenodictyon excelsum* forest (**Latikarma ban**) (270 m asl) **	Morang	**Trees**: *Hymenodictyon excelsum, Shorea robusta* **Shrubs**: *Murraya koenigii, Clerodendron viscosum*	Pure stand of this species is found along the flood plains of Chure in eastern Nepal.
3	*Syzygium cumini* forest (**Jamun ban**) (186–237 m asl) *	Kailali (Masuriya), Karnali flood plain	**Trees**: *Lagestroemia parviflora, Terminalia alata, Dalbergia sissoo* **Shrubs**: *Clerodendron viscosum, Colebrookea oppositifolia*	Primary evergreen forest of *Syzygium cumini* is found in western Terai. It replaces *Shorea robusta* forest along the large riversides in moist and shady areas, *Syzygium cumini* is also one of the major associated species of other forest types- *Shorea robusta* forest, *Terminalia* forest, *Terminalia anogeissiana* forest.
4	*Terminalia anogeissiana* forest (**Banjhi ban**) (1073 m asl) *	Palpa (Majhuwa, Kanchakhola) Surkhet (Pokharikada)	**Trees**: *Terminalia anogeissiana, Shorea robusta, Terminalia alata* **Shrubs**: *Woodfordia fruticosa, Maesa montana*	The species occurs from Terai to about 1700 m, usually in Sal forest but it is also a common constituent of rather dry forest in the Chure, particularly in western Nepal, where it is sometimes dominant and form a distinct ecosystem.
5	*Terminalia* forest (**Sajh ban**) (386–468 m asl) ***	Kanchanpur (Salghad), Kailali (Godawari)	**Trees**: *Terminalia alata, Terminalia chebula, Shorea robusta, Mallotus philippensis, Terminalia anogeissiana, Trichilla connaroides, Buchanania latifolia* **Shrubs**: *Woodfordia fruticosa, Justicia adhatoda, Colebrookea oppositifolia*	These types of forests are often mixed with tropical *Shorea robusta* forest, but in some places they form the pure stand with more than 60 % coverage. *Terminalia alata* is the predominant species of this type.
6	*Senegalia catechu* forest (**Khair ban**) (134–185 m asl) ****	Sarlahi (Bagmati- 11, Lopchan Tol), Dhanusha (Bhatighari CF Puspabanpur), Siraha (Baba Tal, Bandipur-3)	**Trees**: *Senegalia catechu, Albizia procera, Syzygium cumini, Mallotus philippensis, Azadirchata indica, Lagestroemia parviflora, Terminalia alata, Dalbergia latifolia, Bridelia retusa* **Shrubs**: *Murraya koenigii, Urena lobata*	Mature secondary forest of these types is very much limited due to deforestation, however in some areas of eastern Nepal, primary forest of *Senegalia catechu,* can be observed associated with *Dalbergia sissoo* along the river banks.
7	*Albizia* forest (**Siris ban**) (144–700 m asl) **	Sunsari (Chatara)	**Trees**: *Albizia procera, Albizia lebbeck, Adina cordifolia, Cassia fistula, Alstonia scholaris, Wendlandia exserta* **Shrubs**: *Murraya koengii, Clerodendron viscosum*	*Albizia procera,* a semi-deciduous tree and *A. lebbeck* occurs in dry open forest and at the Sal forest margins in Chure-Terai region of Central and Eastern Nepal. *Albizia procera* associates with *Albizia julibrissin, Albizia chinensis* and *Erythrina stricta* in south facing slopes of outer foothills at the edge of abandoned land terrace.

Table 1. *Cont.*

SN	Ecosystem Type and Threat Level [#]	Representative Location(s) in Chure	Characteristic Vegetation	Description
8	*Dalbergia sissoo–Senegalia catechu* forest **(Sisoo–Khair ban)** (195–346 m asl) ***	Dang (Lamahi), Kailali (Godawari, Malakheti, Geta, Shreepur)	**Trees**: *Senegalia catechu, Terminalia alata, Dalbergia sissoo, Syzygium cumini, Aegle marmelos, Shorea robusta, Sapium insigne, Terminalia anogeissiana, Lagerstroemia parviflora* **Shrubs**: *Murraya koenigii, Randia spinosa, Urena lobata, Colebrookea oppositifolia*	Forest as a discrete patch close to the river edge on newly formed gravels or midstream islands created by floods from Chure rivers. Forest types of Sisoo-Khair abundance of various heights, resulting in a discontinuous canopy with poor understory.
9	Tropical deciduous riverine forest **(Usna Pradeshya Nadi Tatiye Pathjhar ban)** (256–321 m asl) **	Kailali (Godawori), Bardiya (Padanaha, Chepang), Ilam	**Trees**: *Bombax ceiba, Tetrameles nudiflora, Sapium insigne, Holopetela integrifolia, Adina cordifolia, Terminalia alata, Dalbergia sissoo, Senegalia catechu.* **Shrubs**: *Murraya koengii, Colebrookea oppositifolia*	It is found along the streams of Bhabar and Dun valleys. In west and central Nepal major component of this forest types are *Bombax ceiba, Adina cordifolia* and *Sapium insigne,* however in east Nepal the species composition is different, where the area is dominated by deciduous *Tetrameles nudiflora* and other associated species like *Alangium salviifolium* and *Toona ciliata.*
10	Tropical mixed broadleaved forest **(Usna Pradeshiye Misrit Chaudapate ban)** (151–1026 m asl) ***	Bara (3 no. Khola), Makawanpur (Gadhi), Morang (Thakaldada), Jhapa (Kankai), Ilam (Jorkalas), Kanchanpur (Daiji Bedkot, Libna), Kailali (Mohaniyal), Dang (Koilabash thulichaur, Suraikhola), Palpa (Majhuwa, Kanchakhola), Chitwan (Kuwapani, Shaktikhor)	**Trees**: *Shorea robusta, Terminala alata, Terminalia chebula, Terminalia bellerica, Adina cordifolia, Semecarpus anacardium, Terminalia chebula, Cleistocalyx operculatus, Diploknema butyracea, Lagerstroemia parviflora, Litsea monopetala, Mallotus philippensis* **Shrubs**: *Murraya koengii, Woodfordia fruticosa, Phoenix humilis*	These forest types are common from east to west where Sal alone cannot dominate the area entirely. This type of forest shows heterogeneous species distribution representing both tall and short trees with various canopy structures. They also offer a wide array of microhabitat conditions, therefore high species diversity and productivity in these forests are seen.
11	*Schima wallichii–Shorea robusta* forest **(Chilaune-Sal ban)** (665–730 m asl) **	Ilam, Morang	**Trees**: *Shorea robusta, Schima wallichii, Duabanga grandiflora, Terminalia alata, Lagestroemia parviflora, Syzygium cumini* **Shrubs**: *Woodfordia fruticosa, Colebrookea oppositifolia*	Disturbed forest in the Chure region of Ilam, may *Schima wallichii* soon replace *Shorea robusta* if disturbances (mostly logging) continue.
12	*Pinus roxburghii–Shorea robusta* forest **(Khote Sallo-Sal ban)** (440 m asl) ***	Bara (3 no. Khola)	**Trees**: *Shorea robusta, Pinus roxburghii, Semecarpus anacardium, Albizia lebbeck* **Shrubs**: *Woodfordia fruticosa, Inula cappa, Berberis asiatica*	These types of associations are developing in Chure, Observed in Bara, and probably occurs also in Makwanpur.
13	*Pinus roxburghii* forest **(Khote sallo ban)** (435–1200 m asl) **	Kailali (Khanidanda, Chure VDC), Kanchapur (Bedkot), Bara (3 no. Khola) Daldeldhura	**Trees**: *Pinus roxburghii, Myrica esculenta, Quercus lanata, Semecarpus anacardium, Lagestroemia parviflora* **Shrubs**: *Rubus ellipticus Phoenix humilis, Berberis asiatica, Inula cappa*	Pine forests are found in Chure region of western and Central Nepal in dry north facing slopes. It reduces the growth of other native tree species to its range by forming thick mat of fallen dwarf shoots on the ground. A Forest type is characterized by continuous canopy and very poor understory.
14	Bamboo thickets **(Bans ban)** (113 m asl) *	Siraha (Baba Tal)	Monospecific thickets; no other vegetation- ground is completely covered with the litter from the bamboos and there are no other shrubs, herbs, ferns or bryophytes.	Probably colonized after disturbance (landslide, forest fire); will be persisted for an unknown period until their dieback following flowering, which occurs at unknown intervals.

[#] Threat level: Low = *, Moderate = **, High = ***, Very High = ****. The single asterisk was given when there was no or at least one evidence of disturbance. Likewise, two asterisks were given when there was two evidence of disturbance, three asterisks for three evidence of disturbance, and four asterisks were given for more than three evidence of disturbance. The bold names within parentheses are Nepali names for particular forest type.

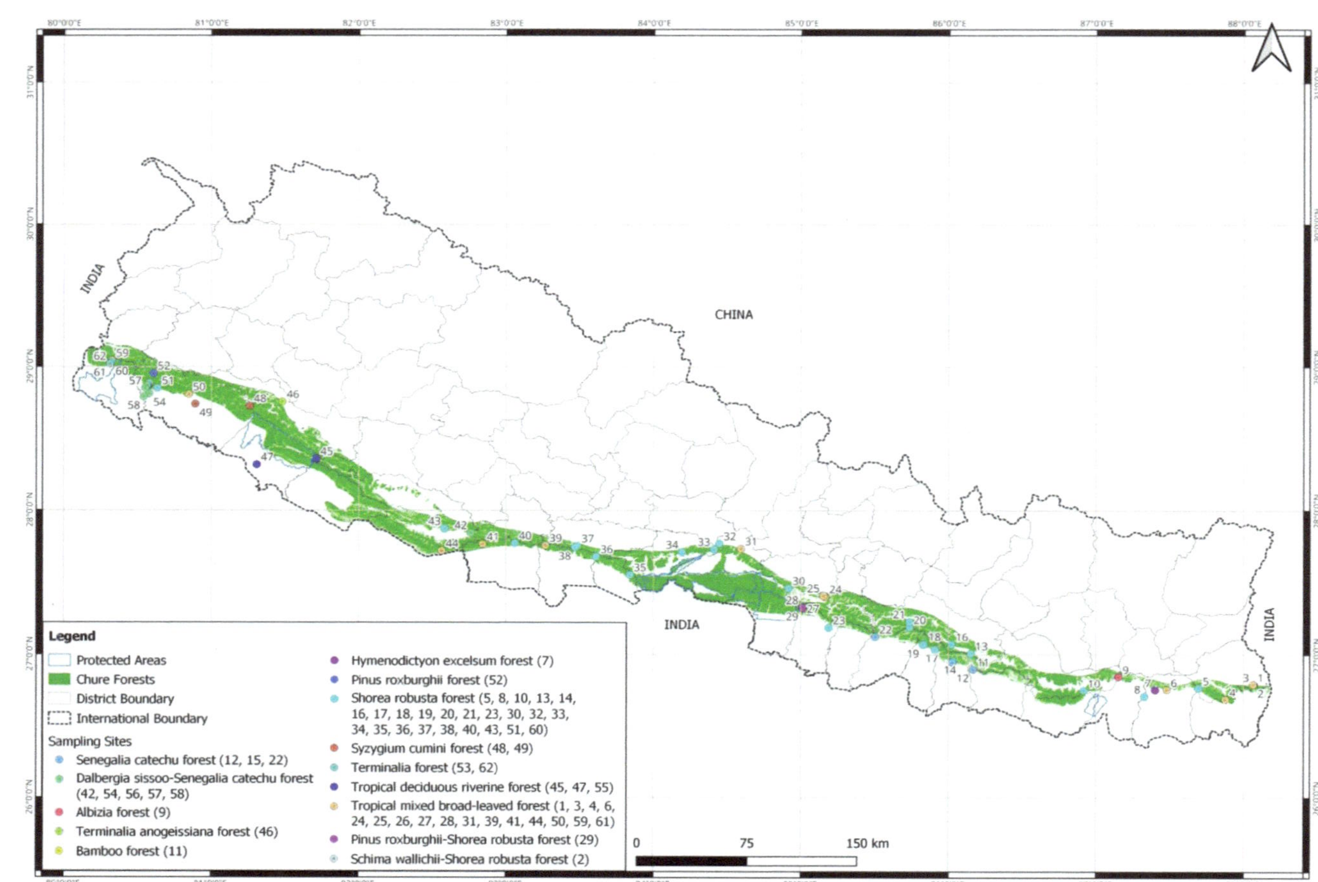

Figure 2. Forest ecosystem types in Chure. The numbers in the parentheses represent the sampling site(s) from where the particular ecosystem was reported. The polygons denote the protected areas.

Figure 3. *Cont.*

Figure 3. Phtoplates showing some of the forest ecosystem types in Chure. (**A**) *Shorea robusta* forest ecosystem, (**B**) *Syzygium cumini* forest ecosystem, (**C**) *Terminalia anogeissiana* forest ecosystem, (**D**) *Terminalia* forest ecosystem, (**E**) *Senegalia catechu* forest ecosystem, (**F**) *Dalbergia sissoo–Senegalia catechu* forest ecosystem, (**G**) Tropical deciduous riverine forest ecosystem, (**H**) Tropical mixed broadleaved forest ecosystem, (**I**) *Schima wallichii–Shorea robusta* forest ecosystem, (**J**) *Pinus roxburghii–Shorea robusta* forest ecosystem, (**K**) *Pinus roxburghii* forest ecosystem, (**L**) Bamboo thickets.

Table 2. Forest ecosystems in Chure.

SN	Type of Ecosystem	Region	Reported in BPP Report (and Referred in Chure Master Plan)	Reported as New
1	*Shorea robusta* forest	East, Central, West	√	
2	*Hymenodictyon excelsum* forest	East		√
3	*Syzygium cumini* forest	West		√
4	*Terminalia anogeissiana* forest	West		√
5	*Terminalia* forest	West	√	
6	*Senegalia catechu* forest	East	√	
7	*Albizia* forest	East	√	
8	*Dalbergia sissoo–Senegalia catechu* forest	West	√	
9	Tropical deciduous riverine forest	East and West	√	
10	Tropical mixed broadleaved forest	East, Central, West	√	
11	*Schima wallichii–Shorea robusta* forest	East		√
12	*Pinus roxburghii–Shorea robusta* forest	Central		√
13	*Pinus roxburghii* forest	Central, West		√
14	Bamboo thickets	East		√

3.3. Forest Regeneration

Out of 14 forest ecosystem types reported from Chure, six forest types namely *Shorea robusta*, *Hymenodictyon excelsum*, *Syzygium cumini*, Tropical mixed broadleaved forests, *Schima wallichii–Shorea robusta*, and Bamboo thickets are producing a good number of seedlings (>60 individuals per plot) and naturally regenerating well. Adequate regeneration of these species was also reported by DFRS [16]. This might be due to the fact that these forests are relatively less disturbed. Similarly, four forest ecosystems showed moderate germination showing less than 60 and more than 20 individuals within the plot studied. Four forest types namely *Terminalia anogeissiana*, *Senegalia catechu*, *Dalbergia sissoo–Senegalia catechu*, and *Albizia* forests showed poor natural regeneration representing less number of seedlings in the plot (Figure 4). The poor representation of seedlings in the plot of these forest types might be due to high disturbances like grazing, deforestation, fire, flood, and invasive species.

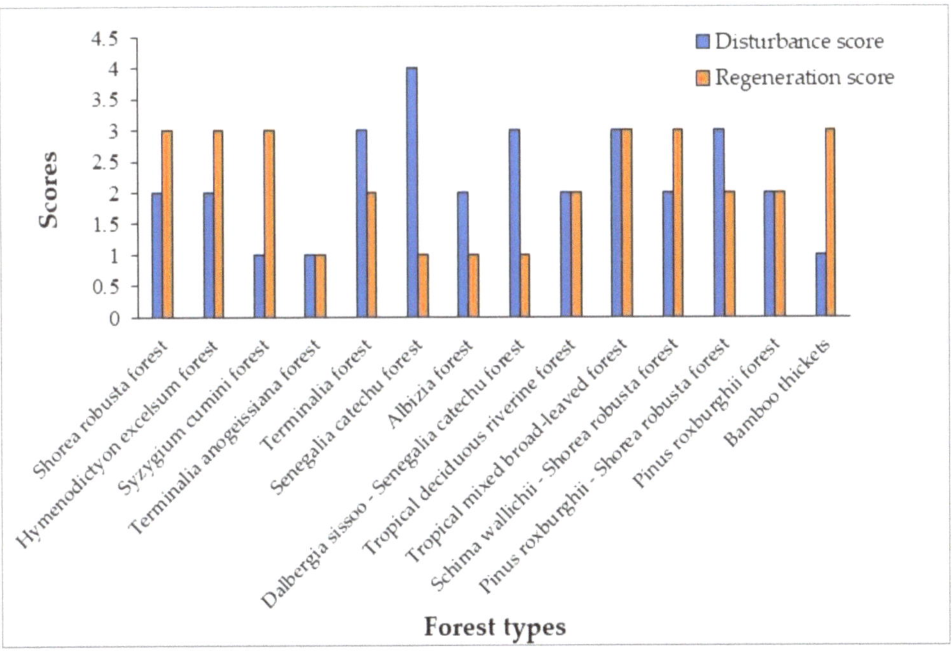

Figure 4. Disturbance and regeneration score according to forest ecosystem types in Chure landscape.

The forest regeneration status is satisfactory in Chure as most of the forests showed moderate to high regeneration. This is because seventy-two percent of our sampling plots were located in community forests meaning that the user groups put some restrictions on resource extraction. However, still, some of the forests showed poor regeneration. Moreover, our sampling design purposively selected relatively good-quality less disturbed forests. Therefore, the overall forest quality of Chure could be much lower than reported in our study (see Figure 5a).

3.4. Floral and Faunal Biodiversity of Chure

The Chure harbors about 1000 plant species of different life forms [16,29,47]. The floral diversity is comparable with that of the Indian part of Chure [54]. It includes more than 281 tree species that constitute about 40% of total tree species recorded in Nepal, 186 shrubs, and 322 herbs including Pteridophytes [16]. Of the total 293 endemic plants of Nepal [55], some nine species are found in the Chure region including *Begonia minicarpa* H. Hara (Locality-Sunsari, 630 m); *B. tribenensis* C.R. Rao (Sunsari, 130 m); *Eriocaulonex sertum* Satake

(Jhapa, 200–300 m); *E. obclavatum* Satake (Jhapa, 200–300 m); *Isodondhan kutanus* Murata (Dhankuta, 1200 m); *Eria nepalensis* D.M. Bajracharya & K.K. Shrestha (Chitwan, 200 m); *Malaxis tamurensis* Tuyama (Dhankuta, 1200 m); *Ophiorrhiza nepalensis* Deb & Mondal (Ilam, 450 m); *Salix plectiles* Kimura (E. Nepal, 200 m) [29,55,56]. However, a detailed floristic inventory of the Chure is lacking.

Figure 5. One of the most degraded sites in Chure (**a**), young and fragile deposits in Chure as nesting sites (**b**), endangered *Cycas pectinata* (**c**), freshwater swamp forests of *Cephalanthus tetrandra* (**d**).

Chure is also a pristine habitat for at least 41 species of mammals, 99 species of herpetofauna (24 species of amphibians and 75 species of reptiles), and 279 species of butterflies [17,19]. Most of the species found in the region are also listed in the CITES Appendices and IUCN Red data book, e.g., tiger, elephant, and rhinoceros. Megafauna like tiger, elephant, rhinoceros, wild buffalo, bison, and many others are flagship species to the Chure and lowland ecosystems. As the landscape is rich in biodiversity, the Government of Nepal has established networks of protected areas in Terai-Chure to conserve mostly the major faunal species. Of these protected areas, Chitwan National Park, Parsa National Park, Bardiya National Park, and Banke National Park cover parts of Chure region. Likewise, four Ramsar sites in region namely Koshi Tappu, Beesshhajari and Associated Lakes, Jagadishpur Reservoir, and Ghodaghodi Lake provide important habitat for wildlife. Ramdhuni (Sunsari), Rauta (Udaypur), Dhanushadham (Dhanusha), Barandabhar (Chitwan), Khata (Bardiya), Kakrebihar (Surkhet), Basanta (Kailai) and Laljhadi-Mohana (Kanchanpur) are protection forests in Chure-Terai [34]. Young and fragile deposits in Chure provide important nesting sites for birds (Figure 5b).

3.5. Biodiversity Hotspots

The whole range of Chure landscape in Nepal from the east to the west is a biodiversity hotspot [29]. Of the total 27 Important Bird Areas of Nepal, 13 are located in Chure-Terai [57]. They are Barandabhar Forest and Wetlands, Bardia National Park, Chitwan National Park; Dang Deukhuri Foothill Forests and West Rapti Wetlands, Dharan Forests, Ghodaghodi Lake, Jagdishpur Reservoir, Koshi Tappu Wildlife Reserve and Barrage, Farmlands in

Lumbini, Nawalparasi Forests, Parsa Wildlife Reserve, Shukla Phanta Wildlife Reserve, and Urlabari Forest Groves. Likewise, Chure forests are important habitats for species under different conservation status. We recorded the endangered *Cycas pectinata* species (a member of a group called Cycads which are an ancient group of seed plants that originated over 280 million years ago [58] in its natural habitats at central (Makawanpur) and eastern Nepal (Andha Rajarani area of Ilam, Bhatighari Communnity Forest (CF) of Danusha) (Figure 5c). Likewise, Bashyal et al. [59] reported several important sites for *Cycas pectinata* in central Nepal. Bhuju and Joshi [27] identified important sites for tree fern (*Cyathea spinulosa)* from eastern Nepal (Madi in Morang and Bajho and Mahmai in Ilam and Bajhoand Chisapani in Ilam). Likewise, Chure forests in far-west Nepal constitute a good population of mature *Pterocarpus marsupium* and *Dalbergia latifolia* forest (Bedkot Tal, Kanchanpur). There are also good natural forest patches of *Senegalia catechu*, a threatened species, in Chure forests of eastern Nepal (Sarlahi-Lopchan Tol, Dhanusha-Bhatighari CF Puspabanpur, Siraha-Baba Tal, Bandipur- 3) (Table 1). These primary forests are important from conservation point of view and the areas can be termed Important Plant Areas (IPAs). We propose three additional IPAs in Chure landscape to the list of Hamilton and Radford [60] (Table 3). Apart from protected areas and Ramsar sites in Chure-Terai, the Dang-Deukhuri Foothill Forests and West Rapti Wetlands, Dharan Forests, Nawalparasi forests, Farmlands in Lumbini and Urlabari Forest Groves are Important Bird Areas [57].

Table 3. Important Plant Areas in Chure.

IPA Complex	Number of Site	District(s)
Lower Mahakali–Seti	1 + 1	Dadeldhura, Kanchanpur
Lower Bheri–Rapti	2	Salyan and Surkhet
Terai Arc Landscape	8	Kailali, Bardia, Banke, Dang, Palpa, Nawalparasi, Chitwan, Parsa
Rapti–Lumbini	2	Pyuthan and Argahkhanchi
Narayani	2	Makwanpur and Bara
Lower Janakpur	2 + 1	Sarlahi, Sindhuli, Dhanusa
Udayapur	1	Udayapur
Morang	1	Morang
Lower Kangchenjungha	1 + 1	Ilam, Jhapa

Modified from Hamilton and Radford [60]. Added one site each in Kanchanpur, Dhanusa and also suggested one site in Morang.

We explored some of the important wetlands of Chure landscape during the field study (Chuli Pokhari, Ilam; Rajarani Tal, Morang; Baba Tal, Siraha; and Bedkot Tal, Kanchanpur). These wetlands and catchment areas are exceptionally rich in biodiversity. For example, we found good population of *Cephalanthus tetrandra* in the natural habitat in the Rajarani Tal of Morang district within Chure landscape (Figure 5d). The freshwater swamp forests of *Cephalanthus tetrandra* trees are believed to be rare in South East Asia. Mikhama and Sirisant [61] have reported *C. tetrandra* freshwater swamp forests from Don Daeng village, Nakhon Phanom province as the only one of its kind in Northeast Thailand, and they have highlighted the ecological role of *C. tetrandra* forest and the active involvement of local people for managing and conserving the important wetland tree species. Furthermore, wetlands play important role in maintaining hydro-climatic balance, and control of flood and landslide both upstream and downstream along the Chure landscape. Further inventory of wetlands in Chure would provide important information for identifying biodiversity hotspots and designing biodiversity conservation programs accordingly.

The human-dominated landscapes within Chure such as in Makwanpur and Chitwan which are inhabited by the Chepang ethnic group are also important for the preservation of biocultural diversity as these landscapes hold Chiuri (*Diploknema butyracea*), a cultural keystone species for Chepang [62].

3.6. Threatened Plant Species

Forest ecosystems in Chure provide important habitats for twenty-six plant species with different conservation status. This further signifies the conservation importance of Chure from the biodiversity point of view (Table 4). Of 26 species documented, five species are Endangered (IUCN); four species each fall under the protected plant list of the Government of Nepal, CITES appendix II; and Vulnerable category of CAMP; three species are in IUCN rare category, two species are in Endangered category of CAMP and one species each is in Threatened (IUCN), Commercially threatened (IUCN) and CITES III list.

Table 4. Plant species with different conservation status in Chure.

SN	Species	Category
1	*Alstonia scholaris* (L.) R.Br.	Rare (IUCN category)
2	*Asparagus racemosus* Willd.	Vulnerable (Conservation Assessment and Management Planning, CAMP *)
3	*Bombax ceiba* L.	Nationally protected (Under the National list of timber trees banned for felling, transportation or export)
4	*Butea monosperma* (Lam.) Kuntze	Endangered (IUCN category)
5	*Choerospondias axillaris* (Roxb.) B.L. Burtt & A.W. Hill	Rare (IUCN category)
6	*Cinnamomum glaucescens* (Nees.) B.L.Burtt & A.W.Hill	Protected
7	*Crateva unilocularis* Buch.-Ham.	Rare
8	*Curculigo orchioides* Gaertn.	Vulnerable (CAMP)
9	*Cycas pectinata* Griff.	CITES Appendix II; Endangered (IUCN category)
10	*Dalbergia latifolia* Roxb.	Nationally protected (Under the National list of timber trees banned for felling, transportation or export)
11	*Dendrobium fimbriatum* Hook.	CITES Appendix II
12	*Dioscorea deltoidea* Wall. Ex Griseb.	Commercially threatened
13	*Magnolia champaca* (L.) Baill. Ex Pierre	Endangered (IUCN category)
14	*Gnetum montanum* Markgr.	CITES Appendix III; Endangered (IUCN category)
15	*Operculina turpethum* (L.) Silva Manso	Endangered (CAMP)
16	*Oroxylum indicum* (L.) Kurz	Vulnerable (IUCN category)
17	*Piper longum* L.	Vulnerable (CAMP)
18	*Pterocarpus marsupium* Roxb.	Nationally protected (Under the National list of timber trees banned for felling, transportation or export)
19	*Rauvolfia serpentina* (L.) Benth. Ex Kurz	Endangered (IUCN category)
20	*Rhynchostylis retusa* (L.) Blume	CITES Appendix II
21	*Rubia manjith* Roxb. Ex Fleming	Vulnerable (CAMP)
22	*Senegalia catechu* (L.f.) Willd.	Threatened (IUCN category)
23	*Shorea robusta* Gaertn.	Nationally protected (Under the National list of timber trees banned for felling, transportation or export)
24	*Swertia angustifolia* Buch.-Ham. Ex D.Don	Endangered (CAMP)
25	*Tinospora sinensis* (Lour.) Merr.	Vulnerable (CAMP)
26	*Vanda tessellata* (Roxb.) Hook. ex G.Don	CITES Appendix II

* Bhattarai et al. [63].

4. Conclusions and Recommendations

The forest ecosystems in Chure are diverse and dynamic. We revisited the forest ecosystems outside protected areas in Chure and found that 14 forest ecosystem types are available in the Chure landscape of Nepal. The reference ecosystem types in Chure were taken from the Biodiversity Profiles Project [15] where 11 forest ecosystems are reported to occur outside protected areas in Chure. Our study further shows that the BPP reported *Alnus nitida* riverine forest in the west does not occur in Chure. The present study reported *Hymenodictyon excelsum* Forest, *Syzygium cumini* Forest, *Terminalia anogeissiana* Forest, *Schima wallichii–Shorea robusta* Forest, *Pinus roxburghii–Shorea robusta* Forest, *Pinus roxburghii* Forest, and Bamboo thickets as new forest ecosystems in Chure because they are not reported previously. These newly formed associations (forest types) are at the preliminary stages of forest ecosystem development, as these types are not common in east-west of Chure landscapes, but slowly evolving. As some of the forest ecosystems such as *Hymenodictyon excelsum* Forest and *Dalbergia sissoo–Senegalia catechu* Forest are found in Terai, mostly on Chure flood plains, these types are included in the total count as they form the contiguous ecosystems in Terai-Chure.

Though our study only located the forest ecosystems, we emphasize that this study complements what is existed for Chure at present and the ongoing national-level ecosystem mapping project of the Ministry of Forests and Environment. The present findings are also helpful to design the conservation programs targeting forest ecosystems in Chure. We affirm with caution that our study is comprehensive so far, but we do not claim that the study is complete as many remote and inaccessible areas were not covered during the field survey. Based on our study, we recommend some priority activities for biodiversity conservation and ecosystem restoration to be implemented in the Chure landscape as below.

1. Our study has identified biodiversity hotspots based on species richness and the occurrence of different plant species under different conservation categories. These areas should be given priority for conservation. Special status could be suggested for the conservation of such species, for example, the Andha Rajarani area in Ilam can be given the status of the park or botanical garden as it is one of the diverse areas for biodiversity.
2. Some of the hotspots identified in this survey are in the community forests, so the local people can be informed about the conservation values of these ecosystems and they can make aware and educated. The CFs operational plan could be revised to incorporate the conservation values of these ecosystems.
3. As ecosystems are dynamic in nature their existence could be of short-term as well. Therefore, long-term monitoring of vulnerable ecosystems such as *Senegalia catechu*, *Dalbergia sissoo–Senegalia catechu*, and Bamboo ecosystems is important. Rapidly changing land use patterns and climate change may put additional pressure on Chure ecosystems.
4. Restoration of the degraded ecosystems such as *Shorea robusta* Forest Ecosystems, *Senegalia catechu* Forest Ecosystems, and *Dalbergia sissoo–Senegalia catechu* Forest Ecosystems should be given high priority. At present, the restoration activities are not ecosystem focused and there is no active participation of the communities. Local governments should be informed about the ecosystem types and their conservation values within their jurisdiction.

Supplementary Materials: The following supporting information can be downloaded at: https://www.mdpi.com/article/10.3390/f14010100/s1, Table S1: Location of the forest ecosystem type and GPS coordinates, Table S2: Forest ecosystems in Chure and Terai.

Author Contributions: Conceptualization, Y.U., A.T., S.K., R.K.P.Y., S.S., S.G., K.P. and M.D.; methodology, Y.U., A.T., S.K. and A.C.; formal analysis, Y.U., A.T., S.K. and A.C.; writing—original draft preparation, Y.U., A.T. and S.K.; funding acquisition, Y.U. All authors have read and agreed to the published version of the manuscript.

Funding: This research was funded by the President Chure Terai-Madhesh Conservation Development Board, Nepal. Article publication fee was supported by the University Grants Commission Nepal.

Data Availability Statement: The data included in this study are available upon request from the corresponding authors.

Acknowledgments: We are thankful to Pramod Kumar Jha, Krishna Prasad Oli, and Hemlal Aryal for constructive suggestions to improve the manuscript.

Conflicts of Interest: The authors declare no conflict of interest.

References

1. Spies, T.A. Forest structure: A key to the ecosystem. *Northwest Sci.* **1998**, *72*, 34–39.
2. Pommerening, A. Evaluating structural indices by reversing forest structural analysis. *For. Ecol. Manag.* **2006**, *224*, 266–277. [CrossRef]
3. Brown, C.; Law, R.; Illian, J.B.; Burslem, D.F.R. Linking ecological processes with spatial and non-spatial patterns in plant communities. *J. Ecol.* **2011**, *99*, 1402–1414. [CrossRef]
4. Brian, J.E.; Geoffrey, B.W.; James, H.B. Extensions and evaluations of a general quantitative theory of forest structure and dynamics. *Proc. Natl. Acad. Sci. USA* **2009**, *106*, 7046–7051.
5. Gamfeldt, L.; Snäll, T.; Bagchi, R.; Jonsson, M.; Gustafsson, L.; Kjellander, P.; Ruiz-Jaen, M.C.; Fröberg, M.; Stendahl, J.; Philipson, C.D.; et al. Higher levels of multiple ecosystem services are found in forests with more tree species. *Nat. Commun.* **2013**, *4*, 1340. [CrossRef] [PubMed]
6. Pretzsch, H.; Zenner, E.K. Toward managing mixed-species stands: From parametrization to prescription. *For. Ecosyst.* **2017**, *4*, 19. [CrossRef]
7. Perry, D. *Forest Ecosystems*; The Johns Hopkins University Press: Baltimore, MD, USA, 1994.
8. FAO. *Forest Resource Assessment*; Food and Agriculture Organization of the United Nations: Rome, Italy, 2000.
9. Comer, P.; Faber-Langendoen, D.; Evans, R.; Gawler, S.; Josse, C.; Kittel, G.; Menard, S.; Pyne, M.; Reid, M.; Schulz, K.; et al. *Ecological Systems of the United States: A Working Classification of U.S. Terrestrial Systems*; NatureServe: Arlington, VA, USA, 2003; 75p.
10. Kulhavý, J.; Suchomel, J.; Menšík, L. *Forest Ecology*; Mendel University in Brno: Brno, Czech Republic, 2013.
11. Franklin, J.F.; Spies, T.A.; Van Pelt, R.; Carey, A.B.; Thornburgh, D.A.; Berg, D.R.; Lindenmayer, D.B.; Harmon, M.E.; Keeton, W.S.; Shaw, D.C.; et al. Disturbances and structural development of natural forest ecosystems with silvicultural implications, using Douglas-fir forests as an example. *For. Ecol. Manag.* **2002**, *155*, 399–423. [CrossRef]
12. Keith, D.A.; Rodríguez, J.P.; Rodríguez-Clark, K.M.; Nicholson, E.; Aapala, K.; Alonso, A.; Asmussen, M.; Bachman, S.; Basset, A.; Barrow, E.G.; et al. Scientific Foundations for an IUCN Red List of Ecosystems. *PLoS ONE* **2013**, *8*, e62111. [CrossRef]
13. Upreti, B.N. *The Physiography and Geology of Nepal and Landslide Hazards. Landslide Problem Mitigation to the Hindukush-Himalayas*; ICIMOD: Lalitpur, Nepal, 2001; Volume 312.
14. Ghimire, S.K.; Higaki, D.; Bhattarai, T.P. Estimation of soil erosion rates and eroded sediment in a degraded catchment of the Siwalik Hills, Nepal. *Land* **2013**, *2*, 370–391. [CrossRef]
15. Biodiversity Profiles Project (BPP). *Biodiversity Profile of Terai and Siwalik Physiographic Zones*; Biodiversity Profiles Project Publication No. 12; Department of National Parks and Wildlife Conservation, Ministry of Forests and Soil Conservation, His Majesty's Government of Nepal: Kathmandu, Nepal, 1995.
16. DFRS. *Chure Forests of Nepal*; Forest Resource Assessment (FRA), Department of Forest Research and Survey (DFRS), Ministry of Forest and Soil Conservation (MoFSC), Government of Nepal: Kathmandu, Nepal, 2014.
17. PCTMCDB. *President Chure-Terai Madhesh Conservation and Management Master Plan*; President Chure-Terai Madhesh Conservation Development Board: Lalitpur, Nepal, 2017.
18. Singh, B.K. Land tenure and conservation in Chure. *J. For. Livelihood* **2017**, *15*, 87–102. [CrossRef]
19. Subedi, N.; Bhattarai, S.; Pandey, M.R.; Kadariya, R.; Thapa, S.K.; Gurung, A.; Prasai, A.; Lamichhane, S.; Regmi, R.; Dhungana, M.; et al. *Report on Faunal Diversity in Chure Region of Nepal*; President Chure-Terai Madhesh Conservation Development Board and National Trust for Nature Conservation: Kathmandu, Nepal, 2021.
20. Jackson, J.K. *Manual of Afforestation in Nepal. Volume 1*; Forest Research and Survey Centre, Ministry of Forests and Soil Conservation: Kathmandu, Nepal, 1994.
21. Schweik, C.M.; Adhikari, K.; Pandit, K.N. Land-cover change and forest institutions: A comparison of two sub-basins in the southern Siwalik hills of Nepal. *Mt. Res. Dev.* **1997**, *17*, 99–116. [CrossRef]
22. Acharya, R.P.; Maraseni, T.N.; Cockfield, G. Local users and other stakeholders' perceptions of the identification and prioritization of ecosystem services in fragile mountains: A case study of Chure Region of Nepal. *Forests* **2019**, *10*, 421. [CrossRef]
23. NPC. *The Fifteenth Plan*; Government of Nepal, National Planning Commission: Kathmandu, Nepal, 2020.
24. Bhuju, D.R. *Nepal's Last Hope for Landscape Level Conservation*; Habitat Himalaya, a Resources Himalaya Factfile: Kathmandu, Nepal, 2000; Volume 7.
25. Pokhrel, K.P. Conservation of Chure Forest in Nepal: Issues, Challenges and Options. *Econ. J. Nepal* **2012**, *35*, 262–278.

26. FAO. *Building a Resilient Chure Region in Nepal*; A project document prepared by FAO; Nepal for Ministry of Forests and Environment: Kathmandu, Nepal, 2019.

27. Bhuju, D.R.; Joshi, G.P. Records of *Cyathea spinulosa* Wallich ex Hooker (Cyatheaceae) and *Cycas pectinata* Griff. (Cycadaceae) from the Churiya hills of eastern Nepal. *Nepal J. Sci. Technol.* **2009**, *10*, 69–72. [CrossRef]

28. Bhandari, P.; Mohan, K.C.; Shrestha, S.; Aryal, A.; Shrestha, U.B. Assessments of ecosystem service indicators and stakeholder's willingness to pay for selected ecosystem services in the Chure region of Nepal. *Appl. Geogr.* **2016**, *69*, 25–34. [CrossRef]

29. Chaudhary, R.P.; Subedi, C.K. Chure-Tarai Madhesh Landscape, Nepal from biodiversity research perspective. *Plant Arch.* **2019**, *19*, 351–359.

30. Gurung, H.; Khanal, N. Landscape processes in the Chure range Central Nepal. *Himal. Rev.* **1986**, *17–19*, 1–39.

31. Khadka, D.; Aryal, A.; Bhatta, K.P.; Dhakal, B.P.; Baral, H. Agroforestry systems and their contribution to supplying forest products to communities in the Chure range, Central Nepal. *Forests* **2021**, *12*, 358. [CrossRef]

32. Biodiversity Profiles Project (BPP). *An Assessment of the Representation of the Terrestrial Ecosystems within the Protected Areas System of Nepal*; Biodiversity Profiles Project Publication No. 15; Department of National Parks and Wildlife Conservation, Ministry of Forests and Soil Conservation, His Majesty's Government of Nepal: Kathmandu, Nepal, 1996.

33. DFRS. *District Forest Cover Maps of Nepal*; Forest Resource Assessment (FRA), Department of Forest Research and Survey (DFRS), Ministry of Forest and Soil Conservation (MoFSC), Government of Nepal: Kathmandu, Nepal, 2015.

34. Chaudhary, R.P.; Uprety, Y.; Devkota, S.; Adhikari, S.; Rai, S.K.; Joshi, S.P. Plant Biodiversity in Nepal: Status, Conservation Approaches, and Legal Instruments under New Federal Structure. In *Plant Diversity in Nepal*; Siwakoti, M., Jha, P.K., Rajbhandary, S., Rai, S.K., Eds.; Botanical Society of Nepal: Kathmandu, Nepal, 2020; pp. 167–206.

35. MoFSC. *Strategy and Action Plan 2015–2025, Terai Arc Landscape, Nepal*; Ministry of Forests and Soil Conservation: Kathmandu, Nepal, 2015.

36. MoFSC. *Strategy and Action Plan 2016–2025, Chitwan-Annapurna Landscape, Nepal*; Ministry of Forests and Soil Conservation: Kathmandu, Nepal, 2015.

37. CBS. (Central Bureau of Statistics). *Population Census of Nepal*; National Planning Commission: Kathmandu, Nepal, 2021.

38. Dobremez, J.F. *Le Nepal. Ecologique et Biogeographie*; CNRS: Paris, France, 1976.

39. Miehe, G.; Pendry, C.A.; Chaudhary, R.P. *Nepal: An Introduction to the Natural History, Ecology and Human Environment of the Himalayas*; Royal Botanical Garden: Edinburgh, UK, 2015.

40. FRTC. *Vegetation Types of Nepal: A Report Based on Review of Literature and Expert Knowledge*; EFTMP Technical Working Document, No. 2; Ecosystem and Forest Types Mapping Program (EFTMP), Forest Research and Training Centre (FRTC): Kathmandu, Nepal, 2021.

41. Ballabha, R.; Tiwari, J.K.; Tiwari, P. Regeneration of tree species in the sub-tropical forest of Alaknanda Valley, Garhwal Himalaya, India. *For. Sci. Pract.* **2013**, *15*, 89–97. [CrossRef]

42. Chaudhary, R.P.; Uprety, Y.; Rimal, S.K. Deforestation in Nepal: Causes, Consequences, and Responses. In *Biological and Environmental Hazards, Risks, and Disasters*; Shroder, J.F., Sivanpillai, R., Eds.; Elsevier: Amsterdam, The Netherlands, 2016; pp. 335–372; ISBN 978-0123-9-4847-2.

43. Hara, H.; Charter, A.H.; Williams, L.J.H. *An Enumeration of the Flowering Plants of Nepal (Vol. iii)*; British Natural History Museum: London, UK, 1982.

44. Hara, H.; Williams, L.H.J. *An Enumeration of the Flowering Plants of Nepal (Vol. ii)*; British Natural History Museum: London, UK, 1979.

45. Polunin, O.; Stainton, A. *Flowers of the Himalaya*; Oxford University Press: New Delhi, India, 1984.

46. Press, J.R.; Shrestha, K.K.; Sutton, D.A. *Annotated Checklist of Flowering Plants of Nepal*; British Natural History Museum: London, UK, 2000.

47. Rajbhandari, K.R.; Thapa Magar, M.S.; Kandel, D.R.; Khanal, C. *Plant Resources of Kailali, West Nepal*; District Plant Resources Office Kailali: Dhangadhi, Nepal, 2016.

48. Shrestha, K.; Bhandari, P.; Bhattarai, S. *Plants of Nepal (Gymnosperms and Angiosperms)*; Heritage Publishers and Distributors Pvt. Ltd.: Kathmandu, Nepal, 2022.

49. Jackson, J.K. *Manual of Afforestation in Nepal. Volume 2*; Forest Research and Survey Centre, Ministry of Forests and Soil Conservation: Kathmandu, Nepal, 1994.

50. TISC. *Forest and Vegetation Types of Nepal*; TISC Document Series No. 105; Department of Forest, HMG/NARMSAP: Kathmandu, Nepal, 2002.

51. Shrestha, T.B. Classification of Nepalese forests and their distribution in protected areas. *Initiation* **2008**, *2*, 1–9. [CrossRef]

52. Smith, M.D.; Knapp, A.K.; Collins, S.L. A framework for assessing ecosystem dynamics in response to chronic resource alterations induced by global change. *Ecology* **2009**, *12*, 3279–3289. [CrossRef] [PubMed]

53. Luo, Y.; Melillo, J.; Niu, S.; Beier, C.; Clark, J.S.; Classen, A.T.; Davidson, E.; Dukes, J.S.; Evans, R.D.; Field, C.B.; et al. Coordinated approaches to quantify long-term ecosystem dynamics in response to global change. *Glob. Chang. Biol.* **2011**, *17*, 843–854. [CrossRef]

54. Haq, S.M.; Waheed, M.; Bussmann, R.W.; Arshad, F. Vegetation composition and ecological characteristics of the forest in the Shawilks Mountain Range from Western Himalayas. *Acta Ecol. Sin.* **2022**; *in press*. [CrossRef]

55. Rajbhandari, K.R.; Rai, S.K.; Joshi, M.D.; Khatri, S.; Bhatt, G.D.; Chhetri, R. Endemic Flowering Plants of Nepal: Status and Distribution. In *Integrating Biological Resources for Prosperity*; Siwakoti, M., Mandal, T.N., Rai, S.K., Rai, S.K., Gautam, T.P., Aryal, H.P., Limbu, K.P., Eds.; Botanical Society of Nepal: Kathmandu, Nepal; Nepal Biological Society: Biratnagar, Nepal; Department of Plant Resources: Kathmandu, Nepal, 2021.

56. Tiwari, A.; Uprety, Y.; Rana, S.K. Plant endemism in the Nepal Himalayas and phytogeographical implications. *Plant Divers.* **2019**, *41*, 174–182. [CrossRef] [PubMed]

57. Inskipp, C.; Baral, H.S. *Important Bird Areas in Nepal: Key Sites for Bird Conservation*; Bird Conservation Nepal: Kathmandu, Nepal, 2005.

58. Hossain, M.K.; Hossain, M.A.; Hossain, S.; Rahman, M.R.; Hossain, M.I.; Nath, S.K.; Siddiqui, M.B.N. Status and conservation needs of *Cycas pectinata* Buch.-Ham. in its natural habitat at Baroiyadhala National Park, Bangladesh. *J. Threate. Taxa* **2021**, *13*, 19070–19078. [CrossRef]

59. Bashyal, A.; Bhattarai, S.; Gautam, J.; Tamang, R.; Bhatta, B. Status, Distribution and Habitat Suitability Mapping of *Cycas pectinata* in Chure Range of Makawanpur, Central Nepal. *J. Plant Resour.* **2019**, *18*, 183–189.

60. Hamilton, A.C.; Radford, E.A. *Identification and Conservation of Important Plant Areas for Medicinal Plants in the Himalaya*; Plantlife International: Salisbury, UK; Ethnobotanical Society of Nepal: Kathmandu, Nepal, 2007.

61. Mikhama, K.; Sirisant, P. The Last and Largest of *Cephalanthus tetrandra* Freshwater Swamp Forest in Northeast Thailand: Natural Resource Appreciation and Management of Local Community. *Int. J. Agric. Technol.* **2016**, *2*, 429–438.

62. Uprety, Y.; Asselin, H. Biocultural importance of the Chiuri [*Diploknema butyracea* (Roxb.) H. J. Lam] tree for the Chepang communities of Central Nepal. 2023; *in preparation*.

63. Bhattarai, N.K.; Tandon, V.; Ved. D.K. Highlights and outcomes of the Conservation Assessment and Management Planning (CAMP) Workshop, Pokhara, Nepal. In *Sharing Local and National Experience in Conservation of Medicinal and Aromatic Plants in South Asia*; Bhattarai, N., Karki, M., Eds.; MAPPA/IDRC: New Delhi, India; MOFSC: Kathmandu, Nepal; IOF: Pokhara, Nepal, 2002; pp. 46–53.

Article

Criteria and Indicators to Define Priority Areas for Biodiversity Conservation in Vietnam

Xuan Dinh Vu [1], Elmar Csaplovics [2], Christopher Marrs [2] and Trung Thanh Nguyen [3,*]

[1] College of Land Management and Rural Development, Vietnam National University of Forestry, Hanoi 13417, Vietnam
[2] Institute of Photogrammetry and Remote Sensing, Technical University of Dresden, 01069 Dresden, Germany
[3] Institute of Environmental Economics and World Trade, Leibniz University of Hannover, 30167 Hannover, Germany
* Correspondence: thanh.nguyen@iuw.uni-hannover.de

Abstract: Balancing biodiversity conservation with land use for agricultural production is a major societal challenge. Conservation activities must be prioritized since funds and resources for conservation are insufficient in the context of current threats, and conservation competes with other societal priorities. In order to contribute to conservation priority-setting literature, we applied an environmental model, Pressure–State–Response (PSR), to develop a set of criteria for identifying priority areas for biodiversity conservation in Vietnam. Our empirical data have been compiled from 185 respondents and categorized into three groups: Governmental Administration and Organizations, Universities and Research Institutions, and Protected Areas. The Analytic Hierarchy Process (AHP) theory was used to identify the weight of all criteria. Our results show that the priority levels for biodiversity conservation identified by these three factors are 41% for "Pressure", 26% for "State", and 33% for "Response". Based on these three factors, seven criteria and seventeen indicators were developed to determine priority areas for biodiversity conservation. Besides, our study also reveals that the groups of Governmental Administration and organizations and Protected Areas put a focus on the "Pressure" factor, while the group of Universities and Research Institutions emphasized the importance of the "Response" factor in the evaluation process. We suggest that these criteria and indicators be used to identify priority areas for biodiversity conservation in Vietnam.

Keywords: analytic hierarchy process; biodiversity conservation; condition–pressure–response model; priority areas; Vietnam

Citation: Vu, X.D.; Csaplovics, E.; Marrs, C.; Nguyen, T.T. Criteria and Indicators to Define Priority Areas for Biodiversity Conservation in Vietnam. *Forests* **2022**, *13*, 1341. https://doi.org/10.3390/f13091341

Academic Editors: Konstantinos Poirazidis and Panteleimon Xofis

Received: 11 July 2022
Accepted: 18 August 2022
Published: 23 August 2022

Publisher's Note: MDPI stays neutral with regard to jurisdictional claims in published maps and institutional affiliations.

1. Introduction

Humans and their wellbeing, health, and livelihood have benefitted significantly from biodiversity [1–4]. However, biodiversity conservation remains one of the greatest challenges facing the modern world. It is estimated that humans have caused the extinction of between 5 and 20% of all species around the world [5], and recent extinction rates are between 100 to 1000 times their pre-human levels [5,6]. Coping with land cover conflicts caused by different stakeholder interests, biodiversity conservation often has to take a step back in relation to other interests [7]. Increasing demands on land are offset by increasingly scarce land resources. Thus, it is of crucial importance for the conservation of biodiversity to have an objective framework for the selection of areas at hand that allows high-priority areas to be identified. Priority areas should cover the most critical areas needed for biodiversity [8,9]. Identifying priority areas is one of the crucial tasks in the process of establishing protected areas since humans are not able to protect all places on Earth that contribute to biodiversity conservation [10]. However, the identification of priority areas for conservation requires the integration of biodiversity data together with socio-economic data on human pressures and responses.

There has been an increasing need for methods that define biodiversity conservation priorities to demarcate where the need for conservation action is most urgent and where the benefits of conservation strategies might be maximized [11]. Previous reviews on the criteria to identify priority areas for biodiversity conservation have mainly focused on an extensive list of relevant ecological and biological criteria [12–14]. From a perspective of a geographic scale for investigation, the establishment of biodiversity conservation priorities can be classified into three categories [11]. At the local scale, researchers and conservationists use criteria relating to genetic diversity and indicator species to provide a focus for establishing conservation priorities [11]. At the regional scale, the Habitat Conservation Planning (HCP) practice is applied to make use of information on the home range and state of organisms to designate habitat reserves. At the regional to a global scale, priority areas for biodiversity conservation are identified by using criteria such as species richness, rarity, endemism, representativeness, and complementarity to drive the conservation effort [15–17]. Nevertheless, there remains a surprising lack of empirically substantiated research that attempts to integrate both biological and socio-economic aspects into the criteria for identifying priority areas for biodiversity conservation.

Vietnam is one of the most important hotspots for biodiversity in the world [18,19]. Previous studies indicated substantial values of biodiversity in Vietnam such as climate change mitigation and adaptation, poverty reduction, education and cultural values [20,21]. However, the rate of biodiversity loss in Vietnam is alarming [16,22]. Like many other countries, much of the conservation effort in Vietnam is put into the formulation of a protected area system as an essential strategy to protect the remaining biodiversity. Currently, establishing a protected area in Vietnam requires feasibility studies to be undertaken to provide information on location, demarcation, area, and biodiversity value. Although the number of protected areas is predicted to increase in the coming years [23], there are still inherent obstacles to identify priority areas for biodiversity conservation in this country, such as limited comprehensive data, lack of time and resources for surveys and assessments, and a deficit of reliable methods. This is the main motivation for our study aimed at developing criteria and indicators for prioritizing areas for biodiversity conservation in Vietnam.

The remaining part of the paper is structured as follows. Section 2 briefly describes the environmental model "Pressure-State-Response". Section 3 provides background information on Vietnam and reviews the literature. Section 4 describes the data and methods. Section 5 presents, and Section 6 discusses, the results. Section 7 concludes.

2. Analytical Framework: "Pressure-State-Response" Model

The Pressure–State–Response (PSR) was first developed in 1993 by the Organization for Economic Co-operation and Development [24]. This framework includes three factors: Pressure (P), State (S), and Response (R), and is based on a concept of causality: human activities create pressure on the environment that changes the quality and quantity of resources (state), and then society responds to these changes with adaptive, preventive and mitigation actions [25].

The PSR (Figure 1) presents the linkages between the pressures exerted on biodiversity conservation caused by human activities (pressure box), the change in quality and quantity of biodiversity (state box), and the response to these changes as society tries to reduce the pressure and conserve the biodiversity resources (response box). The interchanges among these boxes form a continuous feedback mechanism that can be monitored and used for the assessment of biodiversity resources. Therefore, the factor *"Pressure"* describes developments in physical and biological agents, the use of resources, and the use of land. The pressures exerted by society are transported and transformed into a variety of natural processes to manifest themselves in changes in environmental conditions. The factor *"State"* describes the quantity and quality of physical, biological, and chemical phenomena in a particular area, while the factor *"Response"* refers to responses by groups and individuals in society and government attempts to prevent, compensate, ameliorate, or adapt to changes in the state of the environment [26].

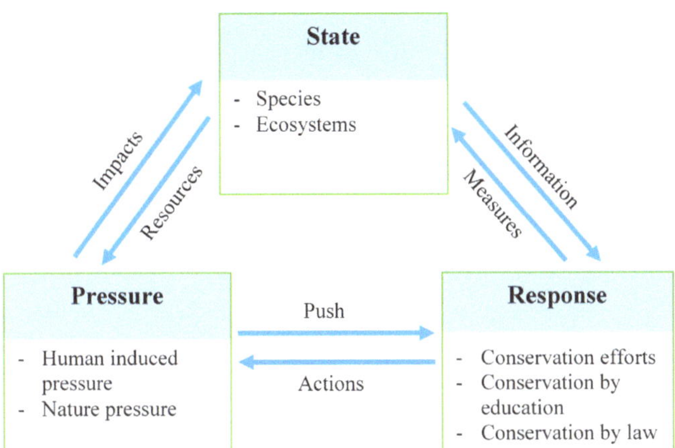

Figure 1. Adapted PSR model for evaluating biodiversity conservation.

This framework has been used to develop criteria and indicators in many fields, for example, in sustainable development [25], which formulate environmental indicators and land quality indicators, and establish state-of-the environment reporting and national environmental performance reviews [27]. Indeed, the PSR model provides a useful tool to formalize environmental problems due to its intuitive structure–human pressure on environmental state and political responses to adopt solutions [28]. The model focused on where ecological system dynamics depend exclusively on human activities [28]. This framework has been adopted by many OECD countries and by the World Bank for environmental reporting [27].

In our study, the PRS model is used to serve as a conceptual framework for analyzing the links between socio-economic activities and biodiversity change. From a theoretical perspective, this study contributes to closing gaps in our understanding of the interrelations between these three factors, namely Pressure, State, and Response, for analyzing problems of biodiversity conservation and identifying criteria for defining priority areas.

3. Literature Review and Research Context

3.1. Literature Review: Determining Priority Areas for Biodiversity Conservation

Biodiversity conservation has been one of the critical environmental issues, which aims to preserve the varieties of species and communities as well as the genetic and functional diversity of species [22,29]. According to BirdLife International, the priority areas of biodiversity conservation are Endemic Bird Areas (EBAs) [30]. Although the critical areas for biodiversity conservation are either EBAs or the diversity of all species and communities, the functions of biodiversity conservation are entirely to preserve species diversity, ecosystem diversity, soil and water conservation functions, and prevent potential threats [31].

Due to a lack of available resources for biodiversity conservation, humans cannot preserve all places on the planet that contribute to biodiversity. Thus, selecting priority areas plays a crucial role in maximizing the effectiveness of conservation and saves resources for other goals. To systemize the setting of priority areas for biodiversity conservation, a combination of criteria and scoring and ranking procedures have developed over the last couple of decades [8,32]. In these processes, multiple criteria such as diversity, rarity, naturalness, and size, among others, have been determined and given scores based on literature reviews and participation techniques [31,33,34]. These ratings have been then combined for each selected area. The areas have been ranked, and the highest priority has been given to the areas with the highest top scores [8]. Although several conservation organizations have proposed criteria to define priority areas for biodiversity conservation, they only focused on those main criteria that they have a great interest in. While BirdLife

International puts stress on the state of species and ecosystems, the Alliance for Zero Extinction emphasized the importance of endemic and threatened species to zone priority areas for biodiversity conservation [22]. However, there remains a surprising lack of studies related to the synthesis of a systematic set of criteria and indicators that support identifying priority areas for biodiversity conservation. To contribute to literature on methods of defining priority areas for biodiversity conservation, this study used the environmental PSR model as a conceptual framework to monitor biodiversity based on three key factors, including state, pressure, and response. One of the strengths of the PSR model is to show the relationships among human activities, biodiversity, and management solutions to assess the influence levels on biodiversity conservation [24,35–37]. Since the criteria were categorized into three factors of the PSR model, the number of pairs of criteria could be reduced to efficiently apply the method of pairwise comparison. Besides, the application of the PSR model also helps to have an insight into negative and positive aspects that influence biodiversity conservation.

The assessment of biodiversity conservation is an important task where conservationists and policymakers have to choose criteria to define potential areas for biodiversity conservation carefully. Previous studies have shown that the Analytic Hierarchy Process (AHP) is a helpful tool for handling complicated decision making and supporting the decision-maker in determining priorities and making the best decision in multiple-use planning of forest resources [38–40] as well as in environmental planning processes [39,41–43]. The pairwise comparison of the AHP method helps to reduce the difficulties of complex decisions and captures both subjective and objective aspects of a choice [44]. The AHP approach is the most suitable tool to determine the weights of assessment factors that significantly impact decision-making processes [42,45–47]. Although the AHP theory was first developed in the late 1970s and has been used as a decision support tool in various fields, few studies have applied it in the fields of forestry, agriculture, and natural resources [48]. Some examples of such applications include the decision making for forest planning [39,40,49,50]; selection of risk factors for forest protection [51–54]; forest management [55–61]; and suitability analysis of land use [45,62–66]. However, previous studies have not pointed and compared the influence levels of criteria on biodiversity conservation. From a technical perspective, our contribution is to calculate the weights of criteria that are used to show the influence levels of criteria on defining priority areas for biodiversity conservation.

3.2. Research Context: Vietnam

The S-shaped country of Vietnam ranges along the latitude from the 23° to 8°30′ N and includes much hilly and mountainous terrain [67]. The country occupies an area of around 329,500 km^2 and is bordered by China, Laos, and Cambodia on the north, northwest, and southwest, respectively. The rest of the country borders the East Vietnam Sea and is 3260 km long (Figure 2). The country comprises eight different eco-regions, including Northwest, Northeast, Red River Delta, North Central Coast, South Central Coast, Central Highlands, Southeast, and Mekong River Delta. The estimated population was 95.5 million people in 2017, with 85.8% belonging to the majority Kinh group, and the rest belonging to many different ethnic minorities such as Hmong, Dao, Tay, Muong, Thai, and Nung.

A long coastline and a wide-ranging latitude and altitude with a variety of hydrological conditions, climatic, soil, and terrain are the main characteristics that have created a high diversity of genes, species, and ecosystems in Vietnam [68,69]. Vietnam is one of the 16 countries with the highest biodiversity in the world and is one of the priority countries for global conservation, with about 10% of species worldwide in only 1% of the world's land area [70]. Vietnam is home to 59 Important Bird Areas [71], and 110 Key Biodiversity Areas [19]. The natural protected areas system comprises 167 protected areas with 34 national parks, 56 nature reserves, 14 species and habitat conservation areas, 54 landscape protection areas, and nine areas of empirical scientific research [72,73]. It is estimated that there are more than 20,000 plant species, 5500 insects, 3000 fishes, more than 1000 birds, and more than 300 mammals found throughout Vietnam [74]. Thus, PAs protection and management are of vital significance for biodiversity conservation in Vietnam [75]. The increase of forest cover has been seen from 28% in 1990 to about 41% in 2015. However, new plantations were established (account for 2.1 million hectares), covering most of the increase, while more than 60% of natural forests are assessed as poor or regenerating [76]. Forest degradation and deforestation in Vietnam continues in the top countries of tree cover loss [76,77]. This led to 13 million ha or 40% of the country's land area being classified as unproductive or barren land. The loss of biodiversity in Vietnam has been very critical, with many species on the brink of extinction because the natural resources have been exploited by humans [69,78]. This alarming degradation of biodiversity has occurred throughout the country [16,22] in all three types of biodiversity: species, ecosystems, and genetic [74]. About 10% of plant species are listed as endemic; separately, orchid endemism is 19.2% [74].

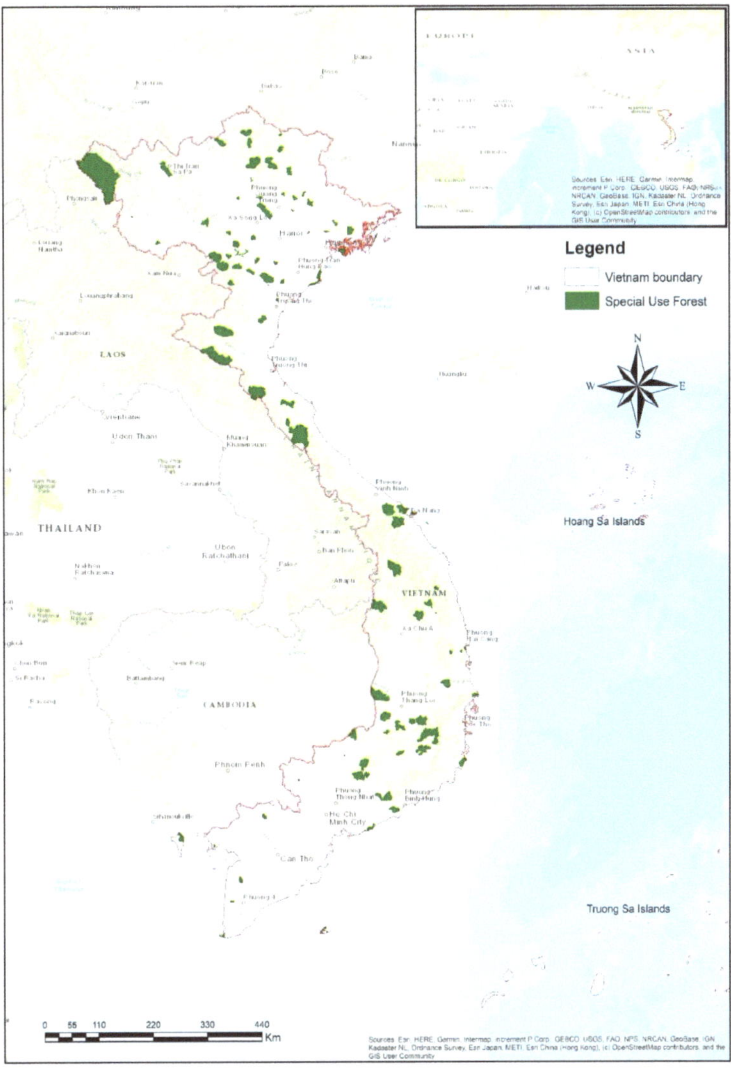

Figure 2. Map of Vietnam and the distribution of special-use forests for biodiversity conservation (Source: Modified from MARS and ESRI).

According to Global Biodiversity, the direct key causes of biodiversity loss in Vietnam include land conversion, infrastructure development, invasive species, overexploitation, pollution, climate change, natural disaster, extreme weather, population growth, forest fire, deterioration of natural ecosystem, replacement of exotic crop plants and domestic animals [69,74]. These direct causes are fueled by socio-economic factors at various scales, such as population growth and poverty [67,74,79]. Human activities are considered the main cause of deforestation and biodiversity loss [22,67,74,80–84]. Previous studies have shown that drivers of deforestation and degradation in Vietnam are highly complex and can form the networks of economic and political interests [85]. There are four main direct causes of deforestation in Vietnam including conversion to agriculture; infrastructure development; unsustainable logging (notably illegal logging); and forest fires [86,87]. The studies have indicated that biodiversity conservation in a protected area is influenced by various factors related to the establishment and management of the protected areas, local communities living next to the protected areas, and policy on protected areas' national management and financial facilities. These factors are socioeconomic and cultural factors, and are related to the management of local communities that neighbor protected areas, which are as important as resources within the protected areas [88]. This means that any conservation efforts need to be combined with economic development plans to provide livelihood opportunities for local communities to reduce the pressure on biodiversity. Many efforts have been made to protect the remaining biodiversity and halt the loss of species in Vietnam [23]. Current conservation legislation in Vietnam is focused on biodiversity, with limited consideration of socio-economic issues [89].

Vietnam is located in the Indo-Burma region, ranked as one of the top 10 biodiversity hotspots and as the top five for being threatened in the world [90]. The protected areas have played an essential role in remaining, preserving flora and fauna diversity [91–95]. Recognizing the value and significance of protected areas for biodiversity, Vietnam is on the way to achieving the National Biodiversity Strategy to 2020 and the vision for 2030 to cover 9% of the country's territory as protected areas [96]. This shows a need to define the priority areas for conservation (Nhan et al., 2015) and the establishment of new protected areas [95].

Boundary marking has been shown as one of the most substantial factors relating to the establishment and administrative effectiveness of protected areas [93]. Most protected areas in Vietnam were formulated with restricted borders that often stay inside the administrative boundaries of provinces [97]. While many specific landscapes, habitats, and ecosystems are the targets of conservation [92], they have existed outside the protected areas [98]. It is necessary to re-assess the value of biodiversity within the protected areas of Vietnam to identify the functional zones for the efficiency of governance, propose the establishment of new protected areas or dissolve the low-value ones [97].

4. Data and Methods

4.1. Data Collection

This study is based on a national survey and expert interviews. We conducted a national survey in Vietnam from March to July 2017 using a questionnaire for both direct (face to face) and indirect (via emails and phone calls) interviews (Appendix S1). We used the stratified sampling method to select the respondents. First, we collected the list of employees who worked in the fields of forest protection and biodiversity conservation. The employees were classified into the following three groups: Governmental Administration and Organizations, Universities and Research Institutions, and Protected Areas. We then collected a random sample of respondents from each group. The stratified sampling reduces errors relative to simple random sampling and ensures that observations and interviews from all relevant groups are included in the sample [99]. We had intended to conduct face-to-face interviews with all. However, some respondents were not available for face-to-face interviews, and thus the questionnaire was sent to them via email. In total, we interviewed 185 respondents from all groups (Appendix S2), including 128 face-to-face interviews and

57 email interviews. The survey questions focused on the criteria, indicators, and their importance levels that belong to the three factors of the PSR model: Pressure, State, and Response. The face-to-face interviews lasted between 1 and 2 h and were conducted in Vietnamese. The research results were based on the first author's PhD dissertation, which was accepted by an ethnic committee from the Technical University Dresden, Germany.

4.2. Data Analysis

All data from the survey were cross-checked following the triangulation method [100] to identify reliable information. Three processes were performed to identify the final criteria system, including: (1) synthesizing from the relating studies, (2) interviewing and consulting experts of conservation and biodiversity, and (3) organizing an academic seminar to identify a final criteria system. Then, we analyzed the collected data to identify and categorize the criteria and sub-criteria according to the environmental PSR model. In addition, the statistics of pairwise comparison and Analytic Hierarchy Process (AHP) were used to measure and compare the influence levels of criteria and sub-criteria in defining priority areas for biodiversity conservation.

4.2.1. Statistics of Pairwise Comparison

Since the 1950s, numerous methodologies of psychology called multidimensional scaling (MDS) have been studied and applied in analyzing the similarity and preferential choice data. Nevertheless, the solution has not yet been found to deal with the problem of gaining the perfect voting in multidimensionality [101]. The majority rule, called the Condorcet Winner, chooses the winner, which is preferred in every one-to-one comparison with the other choices [102].

As shown in the heading of the questionnaire, the data of the pairwise comparison was not gathered directly. The pairwise value was calculated by comparing two criteria in one pairwise of each respondent and synthesizing as in Appendix S3. To examine their importance, the scale of integers ranging from 1 to 9 was applied [44,103]. For example, A and B are two criteria in one pairwise comparison. These are three possible situations: A greater than B (A > B), A equal to B (A = B), and A less than B (A < B). The intensity of importance is 1 represented for the second case (A = B). Consequently, each pairwise can be presented by 8 cases of A > B, one case of A = B, and 8 cases of A < B.

The total of assessments was synthesized for each pairwise from all the respondents (183 people after removing 2 cases of outliers). Twenty-eight pairwise comparisons and 17 instances of them are described entirely in Appendix S3. According to the majority rule (Condorcet Winner), the total of respondents selecting the same situation of A > B, A = B, and A < B for each pairwise comparison was calculated for comparison.

The acceptable risk was demonstrated in the formula of error estimation from Cochran [104]. The chance is commonly called the margin of error, which has been used by researchers as the limit for the willingness to accept [104]. The acceptable margin of error is 5% and 3% for categorical data and continuous data, respectively [105]. The appropriate precision for prevalence is 5% by experience [106–108]. Therefore, in this study, a percentage of difference was used to show the reliability of comparison among the number of respondents who chose A > B, A = B, or A < B. The rule was used to identify the appropriate level of one pairwise comparison as follows:

- Five percent of the difference was used to select the majority to belong to A > B, A = B, or A < B.
- If the number of A > B and A < B is similar or higher than under 5% out of total respondents, the situation of A = B is the priority option.
- If the highest number of three situations (A > B, A = B, and A < B) are higher than others above 5% out of total respondents, this situation is the opinion of the majority.
- If A > B or A < B accounts for the majority, the case of the statistical model is used in this situation.

- If the number of A > B is similar to the number of A < B, and they are more significant than the A = B, the situation of A = B is the collective opinion of all respondents. As such, A and B have equal importance.
- If once the number of either A > B or A < B is less than the number of A = B, and the other is similar to the number of A = B, the trend of majority opinions inclines to the number identical to the A = B.

There are 27 pairwise comparisons for 3 factors, 7 criteria and 17 sub-criteria (Appendix S3) that were used to identify the level of importance of each criterion in the pairwise comparisons. A and B refers to two criteria in each pairwise comparison. Three situations of A > B, A = B, and A < B were calculated regarding the total number of respondents. Various pairwise comparisons were significantly different among of three situations such as Nature–Human, Location–Hydrology, Location–Forest Type, Topography–Forest Type, Hydrology–Forest Type, Climate Change–Nature Disaster, Distribution–Livelihood, Density–Population, Density–Livelihood, and Forest Management Types–Size of Forest Area. The rest of the gained quantities are similar in three situations, hence the percentage of difference value was used to judge which one is greater or whether they are equal together.

The critical levels of two criteria are alike for seven pairwise comparisons, and they are; State–Response, Species–Ecosystem, Conservation–Law, Education–Law, Hydrology–Climate, Distribution–Density, and Density–Population. Most respondents chose the case of A = B in comparison to A > B or A < B. It also points out a unique case of pairwise comparisons that obtained a similar number of respondents that chose the situations of A = B and A < B. It is the pairwise comparison of distribution and quantity with the same amount of responses accounting for 37% of the situations and 26% of the rest (A > B). It can be seen that the trend of the majority is tilted towards the status of A < B.

4.2.2. Analytic Hierarchy Process (AHP)

The Analytic Hierarchy Process (AHP) is a multi-criteria decision-making approach firstly developed by Saaty [46]. The AHP is attractive to many researchers due to the effective mathematical properties of the method [109]. Another advantage of the AHP method is that it enables users to determine the weights of the parameters in the solution of a multi-criteria problem [62]. Solving a problem using AHP is conducted using the weights or priorities of the criteria subjected to pairwise comparison. Weights or priorities are determined by normalizing the pairwise comparison matrix [62].

While performing pairwise comparisons of criteria in the AHP method, a certain level of inconsistency can occur [62]. Thus, the logical consistency of pairwise comparisons must be checked [110]. To measure the consistency of pairwise comparison judgments, the consistency ratio proposed by Saaty [44] is used. A consistency ratio is calculated for the pairwise comparison matrix. In our study, the AHP-based weights of criteria are used to synthesize the mapping data of the criteria to identify and to consider when making a spatial decision, which will be calculated through the overlay equation integrated into MCDA and GIS [31,47,62]. Equation (1) (below) was used to synthesize biodiversity conservation value for study areas.

$$C_k = \sum_{i=1}^{I} W_i X_i^k \tag{1}$$

where C_k is the biodiversity conservation value at the k_{th} intersection region; X_i^k is the score contained within GIS layer of i_{th} at the k_{th} intersection region; W_i is the weight of i_{th} indicator, which can be changed based on the critical level of each indicator.

The steps of estimating a biodiversity conservation index was implemented as follows:

(a) Criteria and their factors of biodiversity conservation were chosen from the literature review and the interviews.
(b) The grade of each factor was transformed from the measured data through the fuzzy set.

(c) The weights of each factor were assigned by the AHP method based on Saaty's scale and the pair-wise comparison matrix (Table 1).

(d) Biodiversity conservation index was then calculated by a simple linear priority function as in Equation (1).

Table 1. Scale for pair-wise AHP comparisons.

Intensity of Importance	Description
1 2	Equal importance
3 4	Moderate importance
5 6	Strong or essential importance
7 8	Very strong or demonstrated importance
9	Extreme importance

Source: [103,111].

In order to use the results calculated by the AHP method, a critical aspect of the AHP is to check the consistency [44,111,112]. Saaty [44] proposed the consistency ratio (CR) to identify the consistencies of the pairwise comparison matrices. The test of consistency must be done when the number of criteria used in a pairwise comparison matrix is higher than 2. When the number increases, the pairwise comparisons climb significantly. This makes inconsistencies arise, and it becomes complicated to check the consistencies.

A pairwise comparison matrix considered as consistent or inconsistent depends on the test of the Consistency Ratio (3). The test can pass when the Consistency Ratio is less than 0.1.

$$\mathbf{CI} = \frac{\lambda - \mathbf{n}}{\mathbf{n} - 1} \tag{2}$$

$$\mathbf{CR} = \frac{\mathbf{CI}}{\mathbf{RI}} \tag{3}$$

where **CR** is the Consistency Ratio as in Equation (3); **CI** is the Consistency Index as in Equation (2); **RI** is the average Random Index based on the Matrix Size (Table 2), **n** is the number of criteria used in a pairwise comparison matrix (**n** $\leq$ 10), and λ is the average of the elements of consistency vector.

Table 2. The average values of the Random Index.

n	1	2	3	4	5	6	7	8	9	10
RI	0	0	0.52	0.89	1.11	1.25	1.35	1.4	1.45	1.49

Source: [111].

The procedure for checking the consistency includes the four following steps:

- Step 1: Identify the λ of the pairwise comparison matrix.
- Step 2: Apply Equation (2) to calculate the Consistency Index (**CI**)
- Step 3: Apply Equation (3) to estimate the Consistency Ratio (**CR**).
- Step 4: The judgment of the consistency of the pairwise comparison matrix is performed through the comparison between the **CR** value and the consistency threshold (0.1). The pairwise comparison matrix is identified to be acceptable when **CR** < 0.1.

5. Results

5.1. Criteria and Indicators for Defining Priority Areas

In many previous studies, the indicators for setting priorities for conservation focus on plant and animal species such as species richness, rarity, endemism, representativeness, and complementarity to drive the conservation effort [15–17]. Some studies emphasize the importance of human population pressure [113–115] or human efforts to protect habitat [115,116], where deforestation and forest degradation have happened [117]. In our study, we integrate conservation and social aspects in the criteria for setting priorities for biodiversity conservation.

Table 3 presents the criteria set to define priority areas for biodiversity conservation in Vietnam. Based on three factors of the PSR model, seven criteria and 17 indicators were identified to support prioritizing areas for conservation in the context of Vietnam.

Table 3. Criteria and indicators for defining priority area for biodiversity conservation.

Factors of PSR Model	Criteria	Indicator
Pressure	Human-induced Pressure	Distribution of population
		Density of population
		Population
		Livelihood of locals
	Natural Pressure	Climate change
		Natural disaster
State	State of Species	Richness
		Rarity
	State of Ecosystem	Location
		Topography
		Hydrology
		Climate
		Forest type
Response	Conservation efforts	Forest management types
		Size of forest area
	Education	Conservation through strengthening education
	Law	Conservation through exerting law

a. Pressure

The current literature has demonstrated that analyzing the pressure on biodiversity, its trends, and origins have become even more urgent since the loss of biodiversity is at such an alarming rate in many countries [118]. In the study, pressures include natural and human-induced factors that cause environmental change. Natural pressures are derived from unexpected natural changes such as climate change and natural disasters. Usually, these changes are unwanted and seen as negative (damage, degradation). Therefore, two main indicators of natural pressure were used, including climate change and natural disasters (flooding, drought, or earthquake). Natural pressures are unpredictable, and so human society is struggling to find solutions to minimize the impact of natural pressures. Human-induced pressures are consequences of human activities (land-use change, logging, hunting, extraction, and use of resources) that have the potential to cause or contribute to adverse effects [74,119,120]. Previous studies have indicated that 6 million square kilometers (32.8%) of protected land in the world is under intense human pressure.

Our study suggests four indicators of human-induced pressure, namely population numbers, distribution of population, the density of population, and the livelihood of locals. Each of these pressures can have different effects, some of which emerge in the short-term (e.g., land use, deforestation), while others are long-term (e.g., climate change) [74,119].

b. State

The State of biodiversity is represented by the number of biological features (measured within species, between species and ecosystems), of physical and chemical features of ecosystems, and/or of environmental functions, vulnerable to pressures in a certain area [119]. Many studies have shown that habitats or environmental conditions are considered as indicators of the existence of species [121–123]. Prediction of species distribution based on the existence of their habitat is used to identify the priority areas for conservation as well as for field surveys [122]. Therefore, one of the most important methods to conserve species is the protection of their habitats [22]. It has shown that the occurrence of rare and sensitive species is determined within their range of appropriate habitats [30].

In our study, the State refers to the number of species, the status of the forest ecosystem and wildlife resources. Due to the pressure on the environment, the State of the environment changes. These changes then have impacts on the functions of the environment, such as human and ecosystem health, resource availability, losses of manufactured capital, and biodiversity [26]. Previous studies have shown that the State may refer either to natural systems alone [124,125] or to both natural and socio-economic systems [126]. Depending on the systems chosen for description, indicators of State can be very different from one study to another [119]. For species, we focused on two main indicators: richness and rarity. The state of the ecosystem consists of five indicators: location, topography, hydrology, climate, and forest type.

Many studies have shown that the increase of species richness in one region depends on the stability level of the forest area at that time [67,127–139]. Forest is considered a significant factor in biodiversity conservation since it provides appropriate habits for many species. It means that the level of species richness is higher in the area covered by forest for a longer period. The diversity of the plant life significantly depends on the disturbance in the past [123]. Therefore, the priority levels of richness for biodiversity conservation are determined by monitoring the forest cover.

c. Responses

Responses may seek to control Pressures (prevention, mitigation), to maintain or restore the State of the environment, to help to accommodate impacts (adaptation) or even deliberate "do nothing" strategies [26,119,140]. For applications regarding biodiversity, Responses are the measures taken to address drivers, pressures, state, or impacts. They include measures to protect and conserve biodiversity (in situ and ex situ), and include, for example, measures to promote the equitable sharing of the monetary or non-monetary gains arising from the utilization of genetic resources.

Most of the indicators developed for Responses concern political actions of protection, mitigation, conservation, or promotion [26,37]. Other indicators refer to Responses as being a mixed result of both effective top-down political action and bottom-up social awareness [119]. Some societal responses may be considered as negative driving forces because they aim at redirecting prevailing trends in consumption and production patterns. Other responses aim at raising the efficiency of products and processes [26].

In this study, we focused on responses related to conservation efforts, education, and law. Conservation efforts are measured by forest management types and the size of forest areas. In addition, conservation activities delivered through strengthening education and enforcing the law are also evaluated as responses to reduce the pressure on biodiversity.

5.2. Weights of Criteria Based on All Respondents

The weights of the factors, criteria and indicators for identifying the priority areas of biodiversity conservation in Vietnam were calculated using the data from all respondents. The values of the Consistency Ratio (CR) in Table 4 of the pairwise comparison matrices were lower than the Consistency Ranking (10%). Thus, the matrices were consistent, and the calculated weights were appropriate to use. The assignment of percentage values of factors, criteria, and indicators are computed and are shown in Table 4.

Table 4. Weights of criteria based on all respondents and groups for identifying the priority areas of biodiversity conservation in Vietnam.

Factors	Criteria	Sub-Criteria	All (%)		PAs (%)		URIs (%)		GOs (%)	
Pressure	Nature	Climate change	7.7		6.2		5.8		7.7	
		Natural disaster	2.6		2.1		2.9		2.6	
		Sum	10.3		8.2		8.7		10.3	
	Human	Distribution	4.6		7.1		2.9		4.3	
		Density	6.0		5.5		3.4		7.2	
		Population	7.1		6.2		4.1		7.2	
		Livelihood	13.1		14.0		6.9		12.1	
		Sum	CR = 1.7%	30.8	CR = 4.5%	32.8	CR = 2.3%	17.3	CR = 2.3%	30.8
	Total		41.0		41.0		26.0		41.1	
State	Species	Richness	2.9		4.4		5.5		4.3	
		Rarity	5.8		4.4		10.9		4.3	
		Sum	8.7		8.8		16.4		8.6	
	Ecosystem	Location	2.5		2.7		2.4		2.6	
		Topography	3.6		4.3		4.0		2.6	
		Hydrology	1.7		1.6		1.8		1.4	
		Climate	3.6		2.6		4.7		5.7	
		Forest type	6.1		6.2		3.5		5.1	
		Sum	CR = 2.0%	17.5	CR = 6.7%	17.4	CR = 2.6%	16.4	CR = 5.8%	17.4
	Total		26.2		26.2		32.8		26.0	
Response	Conservation	Forest management types	6.4		6.4		7.2		7.7	
		Size of forest area	2.1		2.1		3.6		2.6	
		Sum	8.6		8.6		10.8		10.2	
	Education		13.5		13.5		16.9		6.5	
	Law		10.6		10.7		13.5		16.2	
	Total		CR = 5.2%	32.7	CR = 5.2%	32.8	CR = 5.2%	41.2	CR = 5.2%	32.9
Total			CR = 5.2%	100.0	CR = 5.2%	100.0	CR = 5.2%	100.0	CR = 5.2%	100.0

All—All respondents; PAs—Protected Areas group; URIs—Universities and Research Institutes Group; GOs—Government Organizations Group.

The first level is the ultimate purpose (100%) of assessing the influences of the criteria on biodiversity conservation in Vietnam. The second level comprises the three factors used in the environmental model to measure biodiversity conservation. The factors account for 26%, 41%, and 33% of "Condition", "Pressure", and "Response", respectively. The third level witnesses a huge difference among seven criteria, with 31% of "Human", then it is 17% of "Ecosystem", the rest is distributed evenly. With the exception of 24% for "Education" and "Law", the fourth level illustrates 17 indicators of five criteria. The "Livelihood" factor is highest with 13.12%. The mediate group consists of "Types of forest management", "Rarity", "Forest type", "Climate change", "Density", and "Quantity", fluctuating from 5.8% to 7. 71%. Finally, the remaining eight of the seventeen indicators account for just 23.84%.

5.3. Weights of Criteria Based on the Groups

5.3.1. The Protected Areas Group

The respondents working at Protected Areas (PAs) are the key participants of the survey. Their attitudes and experiences help to assess precisely the role of the criteria in biodiversity conservation. The respondents of PAs, located mostly in the North of Vietnam, account for 34.59% of all respondents, of which 51 people (26.27%) were interviewed in person.

The data of the respondents at PAs are filtered separately to calculate the weights of the entire presented criteria in the questionnaire. The consistent tests were applied to all pairwise comparison matrices. Their consistency ratio was all within the consistency threshold (Table 4). The weights were synthesized and described in detail in Table 4. It can be seen that the distribution of weights in PAs is relatively similar to the distribution calculated by total respondents. Exceptionally, the "Richness" and "Rarity" indicators obtained the same percentage (50%) for each instead of 33% and 67% in synthesizing all respondents, respectively. It expresses the required functions and characteristics in establishing protected areas in Vietnam that they are places not only to preserve rare species, but also to remain and enhance the diversity of species.

5.3.2. The Government Administration and Organizations Group

Government administration and organizations represent the communities and the state to implement the law and the policy of the country. Although the group only accounts for 10.26% of all respondents, they are as a group representing the opinion of the Vietnamese government. The data of responses of the organizations were synthesized and calculated separately to gain the weights of all criteria (Table 4).

The pairwise comparison matrices are considered consistent as they passed the consistency test with the Consistency Ratio (CR) lower than 10% (Table 4). The results of the group are pretty similar to the group of Protected Areas and all respondents within the second level. The criteria in the third level assessed are reasonably different. The distribution of weight among "Education", "Law", and "Conservation" changed. The biodiversity conservation by "Law" is the most crucial, accounting for nearly 50% of weight, while the results of other groups, as well as all respondents, show that almost all of their consideration focused on "Education" criterion. It shows that "Law" has been the most interested of the Government Organizations in responding to reduce the loss of biodiversity, which is to be expected when considering the functions and characteristics of Government Organizations.

5.3.3. Universities and Research Institutes Group

The group with the highest number of respondents is the Universities and Research Institutes group with 50.41%. It includes six universities and nine research institutes.

The pairwise comparison matrices were established using the data of 92 respondents (49.86%) from the Universities and Research Institutes group. The weight set calculated by the respondents of universities and research institutes is presented in Table 4.

There is a significant change in the assignment of the weights when it is compared with the case of all respondents. In the second level, the "Response" factor with 41% of weight replaced the top position of "Pressure" and pushed it to bottom with 26% of the weight. The third level witnesses the adjustment of the ratio between "Species" and "Ecosystem" when they are equal in terms of their importance. Instead, the "Ecosystem" criterion is assessed as more important than "Species" by all respondents. The ratio of the indicators in the fourth level is more or less unchanged. Remarkably, the "Climate" indicator climbed to the peak, accounting for 28% impact on the "Ecosystem" criterion, while the "Forest type" indicator fell from 34% to 22% in the importance scale in the assessment of the Universities and Research Institutes group.

The Universities and Research Institutes group represents the people who are working on the training and research field of forestry, biodiversity, and conservation in Vietnam. The assessment results in defining priority areas for biodiversity conservation reveal that the highest interest focuses on disseminating information and education to prevent and reduce the pressure on biodiversity.

5.3.4. Comparison between Weights of the Groups

The survey received the opinions of respondents regarding the importance levels of the established criteria. To use the AHP theory for identifying the weights, the value of each pairwise comparison was the difference in importance level calculated between two criteria in one pairwise. The relationships among the criteria formulated twenty-seven pairwise comparisons.

They were grouped into three different levels, including the factors of the second level, the criteria of the third, and the indicators of the fourth. These were nine pairwise comparison matrixes with one matrix for the second level, three matrices for the third, and five matrices for the fourth. The Consistency Ratio (CR) was used to identify the consistencies of the pairwise comparison matrices (Table 4).

The statistics of weight set in the groups of respondents helped in the analysis and assessment of each factor, criterion, and indicator. Although the different values of weight appeared in a few sectors of each group, those values were not the opinion of the majority. The synthesis of the weights for each group in Table 4 illustrates a part of the common trend of the different fields, such as research, training, planning, policy, decision-making, and implementation. The Protected Areas group assessed the importance level of the factors in the environmental model in a similar way to that of the result based on the assessment of all respondents. The results from the Universities and Research Institutes group showed the most important factor, which was the "Response" factor with 41%. In contrast, the weight of the "Pressure" factor accounted for the most significant percentage at 41% and was calculated based on the data selected from the Protected Areas group, Government Organizations group, and all respondents.

6. Discussion

Prioritization exercises are valuable tools to structure a balance between conservation measures and sustainable land use management. In the last decades, many efforts have been made to develop science-based methodologies to select priority areas for biodiversity conservation [141–143]. Previous studies have documented a range of priority-setting approaches covering a broad spectrum from mathematical to intuitive [142,144,145]. Much of the conservation priority-setting literature concerns the establishment of criteria to identify priority areas for biodiversity conservation [7,13,14,146]. Some studies to formalize the process of setting conservation priorities have focused on ecological and biological criteria to aid systematic selection of areas for biodiversity conservation [7,12–14,146]. According to Asaad, Lundquist, Erdmann and Costello [146], eight ecological and biological criteria were developed to identify suitable locations for biodiversity conservation. Among these, four habitat-based criteria are captured, including uniqueness and rarity of habitats, fragility and susceptibility of habitats, importance for ecological integrity, and representativeness of all habitats [146].

Previous studies have shown that the selection of most priority area networks focuses primarily on species richness, especially that of rare or endangered species [16] and occurrence locations [147,148]. Studies have used either species or ecosystems as biodiversity surrogates to identify priority areas for biodiversity conservation that ensure both species and ecosystems conservation [141,149,150]. Asaad et al. [151] have developed an alternative approach to delineate areas of importance for biodiversity conservation that uses a range of ecological criteria, multiple sources of data, and wide-ranging species taxonomic groups. The study suggested five different criteria to assess the value of critical habitat, species diversity, and charismatic threatened and endemic species.

Despite the range of conservation priority-setting studies investigated, there remain significant challenges in implementing biodiversity conservation that reconcile criteria for identifying priority areas and representative networks for biodiversity protection [146]. The identification of priority areas demands the integration of biophysical data on ecosystems together with social data on human pressures and planning opportunities [11]. However, previous studies have tended to focus on ecological and biological criteria, with little effort to understand the social, economic, and political dimensions within which the protected areas are placed [11]. Our contribution to the conservation priority-setting literature is to apply a multiple criteria analysis based on the PSR model for the establishment of seven criteria and 17 indicators that integrates both biological and socio-economic aspects to set priorities for biodiversity conservation in Vietnam.

One value addition of the research includes the application of the analytical hierarchy process (AHP) to determine the weight of each criterion from a multi-stakeholder perspective that helps to measure its importance level in defining priority areas for biodiversity conservation. In addition, a comprehensive assessment of respondents' opinions provides an insight into different preferences within three main groups including governmental, conservational, research-specific, with regard to weighting the influence of factors and criteria for defining high-priority areas for biodiversity conservation. Our study also reveals that the groups of Governmental Administration and Protected Areas put a focus on the "Pressure" factor while the group of Research Institutions emphasized the importance of "Response" factor in the evaluation process. These findings allow for a better understanding of different approaches to managing and developing protected areas and designate new priority areas based on respective multi-criteria decision-making (prioritization) in Vietnam.

7. Conclusions

Identification of priority areas is an important step in conservation planning to maximize the benefits of conservation strategies. However, the formulation of a criteria system that integrates both biological and socio-economic aspects is still a challenging task for conservationists and practitioners. In this study, we used the Pressure–State–Response model to develop criteria and indicators for defining priority areas for biodiversity conservation. We use empirical data from 185 respondents categorized into three groups: Governmental Administration and Organizations, Universities and Research Institutions, and Protected Areas, collected in a 2017 survey in Vietnam. We apply Analytic Hierarchy Process (AHP) theory to calculate the weight of the criteria and indicators based on information from all respondents and the groups of respondents.

Our results have suggested seven criteria and 17 indicators that integrate biological and socio-economic factors to set priorities for biodiversity conservation in Vietnam. The results have slao shown that the priority levels for biodiversity conservation could be identified by three main factors: Pressure, State, and Response, with the value of the weight of 41%, 26%, and 33%, respectively. In addition, our study revealed that the groups of Governmental Administration and Protected Areas put a focus on the "Pressure" factor while the group of Research Institutions emphasized the importance of the "Response" factor in the evaluation process.

Based on these findings, we emphasize the importance of setting priorities for biodiversity conservation through criteria and indicators. A criteria system integrating biological and social aspects could provide a useful tool to define priority areas for biodiversity conservation. However, further research is needed to apply these suggested criteria and indicators to identify priority areas for biodiversity conservation in practice and to examine the reliability of these criteria and indicators.

Supplementary Materials: The following supporting information can be downloaded at: https://www.mdpi.com/article/10.3390/f13091341/s1, Appendix S1: Questionnaire; Appendix S2: List of interviewees; Appendix S3: Synthesizing the numbers of respondents in pairwise.

Author Contributions: Conceptualization, X.D.V.; methodology, X.D.V.; software, X.D.V.; validation, X.D.V., E.C., T.T.N. and C.M.; formal analysis, X.D.V.; investigation, X.D.V.; resources, X.D.V.; data curation, X.D.V.; writing—original draft preparation, X.D.V.; writing—review and editing, E.C., T.T.N. and C.M.; visualization, X.D.V.; supervision, E.C., T.T.N. and C.M.; project administration, X.D.V. All authors have read and agreed to the published version of the manuscript.

Funding: This research was funded by the Vietnamese Government Scholarship (3979/QĐ-BGDĐT). The publication of this article was funded by the Open Access Fund of the Leibniz University Hannover.

Data Availability Statement: Not applicable.

Acknowledgments: This research was supported by the Chair Group of Remote Sensing of the Institute of Photogrammetry and Remote Sensing, Technical University of Dresden, Dresden, Germany, and the Vietnam National University of Forestry. We would like to express our gratitude to the experts, researchers, and officials involved in the study, particularly those from the protected areas in Vietnam, IUCN Vietnam, Biodiversity Conservation Agency (MONRE), Department of Nature Conservation (MARD), Forest Protection Department (MARD), and the Vietnam National University of Forestry for providing valuable data. My special thanks to forest rangers and communities at Cuc Phuong National Park, Pu Luong Nature Reserve, and Ngoc Son Ngo Luong Nature Reserve.

Conflicts of Interest: The authors declare no conflict of interest.

References

1. Mittermeier, R.A.; Turner, W.R.; Larsen, F.W.; Brooks, T.M.; Gascon, C. Global biodiversity conservation: The critical role of hotspots. In *Biodiversity Hotspots*; Springer: Berlin/Heidelberg, Germany, 2011; pp. 3–22.
2. De Groot, R.; Brander, L.; Van Der Ploeg, S.; Costanza, R.; Bernard, F.; Braat, L.; Christie, M.; Crossman, N.; Ghermandi, A.; Hein, L. Global estimates of the value of ecosystems and their services in monetary units. *Ecosyst. Serv.* **2012**, *1*, 50–61. [CrossRef]
3. Christie, M.; Fazey, I.; Cooper, R.; Hyde, T.; Kenter, J.O. An evaluation of monetary and non-monetary techniques for assessing the importance of biodiversity and ecosystem services to people in countries with developing economies. *Ecol. Econ.* **2012**, *83*, 67–78. [CrossRef]
4. Costanza, R.; De Groot, R.; Sutton, P.; Van der Ploeg, S.; Anderson, S.J.; Kubiszewski, I.; Farber, S.; Turner, R.K. Changes in the global value of ecosystem services. *Glob. Environ. Change* **2014**, *26*, 152–158. [CrossRef]
5. Chapin, F.S., III; Zavaleta, E.S.; Eviner, V.T.; Naylor, R.L.; Vitousek, P.M.; Reynolds, H.L.; Hooper, D.U.; Lavorel, S.; Sala, O.E.; Hobbie, S.E.; et al. Consequences of changing biodiversity. *Nature* **2000**, *405*, 234. [CrossRef]
6. Pimm, S.L.; Russell, G.J.; Gittleman, J.L.; Brooks, T.M. The future of biodiversity. *Science* **1995**, *269*, 347–350. [CrossRef] [PubMed]
7. Wilson, K.A.; Carwardine, J.; Possingham, H.P. Setting conservation priorities. *Ann. N. Y. Acad. Sci.* **2009**, *1162*, 237–264. [CrossRef]
8. Margules, C.R.; Pressey, R.L.; Williams, P.H. Representing biodiversity: Data and procedures for identifying priority areas for conservation. *J. Biosci.* **2002**, *27*, 309–326. [CrossRef]
9. Mehri, A.; Salmanmahiny, A.; Mirkarimi, S.H.; Rezaei, H.R. Use of optimization algorithms to prioritize protected areas in Mazandaran Province of Iran. *J. Nat. Conserv.* **2014**, *22*, 462–470. [CrossRef]
10. Lu, Z.; Xu, W.-H.; Ouyang, Z.-Y.; Zhu, C.-Q. Determination of priority nature conservation areas and human disturbances in the Yangtze River Basin, China. *J. Nat. Conserv.* **2014**, *22*, 326–336. [CrossRef]
11. Balram, S.; Dragićević, S.; Meredith, T. A collaborative GIS method for integrating local and technical knowledge in establishing biodiversity conservation priorities. *Biodivers. Conserv.* **2004**, *13*, 1195–1208. [CrossRef]
12. Day, J.C.; Roff, J. *Planning for Representative Marine Protected Areas: A Framework for Canada's Oceans*; World Wildlife Fund: Toronto, Canada, 2000.
13. Roberts, C.M.; Andelman, S.; Branch, G.; Bustamante, R.H.; Carlos Castilla, J.; Dugan, J.; Halpern, B.S.; Lafferty, K.D.; Leslie, H.; Lubchenco, J. Ecological criteria for evaluating candidate sites for marine reserves. *Ecol. Appl.* **2003**, *13*, 199–214. [CrossRef]
14. Gilman, E.; Dunn, D.; Read, A.; Hyrenbach, K.D.; Warner, R. Designing criteria suites to identify discrete and networked sites of high value across manifestations of biodiversity. *Biodivers. Conserv.* **2011**, *20*, 3363–3383. [CrossRef]
15. Kier, G.; Barthlott, W. Measuring and mapping endemism and species richness: A new methodological approach and its application on the flora of Africa. *Biodivers. Conserv.* **2001**, *10*, 1513–1529. [CrossRef]
16. Myers, N.; Mittermeier, R.A.; Mittermeier, C.G.; da Fonseca, G.A.B.; Kent, J. Biodiversity hotspots for conservation priorities. *Nature* **2000**, *403*, 853–858. [CrossRef]
17. Woodhouse, S.; Lovett, A.; Dolman, P.; Fuller, R. Using a GIS to select priority areas for conservation. *Comput. Environ. Urban Syst.* **2000**, *24*, 79–93. [CrossRef]
18. Marchese, C. Biodiversity hotspots: A shortcut for a more complicated concept. *Glob. Ecol. Conserv.* **2015**, *3*, 297–309. [CrossRef]
19. Mittermeier, R.A.; Gil, P.R.; Hoffmann, M.; Pilgrim, J.; Brooks, T.; Mittermeier, C.G.; Lamoreux, J.; Fonseca, G.A. *Earth's Biologically Richest and Most Endangered Terrestrial Ecoregions*; Conservation International: Arlington, VA, USA, 2004.
20. Tran, H.T.; Tinh, B.D. *Asian Cities Climate Resilience Working Paper Series 4: Cost-Benefit Analysis of Mangrove Restoration in Thi Nai Lagoon, Quy Nhon City, Vietnam*; International Institute for Environment and Development (IIED): Quy Nhon, Vietnam, 2013.
21. Dang, L.H.; Tyl, N. *Willingness to Pay for the Preservation of Lo Go-Xa Mat National Park in Vietnam*; Economy and Environment Program for Southeast Asia (EEPSEA): Tay Ninh, Vietnam, 2009.
22. Gordon, E.A.; Franco, O.E.; Tyrrell, M.L. *Protecting Biodiversity: A Guide to Criteria Used by Global Conservation Organizations*; Yale School of Forestry & Environmental Studies: New Haven, CT, USA, 2005.
23. MARD. *Report on Planning for Special Use Forest to 2020*; Vietnam Ministry of Agriculture and Rural Development: HaNoi, Vietnam, 2014.
24. OECD. *Core Set of Indicators for Environmental Performance Reviews: A Synthesis Report by the Group on the State of the Environment*; Organisation for Economic Co-operation and Development: Paris, France, 1993; pp. 1–39.

25. Martins, J.H.; Camanho, A.S.; Gaspar, M.B. A review of the application of driving forces–Pressure–State–Impact–Response framework to fisheries management. *Ocean. Coast. Manag.* **2012**, *69*, 273–281. [CrossRef]
26. Gabrielsen, P.; Bosch, P. *Environmental Indicators: Typology and Use in Reporting*; European Environment Agency: Copenhagen, Denmark, 2003.
27. Dumanski, J.; Pieri, C. *Application of the Pressure-State-Response Framework for the Land Quality Indicators (LQI) Programme*; FAO: Rome, Italy, 1996.
28. Levrel, H.; Kerbiriou, C.; Couvet, D.; Weber, J. OECD pressure–state–response indicators for managing biodiversity: A realistic perspective for a French biosphere reserve. *Biodivers. Conserv.* **2009**, *18*, 1719. [CrossRef]
29. Saunders, D.; Margules, C.; Hill, B. *Environmental Indicators for National State of the Environment Reporting: Biodiversity*; CSIRO Wildlife Ecology: Canberra, Australia, 1998.
30. Miller, R.I. *Mapping the Diversity of Nature*; Springer Science + Business Media: Berlin/Heidelberg, Germany, 1994.
31. Phua, M.-H.; Minowa, M. A GIS-based multi-criteria decision making approach to forest conservation planning at a landscape scale: A case study in the Kinabalu Area, Sabah, Malaysia. *Landsc. Urban Plan.* **2005**, *71*, 207–222. [CrossRef]
32. Smith, P.G.R.; Theberge, J.B. A review of criteria for evaluating natural areas. *Environ. Manag.* **1986**, *10*, 715–734. [CrossRef]
33. Boteva, D.; Griffiths, G.; Dimopoulos, P. Evaluation and mapping of the conservation significance of habitats using GIS: An example from Crete, Greece. *J. Nat. Conserv.* **2004**, *12*, 237–250. [CrossRef]
34. de Oliveira AvernaValente, R.; Vettorazzi, C.A. Definition of priority areas for forest conservation through the ordered weighted averaging method. *For. Ecol. Manag.* **2008**, *256*, 1408–1417. [CrossRef]
35. Lee, W.; McGlone, M.; Wright, E. Biodiversity inventory and monitoring: A review of national and international systems and a proposed framework for future biodiversity monitoring by the Department of Conservation. In *Landcare Research Contract Report*; LC0405/122; New Zealand, 2005; *(unpublished)*.
36. Long, T.T.; Dang, N.X.; Rastall, R. *Giám sát Đa dạng Sinh học có sự Tham gia: Hướng dẫn phương pháp*; Netherlands Development Organisation (SNV): Hanoi, Vietnam, 2016.
37. OECD. *Environmental Indicators–Development, Measurement and Use*; OECD Environment Directorate: Paris, France, 2003.
38. Kangas, J. Multiple-use planning of forest resources by using the analytic hierarchy process. *Scand. J. For. Res.* **1992**, *7*, 259–268. [CrossRef]
39. Kangas, J.; Kuusipalo, J. Integrating biodiversity into forest management planning and decision-making. *For. Ecol. Manag.* **1993**, *61*, 1–15. [CrossRef]
40. Mendoza, G.A.; Sprouse, W. Forest planning and decision making under fuzzy environments: An overview and illustration. *For. Sci.* **1989**, *35*, 481–502.
41. Anselin, A.; Meire, P.M.; Anselin, L. Multicriteria techniques in ecological evaluation: An example using the analytical hierarchy process. *Biol. Conserv.* **1989**, *49*, 215–229. [CrossRef]
42. Saaty, T.L.; Gholamnezhad, H. *High-Level Nuclear Waste Management: Analysis of Options*; SAGE Publications: New York, NY, USA, 1982; Volume 9, pp. 181–196.
43. Varis, O. The Analysis of Preferences in Complex Environmental. *J. Environ. Manag.* **1989**, *28*, 283–294.
44. Saaty, T.L. *The Analytic Hierarchy Process*; McGraw-Hill: New York, NY, USA, 1980; p. 324.
45. Malczewski, J. GIS-based land-use suitability analysis: A critical overview. *Prog. Plan.* **2004**, *62*, 3–65. [CrossRef]
46. Saaty, T.L. A scaling method for priorities in hierarchical structures. *J. Math. Psychol.* **1977**, *15*, 234–281. [CrossRef]
47. Jiuquan, Z.; Su, Y.; Wu, J.; Liang, H. GIS based land suitability assessment for tobacco production using AHP and fuzzy set in Shandong province of China. *Comput. Electron. Agric.* **2015**, *114*, 202–211. [CrossRef]
48. Schmoldt, D.L.; Kangas, J.; Mendoza, G.A.; Pesonen, M. *The Analytic Hierarchy Process in Natural Resource and Environmental Decision Making*; Springer Science & Business Media: Berlin/Heidelberg, Germany, 2001; Volume 3.
49. Pukkala, T. Multi-Objective Forest Planning. *Manag. For. Ecosyst.* **2003**, *6*, 216. [CrossRef]
50. Pukkala, T.; Kangas, J. A method for integrating risk and attitude toward risk into forest planning. *For. Sci.* **1996**, *42*, 198–205.
51. Jung, J.; Kim, C.; Jayakumar, S.; Kim, S.; Han, S.; Kim, D.H.; Heo, J. Forest fire risk mapping of Kolli Hills, India, considering subjectivity and inconsistency issues. *Nat. Hazards* **2013**, *65*, 2129–2146. [CrossRef]
52. Mahdavi, A.; Shamsi, S.R.F.; Nazari, R. Forests and rangelands' wildfire risk zoning using GIS and AHP techniques. *Casp. J. Environ. Sci.* **2012**, *10*, 43–52.
53. Reynolds, K.M.; Holsten, E.H. Relative importance of risk factors for spruce beetle outbreaks. *Can. J. For. Res.* **1994**, *24*, 2089–2095. [CrossRef]
54. Vadrevu, K.P.; Eaturu, A.; Badarinath, K.V.S. Fire risk evaluation using multicriteria analysis—A case study. *Environ. Monit. Assess.* **2010**, *166*, 223–239. [CrossRef]
55. Jalilova, G.; Khadka, C.; Vacik, H. Developing criteria and indicators for evaluating sustainable forest management: A case study in Kyrgyzstan. *For. Policy Econ.* **2012**, *21*, 32–43. [CrossRef]
56. Kaya, T.; Kahraman, C. Fuzzy multiple criteria forestry decision making based on an integrated VIKOR and AHP approach. *Expert Syst. Appl.* **2011**, *38*, 7326–7333. [CrossRef]
57. Mendoza, G.A.; Prabhu, R. Multiple criteria decision making approaches to assessing forest sustainability using criteria and indicators: A case study. *For. Ecol. Manag.* **2000**, *131*, 107–126. [CrossRef]

58. Peterson, D.L.; Silsbee, D.G.; Schmoldt, D.L. A case study of resources management planning with multiple objectives and projects. *Environ. Manag.* **1994**, *18*, 729–742. [CrossRef]
59. Schmoldt, D.L.; Peterson, D.L. Analytical Group Decision Making in Natural Resources: Methodology and Application. *For. Sci.* **2000**, *46*, 62–75.
60. Segura, M.; Ray, D.; Maroto, C. Decision support systems for forest management: A comparative analysis and assessment. *Comput. Electron. Agric.* **2014**, *101*, 55–67. [CrossRef]
61. Schmoldt, D.L.; Peterson, D.L.; Silsbee, D.G. Developing inventory and monitoring programs based on multiple objectives. *Environ. Manag.* **1994**, *18*, 707–727. [CrossRef]
62. Akıncı, H.; Özalp, A.Y.; Turgut, B. Agricultural land use suitability analysis using GIS and AHP technique. *Comput. Electron. Agric.* **2013**, *97*, 71–82. [CrossRef]
63. Banai-Kashani, R. A new method for site suitability analysis: The analytic hierarchy process. *Environ. Manag.* **1989**, *13*, 685–693. [CrossRef]
64. Hutchinson, C.F.; Toledano, J. Guidelines for demonstrating geographical information systems based on participatory development. *Int. J. Geogr. Inf. Syst.* **1993**, *7*, 453–461. [CrossRef]
65. Pourebrahim, S.; Hadipour, M.; Mokhtar, M.B. Integration of spatial suitability analysis for land use planning in coastal areas; case of Kuala Langat District, Selangor, Malaysia. *Landsc. Urban Plan.* **2011**, *101*, 84–97. [CrossRef]
66. Xiang, W.-N.; Whitley, D.L. Weighting Land Suitability Factors by the PLUS Method. *Environ. Plan. B Plan. Des.* **1994**, *21*, 273–304. [CrossRef]
67. Sterling, E.J.; Hurley, M.M. Conserving Biodiversity in Vietnam: Applying Biogeography to Conservation Research. *Proc. Calif. Acad. Sci.* **2005**, *56*, 98–118.
68. Thao, L.B.; Phuong, N.Đ. *Vietnam: The Country and Its Geographical Regions*; The Gioi: Hanoi, Vietnam, 1997.
69. MONRE. *The Third National Communication of Vietnam to the United Nations Framework Convention on Climate Change*; Vietnam Publishing House of Natural Resources; Environment and Cartography: Hanoi, Vietnam, 2019; Volume 3, p. 125.
70. PARC. *Policy Summary: Building Protected Area System in Vietnam*; Project, P., Ed.; United Nations Development Programme: Hanoi, Vietnam, 2010; pp. 1–20.
71. BirdLife. Vietnam IBA Status. Available online: http://www.birdlife.org/datazone/country/vietnam/ibas (accessed on 1 June 2013).
72. MARD. *Decree 1588 on National Forest Status in 2020*; MARD: Hanoi, Vietnam, 2020.
73. VNFOREST. *Special-Use Forests in Vietnam*; Forestry, V.A.O., Ed.; NXB Nông nghiệp: Hanoi, Vietnam, 2021; p. 294.
74. Loc, P.; Yen, M.; Averyanov, L. Biodiversity in Vietnam. In *Global Biodiversity*; Apple Academic Press: Palm Bay, FL, USA, 2018; pp. 473–502.
75. Bruner, A.G.; Gullison, R.E.; Rice, R.E.; da Fonseca, G.A. Effectiveness of parks in protecting tropical biodiversity. *Science* **2001**, *291*, 125–128. [CrossRef]
76. MARD. *Vietnam's Modified Submission on Refreence Levels for REDD+ Results Based Payments under UNFCCC*; MARD: Hanoi, Vietnam, 2016; p. 85.
77. Hansen, M.C.; Potapov, P.V.; Moore, R.; Hancher, M.; Turubanova, S.A.; Tyukavina, A.; Thau, D.; Stehman, S.V.; Goetz, S.J.; Loveland, T.R. High-resolution global maps of 21st-century forest cover change. *Science* **2013**, *342*, 850–853. [CrossRef]
78. Ghehi, N.K.; MalekMohammadi, B.; Jafari, H. Integrating habitat risk assessment and connectivity analysis in ranking habitat patches for conservation in protected areas. *J. Nat. Conserv.* **2020**, *56*, 125867. [CrossRef]
79. Nguyen, T.T.; Nghiem, N. Optimal forest rotation for carbon sequestration and biodiversity conservation by farm income levels. *For. Policy Econ.* **2016**, *73*, 185–194. [CrossRef]
80. Dudley, N.; Parrish, J. *Closing the Gap: Creating Ecologically Representative Protected Area Systems*; Secretariat of the Convention on Biological Diversity: Montreal, Canada, 2006.
81. Joppa, L.N. Population change in and around Protected Areas. *J. Ecol. Anthropol.* **2012**, *15*, 58–64. [CrossRef]
82. Joppa, L.N.; Loarie, S.R.; Pimm, S.L. On population growth near protected areas. *PLoS ONE* **2009**, *4*, e4279. [CrossRef] [PubMed]
83. Luck, G.W. A review of the relationships between human population density and biodiversity. *Biol. Rev.* **2007**, *82*, 607–645. [CrossRef] [PubMed]
84. Naughton-Treves, L.; Holland, M.B.; Brandon, K. The Role of Protected Areas in Conserving Biodiversity and Sustaining Local Livelihoods. *Annu. Rev. Environ. Resour.* **2005**, *30*, 219–252. [CrossRef]
85. Thuy, P.T.; Moeliono, M.; Hien, N.T.; Tho, N.H.; Hien, V.T. *The Context of REDD+ in Vietnam: Drivers, Agents and Institutions*; CIFOR: Bogor Regency, Indonesia, 2012; Volume 75.
86. Vu, T.D.; Takeuchi, W.; Van, N.A. Carbon stock calculating and forest change assessment toward REDD+ activities for the mangrove forest in Vietnam. *Trans. Jpn. Soc. Aeronaut. Space Sci. Aerosp. Technol. Jpn.* **2014**, *12*, Pn_23–Pn_31. [CrossRef]
87. Yang, A.; Nguyen, D.T.; Vu, T.P.; Le Quang, T.; Pham, T.T.; Larson, A.M.; Ashwin, R. *Analyzing Multilevel Governance in Vietnam: Lessons for REDD+ from the Study of Land-Use Change and Benefit Sharing in Nghe An and Dien Bien Provinces*; JSTOR: New York, NY, USA, 2016.
88. Sam, H.V.; Tung, D.Q.; Ngoc, D.T.B. *Current Status Assessment of Vietnamese Protected Areas in International Context to Support Vietnam's National Forest Plan 2021–2030*; WWF: Hanoi, Vietnam, 2021; p. 63.
89. MONRE. *Biology Diversity and Conservation*; Vietnam Ministry of Natural Resources and Environment: Hanoi, Vietnam, 2004.

90. CEPF. *Indo-Burma Biodiversity Hotspot*; Critical Ecosystem Partnership Fund: Arlington, VA, USA, 2012; p. 360.
91. Csaplovics, E.; Wagenknecht, S.; Seiler, U. *Spatial Information Systems for Transnational Environmental Management of Protected Areas and Regions in the Central European Space Selected Results and Outputs of the Interreg IIIB Project SISTEMaPARC*; Rhombos-Verlag: Berlin, Germany, 2008.
92. Dudley, N. *Guidelines for Applying Protected Area Management Categories*; International Union for Conservation of Nature and Natural Resources (IUCN): Gland, Switzerland, 2008.
93. Leverington, F.; Lemos, K.; Pavese, H.; Lisle, A.; Hockings, M. A Global Analysis of Protected Area Management Effectiveness. *Environ. Manag.* **2010**, *56*, 685–698. [CrossRef]
94. Nhan, H.T.T.; Tinh, T.K.; Hung, P.V. *Thực Trạng Quản lý Khu Bảo tồn Thiên Nhiên tại Việt Nam*; Vietnam National University: Hanoi, Vietnam, 2015.
95. Watson, J.E.M.; Dudley, N.; Segan, D.B.; Hockings, M. The performance and potential of protected areas. *Nature* **2014**, *515*, 67–73. [CrossRef]
96. MONRE. *Vietnam National Biodiversity Strategy to 2020, Vision to 2030*; Vietnam Ministry of Natural Resources and Environment: Hanoi, Vietnam, 2015; p. 176.
97. Quang, H.D. *Những vấn đề quản lý Rừng đặc dụng ở Việt Nam*; Bidoup Nui Ba National Park: Lâm Đồng, Vietnam, 2011.
98. Rodrigues, A.S.L.; Andelman, S.J.; Bakarr, M.I.; Boitani, L.; Brooks, T.; Cowling, R.M.; Fishpool, L.D.C.; Da Fonseca, G.A.B.; Gaston, K.J.; Hoffmann, M.; et al. Effectiveness of the global protected area network in representing species diversity. *Nature* **2004**, *428*, 9–12. [CrossRef] [PubMed]
99. Salkind, N.J. *Encyclopedia of Research Design*; Sage: Newcastle upon Tyne, UK, 2010; Volume 1.
100. Hussein, A. The use of triangulation in social sciences research: Can qualitative and quantitative methods be combined. *J. Comp. Soc. Work.* **2009**, *1*, 1–12. [CrossRef]
101. Poole, K.T. *Spatial Models of Parliamentary Voting*; Cambridge University Press: Cambridge, UK, 2005; p. 252.
102. Poole, K.T.; Iii, S.M.; Grofman, B. A Unified Theory of Voting: Directional and Proximity Spatial Models. *Am. Political Sci. Rev.* **2000**, *94*, 953. [CrossRef]
103. Saaty, T.L.; Vargas, L.G. *The Logic of Priorities: Applications of the Analytic Hierarchy Process in Business, Energy, Health and Transportation*; RWS Publications: Pittsburgh, PA, USA, 1991; Volume 3.
104. Cochran, W.G. *Sampling Techniques*, 5th ed.; John Wiley & Sons: New York, NY, USA, 1977.
105. Krejcie, R.V.; Morgan, D.W. Determining Sample Size for Research Activities. *Educ. Psychol. Meas.* **1970**, *3*, 607–610. [CrossRef]
106. Bartlett, J.E.; Kotrlik, J.W.; Higgins, C.C. Organizational research: Determining appropriate sample size in survey research. *Inf. Technol. Learn. Perform. J.* **2001**, *19*, 43–50. [CrossRef]
107. Mandallaz, D. *Sampling Techniques for Forest Inventories*; CRC Press: Boca Raton, FL, USA, 2007; p. 256.
108. Naing, L.; Winn, T.; Rusli, B.N. Practical Issues in Calculating the Sample Size for Prevalence Studies. *Arch. Orofac. Sci.* **2006**, *1*, 9–14. [CrossRef]
109. Triantaphyllou, E.; Mann, S.H. Using the analytic hierarchy process for decision making in engineering applications: Some challenges. *Int. J. Ind. Eng. Appl. Pract.* **1995**, *2*, 35–44.
110. Öztürk, D.; Batuk, F. Analytic Hierarchy Process for Spatial Decision Making. *Sigma* **2010**, *28*, 124–137.
111. Kou, G.; Ergu, D.; Peng, Y.; Shi, Y. *Data Processing for the AHP/ANP*; Springer Science & Business Media: Berlin/Heidelberg, Germany, 2012; Volume 1.
112. Saaty, T.L. Some Mathematical Concepts of the Analytic Hierarchy Process. *Behaviormetrika* **1991**, *18*, 1–9. [CrossRef]
113. Cincotta, R.P.; Wisnewski, J.; Engelman, R. Human population in the biodiversity hotspots. *Nature* **2000**, *404*, 990–992. [CrossRef]
114. Sanderson, E.W.; Jaiteh, M.; Levy, M.A.; Redford, K.H.; Wannebo, A.V.; Woolmer, G. The human footprint and the last of the wild: The human footprint is a global map of human influence on the land surface, which suggests that human beings are stewards of nature, whether we like it or not. *BioScience* **2002**, *52*, 891–904. [CrossRef]
115. Shi, H.; Singh, A.; Kant, S.; Zhu, Z.; Waller, E. Integrating habitat status, human population pressure, and protection status into biodiversity conservation priority setting. *Conserv. Biol.* **2005**, *19*, 1273–1285. [CrossRef]
116. Kareiva, P.; Marvier, M. Conserving biodiversity coldspots: Recent calls to direct conservation funding to the world's biodiversity hotspots may be bad investment advice. *Am. Sci.* **2003**, *91*, 344–351. [CrossRef]
117. Van Khuc, Q.; Tran, B.Q.; Meyfroidt, P.; Paschke, M.W. Drivers of deforestation and forest degradation in Vietnam: An exploratory analysis at the national level. *For. Policy Econ.* **2018**, *90*, 128–141. [CrossRef]
118. Johnson, C.N.; Balmford, A.; Brook, B.W.; Buettel, J.C.; Galetti, M.; Guangchun, L.; Wilmshurst, J.M. Biodiversity losses and conservation responses in the Anthropocene. *Science* **2017**, *356*, 270–275. [CrossRef]
119. Maxim, L.; Spangenberg, J.H.; O'Connor, M. An analysis of risks for biodiversity under the DPSIR framework. *Ecol. Econ.* **2009**, *69*, 12–23. [CrossRef]
120. VNFOREST. *Special-Use and Protection Forest in Vietnam 2017–2018*; VNFOREST: Hanoi, Vietnam, 2019; p. 80.
121. Feddes, R.A.; Dam, J.C.V. *Unsaturated-Zone Modeling: Progress, Challenges and Applications*; Springer Science & Business Media: Berlin/Heidelberg, Germany, 2004; p. 364.
122. Franklin, S.E. *Remote Sensing for Biodiversity and Wildlife Management*; McGraw-Hill Education: New York, NY, USA, 2010.
123. Moore, P.D. *Tropical Forests*; Infobase Publishing: New York, NY, USA, 2008; p. 267.

124. Bowen, R.E.; Riley, C. Socio-economic indicators and integrated coastal management. *Ocean. Coast. Manag.* **2003**, *46*, 299–312. [CrossRef]

125. Giupponi, C.; Vladimirova, I. Ag-PIE: A GIS-based screening model for assessing agricultural pressures and impacts on water quality on a European scale. *Sci. Total Environ.* **2006**, *359*, 57–75. [CrossRef]

126. Rogers, S.I.; Greenaway, B. A UK perspective on the development of marine ecosystem indicators. *Mar. Pollut. Bull.* **2005**, *50*, 9–19. [CrossRef]

127. Fargione, J.; Tilman, D.; Dybzinski, R.; Lambers, J.H.R.; Clark, C.; Harpole, W.S.; Knops, J.M.H.; Reich, P.B.; Loreau, M. From selection to complementarity: Shifts in the causes of biodiversity–productivity relationships in a long-term biodiversity experiment. *Proc. R. Soc. Lond. B Biol. Sci.* **2007**, *274*, 871–876. [CrossRef]

128. Gould, W. Remote sensing of vegetation, plant species richness, and regional biodiversity hotspots. *Ecol. Appl.* **2000**, *10*, 1861–1870. [CrossRef]

129. Griffiths, G.H.; Lee, J.; Eversham, B.C. Landscape pattern and species richness: Regional scale analysis from remote sensing. *Int. J. Remote Sens.* **2000**, *21*, 2685–2704. [CrossRef]

130. Kerr, J.T.; Southwood, T.R.E.; Cihlar, J. Remotely sensed habitat diversity predicts butterfly species richness and community similarity in Canada. *Proc. Natl. Acad. Sci. USA* **2001**, *98*, 11365–11370. [CrossRef] [PubMed]

131. Kuusipalo, J. Diversity pattern of the forest understorey vegetation in relation to some site characteristics. *Silva Fenn* **1984**. [CrossRef]

132. Myers, N. Threatened biotas: "Hot spots" in tropical forests. *Environmentalist* **1988**, *8*, 187–208. [CrossRef]

133. Oindo, B.O.; Skidmore, A.K. Interannual variability of NDVI and species richness in Kenya. *Int. J. Remote Sens.* **2002**, *23*, 285–298. [CrossRef]

134. Phillips, L.B.; Hansen, A.J.; Flather, C.H. Evaluating the species energy relationship with the newest measures of ecosystem energy: NDVI versus MODIS primary production. *Remote Sens. Environ.* **2008**, *112*, 4381–4392. [CrossRef]

135. Reich, P.B.; Tilman, D.; Isbell, F.; Mueller, K.; Hobbie, S.E.; Flynn, D.F.B.; Eisenhauer, N. Impacts of biodiversity loss escalate through time as redundancy fades. *Science* **2012**, *336*, 589–592. [CrossRef]

136. Rocchini, D.; Perry, G.L.W.; Salerno, M.; Maccherini, S.; Chiarucci, A. Landscape change and the dynamics of open formations in a natural reserve. *Landsc. Urban Plan.* **2006**, *77*, 167–177. [CrossRef]

137. Tilman, D. *Resource Competition and Community Structure*; Princeton University Press: Princeton, NJ, USA, 1982.

138. Wang, R.; Gamon, J.A.; Montgomery, R.A.; Townsend, P.A.; Zygielbaum, A.I.; Bitan, K.; Tilman, D.; Cavender-Bares, J. Seasonal variation in the NDVI-species richness relationship in a prairie grassland experiment (cedar creek). *Remote Sens.* **2016**, *8*, 128. [CrossRef]

139. Wickham, J.D.; Wu, J.; Bradford, D.F. A conceptual framework for selecting and analyzing stressor data to study species richness at large spatial scales. *Environ. Manag.* **1997**, *21*, 247–257. [CrossRef] [PubMed]

140. Smeets, E.; Weterings, R. *Environmental Indicators: Typology and Overview*; European Environment Agency: Copenhagen, Denmark, 1999; Volume 19.

141. Cuesta, F.; Peralvo, M.; Merino-Viteri, A.; Bustamante, M.; Baquero, F.; Freile, J.F.; Muriel, P.; Torres-Carvajal, O. Priority areas for biodiversity conservation in mainland Ecuador. *Neotrop. Biodivers.* **2017**, *3*, 93–106. [CrossRef]

142. Sarkar, S.; Pressey, R.L.; Faith, D.P.; Margules, C.R.; Fuller, T.; Stoms, D.M.; Moffett, A.; Wilson, K.A.; Williams, K.J.; Williams, P.H. Biodiversity conservation planning tools: Present status and challenges for the future. *Annu. Rev. Environ. Resour.* **2006**, *31*, 123–159. [CrossRef]

143. Moilanen, A.; Wilson, K.; Possingham, H. *Spatial Conservation Prioritization: Quantitative Methods and Computational Tools*; Oxford University Press: Oxford, UK, 2009.

144. Brooks, T.M.; Mittermeier, R.A.; Da Fonseca, G.A.; Gerlach, J.; Hoffmann, M.; Lamoreux, J.F.; Mittermeier, C.G.; Pilgrim, J.D.; Rodrigues, A.S. Global biodiversity conservation priorities. *Science* **2006**, *313*, 58–61. [CrossRef]

145. Costello, C.; Polasky, S. Dynamic reserve site selection. *Resour. Energy Econ.* **2004**, *26*, 157–174. [CrossRef]

146. Asaad, I.; Lundquist, C.J.; Erdmann, M.V.; Costello, M.J. Ecological criteria to identify areas for biodiversity conservation. *Biol. Conserv.* **2017**, *213*, 309–316. [CrossRef]

147. Bonn, A.; Gaston, K.J. Capturing biodiversity: Selecting priority areas for conservation using different criteria. *Biodivers. Conserv.* **2005**, *14*, 1083–1100. [CrossRef]

148. Cabeza, M.; Moilanen, A. Design of reserve networks and the persistence of biodiversity. *Trends Ecol. Evol.* **2001**, *16*, 242–248. [CrossRef]

149. Lessmann, J.; Munoz, J.; Bonaccorso, E. Maximizing species conservation in continental E cuador: A case of systematic conservation planning for biodiverse regions. *Ecol. Evol.* **2014**, *4*, 2410–2422. [CrossRef]

150. Sierra, R.; Campos, F.; Chamberlin, J. Assessing biodiversity conservation priorities: Ecosystem risk and representativeness in continental Ecuador. *Landsc. Urban Plan.* **2002**, *59*, 95–110. [CrossRef]

151. Asaad, I.; Lundquist, C.J.; Erdmann, M.V.; Costello, M.J. Delineating priority areas for marine biodiversity conservation in the Coral Triangle. *Biol. Conserv.* **2018**, *222*, 198–211. [CrossRef]

Article

Assessing Forest Biodiversity: A Novel Index to Consider Ecosystem, Species, and Genetic Diversity

Jana-Sophie Ette *, Markus Sallmannshofer and Thomas Geburek

Austrian Research Centre for Forests, Department for Forest Growth, Silviculture, and Genetics, Seckendorff-Gudent Weg 8, 1130 Vienna, Austria
* Correspondence: sophie.ette@yahoo.de

Abstract: Rates of biodiversity loss remain high, threatening the life support system upon which all human life depends. In a case study, a novel biodiversity composite index (BCI) in line with the Convention on Biological Diversity is established in Tyrol, Austria, based on available national forest inventory and forest typing data. Indicators are referenced by ecological modeling, protected areas, and unmanaged forests using a machine learning approach. Our case study displays an average biodiversity rating of 57% out of 100% for Tyrolean forests. The respective rating for ecosystem diversity is 49%; for genetic diversity, 53%; and for species diversity, 71%. Coniferous forest types are in a more favorable state of preservation than deciduous and mixed forests. The BCI approach is transferable to Central European areas with forest typing. Our objective is to support the conservation of biodiversity and provide guidance to regional forest policy. BCI is useful to set restoration priorities, reach conservation targets, raise effectiveness of financial resources spent on biodiversity conservation, and enhance Sustainable Forest Management.

Keywords: convention on biological diversity; national forest inventory; dynamic forest typing; machine learning; sustainable forest management; temperate forests

Citation: Ette, J.-S.; Sallmannshofer, M.; Geburek, T. Assessing Forest Biodiversity: A Novel Index to Consider Ecosystem, Species, and Genetic Diversity. *Forests* **2023**, *14*, 709. https://doi.org/10.3390/f14040709

Academic Editors: Konstantinos Poirazidis and Panteleimon Xofis

Received: 27 December 2022
Revised: 26 February 2023
Accepted: 2 March 2023
Published: 30 March 2023

1. Introduction

Biodiversity loss is one of the greatest ecological challenges of our time [1] Biodiversity plays a crucial role in biological processes, provision of ecosystem services, and stability of forest ecosystems [2–5]. With current rates of biodiversity loss [6,7], forest multifunctionality and productivity are decreasing at an accelerating rate [8].

Evaluating biodiversity is a highly complex task [9,10]. Additionally, biodiversity indicators are still criticized for poor indicator–indicandum relationships [11–14]. Following Heink and Kowarik [15], an *indicator* is of major relevance for a given issue, e.g., assessment of a specific impact for conservation policy (tree diameter and age classes), while an *indicandum* is the indicated phenomenon (old-growth forests). Although the relationship to the indicandum may not be fully understood yet, we will refer to these metrics as "indicators" in the following.

Due to weak correlations with the indicandum, *indicator species concepts* have not been successful [12,13], while concepts for forest genetic monitoring are missing in Europe [16]. Policymakers, forest managers, and scientists are facing severe knowledge gaps while having to decide which and how to choose and aggregate biodiversity indicators [17–20] as well as defining baselines.

Structures, processes, and taxonomic groups are currently used as *ecological indicators* [15]. Our study applies metrics of structural diversity relevant to forest biodiversity based on scientific evidence. *Structural diversity concepts* indicate potential habitat quality, niche differentiation, structural complexity [7], and other sources of forest biodiversity [18], e.g., for umbrella species [21] and bird species [22]. There is broad scientific evidence for positive relationships between measures of forest structural variety and elements of biodiversity [23–25].

On large spatial and temporal scales, the availability of reliable data sets is a limiting factor for biodiversity assessments and monitoring [9,10]. Without sound biodiversity monitoring and reporting systems, natural resources get overexploited or marginalized in decision-making [26]. Gaps in biodiversity monitoring may contribute to the lack of success in biodiversity policy implementation [16]. This may be one of the main reasons why, despite international conventions and large financial efforts [27], current rates of biodiversity loss remain high, threatening the life support system upon which all human life depends [28].

There are three biodiversity indicator sets internationally accepted, developed by the European Environment Agency, Biodiversity Indicators Partnership, and Ministerial Conference on the Protection of Forests in Europe. All of them cannot be used to judge, compare, or predict consequences of forest management for forest biodiversity at the regional level.

Understanding ecological impacts of forest management practices on biodiversity and associated ecosystem processes is essential for developing Sustainable Forest Management approaches [29,30]. Some forest ecosystem services can work in synergy whereas others, such as biodiversity and intensive timber production, are hardly compatible [31]. This policy–policy conflict is one of the most acknowledged trade-offs related to forest management [32–34]. Sustainable Forest Management is characterized by taking consequences of operational decisions for biodiversity into consideration [35], which is very difficult to achieve for forest enterprises. Unambiguous and practical concepts to define and measure forest biodiversity relevant to scale and purpose are needed [36,37].

Selecting appropriate indicators is particularly challenging using forest inventory data which originally were designed for forest resource management purposes [18]. Main impacts of forest management on forest biodiversity are changes in forest structure, species composition [38,39], and forest genetic resources. It is therefore reasonable to monitor changes in these determinants [40,41].

Large-scale forest inventories have rarely been used for biodiversity assessments [42,43]. However, forest inventories proved their potential to overcome data deficits on large spatial and temporal scales [21,25,41,44,45]. Major advantages of inventory-based biodiversity assessments are the repeated measurements which detect temporal changes [10] at low additional costs [45,46] for a high number of attributes, forest types, sample sizes, and scales [10,41]. In the long term, changes in biodiversity may even be related to forest management [41] and forest policy measures, which makes it highly reasonable to choose biodiversity indicators based on existing forest inventory data. Forest typing models cannot solely be used for tree species selection under various climate warming scenarios. An Austrian case study demonstrates the great potential of forest typing models and machine learning for conservation planning and policy guidance.

In this study, a novel biodiversity composite index (BCI) to assess forest biodiversity of the federal autonomous province of Tyrol, Austria, is presented. BCI was created in the Interreg-project "BioΔ4" and was designed to be transferable to neighboring Central European regions in, e.g., Austria, Italy, and Germany. The basic assumption of BCI is that forests of high naturalness can maintain biodiversity best on large temporal and spatial scales. BCI targets heterogeneity and levels of diversity evolving naturally (or nature identical) at a forest stand to conserve overall forest biodiversity on a landscape scale. BCI logic structure is in line with the Convention on Biological Diversity (CBD). It follows the Convention's internationally accepted definition of biodiversity, stating that "biodiversity is the variability among living organisms from all sources including, inter alia, terrestrial, marine and other aquatic ecosystems and the ecological complexes of which they are part; this includes diversity within species, between species, and of ecosystems" (CBD, 1992). In line with the CBD, we define ecosystems as "a dynamic complex of plant, animal and micro-organism communities and their non-living environment interacting as a functional unit." In our case study, BCI "ecosystem diversity" indicates the variations in forest ecosystems within the geographical location Tyrol.

Our objective was to create a stand-scale biodiversity index assembled from indicators which (1) are based on available data sets, (2) are based on high scientific evidence relevant to biodiversity, and (3) equally consider ecosystem, species, and genetic diversity. The new approach can be repeated cost-efficiently in each forest inventory period.

BCI provides quantitative aggregation and simplification of ecological information which can help policy makers to implement biodiversity policies and distribute conservation funding, e.g., for ecosystem restoration. Ranking of forest types on four levels and high-resolution spatial maps of forest diversity with BCI can support decision-making in biodiversity conservation (e.g., target forest types, target regions, ecosystems, species or genetic restoration, conservation priorities, etc.) and evaluate effectiveness of financial resources spent on ecosystem restoration and Sustainable Forest Management (e.g., cost-benefit-analysis). As a minimum requirement, we recommend a future positive BCI trend on all levels as a quantitative goal for regional to national forest policy in order to halt the loss of biological diversity and meet strategic CBD targets.

2. Material

2.1. Forests of Tyrol

Tyrol has a size of 12,684 km^2 and is located in the Eastern Alps. The territory is separated into two parts, namely North Tyrol and East Tyrol (Figure 1). It ranges from 500 to 3800 m above sea level and shows an inner alpine mountainous climate with subcontinental traits. The 520,000 ha of alpine coniferous forests are characterized by dense vegetation in combination with cold climate leading to acidic, thick organic soil horizons [47]. Total stock levels are about 114 M. m^3 (328 m^3/ha) with annual growth rates of 2.2 M. m^3 [48]. With 57.6% tree species abundance, Norway spruce *(Picea abies)* is predominant [48].

Figure 1. Study area. Maps of the study area Tyrol, which is located in Austria (47°41′47.30″ N, 13°20′44.64″ E), Central Europe. The map was taken from https://geology.com/world/austria-satellite-image.shtml (accessed 6 March 2023).

Designated protective forests (e.g., forests protecting infrastructure and settlements from natural hazards) can be found on 48% of the total forest area [48]. Forest regeneration deficits in Tyrolean protective forests have repeatedly been reported [48,49]. Severe game impact on forest regeneration can be found on 57% of the forest area [48]. Dead wood levels account for 10.8% of the living stand volume [48].

2.2. Data Sets

This case study combines field-based measures and lidar-derived approaches using data sets provided by the Austrian Research Centre for Forests (AFI (Austrian Forest Inventory), AUPICMAP study (Geographic-genetic map of the Austrian Norway spruce population), Austrian Planting Statistics, and Nature Forest Reserves), and by the Tyrolean Regional Government (Forest typing project, vegetation surveys, TIRIS (Tyrolean Spatial Information System)). Data processing is done in R, QGIS, and python (Table 1). Reference values for the dead wood levels are supplied by protected areas, e.g., the National parks "Hohe Tauern" and "Berchtesgaden". For other biodiversity indicators, reference values can be found in earlier scientific studies [48,50–53].

Table 1. Data provision and processing. Twelve biodiversity indicators are established based on data sets provided by the Austrian Research Centre for Forests, the Tyrolean Regional Government, and national park managements.

Indicator	Method	Data Set	Reference Data	Processing
Tree species diversity	Grabherr et al. (1998) [50]	AFI	Forest typing	R
Ground vegetation	Grabherr et al. (1998) [50]	Vegetation surveys	Grabherr et al. (1998) [50]	R
Surface soil quality	case study Tyrol	AFI	Hotter et al. (2013) [53]	R
Game impact	case study Tyrol	AFI	-	R
Tree layer structure	Grabherr et al. (1998) [50]	AFI	Grabherr et al. (1998) [50]	R
Developmental level	case study Tyrol	AFI	Grabherr et al. (1998) [50]	R
Dead wood	case study Tyrol, Grabherr et al. (1998) [50]	AFI	AFI, protected area management	R, QGIS
Structural features	case study Tyrol, Grabherr et al. (1998) [50]	AFI	Grabherr et al. (1998) [50]	R
Forest gap structure	case study Tyrol	TIRIS	Grabherr et al. (1998) [50]	QGIS
Autochthony	Geburek and Schweinzer (2012) [52]	AFI	AUPICMAP	Python, QGIS
Management constraints	case study Tyrol	TIRIS	Raab et al. (2002) [51]	QGIS
Genetic features	case study Tyrol	AFI, Austrian Planting Statistics	-	R

AFI = Austrian Forest Inventory. TIRIS = Tyrolean Spatial Information System. AUPICMAP = Geographic-genetic map of the Austrian Norway spruce population.

Biodiversity assessment is performed on 1162 Austrian Forest Inventory subplots. The AFI uses a permanent foursome grid sampling with a grid size of 3.89 km (1 AFI plot ≙ 4 AFI sub plots). Biodiversity indicators are assessed on the AFI subplot level. A detailed AFI field sampling manual, calculation methods, and theoretical background can be found in Hauk and Schadauer [54]. High-resolution forest typing of Tyrol based on ecological modeling was performed in 2019. Considering terrain models, geological models, climate models, expert knowledge, and field data [53], ecological modeling demarcates forest types on small scales (Figure 2A).

2.3. Assignment of AFI Plots to Forest Typing

Firstly, forest typing data is spatially overlaid with TIRIS, AFI, AUPICMAP, and reference area data in QGIS version 3.16 LTR (Figure 2B). Secondly, all AFI subplots outside the forest typing objects are excluded from analysis. Thirdly, if AFI subplots lay outside of the forest typing objects but contain field data; they are assigned manually to the forest type with the closest air-line distance by photo referencing (Figure 3). Biodiversity assessment of Tyrol is based on 1162 AFI subplots and 1521 vegetation survey plots. Forest inventory data and vegetation surveys are assigned to 82 forest types [53] and 223,628 QGIS objects. A total of 347 AFI subplots were excluded in the case study.

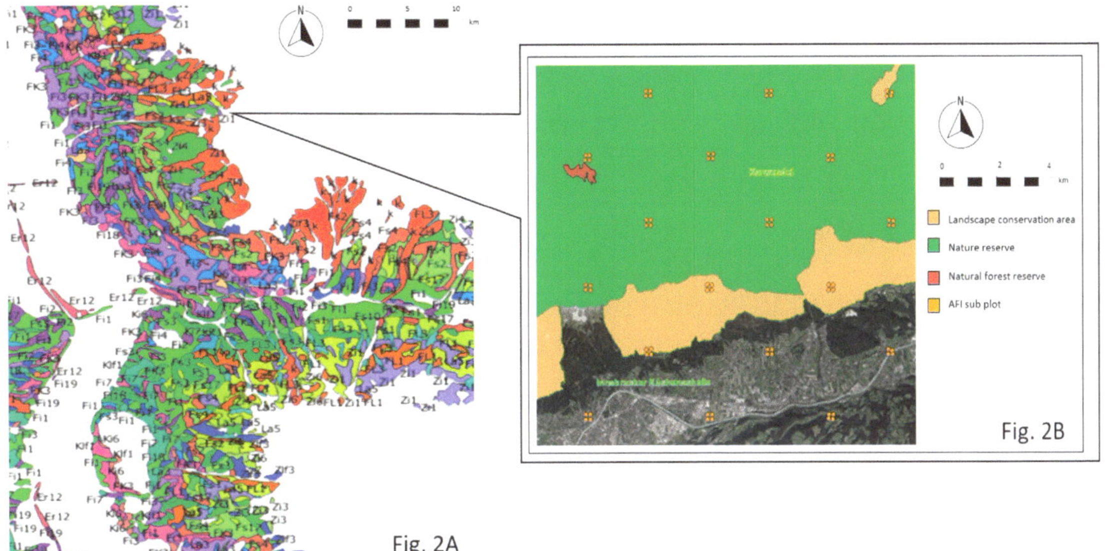

Figure 2. Processing inventory data with QGIS (47°24′59.99″ N, 11°27′59.99″ E). (**A**) Excerpt of the QGIS forest typing [53]. Different colors and abbreviations indicate the forest types. (**B**) Forest typing data (e.g., beige and green objects) is overlaid with TIRIS (e.g., orthophoto), AFI (orange dots), and reference area (red polygon) data.

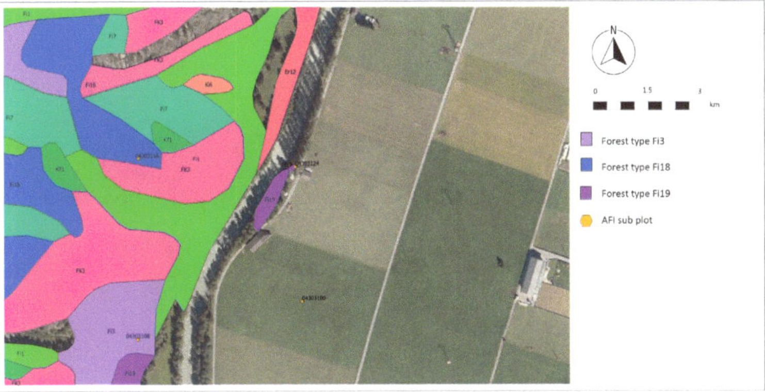

Figure 3. Assigning AFI plots to forest types (47°24′59.99″ N, 1°27′59.99″ E). AFI plots (orange dots, numerical codes) are assigned to forest types (colored areas, alpha-numerical codes) in QGIS by position (e.g., 04303108 to Fi3) or photo referencing (e.g., 04303124 to Fi19). AFI subplots without AFI data located outside of forest type objects are excluded from analysis (e.g., 04303100).

3. Methods

Following McElhinny et al. [55], we collected all data sets available for Tyrol and the neighboring countries, quantified all stand attributes, identified a logical structure, defined a set of indicators according to the CBD definition of biodiversity, and combined these attributes into an additive biodiversity index.

Assessments of BCI can be done using one out of four levels, namely species, ecosystem, genetic, and biodiversity. In line with Grabherr et al. [50], indicators can assume ratings between zero (lowest) and 9.0 (highest) points. Following McElhinny et al. [55], outcomes are expressed as percentage (0%–100%) to ease interpretation.

On the one hand, rare but ecologically highly valuable traits *(bonus indicators)* may compensate for a lower level of common forest traits *(biodiversity indicators)*. On the other hand, missing but rarely occurring forest traits are not rated disadvantageous and BCI does not benchmark against a particular scale of temporal variation [56]. Among available data sets, we favored quantitative and high-resolution measurements of high scientific value and large temporal scales in the choice between biodiversity and bonus indicators (e.g., "management constraints" is a biodiversity indicator, "planting intensity" is a bonus indicator).

3.1. Ecosystem Diversity

Ecosystem, species, and genetic diversity assessment considers three biodiversity indicators and one bonus indicator, respectively. If at least two out of four (>50%) indicators are rated, AFI subplots are included. The indication of 100% is always adapted to the maximum number of points possible under the current number of indicators at the AFI subplot (e.g., 100% = 27.0 points if *three biodiversity indicators* could be rated with up to 9.0 points; or 100% = 18.0 points, if *two biodiversity indicators* could be rated with up to 9.0 points).

$$ecosystem\,diversity = BI_{layer} + BI_{devel.\ level} + BI_{deadwood} + Bonus_{structure}$$

BI_{layer} assesses the deviation of the actual tree layer structure (AFI) from an expected, site-specific layer structure (forest typing). $BI_{devel.level}$ rewards differentiation of successional stages on small scales and late forest successional phases. $BI_{deadwood}$ considers dead wood quantity ($DW_{quantity}$) and quality ($DW_{quality}$). Dead wood quantity is assessed by comparing actual quantities (AFI) to reference values in protected areas and within the AFI data set (Figure 4A,B). $Bonus_{structure}$ rewards shrub layers established naturally in certain forest types, late stand ages, and large tree diameter breast heights.

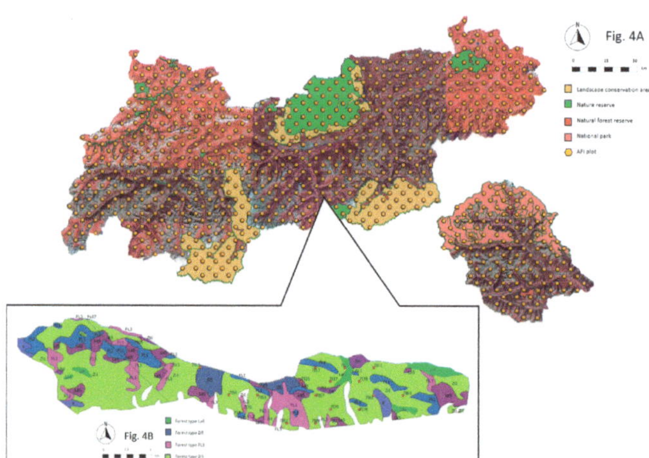

Figure 4. Dead wood reference levels in protected areas (47°15′13.468″ N, 11°36′5.353″ E). (**A**) AFI plots (orange dots) within protected areas (green, beige, and salmon-colored polygons) and protected area inventories are used as reference levels for dead wood quantity $\left(DW_{quantity}\right)$. (**B**) In addition, nature forest reserves and protected area inventories are surveyed. Within a forest type, the respectively highest dead wood quantity out of all inventory data sets is compared to the actual subplot level.

3.2. Species Diversity

$BI_{treespecies}$ is based on a target-performance comparison between actual (AFI) and potential (forest typing data) tree species composition. $BI_{vegetation}$ evaluates the naturalness of species composition of the ground vegetation and their ecosystem disturbance indicating value. BI_{soil} assesses if the actual humus form (AFI) deviates from the expected ones (forest typing). $Bonus_{game}$ rewards extensive game impact on forest regeneration.

$$speciesdiversity = BI_{treespecies} + BI_{vegetation} + BI_{soil} + Bonus_{game}$$

3.3. Genetic Diversity

BI_{gap} characterizes forest gap structure by calculating a surface balance between forest and non-forest area (Figure 5). $BI_{autochthony}$ evaluates genetic diversity of the predominant tree species, Norway spruce, by computing intraspecific haplotype distance to reference populations. $BI_{management}$ considers inclination and distance to forest road systems of a forest site to estimate probability of extensive forest management. $Bonus_{regeneration}$ evaluates the probability of tree species to contain a native gene pool by examining their planting intensity. $Bonus_{phenology}$ uses varying branching types of Norway spruce as a proxy for detecting genetically allochthonous plant material.

$$geneticdiversit = BI_{gap} + BI_{autochthony} + BI_{management} + Bonus_{genetic}$$

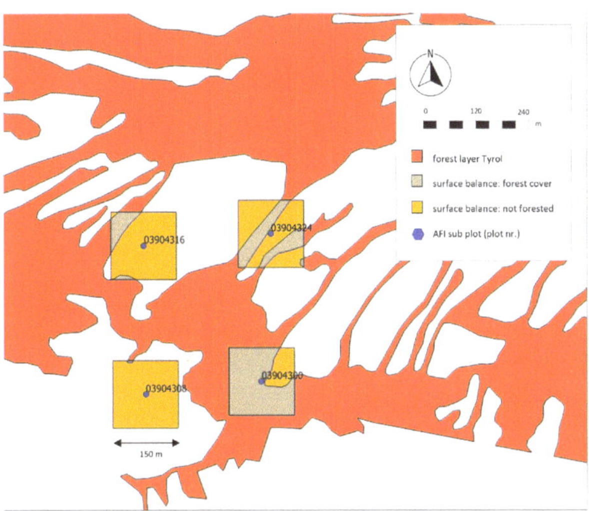

Figure 5. Assessment of surface balance (47°15′34.7724″ N, 11°24′1.3500″ E). Squares with side length 150 m are used to compute surface balance around AFI subplots (e.g., plot nr. 03904308) between forest (light orange polygons) and non-forest area (grey polygons) in QGIS.

3.4. Biodiversity

BCI considers nine biodiversity indicators and three bonus indicators (Figure 6). Indicator ratings (0–9.0 points) are aggregated on the AFI subplot level. BCI is computed by addition of indicators and levels of diversity without weighting, which makes the concept transferable and easy to adapt. If at least six out of twelve (>50%) indicators are rated, the AFI subplots are included. Rare forest types containing less than three AFI subplots ($n = 14$) are not considered in BCI assessment.

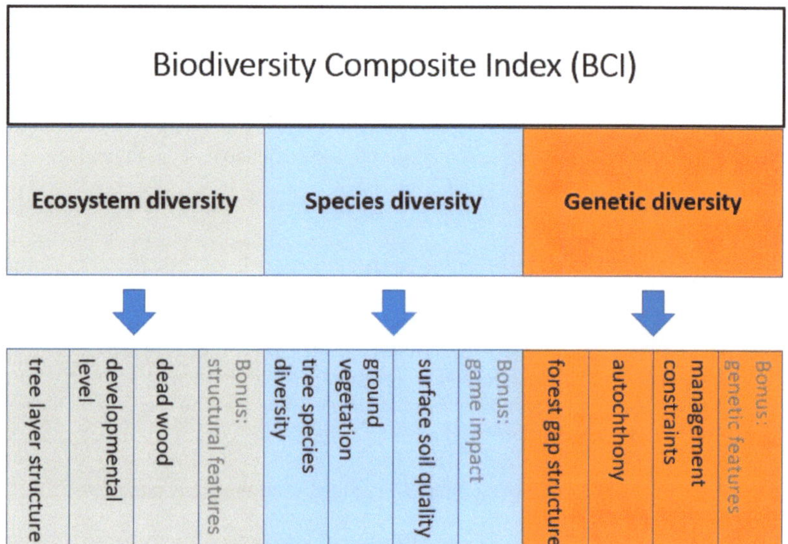

Figure 6. BCI index structure. The modular structure equally considers ecosystem, species, and genetic diversity with three biodiversity indicators (black font color) and one bonus indicator (gray font color), respectively. Indicators are aggregated without weighting. Abbreviations are listed in the Appendix A (Table A1).

From a methodological point of view, BCI can be seen as an enhancement of the studies of Grabherr et al. [50] and Geburek et al. [57]. The framework follows classic niche theories [58–60] which explain co-existence of species with unique species traits and ecological niches varying in space and time. Consequently, species cannot be interchanged easily in a community. In the sense of Whittaker [61], BCI targets high beta-diversity levels to conserve overall forest biodiversity.

The choice of indicators relevant to biodiversity needs to be legitimated [15]. Scientific evidence for the relation between the diversity metric (indicator) and indicandum is provided in the Appendix A (Table A2), and detailed description of indicator evaluation can be found in the Electronic Supplementary Material. Following Virkkala [62], Brin et al. [63], and Gao et al. [14], indicators are selected from data sets available according to logical inference and by referring to other studies of high statistical validity.

3.5. Predictive Modeling with R Randomforest

In line with Bitterlich [64], Lappi and Bailey [65], and Sterba [66], evaluation outcomes are aggregated on the stratum level. With the help of the training data set (AFI subplots; n = 1162 data points with BCI ratings) and machine learning, R randomForest predicts biodiversity levels of 223,628 forest patches (QGIS polygons).

We applied the bagging classification algorithm randomForest in R, which is a group of regression trees made from random selection of samples of the training data [67]. Every random forest in this study is composed of 500 regression trees. For every regression tree in the forest, a training set is drawn from the sample plots, using bootstrap aggregating (bagging). The decision tree is built by rule-based splitting of the bagging sample into subsets, maximizing the variance between the subsets [68]. At each split in the learning process, a random subset of impact variables is used [69]. The splitting process is repeated recursively on each derived subset until (i) the subset has identical values with the target variable or (ii) the splitting no longer adds value to the prediction [70]. The mean value of the target variable within a final subset (leaf of the decision tree) is used as the conditional prediction of the target variable for a corresponding combination of impact variables [68].

For the application of a high-precision data-mining machine learning algorithm, we created polygon centroids of all forest type areas in QGIS as a prediction data set (Figure 7A). Predictive model performance is improved by adding the variables forest type (forest typing), altitude (Copernicus V1.1 DEM), geographical coordinates (TIRIS), and forest type groups (Appendix A Table A3) to the training and prediction data set. Model fit is controlled by additionally repredicting the training set and comparing prediction with R randomForest training data. The standard deviance between training and prediction data is 0%–19%. High deviances of 19% occur seldom in case extraordinary low values in the training set are repredicted.

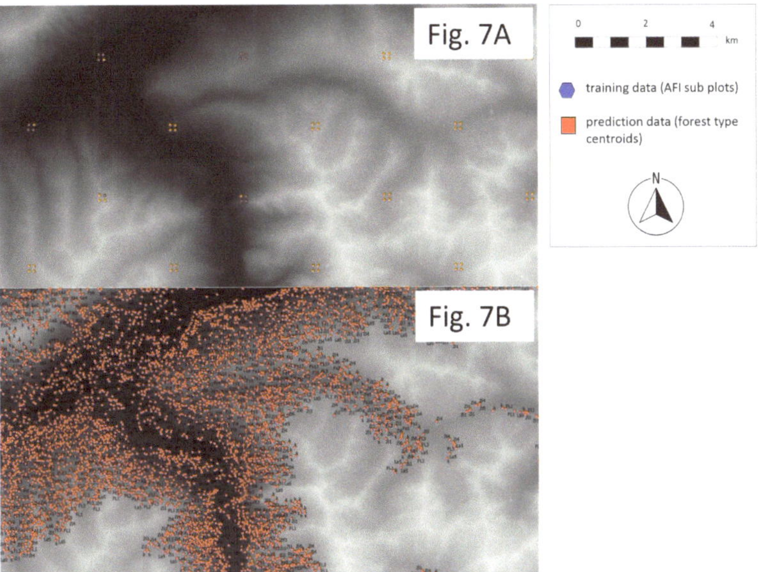

Figure 7. The training and prediction data set in R randomForest (47°15′34.7724″ N, 11°24′1.3500″ E). (**A**) The training set of AFI subplots. (**B**) The prediction set of forest typing polygon centroids for modeling ecosystem, species, genetic diversity, and biodiversity with R randomForest.

In the next step, we assigned centroid values of the prediction set to the polygons to create maps of Tyrol (Figure 7B). Overall, we applied four prediction models, as BCI indicators can be aggregated on the level of species, ecosystem, genetic, and biodiversity. For our case study, we considered the area-weighted mean of the forest area objects in QGIS in high resolution.

To illustrate the prediction outcomes, forest area coloring was done in five classes (0%–20%, 20%–40%, 40%–60%, 60%–80%, 80%–100%) in QGIS. For the additional creation of spatial. jpg maps of Tyrolean forest diversity for the Tyrolean Regional Government, we applied cube spline interpolation in SAGA GIS. Before running the final models, we tested the model approach several times, performing probability checks using solely data of the smallest political district of Tyrol ('Innsbruck', forest area 37 km^2).

4. Results

BCI spatial area assessments can be interpreted on the level of diversity of species and ecosystem, genetic diversity, and biodiversity. Our study displays an average biodiversity rating of 57% (area-weighted mean of forest area) for Tyrol. The respective rating of ecosystem diversity is 49%; of genetic diversity, 53%; and of species, 71% (Figure 8).

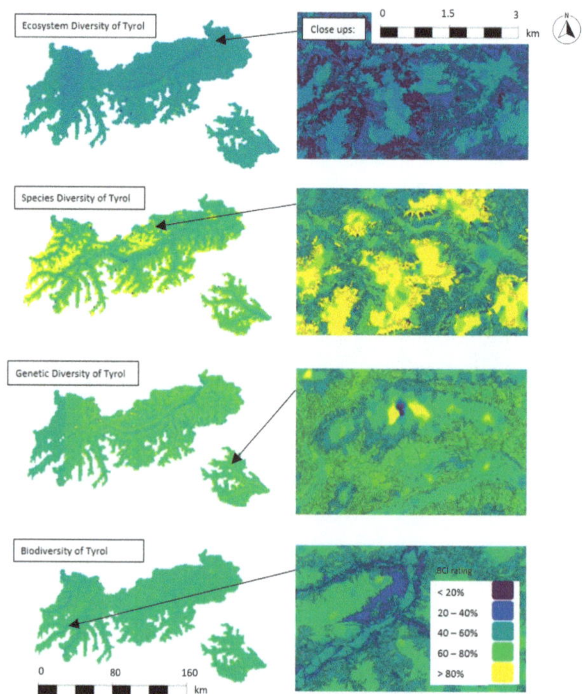

Figure 8. Maps of forest ecosystem, species, genetics, and biodiversity (47°15′13.468″ N, 11°36′5.353″ E). High-resolution maps of forest ecosystem, species, genetics, and biodiversity of Tyrolean forests. Outcomes are displayed in five classes (0%–20%, 21%–40%, 41%–60%, 61%–80%, 81%–100%).

BCI outcomes display high spatial heterogeneity on small scale. The divergence between valleys (e.g., "Inntal") and higher alpine areas in all models is evident through their darker coloring (i.e., lower BCI ratings). This effect is most pronounced for species diversity (Figure 8, left side).

Indicators with high average ratings in the biodiversity assessment are "autochthony" (94%), "tree layer structure" (94%), and "game impact" (91%). Indicators with low average ratings are "structural features" (19%), "management constraints" (36%), and "tree species diversity" (46%).

Indicators available most frequently at the 1507 AFI subplots studied are "forest gap structure" (1507 AFI sub plots), "forest vegetation" (1454 plots), and "structural features" (1004 plots). Indicators with low availability are "autochthony" (520 plots), "surface soil quality" (836 plots), and "tree layer structure" (943 plots).

High-altitude areas received higher BCI ratings than low elevation areas, which is in line with forest type evaluation. Surveying model outcomes on the level of the forest type (Table 2), it can be concluded that coniferous forests in Tyrol are in a more favorable state of preservation and can maintain biodiversity with higher probability than broad-leaved and mixed forests.

Table 2. BCI evaluation outcomes on the forest type level.

Forest Type	Code	Species Diversity [%]	Ecosystem Diversity [%]	Genetic Diversity [%]	Biodiversity (BCI) [%]
Subalpine dry silicate larch-spruce forest	Fs4	77	65	69	70
Overlying humus-carbonate larch-stone pine forest	Zi2	66		87	69

Table 2. *Cont.*

Forest Type	Code	Species Diversity [%]	Ecosystem Diversity [%]	Genetic Diversity [%]	Biodiversity (BCI) [%]
Subalpine coniferous avalanche sites	FL3	100	48	62	69
Subalpine cool silicate steep slope (green alder-larch) spruce forest	Fs3	79	57	74	68
Cool silicate steep slope spruce-larch fir forest	FT12	77	54	70	67
Montane sunny rock sites on carbonate	FK2	100	63	67	65
Montane dry silicate (pine) spruce forest	Fi4	65	59	73	65
Warm silicate larch-stone pine forest	Zi4	81		56	65
Poor silicate larch stone pine forest	Zi1	80		57	64
Subalpine basic larch-spruce forest	Fs5	84	55	57	64
Dry carbonate pine forest	Ki1	86	48	62	64
Subalpine poor silicate (larch) spruce forest	Fs1	82	48	61	63
Marl steep slope spruce-fir-beech forest	Ftb20	81	52	59	63
Montane poor silicate (larch) spruce forest	Fi2	79	52	59	63
Mountain pine, green alder, hardwood scrub, scrubby areas	k	100	38	54	63
Montane poor carbonate larch-spruce forest	Fi6	81	50	57	61
Lawinar silicate (green alder) larch-spruce forest	Fs10	83	50	51	61
Moderately dry carbonate pine-spruce-beech forest	Fkb1	71	48	67	61
Cool carbonate steep slope larch-pine forest	Ki18	72	43	68	61
Floodplain sites of the montane level	Er12	93	40	49	61
Fresh silicate fir-spruce forest of the intermediate Alps	FT10	79	46	55	60
Subalpine fresh silicate spruce forest	Fs17	82	47	51	60
Fresh alkaline spruce-fir forest	FT9	84	44	49	60
Moderately fresh silicate fir-spruce forest	Fi22	65	60	50	59
Subalpine warm silicate larch-spruce forest	Fs2	82	45	53	58
Poor silicate spruce-fir forest	FT2	71	44	63	58
Montane warm carbonate spruce forest	Fi8	56	51	63	58
Montane fresh basic spruce forest	Fi5	58	50	50	58
Montane fresh silicate (larch) spruce forest	Fi1	69	47	53	58
Montane warm silicate (larch) spruce forest	Fi3	72	52	50	58
Subalpine fresh carbonate spruce forest	Fs6	44	53	68	57
Moderately fresh silicate spruce-fir-beech forest	Ftb4	68	52	55	57
Moderately fresh carbonate spruce-fir-beech forest	Ftb7	68	49	57	57
Subalpine dry carbonate (larch) spruce forest	Fs7	26	64	59	57
Fresh carbonate spruce-fir-beech forest	Ftb6	75	45	53	56
Warm carbonate beech forest	Bu3	66	48	55	56
Fresh silicate spruce-fir forest	FT1	69	47	54	56
High montane carbonate spruce-fir beech forest	Ftb13	66	50	61	56
Rich basic spruce-fir forest	FT6	55	57	47	56
Rich silicate spruce-fir forest	FT5	63	49	55	56
Overlying humus carbonate spruce-fir forest	FT20	45	58	58	56
Fresh silicate beech forest with conifers	TB2	78		44	55
Rich loam-deciduous beech forest	LhB1	72		48	55
Fresh basic spruce-fir beech forest	Ftb8	76	50	52	55
Moderately fresh carbonate spruce-fir forest	FT15	35	54	64	55
Moist acid spruce-fir forest	FT8	68	47	48	55
Warm basic (larch) spruce forest	Fi7	66	50	41	54
Montane poor carbonate spruce-fir forest	Fi23	63	49	48	53
High montane carbonate spruce-fir forest	FT19	39	46	67	53
Fresh loam (beech) spruce-fir forest	FT16	63	51	54	53
Overlying humus carbonate spruce-fir -beech forest	Ftb16	63	48	61	53
Rich clay spruce-fir-beech forest	Ftb10	74	51	43	53
Rich silicate ash-lime mixed forest	Lh3	61		56	52
Fresh basic deciduous beech forest	Bu1	58	51	47	52
Fresh clay spruce-fir-beech forest	Ftb1	62	43	49	52
Moderately fresh carbonate and clay beech forest	Bu17	55	51	44	50

Table 2. *Cont.*

Forest Type	Code	Species Diversity [%]	Ecosystem Diversity [%]	Genetic Diversity [%]	Biodiversity (BCI) [%]
Rich silicate spruce-fir beech forest	Ftb11	40	55	51	50
Montane rich silicate spruce forest	Fi19	60	47	43	50
Fresh silicate spruce-fir-beech forest of the Northern Alps	Ftb2	49	49	49	49
Fresh clay beech forests with conifers	TB1	66		43	49
Warm carbonate oak-ash-lime forest	Lh2	52		48	48
Moist basic (gray alder) maple-ash mixed forest	Lh5	54		41	47
Silicate hardwood spruce-fir forest	LhT1	66		35	46
Colline grey alder riparian forest	Er3	49	38	42	45
Montane grey alder riparian forest	Er2	53	53	21	42
Fresh silicate lime-ash-pedunculate oak forest	Ei1	24	46	44	38

☐ Deciduous ☐ Mixed ☐ Coniferous ☐ Shrub

High ratings indicate that the forest type is in a more favorable state of preservation and can maintain a certain aspect of biodiversity with higher probability. On the contrary, low ratings display a less favorable state of preservation. They may indicate the need for active management to conserve certain aspects of forest biodiversity.

High ecosystem diversity ratings are displayed by the forest types of "Overlay humus-carbonate larch-stone pine forest" (68%), "Subalpine dry silicate larch-spruce forest" (65%), and "Subalpine dry carbonate (larch) spruce forest" (64%). In contrast, models indicate low ecosystem diversity ratings in "Moist basic (gray alder) maple-ash mixed forest" (40%), "Colline grey alder riparian forest" (38%), and "Silicate hardwood spruce-fir forest" (37%).

In the case study, forest types of high species diversity ratings are "Subalpine coniferous avalanche sites" (100%), "Mountain pine, green alder, hardwood scrub, scrubby areas" (98%), and "Subalpine basic larch-spruce forest" (89%). Low species diversity ratings are assigned to "Warm carbonate oak-ash-lime forest" (44%), "Rich loam-deciduous beech forest" (43%), and "Fresh silicate lime-ash-pedunculate oak forest" (27%).

Our models indicate high genetic diversity ratings for "Subalpine dry silicate larch-spruce forest" (82%), "Cool carbonate steep slope larch-pine forest" (82%), and "Overlay humus-carbonate larch-stone pine forest" (78%). Low genetic diversity ratings are found in "Colline grey alder riparian forest" (49%), "Fresh clay beech forest with conifers" (49%), and "Montane grey alder riparian forest" (40%). Highest probability for autochthony of the Norway spruce populations is detected in the Central and Eastern parts of Northern Tyrol. For detailed outcomes, please consider the Supplementary Material.

Overall, high biodiversity ratings can be found in "Subalpine dry silicate larch-spruce forest" (74%), "Subalpine coniferous avalanche sites" (72%), and "Overlay humus-carbonate larch-stone pine forest" (72%). On the contrary, low biodiversity ratings are in "Colline grey alder riparian forest" (46%), "Montane grey alder riparian forest" (45%), and "Fresh silicate lime-ash-pedunculate oak forest" (42%).

5. Discussion

5.1. Approach and Biodiversity Indicator Choice

The BCI approach differs substantially from the way other authors identified, weighted, and scored indicators. As we chose indicators based on inventory data availability and scientific literature, we forwent performing a principal component analysis to test for redundancy, such as in McElhinny et al. [55] and Storch et al. [41]. In line with LaRue et al. [71] and Ette et al. [72], we expect the BCI indicators to be intercorrelated and neither ecologically nor statistically independent.

Some indicators can be a proxy for more than one level of biodiversity which, based on scientific knowledge, might seem difficult to assign, e.g., on the one hand, the availability of about 25 m^3/ha of dead wood is an important quantitative threshold value for many endangered species [73,74]. On the other hand, general positive correlations between dead wood volume and wood-living fungi species, dead wood volume, and saproxylic species diversity, and between dead wood diversity and saproxylic species diversity are found for Europe in a meta-study [14]. However, this does not endanger assessment quality. In our case study, dead wood quantity is an ecological diversity indicator. Ratings are rising linearly with the share of reference levels (see Supplementary Material). If species diversity were targeted instead, ratings other than linear ones might be more appropriate.

Weighting as a final step in aggregation would have a major impact on the results. Nevertheless, respecting the limited knowledge about ecological communities, biological interactions, and genetic diversity in forests, putting weights to biodiversity indicators reveals more about the study authors and scientific community than substantially reaching an assessment that is closer to the true status of biological diversity. We agree with Okland [75] and Storch et al. [41] that indicator weighting is only reasonable for monitoring certain taxonomic groups with known correlations to specific habitat quality requirements. In line with McElhinny et al. [55], we expected weighting to probably subjoin more subjectivity to the BCI without providing additional insights.

5.2. Compare Study Outcomes

Spatial comparison within Tyrol shows that forest areas of high elevation tend to have higher BCI ratings compared to valleys in all models. Coniferous forests are in a more favorable state of preservation and can maintain biodiversity with higher probability than broad-leaved and mixed forests. In Austria, natural or semi-natural forests are mainly stocked in the subalpine, inner parts of the Alps and are characterized by a dominance of coniferous tree species [50]. In Tyrol, only 13% of the area is suitable for permanent settlement [76], which puts high pressure on ecosystems of low elevations such as broad-leaved and mixed forests. This effect is most pronounced in a species diversity model which also shows highest assessment heterogeneity on small spatial scales. BCI can be used to regionally define conservation targets, e.g., ecosystem restoration of forest types ('Silicate hardwood spruce-fir forest") in regions with below-average BCI performance (e.g., low elevation sites), or to regionally promote a particular level of biodiversity in a specific area (e.g., measures for ecosystem diversity such as retention trees).

It is not possible for us to directly compare our case study outcomes with other biodiversity assessments [10,41,55,77], due to unavailable indicator values in Tyrol (e.g., bark diversity, hollow trees, litter dry weight, litter decomposition, tree age, vegetation cover), different scales [57], and different study purposes [77]. However, there is partial agreement in the choice of indicators such as perennial species richness [55,77], natural regeneration [41,57,77], standing and lying dead wood [41,55,57,77], old growth trees [10,41,55,57,77], genetic diversity of Norway spruce [57], forest fragmentation [57,77], and tree species frequency [10]. Benchmarking based on vegetation types can also be found in Parkes et al. [77]. The choice of indicators in this study largely corresponds to a meta-study of Gao et al. [14], who demonstrated that the biodiversity indicators chosen most frequently in 142 European ecological studies are dead wood volume, age of canopy trees, vascular plant species, tree canopy cover, decay classes, and dead wood diversity.

5.3. Advantages and Disadvantages

Major advantages of the BCI approach are easy transferability, cost-efficient long-term monitoring of forest policy measures, and the logical indicator structure in line with the CBD. BCI can be used as a conservation planning tool to halt biodiversity loss on the national scale. With our state-of-the-art data pre- and postprocessing, BCI sticks very close to the policy-relevant definition of biodiversity in the CBD, which 183 member countries agreed on in 1992. We provide a new option to assess biodiversity based on available national forest inventory and forest typing data. Outcomes can be interpreted on three levels (diversity of ecosystem, species diversity, and genetic diversity) and aggregated to assess forest biodiversity in high resolution on varying spatial scales. By not weighting indicators, the framework remains easy to adapt to neighboring regions in Central Europe.

Quantitative aggregation and simplification of ecological spatial information may help policy makers and conservationists to implement biodiversity policies and assign conservation funding, e.g., for ecosystem restoration. The ranking of forest types and high-resolution spatial maps of forest diversity can support decision-making in biodiversity conservation (e.g., target forest types, target regions, conservation priorities, and ecosystem-, species-, or genetic restoration measures) and retrospectively evaluate effectiveness of financial resources spent on ecosystem restoration and nature conservation. Additionally, effects of different forest management measures on biodiversity can be assessed per forest type and used to advance Sustainable Forest Management. Within one forest inventory period, performing cost-benefit analyses of, e.g., biodiversity conservation efforts, forest management practices, forest road building, regional forest policy funding, and Sustainable Forest Management measures will be made available. Quantifying forest biodiversity with BCI allows targeted management of a landscape's biodiversity and distributed biodiversity values. BCI can be used as a measurable, objective, and quantitative guidance for regional forest and conservation policy using the first BCI assessment as a baseline minimum.

However, the BCI concept could not overcome all weaknesses of forest inventory-based approaches described in Storch et al. [41], e. g., large-scale forest inventory design may not capture small areas like nature reserves well enough and very rare forest types must be excluded from the analysis. Plot measures may not be representative for the forest stand and most biodiversity aspects can only be addressed through surrogates. Additionally, most genetic diversity indicators focus on the major tree species of Tyrol, Norway spruce, as data for other species are not available. The indicators "autochthony", "tree layer structure", and "game impact" display high ratings in the BCI assessment. For upcoming BCI assessments, these indicator evaluations should be revised based on experience gained from the Tyrolean case study. Applying BCI, error propagation of forest typing models can possibly occur. Nevertheless, by using ecological modeling, referencing indicators by forest type, employing GIS data such as orthophotos, and machine learning, we were able to advance reliability and spatial resolution of forest biodiversity assessments.

6. Conclusions

Assessing biodiversity is highly complex. The intention of BCI is to aggregate and simplify ecological information in a surrogate approach, advance forest-inventory based assessments, and monitor all levels of forest biodiversity in line with the CBD.

In the case study, average ecosystem diversity is 49%, species diversity is 71%, genetic diversity is 53%, and biodiversity is 57%. In Tyrol, coniferous forests are in a better state of preservation and can maintain biodiversity with higher probability than broad- leaved and mixed forests. These findings, next to rankings of forest types and high-resolution spatial maps of forest biodiversity, can be used to advance land use policies, forest management, nature conservation, and landscape planning in Austria, e.g., by cost-benefit analysis. The approach is transferable to neighboring regions with forest-typing, e.g., in Germany, Italy, and Austria.

For Tyrol, we highly recommend a second BCI assessment within six years to solve the baseline problem, monitor temporal and spatial changes, detect trends in forest biodiversity, and evaluate effects of forest management and biodiversity conservation. BCI can give objective guidance and feedback to forest policy to counteract the biodiversity crisis. We recommend a future positive BCI trend on all levels as a quantitative goal for regional forest policy to meet strategic CBD targets.

Supplementary Materials: The following supporting information can be downloaded at: https://www.mdpi.com/article/10.3390/f14040709/s1.

Author Contributions: Conceptualization, J.-S.E.; Methodology, J.-S.E.; Software, J.-S.E. and M.S.; Validation, J.-S.E.; Formal analysis, J.-S.E.; Investigation, J.-S.E.; Resources, T.G.; Data curation, J.-S.E.; Writing—original draft, J.-S.E.; Writing—review & editing, J.-S.E.; Visualization, J.-S.E.; Supervision, T.G.; Project administration, J.-S.E. and T.G.; Funding acquisition, J.-S.E. and T.G. All authors have read and agreed to the published version of the manuscript.

Funding: Parts of the study were funded by the European Union (Interreg-project). Open access funding was provided by the Austrian Research Centre for Forests (BFW). Besides, no funding has been received.

Institutional Review Board Statement: Not applicable.

Informed Consent Statement: Not applicable.

Data Availability Statement: Data is not publicly available due to third party policies. Restrictions apply to the availability of forest inventory data. Data was obtained from the protected area managements, the Tyrolean Regional Government, and the Austrian Research Centre for Forests.

Conflicts of Interest: The authors declare no conflict of interest.

List of Acronyms

AFI	Austrian Forest Inventory
AUPICMAP	Geographic-genetic map of the Austrian Norway spruce population
BCI	Biodiversity composite index
BI	Biodiversity Indicator
BONUS	Bonus Indicator
CBD	Convention on Biological Diversity
TIRIS	Tyrolean Spatial Information System
QGIS	Quantum-Geographic Information System

Appendix A

Table A1. Abbreviations of indicators.

Ecosystem Diversity	
BI_{layer}	Biodiversity indicator: Tree layer structure
$BI_{devel.level}$	Biodiversity indicator: Developmental level
$BI_{deadwood}$	Biodiversity indicator: Dead wood
$DW_{quantity}$	Biodiversity indicator: Dead wood I: Quantity
$DW_{quality}$	Biodiversity indicator: Dead wood II: Quality
$Bonus_{structure}$	Bonus indicator: Structural features
$Bonus_{shrub}$	Bonus indicator: Structural features I: Shrub cover
$Bonus_{standage}$	Bonus indicator: Structural features II: Stand age
$Bonus_{dbh}$	Bonus indicator: Structural features III: Diameter breast height
Species diversity	
$BI_{treespecies}$	Biodiversity indicator: Tree species diversity
$BI_{vegetation}$	Biodiversity indicator: Ground vegetation
BI_{soil}	Biodiversity indicator: Surface soil quality
$Bonus_{game}$	Bonus indicator: Game impact
Genetic diversity	
BI_{gap}	Biodiversity indicator: Forest gap structure
$BI_{autochthony}$	Biodiversity indicator: Autochthony
$BI_{management}$	Biodiversity indicator: Management constraints
$m.constraint_{inclination}$	Biodiversity indicator: Management constraints I: Inclination
$m.constraint_{distance}$	Biodiversity indicator: Management constraints II: Distance to forest road
$Bonus_{genetic}$	Bonus indicator: Genetic features
$Bonus_{regeneration}$	Bonus indicator: Genetic features I: Natural Regeneration
$Bonus_{phenology}$	Bonus indicator: Genetic features II: Phenology

Table A2. Indicative value of diversity indicators. Surrogates for forest biodiversity and scientific evidence for their indicative value.

Indicators	Scientific Evidence
	Ecosystem diversity indication
tree layer structure (s. s., naturalness of tree layer composition)	Structural spatial diversity increases resource partitioning of light use among species [78–80] and indicates the number of niches occurring vertically and horizontally within the canopy [22]. Greater overlap of crowns indicates a greater use of niche space for light in the canopy [37,81] and is a measure of ecological niche space size [71]. Tree layer structure is a proxy for forest management intensity [50]. Heterogenous vegetation heights are associated with greater ecosystem function [71].
developmental level (s. s., diversity of the developmental stages)	Variation of tree dimension can be used as a proxy for habitat quality or biotope trees [10] and related macro- and microhabitats [82]. Forest age differentiation indicates high niche supply and affects community composition [83]. Late successional stages are proxies for ecosystem productivity [84], biotic resistance to invasion [85], and light absorption [80]. The developmental level can be a hint towards management intensity [50]. The indicator approach is based on mosaic cycle concepts [86,87] and niche theory [88,89].
dead wood (s. s., dead wood quantity and quality)	Dead wood promotes forest biodiversity [74,90–96]. It provides habitat, shelter, growth substratum, and nutrition for various organisms, e.g., bryophytes, saproxylic insects, and fungi [96–99]. Coarse woody debris supports numerous forest ecosystem functions [100], e.g., nutrient cycling [101,102]. Occurrence of coarse woody debris may indicate forest ecosystem processes such as mortality, ingrowth, competition [103], and ecosystem disturbance [104].
structural features (s. s., shrub cover stand age, and diameter breast height)	Structural diversity is a proxy for structural complexity, potential habitat variability, and niche differentiation for umbrella species [21,105]. The occurrence of shrub species can be an important contribution to maintain forest biodiversity [106]. Shrub and tree height is a proxy for vertical stratification of niche space [71], e.g., for birds [22]. Mean canopy height indicates the number of niches filled within the ecosystem volume [107]. Canopy tree age was found to correlate positively with epiphytic lichen diversity [14]. Large tree diameters indicate high potential for tree-related habitats [108].
	Species diversity indication
tree species diversity (s. s., naturalness of tree species composition)	There is high scientific evidence for a positive correlation between tree species diversity and the number of bird [109], ground beetle [110,111], arthropod [112], and ground vegetation species [110,113]. Tree species abundance can be used as a proxy for species diversity of, e.g., saproxylic beetles, bryophytes, lichens, fungi, and arthropods [114–117].
ground vegetation (s. s., naturalness of plant species composition, disturbance indication)	Plant species diversity indicates partitioning of resource use between species [118]. Native plant species diversity is a proxy for the number of different niche spaces filled by native plant species [119].
surface soil quality (s. s., divergence from the expected humus form)	Most species diversity of Europe can be found in the soil ecosystems [16]. Humus form is one of the regulating factors for the composition of species communities [120–123]. The diversity of zoological groups linearly correlates to soil pH value and humus type [124]. There is high evidence for the relevance of humus type for forest biodiversity and overall species diversity [2,120,124–126]. Slight changes in physico-chemical components can lead to great changes in soil biota communities [127].
Game impact (s. s., extensive game impact on forest regeneration)	There is broad consensus on the relevance of tree browsing for forest biodiversity [128–130]. Severe herbivore impact leads to tree species segregation, lacking regeneration, and disturbed forest succession [131].
	Genetic diversity indication
forest gap structure (s. s., surface balance forest–non forest area)	Forest fragmentation is a serious threat to genetic diversity [132–135]. Fragmentation subdivides populations into small units and imposes barriers to migration, which is an important driver for extinction [136]. Fragmentation can erode neutral and adaptive genetic diversity and lowers effective population sizes and genetic variability [137,138]. It promotes genetic drift and inbreeding depression [139]. Habitat fragmentation may affect adaptive potential of populations and their fitness level negatively [135,140]. Susceptibility to habitat fragmentation and habitat split is highly species-specific [138]. Dispersal ability, migration, habitat availability, and range of environmental tolerance is decisive for genetic consequences for species, populations, and individuals [138,141,142]. Allelic richness is most vulnerable to habitat fragmentation with rare gene expressions preferentially being eliminated [135,139].
autochthony (s. s., genetic distance between populations of Norway spruce)	Autochthonous populations show small-scale genetic differentiation and local adaption of tree species [143,144], promoting tree population differentiation [145,146]. Negative effects of allochthonous seed sources are maladaptation to the local environment, intraspecific hybridization (introgression), cryptic invasion, and other unintended effects on associated species which can be seen as environmental risks [144,147]. Genetic variability of introduced forest reproductive material tends to be considerably lower compared to local populations [148]. Artificial transfer of genetic information, e.g., by using forest reproductive material, tends to degrade forest genetic structures [149].
Management constraints (s. s., inclination, and distance to forest road)	Main drivers of extinction are of anthropogenic origin [150,151]. Forest management may affect forest genetic resources through changes in genetic drift, mating systems, fertility, and species migration [147,152]. It can lead to the loss of rare and localized alleles [153,154]. Silviculture influences the major evolutionary forces of selection, genetic drift, and gene flow [136,155–157]. Forest management can affect mating systems, genetic variation and population structure of forest trees [158,159], lowers effective population sizes [155,160], and impacts the adaptive potential of forests [159].

Table A2. *Cont.*

Indicators	Scientific Evidence
Genetic features (s. s., tree planting intensity, phenology of Norway spruce, and crown structure)	Choice of reproductive forest material has probably the most significant impact on the genetic diversity of forest trees in Europe [161]. Possible negative effects of long-distance seed transfer on genetic diversity are described in Kremer [147] and Carnus et al. [162]. Hybridization of local and non-local genotypes may affect genetic population structure negatively through outbreeding depression, introgression, demographic invasion, introduction of diseases, and genetic erosion [144,147]. Branching types of Norway spruce may be used as a hint to detect genetically allochthonous plant material [163]. Lower stand density affects pollen and seed dispersal positively and promotes pollen dispersal [164] and pollen densities [165,166]. Decreasing tree density is likely to increase wind turbulences, and pollen and seed long-distance dispersal [164,167,168].

Table A3. Assignment to forest type groups. Assignment of forest types [53] to forest type groups [50].

Forest Type Groups [50]	Forest Types Assigned [53]
Mountain pine and scrub forest communities	Bu10, Ge1, Ge8, Ge9, Lat2, Lat3, Lat4, Lh8, k
Carbonate-rich subalpine pine and larch forests	Fs5, Ki20, La1, La2, La4, La6, Lat1, Zi2, Zi3, Zi6
Carbonate-rich montane mixed spruce-coniferous forests	Fi5, Fi6, Fi7, Fi8, Fi10, Fi13, Fi14, Fi18, Fi20, Fi23, Fi25, Fs13, FT3, FT9, FT13, FT15, FT18, FT19, FT20, LhT2
Silicate-rich spruce-fir forests	Fi2, Fi3, Fi4, Fi9, Fi11, Fi19, Fi22, FT1, FT2, FT5, FT8, FT10, FT12
Moist coniferous and birch forests (including bog edge forests)	Fi16, Fi17, FT7, FT22, Ki21, Fs11, Fs14, Fs18
Mixed pine forests on carbonate	Ki1, Ki2, Ki3, Ki17, Ki18, Ki19, LhK3
Silicate (spruce-fir) beech forests	Bu5, FT17, Ftb2, Ftb4, Ftb11, Ftb12, LhT1, TB2
Mixed maple and ash forests	Bu4, Ei1, Lh4, Lh6, Lh9, Lh16, Lh17, Lh18
Lime and mixed lime forests	Ei3, Ei4, Lh1, Lh2, Lh3, Lh11, Lh13, LhB2
Base-rich dry beech forests	Bu3, Bu7, Fkb1, Ftb20
Brown soil (spruce-fir) beech forests	Ftb3
Downy oak forests	MH2
Oak and oak-pine forests	Ei2, Ei7, Ei12, Ki4, Ki15
Willow communities	Er4, Er11, Er12
Hard riparian forests	Fi21, Ki9, Lh12, Lh14
Stream-accompanying alder-ash forests	Er2, Er3, Er7, Er8
Fresh carbonate (spruce-fir) beech forests	Bu1, Bu17, FT16, Ftb1, Ftb6, Ftb7, Ftb10, Ftb13, Ftb 14, Ftb16, LhB1, LhB3, TB1
High-altitude beech forests with maple	Bu11, Bu20, Ftb8
Subalpine coniferous forests on silicate	Fi12, Fs1, Fs2, Fs3, Fs4, Fs10, Fs12, Fs17, La5, La7, Zi1, Zi4, Zi5
Gray alder forests	Er1, Er5, Lh5, Lh21, Er13
Pine forests on silicate	FT21, Ki6, Ki7, Ki16, La3
Carbonate-rich subalpine coniferous mixed spruce forests	Fi1, Fs6, Fs7, Fs8, Fs9, FT6
Block forest, rubble, and rock sites on carbonate (newly established)	FK1, FK2, Fkb2, FL2, Lh10, Klf2, LhK2, Ki23, Zlf4
Block forest, rubble, and rock sites on silicate (newly established)	FK3, Fkb3, FL1, Klf1, LhK1, Zlf3, Ki5, Lh7, FL3

References

1. Rockström, J.; Steffen, W.; Noone, K.; Persson, Å.; Chapin, F.S. Planetary Boundaries: Exploring the Safe Operating Space for Humanity. Ecology and Society 14, 32. Available online: http://www.ecologyandsociety.org/vol14/iss2/art32/ (accessed on 1 March 2023).
2. Hooper, D.U.; Chapin, F.S.; Ewel, J.J.; Hector, A.; Inchausti, P.; Lavorel, S.; Lawton, J.H.; Lodge, D.M.; Loreau, M.; Naeem, S.; et al. Effects of Biodiversity on Ecosystem Functioning: A Consensus of Current Knowledge. *Ecol. Monogr.* **2005**, *75*, 3–35. [CrossRef]
3. Balvanera, P.; Pfisterer, A.B.; Buchmann, N.; He, J.-S.; Nakashizuka, T.; Raffaelli, D.; Schmid, B. Quantifying the Evidence for Biodiversity Effects on Ecosystem Functioning and Services. *Ecol. Lett.* **2006**, *9*, 1146–1156. [CrossRef]
4. Mace, G.M.; Norris, K.; Fitter, A.H. Biodiversity and Ecosystem Services: A Multilayered Relationship. *Trends Ecol. Evol.* **2012**, *27*, 19–26. [CrossRef] [PubMed]
5. Gamfeldt, L.; Snäll, T.; Bagchi, R.; Jonsson, M.; Gustafsson, L.; Kjellander, P.; Ruiz-Jaen, M.C.; Froberg, M.; Stendahl, J.; Philipson, C.D.; et al. Higher Levels of Multiple Ecosystem Services Are Found in Forests with More Tree Species. *Nat. Commun.* **2013**, *4*, 1340. [CrossRef]

6.   Reich, P.B.; Tilman, D.; Isbell, F.; Mueller, K.; Hobbie, S.E.; Flynn, D.F.B.; Eisenhauer, N. Impacts of Biodiversity Loss Escalate Through Time as Redundancy Fades. *Science* **2012**, *336*, 589–592. [CrossRef] [PubMed]
7.   Cardinale, B.J.; Duffy, J.E.; Gonzalez, A.; Hooper, D.U.; Perrings, C. Biodiversity Loss and Its Impact on Humanity. *Nature* **2012**, *486*, 59–67. [CrossRef] [PubMed]
8.   Liang, J.; Crowther, T.W.; Picard, N.; Wiser, S.; Zhou, M.; Alberti, G.; Schulze, E.-D.; McGuire, A.D.; Bozzato, F.; Pretzsch, H.; et al. Positive Biodiversity-Productivity Relationship Predominant in Global Forests. *Science* **2016**, *354*, aaf8957. [CrossRef]
9.   Purvis, A.; Hector, A. Getting the Measure of Biodiversity. *Nature* **2000**, *405*, 212–219. [CrossRef]
10.   Heym, M.; Uhl, E.; Moshammer, R.; Dieler, J.; Stimm, K.; Pretzsch, H. Utilising Forest Inventory Data for Biodiversity Assessment. *Ecol. Indic.* **2020**, *121*, 107196. [CrossRef]
11.   Ferris, R.; Humphrey, J.W. A Review of Potential Biodiversity Indicators for Application in British Forests. *Forestry* **1999**, *72*, 313–328. [CrossRef]
12.   Margules, C.; Pressey, R.L.; Williams, P.H. Representing Biodiversity: Data and Procedures for Identifying Priority Areas for Conservation. *J. Biosci.* **2002**, *27*, 309–326. [CrossRef] [PubMed]
13.   Duelli, P.; Obrist, M.K. Biodiversity Indicators: The Choice of Values and Measures. *Agric. Ecosyst. Environ.* **2003**, *98*, 87–98. [CrossRef]
14.   Gao, T.; Nielsen, A.B.; Hedblom, M. Reviewing the Strength of Evidence of Biodiversity Indicators for Forest Ecosystems in Europe. *Ecol. Indic.* **2015**, *57*, 420–434. [CrossRef]
15.   Heink, U.; Kowarik, I. What Criteria Should Be Used to Select Biodiversity Indicators? *Biodivers. Conserv.* **2010**, *19*, 3769–3797. [CrossRef]
16.   Ette, J.S.; Geburek, T. Why European Biodiversity Reporting Is Not Reliable. *AMBIO* **2021**, *50*, 929–941. [CrossRef]
17.   Yoccoz, N.G.; Nichols, J.D.; Boulinier, T. Monitoring of Biological Diversity in Space and Time. *Trends Ecol. Evol.* **2001**, *16*, 446–453. [CrossRef]
18.   McElhinny, C.; Gibbons, P.; Brack, C.; Bauhus, J. Forest and Woodland Stand Structural Complexity: Its Definition and Measurement. *For. Ecol. Manag.* **2005**, *218*, 1–24. [CrossRef]
19.   Katzner, T.; Milner-Gulland, E.J.; Bragin, E. Using Modelling to Improve Monitoring of Structured Populations: Are We Collecting the Right Data? *Conserv. Biol.* **2007**, *21*, 241–252. [CrossRef]
20.   Jones, J.P.; Collen, B.; Atkinson, G.; Baxter, P.W.; Bubb, P.; Illian, J.B.; Katzner, T.; Keane, A.; Loh, J.; Mcdonald-Madden, E.; et al. The Why, What, and How of Global Biodiversity Indicators beyond the 2010 Target. *Conserv. Biol.* **2010**, *25*, 450–457. [CrossRef]
21.   Müller, J.; Pöllath, J.; Moshammer, R.; Schröder, B. Predicting the Occurrence of Middle Spotted Woodpecker *Dendrocopos Medius* on a Regional Scale, Using Forest Inventory Data. *For. Ecol. Manag.* **2009**, *257*, 502–509. [CrossRef]
22.   MacArthur, R.H.; MacArthur, J.W. On Bird Species Diversity. *Ecology* **1961**, *42*, 594–598. [CrossRef]
23.   Begon, M.; Harper, J.L.; Townsend, C.R. *Ökologie, Individuen, Populationen und Lebensgemeinschaften [Ecology, Indi-Viduals, Populations, and Communities]*; Birkhäuser Publishers: Basel, Switzerland, 1991.
24.   McNally, R.; Parkinson, A.; Horrocks, G.; Conole, L.; Tzaros, C. Relationships between Terrestrial Vertebrate Diversity, Abundance, and Availability of Coarse Woody Debris on South-Eastern Australian Floodplains. *Biol. Conserv.* **2001**, *99*, 191–205. [CrossRef]
25.   Winter, S.; Chirici, G.; McRoberts, R.E.; Hauk, E.; Tomppo, E. Possibilities for Harmonizing National Forest Inventory Data for Use in Forest Biodiversity Assessments. *Int. J. Environ. Res. Public Health* **2008**, *81*, 33–44. [CrossRef]
26.   Norton, B.G. Improving ecological communication: The role of ecologists in environmental policy formation. *Ecol. Appl.* **1998**, *8*, 350–364. [CrossRef]
27.   Waldron, A.; Miller, D.C.; Redding, D.; Mooers, A.; Kuhn, T.S.; Nibbelink, N.; Roberts, J.T.; Tobias, J.A.; Gittleman, J.L. Reductions in Global Biodiversity Loss Predicted from Conservation Spending. *Nature* **2017**, *551*, 364–367. [CrossRef]
28.   Science for Environment Policy: Ecosystem Services and Biodiversity; Bristol, UK. 2015. Available online: https://ec.europa.eu/environment/integration/research/newsalert/pdf/ecosystem_services_biodiversity_IR11_en.pdf (accessed on 1 March 2023).
29.   Kusumoto, B.; Shiono, T.; Miyoshi, M.; Maeshiro, R.; Fujii, S.; Kuuluvainen, T.; Kubota, Y. Functional Response of Plant Communities to Clearcutting: Management Impacts Differ between Forest Vegetation Zones. *J. Appl. Ecol.* **2015**, *52*, 171–180. [CrossRef]
30.   Henneron, L.; Aubert, M.; Bureau, F.; Dumas, Y.; Ningre, F.; Perret, S.; Richter, C.; Balandier, P.; Chauvat, M. Forest Management Adaptation to Climate Change: A Cornelian Dilemma between Drought Resistance and Soil Macro-Detritivore Functional Diversity. *J. Appl. Ecol.* **2015**, *52*, 913–927. [CrossRef]
31.   Pohjanmies, T.; Eyvindson, K.; Triviño, M.; Mönkkönen, M. More Is More? Forest Management Allocation at Different Spatial Scales to Mitigate Conflicts between Ecosystem Services. *Landsc. Ecol.* **2017**, *32*, 2337–2349. [CrossRef]
32.   Boscolo, M.; Vincent, J.R. Nonconvexities in the Production of Timber, Biodiversity, and Carbon Sequestration. *J. Environ. Econ. Manag.* **2003**, *46*, 251–268. [CrossRef]
33.   Duncker, P.S.; Barreiro, S.M.; Hengeveld, G.M.; Lind, T.; Mason, W.L.; Ambrozy, S.; Spiecker, H. Classification of Forest Management Approaches: A New Conceptual Framework and Its Applicability to European Forestry. *Ecol. Soc.* **2012**, *17*, 51. [CrossRef]
34.   Eyvindson, K.; Repo, A.; Mönkkönen, M. Mitigating Forest Biodiversity and Ecosystem Service Losses in the Era of Biobased Economy. *For. Policy Econ.* **2018**, *92*, 119–127. [CrossRef]

35. UNECE; FAO. *State of Europe's Forests 2020. Status and Trends in Sustainable Forest Management in Europe*; Liaison Unit Bratislava: Bratislava, Slovakia, 2020.

36. Sarkar, S.; Margules, C. Operationalizing Biodiversity for Conservation Planning. *J. Biosci.* **2002**, *27*, 299–308. [CrossRef] [PubMed]

37. Williams, J. Metrics for Assessing the Biodiversity Values of Farming Systems and Agricultural Landscapes. *Pac. Conserv. Biol.* **2004**, *10*, 145–163. [CrossRef]

38. Lindenmayer, D.B.; Margules, C.R.; Botkin, D.B. Indicators of Biodiversity for Ecologically Sustainable Forest Management. *Conserv. Biol.* **2000**, *14*, 941–950. [CrossRef]

39. Kuuluvainen, T. Forest Management and Biodiversity Conservation Based on Natural Ecosystem Dynamics in Northern Europe: The Complexity Challenge. *AMBIO* **2009**, *38*, 309–315. [CrossRef] [PubMed]

40. Taboada, Á.; Tárrega, R.; Calvo, L.; Marcos, E.; Marcos, J.A.; Salgado, J.M. Plant and Carabid Beetle Species Diversity in Relation to Forest Type and Structural Heterogeneity. *Eur. J. For. Res.* **2008**, *129*, 31–45. [CrossRef]

41. Storch, F.; Dormann, C.F.; Bauhus, J. Quantifying forest structural diversity based on large-scale inventory data: A new approach to support biodiversity monitoring. *J. Ecosyst. Ecography* **2018**, *5*, 34. [CrossRef]

42. Kändler, G. Bestockungsinventur Auf den Stichproben der Bodenzustandserhebung 2006 [Stocking Inventory on the Soil Condition Survey Samples 2006]. *FVA Einblick* **2006**, *2*, 11–12.

43. Polley, H. *Monitoring in Wäldern: Die Bundeswaldinventur und Verknüpfungen für Naturschutzfragen [Forest Monitoring: National Forest Inventory and Links for Nature Conservation Issues]*; Thünen Institute for Forest Ecosystems: Braunschweig, Germany; Naturschutz und Biologische Vielfalt: Vienna, Austria, 2010; Volume 83, pp. 65–78.

44. Chirici, G.; Winter, S.; McRoberts, R.E. *National Forest Inventories: Contributions to Forest Biodiversity Assessments*; Springer Publishers: Dordrecht, The Netherlands, 2011. [CrossRef]

45. Corona, P.; Chirici, G.; McRoberts, R.E.; Winter, S.; Barbati, A. Contribution of Large-Scale Forest Inventories to Biodiversity Assessment and Monitoring. *For. Ecol. Manag.* **2011**, *262*, 2061–2069. [CrossRef]

46. Corona, P.; Köhl, M.; Marchetti, M. *Advances in Forest Inventory for Sustainable Forest Management and Biodiversity Monitoring*; Springer Publishers: Dordrecht, The Netherlands, 2003. [CrossRef]

47. Spiecker, H. Silvicultural Management in Maintaining Biodiversity and Resistance of Forests in Europe—Temperate Zone. *J. Environ. Manag.* **2003**, *67*, 55–65. [CrossRef]

48. BFW. *Ergebnisse der österreichischen Waldinventur (ÖWI) 2007–2009. [Results of the Austrian Forest Inventory (ÖWI) 2007–2009]*; Austrian Research Centre for Forests: Vienna, Austria, 2011. Available online: https://bfw.ac.at/030/pdf/1818_pi24.pdf (accessed on 1 March 2023).

49. Government, T.R. Waldzustandsinventur [Forest Condition Inventory]. Available online: https://www.tirol.gv.at/umwelt/wald/waldzustand/waldberichte/ (accessed on 1 March 2023).

50. Grabherr, G.; Koch, G.; Kirchmeir, H.; Reiter, K. *Hemerobie österreichischer Waldökosysteme [Hemeroby of Austrian Forest Ecosystems]*; Wagner University Press: Innsbruck, Austria, 1998.

51. Raab, S.; Feller, S.; Uhl, E.; Schäfer, A.; Ohrner, G. Aktuelle Holzernteverfahren Am Hang [Temporary Forest Harvesting Techniques on Slopes]; LWF Wissen: 2002; Volume 36. Available online: https://www.lwf.bayern.de/service/publikationen/lwf_wissen/064166/index.php (accessed on 1 March 2023).

52. Geburek, T.; Schweinzer, K. Simulationsstudie Fichte: Daten Des AUPICMAP Projekts [A Simulation Study on Norway Spruce: Data of the AUPICMAP Project]. Austrian Research Centre for Forests: Vienna, Austria, unpublished.

53. Hotter, M.; Simon, A.; Vacik, H.; Wallner, M.; Simon, A. Waldtypisierung Tirol [Forest Typing Tyrol]; Tyrolean Regional Government: Innsbruck, Austria. Available online: https://www.tirol.gv.at/umwelt/wald/schutzwald/waldtypisierung/waldtypenhandbuch/ (accessed on 1 March 2023).

54. Hauk, E.; Schadauer, K. Instruktionen für die Feldarbeiten der österreichischen Waldinventur 2007–2009 [Field Work Manual of the Austrian Forest Inventory 2007–2009]. Available online: https://www.bfw.gv.at/instruktion-feldarbeit-oesterreichische-waldinventur/ (accessed on 1 March 2023).

55. McElhinny, C.; Gibbons, P.; Brack, C. An Objective and Quantitative Methodology for Constructing an Index of Stand Structural Complexity. *For. Ecol. Manag.* **2006**, *235*, 54–71. [CrossRef]

56. Landres, P.B.; Morgan, P.; Swanson, F.J. Overview of the Use of Natural Variability Concepts in Managing Ecological Systems. *Ecol. Appl.* **1999**, *9*, 1179–1188. [CrossRef]

57. Geburek, T.; Milasowszky, N.; Frank, G.; Konrad, H.; Schadauer, K. The Austrian Forest Biodiversity Index: All in One. *Ecol. Indic.* **2010**, *10*, 753–761. [CrossRef]

58. MacArthur, R.H. *Geographical Ecology: Patterns in the Distribution of Species*; Princeton University Press: Princeton, NJ, USA, 1972.

59. Tilman, D. *Resource Competition and Community Structure*; Princeton University Press: Princeton, NJ, USA, 1982.

60. Gause, G.F. *The Struggle for Existence*; Dover Publication: London, UK, 2019.

61. Whittaker, R.H. Vegetation of the Siskiyou Mountains, Oregon and California. *Ecol. Monogr.* **1960**, *30*, 279–338. [CrossRef]

62. Virkkala, R. Why study woodpeckers? The significance of woodpeckers in forest ecosystems. *Ann. Zool. Fennici.* **2006**, *43*, 82–85. Available online: https://www.jstor.org/stable/23735920 (accessed on 1 March 2023).

63. Brin, A.; Meredieu, C.; Piou, D.; Brustel, H.; Jactel, H. Changes in Quantitative Patterns of Dead Wood in Maritime Pine Plantations over Time. *For. Ecol. Manag.* **2008**, *256*, 913–921. [CrossRef]

64. Bitterlich, W. Die Winkelzählprobe [Angle Count Sampling]. *Forstwiss. Cent.* **1952**, *71*, 215–225. [CrossRef]

65. Lappi, J.; Bailey, R.L. Estimation of the Diameter Increment Function or Other Tree Relations Using Angle-Count Samples. *For. Sci.* **1987**, *33*, 725–739. [CrossRef]
66. Sterba, H. Diversity Indices Based on Angle Count Sampling and Their Interrelationships When Used in Forest Inventories. *Forestry* **2008**, *8*, 587–597. [CrossRef]
67. Breiman, N.L. Random Forests. *Mach. Learn.* **2001**, *45*, 5–32. [CrossRef]
68. Venables, W.N.; Ripley, B.D. *Modern Applied Statistics with S*; Springer Publishing: New York, NY, USA, 2002.
69. Ho, T.K. The Random Subspace Method for Constructing Decision Forests. *IEEE Trans. Pattern Anal. Mach. Intell.* **1998**, *20*, 832–844. [CrossRef]
70. Quinlan, J.R. Induction of Decision Trees. *Mach. Learn.* **1986**, *1*, 81–106. [CrossRef]
71. LaRue, E.; Hardiman, B.; Elliott, J.; Fei, S. Structural Diversity as a Predictor of Ecosystem Function. *Environ. Res. Lett.* **2019**, *14*, 114011. [CrossRef]
72. Ette, J.S.; Ritter, T.; Vospernik, S. Insights in Forest Structural Diversity Indices with Machine Learning: What Is Indicated? *Biodivers. Conserv.* **2023**. [CrossRef]
73. Bütler, R.; Schläpfer, R. Wie viel Totholz braucht der Wald [How much dead wood does a forest need]? *Schweiz. Z. Forstwes.* **2004**, *155*, 31–37. [CrossRef]
74. Müller, J.; Bussler, H.; Utschick, H. Wie viel Totholz braucht der Wald? Ein wissenschaftsbasiertes Konzept gegen Artenschwund in den Totholzzönosen [How much deadwood does the forest need? A science-based concept against species loss in coenoses of dead wood]. *Nat. Landsch.* **2007**, *39*, 165–170.
75. Okland, B. Unlogged Forests: Important Sites for Preserving the Diversity of Mycetophilids (Diptera: Sciaroidea). *Biol. Conserv.* **1996**, *76*, 297–310. [CrossRef]
76. Government, T.R. Schutzwald in Tirol: Landesschutzwaldkonzept 2000 [Protective Forests of Tyrol: State Protection Forest Concept 2000]; Tyrolean Regional Government: Innsbruck, Austria. Available online: https://www.tirol.gv.at/umwelt/wald/schutzwald/landesschutzwaldkonzept/ (accessed on 1 March 2023).
77. Parkes, D.; Greame, N.; Chea, D. Assessing the Quality of Native Vegetation: The 'Habitat Hectares' Approach. *Ecol. Manag. Restor.* **2003**, *4*, 29–38. [CrossRef]
78. Kohyama, T. Size-Structured Tree Populations in Gap-Dynamic Forest: The Forest Architecture Hypothesis for the Stable Coexistence of Species. *J. Ecol.* **1993**, *81*, 131–143. [CrossRef]
79. Yachi, S.; Loreau, J. Does complementary resource use enhance ecosystem function? A model of light competition in plant communities. *Ecol. Lett.* **2006**, *10*, 54–62. [CrossRef]
80. Atkins, J.W.; Fahey, R.T.; Hardiman, B.H.; Gough, C.M. Forest Canopy Structural Complexity and Light Absorption Rela-Tionships at the Subcontinental Scale. *J. Geophys. Res. Biogeosci.* **2018**, *123*, 1387–1405. [CrossRef]
81. Zheng, L.T.; Chen, H.Y.H.; Yan, E.R. Tree Species Diversity Promotes Litterfall Productivity through Crown Complementarity in Subtropical Forests. *J. Ecol.* **2019**, *107*, 1852–1861. [CrossRef]
82. Larrieu, L.; Cabanettes, A.; Brin, A.; Bouget, C.; Deconchat, M. Tree Microhabitats at the Stand Scale in Montane Beech–Fir Forests: Practical Information for Taxa Conservation in Forestry. *Eur. J. For. Res.* **2013**, *133*, 355–367. [CrossRef]
83. Gossner, M.M.; Schall, P.; Ammer, C.; Ammer, U.; Engel, K.; Schubert, H.; Simon, U.; Utschick, H.; Weisser, W.W. Forest Management Intensity Measures as Alternative to Stand Properties for Quantifying Effects on Biodiversity. *Ecosphere* **2014**, *5*, 113. [CrossRef]
84. Hardiman, B.S.; Bohrer, G.; Gough, C.M.; Vogel, C.S.; Curtis, P.S. The Role of Canopy Structural Complexity in Wood Net Primary Production of a Maturing Northern Deciduous Forest. *Ecology* **2011**, *92*, 1818–1827. [CrossRef]
85. Iannone, B.V.; Potter, K.M.; Hamil, K.A.D.; Huang, W.; Zhang, H.; Guo, Q.; Oswalt, C.M.; Woodall, C.W.; Fei, S. Evidence of Biotic Resistance to Invasions in Forests of the Eastern USA Landscape. *Ecology* **2016**, *31*, 85–99. [CrossRef]
86. Remmert, H. Sukzessionen im Klimax-System [Successions in the climax system]. *Verh. Ges. Okol.* **1987**, *16*, 27–34.
87. Remmert, H. Das Mosaik-Zyklus-Konzept und seine Bedeutung für den Naturschutz: Eine Übersicht [The mosaic cycle concept and its relevance to conservation: An overview]. *Congr. Rep. Lauf. Semin.* **1989**, *5*, 5–15.
88. Rosenzweig, M.L. *Species Diversity in Space and Time*; Cambridge University Press: Cambridge, UK, 1995.
89. Chase, J.M.; Leibold, M.A. *Ecological Niches: Linking Classical and Contemporary Approaches*; University of Chicago Press: Chicago, IL, USA, 2003. [CrossRef]
90. Ohlson, M.; Söderström, L.; Hörnberg, G.; Zackrisson, O.; Hermansson, J. Habitat Qualities versus Long-Term Continuity as Determinants of Biodiversity in Boreal Old-Growth Swamp Forests. *Biol. Conserv.* **1997**, *81*, 221–231. [CrossRef]
91. Siitonen, J.; Martikainen, P.; Punttila, P.; Rauh, J. Coarse Woody Debris and Stand Characteristics in Mature, Managed and Boreal Mesic Forests in Southern Finland. *For. Ecol. Manag.* **2000**, *128*, 211–225. [CrossRef]
92. Kappes, H.; Topp, W. Emergence of Coleoptera from Deadwood in a Managed Broadleaved Forest in Central Europe. *Biodivers. Conserv.* **2004**, *13*, 1905–1924. [CrossRef]
93. Persiani, A.M.; Audisio, P.; Lunghini, D.; Maggi, O.; Granito, V.M.; Biscaccianti, A.B.; Chiavetta, U.; Marchetti, M. Linking Taxonomical and Functional Biodiversity of Saproxylic Fungi and Beetles in Broad-Leaved Forests in Southern Italy with Varying Management Histories. *Plant Biosyst.* **2010**, *144*, 250–261. [CrossRef]
94. Rondeux, J.; Sanchez, C. Review of Indicators and Field Methods for Monitoring Biodiversity within National Forest Inventories: Core Variable Dead Wood. *Environ. Monit. Assess.* **2009**, *164*, 617–630. [CrossRef] [PubMed]

95.   Brin, A.; Bouget, C.; Brustel, H.; Jactel, H. Diameter of Downed Woody Debris Does Matter for Saproxylic Beetle Assemblages in Temperate Oak and Pine Forests. *J. Insect Conserv. Divers.* **2011**, *15*, 653–669. [CrossRef]
96.   Lassauce, A.; Paillet, Y.; Jactel, H.; Bouget, C. Deadwood as a Surrogate for Forest Biodiversity: Meta-Analysis of Correlations between Deadwood Volume and Species Richness of Saproxylic Organisms. *Ecol. Indic.* **2011**, *11*, 1027–1039. [CrossRef]
97.   Harmon, M.E.; Franklin, J.F.; Swanson, F.J.; Sollins, P.; Gregory, S.V.; Lattin, J.D.; Anderson, N.H.; Cline, S.P.; Aumen, N.G.; Sedell, J.R.; et al. Ecology of Coarse Woody Debris in Temperate Ecosystems. *Adv. Ecol. Res.* **1986**, *15*, 133–302. [CrossRef]
98.   Blasi, C.; Marchetti, M.; Chiavetta, U.; Aleffi, M.; Audisio, P.; Azzella, M.M.; Brunialti, G.; Capotorti, G.; Del Vico, E.; Lattanzi, E.; et al. Multi-taxon and Forest Structure Sampling for Identification of Indicators and Monitoring of Old-growth Forest. *Plant Biosyst.* **2010**, *144*, 160–170. [CrossRef]
99.   Stokland, J.N.; Siitonen, J.; Jonsson, B.G. *Biodiversity in Dead Wood*; Cambridge University Press: Cambridge, UK, 2012.
100.   Cornwell, W.K.; Cornelissen, J.H.C.; Allison, S.D.; Bauhus, J.; Eggleton, P.; Preston, C.M.; Scarff, F.; Weedon, J.T.; Wirth, C.; Zanne, A.E. Plant Traits and Wood Fates across the Globe: Rotted, Burned, or Consumed? *Glob. Chang. Biol.* **2009**, *15*, 2431–2449. [CrossRef]
101.   Litton, C.M.; Raich, J.W.; Ryan, M.G. Carbon Allocation in Forest Ecosystems. *Glob. Chang. Biol.* **2007**, *13*, 2089–2109. [CrossRef]
102.   Kahl, T.; Mund, M.; Bauhus, J.; Schulze, E.D. Dissolved Organic Carbon from European Beech Logs: Patterns of Input to and Retention by Surface Soil. *Ecoscience* **2012**, *19*, 364–373. [CrossRef]
103.   Svensson, J.S.; Jeglum, J.K. Structure and Dynamics of an Undisturbed Old-Growth Norway Spruce Forest on the Rising Bothnian Coastline. *For. Ecol. Manag.* **2001**, *15*, 67–79. [CrossRef]
104.   Schliemann, S.A.; Bockheim, J.G. Methods for Studying Treefall Gaps: A Review. *For. Ecol. Manag.* **2011**, *7*, 1143–1151. [CrossRef]
105.   Zahner, V.; Sikora, L.; Pasinelli, G. Heart Rot as a Key Factor for Cavity Tree Selection in the Black Woodpecker. *For. Ecol. Manag.* **2012**, *271*, 98–103. [CrossRef]
106.   Paul, M.; Hinrichs, T.; Janßen, A.; Schmitt, H.P.; Soppa, B. *Forstliche Genressourcen in Deutschland: Konzepte zur Erhaltung und nachhaltigen Nutzung forstlicher Genressourcen in der Bundesrepublik Deutschland [Forest Genetic Re-sources of Germany: Concepts for Maintenance and Sustainable Use of Forest Genetic Resources in Germany]*; Bundesministerium für Ernährung, Landwirtschaft und Verbraucherschutz: Bonn, Germany, 2000.
107.   Currie, D.J.; Mittelbach, G.G.; Cornell, H.V.; Field, R.; Guégan, F.; Hawkins, B.A.; Kaufman, D.M.; Kerr, J.T.; Oberdorff, T.; O'Brien, E.; et al. Predictions and Tests of Climate-based Hypotheses of Broad-scale Variation in Taxonomic Richness. *Ecol. Lett.* **2004**, *7*, 1121–1134. [CrossRef]
108.   Hilmo, O.; Holien, H.; Hytteborn, H.; Ely-Aastrup, H. Richness of Epiphytic Lichens in Differently Aged *Picea abies* Plantations Situated in the Oceanic Region of Central Norway. *Lichenologist* **2009**, *41*, 97–108. [CrossRef]
109.   Baguette, M.; Deceuninck, B.; Muller, Y. Effects of spruce afforestation on bird community dynamics in a native broad-leaved forest area. *Acta Oecologica* **1994**, *15*, 275–288. Available online: http://hdl.handle.net/2078.1/48606 (accessed on 1 March 2023).
110.   Fahy, O.; Gormally, M. A Comparison of Plant and Carabid Beetle Communities in Irish Oak Woodland with a Nearby Conifer Plantation and Clear-Felled Site. *For. Ecol. Manag.* **1998**, *110*, 263–273. [CrossRef]
111.   Magura, T.B.; Tothmeresz, M.; Bordan, Z. Effects of nature management practices on carabid assemblages (Coleoptera: Carabidae) in a non-native plantation. *Biol. Conserv.* **2000**, *93*, 95–102. [CrossRef]
112.   Chey, V.K.; Holloway, J.D.; Speight, M.R. Diversity of Moths in Forest Plantations and Natural Forests in Sabah. *Bull. Entomol. Res.* **1997**, *87*, 371–385. [CrossRef]
113.   Humphrey, J.W.; Ferris, R.; Jukes, M.R.; Peace, A.J. The Potential Contribution of Conifers Plantations to the UK Biodiversity Action Plan. *Bot. J. Scotl.* **2002**, *54*, 49–62. [CrossRef]
114.   Uliczka, H.; Angelstam, P. Occurrence of Epiphytic Macrolichens in Relation to Tree Species and Age in Managed Boreal Forest. *Ecography* **1999**, *22*, 396–405. [CrossRef]
115.   Brändle, M.; Brandl, R. Species Richness of Insects and Mites on Trees: Expanding Southwood. *J. Anim. Ecol.* **2001**, *70*, 41–504. [CrossRef]
116.   Berglund, H.; O'Hara, R.B.; Jonsson, B.G. Quantifying Habitat Requirements of Tree-Living Species in Fragmented Boreal Forests with Bayesian Methods. *Conserv. Biol.* **2009**, *23*, 1127–1137. [CrossRef]
117.   Ulyshen, M.D. Arthropod Vertical Stratification in Temperate Deciduous Forests: Implications for Conservation-Oriented Management. *For. Ecol. Manag.* **2011**, *261*, 1479–1489. [CrossRef]
118.   Silvertown, J. Plant coexistence and the niche. *Trends Ecol. Evol.* **2004**, *19*, 605–611. [CrossRef]
119.   Turnbull, L.A.; Isbell, F.; Purves, D.W.; Loreau, M.; Hector, A. Understanding the Value of Plant Diversity for Ecosystem Function through Niche Theory. *Philos. Trans. R. Soc. B* **2016**, *283*, 20160536. [CrossRef]
120.   Ponge, J.F. Humus Forms in Terrestrial Ecosystems: A Framework to Biodiversity. *Soil Biol. Biochem.* **2003**, *35*, 935–945. [CrossRef]
121.   Paquin, P.; Coderre, D. Changes in Soil Macroarthropod Communities in Relation to Forest Maturation through Three Successional Stages in the Canadian Boreal Forest. *Oecologia* **1997**, *112*, 104–111. [CrossRef]
122.   Peltier, A.; Ponge, J.-F.; Jordana, R.; Arino, A. Humus Forms in Mediterranean Scrublands with Aleppo Pine. *Soil Sci. Soc. Am. J.* **2001**, *65*, 884–896. [CrossRef]
123.   Ponge, J.F. Biocenoses of Collembola in atlantic temperate grass-woodland ecosystems. *Pedobiologia* **1993**, *37*, 223–244.
124.   Salmon, S.; Artuso, N.; Frizzera, L.; Zampedri, R. Relationships between Soil Fauna Communities and Humus Forms: Response to Forest Dynamics and Solar Radiation. *Soil Biol. Biochem.* **2008**, *40*, 1707–1715. [CrossRef]

125. Schäfer, M.; Schauermann, J. The Soil Fauna of Beech Forests: Comparison between a Mull and a Moder Soil. *Pedobiologia* **1991**, *34*, 299–314.
126. Salmon, S.; Mantel, J.; Frizzera, L.; Zanella, A. Changes in humus forms and soil animal communities in two developmental phases of Norway spruce on an acidic substrate. *For. Ecol. Manag.* **2006**, *237*, 47–56. [CrossRef]
127. Cassagne, N.; Bal-Serin, M.C.; Gers, C.; Gauquelin, T. Changes in Humus Properties and Collembolan Communities Following the Replanting of Beech Forests with Spruce. *Pedobiologia* **2004**, *48*, 267–276. [CrossRef]
128. Gill, R.M.A. A Review of Damage by Mammals on North Temperate Forests III: Impact on Trees and Forests. *Forestry* **1992**, *65*, 363–388. [CrossRef]
129. Pastor, J.; Moen, R.A.; Cohen, Y. Spatial Heterogeneities, Carrying Capacity, and Feedbacks in Animal-Landscape Interactions. *J. Mammal.* **1997**, *78*, 1040–1052. [CrossRef]
130. Reimoser, F. Steering the Impacts of Ungulates on Temperate Forest. *J. Nat. Conserv.* **2003**, *10*, 243–252. [CrossRef]
131. Reimoser, F.; Reimoser, S.; Klansek, E. *Wildlebensräume: Habitatqualität, Wildschadensanfälligkeit, Bejagbarkeit [Wild-Life Areas: Habitat Quality, Susceptability for Game Damage, Huntability]*; Zentralstelle Österreichischer Landesjagdverbände: Vienna, Austria, 2006. Available online: https://wildlife.reimoser.info/special.php (accessed on 1 March 2023).
132. Fahrig, L. Effects of habitat fragmentation on biodiversity. *Annu. Rev. Ecol. Evol. Syst.* **2003**, *34*, 487–515. [CrossRef]
133. Henle, K.; Lindenmayer, D.B.; Margules, C.R.; Saunders, D.A.; Wissel, C. Species Survival in Fragmented Landscapes: Where Are We Now? *Biodivers. Conserv.* **2004**, *13*, 1–8. [CrossRef]
134. Haddad, N.M.; Brudvig, L.A.; Clobert, J.; Davies, K.F.; Gonzalez, A.; Holt, R.D.; Lovejoy, T.E.; Sexton, J.O.; Austin, M.P.; Collins, C.D.; et al. Habitat Fragmentation and Its Lasting Impact on Earth's Ecosystems. *Sci. Adv.* **2015**, *1*, 1500052. [CrossRef] [PubMed]
135. Dobeš, C.; Konrad, H.; Geburek, T. Potential Population Genetic Consequences of Habitat Fragmentation in Central European Forest Trees and Associated Understorey Species—An Introductory Survey. *Diversity* **2017**, *9*, 9. [CrossRef]
136. Ledig, F.T. Human impacts on genetic diversity in forest ecosystems. *Oikos* **1992**, *63*, 87–108. [CrossRef]
137. Johansson, M.; Primmer, C.R.; Merila, J. Does Habitat Fragmentation Reduce Fitness and Adaptability? A Case Study of the Common Frog (Rana Temporaria). *Mol. Ecol.* **2007**, *16*, 2693–2700. [CrossRef] [PubMed]
138. Dixo, M.; Metzger, J.P.; Morgante, J.S.; Zamudio, K.R. Habitat Fragmentation Reduces Genetic Diversity and Connectivity among Toad Populations in the Brazilian Atlantic Coastal Forest. *Biol. Conserv.* **2009**, *142*, 1560–1569. [CrossRef]
139. Amos, W.; Harwood, J. Factors Affecting Levels of Genetic Diversity in Natural Populations. *Philos. Trans. R. Soc. B* **1998**, *353*, 177–186. [CrossRef]
140. Sgro, C.M.; Lowe, A.J.; Hoffmann, A.A. Building Evolutionary Resilience for Conserving Biodiversity under Climate Change. *Evol. Appl.* **2011**, *4*, 326–337. [CrossRef]
141. Andersen, L.W.; Fog, K.; Damgaard, C. Habitat Fragmentation Causes Bottlenecks and Inbreeding in the European Tree Frog (*Hyla arborea*). *Proc. Biol. Sci.* **2004**, *271*, 1293–1302. [CrossRef]
142. Peakall, R.; Lindenmayer, D. Genetic Insights into Population Recovery Following Experimental Perturbation in a Fragmented Landscape. *Biol. Conserv.* **2006**, *132*, 520–532. [CrossRef]
143. Kleinschmit, J.R.G.; Kownatzki, D.; Gregorius, H.R. Adaptational characteristics of autochthonous populations—Consequences for provenance delineation. *For. Ecol. Manag.* **2004**, *197*, 213–224. [CrossRef]
144. Mijnsbrugge, K.; Bischoff, A.; Smith, B. A Question of Origin: Where and How to Collect Seed for Ecological Restoration. *Basic Appl. Ecol.* **2010**, *11*, 300–311. [CrossRef]
145. Worrell, R. A Comparison between European Continental and British Provenances of Some British Native Trees: Growth, Survival and Stem Form. *For. Int. J. For. Res.* **1992**, *65*, 253–280. [CrossRef]
146. Jones, A.T.; Hayes, M.J.; Sackville Hamilton, N. 2001 The Effect of Provenance on the Performance of *Crataegus monogyna* in hedges. *J. Appl. Ecol.* **2001**, *38*, 952–962. [CrossRef]
147. Kremer, A. 2001 Genetic Risks in Forestry. In *Risk Management and Sustainable Forestry*; EFI Proceedings No. 45; EFI: Joensuu, Finland, 2001.
148. Laikre, L.; Ryman, N. Effects on Intraspecific Biodiversity from Harvesting and Enhancing Natural Populations. *AMBIO* **1996**, *25*, 504–509. Available online: http://www.jstor.org/stable/4314530 (accessed on 1 March 2023).
149. Jansen, S.; Konrad, H.; Geburek, T. The Extent of Historic Translocation of Norway Spruce Forest Reproductive Material in Europe. *Ann. For. Sci.* **2017**, *74*, 56–72. [CrossRef]
150. Sala, O.E.; Chapin, F.S.; Armesto, J.J.; Berlow, E.; Bloomfield, J.; Dirzo, R.; Huber-Sanwald, E.; Huenneke, L.F.; Jackson, R.B.; Kinzig, A.; et al. Global Biodiversity Scenarios for the Year 2100. *Science* **2000**, *287*, 1770–1774. [CrossRef]
151. Newbold, T.; Hudson, L.N.; Hill, S.L.L.; Contu, S.; Lysenko, I.; Senior, R.A.; Börger, L.; Bennett, D.J.; Choimes, A.; Collen, B.; et al. Global Effects of Land Use on Local Terrestrial Biodiversity. *Nature* **2015**, *520*, 45–50. [CrossRef]
152. Buiteveld, J.; Vendramin, G.G.; Leonardi, S.; Kamer, K.; Geburek, T. Genetic Diversity and Differentiation in European Beech (*Fagus sylvatica* L.) Stands Varying in Management History. *For. Ecol. Manag.* **2007**, *247*, 98–106. [CrossRef]
153. Buchert, G.P.; Rajora, O.P.; Hood, J.V. Effects of Harvesting on Genetic Diversity in Old-Growth Eastern White Pine in Ontario, Canada. *Conserv. Biol.* **1997**, *11*, 747–758. [CrossRef]
154. Rajora, O.P.; Rahnam, M.H.; Buchert, G.P.; Dancik, B.P. Microsatellite DNA Analysis of Genetic Effects of Harvesting in Old-Growth Eastern White Pine (*Pinus strobus*) in Ontario, Canada. *Mol. Ecol.* **2000**, *9*, 330–348. [CrossRef]
155. Finkeldey, R.; Ziehe, M. Genetic implications of silvicultural regimes. *For. Ecol. Manag.* **2004**, *197*, 231–244. [CrossRef]

156. Ratnam, W.; Rajora, O.P.; Finkeldey, R.; Aravanopoulos, F.; Bouvet, J.-M.; Vaillancourt, R.E.; Kanashiro, M.; Fady, B.; Tomita, M.; Vinson, C. Genetic Effects of Forest Management Practices: Global Synthesis and Perspectives. *For. Ecol. Manag.* **2014**, *333*, 52–65. [CrossRef]
157. Kavaliauskas, D.; Fussi, B.; Westergren, M.; Aravanopoulos, F.; Finzgar, D.; Baier, R.; Alizoti, P.; Bozic, G.; Avramidou, E.; Konnert, M.; et al. The Interplay between Forest Management Practices, Genetic Monitoring, and Other Long-Term Monitoring Systems. *Forests* **2018**, *9*, 133. [CrossRef]
158. Paffetti, D.; Travaglini, D.; Buonamici, A.; Nocentini, S.; Vendramin, G.G.; Giannini, R.; Vettori, C. The Influence of Forest Management on Beech (*Fagus sylvatica* L.) Stand Structure and Genetic Diversity. *For. Ecol. Manag.* **2012**, *284*, 34–44. [CrossRef]
159. Aravanopoulos, F.A. Do Silviculture and Forest Management Affect the Genetic Diversity and Structure of Long-Impacted Forest Tree Populations? *Forests* **2018**, *9*, 355. [CrossRef]
160. Oliver, T.H.; Heard, M.S.; Isaac, N.J.B.; Roy, D.B.; Procter, D.; Eigenbrod, F.; Freckleton, R.P.; Hector, A.L.; Orme, C.D.L.; Petchey, O.; et al. Biodiversity and resilience of ecosystem functions. *Trends Ecol. Evol.* **2015**, *30*, 673–684. [CrossRef]
161. Geburek, T.; Konrad, H. Why the Conservation of Genetic Resources Has Not Worked. *Conserv. Biol.* **2008**, *22*, 267–274. [CrossRef]
162. Carnus, J.M.; Parrotta, J.; Brockerhoff, E.; Arbez, M.; Jactel, H. Planted Forests and Biodiversity. *J. For.* **2006**, *104*, 65–77. [CrossRef]
163. Geburek, T.; Robitschek, K.; Milasowszky, N. A Tree of Many Faces: Why Are There Different Crown Types in Norway Spruce (*Picea abies* [L.] Karst.)? *Flora* **2008**, *203*, 126–133. [CrossRef]
164. Hardy, J.O. How Fat Is the Tail? *Heredity* **2009**, *103*, 437–438. [CrossRef]
165. Smouse, P.E.; Sork, V.L. Measuring Pollen Flow in Forest Trees: An Exposition of Alternative Approaches. *For. Ecol. Manag.* **2004**, *197*, 21–38. [CrossRef]
166. Bacles, C.F.E.; Burczyk, J.; Lowe, A.J.; Ennos, R.A. Historical and Contemporary Mating Patterns in Remnant Populations of the Forest Tree *Fraxinus excelsior* L. *Evolution* **2005**, *59*, 979–990. [CrossRef] [PubMed]
167. Digiovanni, F.; Kevan, P.G. Factors Affecting Pollen Dynamics and Its Importance to Pollen Contamination: A Review. *Ca-Nadian J. For. Res.* **1991**, *21*, 1155–1170. [CrossRef]
168. Salzer, K. Wind- and Bird-Mediated Gene Flow in Pinus Cembra: Effects on Spatial Genetic Structure and Potential Close-Relative Inbreeding. Ph.D. Thesis, University of Zurich, Zurich, Switzerland, 2011.

forests

Article

Effect of Land Use and Land Cover Change on Plant Diversity in the Ghodaghodi Lake Complex, Nepal

Manoj Naunyal [1], Bidur Khadka [1,*] and James T. Anderson [2]

[1] Kathmandu Forestry College, Amarawati Marga, Koteshwor, Kathmandu P.O. Box 1276, Nepal

[2] James C. Kennedy Waterfowl and Wetlands Conservation Center, Belle W. Baruch Institute of Coastal Ecology and Forest Science, Clemson University, P.O. Box 596, Georgetown, SC 29442, USA

* Correspondence: bidurkhadka2005@gmail.com

Abstract: The Ghodaghodi Lake Complex is a Ramsar site, Nepal's first bird sanctuary, and has significant ecological and economic values. The lake complex is in the western part of the lowland of the Terai region. Numerous studies indicate a relation between the normalized difference vegetation index (NDVI), land use, and land cover with plant diversity. However, the association between terrestrial plant diversity and NDVI in the Ghodaghodi Lake Complex is unknown but has important implications due to potential land use changes. We aimed to understand the relationship between plant diversity and NDVI in the Ghodaghodi Lake Complex. We performed a vegetation survey using a simple random sampling methodology. Shannon–Wiener's diversity index (H') was calculated from the field data, and Landsat images were used to compare land use and land cover changes and calculate NDVI values for 2000 and 2022. The image classification shows that forest cover in April and December 2000 was 71.1% and 58.5%, respectively, and was the dominant land cover in the study area. In contrast, agriculture occupied 18.8% and 27.3% in April and December 2000, respectively, and was the primary land use. Forests covered the most land in April (64.8%) and December (65.3%) of 2022. Likewise, agriculture was a widespread land use. We found a significant correlation (r = 0.80, *p* < 0.05) between the NDVI and plant species diversity, as the NDVI explained 65% of plant species diversity. There was a decrease in forest cover from 2000 to 2022. The strong correlation between the NDVI and vegetation species diversity shows that the NDVI can be a substitute for plant diversity. Our findings show that increased NDVI corresponds to increased plant species diversity and that the lake complex had more plant diversity in 2022 than in 2000, despite a decrease in forested lands.

Keywords: correlation; Ghodaghodi Lake; land cover; land dynamic; land use; NDVI; Ramsar site; Shannon's diversity index

Citation: Naunyal, M.; Khadka, B.; Anderson, J.T. Effect of Land Use and Land Cover Change on Plant Diversity in the Ghodaghodi Lake Complex, Nepal. *Forests* **2023**, *14*, 529. https://doi.org/10.3390/f14030529

Academic Editors: Konstantinos Poirazidis and Panteleimon Xofis

Received: 26 January 2023
Revised: 2 March 2023
Accepted: 6 March 2023
Published: 8 March 2023

1. Introduction

Land use and cover are often used interchangeably, but their meanings differ [1]. Land use refers to human activities of how it is worked upon and managed, such as residential, industrial zones, or agriculture. In contrast, land cover is a physical characteristic of Earth's surface, such as forests, agriculture, wetlands, or lakes, solely created by human activities [2]. Land use and land cover change (LULCC) analyses focus on how the land has been used, what types of changes are predicted in the future, and the various driving forces and processes behind these changes [3]. The main driver of LULCC is anthropogenic activities, such as socio-economic activities. Due to rapid human settlement, the land use/cover effect has become an emerging issue worldwide. LULCC responds to social and ecological processes in the landscape.

Vegetation species diversity measures the health of plant communities and ecosystems and influences other biotic communities [4–8] and ecological processes [9–13]. Anthropogenic activities and land use affect plant diversity [14,15]. Therefore, LULCC is essential in assessing plant diversity. Land use and land cover are correlated; thus, a change in one factor affects the other [2]. LULCC impacts biodiversity, water and radiation budgets, trace gas emissions, and other climatic processes [16]. LULCC affects the local, regional, and global climate.

Vegetation species diversity and spectral information from the image are essential in assessing biodiversity. The biodiversity assessment is based on the spectral resolution of the data [17]. Landsat, the remote sensing system, analyzes spectral information collected from visible, near-infrared, and the middle region related to plant properties [18]. Near-infrared and red bands are used for estimating tree diversity [19–21]. Plant diversity and normalized difference vegetation index (NDVI) are also strongly correlated in the savannah biomes [22]. NDVI is sensitive to critical environmental factors, such as rainfall, that affect biodiversity [23]. The NDVI analyzes energy in an ecosystem as primary productivity that defines spectral patterns in plant diversity [20,24]. NDVI measures the energy entering the ecosystem through primary productivity [25]. Because of this relationship between primary productivity and NDVI, the NDVI predicts a change in species richness [19]. Gould [19] studied that this change in the NDVI corresponds with a change in species richness in the arctic landscape of Canada. A positive correlation between the NDVI and vegetation species diversity was presented by Gould [19] and Levin et al. [25]. However, a study in Argentina indicated a poor relationship between the NDVI and secondary productivity [26].

Historical spatial attribute data allow for assessing changes in plant species diversity over time [27]. Walter [28] presented a positive correlation between plant species richness and NDVI in California, USA. Zhang et al. found that the NDVI and net primary productivity (NPP) are correlated. Because of this relationship between the NDVI and NPP and NPP and species richness, researchers showed a connection between the NDVI and species richness [29]. Wang et al. [30] studied a positive correlation between the NDVI and plant diversity in grasslands. Similar results were found by Madonsela et al. [22] in savannahs, Pouteau et al. [31] in rainforests, and Levin et al. [25] in mountainous regions. NDVI explained up to 87% variation in species diversity for a particular vegetation type, landscape, or region [32]. The normalized difference vegetation index (NDVI) ranges from -1 to 1. The negative value represents the surface covered with clouds and water, the positive value is vegetation, and zero shows bare land [33]. An area rich in vegetation reflects a higher NDVI, while the negative value of NDVI represents no vegetation [34]. The NDVI is less sensitive to soil differences, so it is less sensitive to solar elevation. Still, it is susceptible to green vegetation [35].

Due to plant diversity's role in promoting the biodiversity of other taxa and the ecosystem services they provide [4,6,7,11,13], we thought it prudent to study plant diversity in the Ghodaghodi Lake Complex due to potential perturbations on site. The study's general objective is to show the effect of LULCC on plant diversity in the Ghodaghodi Lake Complex, Nepal. The study's goals are to (1) assess the LULCC, (2) show the relation between the NDVI and vegetation species diversity, and (3) assess the terrestrial plant biodiversity in the Ghodaghodi Lake Complex. We assume that there is a linear relationship between the NDVI and plant diversity.

2. Materials and Methods

2.1. Study Area

The Ghodaghodi Lake Complex (GLC) lies in the Kailali district of the Sudur Pashchim Province of Nepal (Figure 1). It is in the southwestern part of Terai, having spatial extents between latitude 28°41′17″ N and longitude 80°56′47″ E, and the elevation is 205 m (about 672.57 ft). The wetlands cover approximately 2500 ha [36]. The Ramsar site comprises 14 large and small lakes and ponds separated by a hillock [37]. On the lower slope of Siwalik, the lake complex is surrounded by tropical deciduous forests [38]. Ghodaghodi (138 ha), Nakharodi (70 ha), and Baishawa (10 ha) are the major lakes of the complex [37]. The GLC is a Ramsar site, a key biodiversity area, and Nepal's first bird sanctuary. It supports a significant population of fishing cats (*Prionailurus viverrinus*) (vulnerable), mugger crocodiles (*Crocodylus palustris*) (vulnerable), a national biodiversity indicator species, *Cotton Pygmy*

Goose (least concern), 319 birds, and 29 fish species [39]. The Basanta forest corridor enables wildlife movement from Siwalik to the Western Terai Complex: Bardia and Suklaphanta National Parks and Dudhwa Tiger Reserve [38]. The GLC provides suitable habitat for the red-crowned roofed turtle (*Kachuga kachuga*) (critically endangered), tiger (*Panthera tigris tigris*) (endangered), three-striped roof turtle (*Kachuga dhongoka*), smooth-coated otter (*Lutrogale perspicillata*) (vulnerable), common otter (*Lutralutra*), swamp deer (*Cervus duvaucelii*), lesser adjutant stork (*Leptoptilos javanicus*), orchid (*Aerideso dorata*) (endangered), lotus (*Nelumbo nucifera*) (threatened), and wild rice (*Hygroryza aristata*) (threatened) [40].

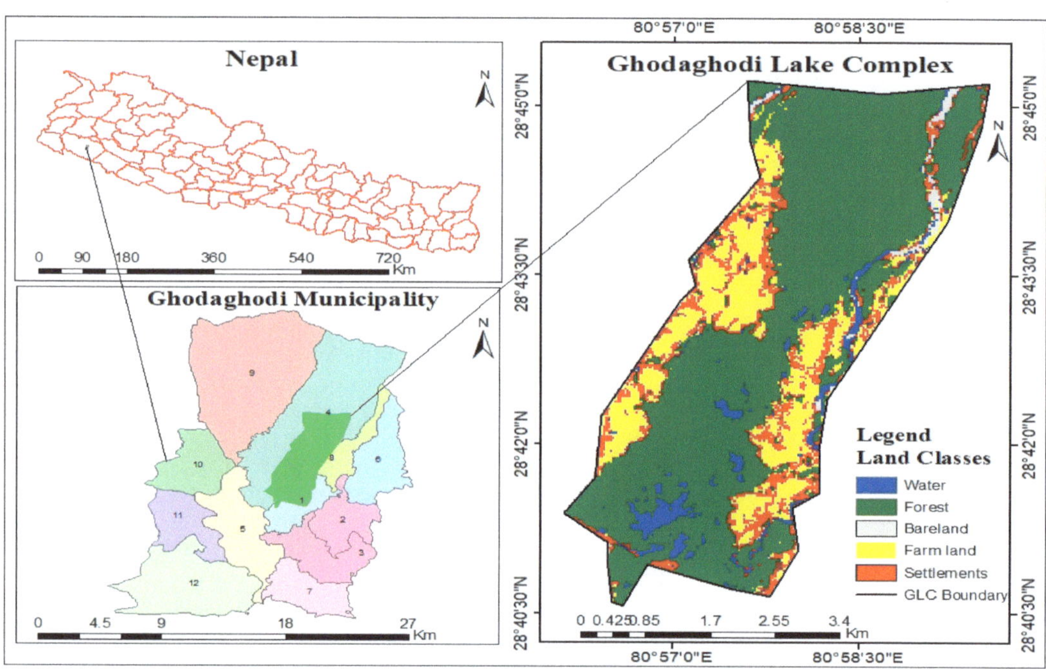

Figure 1. Location of Ghodaghodi Lake Complex (GLC) in the map of Nepal and its land use.

Tharus are the indigenous and dominant people. The site received an annual rainfall of 1385 mm (about 4.54 feet) with an average monthly maximum temperature ranging from 18.8 degrees Celsius to 34.6 degrees and a minimum temperature ranging from 5.5 to 22.9 degrees Celsius in Tikapur (District Profile, 2021, unpublished report). The Chure Rural Municipality borders the Ghodaghodi Lake Complex in the north, Bardagoriya Rural Municipality in the east, Gauriganga Municipality in the west, and Vajani Municipality in the south (Division Forest Office, Pahalmanpur, Kailali, 2022, unpublished report). Ward numbers 1, 4, and 8 are immediate to the Ghodaghodi Lake area, with 39 settlements, of which 14 are in Ward 1, 13 are in Ward 4, and 12 are in Ward 8 (Divisional Forest Office, Pahalmanpur, Kailali, 2022). Tharu have been the significant inhabitants of the area for over two centuries [41]. Migrants from the mountainous region might create massive pressure on wetland resources, causing deforestation, encroachment in and around wetlands, and uncontrolled fishing and illegal hunting. Changes in demographic conditions affect the traditional use of resources. According to the Division Forest Office, Pahalmanpur, Kailali, "The lake complex is surrounded by 11 community forests and two main rivers, Doda to the east and Kauwa to the west".

2.2. Data Collection

Sample plots were established through simple random sampling in the study area. The boundary points of GLC were obtained from the Division Forest Office, Pahalmanpur, Kailali. The GLC was overlain with an 850 m × 850 m grid, resulting in 722,500 m^2 plots. Thirty-one random sampling points were selected. At each sampling point, we set up four plots: a 12.61 m radius circular plot for tree species (diameter at breast height (DBH) $\geq$ 30 cm), 5.64 m for pole species (DBH = 10–29.9 cm), 2.82 m (DBH < 10 cm (3.94 in), height $\geq$ 1 m) for saplings, and 1.78 m (30 cm (11.81 in) $\leq$ height < 1 m) for short saplings [42]. After the plots were set up, each plot's total number of species was recorded (i.e., species richness and number of species per plot). Secondary data were collected from related sources, articles, and Landsat 7 (ETM+) and 9 (OLI2/TIRS2). Landsat 7 (ETM+) capturing on 1 April 2000 and 29 December 2000, and Landsat 9 (OLI2/TIRS2 images (path 144 and row 40)) capturing on 30 April 2022 and 24 November 2022, were downloaded from the USGS (https://earthexplorer.usgs.gov/) accessed on 15 August 2022. Each Landsat image represents the cold and hot seasons in the study area. These Landsat images were radiometrically and geometrically corrected. These images were processed using ERDAS Imagine 2015 and ArcGIS 10.8.

2.3. Image Processing

ERDAS Imagine 2015 software combined multiple images into one image to have the same degree of extent [43]. It also resampled bands with different spatial resolutions into target spatial resolutions. After stacking the layer, the image was subset from Landsat images using the area of interest file (AOI) with a defined study area by applying the sub-set tool in ERDAS Imagine 2015. The Landsat 7 (ETM+) image (2000) has a scanned line error. We removed the error by overlapping images and using the focal analysis tool in ERDAS Imagine 2015. Image improvement was conducted to improve the quality, information content of original data, and visual interpretation of different objects or features in the scene. ERDAS Imagine 2015 image enhancement tools, such as General Contrast, Radiometric Correction, and Noise Correction, were used. Various indices have been developed to extract features of interest from satellite imagery for better classifications [44].

Following supervised classification based on maximum likelihood, five land use–land cover classes were identified in the GLC (i.e., forest, agricultural land, water body, settlement, and bare land). Training samples (i.e., area of interest (AOI)) were provided based on field knowledge [45]. Accuracy assessment plays a significant role in image classification. It compares the classified image to ground truth data [46]. The ground truth points were calculated from the field. The classification was evaluated using Kappa accuracy on a scale between 0 and 1, where 1 stands for complete agreement [47].

$$\text{Kappa} = K^\cap = \frac{\text{observed accuracy} - \text{chance agreements}}{1 - \text{chance agreements}} \tag{1}$$

$$\text{Kappa} = K^\cap = \frac{N \sum_{i=1}^{i} X_{ii} - \sum_{i=1}^{r} (X_{i+} \times X_{+i})}{N^2 - \sum_{i=1}^{r} (X_{i+} \times X_{+i})} \tag{2}$$

where

r = number of rows in the error matrix;
X_{ii} = the number of observations in row i and column i (on the same diagonal);
X_{i+} = total observations in row i;
$sumX_{+i}$ = total observations in column i;
N = total number of observations included in the matrix.

The change assessment was calculated for 2000 and 2022. The data on land classification class changes were assessed by calculating the percentage in the respective years, and the annual change rate was calculated using the following formula [48].

$$\text{Annual change rate} = \left[\left(\frac{b}{a}\right)^{1/n} - 1\right] \times 100 \tag{3}$$

where

a = base year data;
b = end-year data;
n = number of years.

NDVI is commonly used to distinguish vegetation from other features. It measures the amount and vigor of green foliage in an area of land and is a standardized way to measure healthy vegetation [49]. It is a dimensionless index [50]. NDVI is sensitive to visible and near-infrared light reflected by vegetation. The green plant reflects near-infrared and green light because of chlorophyll, while red and blue light are absorbed. Landsat has bands with NIR and red. NDVI measures the difference between near-infrared and red light because near-infrared is strongly reflected by vegetation, while red light is absorbed. So, a high NDVI value corresponds to dense vegetation.

The NDVI is calculated as follows:

$$\text{NDVI} = \frac{(\text{NIR} - \text{RED})}{(\text{NIR} + \text{RED})} \tag{4}$$

where RED and NIR reflect red and near-infrared bands, respectively [51].

2.4. Statistical Analysis

Shannon–Wiener's Diversity Index (H') was calculated for the 31 sampling plots. We used H' as it accounts for species richness and evenness and is not affected by sample size [52,53]. A *t*-test was performed between the NDVI of 2000 and 2022. A simple linear regression test was also performed between the NDVI and plant diversity. We considered tests to be significant at $p = 0.05$.

$$\text{H}' = -\sum_{i=1}^{s}(\text{pi}\ln\text{pi}) \tag{5}$$

where

H' = Shannon–Wiener's Diversity Index;
pi = proportion of individuals in the ith species, i.e., (ni/N);
ni = importance value index of the species;
N = importance value index of all of the species.

3. Results

3.1. Comparing the NDVI of 2000 and 2022 of the Lake Complex

We found vast differences in the NDVI between 2000 and 2022 (Figure 2). We found that the NDVI was higher in 2022 (mean = 0.47, SE = 0.0084, SD = 0.0467) than in 2000 (mean = 0.37, SE = 0.0144, SD = 0.0804) ($p < 0.00001$). The NDVI of 31 sampling plots was compared, which showed that in 2000, Plot 12 had the lowest NDVI (0.11), and Plot 15 had the highest NDVI (0.51), while in 2022, Plot 12 had an NDVI of 0.45 and plant diversity of 2.52 and Plot 15 had an NDVI of 0.56 and plant diversity of 3.1. This relation shows that an increase in the NDVI corresponds with an increase in plant diversity. The NDVI showed a stronger indication of plant diversity in 2022 than in 2000. In 2022, the NDVI ranged from 0.58 to 0.37.

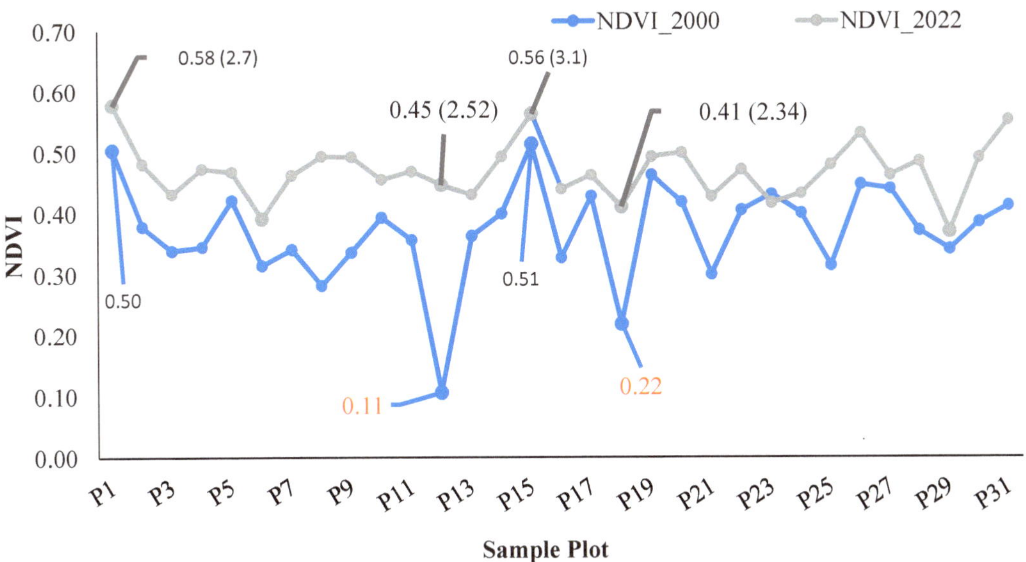

Figure 2. Comparing normalized difference vegetation index (NDVI) between 2000 and 2022 in Ghodaghodi Lake Complex, Nepal.

3.2. Land Use and Land Cover Change in the Ghodaghodi Lake Complex

Landsat 7 (ETM+) and Landsat 9 (OLI 2/TIRS 2) images were used for the land use–land cover classification. The Landsat images showed major changes in the forest and agricultural land in the GLC (Tables 1–3). The image classification of April 2000 showed that about 71.05% of the land was covered by forest, the main land cover in the study area, while agriculture occupied 18.81% (Figures 3 and 4). Water bodies, settlements, and bare ground occupied 3.57%, 2.02%, and 4.55%, respectively. In December, about 54.48% of the land was covered by forest; agriculture, water, settlements, and bare land occupied 27.28%, 9.37%, 1.78%, and 7.19%, respectively. Similarly, the image classification of April 2022 shows that forest cover was the dominant land cover (64.76%) (Figures 3 and 5). Likewise, agriculture was still the primary land use covering 19.17%, while water bodies, settlements, and barren land covered around 5.75%, 9.01%, and 1.24%, respectively. In November, about 65.3% of the land was covered by forests; agriculture, water, settlements, and bare ground occupied 18.6%, 5%, 6.9%, and 4.2%, respectively.

Table 1. The change in land use and land cover (LULC) in Ghodaghodi Lake Complex, Nepal, between April and December 2000.

Site	LULC Classes	LULC 2000 (April)		LULC 2000 (December)	
		Area (ha)	Area (%)	Area (ha)	Area (%)
1	Forest	1646.78	71.05	1262.53	54.48
2	Agricultural land	435.99	18.81	632.21	27.28
3	Water bodies	82.72	3.57	214.80	9.27
4	Settlement	46.72	2.02	41.24	1.78
5	Bare land	105.53	4.55	166.62	7.19
	Grand total	2317.74	100	2317.39	100

Table 2. Land use and land cover (LULC) change in Ghodaghodi Lake Complex, Nepal, between April and November 2022.

Site	LULC Classes	LULC 2022 (April)		LULC 2022 (November)	
		Area (ha)	Area (%)	Area (ha)	Area (%)
1	Forest	1500.85	64.762	1513.14	65.29
2	Agricultural land	444.28	19.171	431.36	18.61
3	Water bodies	133.31	5.752	116.74	5.04
4	Settlement	210.36	9.077	159.95	6.90
5	Bare land	28.69	1.238	96.53	4.16
	Grand total	2317.48	100	2317.72	100

Table 3. Current land use and land cover (LULC) change in Ghodaghodi Lake Complex, Nepal, between 2000 (April) and 2022 (April).

Site	LULC Classes	LULC 2000		LULC 2022	
		Area (ha)	Area (%)	Area (ha)	Area (%)
1	Forest	1646.78	71.05	1500.85	64.762
2	Agricultural land	435.99	18.81	444.28	19.171
3	Water bodies	82.72	3.57	133.31	5.752
4	Settlement	46.72	2.02	210.36	9.077
5	Bare land	105.53	4.55	28.69	1.238
	Grand total	2317.74	100	2317.48	100

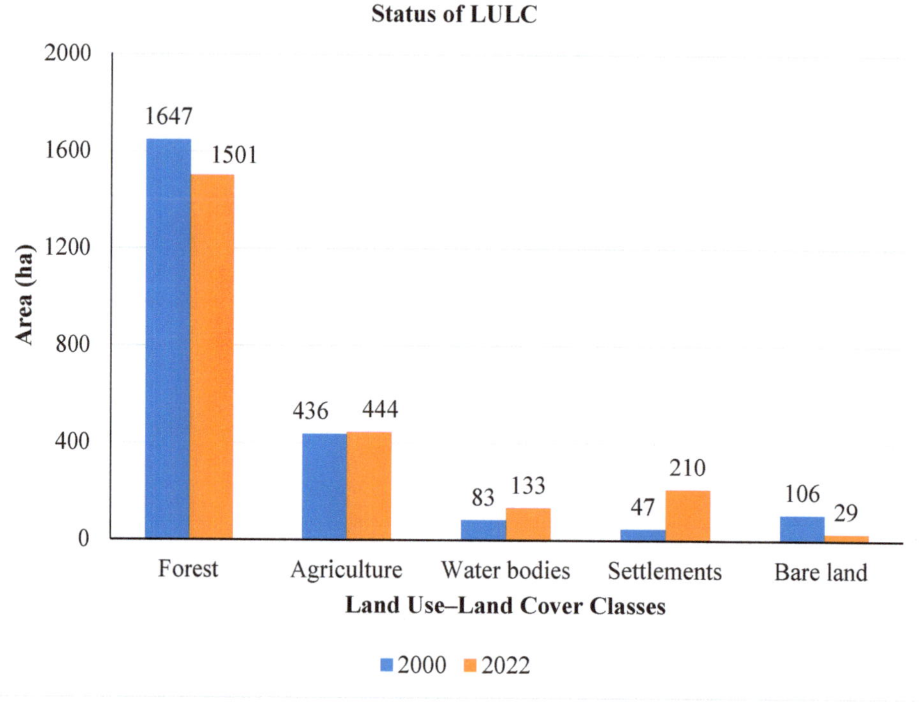

Figure 3. Histogram of land use and land cover (LULC) change in Ghodaghodi Lake Complex (GLC), Nepal, 2000 (April) and 2022 (April).

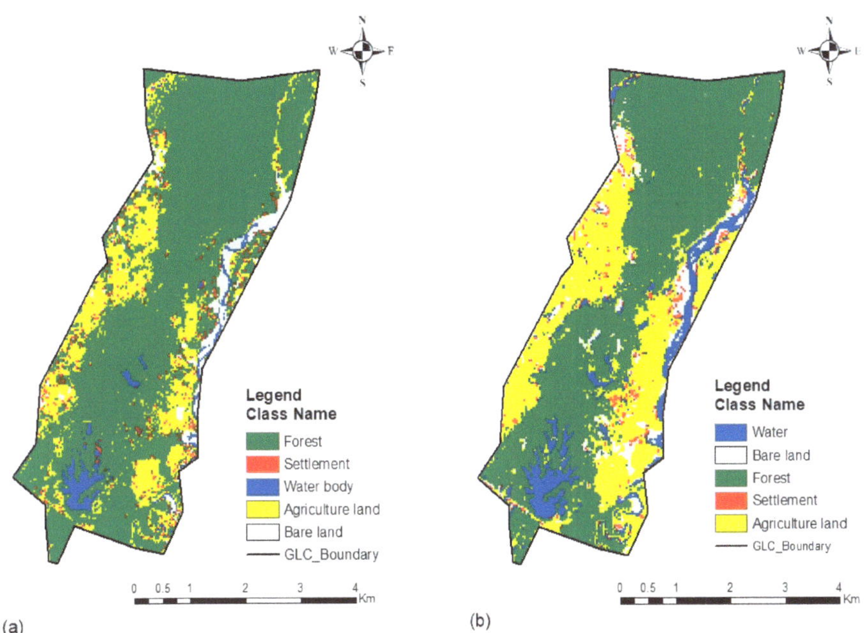

Figure 4. Classified land use and land cover (LULC) map of Ghodaghodi Lake Complex, Nepal, 2000 data. (**a**) LULC map in April, (**b**) LULC map in December.

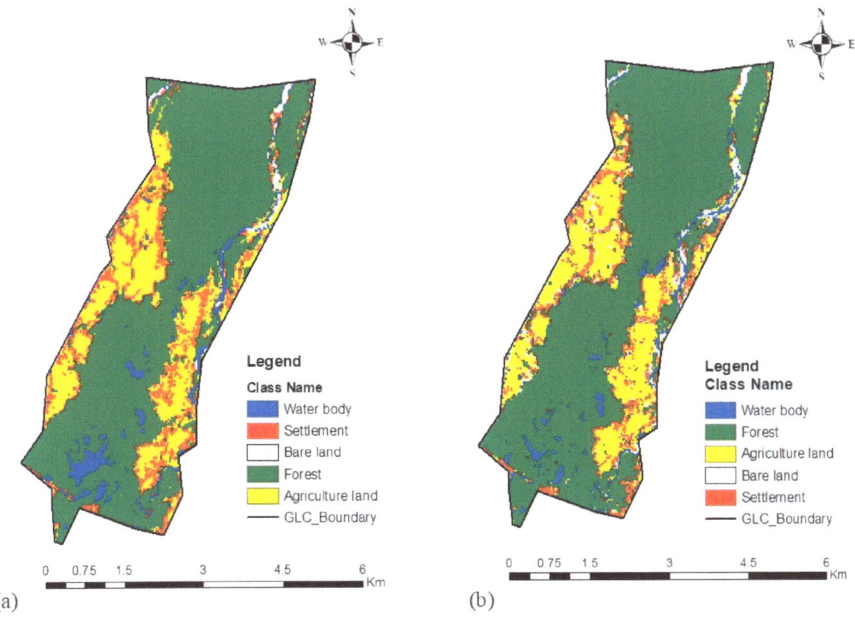

Figure 5. Classified land use and land cover (LULC) map of Ghodaghodi Lake Complex, Nepal, 2022 data. (**a**) LULC map in April, (**b**) LULC map in November.

Forest cover changed into other land uses and cover from 2000 to 2022 (Table 3). While water bodies were changed into forests of 33 ha, similarly, settlements were converted into forest areas of 10 ha (Figures 6 and 7). In contrast, forest cover was transformed into water bodies, settlements, forests, bare land, and agricultural land of 49 ha, 99 ha, 1357 ha, 7 ha, and 133 ha, respectively (Figure 7). Barren land was converted into forests of 23 ha, and agricultural land was changed into forests of 78 ha (Figure 7).

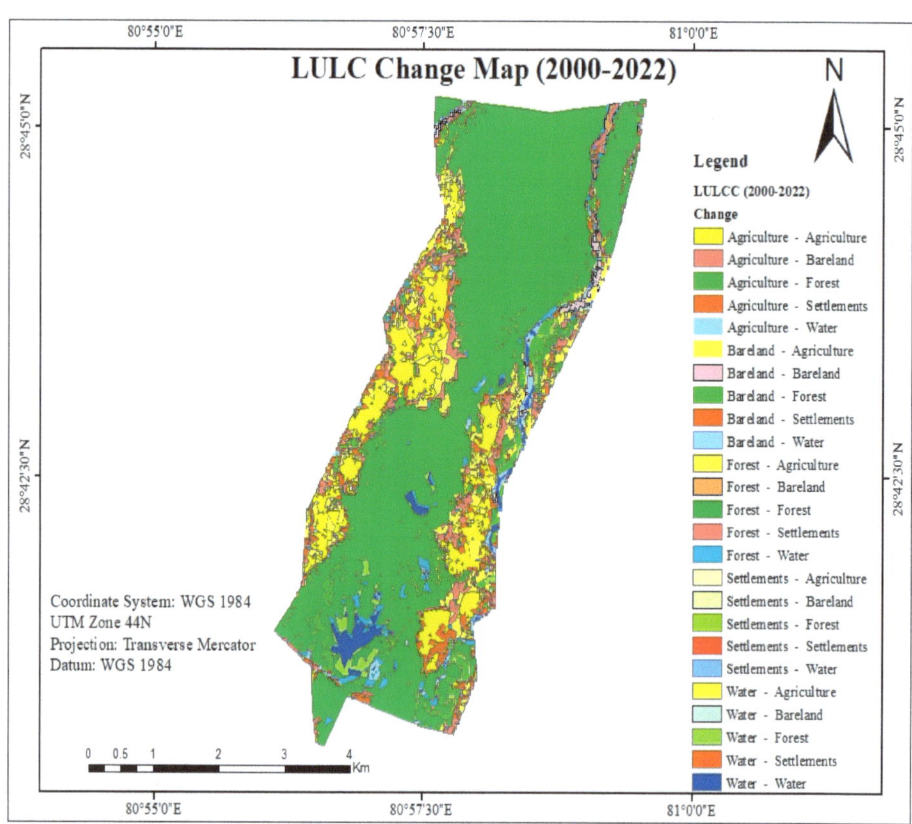

Figure 6. Change detection of land use and land cover (LULC) change in Ghodaghodi Lake Complex, Nepal, between 2000 (April) and 2022 (April).

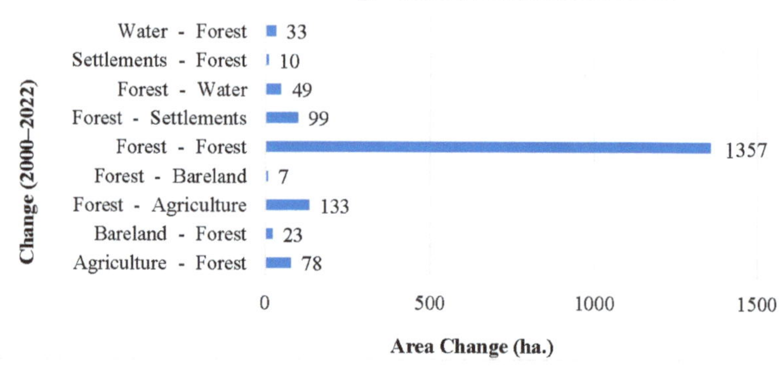

Figure 7. Histogram showing changes in land use cover between 2000 and 2022 in the Ghodaghodi Lake Complex, Nepal.

3.3. Accuracy Assessment of the Classified Map of 2000 and 2022

Accuracy assessment is integral to any classification project. It compares the classified image to another source considered to be accurate or ground truth data [46]. One hundred and fifty ground truth points (fifty from each class) were collected from different land use–land cover classes. The classification was evaluated using Kappa accuracy. Accuracy was 88.67% and 90.67% for the classified map of 2000 (Tables 4 and 5) and 86.00% and 89.33% for 2022 (Tables 6 and 7).

Table 4. Accuracy Assessment of Classified Map of April 2000.

Class Name	Water Body	Bare Land	Forest	Settlement	Agricultural Land	TU	UA
Water Body	28	1	1	0	0	30	93.33
Bare Land	0	27	0	0	3	30	90
Forest	0	0	28	0	2	30	93.33
Settlement	0	0	1	26	3	30	86.67
Agricultural Land	2	2	2	0	24	30	80
TP	30	30	32	26	32	150	OA = 88.67
PA	93.33	90	87.5	100	75	Kappa = 0.858	

TU: total user, TP: total producer, UA: users' accuracy, PA: producers' accuracy, OA: overall accuracy.

Table 5. Accuracy Assessment of Classified Map of December 2000.

Class Name	Water Body	Bare Land	Forest	Settlement	Agricultural Land	TU	UA
Water Body	26	0	3	1	0	30	86.67
Bare Land	0	26	2	0	2	30	86.67
Forest	0	0	30	0	0	30	100
Settlement	0	1	1	25	3	30	83.33
Agricultural Land	0	0	1	0	29	30	96.67
TP	26	27	37	26	34	150	OA = 90.67
PA	100	96.29	81.08	96.15	85.29	Kappa = 0.883	

TU: total user, TP: total producer, UA: users' accuracy, PA: producers' accuracy, OA: overall accuracy.

Table 6. Accuracy Assessment of Classified Map of April 2022.

Class Name	Water Body	Bare Land	Forest	Settlement	Agricultural Land	TA	UA
Water Body	24	2	3	0	1	30	80
Bare Land	1	21	1	4	3	30	70
Forest	0	0	30	0	0	30	100
Settlement	1	2	0	24	3	30	80
Agricultural Land	0	0	0	0	30	30	100
TP	26	25	34	28	37	150	OA = 86.00
PA	92.3	84	88.23	85.71	81.08	Kappa = 0.825	

TU: total user, TP: total producer, UA: users' accuracy, PA: producers' accuracy, OA: overall accuracy.

Table 7. Accuracy Assessment of Classified Map of November 2022.

Class Name	Water Body	Bare Land	Forest	Settlement	Agricultural Land	TU	UA
Water Body	25	2	3	0	0	30	83.33
Bare Land	0	28	0	0	2	30	93.33
Forest	0	0	30	0	0	30	100
Settlement	0	0	0	23	7	30	76.67
Agricultural Land	0	0	1	1	28	30	93.33
TP	25	30	34	24	37	150	OA = 89.33
PA	100	93.33	88.23	95.83	75.67	Kappa = 0.867	

TU: total user, TP: total producer, UA: users' accuracy, PA: producers' accuracy, OA: overall accuracy.

3.4. Diversity Dynamic Depicted Using Remote Sensing

Shannon's Diversity Index was calculated from the field data. The diversity index ranged from 2.07 to 3.05 (Table 8). Similarly, Plot 15 and Plot 6 have the highest and lowest diversity index of 3.05 and 2.07, respectively (Table 8). A regression test was performed between Shannon's Diversity Index and NDVI. We found that plant species diversity depends on NDVI (r = 0.80, p < 0.001). NDVI explained about 65% (r^2 = 0.6522) of the variety in plant species (Figure 8).

Table 8. Shannon's Diversity Index Calculated from sample plots.

Sample Plots	Shannon's Diversity Index
Plot 1	2.90
Plot 2	2.77
Plot 3	2.58
Plot 4	2.74
Plot 5	2.69
Plot 6	2.07
Plot 7	2.93
Plot 8	3.03
Plot 9	2.75
Plot 10	2.70
Plot 11	2.68
Plot 12	2.52
Plot 13	2.49
Plot 14	2.94
Plot 15	3.05
Plot 16	2.48
Plot 17	2.65
Plot 18	2.34
Plot 19	2.88
Plot 20	2.89
Plot 21	2.52
Plot 22	2.83
Plot 23	2.71
Plot 24	2.53
Plot 25	2.77
Plot 26	2.80
Plot 27	2.55
Plot 28	2.74
Plot 29	2.33
Plot 30	2.63
Plot 31	2.88

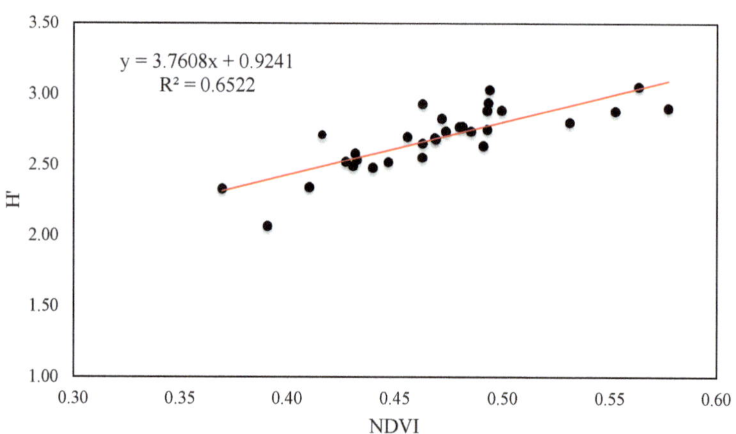

Figure 8. Relationship between normalized difference vegetation index (NDVI) and Shannon–Wiener diversity index (H') for vegetation in Ghodaghodi Lake Complex, Nepal.

4. Discussion

The results present the change in land cover over the past 22 years. The forest cover decreased by 6.29% over 22 years. Similarly, the agricultural area increased by 0.36%, water bodies increased by 2.18%, settlements increased by 7.06%, and bare land decreased by 3.31%. Although there was a decrease in forest cover, water bodies, primarily impoundments and dugouts, grew.

Khanal [54] found that forest cover decreased by 75%, 70%, 65%, and 64% in 1977, 1990, 1999, and 2008, respectively, in three village development committees (VDCs) of the lake complex (i.e., Darakh, Sadapani, and Ramshikharjhala). The loss in forest cover in the VDC lake complex was highest between 1990 and 1999 [54]. Anthropogenic activities, such as continuous grazing, deforestation, roads, encroachment, and illegal forest product extraction, were the significant causes of forest cover loss [55,56]. Others have also indicated the role of road construction on forest loss [54]. However, forest loss associated with roads is only part of the story. Roads influence vegetative communities [57], fish and wildlife assemblages [58–61], soil chemistry [62], stream sedimentation [63], stream morphology [64], water quality [65], and benthic macroinvertebrates [63,65]. Khanal [54] found that southern parts suffer more shrinkage of forest cover than northern parts. This notable change in northern forest cover was due to the natural disaster of floods and landslides [54]. The southern plain is dominated by the highly productive Sal (*Shorea robusta*) forests and has a higher population density than the northern part. Continuous grazing, illicit tree cutting, and encroachment result in more forest cover loss [54]. Forest cover loss was the most intensive between 1990 and 1999 because of the construction of roads [54]. However, the intensity of deforestation was low but still sustained. The active involvement of local communities, ethnic groups, community forest user groups, and youth groups helped to conserve the lake area.

Similarly, the relationship between plant diversity and NDVI was also assessed. We found a strong correlation between the NDVI and plant diversity from this relationship. Chapungu et al. [27] found that the NDVI explained about 62% of the vegetation index. Our results support previous research suggesting that the NDVI can substitute for vegetative species diversity [19,25,27]. Chapungu et al. [27] used the NDVI as a proxy for plant diversity to cover the absence of long-term historical data on plant diversity. Wang et al. [66] presented the linear relationship of NDVI with biomass but a log relationship with vegetation percentage cover. In the case of sparse canopies (crown cover below 60%), the NDVI is more sensitive, while the NDVI is less sensitive to dense canopies (crown cover above 60%) [66]. Mahananda et al. [67] modeled three categories of forest (i.e., evergreen forest, semi-evergreen forest, and deciduous forest) based on the relation between the NDVI and plant species diversity. Around 2000, heavy road expansion resulted in a significant loss in forest cover [54], affecting the NDVI and likely biota [57,60,61,63] and environmental quality [62–65]. However, in 2022, because of the active involvement of local communities in forest conservation and protection, the NDVI improved. Gould [19] and Levin et al. [25] also present a positive correlation between the NDVI and species richness. Here, the NDVI and plant species diversity correspond with each other.

5. Conclusions

The land use and land cover change map showed less forest area in 2022 than in 2000. There was an increase in water bodies, and most bare lands were converted to forests. The major change in forest cover might be due to various road construction activities, migrants from the surrounding hills, various illegal activities, and factors such as climate change, forest fires, and landslides. Although there has been a decrease in forest cover, the lake complex has more plant diversity than in 2000. This change might be due to the involvement of local people in community forest management. The lake complex was declared a Ramsar site in 2003, bringing international attention to the conservation of wetlands. Remote sensing attributes play an essential role in assessing plant diversity. The strong correlation between the NDVI and vegetation species diversity shows that the NDVI and plant diversity are surrogates for each other. From this relationship, we conclude that an increase in plant diversity corresponds to a rise in the NDVI and vice versa.

Author Contributions: M.N., field data collection, methodology, data analysis, and writing the original draft. B.K., methodology, review and editing, and supervision. J.T.A., editing and supervision. All authors have read and agreed to the published version of the manuscript.

Funding: This research was partially funded by Ministry of Industry, Tourism, Forest and Environment, Sudurpaschim province, Nepal. The James C. Kennedy Waterfowl and Wetlands Conservation Center at Clemson University funded the APC.

Data Availability Statement: The data supporting the findings of this study are available from the first author upon request.

Acknowledgments: The author acknowledges the colleagues from Kathmandu Forestry College as well as Assistant Forest Officer from Division Forest Office, Pahalmanpur Kailali, Bishow Raj Pandit for constant support.

Conflicts of Interest: The authors declare no conflict of interest.

References

1. Dimyati, M.; Mizuno, K.; Kobayashi, S.; Kitamura, T. An analysis of land use/cover change in Indonesia. *Int. J. Remote Sens.* **1996**, *17*, 931–944. [CrossRef]
2. Rawat, J.S.; Kumar, M. Monitoring land use/cover change using remote sensing and GIS techniques: A case study of Hawalbagh block, district Almora, Uttarakhand, India. *Egypt. J. Remote Sens. Space Sci.* **2015**, *18*, 77–84. [CrossRef]
3. Maitima, J.M.; Mugatha, S.M.; Reid, R.S.; Gachimbi, L.N.; Majule, A.; Lyaruu, H.; Pomery, D.; Mathai, S.; Mugisha, S. The linkages between land use change, land degradation and biodiversity across East Africa. *Ajol. Info* **2009**, *3*, 310–325. Available online: https://www.ajol.info/index.php/ajest/article/view/56259 (accessed on 16 August 2022).
4. Anderson, J.T.; Zilli, F.L.; Montalto, L.; Marchese, M.R.; McKinney, M.; Park, Y.L. Sampling and Processing Aquatic and Terrestrial Invertebrates in Wetlands. In *Wetland Techniques*; Anderson, J.T., Davis, C.A., Eds.; Springer: New York, NY, USA, 2013; Volume 2, pp. 142–195.
5. Balcombe, C.K.; Anderson, J.T.; Fortney, R.H.; Kordek, W.S. Wildlife use of mitigation and reference wetlands in West Virginia. *Ecol. Eng.* **2005**, *25*, 85–99. [CrossRef]
6. Clipp, H.L.; Anderson, J.T. Environmental and anthropogenic factors influencing salamanders in riparian forests: A review. *Forests* **2014**, *5*, 2679–2702. [CrossRef]
7. Edalgo, J.A.; JT Anderson, J.T. Effects of prebaiting on small mammal trapping success in a Morrow's honeysuckle–dominated area. *J. Wildl. Manag.* **2007**, *71*, 246–250. Available online: https://www.jstor.org/stable/4495169 (accessed on 25 January 2023). [CrossRef]
8. Veselka, W.V.; Anderson, J.T.; Kordek, W.S. Using dual classifications in the development of avian wetlands indices of biological integrity for wetlands in West Virginia, USA. *Environ. Monit. Assess.* **2010**, *164*, 533–548. [CrossRef]
9. Gingerich, R.T.; Anderson, J.T. Decomposition trends of five plant litter types in mitigated and reference wetlands in West Virginia, USA. *Wetlands* **2011**, *31*, 653–662. [CrossRef]
10. Gingerich, R.T.; Panaccione, D.G.; Anderson, J.T. The role of fungi and invertebrates in litter decomposition in mitigated and reference wetlands. *Limnologica* **2015**, *54*, 23–32. [CrossRef]
11. Hedrick, L.B.; Anderson, J.T.; Welsh, S.A.; Lin, L.S. Sedimentation in mountain streams: A review of methods and measurements. *Nat. Resour.* **2013**, *4*, 92–104. Available online: https://www.scirp.org/journal/paperinformation.aspx?paperid=29185 (accessed on 25 January 2023). [CrossRef]
12. Petty, J.T.; Gingerich, G.; Anderson, J.T.; Ziemkiewicz, P.F. Ecological function of constructed perennial stream channels on reclaimed surface coal mines. *Hydrobiologia* **2013**, *720*, 39–53. [CrossRef]
13. Stella, J.C.; Rodriguez-Gonzalez, P.M.; Dufour, S.; Bendix, J. Riparian vegetation research in Mediterranean-climate regions: Common patterns, ecological processes, and considerations for management. *Hydrobiologia* **2013**, *719*, 291–315. [CrossRef]
14. Vitousek, P.M.; Mooney, H.A.; Lubchenco, J.; Melillo, J.M. Human domination of Earth's ecosystems. *Science* **1997**, *277*, 494–499. [CrossRef]
15. Sala, O.E.; Chapin, F.S.; Armesto, J.J.; Berlow, E.; Bloomfield, J.; Dirzo, R.; Huber-Sanwald, E.; Huenneke, L.F.; Jackson, R.B.; Kinzig, A.; et al. Global biodiversity scenarios for the year 2100. *Science* **2000**, *287*, 1770–1774. [CrossRef] [PubMed]
16. Riebsame, W.E.; Meyer, W.B.; Turner, B.L. Modeling land use and cover as part of global environmental change. *Clim. Chang.* **1994**, *28*, 45–64. [CrossRef]
17. Cho, M.A.; Mathieu, R.; Asner, G.P.; Naidoo, L.; van Aardt, J.; Ramoelo, A.; Debba, P.; Wessels, K.; Main, R.; Smit, I.P.J.; et al. Mapping tree species composition in South African savannas using an integrated airborne spectral and LiDAR system. *Remote Sens. Environ.* **2012**, *125*, 214–226. [CrossRef]

18. Hernandez-Stefanoni, J.L.; Gallardo-Cruz, J.A.; Meave, J.A.; Rocchini, D.; Bello-Pineda, J.; López-Martínez, J.O. Modeling α- and β-diversity in a tropical forest from remotely sensed and spatial data. *Int. J. Appl. Earth Obs. Geoinf.* **2012**, *19*, 359–368. [CrossRef]
19. Gould, W. Remote sensing of vegetation, plant species richness, and regional biodiversity hotspots. *Ecol. Appl.* **2000**, *10*, 1861–1870. [CrossRef]
20. Parviainen, M.; Luoto, M.; Heikkinen, R. NDVI-Based Productivity and Heterogeneity as Indicators of Plant-Species Richness in Boreal Landscapes. 2010. Available online: https://helda.helsinki.fi/bitstream/handle/10138/233102/ber15-3-301.pdf?sequence=1 (accessed on 25 January 2023).
21. Wood, E.M.; Pidgeon, A.M.; Radeloff, V.C.; Keuler, N.S. Image texture predicts avian density and species richness. *PloS ONE* **2013**, *8*, e63211. [CrossRef]
22. Madonsela, S.; Cho, M.A.; Ramoelo, A.; Mutanga, O. Remote sensing of species diversity using Landsat 8 spectral variables. *ISPRS J. Photogramm. Remote Sens.* **2017**, *133*, 116–127. [CrossRef]
23. Pau, S.; Gillespie, T.W.; Wolkovich, E.M. Dissecting NDVI–species richness relationships in Hawaiian dry forests. *J. Biogeogr.* **2012**, *39*, 1678–1686. [CrossRef]
24. Witman, J.D.; Cusson, M.; Archambault, P.; Pershing, A.J.; Mieszkowska, N. The relation between productivity and species diversity in temperate–arctic marine ecosystems. *Ecology* **2008**, *89*, S66–S80. [CrossRef] [PubMed]
25. Levin, N.; Shmida, A.; Levanoni, O.; Tamari, H.; Kark, S. Predicting mountain plant richness and rarity from space using satellite-derived vegetation indices. *Divers. Distrib.* **2007**, *13*, 692–703. [CrossRef]
26. Posse, G.; Cingolani, A.M. A test of the use of NDVI data to predict secondary productivity. *Appl. Veg. Sci.* **2004**, *7*, 201–208. [CrossRef]
27. Chapungu, L.; Nhamo, L.; Gatti, R.C. Estimating biomass of savanna grasslands as a proxy of carbon stock using multispectral remote sensing. *Remote Sens. Appl. Soc. Environ.* **2020**, *17*, 100275. [CrossRef]
28. Walter, R.E.; Stoms, D.M.; Estes, J.E.; Cayocca, K.D. Relationships between biological diversity and multi-temporal vegetation index in California. In Proceedings of the American Society for Photogrammetric Engineering and Remote Sensing. American Congress of Surveying and Mapping, Washington, DC, USA, 2–14 August 1992; pp. 562–571.
29. Skidmore, A.; Oindo, B.; Said, M. Biodiversity assessment by remote sensing. In Proceedings of the 30th International Symposium on Remote Sensing of the Environment: Information for Risk Management and Sustainable Development, Honolulu, HI, USA, 25–30 July 2003; pp. 10–14.
30. Wang, R.; Gamon, J.A.; Cavender-Bares, J.; Townsend, P.A.; Zygielbaum, A.I. The spatial sensitivity of the spectral diversity–biodiversity relationship: An experimental test in a prairie grassland. *Ecol. Appl.* **2018**, *28*, 541–556. [CrossRef]
31. Pouteau, R.; Gillespie, T.W.; Birnbaum, P. Predicting tropical tree species richness from normalized difference vegetation index time series: The devil is not in the detail. *Remote Sens.* **2018**, *10*, 698. [CrossRef]
32. Chitale, V.S.; Behera, M.D.; Roy, P.S. Deciphering plant richness using satellite remote sensing: A study from three biodiversity hotspots. *Biodivers. Conserv.* **2019**, *28*, 2183–2196. [CrossRef]
33. Jones, H.G.; Vaughan, R.A. *Remote Sensing of Vegetation: Principles, Techniques, and Applications*; Oxford University Press: Oxford, UK, 2010.
34. Mpandeli, S.; Nhamo, L.; Moeletsi, M.; Masupha, T.; Magidi, J.; Tshikolomo, K.; Liphadzi, S.; Naidoo, D.; Mabhaudhi, T. Assessing climate change and adaptive capacity at local scale using observed and remotely sensed data. *Weather. Clim. Extrem.* **2019**, *26*, 100240. [CrossRef]
35. Matsushita, B.; Yang, W.; Chen, J.; Onda, Y.; Qiu, G. Sensitivity of the enhanced vegetation index (EVI) and normalized difference vegetation index (NDVI) to topographic effects: A case study in high-density cypress forest. *Sensors* **2007**, *7*, 2636–2651. [CrossRef] [PubMed]
36. Lamsal, P.; Atreya, K.; Ghosh, M.K.; Pant, K.P. Effects of population, land cover change, and climatic variability on wetland resource degradation in a Ramsar listed Ghodaghodi Lake Complex, Nepal. *Environ. Monit. Assess.* **2019**, *191*, 1–16. [CrossRef] [PubMed]
37. Siwakoti, M.; Karki, J.B. Conservation status of Ramsar sites of Nepal Tarai: An overview. *Bot. Orient. J. Plant Sci.* **2009**, *6*, 76–84. [CrossRef]
38. Lamsal, P.; Pant, K.P.; Kumar, L.; Atreya, K.; Atangana, A.R.; Bisht, I.; Chistoserdov, A.; De, P.; Escalante, R.; Vergara, P.M. Diversity, uses, and threats in the GhodaghodiLake complex, a Ramsar site in western lowland Nepal. *Int. Sch. Res. Not.* **2014**. [CrossRef]
39. IUCN. *A Review of the Status and Threats to Wetlands in Nepal*; IUCN Nepal: Kathmandu, Nepal, 2004.
40. Dnpwc, B. *Conserving Biodiversity and Delivering Ecosystem Services at Important Bird Areas in Nepal*; Birdlife International: Cambridge, UK, 2012.
41. Sah, J.P.; Heinen, J.T. Wetland resource use and conservation attitudes among indigenous and migrant peoples in Ghodaghodi Lake area, Nepal. *Environ. Conserv.* **2001**, *28*, 345–356. [CrossRef]
42. Dof. *Community Forest Inventory Guideline*; Department of Forest: Kathmandu, Nepal, 2004.
43. Rahman MT, U.; Tabassum, F.; Rasheduzzaman, M.; Saba, H.; Sarkar, L.; Ferdous, J.; Islam, A.Z. Temporal dynamics of land use/land cover change and its prediction using CA-ANN model for southwestern coastal Bangladesh. *Environ. Monit. Assess.* **2017**, *189*, 565. [CrossRef]

44. Bhatti, S.S.; Tripathi, N.K. Built-up area extraction using Landsat 8 OLI imagery. *GIScience Remote Sens.* **2014**, *51*, 445–467. [CrossRef]

45. Olaniyi, O.; Hlengwa, D.; Anderson, J.T. Assessing the driving forces of Guinea savanna transition using geospatial technology and machine learning in Old Oyo National Park, Nigeria. *Geocarto Int.* **2023**, *37*, 17242–17259. [CrossRef]

46. Didelija, M.; Kulo, N.; Mulahusić, A.; Tuno, N.; Topoljak, J. Segmentation scale parameter influence on the accuracy of detecting illegal landfills on satellite imagery. A case study for Novo Sarajevo. *Ecol. Inform.* **2022**, *70*, 101755. [CrossRef]

47. Cohen, J. Weighted kappa: Nominal scale agreement provision for scaled disagreement or partial credit. *Psychol. Bull.* **1968**, *70*, 213–220. [CrossRef]

48. FAO. *Forest Resources Assessment 1990*; Global Synthesis FAO: Rome, Italy, 1995; Available online: http://www.fao.org/docrep/007/v5695e/v5695e00.htm (accessed on 25 January 2023).

49. Kumar, K.S.; Bhaskar, P.U.; Padmakumari, K. Estimation of land surface temperature to study urban heat island effect using Landsat ETM+ image. *Int. J. Eng. Sci. Technol.* **2012**, *4*, 771–778.

50. Weier, J.; Herring, D. Measuring Vegetation (ndvi & evi). *NASA Earth Obs.* **2000**, *20*. Available online: https://earthobservatory.nasa.gov/features/MeasuringVegetation (accessed on 16 August 2022).

51. Li, F.; Chen, W.; Zeng, Y.; Zhao, Q.; Wu, B. Improving Estimates of Grassland Fractional Vegetation Cover Based on a Pixel Dichotomy Model: A Case Study in Inner Mongolia, China. *Remote Sens.* **2014**, *6*, 4705–4722. [CrossRef]

52. Kent, M.; Coker, P. *Vegetation Description and Analysis: A Practical Approach*; John Wiley and Sons: New York, NY, USA, 1992; p. 363.

53. Shannon, C.E. A mathematical theory of communication. *Bell Syst. Tech. J.* **1948**, *27*, 379–423. [CrossRef]

54. Khanal, S. Change assessment of forest cover in Ghodaghodi lake area in Kailali district of Nepal. *Banko Janakari* **2009**, *19*, 15–19. [CrossRef]

55. Bhandari, B. Wise use of wetlands in Nepal. *Banko Janakari* **2009**, *19*, 10–17. [CrossRef]

56. Diwakar, J.; Barjracharya, S.; Yadav, U. Ecological study of Ghodaghodi lake. *Banko Janakari* **2009**, *19*, 18–23. [CrossRef]

57. Rentch, J.S.; Fortney, R.H.; Stephenson, S.L.; Adams, H.S.; Grafton, W.N.; Anderson, J.T. Vegetation-site relationships of roadside plant communities in West Virginia USA. *J. Appl. Ecol.* **2005**, *42*, 129–138. [CrossRef]

58. Poplar-Jeffers, I.O.; Petty, J.T.; Anderson, J.T.; Kite, J.S.; Strager, M.P.; Fortney, R.H. Culvert replacement and stream habitat restoration: Implications from brook trout management in an Appalachian watershed USA. *Restor. Ecol.* **2009**, *17*, 404–413. [CrossRef]

59. Ward, R.L.; Anderson, J.T.; Petty, J.T. Effects of road crossings on stream and streamside salamanders. *J. Wildl. Manag.* **2008**, *72*, 760–771. [CrossRef]

60. Vance, J.A.; Angus, N.B.; Anderson, J.T. Effects of bridge construction on songbirds and small mammals at Blennerhassett Island, Ohio River, USA. *Environ. Monit. Assess.* **2013**, *185*, 7739–7748. [CrossRef] [PubMed]

61. Vance, J.A.; Angus, N.B.; Anderson, J.T. Riparian and riverine wildlife response to a newly created bridge crossing. *Nat. Resour.* **2012**, *3*, 213–228. [CrossRef]

62. Vance, J.A.; Angus, N.B.; Rentch, J.S.; Anderson, J.T. Vegetation and soil parameters at an island bridge crossing. *Castanea* **2014**, *79*, 59–73. [CrossRef] [PubMed]

63. Hedrick, L.B.; SA Welsh, S.A.; Anderson, J.T. Effects of highway construction on sediment and benthic macroinvertebrates in two tributaries of the Lost River, West Virginia. *J. Freshw. Ecol.* **2007**, *22*, 561–569. [CrossRef]

64. Hedrick, L.B.; Welsh, S.A.; Anderson, J.T. Influences of high-flow events on a stream channel altered by construction of a highway bridge: A case study. *Northeast. Nat.* **2009**, *16*, 375–394. [CrossRef]

65. Chen, Y.; Viadero, R.C., Jr.; Wei, X.; Hedrick, L.B.; Welsh, S.A.; Anderson, J.T.; Lin, L. Effects of highway construction on stream water quality and macroinvertebrate condition in a Mid-Atlantic highlands watershed, USA. *J. Environ. Qual.* **2009**, *38*, 1672–1682. [CrossRef] [PubMed]

66. Wang, R.; Gamon, J.A.; Montgomery, R.A.; Townsend, P.A.; Zygielbaum, A.I.; Bitan, K.; Tilman, D.; Cavender-Bares, J. Seasonal Variation in the NDVI–Species Richness Relationship in a Prairie Grassland Experiment (Cedar Creek). *Remote Sens.* **2016**, *8*, 128. [CrossRef]

67. Mahanand, S.; Behera, M.D.; Roy, P.S. Rapid assessment of plant diversity using MODIS biophysical proxies. *J. Environ. Manag.* **2022**, *311*, 114778. [CrossRef]

forests

Article

Evidence for Alternate Stable States in an Ecuadorian Andean Cloud Forest

Ana Mariscal [1,2], Daniel Churchill Thomas [3,*], Austin Haffenden [4], Rocío Manobanda [2], William Defas [2], Miguel Angel Chinchero [2], José Danilo Simba Larco [2], Edison Jaramillo [2], Bitty A. Roy [5] and Mika Peck [6]

[1] Cambugan Foundation, Quito 170521, Ecuador; amariscal2005@yahoo.com
[2] Instituto Nacional de Biodiversidad, Herbario Nacional del Ecuador (QCNE), Quito 170529, Ecuador; rociomanobanda@yahoo.com (R.M.); williandefas_m@yahoo.com (W.D.); cdtk@hotmail.com (M.A.C.); jdlarco2@hotmail.com (J.D.S.L.); bioedijara78@hotmail.com (E.J.)
[3] BayCEER Institute, 95440 Bayreuth, Germany
[4] School of Environment and Natural Resources, University of Wales Bangor, Gwynedd LL57 2UW, UK; aushaff@zoho.eu
[5] Institute of Ecology and Evolution, University of Oregon, Eugene, OR 97403, USA; bit@uoregon.edu
[6] Ecology, Behaviour and Environment, School of Life Sciences, University of Sussex, Brighton BN1 9RH, UK; m.r.peck@sussex.ac.uk
* Correspondence: daniel.thomas@uni-bayreuth.de

Abstract: Tree diversity inventories were undertaken. The goal of this study was to understand changes in tree community dynamics that may result from common anthropogenic disturbances at the Reserva Los Cedros, a tropical montane cloud forest reserve in northern Andean Ecuador. The reserve shows extremely high alpha and beta tree diversity. We found that all primary forest sites, regardless of age of natural gaps, are quite ecologically resilient, appearing to return to a primary-forest-type community of trees following gap formation. In contrast, forests regenerating from anthropogenic disturbance appear to have multiple possible ecological states. Where anthropogenic disturbance was intense, novel tree communities appear to be assembling, with no indication of return to a primary forest state. Even in ancient primary forests, new forest types may be forming, as we found that seedling community composition did not resemble adult tree communities. We also suggest small watersheds as a useful basic spatial unit for understanding biodiversity patterns in the tropical Andes that confound more traditional Euclidean distance as a basic proxy of dissimilarity. Finally, we highlight the conservation value of Reserva Los Cedros, which has managed to reverse deforestation within its boundaries despite a general trend of extensive deforestation in the surrounding region, to protect a large, contiguous area of highly endangered Andean primary cloud forest.

Keywords: Reserva Los Cedros; tropical forest ecology; tropical Andean biodiversity hotspot; alternate stable states; tropical forest conservation; forest succession

Citation: Mariscal, A.; Thomas, D.C.; Haffenden, A.; Manobanda, R.; Defas, W.; Angel Chinchero, M.; Simba Larco, J.D.; Jaramillo, E.; Roy, B.A.; Peck, M. Evidence for Alternate Stable States in an Ecuadorian Andean Cloud Forest. *Forests* **2022**, *13*, 875. https://doi.org/10.3390/f13060875

Academic Editors: Konstantinos Poirazidis and Panteleimon Xofis

Received: 2 May 2022
Accepted: 25 May 2022
Published: 3 June 2022

Publisher's Note: MDPI stays neutral with regard to jurisdictional claims in published maps and institutional affiliations.

1. Introduction

Succession has been a central topic of discussion in ecology for at least a century [1–11]. As with many emergent properties of ecosystems, succession in ecosystems is endlessly controversial: the very existence of stable equilibria states and predictable successional (seral) stages have long been both called into question and defended [2,9,12,13]. In recent times, increasing attention has been given to models that lie within a community assembly and/or coexistence framework to explain and predict community composition in plants [14,15]. Rather than attempting to broadly predict changes in dominant species associations in the form of seral stages, community-assembly and coexistence models give greater attention to the importance of individuals, which sum to explain differences in community compositions. These models emphasize dispersal, priority, and stochastic effects on individuals, and in the case of niche-based or trait-based models, a further discussion of selection or "filtering" by biotic and abiotic factors [16–21]. There is no inherent contradiction among these two broad approaches to community modeling: successional and assembly/coexistence approaches are highly complementary [15,22], and are sometimes packaged into a general "dynamics" framework [23]. This cultural shift in community ecology is perhaps due to a desire by ecologists to move away from the highly idiosyncratic, localized knowledge of a site that is often required to successfully predict successional patterns [12,24], and instead use more abstract and more universal ecological models.

This shift may also perhaps be partly due to the chaotic times in which we find ourselves. In the current era, most ecosystem types are experiencing historical rates of species loss [25,26], and are already undergoing consequences of climate change [27–33]. Ecosystems are also continually receiving new species from human activity [34] and face direct modification or even wholesale elimination due to land-use change [35–37]. Few anthropogenically disturbed ecosystems seem to have been successfully "returned" to stable historical ecological states [34], let alone to states that resemble some type of prehistoric ecological stability. In our destabilized era, the search for ancient species associations that require multiple uninterrupted, decadal journeys through intermediate ecological stages can sometimes seem quixotic.

However, the discussion of succession in forests has taken on particular and new urgency in the current era of accelerating climate change. Forests and their soils fix a significant portion of carbon every year, which has probably been essential in slowing negative effects of anthropogenic atmospheric carbon release [38]. Forests and their soils are also sitting reservoirs of vast amounts of sequestered carbon in soils that are readily lost through disturbance [39]. After many decades of underappreciation as essentially carbon-neutral ecosystems [40], a vibrant discussion has arisen around the role of primary forests not only as immense storehouses of carbon, but also as continuing carbon sinks [41,42]. New, proforestation-centered prescriptions for climate mitigation have therefore gained support: primary or late-stage successional forests should be protected as reserves of existing fixed carbon and potentially high-functioning sinks for further carbon fixation, and management of secondary forests should enhance characteristics that maximize carbon sequestration [43–46]. Such characteristics are often exemplified by old forests [47,48]. Tropical forests are the most volatile portion of the global forest carbon sink, and their role as a sink for atmospheric CO_2, rather than a source, will depend greatly on how they are managed and protected in the coming years [49]. Tropical montane forests are likely more significant carbon stores than was historically thought, given higher-than-predicted productivity and their large surface areas, resulting from complex, high-angle topography [50,51]. Thus, predictive models of succession may now become essential tools in mitigating the climate crisis: an understanding of seral stages (if indeed they exist!) in forests will be necessary for reaching maximal carbon storage and protecting existing carbon reserves.

The current study was undertaken at Reserva Los Cedros, a midelevation cloud forest reserve located in the heart of the tropical Andean biodiversity hotspot [52]. While the Tropical Andes may not display the same levels of alpha tree diversity often seen in the forests of the lower Amazon basin [53,54], the Tropical Andes have long been recognized for high biodiversity across multiple taxonomic groups [55–63]. This biodiversity is characterized by endemism at very fine spatial scales [64,65], making it an extremely important conservation priority [66]. The mechanisms for the high biodiversity and endemism in the tropics at large, including montane forests, have been debated for decades and perhaps centuries [67,68]. High solar energy and constant, plentiful rainfall are characteristic in many areas of the neotropics, and these traits are often associated with high biodiversity [69]. Additionally, the neotropics may have higher speciation rates, acting as evolutionary "cradles", and/or have higher retention of taxa, thus acting as refugia or "museums", allowing for greater accumulation of species over geological time [70,71]. The tropical Andes augment these general biodiversity trends in the tropics with characteristics that promote the role of endemism even further: complex topography creates diversified habitats and environmental gradients, as well as increased niches [58,67,72], increasing the potential for sympatric speciation, and also adding dispersal barriers to increase the potential for fine-scale allopatric speciation [73]. The insular or "island" nature of cloud/mountain systems may be particularly enhanced situations for speciation events that result from neutral events, especially founder effects [72]. All of these factors can interact and sum to create the unique patterns of diversity and endemism observed in the Andes.

Here, we examined the dynamics of early (<20 years) forest succession at Reserva Los Cedros following anthropogenic or natural-gap-forming-disturbances. The landscape of the reserve is dominated by primary forest, but contains mixed-age secondary forests of varying land-use histories (see the Results section). In regions where remnants of ancient forest ecosystems—ecosystems without a history of significant modern anthropogenic disturbance—persist on the landscape, there are advantages for both ecological analysis and restoration efforts. First, we have information in the form of local, functioning examples of the biological complexity and reference points for primary-forest equilibria. Second, forest recovery is facilitated by the presence of species that have coevolved under local conditions for many millennia, available to act as nuclei for reforestation efforts [74,75]. The existence of extensive primary forest has proven particularly valuable for the study of succession and community equilibria in Neotropical forests [22,24,76].

We also examined the spatial signature of this fine-scale beta diversity of trees, with the working hypothesis that individual, steep-sided catchments may be the spatial unit of importance for understanding Andean biodiversity. We hypothesized that two characteristics/processes govern the behavior of community similarity in the tropical Andes: (1) large-scale spatial auto-correlation will cause short-distance comparisons to be more similar than farther comparisons, and with distance, all site comparisons will approach complete dissimilarity, a phenomenon known as Tobler's Law [77]; however, (2) we predicted significant "noise" around this general trend of decreasing similarity, because the complex topography of the Andes causes great dissimilarity even among some very localized comparisons, and conversely, causes highly similar habitats to occur sometimes at great distances apart due to similar site conditions. We predicted that in systems with extremely complex topography such as the Andes, the most informative unit for modeling community dissimilarity would not be Euclidean distance (meters), but instead the number of watersheds crossed. If this hypothesis was supported, it would facilitate future understanding of tropical diversity.

Finally, Los Cedros has succeeded at both forest protection and reforestation, despite its location in a region of high deforestation and now mineral exploration. Thus, we examined here the success of Los Cedros in the conservation of a forest in a region of Ecuador that is under intense extractivist pressure, in terms of forest cover change and IUCN red-list species observed.

2. Methods

2.1. Site—Reserva Los Cedros

All fieldwork described was performed at or directly adjacent to Reserva Los Cedros (www.reservaloscedros.org, accessed on 31 April 2022), a protected forest reserve on the western slope of the Andes, in northwestern Ecuador (00°18′031.000 N, 78°46′044.6′00 W), at 1000–2700 m asl. The reserve lies within the Andean Choco bioregion, one of the most biodiverse habitats on the planet [72,78,79]. It is also considered to be within the tropical Andean biodiversity hotspot [66]. The reserve protects 5256 hectares of cloud forest. Definitions of cloud forest vary and can be quite complex [80], but following Foster [81] and Stadtmüller [80], we use the term cloud forest to mean a forest "whose characteristics are tied to the frequent presence of clouds and mist", and consider the terms montane rain forest and cloud forest to be synonyms. Following Grubb et al. [82,83], the ecotones within the reserve are probably most accurately classified as a cloud forest of mostly lower montane rain forest, with some regions of higher montane cloud forest, and some Elfin forest zones in its highest, least-explored areas. The Reserve also shares a border with the 305,000 hectare nationally protected Cotocachi-Cayapas National Park. Rainfall averages 2903 ($\pm$186 mm) per year according to onsite measurements. Humidity is typically high (~100%), and daily temperatures at the site range from 15 °C to 25 °C [84]. Annual fluctuations in temperatures are minimal. Daily precipitation varies according to the time of year, with the wet season (October–May) to dry (June–September) seasons [85].

The rarity of primary forest in the north Andean region is due to deforestation from precolonial times [76,86], and more recently in the 1960s due to land reform efforts that legalized and encouraged homestead-scale settlement of large government and private ("Hacienda") forest holdings [87]. Los Cedros has apparently largely escaped deforestation during both eras. The land within Los Cedros and the surrounding region was inhabited by the poorly understood Yumbo indigenous group until 1690. Their activities probably significantly altered sites, but with unknown effects on the ecology and canopy cover of the forest, and their economy was likely integrated into the forested setting of the midelevation Andean region [88,89]. However, the impacts of indigenous land use in South American forests has been, until recently, greatly underestimated and misunderstood by scholars [76,90]. Given the rainfall and high humidity, it is unlikely that large-scale deforestation resulted from fires at Los Cedros. Additionally, fire scares were not observed in the vicinity of the study, neither within nor outside the study plots. Other than possible small-scale precolonial indigenous activities, the majority of the Los Cedros forest has seen relatively few anthropogenic alterations.

Land acquisitions to build the reserve were made between 1988 and 1995. Records of ownership and land use prior to the establishment of the reserve are generally not available or are not highly reliable, so oral histories of anthropogenically altered sites were collected from Los Cedros staff and the community. Tracts purchased by Los Cedros were usually developed or deforested only in small proportions of their total area, usually for small-scale mixed ("finca") agriculture or for cattle and mule pasture. In some cases, these small-scale clearings were made for assertion of legal rights over land, rather than agricultural production at scale (Jose DeCoux, pers. comm.). Though acquisitions were made during the period of 1988–1995, use of these sites by previous landowners often apparently continued for years. Prior to abandonment to forest regeneration, sites of intensive agricultural use are understood to usually have been in use for longer periods of time compared to the pastured sites, as agricultural sites were often the sites of small homesteads or fincas. In contrast, pasture sites were often cleared merely to establish ownership before sale, or for temporary hosting of cattle herds.

Once under active management by Reserva Los Cedros, all sites began regeneration of the forest at approximately the same time. The approximate time of abandonment of agriculture or grazing for all sites was 6–7 years prior to the initiation of the survey. All former agricultural and pasture sites were selectively grazed by cattle to reduce competition from graminoids for protected tree seedlings. Tree seedlings were not planted, but were allowed to re-establish by encroachment from the adjacent forest during and after selective grazing. This method is known informally by some workers in the region as "reforestation by cattle rotation", and is intended to release naturally regenerated tree seedlings from intensive competition from pasture grasses [91] without the use of herbicides.

Size and growth forms were used to estimate tree age to corroborate oral history of each area of anthropogenic alteration. However, given the informal/incomplete historical record that was available, it was not possible to recover the exact length of human settlement prior to abandonment for most areas. Sites of plots were selected to be comparable among their qualitatively classified use history (see the Survey Methods subsection below).

Satellite data on forest cover from 1990 shows ~96% forest cover of Los Cedros (see results), of which at least 80% is thought to have been primary forest. Non-forest land use was concentrated in the southeastern portion of the reserve, where the vegetation surveys were undertaken.

2.2. Tree Survey and Plant Identification

2.2.1. Selection of Sites and Categorization of Land-Use History

Primary forest was defined as those forests which had no historical record and no physical evidence of canopy removal or other large structural alteration by humans. However, forest ecosystems exist continuously as matrices of various states of nonanthropogenic gap formation and closure [3,92–96], including montane wet forests and cloud forests (Crausbay and Martin, 2016). As such, primary forest was divided into three categories of land cover based on their gap characteristics: Bosque Cerrado (Closed Forest) "BC" = mature forest with no physical signs of a gap-forming disturbance. Canopy is closed. Bosque Secundario (Secondary Forest) "BS" = sites with evidence of a recent gap, now with a closed canopy. Claros del Bosque (Forest Clearings) "CLB" = recent, natural gaps in the forest.

In addition to primary forest, sites with histories of anthropogenic disturbance were placed into two categories: Regeneración de Fincas Agricultura y Ganadería (Regeneration from Agriculture and Pasture) "RG" = abandoned small family farms, with land use mostly consisting of pasture maintained for cattle. Regeneración Caña de Azúcar (Regeneration from Sugar Cane) "RCA" = intensively farmed sites used for sugar cane or corn production.

2.2.2. Survey Methods

Site-selection surveys were undertaken at Reserva Los Cedros for the years of 2005 and 2007. The southern area of Los Cedros was divided into 4 areas of study, which were then further divided into three sub-blocks each, from which one sub-block was randomly selected and searched for natural gaps. Once located, each natural gap was also accompanied by a BS and BC site, at a minimum of 40 m distance between survey sites (Supplementary Materials Figure S1). Two additional smaller blocks were added in the vicinity of previously settled areas to increase coverage of anthropogenic disturbances. Overall, 61 sites from various land-use histories and elevations were sampled (see results.).

Adult trees were defined as trees at the height 1.5 m with a diameter of 10 cm or greater. Material from all trees fitting this description were sampled within a circle plot of 30 m^2 radius from the center point for morphological identification of species where possible. Additionally, within each survey site of BC, RCA, and RG plots, all juvenile trees were examined. Due to time constraints, juvenile trees from BS and CLR plots were not sampled. Juvenile trees were defined as trees with a maximum diameter less than 10 cm, growing from 50 cm up to 2 m of species known to be capable of growing to adult tree size as given above, under ideal circumstances and with sufficient time. Juvenile trees were surveyed within a square subplot of size 5 × 5 m that was centered within the larger 30 m circular plot. Nomenclatures of identifications were based on the *Flora of Ecuador* [97].

2.3. Statistical and Informatic Methods

All analyses were conducted in Python and R. Python version 3.8.10 [98], using Pandas 1.1.3 [99,100] and Matplotlib 3.1.2 [101] for visualization; and R version 3.6.3 [102] with the in-box R plotter engine were used. All analyses were conducted in an Ubuntu 20.04.2 LTS environment. Where Bayesian analyses were used, Python version 3.7.3 was used for compatibility with the PyMC3 package version 3.8 [103]. All Bayesian models used the default NUTS sampler to sample the posterior. The code for all pertinent statistical analyses was run and recorded using a Jupyter Notebook that was stored in the affiliated GitHub repo, and is viewable as a notebook online (https://nbviewer.jupyter.org/github/danchurch/losCedrosTrees/blob/master/anaData/MariscalDataExploration.ipynb, accessed on 31 April 2022).

2.3.1. Species Accumulation Curves and Richness Estimators

Species accumulation curves for the entire area studied were calculated as the number of adult tree species observed per meter-squared of the physically sampled area. Each subplot covered a circle with a diameter of 30 m, or 0.071 ha. Additionally, a permanent tree diversity plot was established concurrently with rapid surveys in 2005. Within this, trees were sampled in a grid format, every 10 m along an east/west axis and every 5 m along a north/south axis, in a rectangular half-hectare area (50 × 100 m). Trees were identified as species where possible, and unidentified trees were grouped into species-level operational taxonomic designations (Supplementary Table S1). In the case of the permanent plot, tree species accumulation was modeled as a function of trees examined. Species accumulation models and species estimators were calculated using the specaccum, specpool, and associated functions with the Vegan package in R version 2.5–6 [104]. Point predictions of diversity for a 1 ha area were generated using the predict function, which when used as a method of specpool model objects, used a Mao Tau rarefaction method [105] to generate predictions.

2.3.2. Tree Community Turnover (Distance Decay or Beta Diversity)

Turnover in tree communities was modeled and visualized using the Bray–Curtis dissimilarity [1,106] as a function of (1) physical distance or (2) watershed crossings mapped at a small scale. The Bray–Curtis dissimilarities and physical distance matrices among subplots were calculated using the SciPy spatial submodule [107]. Bayesian models of community dissimilarity decay were conducted using PyMC3 package in Python, with visualizations created using the companion ArViz package [108].

Turnover by Physical Distance

Tree community turnover by physical distance was calculated using the Bray–Curtis dissimilarity as a function of physical distance (meters). Two Bayesian models were used as alternative formulations of the above two processes of local variability and large-scale spatial autocorrelation.

Asymptotic Model

In one approach, the Bray–Curtis dissimilarity was modeled as an asymptotic function, or "Michaelis–Menten"-type function:

$$y = \frac{x}{K_m + x} \tag{1}$$

where y is the predicted Bray–Curtis dissimilarity, x is the distance between the compared sites, and K_m is the distance at which half of the maximum Bray–Curtis dissimilarity is reached. This model honors both theoretical constraints of complete similarity at proximal sites and maximum dissimilarity. Additionally, we modeled variation around the mean as a linear variable, reducing as all distant comparisons approached complete dissimilarity (Bray–Curtis dissimilarity = 1). This shrinking variance term ε was exponentiated to avoid negative values. As a Bayesian model, this was formulated in probabilistic terms, with priors, as follows:

$$y \sim N\left(\frac{x}{K_m + x}, \varepsilon\right), \; \varepsilon = e^{(\delta x + \gamma)} \tag{2}$$

$$K_m \sim N(\mu = 200, \sigma = 10) \tag{3}$$

$$\delta \sim N(\mu = -0.0006, \sigma = 0.005) \tag{4}$$

$$\gamma \sim N(\mu = 1.5, \sigma = 0.5) \tag{5}$$

Prior distribution for K_m was loosely estimated based on Draper et al. [53], who showed that in many Amazonian forest ecosystem types, the decay of tree community similarity occurs rapidly, losing more than half their similarity within 200 m. Owing to the novel formulation, priors for delta and gamma of Equation (2) were not found in the existing literature, and were assigned weak priors, with means intended to reflect the initial wide variation and its subsequent tightening around the Bray–Curtis dissimilarity = 1.

Linear Model

In a second formulation, the decay in tree community similarity was modeled as a simple linear equation. Variance around the mean Bray–Curtis dissimilarity was still allowed to vary negatively with distance, beginning with a large initial spread to encompass both highly similar and dissimilar sites at close distances. Additionally, a skewed mean distribution [109,110] was used to describe the variance around the mean function, given the long skew that was observable in the data at most distances. As with the asymptotic model, the variance around the mean function of the Bray–Curtis dissimilarity was described as an exponentiated linear function.

$$y \sim skewNormal(\alpha + \beta x, \varepsilon, \alpha), \ \varepsilon = e^{(\delta x + \gamma)} \tag{6}$$

$$A \sim N(\mu = 0.9, \sigma = 0.2) \tag{7}$$

$$B \sim N(\mu = 0.0001, \sigma = 0.00001) \tag{8}$$

$$A \sim N(\mu = -2.0, \sigma = 0.5) \tag{9}$$

$$\Delta \sim N(\mu = -0.0006, \sigma = 0.005) \tag{10}$$

$$\Gamma \sim N(\mu = -1.5, \sigma = 0.5) \tag{11}$$

Prior distribution for β was again loosely estimated based on Draper et al. [53], using data from community turnover of Amazonian terra firme forests. All other priors were assigned as weakly informative priors.

Model Comparison

Comparisons of performance between these two models were conducted via posterior predictive checks on the data and variance explained (Bayesian R^2) using the sample_posterior_predictive command in the PyMC3 package and the r2_score command in the ArViz package. Bayesian p-values to indicate a balanced performance were calculated using the distribution of residual differences between the model predictions and observed data. To confirm Bayesian linear model results, a classic linear least-squares regression was also applied to the data, using a Wald test with a null hypothesis that the slope of community turnover was zero, as implemented with the linregress function in the SciPy Stats module. Variance defaults (bivariate normal) for the function were not modified.

Overall Turnover by Watershed Crossings

To better understand changes in community as a function of topographic complexity, we delineated small watersheds in the area of the study area. The original digital elevation model used was the ASTER Global Digital Elevation Map version 2 [111]. Every subplot was assigned to a microwatershed and a distance matrix of watershed crossings. Watersheds were delineated using the pysheds package in Python [112]. The distance matrix of microwatershed crossings was calculated using Dijkstra's algorithm for shortest paths [113] in the NetworkX package version 2.4 [114]. Methods for specific watershed delineations are further explained in the statistical Jupyter Notebook (https://nbviewer.jupyter.org/github/danchurch/losCedrosTrees/blob/master/anaData/MariscalDataExploration.ipynb#watersheds, accessed on 31 April 2022).

2.3.3. Ordination by Historical Land Use/Habitat

Structuring in adult tree communities as a function of historical land use/habitat was visualized using the Bray–Curtis dissimilarity and nonmetric multidimensional scaling (NMDS) (Legendre and Legendre, 2012) with the metaMDS function in the Vegan package in R.

2.3.4. GIS Data and Additional Environmental Data

Except where otherwise noted, geospatial data were explored and visualized using tools from GeoPandas (version 0.8.1, https://geopandas.org, accessed on 31 April

2022), Rasterio (version 1.1.8, https://rasterio.readthedocs.io, accessed on 31 April 2022), and QGIS (QGIS v3.4.11-Madeira, OSGEO v3.7.1). Several environmental variables were generated using ASTER Global Digital Elevation Map version 2, including slope, aspect, elevation, eastern and northern exposures, and distance to nearest stream of all subplots. Stream data were digitized using Map CT-NII-C3-d of Bosque Protector Los Cedros from the Ecuadorian Instituto Geográfica Militar. Rasters of slope and aspect were generated with the Raster Terrain Analysis tool suite in QGIS.

2.3.5. Hierarchical Clustering of Sites

To understand the current types of forests present at Los Cedros, adult tree community data from all sites were partitioned into clusters using Ward's minimum variance clustering [106] using the Bray–Curtis dissimilarity, as implemented in the hclust command in the Vegan package in R. The number of clusters was decided by eye from the resulting dendrogram while taking into consideration the branch lengths and meaningful ecological groups.

2.3.6. Prediction of Current Ecological State by Land Use History and Elevation

The current ecological state, as defined based on the cluster analysis, was modeled as a function of historical land-use/habitat data and elevation. Each of the two predictors (height, land use/habitat) was considered individually, and then a combined model was created using both elevation and historical land use/habitat. All models were Bayesian and were written in Python using the PyMC3 package.

For land use/habitat, a multinomial logistic regression model was constructed with a softmax link function and the following priors:

$$\alpha \sim N(\mu = 0, \sigma = 5) \tag{12}$$

$$\beta \sim N(\mu = 0, \sigma = 5) \tag{13}$$

$$\theta = \text{softmax}(\alpha + \beta x) \tag{14}$$

$$y \sim \text{categorical}(\theta \mid y_{\text{observed}}) \tag{15}$$

where y is a vector of predicted probabilities for forest type (cluster) of a site, distributed as a categorical random variable of α, β, and x, and conditioned on our observed current forest type (i.e., cluster number); x is the observed historical land use/habitat in a dummy variable matrix format; α is the y-intercept for each current forest type, and β is the vector of slope coefficients for each of five types of historical land use/habitat for each current forest type. α and β were assigned weak, normally distributed priors centered on zero.

Elevation: Hierarchical clustering of tree communities and NMS ordinations both indicated two distinct "natural" forest types occurring within the extent of the survey, which were designated as forest types III and IV (see the Results section). A logistic regression model was constructed to model differences among forest types III and IV from the hierarchical cluster analysis. To further understand differences between these two forest types, exploratory PERMANOVA models were used to test grouping by all available environmental predictors (see the Jupyter Notebook). Following this, the forest type was modeled as a function of elevation using a Bayesian logistic regression model. Priors and posterior were modeled as follows:

$$\alpha \sim N(\mu = 0, \sigma = 10) \tag{16}$$

$$\beta \sim N(\mu = 0, \sigma = 10) \tag{17}$$

$$\theta = \text{logistic}(\alpha + \beta x) \tag{18}$$

$$y \sim \text{Bernoulli}(\theta \mid y_{\text{observed}}) \tag{19}$$

where α and β are the intercept and slope controlling the boundary decision between type III and type IV forests in terms of elevation; θ is the prior probability that a forest site will be a type III forest given its elevation; and y is a Bernoulli distribution with θ as the mean conditioned by the observed forest type.

For the combined elevation and historical land-use/habitat model, the elevation and historical land-use/habitat predictors were combined into one linear model:

$$\alpha \sim N(\mu = 0, \sigma = 5) \tag{20}$$

$$\beta \sim N(\mu = 0, \sigma = 5) \tag{21}$$

$$\Theta = \text{softmax}(\alpha + \beta x) \tag{22}$$

$$y \sim \text{categorical}(\theta \mid y_{\text{observed}}) \tag{23}$$

where x is the observed historical land use/habitat in a dummy variable matrix format, with an additional column vector giving the elevation of each site; α is the y-intercept for each row of x; and β is the vector of slope coefficients for each column of x for each current forest type (= row of x). α and β were assigned weak, normally distributed priors centered on zero; y is a vector of predicted probabilities for forest type (cluster) of a site distributed as a categorical random variable of α, β, and x, and conditioned on our observed current forest type (i.e., cluster number).

2.3.7. Indicator Species Analysis

Indicator species analyses were conducted to ascertain representative species for both historical land-use/habitat types and current forest type (cluster analysis results) using the indicspecies package in R [115] with the multipatt function and Pearson's phi coefficient correlation (function "r.g").

2.3.8. Spatial Analyses

To determine important spatial scales on which the tree community was changing, and to give shape to possible spatial structuring of the tree community, Moran's eigenvector maps (MEMs) were constructed [116] by following examples as found in documentation for the ADEspatial package (https://cran.r-project.org/web/packages/adespatial/vignettes/tutorial.html, accessed on 1 August 2020). A spatially weighted neighborhood matrix of sample sites was constructed using a Gabriel connectivity matrix with an inverse distance weighting. Abundances of our adult tree community matrix were Hellinger-transformed [106], and new PCA axes of the transformed were determined with the dudi.pca command in the ade4 package. Important MEMs were then selected based on their statistical significance as explanatory terms in a linear model, with the four most important tree community PCA axes as the response variables. Important MEMS were determined using a forward model-selection process, as implemented with the mem.select command in the adespatial package. The default cutoff of alpha = 0.05 was used to decide the statistical significance of MEMs. All available environmental variables were checked for correlations with MEMs using standard linear regression, as implemented in the linregress function in the Python SciPy Stats package. Correction for multiple testing was done with a Benjamini–Hochberg correction, as implemented in the multipletests command in the Python Statsmodels package. For exploratory purposes, the false discovery rate was set at FDR = 0.1.

2.3.9. Juvenile Communities

Juvenile tree community data were available for a subset of historical land-use/habitat types: "Bosque Cerrado (Closed Forest)" (BC), "Regeneración Caña de Azúcar (Regeneration from Sugar Cane)" (RCA), and "Regeneración de Fincas Agricultura y Ganadería (Regeneration from Agriculture and Pasture)" (RG). Juvenile data were not taken for "Claro del Bosque (Forest Clearings)" (CLB) or "Bosque Secondario (Secondary Forest)" (BS). Due to the difficulties associated with identifying poorly understood herbaceous taxa and/or juvenile trees, these data required extensive additional cleaning and verification following the botanical identifications. For the purposes of this study, juvenile trees were defined as undersized plants that would, in time and with proper growth conditions, reach the stature of the mature trees as defined above; namely, woody plants with a diameter-at −1.5 m of 10 cm or greater. This was as compared to herbaceous plants or small woody plants (such as lianas or subshrubs), which would never or rarely reach a 10 cm girth at 1.5 m height.

Samples that did not receive sufficient identification to confidently conclude that the sample was indeed a juvenile tree were discarded from the analysis. This selection was done using automated and manual queries of growth form data requested from the TRY database [117], manual checks of the Encyclopedia of Life [118], and the Smithsonian Tropical Research collections site (https://stricollections.org/portal/index.php, accessed on 1 October 2020). The Gentry manual [119] was also frequently consulted. Exact species information was often not available, and the designation of juvenile tree status was made using genus growth form data where possible. Juvenile tree communities were compared to adult trees by subsetting adult tree community data to just those sites with juvenile data and combining community matrices from both. These were then ordinated using the Bray–Curtis dissimilarity and nonmetric multidimensional scaling (NMS) via the metaMDS function in the Vegan package in R.

2.3.10. Deforestation in the Region

To understand the extractive pressure in the region of Los Cedros, as well as the efficacy of Los Cedros as a conservation project, we examined land-cover changes from 1990 to 2018. Data on changing land cover were accessed from the Ecuadorian Ministerio del Ambiente's ("MAE") interactive map and GIS catalogue, using the "Cobertura vegetal" layers from 1990 and 2018 (available at https://gis-sigde.maps.arcgis.com/apps/webappviewer/idex.html?id=8b53f9388c034b5e8e3147f03583d7ec&fbclid=IwAR2XobS46Szpz4A7IGroPuLCZh5GSJC, accessed on 10 April 2021). Vector data from the MAE were rasterized to a pixel resolution of 30 m^2 using the gdal_rasterize program in the GDAL tool suite of the OSGeo project [120] and visualized in Matplotlib using the rasterio package. Background deforestation rates were constrained to the Cotacachi canton of the Imbabura province, in which Los Cedros is located. Deforestation rates in the similarly-sized, nearby Bosque Protector el Chontal were quantified for comparison with Los Cedros. BP Chontal is managed by the Cattlemen's Association of the nearby town of Brillasol (pers comm. Jose DeCoux). Administrative boundaries of BP Los Cedros and BP Chontal were supplied using the official boundary layer "Bosque y Vegetación protectora", also available at MAE's interactive map website.

3. Results

Pertinent posterior distributions are described here. Full results of the posterior distributions of all models are available in the Jupyter Notebook.

3.1. Species Accumulation Curves and Richness Estimators

A total of 343 adult tree species were directly observed in plots. Estimates of total adult tree species in the study area range from 404 to 566 species (Figure 1A, Table 1). In the permanent 0.5 ha plot, 43 species of tree were observed, distributed among 36 genera and 25 plant families, with a range of predicted total species from 51 to 72 species in the half-hectare (Figure 1B, Table 1). On average, 1 ha in the southern portion of Los Cedros was predicted to host 169 species of adult tree. Chao estimates, first- and second-order jackknife, and bootstrap estimates for the total study area and for the permanent 0.5 ha plot are given in Table 1.

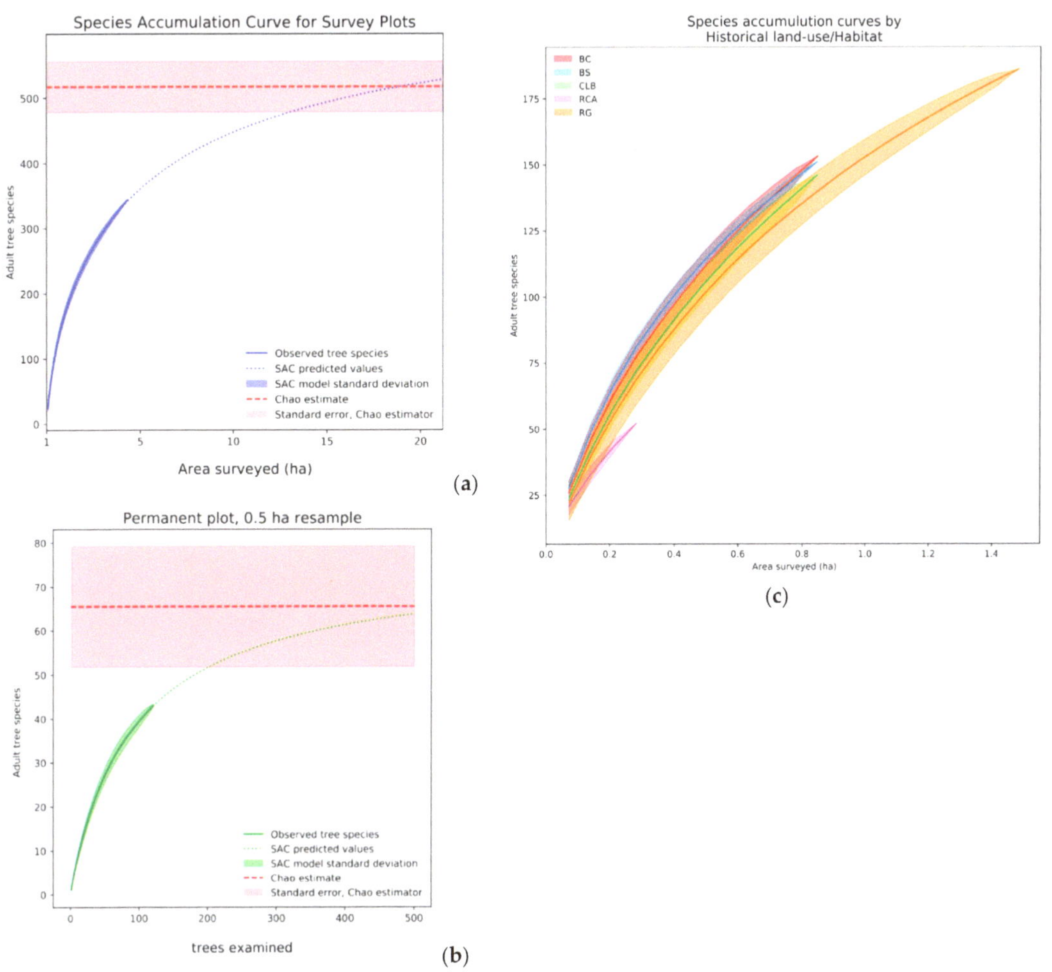

Figure 1. Species accumulation curves (SACs): (**a**) SAC for entirety of study, with all land uses combined. Dotted line is the estimated number of species using a Chao estimator, with one standard error. (**b**) SAC for single 0.5 hectare permanent plot. Dotted line is the estimated number of species using a Chao estimator, with one standard error. (**c**) SAC for entirety of study, itemized by land-use/habitat type. See Section 2.2.1 for explanation of land-use/habitat abbreviations.

Table 1. Estimates of adult tree species diversity in survey area and permanent plot, Using Chao, First- and Second-order jackknife, and bootstrap methods. Uncertainties are given as standard error (".Se").

	Sample Size	Species Observed	Chao	Chao.Se	Jack1	Jack1.Se	Jack2	Boot	Boot.Se
Survey Plots	61	343	516.4	39.2	483.7	21.6	566.8	404.4	10.5
Permanent Plot (Trees)	121	43	65.4	13.8	61.8	4.3	72.7	51.4	2.2

When examined by land-use/habitat type, more adult tree species were observed in reforested pasture sites, possibly due to deeper sampling, but natural-disturbance sites (BS, BC, and CLB) were comparable when rarified to a common sample size (Figure 1C, Supplementary Table S2). Sites with a history of intensive agricultural use were the least diverse, though the depth of sampling was lowest in this group. Species estimators by land-use history are given in Supplementary Table S2.

3.2. Tree Community Turnover (Distance Decay)

3.2.1. Turnover by Physical Distance

1. Asymptotic model

The asymptote model was visualized with the 50% and 95% highest posterior densities ("HPD") for the predicted Bray–Curtis values, as shown in Figure 2A. The posterior distribution of our asymptotic model explained a large amount of the variance in the adult tree community data ($R^2 = 0.53 \pm 0.11$). The distribution of posterior predictive values was nearly symmetric around the observed Bray–Curtis values (Bayesian *p*-value = 0.49, Supplementary Figure S2). The posterior distribution of K_m was centered on 153 m ($\pm$ 6.13 m), a shift of -47 m from the prior estimate of 200 m, with no overlap between the posterior and prior (Supplementary Figure S3).

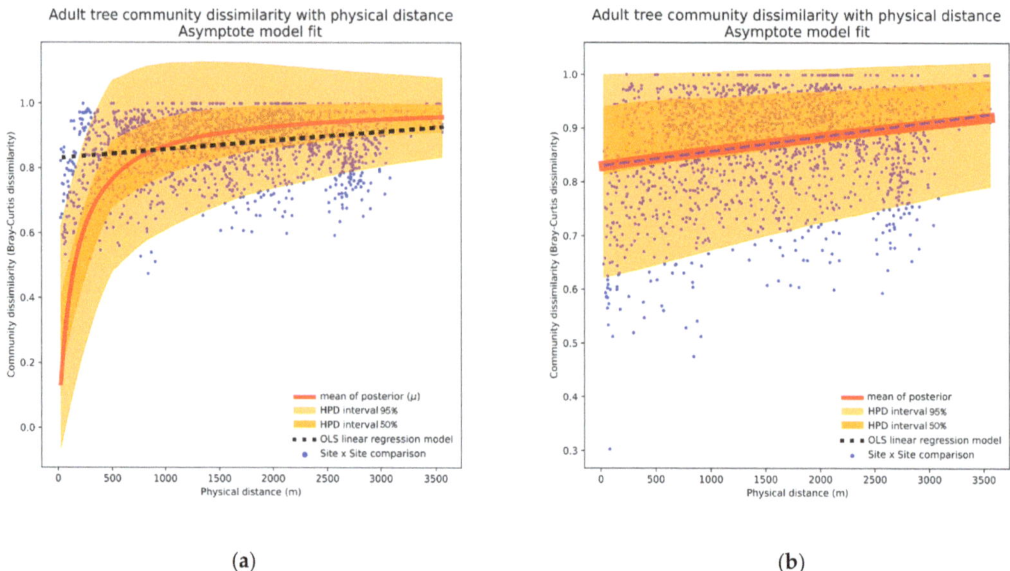

(**a**)   (**b**)

Figure 2. Community turnover and community dissimilarity models. Blue points represent Bray–Curtis dissimilarities between tree communities among all 61 sites, with increasing physical distance between sites being compared. Two approaches are considered, an asymptotic model and a simple linear model. (**a**) Bayesian, asymptotic (Michaelis–Menten) model fit. In this model, the function of the mean community dissimilarity value of zero was enforced when comparing sites that were nearly zero meters apart, and asymptotically approached complete dissimilarity (Bray–Curtis dissimilarity = 1). For comparison, the classical ordinary-least-squares fit is also given (black dashed line). (**b**) Bayesian linear model fit.

2. Linear model

The Bayesian linear model was visualized with the 50% and 95% highest posterior densities ("HPD") for the predicted Bray–Curtis values, as shown in Figure 2B. As constructed, the posterior distribution of our linear model explained only a small amount of the variance in the adult tree community data ($R^2 = 0.06 \pm 0.01$). The posterior predictive values tended to be lower than observed Bray–Curtis values (Bayesian *p*–value = 0.41, Supplementary Figure S2). The frequentist, least-squares linear regression model highly resembled the Bayesian model, also reporting very little variance explained ($R^2 = 0.06 \pm 0.01$, $p = < 0.01$, Figure 2B).

3.2.2. Overall Turnover by Watershed Crossings

When plotted as a function of watershed crossings (Figure 3), the widest range of Bray–Curtis values occurred among sites within the same watershed (distance class 0), as did the lowest mean Bray–Curtis score (mean Bray–Curtis dissimilarity = 0.82, $\pm$ 0.11). Following this, other mean Bray–Curtis values of comparisons of other distance class did not vary heavily, and were mostly statistically indistinguishable (Kruskal–Wallis H-test (H (5, 1825) = 73.27, $p < 0.01$; and Tukey HSD (p–adj = 0.001); Tables 2 and 3, Figure 3). After the first watershed crossing, further comparisons were all approximately equally dissimilar, with a mean of Bray–Curtis = 0.88 ($\pm$0.086).

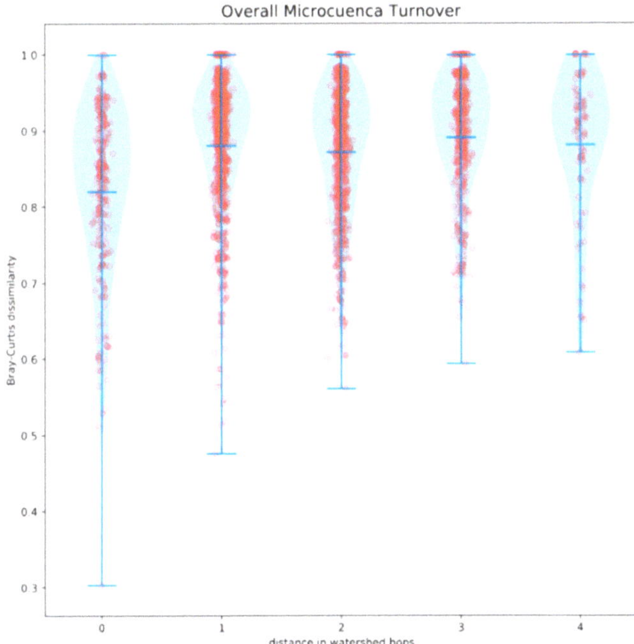

Figure 3. Binning of community turnover into watershed crossings ("hops"). Red points represent Bray–Curtis dissimilarities between tree communities among all 61 sites with increasing watershed crossings between sites.

Table 2. Mean dissimilarity (Bray-Curtis dissimilarity) of survey sites by number of watersheds crossed.

Watersheds Crossed	Sample Size	Mean BC	Std. dev BC
0	215	0.820	0.114
1	630	0.880	0.089
2	559	0.872	0.085
3	346	0.891	0.079
4	80	0.881	0.096

Table 3. Tukey's honest significant difference of comparisons among watershed distance classes.

Group1	Group2	Meandiff	*p*-Adj	Lower	Upper	Reject
0	1	0.0601	0.001	0.0408	0.0795	TRUE
0	2	0.0516	0.001	0.0319	0.0713	TRUE
0	3	0.0708	0.001	0.0495	0.0921	TRUE
0	4	0.0609	0.001	0.0287	0.093	TRUE
1	2	−0.0085	0.4759	−0.0228	0.0057	FALSE
1	3	0.0106	0.393	−0.0058	0.0271	FALSE
1	4	0.0007	0.9	−0.0284	0.0299	FALSE
2	3	0.0192	0.0158	0.0024	0.036	TRUE
2	4	0.0093	0.9	−0.0201	0.0386	FALSE
3	4	−0.0099	0.9	−0.0404	0.0205	FALSE

3.3. Hierarchical Clustering of Sites and Prediction of Current Ecological State by Land-Use History and Elevation

Tree communities were first sorted into two large groups that aligned well with previous land use: the first grouping contained sites that have historically experienced anthropogenic disturbances of any type, and the second grouping contained sites with no recorded history of anthropogenic disturbance, but all types of natural gap-forming disturbances (Figure 4). Each of these groups then were sorted into two further groups (for a total of four clusters). Cluster I contained all sites considered to be highly disturbed by agricultural use (RCA sites) and some sites with intermediate agricultural use as pasture (RG sites) (Figure 4). Cluster II consisted entirely of RG sites, and also contained the majority of RG sites. Sites with no history of anthropogenic disturbances fell into two clusters: type III and type IV (Figure 4). The hierarchical clustering results were congruent with the NMS ordinations (Figure 5, Supplementary Figure S4). A summary map giving both the current ecological state and the historical land use/habitat is given in Figure 6.

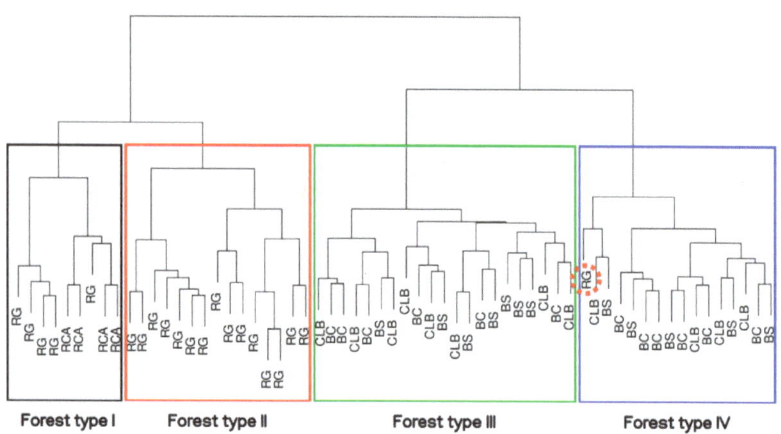

Figure 4. Hierarchical clustering of all sites using species community composition. Tips of the tree represent sites. These four obvious groupings were used to categorize the current ecological state of the forest at each site, or "Forest type". Site circled with a red-dotted line represents site 10.1, which was historically pastured but regenerated to resemble an "old" forest type IV. The history of land use at each site is noted.

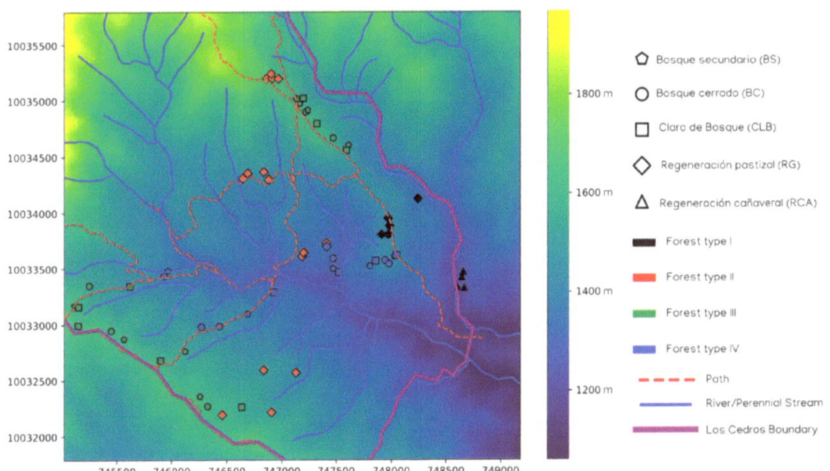

Figure 5. Nonmetric multidimensional scaling (NMS) ordination of adult tree communities using Bray–Curtis dissimilarity. Kruskal's stress = 0.29. Both land-use history (hulls) and current ecological state ("Forest Type", point fill color) are noted. Site circled with a red-dotted line represents site 10.1, which was historically pastured but regenerated to resemble an "old" forest type IV.

Figure 6. Map of sites with land-use history and current ecological state ("forest type"). *X* and *Y* coordinates are Universal Transverse Mercator (zone 17S) coordinates in meters.

A linear model with these ecological states (forest types I–IV) as the response variable using the single variable of previous land-use/habitat as predictors successfully predicted the sites that had been modified for pasture or agriculture, and it also well predicted nonanthropogenically disturbed sites as a type III/IV ecological state ($R^2 = 0.65 \pm 0.04$, Supplementary Figure S5). Posterior predictive checks were correct for 65% of sites, mostly for sites with type I and type II forests. However, this previous land-use/habitat-only model could not distinguish among "natural forest" (III and IV) types.

Among sites with no history of anthropogenic disturbance ("natural forests", types III and IV), elevation strongly predicted the current ecological category of forest: sites with no history of anthropogenic disturbance at elevations of 1503 m (95% credible interval = 1469 m to 1537 m) or higher were categorized almost completely as type III forests, and below this elevation, sites were categorized as type IV forests (Bayesian logistic regression, $R^2 = 0.87$; Figures 6 and 7, Supplementary Figure S6).

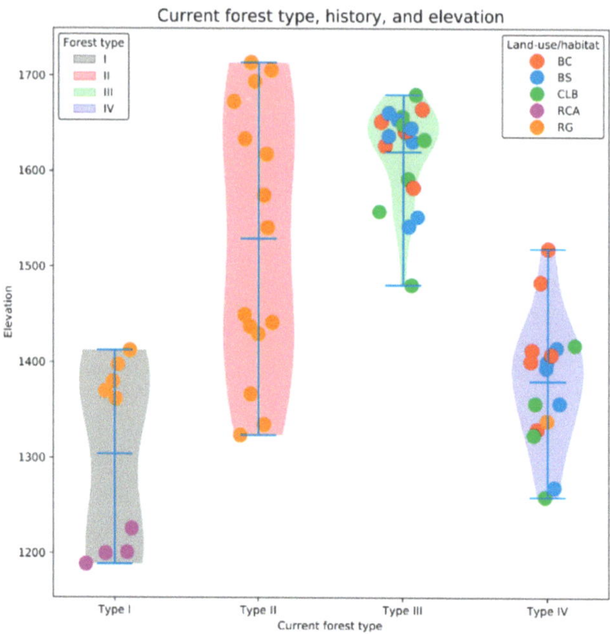

Figure 7. Current ecological state of sites ("forest type") and their elevations. Historical land uses of sites are indicated by color of point fill.

A combined model of previous land use/habitat and elevation explained most of the variance in the total and predicted the current ecological state very well (Bayesian linear model, $R^2 = 0.88 \pm 0.16$; Supplementary Figure S7), and performed much better in the posterior predictions, with 87% of sites correctly predicted as their forest type.

3.4. Indicator Species Analysis

Indicator species were detected for all individual land-use-history/habitat types, all current ecological states (forest types), and for some combinations of site types. The indicator species analysis results are given in Table 4.

Table 4. Indicator species for historical land use/habitat, and indicator species for current ecological state (forest type). "Rpb.G" represents the point-biserial-serial-correlation-coefficient, a measure of the strength of the co-occurrence pattern observed between a tree species and a habitat-type or ecological state. Statistical significance codes: "***" $p < 0.001$; "**" $p < 0.01$; "*" $p < 0.05$.

Indicator Species by Land-Use-History/Habitat-Type			
SGroup BC			
Indicator Species	Rpb.G	*p*-Value	Sig Code
Pseudolmedia rigida (Moraceae)	0.423	0.0317	*
Ficus subandina (Moraceae)	0.406	0.049	*
Group BS			
	Rpb.G	*p*-Value	Sig Code
Psychotria paeonia (Rubiaceae)	0.478	0.0148	*
Persea pseudofasciculata (Lauraceae)	0.432	0.0351	*
Myrcia aff. *aliena* (Myrtaceae)	0.425	0.0377	*
Group CLB			
	Rpb.G	*p*-Value	Sig Code
Endlicheria cf. *chalisea* (Lauraceae)	0.431	0.0268	*
Group RCA			
	Rpb.G	*p*-Value	Sig Code
Cordia colombiana (Boraginaceae)	0.894	0.0001	***
Saurauia sp. 1 (Actinidaceae)	0.694	0.0003	***
Miconia aff. *brevitheca* (Melastomataceae)	0.683	0.0004	***
Ficus caldasiana (Moraceae)	0.667	0.0033	**
Turpinia occidentalis (Staphyleaceae)	0.663	0.0006	***
Clarisia biflora (Moraceae)	0.645	0.0022	**
Nectandra membranacea (Lauraceae)	0.61	0.0027	**
Aegiphila alba (Verbenaceae)	0.594	0.002	**
Ficus andicola (Moraceae)	0.539	0.0191	*
Caryodaphnopsis theobromifolia (Lauraceae)	0.535	0.0099	**
Nectandra aff. *membranacea* (Lauraceae)	0.513	0.0106	*
Group RG			
	Rpb.G	*p*-Value	Sig Code
Cecropia andina (Cecropiaceae)	0.662	0.0018	**
Cecropia ficifolia (Cecropiaceae)	0.58	0.0035	**
Cecropia sp. 2 (Cecropiaceae)	0.427	0.031	*
Meriania tomentosa (Melastomataceae)	0.415	0.0426	*
Group BC + BS			
	Rpb.G	*p*-Value	Sig Code
Aniba aff. *hostmanniana* (Lauraceae)	0.411	0.0471	*
Group RCA + RG			
	Rpb.G	*p*-Value	Sig Code
Solanum lepidotum (Solanaceae)	0.529	0.006	**

Table 4. *Cont.*

Indicator Species by Land-Use-History/Habitat-Type			
Cestrum megalophyllum (Solanaceae)	0.423	0.0268	*
Group BC + BS + CLB			
	Rpb.G	*p*-Value	Sig Code
Alsophila erinacea (Cyatheaceae)	0.42	0.0423	*
Indicator Species by Current Ecological State (Forest Type)			
Group I			
	Rpb.G	*p*-Value	Sig Code
Cordia colombiana (Boraginaceae)	0.661	0.0001	***
Meriania tomentosa (Melastomataceae)	0.641	0.0001	***
Saurauia sp. 1 (Actinidaceae)	0.545	0.0003	***
Cyathea halonata (Cyatheaceae)	0.458	0.0033	**
Senna dariensis (Fab. Caesalpiniaceae)	0.449	0.0046	**
Ficus caldasiana (Moraceae)	0.42	0.0206	*
Leandra subseriata (Melastomataceae)	0.37	0.015	*
Miconia aff. *brevitheca* (Melastomataceae)	0.368	0.027	*
Piper fuliginosum (Piperaceae)	0.356	0.05	*
Turpinia occidentalis (Staphyleaceae)	0.346	0.0426	*
Group II			
	Rpb.G	*p*-Value	Sig Code
Cecropia andina (Cecropiaceae)	0.811	0.0001	***
Cecropia sp. 2 (Cecropiaceae)	0.5	0.0009	***
Melastomataceae sp. 1	0.468	0.0051	**
Dussia lehmannii (Fab. Faboideae)	0.334	0.0444	*
Group III			
	Rpb.G	*p*-Value	Sig Code
Otoba gordoniifolia (Myristicaceae)	0.524	0.0007	***
Alsophila erinacea (Cyatheaceae)	0.515	0.0015	**
Persea aff. *pseudofasciculata* (Lauraceae)	0.494	0.0022	**
Vismia lauriformis (Clusiaceae)	0.44	0.0065	**
Conostegia aff. *centronioides* (Melastomataceae)	0.418	0.0083	**
Wettinia aff. *oxycarpa* (Arecaceae)	0.403	0.0115	*
Persea pseudofasciculata (Lauraceae)	0.386	0.0217	*
Psychotria paeonia (Rubiaceae)	0.373	0.0264	*
Ficus dulciaria (Rubiaceae)	0.354	0.03	*
Group IV			
	Rpb.G	*p*-Value	Sig Code
Dacryodes cupularis (Burseraceae)	0.754	0.0001	***
Protium ecuadorense (Burseraceae)	0.638	0.0001	***
Garcinia macrophylla (Clusiaceae)	0.613	0.0001	***

Table 4. *Cont.*

Indicator Species by Land-Use-History/Habitat-Type			
.*Beilschmiedia* aff. *costaricensis* (Lauraceae)	0.448	0.0018	**
Conostegia superba (Melastomataceae)	0.444	0.0024	**
Ocotea stenoneura (Lauraceae)	0.369	0.0454	*
Gustavia dodsonii (Lecythidaceae)	0.362	0.029	*
Styrax weberbaueri (Styracaceae)	0.361	0.0408	*
Pseudolmedia rigida (Moraceae)	0.355	0.0263	*
Group I + II			
	Rpb.G	*p*-Value	Sig Code
Cecropia ficifolia (Cecropiaceae)	0.52	0.0005	***
Solanum lepidotum (Solanaceae)	0.447	0.0055	**
Cecropia reticulata (Cecropiaceae)	0.371	0.0244	*
Aegiphila alba (Verbenaceae)	0.348	0.0465	*
Urera caracasana (Urticaceae)	0.341	0.037	*
Group I + IV			
	Rpb.G	*p*-Value	Sig Code
Caryodaphnopsis theobromifolia (Lauraceae)	0.416	0.0092	**
Clarisia biflora (Moraceae)	0.403	0.0157	*
Group II + III			
	Rpb.G	*p*-Value	Sig Code
Guatteria megalophylla (Annonaceae)	0.411	0.0133	*
Ficus cuatrecasana (Moraceae)	0.364	0.0299	*
Group III + IV			
	Rpb.G	*p*-Value	Sig Code
Aniba aff. *hostmanniana* (Lauraceae)	0.401	0.0138	*
Capparis sp. (Capparaceae)	0.385	0.0267	*
Hieronyma asperifolia (Euphorbiaceae)	0.383	0.0226	*
Eschweilera integrifolia (Lecythidaceae)	0.379	0.0199	*
Ocotea insularis (Lauraceae)	0.366	0.0295	*
Faramea oblongifolia (Rubiaceae)	0.362	0.0291	*
Helicostylis tovarensis (Moraceae)	0.346	0.0444	*

3.5. Spatial Analyses

A total of 13 Moran's eigenvector maps variables were detected as correlated with changes in tree community (Figure 8; cumulated, adjusted $R^2 = 0.19$, all MEMs $p < 0.02$ as components in forward model selection). Most were substantially correlated with 1–3 available environmental variables or a particular land-use/habitat history (Table 5), with the exception of one highly localized MEM variable, MEM3. The most influential MEM variables in the model, MEM1 and MEM2, presented obvious geographic patterns. MEM1 (contributing $R^2 = 0.026$ to cumulative R^2, or 13.9% of explainable variance) indicated a general difference in tree communities between those of the northeastern and southwestern sides of the Los Cedros River (Supplementary Figure S8). The second most influential MEM variable, MEM2 (contributing $R^2 = 0.026$ to cumulative R^2, or 13.6% of explainable variance), indicated a difference between the highest sites sampled in the study, which were situated along a ridge that runs ultimately to the highest points in the reserve, and lower elevations along this ridge system (Supplementary Figure S9). MEM8 also was correlated with elevation and with distance to nearest stream, likely indicating a difference among "high and dry" sites and lower, wetter sites (Supplementary Figure S10).

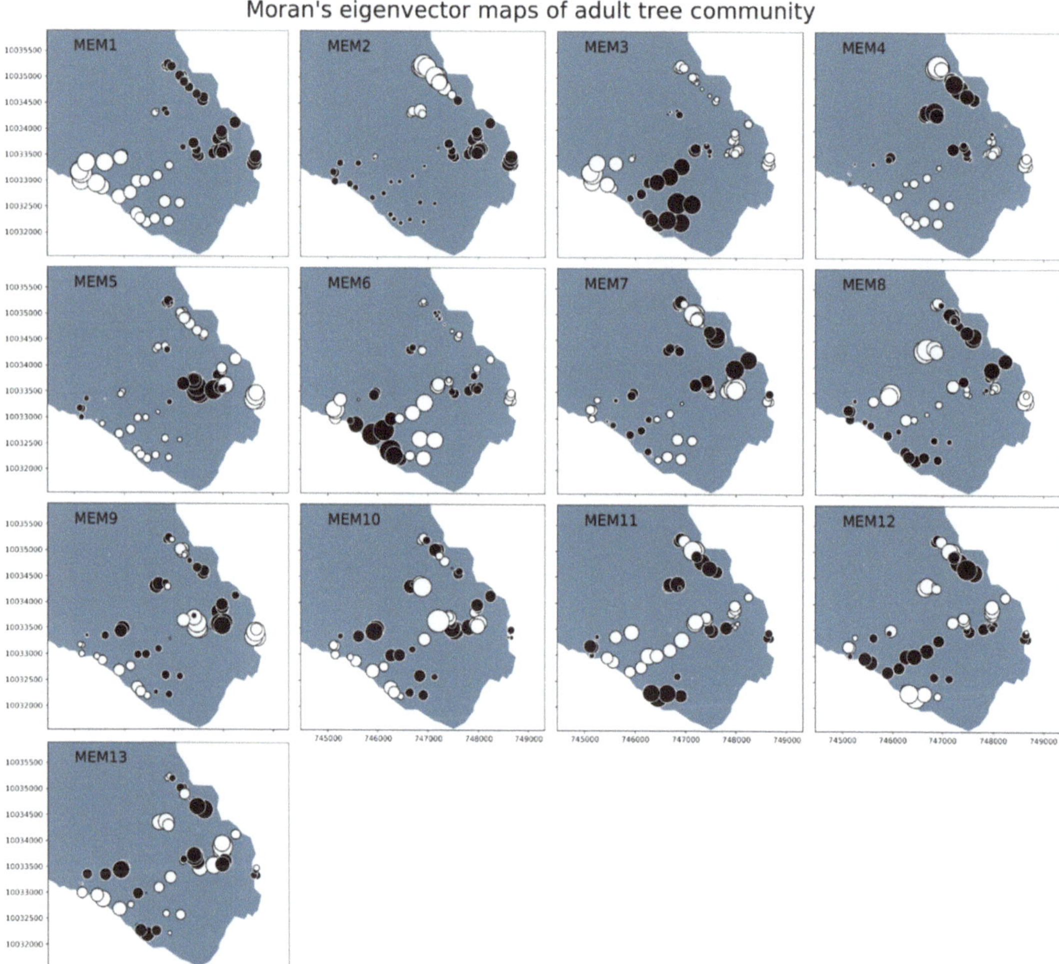

Figure 8. All statistically significant Moran's eigenvector maps (MEMs) detected from tree community data. Within each map, two sites of large size but differing colors were maximally different (i.e., a site with a large white circle had some large difference in species composition with a site with a large black circle). Correlations between these spatial patterns and available environmental data are given in Table 5.

Table 5. Moran's eigenvector maps and their statistically significant environmental correlations.

	Slope	Dem	Aspect	Exposure (Eastern)	Exposure (Northen)	Tostream	BC	BS	CLB	RCA	RG
MEM1		−0.403								0.287	
MEM2	−0.344	−0.636								0.308	
MEM3											
MEM4					0.349						
MEM5										−0.503	
MEM6						0.306					
MEM7											0.385
MEM8		0.404				0.460					
MEM9										−0.451	
MEM10						−0.340					
MEM11						0.282					
MEM12											−0.410
MEM13							0.345				

3.6. Juvenile Communities

Juvenile communities did not cluster within the groups presented by adult communities, neither in terms of historical land use/habitat (Figure 9A) nor current ecological state of adult trees (Figure 9B). Rather, all juvenile communities radiated into new dissimilarity space from their respective adult communities, possibly indicating recruitment of new tree species and/or new species combinations at these sites. Of the 148 species of juvenile tree species observed in the study, 110 species were observed as both juvenile and mature specimens, and an additional 38 species were observed only as juvenile specimens.

Juvenile tree communities changed along the same axes as were observed for their respective adult tree communities. When categorized using the historical land-use/habitat variable, RCA sites and related, similar RG sites were pulled in a negative direction along the NMS1 axis, as shown in Figure 9A, and nonanthropogenically disturbed sites (BC, BS, and CLB) were pulled in a positive direction along the NMS2 axis. When categorized according to the current ecological state of their adult tree communities, the same pattern held (Figure 9B): juvenile tree communities were different and outside of clusters formed by their respective adult communities, but changed along the same axes as their adult communities. Two exceptions to this were observed: the juvenile tree communities of two type I forest sites have come to more closely resemble type III (Figure 9B, sites 1.2 and 1.3). These sites had a land-use/habitat history of pasture conversion (RG), and their adult tree communities clustered into a current ecological state of type I forest, the most anthropogenically disturbed group, and often associated with RCA land-use history (Figure 5). However, their juvenile tree communities now resemble more "natural" forest types. Additionally, the site 1.1 juvenile tree community was so unique as to be an outlier to the rest of the sites studied, and had to be removed for informative examination of the remaining sites (Supplementary Figure S11).

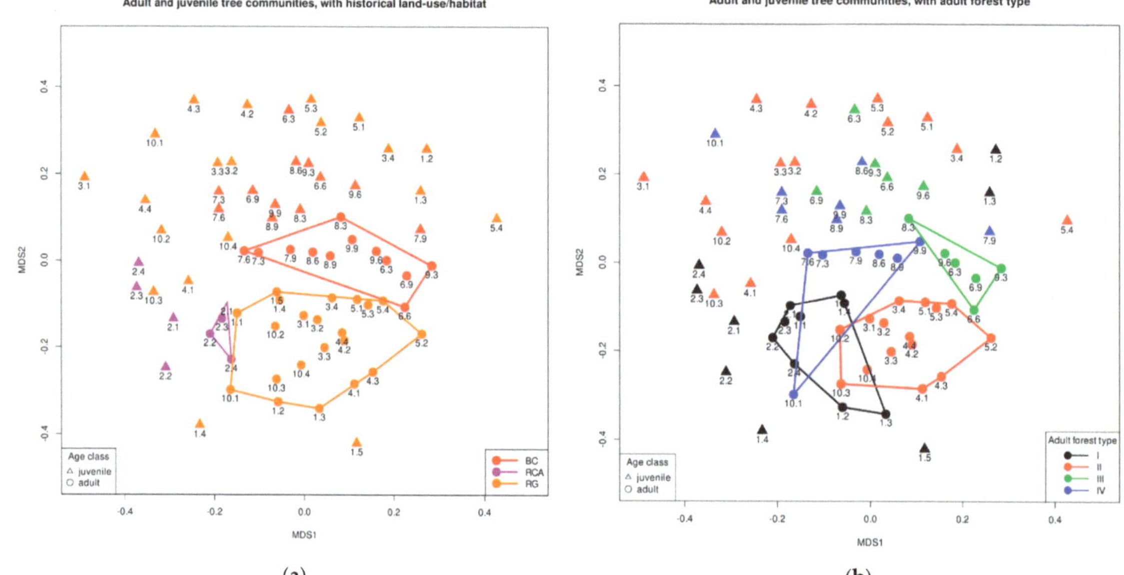

Figure 9. Nonmetric multidimensional scaling (NMS) ordination of combined juvenile and adult tree communities using Bray–Curtis dissimilarity. Only sites with both juvenile and adult data are shown. Each site therefore had two points, a juvenile (triangle) and adult (circle) tree community. (**a**) Ordination colored by historical land use. Hulls are drawn around adult communities. (**b**) Ordination colored by current ecological state ("forest type"). Hulls are drawn around adult communities.

3.7. Deforestation in the Region

Between 1990 and 2018, forest cover in the Cotacachi canton was reduced from an estimated 87,967 ha of native forest cover in 1990 to 71,739 ha in 2018, an 18% reduction in the total forest from 1990 levels, mostly due to conversion to agricultural land (Figure 10). In terms of total land cover, this equated to a shift from approximately 52% of total Cotacachi canton land cover being native forest to 42% of total land cover as native forest. Forest cover in Los Cedros increased from an estimated 5094 ha of native forest cover in 1990 to 5210 ha in 2018, a 2.3% addition to the total forest at 1990 levels, due mostly to the reforestation of former pasture (Figure 11). Forest cover in Bosque Protector Chontal decreased from an estimated 6920 ha of native forest cover in 1990 to 6565 ha in 2018, a 5% reduction in the total forest from 1990 levels, due mostly to conversion of forest to agricultural land uses (Figure 11).

Unlike earlier datasets, the 2018 MAE land cover dataset does not include pasture separately from other forms of agricultural land use, so it was difficult to quantify which types of agricultural land use were most commonly replacing forest in the region. However, based on visitation to the surrounding communities, it is the authors' assessment that much of the recent deforestation was undertaken to create pasture for cattle grazing.

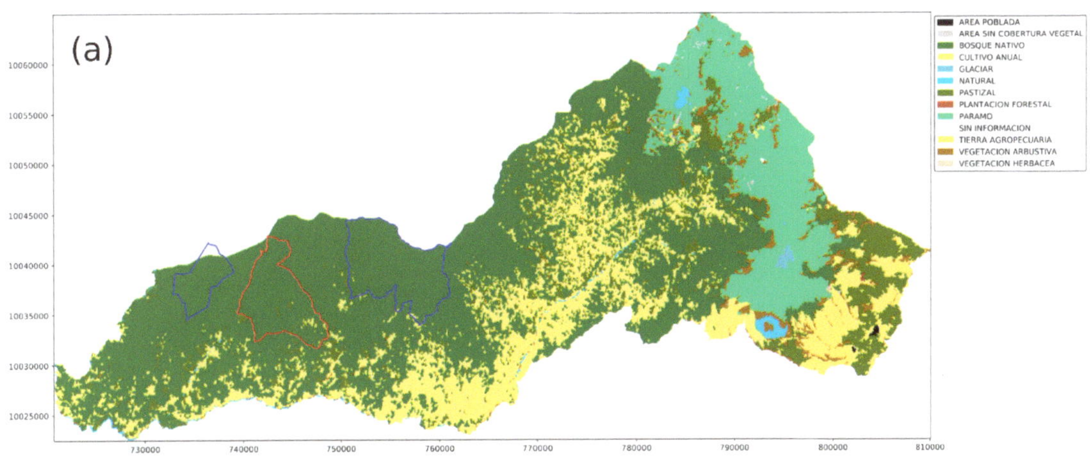

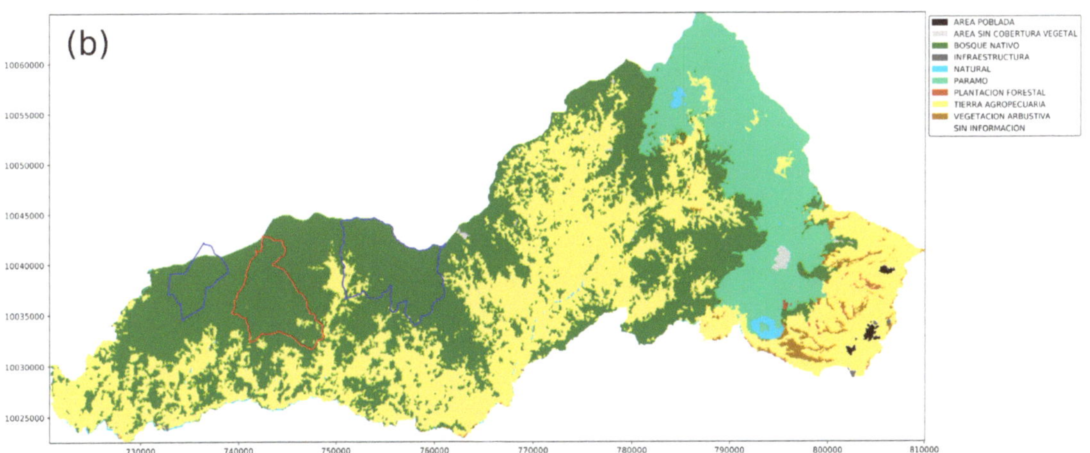

Figure 10. Land-cover changes in the Cotacachi canton. Reserva Los Cedros is outlined in red, and two other nearby protected forests are outlined in blue (B.P. Cebu to the west and B.P. Chontal to the east). Forest cover is shown in dark green, and agricultural land in beige. Geographic coordinate system is PSAD56/UTM zone 17S, in meters. (**a**) 1990; (**b**) 2018.

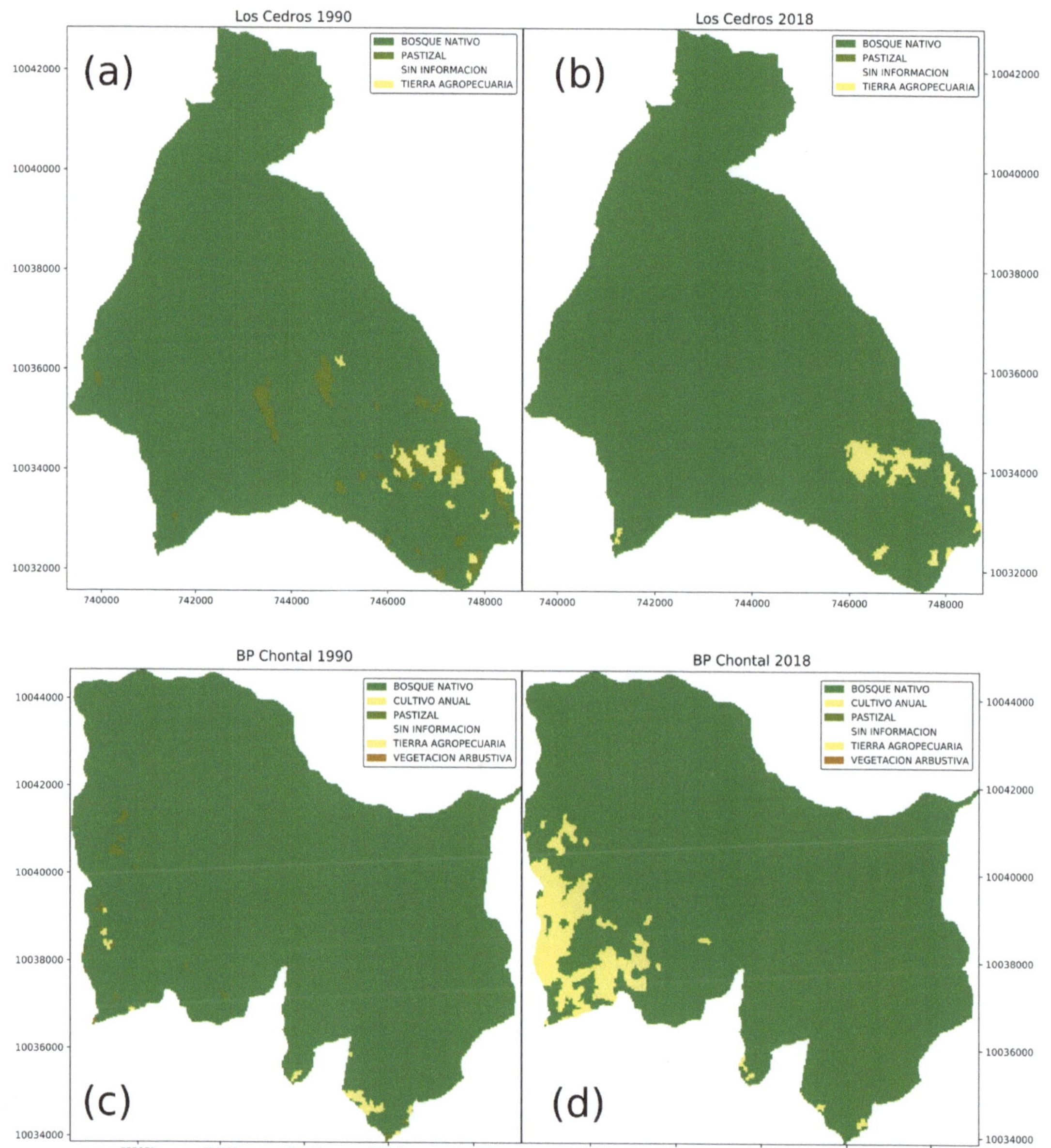

Figure 11. Land-cover changes in Reserva Los Cedros and in nearby protected forest B.P. Chontal. Forest cover is shown in dark green, pasture in tan-green, and agricultural land in beige. Geographic coordinate system is PSAD56/UTM zone 17S, in meters. (**a**) Los Cedros, 1990; (**b**) Los Cedros, 2018; (**c**) B.P. Chontal, 1990; (**d**) B.P. Chontal, 2018.

4. Discussion

4.1. Stable States in the Andean Cloud Forest

In this study, we examined 61 forest sites with known histories of anthropogenic disturbance or natural, gap-forming disturbance, and examined this history as a predictor of current ecological state. We could categorize the southern area of the Los Cedros reserve into four forest types based on adult tree communities (Figure 4), and could well predict these ecological states from past land use/habitat type and elevation. Sites that had no history of anthropogenic disturbance group were readily categorized into either forest type III or IV (Figures 4 and 5). These sites were characterized by endemic, forest-dependent tree species (Table 4). The differences among these "natural" forest types III and IV were strongly predicted by elevation (Figures 6 and 7, Supplementary Figure S6), exhibiting elevation-dependent ecological zonation often observed in montane tropical forests [121,122]. Natural gap-forming disturbances did not change the species compositions of these sites to make them significantly different from other natural forest sites (Figure 5). As such, these natural forest types may represent stable equilibria with basins of attraction that give each some resilience to gap-forming disturbance (Figure 12).

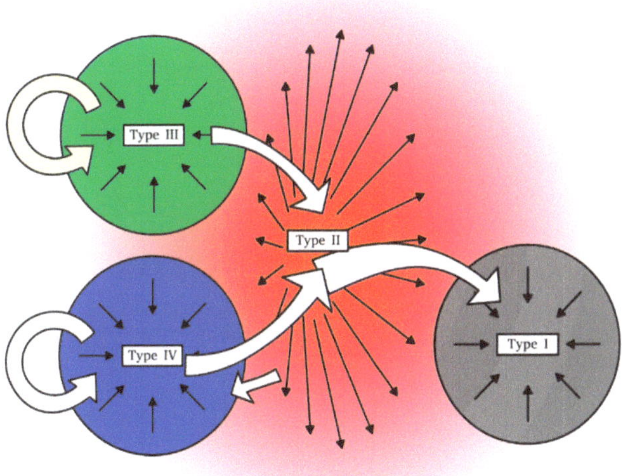

Figure 12. Conceptualization of ecological equilibria at Reserva Los Cedros. Arrows indicate the direction of change between basins that were hypothesized as possible from our observations. Type III and Type IV ecological states observed in our study (see Figure 5) probably represented regional historical basins of attraction that can be "knocked" into a novel forest type (type II) by intermediate agricultural disturbance, and perhaps even into a more modified type I forest by more intensive anthropogenic disturbance. Type II forests may also be capable of transition back to ecological states that resemble historical states, as was directly observed in our study at site 10.1 (see Figure 4).

There were indications that sites that had experienced intermediate disturbance (conversion to pasture followed by reforestation) had entered a less stable, more pluripotent ecological state. Nearly all high and low elevation sites that were converted to pasture prior to reforestation (land-use/habitat type "RG") then developed into a single community type, ("forest type II"). This suggested a homogenizing effect on tree communities due to this kind of disturbance and a common successional response by the forest after conversion to pasture and abandonment, regardless of elevation. In ordinations of community similarity (Figure 5, Supplementary Figure S3), type II forest sites were situated in an intermediate position between all other states. It is therefore possible that the type II forest type represents a low-sloped, convex, unstable equilibrium that can sometimes allow sites to return to a natural forest state. In one case, we observed a site that had undergone anthropogenic disturbance and subsequently developed into a natural forest type (site 10.1, circled in Figures 4 and 5). Additionally, in the case of two pastured (RG) sites, their adult communities resembled the highly affected type I forest, but their juvenile communities currently more closely resembled the "natural" type III forest (Figure 9, sites 1.2 and 1.3). In a different successional direction, several pastured (RG) sites developed into the novel Type I forest, the same forest type into which all the intensive-agriculture sites developed (Figure 5).

The existence of this indeterminate, intermediate ecological state, forest type II, supported the hypothesis that primary cloud forest has some capacity to "repair" highly modified agricultural sites, but also indicated that this is not a certain outcome. This postpasture successional trajectory appeared to be very distinct from that which followed natural, gap-forming disturbances, and may be intermediate to severe agricultural conversions (cane production followed by natural regeneration), which tended to drive sites to another state, forest type I. The adult tree communities of sites that experienced intensive agriculture all developed into forest type I, suggesting that forest type I may be another stable equilibrium or novel ecosystem type [34]. Additionally, all sites that were observed to be in a type I state were low-elevation sites that presumably would have otherwise existed in a type IV state, suggesting vulnerability of these lower-elevation forests to anthropogenic disturbance. However, all type I sites were also colocalized to the same area of the reserve (Figure 6), making it difficult to generalize this pattern to the rest of the study area.

Several forces may be at work in creating these equilibria that we observed. In type III and type IV forests ("natural forest" types), gap-forming incidents appear to keep these sites within their basins of attraction, and the breakdown of these negative feedbacks and ecological conditions that are likely responsible for shifts into forest type I equilibria.

4.1.1. Soil Structure

While soil structure data were not available for this survey, researchers at field sites noted that sites that had undergone intensive agricultural use ("RCA" sites) had more compacted soils. Agricultural use can incur long-lasting legacy effects on soil structural and chemical characteristics, even after abandonment and reforestation [123,124]. Deforestation has been shown to cause long-term changes in soil physical structure and microbial activity in soils, especially among plant-symbiotic microbes [125–127].

4.1.2. Soil Seed Bank Depletion

In addition to soil compaction, several seasons of indiscriminate grazing or corn or sugarcane culture probably greatly depleted the residual seed bank of forest plants [128,129]. When soil-seed banks have been exhausted, recolonization of sites by forest plants must occur through dispersal of seeds to the forest regeneration site.

4.1.3. Large Seeded Plants and Animal Dispersal

Reserva Los Cedros hosts three species of primates: the brown-headed spider monkey (*Ateles fusciceps fusciceps*), the white-headed capuchin (*Cebus capucinus*), and the mantled howler monkey (*Alouatta palliata*), in addition to numerous other frugivorous birds and mammals [52]. These animals likely play an essential role in closing gaps after disturbances. Indeed, it is likely that the differences between our results and those observed at the Maquipucuna Cloud Forest Reserve [130,131], only about 40 km away in a straight line and at similar elevations, resulted from the presence of primates at Los Cedros and their extirpation due to hunting at Maquipucuna [52]. Myster recorded little colonization of pastures, whereas ours were filled with *Cecropia*, which are commonly dispersed by both *Ateles* and *Cebus* monkeys [132,133]. In recent natural gaps ("CLB"), we observed a single indicator species: *Endlicheria* sp. (Table 4). *Endlicheria* tend to have classic bird-dispersed fruits, a small drupe with red color in the cupule [134], but some *Endolicheria* species have been suggested elsewhere to also be primate-dispersed [135]. Lower-elevation natural forests (forest type IV) were characterized by copal trees (*Protium* and *Dacryodes* spp., Table 4), well known locally as primate food sources [136]. In general, type III and IV forest types were generally characterized by plants with larger seeds or fruits, often primate- or otherwise vertebrate-dispersed, such as those in Lecythidaceae and Lauraceae, and

Garcinia (Table 4). This prevalence of trees with larger seeds/fruits is typical of tropical forests, where most trees produce animal-vectored seeds [137]. Absent active transport of seeds by animals, colonization of pastures by forest trees is expected to be very slow [138] and to heavily favor wind-dispersed seeds [139] or rapidly spreading cover shrubs, such as bamboo species [140]. Many ruderal species, conversely, are wind-dispersed [141] and are likely to readily colonize abandoned field sites, especially when aided by human vectors [142]. Given Los Cedros' relatively low elevation among montane tropical forests, it is not surprising that its forests are heavily populated with large-seeded species that likely rely on primate dispersal [143]. Additionally, we found four large-seeded tree species that are commonly associated with primary forests and that are animal fodder, but that were also observed as having a strong presence, even in highly disturbed sites: *Caryodaphnopsis theobromifolia* [144,145], *Clarisia biflora* [146], *Guatteria megalophylla* [147,148], and *Ficus cuatrecasana* [149] (see Table 4, indicator species for group I + IV and group II + III). *C. theobromifolia* is a valuable timber species in the area, and its presence in highly disturbed sites is welcome economically and more evidence for the high regenerative potential in these forests, perhaps due to primate dispersal. A study of forests in Peru that were protected from hunting versus those that were not only found seedlings and juveniles of Caryodaphnopsis in protected forests, and they documented more primate and mammal dispersers in the protected forests [144].

4.1.4. Relative Fluxes of Local vs. Exotic Plant Types

Most of the sites examined here were embedded in a landscape of primary forest, and may have been able to recruit seeds from forest-dependent species even if they experienced seed-bank depletion [150], while still insulated from outside seed sources. Approximately half of the type I forest sites, however, were located on the edge of Los Cedros abutting neighboring farmland, probably allowing for extensive input of small-seeded pioneer tree species such as *Saurauia* spp. The remaining type I sites were located along mule trails that supply the reserve, and also presumably were repeatedly exposed to seeds from outside locations.

The above mechanisms are primarily "gap-filling" mechanisms, and may fail in the face of extensive anthropogenic disturbances such as habitat fragmentation and loss (for example, Cramer et al. [151]). It is probably important to note how dependent these processes may be on large animal dispersers, and especially on primates, as primate populations are in decline in the region [52,152]. The importance of seed-dispersal services by large mammals and spider monkeys in particular can be appreciated by anticipating their loss: Peres et al. [153] have predicted substantial loss of forest biomass in forests if large mammal populations are reduced, especially spider monkeys and tapirs. The importance of the brown-headed spider monkey as a dispersal agent of larger seeds, often representing tree species sought after by loggers, was clearly seen in a study in Western Ecuadorian lowland forest systems [154]. A similar positive feedback may be possible with the loss of forest-dependent birds due to deforestation. Frugivorous, forest-dwelling birds provide a unique seed-dispersal service to forest trees, but are highly interdependent with their habitat; their decline in diversity has been directly linked to vegetative diversity loss and deforestation [155].

Other complex interactions may also be important: the presence of *Cecropia* spp. as an indicator species in forest types with intermediate disturbance (forest type II) and appearance less often in the novel type I forest. In this study, they were also not observed as extremely prevalent in recent natural gaps. This is possibly important, as *Cecropia* trees occupy a somewhat unique position of being an early-successional species while also producing large fruits and canopy structure characteristics useful to both primates and frugivorous birds [136,156,157]. *Cecropia* trees may attract primate dispersers and forest-dependent birds back to a disturbed site, which therefore bring with them the heavier seeds of other forest-dependent, heavy-seeded species. At Los Cedros, the smallest primate dispersers, capuchin monkeys, are regularly observed in *Cecropia* trees. *Cecropia* spp. may therefore act as another gap-repairing negative-feedback mechanism, one that plays out in cases of forest gaps of larger size or severity than the smaller natural gaps examined here. Thus, the success or failure of *Cecropia* may play a deciding role that allows sites to return from the ecological state we observed here as forest type II (resulting from intermediate disturbance), and potentially back to primary-like forest states. Note again, however, that this mechanism would depend upon the presence of large animal vectors, especially primates.

Given the relatively short period of time since disturbance prior to surveys (13–18 years), it was also very difficult to confidently project the stability of the here-proposed candidate equilibria to larger time scales, especially in times of deep ecosystem change due to climate change. Sites were surveyed in 2005, so an additional 16 years had passed since the collection of these data. Indeed, Loughlin et al. [76] suggested that a tropical Andean cloud forest in a nearby ecoregion may have a single long-term ecological equilibrium. However, conditions in the Andes are changing dramatically from those that presumably sustained the long-term resiliency observed by Loughlin. Thus, the candidate alternative states observed here should not be disregarded, as they may mark the beginning of novel forest types in the region [34]. For the moment, visual inspection of these sites in the present day confirmed that unique tree communities continue to exist in type I forest sites compared to surrounding older forests. Updated systematic surveys are now necessary to investigate the stability of the ecosystem types suggested here.

4.2. Juvenile Tree Community

It is difficult to assign cause to the lack of structure observed in juvenile tree communities at Los Cedros (Figure 9). Much of the noise in these juvenile communities presumably stemmed from the incomplete filtering of young plants at each site [158,159]—the environmental pressures that have shaped the adult tree community at each site likely had yet to totally act on the juvenile trees at the time of sampling. Additionally, plots with a history of nonanthropogenic disturbances (BS and CLR plots) were not sampled for juvenile trees, meaning that we could not comment directly on the importance of environmental filtering on seedlings due to "natural" successional patterns.

However, it is also inevitable that the same global environmental changes that are acting on other forests throughout the world [27,160], including cloud forests [81], are at play in the ancient forests of Los Cedros, and are contributing to the reorganization of future tree communities, perhaps uncoupling ancient species associations. The disturbance regime under study (conversion to agriculture, followed by forest regeneration) also introduced new conditions and plant species to Los Cedros. Thus, in these juvenile trees, we may also be observing two sources of disturbance that could shift the forests of Los Cedros out of the basins of attraction of the primary forest state of very different scales: local land-use change and global climate change. In the terminology of Beisner et al. [161], the former may still be considered a state variable change from which it is sometimes within the capacity of the cloud forest to rebound and recover to a primary forest state. The latter, however, is a deep shift in the parameters of the Andean cloud forest ecosystem, which will no doubt change the shape of the possible [162–164]. It is not unreasonable to expect interactions between these two fundamental sources of ecological disturbance [26,162,165,166].

4.3. Beta Diversity and Spatial Heterogeneity in the Andean Cloud Forest

We examined patterns of distance decay in the tree communities of the southern area of Los Cedros. When community turnover was modeled as a function of Euclidean distance, a model in the form of an asymptote function fit well to the observed patterns of community turnover. Our asymptote model suggested turnover at a short distance, reporting that half of the maximum dissimilarity was reached in just ~150 m, and a mean Bray–Curtis dissimilarity >0.8 in comparison with distances larger than 600 m (Figure 2). When small watersheds were used as the basic spatial unit, rather than Euclidean distance, most of the decay in the tree community similarity occurred with the first crossover to a neighboring watershed (Figure 3). Following this, the mean dissimilarity among sites increased slightly but remains uniformly high, and further comparisons were not statistically significantly different, meaning comparisons between sites five watersheds away were not on average more or less similar than comparisons of sites that were only two, three, or four watersheds apart, because so much change in community composition had already occurred just within the first watershed crossing. This was supportive of the colloquial understanding that in the Andes, each small drainage can host an almost entirely distinctive community from its neighbors.

This may also provide more insight into the processes that create the fine-scale of endemism often observed in Andean forests. We hypothesized that this fine-scale, watershed-based community turnover was due to high dispersal limitation and high microsite variability that resulted from the complex, dramatic topography of the Andes. This microsite variation was visible to some degree in our Moran's eigenvector maps and their environmental correlations (Figure 8, Supplementary Figures S9 and S10, Table 5). These MEM maps showed spatially explicit patterns of difference in tree community that were acting at very small distances, close to the scale of the microwatersheds we have delineated, and that correlated with watershed characteristics such as local elevation changes, proximity to water, and hill tops and rivers as possible dispersal barriers. These observed spatioenvironmental patterns may explain up to 19% of variance in the plant community. This fraction may represent much of the environmental filtering that is occurring within the tree community at Los Cedros. Dispersal limitation is harder to test, but the predominance of large-seeded species observed in our ancient forest sites suggested that many important tree species were dispersal-limited to highly local scales and dependent on large animal dispersers to overcome this limitation. This preponderance of heavy-seeded species also might be well approximated by a symmetric dispersal-limited neutral model, a model that can generate significant small-scale spatial patterning in communities even without considering additional microsite environmental conditions [17]. Future studies should incorporate watershed units, as they can explain more in the Andes than simple distance or elevation.

4.4. Conservation Value of Los Cedros

Since the sounding of an alarm by Gentry and Dodson [72,87], Myers et al. [66,167], and numerous others [78], there has been much concern about the future of the Chocó and the tropical Andean biodiversity hotspot by conservation groups. However, there is little evidence that this call for conservation has resulted in sufficient meaningful change for the region of Los Cedros—in fact, quite the opposite. Instead, during this time frame, Cotacachi Canton has lost significant forest cover (Figure 10), as has Ecuador generally [87,168,169]. In stark contrast, Los Cedros—which has onsite, conservation-oriented staff—has well withstood the traditional pressures of timbering and settlement, as evidenced by its increase in forest cover during a time of net forest loss in the Cotacachi Canton. In addition to the historical habitat conversion from timber extraction and settlement, the tropical Andean biodiversity hotspot has now found itself in a new center of metal mining exploration [52,170,171], a new and entirely different extractive pressure on the region. Los Cedros itself is targeted by a junior mining partner (cornerstoneresources.com, accessed on 31 April 2021). Los Cedros has responded with a successful legal challenge that reached Ecuador's highest courts, with implications for all Bosque Protectores in Ecuador [170]. Due to its proactive defensive legal efforts, Los Cedros' conservation effect was thus amplified even beyond the unusually effective physical protection of its forests.

Going forward, however, conservation of primary forest reserves such as Los Cedros will face novel challenges. It is widely understood that the high endemism of the Andean biodiversity hotspot makes it both a conservation priority and an especially difficult conservation challenge. It is less widely appreciated that forests, and especially primary forests, pose a major additional challenge to conservation. In most cases, we have only a crude understanding of the numerous ecological interactions required to maintain a primary ecosystem in its current levels of biodiversity and benefits for human society ("ecosystem services"), otherwise known as the ecological complexity of an ecosystem [172,173]. Ancient ecosystems are unique in large part due to their emergent complexity, and quantifying this complexity is usually difficult, requiring composite measures of numerous indices of forest physical structure, plant community, and other ecological properties [174,175]. In complex ecosystems, the loss of a species is multiplied by disruption of its interactions with other species, such as with trophic cascades [176]. The number of possible interactions among species undergoes quadratic growth as biodiversity increases linearly, so the immense biodiversity of the tropics presents a particular challenge in understanding the local ecological complexity. Primary forest fragments are also presumably subject to the well-known vulnerabilities of "island"—or insular—habitats [177,178]; namely, the stochastic extinction of species without any nearby "mainland" to replenish populations. This is in addition to the uniquely dynamic, weather-dependent boundaries that make all cloud forests insular systems, primary or otherwise, and that may make them very vulnerable to shrinkage from climate change [72,81,164]. Thus, in insular, highly interdependent ancient ecosystems such as primary cloud forests, these stochastic species extinctions may have greater cascading impacts than in simpler ecosystems. Even if completely protected from external changes, primary forest fragments such as Los Cedros may ecologically simplify or "decay" with time if they are not sufficiently understood, buffered, and interconnected. As such, the regional resiliency noted by Loughlin et al. [76] may be subject to novel vulnerabilities that were not so urgent historically. Michaels et al. [179] have suggested that restoration efforts should first attempt to diagnose the ecological conditions and feedbacks that allow a basin of attraction for a desired stable state to persist, and also find the minimum "nucleus" area needed for those processes to proceed and grow. They and others, such as some proponents of rewilding [180], suggest seeding landscapes with species consortiums intended to recreate ecological complexity as much as possible. Los Cedros is far from needing rewilding, but conservation efforts around Los Cedros and other ancient tropical forests will be greatly enriched by such an awareness of protecting and nurturing ecological complexity, beyond simple physical protection of reserve borders.

Given the state of fragmentation and loss of forest in the Choco Andean region, and the poor understanding of its numerous pollination and diaspore dispersal networks, especially the threatened state and habitat requirements of its primate dispersers [152], Los Cedros may be at or close to threshold stable patch size for maintaining its level of biodiversity. Los Cedros may therefore face an important juncture, with a choice of growing or subsiding. Further reduction or anthropogenic disturbance of cloud forest area risks nonlinear, catastrophic changes. Additionally, the fragments of Andean primary cloud forest habitat that persist to date are rare and small enough that there is no economic justification for eliminating or heavily modifying them. Simply as a matter of scale, any economic benefit from the degradation of these forest fragments will not be large enough to justify their loss as reserves of genetic information, providers of ecosystem services, and protectors of species diversity. Other methods for poverty reduction are available [181,182]. However, great opportunity for the future exists in Los Cedros: the forest of Los Cedros need not be seen as a delicate ecological island in a storm, nor is its protection an academic or moral exercise, or a desperate rear-guard conservation strategy. It is instead useful to think in the sense of Michaels [179], supported by the long-term observations of Loughlin et al. [76], and see the Los Cedros forest as a nucleus of primary cloud forest that has prospered for thousands of years, that has weathered the recent waves of deforestation, and that now stands ready to help reseed and restore the new forests of the northern Andes.

Supplementary Materials: The following are available online at https://www.mdpi.com/article/10.3390/f13060875/s1, Figure S1: Sampling design, natural forest sites, Figure S2: Comparison of predictive performance of models of community turnover, Figure S3: Posterior shift for K_m, Figure S4: Ordination of Hierarchical clustering results (forest type), Figure S5: Predictions of current ecological state by historical land-use/habitat, Figure S6: Prediction of primary forest type by elevation, Figure S7: Current forest type, history, and elevation, Figure S8: MEM 1, Figure S9: MEM 2, Figure S10: MEM 8, Figure S11: Adult and juvenile tree communities, with adult forest type, outlier retained, Table S1: Species lists, Table S2: Species diversity estimates for historical land-use/habitat types.

Author Contributions: Conceptualization, A.M., D.C.T., B.A.R. and M.P.; methodology, A.M., D.C.T. and M.P.; formal analysis, D.C.T.; investigation, A.M., R.M., A.H., W.D., M.A.C., J.D.S.L. and E.J.; resources, M.P.; data curation, A.M., B.A.R. and D.C.T.; writing—original draft preparation, D.C.T.; writing—review and editing, A.M., D.C.T., B.A.R. and M.P.; visualization, D.C.T.; supervision, M.P.; project administration, M.P. and A.M.; funding acquisition, M.P. All authors have read and agreed to the published version of the manuscript.

Funding: Field and herbarium work stages were funded by the Darwin Initiative under project number 14–040. Publication costs were funded by the Deutsche Forschungsgemeinschaft (DFG, German Research Foundation, 491183248) through the Open Access Publishing Fund of the University of Bayreuth.

Institutional Review Board Statement: Not applicable.

Informed Consent Statement: Not applicable.

Data Availability Statement: Publicly available datasets were analyzed in this study. This data can be found in the associated github repository (https://github.com/danchurch/losCedrosTrees), and all analyses are recorded as a Jupyter notebook (https://nbviewer.org/github/danchurch/losCedrosTrees/blob/master/anaData/MariscalDataExploration.ipynb).

Acknowledgments: The work would not have been possible without Jose DeCoux and his dedication to the conservation of Reserva Los Cedros. Several field assistants, volunteers, and guides associated with Los Cedros helped with data collection: Laurence Duvauchelle, Gila Roder, Martín Obando, Manuel Moreno, Danny Cumba, Homero Sánchez, Fausto Lomas, and Víctor Lomas. Additional help in the field and herbarium were contributed by: David Neill, Mercedes Asanza, Homero Vargas, Nelson Moyano, Gabriela Montoya, Sonia Sandoval, Iralda Yépez, Byron Nuggerud, Diana Fernández, Efraín Freire, Elsa Toapanta, Carlos Cerón, Walter Palacios, Edison Jiménez, Rosa Masaquiza, Rosa Batallas, Fernando Asanza, Marcia Peñafiel, and Laura Mariscal. Additional PrimeNet project colleagues included: Diego Tirira, Mercedes Gavilanes, María Isabel Estévez, and Xavier Cueva.

Conflicts of Interest: The authors declare no conflict of interest. The funders had no role in the design of the study; in the collection, analyses, or interpretation of data; in the writing of the manuscript, or in the decision to publish the results.

References

1. Bray, J.R.; Curtis, J.T. An Ordination of the Upland Forest Communities of Southern Wisconsin. *Ecol. Monogr.* **1957**, *27*, 325–349. [CrossRef]
2. Bunin, G. Directionality and Community-Level Selection. *Oikos* **2021**, *130*, 489–500. [CrossRef]
3. Chambers, J.Q.; Negron-Juarez, R.I.; Marra, D.M.; Di Vittorio, A.; Tews, J.; Roberts, D.; Ribeiro, G.H.; Trumbore, S.E.; Higuchi, N. The Steady-State Mosaic of Disturbance and Succession across an Old-Growth Central Amazon Forest Landscape. *Proc. Natl. Acad. Sci. USA* **2013**, *110*, 3949–3954. [CrossRef]
4. Chang, C.C.; Turner, B.L. Ecological Succession in a Changing World. *J. Ecol.* **2019**, *107*, 503–509. [CrossRef]
5. Clements, F.E. *Plant Succession: An Analysis of the Development of Vegetation*; Carnegie Institution of Washington: Washington, DC, USA, 1916; ISBN 0-598-48534-1.
6. Connell, J.H.; Slatyer, R.O. Mechanisms of Succession in Natural Communities and Their Role in Community Stability and Organization. *Am. Nat.* **1977**, *111*, 1119–1144. [CrossRef]
7. Cowles, H.C. The Ecological Relations of the Vegetation on the Sand Dunes of Lake Michigan, Part I—Geographical Relations of the Dune Floras. *Bot. Gaz.* **1899**, *27*, 95–117. [CrossRef]
8. Curtis, J.T.; Greene, H.C. A Study of Relic Wisconsin Prairies by the Species-Presence Method. *Ecology* **1949**, *30*, 83–92. [CrossRef]
9. Gleason, H.A. The Individualistic Concept of the Plant Association. *Bull. Torrey Bot. Club* **1926**, *53*, 7–26. [CrossRef]
10. Holling, C.S. Resilience and Stability of Ecological Systems. *Annu. Rev. Ecol. Syst.* **1973**, *4*, 1–23. [CrossRef]
11. Whittaker, R.H. Vegetation of the Great Smoky Mountains. *Ecol. Monogr.* **1956**, *26*, 1–80. [CrossRef]
12. Norden, N.; Angarita, H.A.; Bongers, F.; Martínez-Ramos, M.; Granzow-de la Cerda, I.; Van Breugel, M.; Lebrija-Trejos, E.; Meave, J.A.; Vandermeer, J.; Williamson, G.B.; et al. Successional Dynamics in Neotropical Forests Are as Uncertain as They Are Predictable. *Proc. Natl. Acad. Sci. USA* **2015**, *112*, 8013–8018. [CrossRef] [PubMed]
13. Richards, P.W. *The Tropical Rain Forest*; Cambridge University Press: London, UK, 1952.
14. HilleRisLambers, J.; Adler, P.B.; Harpole, W.S.; Levine, J.M.; Mayfield, M.M. Rethinking Community Assembly through the Lens of Coexistence Theory. *Annu. Rev. Ecol. Evol. Syst.* **2012**, *43*, 227–248. [CrossRef]

15. Young, T.P.; Chase, J.M.; Huddleston, R.T. Community Succession and Assembly Comparing, Contrasting and Combining Paradigms in the Context of Ecological Restoration. *Ecol. Rest.* **2001**, *19*, 5–18. [CrossRef]
16. Fisher, C.K.; Mehta, P. The Transition between the Niche and Neutral Regimes in Ecology. *Proc. Natl. Acad. Sci. USA* **2014**, *111*, 13111–13116. [CrossRef] [PubMed]
17. Hubbell, S.P. Approaching Ecological Complexity from the Perspective of Symmetric Neutral Theory. *Trop. For. Community Ecol.* **2008**, *143*, 159.
18. Keddy, P.A. Assembly and Response Rules: Two Goals for Predictive Community Ecology. *J. Veg. Sci.* **1992**, *3*, 157–164. [CrossRef]
19. Kraft, N.J.B.; Valencia, R.; Ackerly, D.D. Functional Traits and Niche-Based Tree Community Assembly in an Amazonian Forest. *Science* **2008**, *322*, 580–582. [CrossRef]
20. Laughlin, D.C.; Laughlin, D.E. Advances in Modeling Trait-Based Plant Community Assembly. *Trends Plant Sci.* **2013**, *18*, 584–593. [CrossRef]
21. Ulrich, W.; Zaplata, M.K.; Winter, S.; Schaaf, W.; Fischer, A.; Soliveres, S.; Gotelli, N.J. Species Interactions and Random Dispersal Rather than Habitat Filtering Drive Community Assembly during Early Plant Succession. *Oikos* **2016**, *125*, 698–707. [CrossRef]
22. Bhaskar, R.; Dawson, T.E.; Balvanera, P. Community Assembly and Functional Diversity along Succession Post-Management. *Funct. Ecol.* **2014**, *28*, 1256–1265. [CrossRef]
23. Pickett, S.T.A.; Cadenasso, M.L.; Meiners, S.J. Ever since Clements: From Succession to Vegetation Dynamics and Understanding to Intervention. *Appl. Veg. Sci.* **2009**, *12*, 9–21. [CrossRef]
24. Norden, N.; Chazdon, R.L.; Chao, A.; Jiang, Y.-H.; Vílchez-Alvarado, B. Resilience of Tropical Rain Forests: Tree Community Reassembly in Secondary Forests. *Ecol. Lett.* **2009**, *12*, 385–394. [CrossRef]
25. Barnosky, A.D.; Matzke, N.; Tomiya, S.; Wogan, G.O.U.; Swartz, B.; Quental, T.B.; Marshall, C.; McGuire, J.L.; Lindsey, E.L.; Maguire, K.C.; et al. Has the Earth's Sixth Mass Extinction Already Arrived? *Nature* **2011**, *471*, 51–57. [CrossRef]
26. Pimm, S.L.; Jenkins, C.N.; Abell, R.; Brooks, T.M.; Gittleman, J.L.; Joppa, L.N.; Raven, P.H.; Roberts, C.M.; Sexton, J.O. The biodiversity of species and their rates of extinction, distribution, and protection. *Science* **2014**, *344*, 1246752. [CrossRef] [PubMed]
27. Allen, C.D.; Macalady, A.K.; Chenchouni, H.; Bachelet, D.; McDowell, N.; Vennetier, M.; Kitzberger, T.; Rigling, A.; Breshears, D.D.; Hogg, E.H. (Ted); et al. A Global Overview of Drought and Heat-Induced Tree Mortality Reveals Emerging Climate Change Risks for Forests. *For. Ecol. Manag.* **2010**, *259*, 660–684. [CrossRef]
28. Carmody, M.; Waszczak, C.; Idänheimo, N.; Saarinen, T.; Kangasjärvi, J. ROS Signalling in a Destabilised World: A Molecular Understanding of Climate Change. *J. Plant Physiol.* **2016**, *203*, 69–83. [CrossRef]
29. Ciais, P.; Reichstein, M.; Viovy, N.; Granier, A.; Ogée, J.; Allard, V.; Aubinet, M.; Buchmann, N.; Bernhofer, C.; Carrara, A.; et al. Europe-Wide Reduction in Primary Productivity Caused by the Heat and Drought in 2003. *Nature* **2005**, *437*, 529–533. [CrossRef]
30. Day, J.W.; Christian, R.R.; Boesch, D.M.; Yáñez-Arancibia, A.; Morris, J.; Twilley, R.R.; Naylor, L.; Schaffner, L.; Stevenson, C. Consequences of Climate Change on the Ecogeomorphology of Coastal Wetlands. *Estuaries Coasts* **2008**, *31*, 477–491. [CrossRef]
31. Doney, S.C.; Ruckelshaus, M.; Emmett Duffy, J.; Barry, J.P.; Chan, F.; English, C.A.; Galindo, H.M.; Grebmeier, J.M.; Hollowed, A.B.; Knowlton, N.; et al. Climate Change Impacts on Marine Ecosystems. *Annu. Rev. Mar. Sci.* **2012**, *4*, 11–37. [CrossRef]
32. Fadrique, B.; Báez, S.; Duque, Á.; Malizia, A.; Blundo, C.; Carilla, J.; Osinaga-Acosta, O.; Malizia, L.; Silman, M.; Farfán-Ríos, W. Widespread but Heterogeneous Responses of Andean Forests to Climate Change. *Nature* **2018**, *564*, 207. [CrossRef]
33. Van Mantgem, P.J.; Stephenson, N.L.; Byrne, J.C.; Daniels, L.D.; Franklin, J.F.; Fulé, P.Z.; Harmon, M.E.; Larson, A.J.; Smith, J.M.; Taylor, A.H.; et al. Widespread Increase of Tree Mortality Rates in the Western United States. *Science* **2009**, *323*, 521–524. [CrossRef]
34. Hobbs, R.J.; Higgs, E.; Harris, J.A. Novel Ecosystems: Implications for Conservation and Restoration. *Trends Ecol. Evol.* **2009**, *24*, 599–605. [CrossRef] [PubMed]
35. Ellis, E.C.; Goldewijk, K.K.; Siebert, S.; Lightman, D.; Ramankutty, N. Anthropogenic Transformation of the Biomes, 1700 to 2000. *Glob. Ecol. Biogeogr.* **2010**, *19*, 589–606. [CrossRef]
36. Hautier, Y.; Tilman, D.; Isbell, F.; Seabloom, E.W.; Borer, E.T.; Reich, P.B. Anthropogenic Environmental Changes Affect Ecosystem Stability via Biodiversity. *Science* **2015**, *348*, 336–340. [CrossRef] [PubMed]
37. Vitousek, P.M.; Mooney, H.A.; Lubchenco, J.; Melillo, J.M. Human Domination of Earth's Ecosystems. *Science* **1997**, *277*, 494–499. [CrossRef]
38. Pan, Y.; Birdsey, R.A.; Fang, J.; Houghton, R.; Kauppi, P.E.; Kurz, W.A.; Phillips, O.L.; Shvidenko, A.; Lewis, S.L.; Canadell, J.G.; et al. A Large and Persistent Carbon Sink in the World's Forests. *Science* **2011**, *333*, 988–993. [CrossRef]
39. Lal, R. Forest Soils and Carbon Sequestration. *For. Ecol. Manag.* **2005**, *220*, 242–258. [CrossRef]
40. Odum, E.P. The Strategy of Ecosystem Development. In *The Ecological Design and Planning Reader*; Springer: Berlin/Heidelberg, Germany, 2014; pp. 203–216.
41. Gundersen, P.; Thybring, E.E.; Nord-Larsen, T.; Vesterdal, L.; Nadelhoffer, K.J.; Johannsen, V.K. Old-Growth Forest Carbon Sinks Overestimated. *Nature* **2021**, *591*, E21–E23. [CrossRef] [PubMed]
42. Luyssaert, S.; Schulze, E.-D.; Börner, A.; Knohl, A.; Hessenmöller, D.; Law, B.E.; Ciais, P.; Grace, J. Old-Growth Forests as Global Carbon Sinks. *Nature* **2008**, *455*, 213–215. [CrossRef]
43. Lewis, S.L.; Wheeler, C.E.; Mitchard, E.T.A.; Koch, A. Restoring Natural Forests Is the Best Way to Remove Atmospheric Carbon. *Nature* **2019**, *568*, 25–28. [CrossRef]
44. Moomaw, W.R.; Masino, S.A.; Faison, E.K. Intact Forests in the United States: Proforestation Mitigates Climate Change and Serves the Greatest Good. *Front. For. Glob. Chang.* **2019**, *2*, 27. [CrossRef]

45. Sacco, A.D.; Hardwick, K.A.; Blakesley, D.; Brancalion, P.H.S.; Breman, E.; Rebola, L.C.; Chomba, S.; Dixon, K.; Elliott, S.; Ruyonga, G.; et al. Ten Golden Rules for Reforestation to Optimize Carbon Sequestration, Biodiversity Recovery and Livelihood Benefits. *Glob. Chang. Biol.* **2021**, *27*, 1328–1348. [CrossRef]
46. Watson, J.E.M.; Evans, T.; Venter, O.; Williams, B.; Tulloch, A.; Stewart, C.; Thompson, I.; Ray, J.C.; Murray, K.; Salazar, A.; et al. The Exceptional Value of Intact Forest Ecosystems. *Nat. Ecol. Evol.* **2018**, *2*, 599–610. [CrossRef]
47. Harmon, M.E.; Ferrell, W.K.; Franklin, J.F. Effects on Carbon Storage of Conversion of Old-Growth Forests to Young Forests. *Science* **1990**, *247*, 699–702. [CrossRef]
48. Stephenson, N.L.; Das, A.J.; Condit, R.; Russo, S.E.; Baker, P.J.; Beckman, N.G.; Coomes, D.A.; Lines, E.R.; Morris, W.K.; Rüger, N.; et al. Rate of Tree Carbon Accumulation Increases Continuously with Tree Size. *Nature* **2014**, *507*, 90–93. [CrossRef] [PubMed]
49. Mitchard, E.T.A. The Tropical Forest Carbon Cycle and Climate Change. *Nature* **2018**, *559*, 527–534. [CrossRef] [PubMed]
50. Girardin, C.A.J.; Espejob, J.E.S.; Doughty, C.E.; Huasco, W.H.; Metcalfe, D.B.; Durand-Baca, L.; Marthews, T.R.; Aragao, L.E.O.C.; Farfán-Rios, W.; García-Cabrera, K.; et al. Productivity and Carbon Allocation in a Tropical Montane Cloud Forest in the Peruvian Andes. *Plant Ecol. Divers.* **2014**, *7*, 107–123. [CrossRef]
51. Spracklen, D.V.; Righelato, R. Tropical Montane Forests Are a Larger than Expected Global Carbon Store. *Biogeosciences* **2014**, *11*, 2741–2754. [CrossRef]
52. Roy, B.A.; Zorrilla, M.; Endara, L.; Thomas, D.C.; Vandegrift, R.; Rubenstein, J.M.; Policha, T.; Rios-Touma, B.; Read, M. New Mining Concessions Could Severely Decrease Biodiversity and Ecosystem Services in Ecuador. *Trop. Conserv. Sci.* **2018**, *11*, 1940082918780427. [CrossRef]
53. Draper, F.C.; Asner, G.P.; Coronado, E.N.H.; Baker, T.R.; García-Villacorta, R.; Pitman, N.C.A.; Fine, P.V.A.; Phillips, O.L.; Gómez, R.Z.; Guerra, C.A.A.; et al. Dominant Tree Species Drive Beta Diversity Patterns in Western Amazonia. *Ecology* **2019**, *100*, e02636. [CrossRef] [PubMed]
54. Valencia, R.; Balslev, H.; Miño, G.P.Y. High Tree Alpha-Diversity in Amazonian Ecuador. *Biodivers. Conserv.* **1994**, *3*, 21–28. [CrossRef]
55. Brehm, G.; Hebert, P.D.N.; Colwell, R.K.; Adams, M.-O.; Bodner, F.; Friedemann, K.; Möckel, L.; Fiedler, K. Turning Up the Heat on a Hotspot: DNA Barcodes Reveal 80% More Species of Geometrid Moths along an Andean Elevational Gradient. *PLoS ONE* **2016**, *11*, e0150327. [CrossRef] [PubMed]
56. Churchill, S.P. Moss diversity and endemism of the tropical andes. *Ann. Mo. Bot. Gard.* **2009**, *96*, 434–449. [CrossRef]
57. Gentry, A.H.; Dodson, C.H. Diversity and Biogeography of Neotropical Vascular Epiphytes. *Ann. Mo. Bot. Gard.* **1987**, *74*, 205. [CrossRef]
58. Hughes, C.E. The Tropical Andean Plant Diversity Powerhouse. *New Phytol.* **2016**, *210*, 1152–1154. [CrossRef] [PubMed]
59. Küper, W.; Kreft, H.; Nieder, J.; Köster, N.; Barthlott, W. Large-Scale Diversity Patterns of Vascular Epiphytes in Neotropical Montane Rain Forests. *J. Biogeogr.* **2004**, *31*, 1477–1487. [CrossRef]
60. Olson, D.M. The Distribution of Leaf Litter Invertebrates Along a Neotropical Altitudinal Gradient. *J. Trop. Ecol.* **1994**, *10*, 129–150. [CrossRef]
61. Ramírez-Barahona, S.; Luna-Vega, I.; Tejero-Díez, D. Species Richness, Endemism, and Conservation of American Tree Ferns (Cyatheales). *Biodivers. Conserv.* **2011**, *20*, 59–72. [CrossRef]
62. Ríos-Touma, B.; Holzenthal, R.W.; Huisman, J.; Thomson, R.; Rázuri-Gonzales, E. Diversity and Distribution of the Caddisflies (Insecta: Trichoptera) of Ecuador. *PeerJ* **2017**, *5*, e2851. [CrossRef]
63. Smith, S.A.; Oca, A.N.M.D.; Reeder, T.W.; Wiens, J.J. A Phylogenetic Perspective on Elevational Species Richness Patterns in Middle American Treefrogs: Why so Few Species in Lowland Tropical Rainforests? *Evolution* **2007**, *61*, 1188–1207. [CrossRef] [PubMed]
64. Borchsenius, F. Patterns of Plant Species Endemism in Ecuador. *Biodivers. Conserv.* **1997**, *6*, 379–399. [CrossRef]
65. Young, K.R.; Ulloa, C.U.; Luteyn, J.L.; Knapp, S. Plant Evolution and Endemism in Andean South America: An Introduction. *Bot. Rev.* **2002**, *68*, 4–21. [CrossRef]
66. Myers, N.; Mittermeier, R.A.; Mittermeier, C.G.; Da Fonseca, G.A.; Kent, J. Biodiversity Hotspots for Conservation Priorities. *Nature* **2000**, *403*, 853–858. [CrossRef]
67. Janzen, D.H. Why Mountain Passes Are Higher in the Tropics. *Am. Nat.* **1967**, *101*, 233–249. [CrossRef]
68. Von Humboldt, A.; Bonpland, A. *Essay on the Geography of Plants*; University of Chicago Press: Chicago, IL, USA, 1807; ISBN 0-226-36068-7.
69. Hawkins, B.A.; Field, R.; Cornell, H.V.; Currie, D.J.; Guégan, J.-F.; Kaufman, D.M.; Kerr, J.T.; Mittelbach, G.G.; Oberdorff, T.; O'Brien, E.M.; et al. Energy, Water, and Broad-Scale Geographic Patterns of Species Richness. *Ecology* **2003**, *84*, 3105–3117. [CrossRef]
70. Arita, H.T.; Vázquez-Domínguez, E. The Tropics: Cradle, Museum or Casino? A Dynamic Null Model for Latitudinal Gradients of Species Diversity. *Ecol. Lett.* **2008**, *11*, 653–663. [CrossRef] [PubMed]
71. Jablonski, D.; Roy, K.; Valentine, J.W. Out of the Tropics: Evolutionary Dynamics of the Latitudinal Diversity Gradient. *Science* **2006**, *314*, 102–106. [CrossRef]
72. Gentry, A.H. Tropical Forest Biodiversity: Distributional Patterns and Their Conservational Significance. *Oikos* **1992**, *63*, 19. [CrossRef]

73. Särkinen, T.; Pennington, R.T.; Lavin, M.; Simon, M.F.; Hughes, C.E. Evolutionary Islands in the Andes: Persistence and Isolation Explain High Endemism in Andean Dry Tropical Forests. *J. Biogeogr.* **2012**, *39*, 884–900. [CrossRef]

74. Corbin, J.D.; Holl, K.D. Applied Nucleation as a Forest Restoration Strategy. *For. Ecol. Manag.* **2012**, *265*, 37–46. [CrossRef]

75. Zahawi, R.A.; Holl, K.D.; Cole, R.J.; Reid, J.L. Testing Applied Nucleation as a Strategy to Facilitate Tropical Forest Recovery. *J. Appl. Ecol.* **2013**, *50*, 88–96. [CrossRef]

76. Loughlin, N.J.D.; Gosling, W.D.; Mothes, P.; Montoya, E. Ecological Consequences of Post-Columbian Indigenous Depopulation in the Andean–Amazonian Corridor. *Nat. Ecol. Evol.* **2018**, *2*, 1233–1236. [CrossRef]

77. Tobler, W.R. A Computer Movie Simulating Urban Growth in the Detroit Region. *Econ. Geogr.* **1970**, *46*, 234–240. [CrossRef]

78. Londoño-Murcia, M.C.; Tellez-Valdés, O.; Sánchez-Cordero, V. Environmental Heterogeneity of World Wildlife Fund for Nature Ecoregions and Implications for Conservation in Neotropical Biodiversity Hotspots. *Environ. Conserv.* **2010**, *37*, 116–127. [CrossRef]

79. Pérez-Escobar, O.A.; Lucas, E.; Jaramillo, C.; Monro, A.; Morris, S.K.; Bogarín, D.; Greer, D.; Dodsworth, S.; Aguilar-Cano, J.; Sanchez Meseguer, A.; et al. The Origin and Diversification of the Hyperdiverse Flora in the Chocó Biogeographic Region. *Front. Plant Sci.* **2019**, *10*, 1328. [CrossRef] [PubMed]

80. Stadtmüller, T. *Cloud Forests in the Humid Tropics: A Bibliographic Review*; United Nations University Press: Tokyo, Japan; Turrialba, Costa Rica, 1987; ISBN 92-808-0670-X.

81. Foster, P. The Potential Negative Impacts of Global Climate Change on Tropical Montane Cloud Forests. *Earth-Sci. Rev.* **2001**, *55*, 73–106. [CrossRef]

82. Grubb, P.J.; Lloyd, J.R.; Pennington, T.D.; Whitmore, T.C. A Comparison of Montane and Lowland Rain Forest in Ecuador I. The Forest Structure, Physiognomy, and Floristics. *J. Ecol.* **1963**, *51*, 567. [CrossRef]

83. Grubb, P.J.; Whitmore, T.C. A Comparison of Montane and Lowland Rain Forest in Ecuador: II. The Climate and Its Effects on the Distribution and Physiognomy of the Forests. *J. Ecol.* **1966**, *54*, 303. [CrossRef]

84. Policha, T. *Pollination Biology of the Mushroom-Mimicking Orchid Genus Dracula*; University of Oregon: Eugene, OR, USA, 2014.

85. Información meteo e hidro | Productos Y Servicios Inamhi. Available online: https://inamhi.wixsite.com/inamhi/novedades (accessed on 25 March 2022).

86. Wunder, S. *Economics of Deforestation: The Example of Ecuador*; St Antony's Series; Palgrave Macmillan: London, UK, 2000; ISBN 978-0-333-73146-8.

87. Dodson, C.H.; Gentry, A.H. Biological Extinction in Western Ecuador. *Ann. Mo. Bot. Gard.* **1991**, *78*, 273. [CrossRef]

88. Lippi, R.D. Paleotopography and Phosphate Analysis of a Buried Jungle Site in Ecuador. *J. Field Archaeol.* **1988**, *15*, 85–97. [CrossRef]

89. Lippi, R.; Gudiño, A. Inkas and Yumbos at Palmitopamba in Northwestern Ecuador. In *Distant Provinces in the Inka Empire*; University of Iowa Press: Iowa City, IA, USA, 2010; pp. 260–278.

90. Watling, J.; Iriarte, J.; Mayle, F.E.; Schaan, D.; Pessenda, L.C.R.; Loader, N.J.; Street-Perrott, F.A.; Dickau, R.E.; Damasceno, A.; Ranzi, A. Impact of Pre-Columbian "Geoglyph" Builders on Amazonian Forests. *Proc. Natl. Acad. Sci. USA* **2017**, *114*, 1868–1873. [CrossRef] [PubMed]

91. Goosem, S.P.; Tucker, N.I.J.; Queensland; Wet Tropics Management Authority. *Repairing the Rainforest*; Wet Tropics Management Authority: Cairns, Australia, 2013; ISBN 978-1-921591-66-2.

92. Espírito-Santo, F.D.; Keller, M.M.; Linder, E.; Oliveira Junior, R.C.; Pereira, C.; Oliveira, C.G. Gap Formation and Carbon Cycling in the Brazilian Amazon: Measurement Using High-Resolution Optical Remote Sensing and Studies in Large Forest Plots. *Plant Ecol. Divers.* **2014**, *7*, 305–318. [CrossRef]

93. Franklin, J.F.; Spies, T.A.; Pelt, R.V.; Carey, A.B.; Thornburgh, D.A.; Berg, D.R.; Lindenmayer, D.B.; Harmon, M.E.; Keeton, W.S.; Shaw, D.C.; et al. Disturbances and Structural Development of Natural Forest Ecosystems with Silvicultural Implications, Using Douglas-Fir Forests as an Example. *For. Ecol. Manag.* **2002**, *155*, 399–423. [CrossRef]

94. Hunter, M.O.; Keller, M.; Morton, D.; Cook, B.; Lefsky, M.; Ducey, M.; Saleska, S.; de Oliveira, R.C., Jr.; Schietti, J. Structural Dynamics of Tropical Moist Forest Gaps. *PLoS ONE* **2015**, *10*, e0132144. [CrossRef] [PubMed]

95. Brokaw, N.Y.L. The Definition of Treefall Gap and Its Effect on Measures of Forest Dynamics. *Biotropica* **1982**, *14*, 158–160. [CrossRef]

96. Crausbay, S.D.; Martin, P.H. Natural Disturbance, Vegetation Patterns and Ecological Dynamics in Tropical Montane Forests. *J. Trop. Ecol.* **2016**, *32*, 384–403. [CrossRef]

97. Persson, C.; Eriksson, R.; Ståhl, B.; Romoleroux, K. (Eds.) *Flora of Ecuador | University of Gothenburg*; University of Gothenburg: Gothenburg, Sweden, 2011.

98. Python Core Team Python Software Foundation. Available online: https://www.python.org/ (accessed on 27 January 2018).

99. McKinney, W. Data Structures for Statistical Computing in Python. In Proceedings of the 9th Python in Science Conference, Stéfan van der Walt, Jarrod Millman, Austin, TX, USA, 28 June–3 July 2010; pp. 51–56.

100. Reback, J.; McKinney, W.; jbrockmendel; den Bossche, J.V.; Augspurger, T.; Cloud, P.; gfyoung; Sinhrks; Hawkins, S.; Klein, A.; et al. *Pandas 1.1.3*. Zenodo, Online Software Record. 2020. Available online: https://zenodo.org/record/4067057#.YpYM9FRByUk (accessed on 5 October 2020).

101. Hunter, J.D. Matplotlib: A 2D Graphics Environment. *Comput. Sci. Eng.* **2007**, *9*, 90–95. [CrossRef]

102. R Core Team. *R: A Language and Environment for Statistical Computing*; DPLR: Vienna, Austria, 2020.

103. Salvatier, J.; Wiecki, T.V.; Fonnesbeck, C. Probabilistic Programming in Python Using PyMC3. *PeerJ Comput. Sci.* **2016**, *2*, e55. [CrossRef]
104. Oksanen, J.; Guillaume Blanchet, F.; Friendly, M.; Kindt, R.; Legendre, P.; McGlinn, D.; Minchin, P.R.; O'Hara, R.B.; Simpson, G.L.; Solymos, P.M.; et al. *Vegan: Community Ecology Package. R Package Version 2.4-3*; The Comprehensive R Archive Network: Online, 2017.
105. Colwell, R.K.; Chao, A.; Gotelli, N.J.; Lin, S.-Y.; Mao, C.X.; Chazdon, R.L.; Longino, J.T. Models and Estimators Linking Individual-Based and Sample-Based Rarefaction, Extrapolation and Comparison of Assemblages. *J. Plant Ecol.* **2012**, *5*, 3–21. [CrossRef]
106. Legendre, P.; Legendre, L. Developments in Environmental Modelling. In *Numerical Ecology*, 3rd ed.; Elsevier: Amsterdam, The Netherlands, 2012; Volume 24, ISBN 978-0-444-53868-0.
107. Virtanen, P.; Gommers, R.; Oliphant, T.E.; Haberland, M.; Reddy, T.; Cournapeau, D.; Burovski, E.; Peterson, P.; Weckesser, W.; Bright, J.; et al. SciPy 1.0: Fundamental Algorithms for Scientific Computing in Python. *Nat. Methods* **2020**, *17*, 261–272. [CrossRef] [PubMed]
108. Kumar, R.; Carroll, C.; Hartikainen, A.; Martin, O. ArviZ a Unified Library for Exploratory Analysis of Bayesian Models in Python. *J. Open Source Softw.* **2019**, *4*, 1143. [CrossRef]
109. Ashour, S.K.; Abdel-hameed, M.A. Approximate Skew Normal Distribution. *J. Adv. Res.* **2010**, *1*, 341–350. [CrossRef]
110. O'Hagan, A.; Leonard, T. Bayes Estimation Subject to Uncertainty About Parameter Constraints. *Biometrika* **1976**, *63*, 201–203. [CrossRef]
111. Tachikawa, T.; Kaku, M.; Iwasaki, A.; Gesch, D.B.; Oimoen, M.J.; Zhang, Z.; Danielson, J.J.; Krieger, T.; Curtis, B.; Haase, J.; et al. *ASTER Global Digital Elevation Model Version 2—Summary of Validation Results*, 2nd ed.; NASA: Washington, DC, USA, 2011; p. 27.
112. Bartos, M. *Pysheds*. Software Package 2021. Available online: https://github.com/mdbartos/pysheds (accessed on 1 October 2020).
113. Dijkstra, E.W. A Note on Two Problems in Connexion with Graphs. *Numer. Math.* **1959**, *1*, 269–271. [CrossRef]
114. Hagberg, A.; Swart, P.; Chult, D.S. *Exploring Network Structure, Dynamics, and Function Using Networkx*; Los Alamos National Lab.: Los Alamos, NM, USA, 2008.
115. De Cáceres, M.; Legendre, P.; Moretti, M. Improving Indicator Species Analysis by Combining Groups of Sites. *Oikos* **2010**, *119*, 1674–1684. [CrossRef]
116. Dray, S.; Pélissier, R.; Couteron, P.; Fortin, M.-J.; Legendre, P.; Peres-Neto, P.R.; Bellier, E.; Bivand, R.; Blanchet, F.G.; Cáceres, M.D.; et al. Community Ecology in the Age of Multivariate Multiscale Spatial Analysis. *Ecol. Monogr.* **2012**, *82*, 257–275. [CrossRef]
117. Kattge, J.; Bönisch, G.; Díaz, S.; Lavorel, S.; Prentice, I.C.; Leadley, P.; Tautenhahn, S.; Werner, G.D.A.; Aakala, T.; Abedi, M.; et al. TRY Plant Trait Database—Enhanced Coverage and Open Access. *Glob. Chang. Biol.* **2020**, *26*, 119–188. [CrossRef]
118. Parr, C.S.; Wilson, N.; Leary, P.; Schulz, K.; Lans, K.; Walley, L.; Hammock, J.; Goddard, A.; Rice, J.; Studer, M.; et al. The Encyclopedia of Life v2: Providing Global Access to Knowledge About Life on Earth. *Biodivers. Data J.* **2014**, *2*, e1079. [CrossRef]
119. Gentry, A.H. *A Field Guide to the Families and Genera of Woody Plants of Northwest South America (Colombia, Ecuador, Peru), with Supplementary Notes on Herbaceous Taxa*; The University of Chicago Press: Chicago, IL, USA, 1993; ISBN 0-226-28944-3.
120. GDAL/OGR contributors. *GDAL/OGR Geospatial Data Abstraction Software Library*; Open Source Geospatial Foundation: Beaverton, OR, USA, 2021.
121. Grubb, P.J. Control of Forest Growth and Distribution on Wet Tropical Mountains: With Special Reference to Mineral Nutrition. *Annu. Rev. Ecol. Syst.* **1977**, *8*, 83–107. [CrossRef]
122. Leigh, E.G. Structure and Climate in Tropical Rain Forest. *Annu. Rev. Ecol. Syst.* **1975**, *6*, 67–86. [CrossRef]
123. Sohng, J.; Singhakumara, B.M.P.; Ashton, M.S. Effects on Soil Chemistry of Tropical Deforestation for Agriculture and Subsequent Reforestation with Special Reference to Changes in Carbon and Nitrogen. *For. Ecol. Manag.* **2017**, *389*, 331–340. [CrossRef]
124. Zhang, H.; Deng, Q.; Hui, D.; Wu, J.; Xiong, X.; Zhao, J.; Zhao, M.; Chu, G.; Zhou, G.; Zhang, D. Recovery in Soil Carbon Stock but Reduction in Carbon Stabilization after 56-Year Forest Restoration in Degraded Tropical Lands. *For. Ecol. Manag.* **2019**, *441*, 1–8. [CrossRef]
125. Hartmann, M.; Howes, C.G.; VanInsberghe, D.; Yu, H.; Bachar, D.; Christen, R.; Henrik Nilsson, R.; Hallam, S.J.; Mohn, W.W. Significant and Persistent Impact of Timber Harvesting on Soil Microbial Communities in Northern Coniferous Forests. *ISME J.* **2012**, *6*, 2199–2218. [CrossRef]
126. Hartmann, M.; Niklaus, P.A.; Zimmermann, S.; Schmutz, S.; Kremer, J.; Abarenkov, K.; Lüscher, P.; Widmer, F.; Frey, B. Resistance and Resilience of the Forest Soil Microbiome to Logging-Associated Compaction. *ISME J.* **2014**, *8*, 226–244. [CrossRef] [PubMed]
127. Wang, Q.; He, N.; Xu, L.; Zhou, X. Microbial Properties Regulate Spatial Variation in the Differences in Heterotrophic Respiration and Its Temperature Sensitivity between Primary and Secondary Forests from Tropical to Cold-Temperate Zones. *Agric. For. Meteorol.* **2018**, *262*, 81–88. [CrossRef]
128. Bart, D.; Davenport, T. The Influence of Legacy Impacted Seed Banks on Vegetation Recovery in a Post-Agricultural Fen Complex. *Wetl. Ecol Manag.* **2015**, *23*, 405–418. [CrossRef]
129. Young, K.R.; Ewel, J.J.; Brown, B.J. Seed Dynamics during Forest Succession in Costa Rica. *Vegetatio* **1987**, *71*, 157–173. [CrossRef]
130. Myster, R.W. Early Successional Pattern and Process after Sugarcane, Banana, and Pasture Cultivation in Ecuador. *N. Z. J. Bot.* **2007**, *45*, 101–110. [CrossRef]
131. Myster, R.W. Seed Predation, Pathogens and Germination in Primary vs. Secondary Cloud Forest at Maquipucuna Reserve, Ecuador. *J. Trop. Ecol.* **2015**, *31*, 375–378. [CrossRef]

132. Fragaszy, D.M.; Visalberghi, E.; Fedigan, L.M. *The Complete Capuchin: The Biology of the Genus Cebus*; Cambridge University Press: London, UK, 2004; ISBN 0-521-66768-2.
133. Russo, S.E.; Campbell, C.J.; Dew, J.L.; Stevenson, P.R.; Suarez, S.A. A Multi-Forest Comparison of Dietary Preferences and Seed Dispersal by *Ateles* spp. *Int. J. Primatol.* **2005**, *26*, 1017–1037. [CrossRef]
134. Chanderbali, A. Flora Neotropica; publ. for Organization for flora neotropica. In *Lauraceae*; The New York Botanical Garden: New York, NY, USA, 2004; ISBN 978-0-89327-454-2.
135. Bufalo, F.S.; Galetti, M.; Culot, L. Seed Dispersal by Primates and Implications for the Conservation of a Biodiversity Hotspot, the Atlantic Forest of South America. *Int. J. Primatol.* **2016**, *37*, 333–349. [CrossRef]
136. Morelos-Juárez, C.; Tapia, A.; Conde, G.; Peck, M. *Diet of the Critically Endangered Brown-Headed Spider Monkey (Ateles Fusciceps Fusciceps) in the Ecuadorian Chocó: Conflict between Primates and Loggers over Fruiting Tree Species*; PeerJ Inc.: Corte Madera, CA, USA, 2015.
137. Howe, H.F.; Smallwood, J. Ecology of Seed Dispersal. *Annu. Rev. Ecol. Syst.* **1982**, *13*, 201–228. [CrossRef]
138. Cubiña, A.; Aide, T.M. The Effect of Distance from Forest Edge on Seed Rain and Soil Seed Bank in a Tropical Pasture. *Biotropica* **2001**, *33*, 260–267. [CrossRef]
139. Holl, K.D. Factors Limiting Tropical Rain Forest Regeneration in Abandoned Pasture: Seed Rain, Seed Germination, Microclimate, and Soil1. *Biotropica* **1999**, *31*, 229–242. [CrossRef]
140. Mariscal, A.; Tigabu, M.; Savadogo, P.; Odén, P.C. Regeneration Status and Role of Traditional Ecological Knowledge for Cloud Forest Ecosystem Restoration in Ecuador. *Forests* **2022**, *13*, 92. [CrossRef]
141. Grime, J.P. Evidence for the Existence of Three Primary Strategies in Plants and Its Relevance to Ecological and Evolutionary Theory. *Am. Nat.* **1977**, *111*, 1169–1194. [CrossRef]
142. Planchuelo, G.; Kowarik, I.; von der Lippe, M. Endangered Plants in Novel Urban Ecosystems Are Filtered by Strategy Type and Dispersal Syndrome, Not by Spatial Dependence on Natural Remnants. *Front. Ecol. Evol.* **2020**, *8*, 18. [CrossRef]
143. Chapman, H.; Cordeiro, N.J.; Dutton, P.; Wenny, D.; Kitamura, S.; Kaplin, B.; Melo, F.P.L.; Lawes, M.J. Seed-Dispersal Ecology of Tropical Montane Forests. *J. Trop. Ecol.* **2016**, *32*, 437–454. [CrossRef]
144. Nunez-Iturri, G.; Olsson, O.; Howe, H.F. Hunting Reduces Recruitment of Primate-Dispersed Trees in Amazonian Peru. *Biol. Conserv.* **2008**, *141*, 1536–1546. [CrossRef]
145. van der Werff, H. A New Species of Caryodaphnopsis (Lauraceae) from Peru. *Syst. Bot.* **1986**, *11*, 415–418. [CrossRef]
146. Romo, M. Seasonal Variation in Fruit Consumption and Seed Dispersal by Canopy Bats (*Artibeus* spp.) in a Lowland Forest in Peru. *Vida Silv. Neotrop.* **1996**, *5*, 110–119.
147. Acero-Murcia, A.; Almario, L.J.; García, J.; Defler, T.R.; López, R. Diet of the Caquetá Titi (*Plecturocebus caquetensis*) in a Disturbed Forest Fragment in Caquetá, Colombia. *Primate Conserv.* **2018**, *32*, 1–7.
148. Knogge, C.; Heymann, E.W. Seed Dispersal by Sympatric Tamarins, Saguinus Mystax and Saguinus Fuscicollis: Diversity and Characteristics of Plant Species. *Folia Primatol.* **2003**, *74*, 33–47. [CrossRef]
149. Giraldo, P.; Gómez-Posada, C.; Martínez, J.; Kattan, G. Resource Use and Seed Dispersal by Red Howler Monkeys (*Alouatta seniculus*) in a Colombian Andean Forest. *Neotrop. Primates* **2007**, *14*, 55–64. [CrossRef]
150. Willson, M.F.; Crome, F.H.J. Patterns of Seed Rain at the Edge of a Tropical Queensland Rain Forest. *J. Trop. Ecol.* **1989**, *5*, 301–308. [CrossRef]
151. Cramer, J.M.; Mesquita, R.C.G.; Bruce Williamson, G. Forest Fragmentation Differentially Affects Seed Dispersal of Large and Small-Seeded Tropical Trees. *Biol. Conserv.* **2007**, *137*, 415–423. [CrossRef]
152. Peck, M.; Thorn, J.; Mariscal, A.; Baird, A.; Tirira, D.; Kniveton, D. Focusing Conservation Efforts for the Critically Endangered Brown-Headed Spider Monkey (*Ateles fusciceps*) Using Remote Sensing, Modeling, and Playback Survey Methods. *Int. J. Primatol.* **2011**, *32*, 134–148. [CrossRef]
153. Peres, C.A.; Emilio, T.; Schietti, J.; Desmoulière, S.J.M.; Levi, T. Dispersal Limitation Induces Long-Term Biomass Collapse in Overhunted Amazonian Forests. *Proc. Natl. Acad. Sci. USA* **2016**, *113*, 892–897. [CrossRef] [PubMed]
154. Calle-Rendón, B.R.; Peck, M.; Bennett, S.E.; Morelos-Juarez, C.; Alfonso, F. Comparison of Forest Regeneration in Two Sites with Different Primate Abundances in Northwestern Ecuador. *Rev. De Biol. Trop.* **2016**, *64*, 493–506. [CrossRef] [PubMed]
155. Morante-Filho, J.C.; Arroyo-Rodríguez, V.; Pessoa, M.D.; Cazetta, E.; Faria, D. Direct and Cascading Effects of Landscape Structure on Tropical Forest and Non-Forest Frugivorous Birds. *Ecol. Appl.* **2018**, *28*, 2024–2032. [CrossRef] [PubMed]
156. Navarro, A.B.; Bovo, A.A.A.; Alexandrino, E.R.; Oliveira, V.C.; Pizo, M.A.; Ferraz, K.M.P.M.B. Fruit Availability at the Individual and Local Levels Influences Fruit Removal in *Cecropia pachystachya*. *Braz. J. Biol.* **2018**, *79*, 758–759. [CrossRef] [PubMed]
157. Skutch, A.F. The Most Hospitable Tree. *Sci. Mon.* **1945**, *60*, 5–17.
158. Baldeck, C.A.; Harms, K.E.; Yavitt, J.B.; John, R.; Turner, B.L.; Valencia, R.; Navarrete, H.; Bunyavejchewin, S.; Kiratiprayoon, S.; Yaacob, A. Habitat Filtering across Tree Life Stages in Tropical Forest Communities. *Proc. R. Soc. B Biol. Sci.* **2013**, *280*, 20130548. [CrossRef]
159. Ledo, A.; Cayuela, L.; Manso, R.; Condés, S. Recruitment Patterns and Potential Mechanisms of Community Assembly in an Andean Cloud Forest. *J. Veg. Sci.* **2015**, *26*, 876–888. [CrossRef]
160. Seidl, R.; Thom, D.; Kautz, M.; Martin-Benito, D.; Peltoniemi, M.; Vacchiano, G.; Wild, J.; Ascoli, D.; Petr, M.; Honkaniemi, J.; et al. Forest Disturbances under Climate Change. *Nat. Clim. Chang.* **2017**, *7*, 395–402. [CrossRef] [PubMed]
161. Beisner, B.E.; Haydon, D.T.; Cuddington, K. Alternative Stable States in Ecology. *Front. Ecol. Environ.* **2003**, *1*, 376–382. [CrossRef]

162. Clerici, N.; Cote-Navarro, F.; Escobedo, F.J.; Rubiano, K.; Villegas, J.C. Spatio-Temporal and Cumulative Effects of Land Use-Land Cover and Climate Change on Two Ecosystem Services in the Colombian Andes. *Sci. Total Environ.* **2019**, *685*, 1181–1192. [CrossRef] [PubMed]
163. Feeley, K.J.; Silman, M.R.; Bush, M.B.; Farfan, W.; Cabrera, K.G.; Malhi, Y.; Meir, P.; Revilla, N.S.; Quisiyupanqui, M.N.R.; Saatchi, S. Upslope Migration of Andean Trees. *J. Biogeogr.* **2011**, *38*, 783–791. [CrossRef]
164. Rehm, E.M.; Feeley, K.J. The Inability of Tropical Cloud Forest Species to Invade Grasslands above Treeline during Climate Change: Potential Explanations and Consequences. *Ecography* **2015**, *38*, 1167–1175. [CrossRef]
165. Stork, N.E. Re-Assessing Current Extinction Rates. *Biodivers. Conserv.* **2010**, *19*, 357–371. [CrossRef]
166. Tovar, C.; Arnillas, C.A.; Cuesta, F.; Buytaert, W. Diverging Responses of Tropical Andean Biomes under Future Climate Conditions. *PLoS ONE* **2013**, *8*, e63634. [CrossRef]
167. Myers, N. Threatened Biotas:" Hot Spots" in Tropical Forests. *Environmentalist* **1988**, *8*, 187–208. [CrossRef]
168. Mosandl, R.; Günter, S.; Stimm, B.; Weber, M. Ecuador Suffers the Highest Deforestation Rate in South America. In *Gradients in a Tropical Mountain Ecosystem of Ecuador*; Beck, E., Bendix, J., Kottke, I., Makeschin, F., Mosandl, R., Eds.; Ecological Studies; Springer: Berlin/Heidelberg, Germany, 2008; pp. 37–40. ISBN 978-3-540-73526-7.
169. Tapia-Armijos, M.F.; Homeier, J.; Espinosa, C.I.; Leuschner, C.; de la Cruz, M. Deforestation and Forest Fragmentation in South Ecuador since the 1970s—Losing a Hotspot of Biodiversity. *PLoS ONE* **2015**, *10*, e0133701. [CrossRef]
170. Guayasamin, J.M.; Vandegrift, R.; Policha, T.; Encalada, A.C.; Greene, N.; Ríos-Touma, B.; Endara, L.; Cárdenas, R.E.; Larreátegui, F.; Baquero, L.; et al. Tipping Point Towards Biodiversity Conservation? Local and Global Consequences of the Application of 'Rights of Nature' by Ecuador. *Soc. Sci.* **2021**. [CrossRef]
171. Vallejo Galárraga, M.C.; Freslon, W.S. Ecuador: Mineral Policy. In *Encyclopedia of Mineral and Energy Policy*; Tiess, G., Majumder, T., Cameron, P., Eds.; Springer: Berlin/Heidelberg, Germany, 2017; pp. 1–8. ISBN 978-3-642-40871-7.
172. Filotas, E.; Parrott, L.; Burton, P.J.; Chazdon, R.L.; Coates, K.D.; Coll, L.; Haeussler, S.; Martin, K.; Nocentini, S.; Puettmann, K.J.; et al. Viewing Forests through the Lens of Complex Systems Science. *Ecosphere* **2014**, *5*, art1. [CrossRef]
173. Messier, C.; Puettmann, K.; Chazdon, R.; Andersson, K.P.; Angers, V.A.; Brotons, L.; Filotas, E.; Tittler, R.; Parrott, L.; Levin, S.A. From Management to Stewardship: Viewing Forests As Complex Adaptive Systems in an Uncertain World. *Conserv. Lett.* **2015**, *8*, 368–377. [CrossRef]
174. Loehle, C. Challenges of Ecological Complexity. *Ecol. Complex.* **2004**, *1*, 3–6.
175. McElhinny, C.; Gibbons, P.; Brack, C.; Bauhus, J. Forest and Woodland Stand Structural Complexity: Its Definition and Measurement. *For. Ecol. Manag.* **2005**, *218*, 1–24. [CrossRef]
176. Ripple, W.J.; Estes, J.A.; Schmitz, O.J.; Constant, V.; Kaylor, M.J.; Lenz, A.; Motley, J.L.; Self, K.E.; Taylor, D.S.; Wolf, C. What Is a Trophic Cascade? *Trends Ecol. Evol.* **2016**, *31*, 842–849. [CrossRef]
177. MacArthur, R.H.; Wilson, E.O. *The Theory of Island Biogeography*; Princeton University Press: Princeton, NJ, USA, 2001; Volume 1, ISBN 0-691-08836-5.
178. Ross, K.A.; Fox, B.J.; Fox, M.D. Changes to Plant Species Richness in Forest Fragments: Fragment Age, Disturbance and Fire History May Be as Important as Area. *J. Biogeogr.* **2002**, *29*, 749–765. [CrossRef]
179. Michaels, T.K.; Eppinga, M.B.; Bever, J.D. A Nucleation Framework for Transition between Alternate States: Short-Circuiting Barriers to Ecosystem Recovery. *Ecology* **2020**, *101*, e03099. [CrossRef] [PubMed]
180. Fernández, N.; Navarro, L.M.; Pereira, H.M. Rewilding: A Call for Boosting Ecological Complexity in Conservation. *Conserv. Lett.* **2017**, *10*, 276–278. [CrossRef]
181. Balmford, A.; Bruner, A.; Cooper, P.; Costanza, R.; Farber, S.; Green, R.E.; Jenkins, M.; Jefferiss, P.; Jessamy, V.; Madden, J.; et al. Economic Reasons for Conserving Wild Nature. *Science* **2002**, *297*, 950–953. [CrossRef] [PubMed]
182. Kocian, M.; Batker, D.; Harrison-Cox, J. *An Ecological Study of Ecuador's Intag Region: The Environmental Impacts and Potential Rewards of Mining*; Earth Economics: Tacoma, WA, USA, 2011.

Article

Ecological Strategy Spectra for Communities of Different Successional Stages in the Tropical Lowland Rainforest of Hainan Island

Chen Chen [1], Yabo Wen [1], Tengyue Ji [1], Hongxia Zhao [1], Runguo Zang [2,3] and Xinghui Lu [1,*]

[1] College of Agronomy and Agricultural Engineering, Liaocheng University, Liaocheng 252000, China; cc2842855056@163.com (C.C.); wenyabo416@163.com (Y.W.); jity923@163.com (T.J.); zhaohxia1314@163.com (H.Z.)

[2] Key Laboratory of Forest Ecology and Environment of the National Forestry and Grassland Administration, Research Institute of Forest Ecology, Environment, and Protection, Chinese Academy of Forestry, Beijing 100091, China; zangrung@caf.ac.cn

[3] Co-Innovation Center for Sustainable Forestry in Southern China, Nanjing Forestry University, Nanjing 210037, China

[*] Correspondence: luxinghui_0@163.com

Abstract: Plant ecological strategies are shaped by long-term adaptation to the environment and are beneficial to plant survival and reproduction. Research is ongoing to better understand how plants best allocate resources for growth, survival and reproduction, as well as how ecological strategies may shift in plant communities over the course of succession. In this study, 12 forest dynamics plots in three different successional stages were selected for study in the tropical lowland rainforest ecosystem of Hainan Island. For each plot, using Grime's competitor, a stress-tolerator, the ruderal (CSR) scheme and using the CSR ratio tool "StrateFy", an ecological strategy spectrum was constructed using functional trait data obtained by collecting leaf samples from all woody species. The ecological strategy spectra were compared across successional stages to reveal successional dynamics. The results showed: (1) The ecological strategy spectra varied among forest communities belonging to three different successional stages. (2) The community-weighted mean CSR (CWM-CSR) strategies shifted with succession: CWM-S values decreased, while the CWM-C and CWM-R values increased. Overall, shifts in plant functional traits occurred slowly and steadily with succession showing complex and diverse trade-offs and leading to variation among the ecological strategy spectra of different successional stages.

Keywords: ecological strategy; succession; functional traits; forest vegetation; successional dynamics; ecosystem function

Citation: Chen, C.; Wen, Y.; Ji, T.; Zhao, H.; Zang, R.; Lu, X. Ecological Strategy Spectra for Communities of Different Successional Stages in the Tropical Lowland Rainforest of Hainan Island. *Forests* **2022**, *13*, 973. https://doi.org/10.3390/f13070973

Academic Editors: Konstantinos Poirazidis and Panteleimon Xofis

Received: 12 May 2022
Accepted: 20 June 2022
Published: 22 June 2022

Publisher's Note: MDPI stays neutral with regard to jurisdictional claims in published maps and institutional affiliations.

1. Introduction

Ecological strategies reflect how species optimally allocate resources to growth, survival and reproduction, thereby, capturing trade-offs among functional traits [1,2]. After selection by abiotic and biotic environmental factors, a plant's ecological strategy represents its ideal combination of traits [3,4]. To remain competitive within their ecological community, plants may adjust their resource allocations. Species display different combinations of traits [5,6] based on their tolerance of current environmental conditions and ability to cope with resource-poor habitats [7,8].

The study of ecological strategies is an important avenue to understanding biological community assembly and dynamics in response to environmental change [9]. Many famous ecologists have investigated different aspects of species' ecological strategies [10,11]. Additionally, the study of plant functional traits to better our understanding of plant ecological strategies represents a current research hotspot in ecology [12–14].

Grime's competitor, stress-tolerator, ruderal CSR theory is foundational to research on ecological strategies and is also based on plant functional traits [15,16]. Recent developments in CSR theory seek to explain plant ecological strategies in terms of the primary dimension of functional trait variation. Pierce et al. [1] described how three functional traits (of plant leaves) can be used to extrapolate the three main dimensions of functional trait variation and developed a freely available tool, "StrateFy", to implement these calculations.

Plant species are divided into 19 strategy types, including three primary strategies (C, S and R), four secondary strategies (CS, CR, SR and CSR) and twelve tertiary strategies (C/CR, C/CS, C/CSR, CR/CSR, CS/CSR, CS/CSR, R/CR, S/CS, R/CSR, S/CSR, SR/CSR, R/SR and S/SR) [17,18]. Secondary strategy and tertiary strategy are different combinations of the three primary strategies. The three primary strategies (i.e., C, S and R) represent combinations of traits best suited to competition (C), abiotic stress tolerance (S) and ruderal habitats with periodic biomass destruction (R) [1]. A more detailed description refers to CSR classification after Hodgson et al.'s CSR classification [19].

Using CSR theory and community-weighted mean (CWM) trait values, differences in the functional compositions among plant communities can be evaluated. Community-level ecological strategy spectra can then be estimated using the CWM trait values in "StrateFy". These ecological strategy spectra (i.e., the number and relative abundance of species holding different ecological strategy types in the community) can provide a "functional summary" of the vegetation, which can also be used to study how communities vary among successional stages [17].

Grime's CSR theory has been applied to functional analyses, involving global [20], regional [21] and local [22] scales. Previous studies have shown that ecological strategies can be used to explain the distribution of species along environmental gradients [23,24]. Shifts in ecological strategy can reflect the influence of environmental gradients and disturbances of forest dynamics [25]. While there have been many studies of vegetation function and community assembly [7,26], little is known about how community ecological strategy spectra shift with succession.

Succession represents a process of dynamic community construction [27]. Across successional stages, the environmental factors affecting the vegetation are constantly changing [28], as are plant–environment interactions. As a result, the community composition and structure shift over time [29], thus, affecting ecosystem function [30]. Although there are still disagreements about the predictability of community structure and the role of historical contingency [31], many studies have shown that, while species composition during succession is often unpredictable, functional changes are deterministic [32].

Functional analyses of communities may be helpful to better characterize the successional process and related environmental changes [33]. Studying ecological strategies at the community-level may provide insights into the resource balance at each restoration stage, as well as the process of community assembly and ecosystem functioning [34]. However, the knowledge of how ecological strategy spectra of different successional forests in the same region in tropical forests is largely unclear.

In restoring abandoned slash-and-burn farmland in the Bawangling tropical lowland rainforest on Hainan Island, China, communities of different restoration ages have been produced. This region therefore provides an ideal system to evaluate the relationship between restoration age and the ecological strategy spectrum. In this study, forests ecological strategy spectra belonging to different successional stages were determined. Then, the following two questions were discussed: (1) Do the forest ecological strategy spectra change with succession? (2) What is the effect of succession on forest ecological strategy composition?

This approach is valuable for expanding the study of successional forest ecosystem functioning [35]. Succession is hypothesized to affect a plant community's ecological strategy spectrum, with the community-weighted mean S expected to be dominant in the later stages of succession and community-weighted mean C dominant early on. Moreover, plant ecological strategies may become more specialized over the course of succession.

2. Materials and Methods

2.1. Study Area and Sampling Strategy

This study was carried out in the Bawangling Forest Region on Hainan Island (18°52′–19°12′ N, 108°53′–109°20′ E), which occurs at the northern limit of the tropical rainforest in Asia [36]. Bawangling Forest covers an area of about 500 km^2, with an altitudinal range from 100–1654 m. Vegetation in the area varies with altitude; however, this study focused on the tropical lowland rainforest (<800 m above sea level).

Abandoned slash-and-burn farmland, naturally-restored secondary forest and a few undisturbed old-growth forests are distributed here. Information on the history of land-use for the plots was obtained from the management records of the Bawangling National Nature Reserve [37]. Within the region, the annual average temperature is 23.6 °C, and the annual precipitation is 1677 mm. The rainy season occurs from May to October, and the dry season from November to April [36].

According to best practices published by the Center for Tropical Forestry Science (CTFS) [38], twelve forest dynamic monitoring plots of 100 × 100 m were established (twelve plots belong to three successional stages: 30-year-old secondary growth forest, 60-year-old secondary growth forest and the old growth forest) (Table 1). For the convenience of community survey, each plot was divided into 25 quadrats of 20 × 20 m, and cement piles were used to mark the four corners of each quadrat. In these fixed plots, all woody stems with diameter breast height (DBH) > 1 cm were surveyed, and the species name and DBH were recorded.

Table 1. Overview of forest dynamic monitoring plots in communities of various successional stages.

Stages of Succession	Abbreviation	Interference History	Number of Plots
Early succession	E	30-year-old secondary forest	4
Mid-succession	M	60-year-old secondary forest	4
Late succession	O	Old growth forest	4

2.2. Determination of Functional Traits

Ten individuals were sampled from each species (excluding endangered species) in each plot. Random sampling was conducted if there were more than ten individuals in each species. In cases where there were less than ten individual per species, all individuals were sampled. Five to ten mature leaves were collected to measure the leaf functional traits in the field. To ensure that the leaf materials remained fresh, samples were stored in fresh-keeping bags and transported to the laboratory for measurement within 24 h [39]. Two healthy leaves were selected for each individual. These were weighed (to determine the leaf fresh weight [LFW], in mg), then scanned on a flatbed scanner, and the area of each leaf (or leaf area [LA]) was determined using ImageJ. Afterwards, leaf samples were dried in an 80 °C oven to a constant weight. The leaf dry weight (LDW, in mg) was measured, and the specific leaf area (SLA) and leaf dry matter content (LDMC) were calculated [39]. Leaf traits were measured for 434 species in total.

2.3. Data Analysis

Using the CSR ratio tool "StrateFy" [1], one can measure three simple and easy-to-quantify leaf functional traits (the leaf area, specific leaf area and leaf dry matter content) and then use the trade-offs among them to express the degree of C-, S- and R-selection. This enables the classification of plant species according to their CSR strategy, as well as quantitative comparisons among different species or communities [19].

According to the results of the community survey and functional traits determination, we created two matrices of twelve forest dynamic monitoring plots: one is the species abundance matrix, and the other is the three functional traits matrix of leaves (leaf area, specific leaf area and leaf dry matter content) of species. The "dbFD" function in the "FD" package [40] was used to calculate community-weighted mean (CWM) values of leaf area, specific leaf area and leaf dry matter content [41,42]. These were then input into "StrateFy" to obtain the CSR value of each species to determine the classification of the ecological strategies.

Then, we calculated the CWM-CSR value of each plot and the number and relative abundance of species with different ecological strategies in each succession stage. For each plot, the incidence of each ecological strategy was summed over the individuals to obtain the overall "strategy richness". Similarly, the distribution of each strategy in each sample plot was compiled using a three-level classification scheme (CSR). Variation in the CWM-CSR values, strategy richness and strategy distributions among successional stages was assessed using one-way ANOVAs followed by Tukey tests [43]. To compare community types based on these ecological strategy spectra, nonparametric Kruskal–Wallis tests and Wilcoxon tests were implemented.

To further explore whether communities varied in terms of their ecological strategy composition, we established the abundance matrix of each ecological strategy in 12 sample plots. Non-metric multidimensional scaling (NMDS) was utilized. The NMDS was then constrained by community type (i.e., successional stage) to evaluate how successional stage influenced variation in plant ecological strategies. We used an analysis of similarities (ANOSIM) to test the significance of the constrained axes. The "vegan" package in R was used to perform the NMDS (with the "metaMDS" function) and the ANOSIM ("ANOSIM" function). All statistical analyses were conducted in R [44].

The C, S and R values for each species were used to create a "trade-off triangle" in order to compare among target species. For each succession stage, a ternary (or triangle) plot was drawn with each axis representing a strategy (i.e., C, S or R). Individual species were then positioned within the resulting CSR triangle. Ecological strategies are represented by color: pure red indicates the C strategy, pure green indicates the S strategy, and pure blue indicates the R strategy. We copy-and-pasted the "Color values in SigmaPlot format" in "StrateFy" into Sigmaplot to obtain the colors of species. Triangle diagrams were drawn in SigmaPlot.

3. Results

3.1. Types of Community Ecological Strategies in Successional Stages

A total of 434 plant species were identified across all sample plots. We documented 182 species in forest plots of early succession, 247 species in mid-succession and 320 species in late succession. The three successional stages differed in terms of the species diversity. In our study, species were assigned to 16 out of the total 19 ecological strategies by "StrateFy", with two of the three primary strategies being C and S. There were three secondary strategies identified, namely CR, CS and CSR, and eleven tertiary strategies identified: C/CR, C/CS, C/CSR, CR/CSR, CS/CSR, R/CR, R/CSR, S/CS, S/CSR, S/SR and SR/CSR (Table 2).

Table 2. The types of community-level ecological strategies identified in different successional stages.

Early Succession	Mid-Succession	Late Succession
C	C	C
C/CR	C/CR	C/CR
C/CS	C/CS	C/CS
C/CSR	C/CSR	C/CSR
CR	CR	CR
CS	CR/CSR	CR/CSR
CS/CSR	CS	CS
CSR	CS/CSR	CS/CSR
CR/CSR	CSR	CSR
R/CSR	R/CR	R/CSR
S	S	S/CS
S/CS	S/CS	S/CSR
S/CSR	S/CSR	S
S/SR	SR/CSR	
SR/CSR		

(Early succession = E, mid-succession = M and late succession = O).

3.2. Successional Dynamics in Community Ecological Strategy Spectra

The ecological strategy spectra varied among forest communities belonging to three different successional stages. The strategy richness represents the total number of CSR strategy types identified for each successional stage. In total, 16 strategy types (of 19 possible types) were distinguished: 15 in the early successional stage community, 14 in the mid-successional stage community and 13 in the mature forest (Table 2). The strategy richness showed a downward trend with succession.

In early and mid-successional stage communities, species were largely concentrated in the CS area of the ternary plot, and the three dominant strategies were CS, S/CS and CS/CSR. In the early successional stage community, no species adopted the R/CR strategy, while no S/SR and R/CSR strategies were found in the mid-successional stage community. In the late successional stage community, species were concentrated in the CS/CSR area of the ternary plot, with CS/CSR, CS and S/CS being the most common strategies. The R/CR, S/SR and SR/CSR strategies were not identified (Figures 1 and 2).

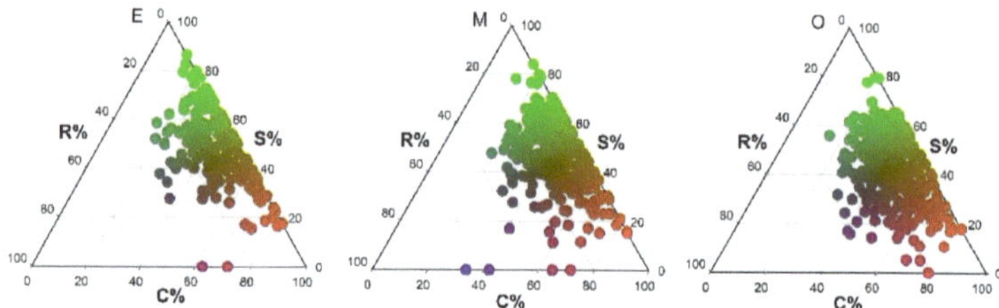

Figure 1. Ternary plots of species ecological strategies for different successional stages in tropical lowland rainforests on Hainan Island (E = early succession, M = mid-succession and O = late succession). C (%), S (%) and R (%) represent the three strategy components C, S and R, respectively. Ecological strategies are represented by color: pure red indicates the C strategy, pure green indicates the S strategy, and pure blue indicates the R strategy. Intermediate (mixed) colors indicate the full range of intermediate strategies (e.g., green/blue = SR strategy).

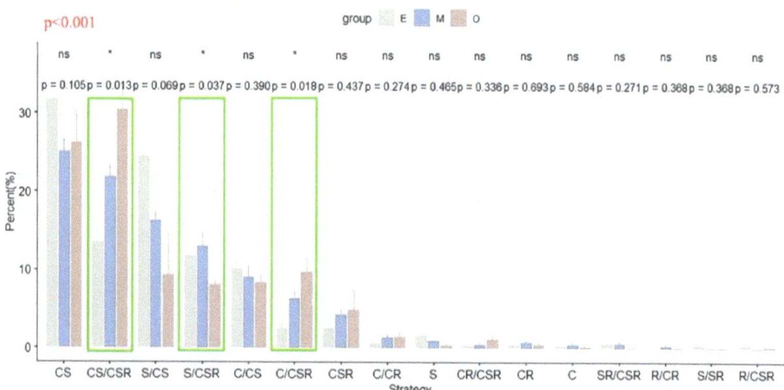

Figure 2. Changes in the ecological strategy spectra (based on CSR theory) across successional stages in the tropical lowland rainforest of Hainan Island (gray, E, early succession; blue, M, mid-succession; and red, O, late succession). Percentages (%) indicate the average number of species in each strategy category ($n = 4$), with error bars denoting the standard errors. * indicates a significant difference among stages ($p < 0.05$), while ns indicates no significant difference ($p > 0.05$); green boxes additionally highlight significant cases. $p < 0.001$ in the top left indicates a significant difference among the ecological strategy spectra of three succession stages.

Based on the variance partitioning analysis, the three successional stages differed significantly in terms of the proportion of S/CSR, C/CSR and CS/CSR strategies adopted by the species within each community (Figure 2). As succession proceeded, the proportion of S, C/CS and S/CS strategies decreased, while the proportion of C/CR, CSR, CR/CSR, C/CSR and CS/CSR strategies increased (Figure 2).

3.3. Effects of Successions on CSR Strategies Composition

The first NMDS axis separated the early and mid-successional stages from the late successional stage, with partial overlap between the early and middle stages (Figure 3). In addition, the ANOSIM (R = 0.6366, p = 0.001 < 0.05) confirmed significant variation among the ecological strategy spectra of different successional periods. The S/CSR, C/CS and CS strategies were common in all communities. However, in the NMDS, the distance between the C/CSR, CS/CSR and C/CR strategies was reduced in the late successional stage community versus the early and mid-successional stage communities, indicating that there were more species holding these strategies in mature forest (i.e., later in succession) (Figure 3).

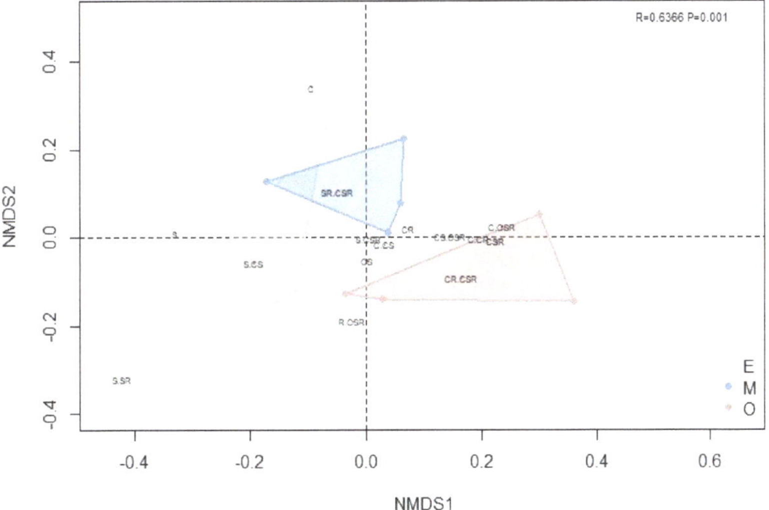

Figure 3. A non–metric multidimensional scaling (NMDS) diagram based on species richness was used to compare forest communities of different successional stages in the tropical lowland rainforest of Hainan Island. Polygons of different colors represent different successional stages (gray, E, early succession; blue, M, mid-succession; and red, O, late succession). Text labels within the plot represent ecological strategies (C = competitors, S = stress tolerators and R = ruderals); please refer to the CSR classification of Hodgson et al. (1999) [19] for a more detailed description. For the NMDS, R = 0.6366 and p = 0.001.

The average CWM-CSR values calculated for each of the four sample plots were used to represent the CSR values for each successional stage. For all three stages, the C and S components contributed more to the CWMs when compared with the R component. The CWM-C, CWM-S and CWM-R values varied among successional stages. With succession, component C increased from 38.19% to 42.22%, while component S declined from a maximum value of 53.90% to 45.35% (Figure 4). Further analysis for component C revealed a significant difference between late versus early/mid- successional stages but no difference between early and mid-successional stages. For component S, the early and mid-stages differed; however, the middle and late stages did not. Similarly, for the R component, the middle and later stages did not differ; however, both were greater than for the early successional stage (Figure 4).

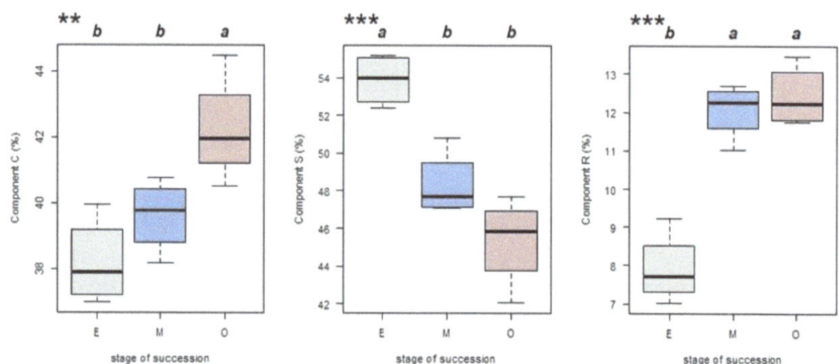

Figure 4. CWM-CSR values (for C, S and R strategy components) for tropical lowland rainforest stands on Hainan Island at different successional stages (gray, E, early succession; blue, M, mid-succession; and red, O, late succession). ** $p < 0.05$ and *** $p < 0.01$ indicate significant differences according to a one-way ANOVA ($n = 4$). Different letters indicate significant differences between stages (Tukey's test; $p < 0.05$).

Ecological strategies were classified into three groups: primary, secondary and tertiary. The proportion of species belonging to each group is shown in Figure 5 for each successional stage. For all three successional stages, the proportion of species increased with the group number. The results showed that most of the species in the three succession stages adopted the ecological strategy with complicated trade-offs. A few species adopted the primary strategies (the C and S ecological strategies).

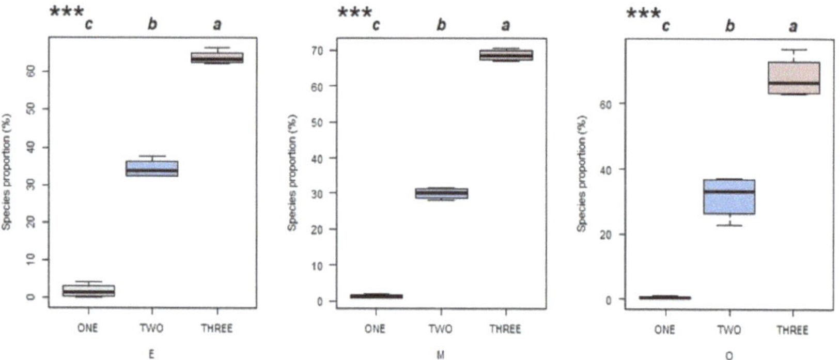

Figure 5. The proportion of species in Hainan tropical lowland rainforest stands belonging to each of three ecological strategies groups for multiple successional stages (E = early succession, M = mid-succession and O = late succession). Primary strategies (in grey) include C and S. Secondary strategies (in blue) include CS, CR and CSR. Tertiary strategies (in red) include C/CR, C/CS, C/CSR, CR/CSR, CS/CSR, R/CR, R/CR, R/CS, S/CS, S/CSR, S/SR and SR/CSR. *** $p < 0.01$ in a one-way ANOVA ($n = 4$). Different letters indicate significant differences between ecological strategies groups (Tukey's test; $p < 0.05$).

4. Discussion

4.1. Dynamic Successional Patterns in Community Ecological Strategy Spectra

This study found Grime's CSR theory allowed the functional interpretation of tropical lowland rainforest communities along a successional gradient [15] as well as the identification of realized functional niches within the communities. Pierce et al. [1] found that the CSR strategies for species characteristic of primary succession from scree vegetation to siliceous alpine grassland, terminating with alpine Nardus pasture, were evident. Our work further confirms that the applicability and effectiveness of the globally-corrected CSR ecological strategy spectrum approach was assessed from the perspective of regional succession [19].

This study found that the ecological strategy spectra varied significantly along the successional gradient. In all three successional stages, most species had intermediate values for components C and S. This is consistent with previous research results [45]. There are few R-selection ecological strategies, likely because our survey does not include herbaceous plants [41]. The proportion of species having each strategy type differed over time. Initially, most species had CS, S/CS or CS/CSR strategies, with only a few having C or CR/CSR strategies.

As succession proceeded, the proportion of species with these three dominant strategies changed, with CS and S/CS becoming less common (31.66% to 25.06% and 24.46% to 16.23%, respectively), while CS/CSR increased in frequency (13.42% to 21.81%). Only a few species possessed C or R/CR strategies. As forests matured (late successional stage), most species had CS/CSR, CS or S/CS strategies, while only a few species had C or R/CSR strategies. The diversity of ecological strategies indicates that plants take various trade-offs to make use of the acquired environmental resources [5,7,26].

However, the environmental driving force behind the changes of these strategies remains unclear. Strengthening these studies will contribute to reveal the relationship between environment and ecological strategies [41,46]. Use of this approach may help to predict patterns of species' functional trade-offs in a specific environment in addition to how community processes respond to the environment [1,41]. The number of strategies tended to decrease over the course of succession. Strategy richness was also negatively correlated with species richness. This is an interesting discovery. More species are thought to have more ecological strategy classifications [47].

However, the late succession stage with the largest number of species in this study showed the least ecological strategy classifications. This may be because, later in succession, after a long period of environmental screening [48], species' ecological strategies have converged [49,50]. The stability of functions shifts during the process of community assembly in secondary (restored) forests [47]. Therefore, although the number of species increases, the number of ecological strategies decreases. The differences in strategy richness (among community types) may be due to variation in the driving forces underlying successional processes, which may be affected by the interaction of multiple environmental factors [46,51].

4.2. Community-Level Differences in Ecological Strategy Composition with Succession

By summarizing the ecological strategy spectra for all succession stages, the CS strategy was found to be the most common strategy across all three successional stages. However, the proportion of species having a CS strategy (secondary strategy) was highest in the early stages of succession and was replaced by the CS/CSR strategy (tertiary strategy) in mature forests (30.40% for CS/CSR vs. 26.19% for CS). For all three successional stages, the proportion of species increased from primary to secondary to tertiary strategy.

Therefore, more species presented with complex versus simple functional trade-offs in the study communities. This is consistent with the research results of tropical forest ecological strategies in these four climatic zones [52]. Primary and secondary strategies were better represented early in succession, while tertiary strategies were more common during the mid-to-late stages of succession. This suggests that, over the course of succession, more complicated trade-offs among traits emerge. As a result, complex combinations of stress tolerance, competitive acquisition and resource allocation traits become more common.

Scholars have found that, if the influence of the plant community species composition and diversity on the community is ignored, the functional traits of individual plants cannot accurately reflect the ecosystem function [5,53]. Instead, the functional traits of plants at the community level can better reflect the ecosystem function [5]. Differences in CWM-CSR strategies may reflect changes in the community functional trait composition caused by succession. The most competitive community (i.e., the highest C-value) was that of the mature forest, while the recently abandoned farmland had the lowest C-value.

The latter community was also the most stress-tolerant (i.e., the highest S-score), in contrast to the mature forest (the lowest degree of S-selection). This contradicts our hypothesis. The mid-successional stage community showed intermediate C- and S-values compared to the early and late stages. Stress tolerance (as S-values) was the highest early during succession and declined over time, likely as the disturbances associated with slash-and-burn cultivation slowly diminished with the passage of time [54]. During recovery from slash-and-burn cultivation, a positive feedback loop also develop [55], whereby vegetation growth provides richer resources to the environment [56], which then, in turn, further promotes plant growth.

As a result, the C-values (i.e., the ability to compete for resources) rise as resources become more abundant [48]. Therefore, when compared to the disturbance of slash-and-burn cultivation, the resource limitations occurring later in succession do not significantly limit community restoration. These resource limitations also play a positive role in the process of secondary succession after slash-and-burn cultivation—for instance, the litter improves the soil environment [57]. In secondary succession, forest habitats and ecosystem functions are restored.

5. Conclusions

This study demonstrated that the CSR framework (based on functional traits) is an effective method to evaluate how succession impacts the functional composition of forests. Over the course of succession in tropical lowland rainforests, both the ecological strategy spectra and CWM-CSR strategies were found to shift. Species with more complex balance and combinations of functional traits have greater survival advantages. This study expands the understanding of ecosystem function in successional forests and provides a perspective on research regarding the ecological strategy of the community building process.

Author Contributions: Conceptualization, X.L. and R.Z.; methodology, X.L. and R.Z.; formal analysis, C.C. and Y.W.; investigation, C.C., Y.W. and T.J.; writing—original draft preparation, C.C.; writing—review and editing, X.L. and H.Z.; visualization, C.C. and Y.W.; project administration, X.L.; funding acquisition, X.L. All authors have read and agreed to the published version of the manuscript.

Funding: This research was funded by the National Natural Science Foundation of China 31901210 and by the Doctoral Research Foundation of Liaocheng University 318051822.

Acknowledgments: We are grateful to the many people who contributed to this study, particularly Xiusen Yang and Yuecai Tang at the Bawangling Nature Reserve for their assistance in specimen identification and work in the field investigation. We would like to thank Emily Drummond at the University of British Columbia for her English editing of the manuscript.

Conflicts of Interest: The authors declare that they have no known competing financial interests or personal relationships that could have appeared to influence the work reported in this paper.

References

1. Pierce, S.; Negreiros, D.; Cerabolini, B.E.L.; Kattge, J.; Diaz, S.; Kleyer, M.; Shipley, B.; Wright, S.J.; Soudzilovskaia, N.A.; Onipchenko, V.G.; et al. A global method for calculating plant CSR ecological strategies applied across biomes world-wide. *Funct. Ecol.* **2017**, *31*, 444–457. [CrossRef]
2. Chauvin, K.M.; Asner, G.P.; Martin, R.E.; Kress, W.J.; Wright, S.J.; Field, C.B. Decoupled dimensions of leaf economic and anti-herbivore defense strategies in a tropical canopy tree community. *Oecologia* **2018**, *186*, 765–782. [CrossRef] [PubMed]
3. DÍAz, S.; Lavorel, S.; McIntyre, S.U.E.; Falczuk, V.; Casanoves, F.; Milchunas, D.G.; Skarpe, C.; Rusch, G.; Sternberg, M.; Noy-Meir, I.; et al. Plant trait responses to grazing—A global synthesis. *Glob. Chang. Biol.* **2007**, *13*, 313–341. [CrossRef]
4. Guo, W.Y.; van Kleunen, M.; Winter, M.; Weigelt, P.; Stein, A.; Pierce, S.; Pergl, J.; Moser, D.; Maurel, N.; Lenzner, B.; et al. The role of adaptive strategies in plant naturalization. *Ecol. Lett.* **2018**, *21*, 1380–1389. [CrossRef]
5. Niu, S.; Classen, A.T.; Luo, Y. Functional traits along a transect. *Funct. Ecol.* **2018**, *32*, 4–9. [CrossRef]
6. GÓMez-Aparicio, L.; GarcÍA-ValdÉS, R.; RuÍZ-Benito, P.; Zavala, M.A. Disentangling the relative importance of climate, size and competition on tree growth in Iberian forests: Implications for forest management under global change. *Glob. Chang. Biol.* **2011**, *17*, 2400–2414. [CrossRef]

7. Díaz, S.; Cabido, M.R. Plant functional types and ecosystem function in relation to global change. *J. Veg. Sci.* **1997**, *8*, 463–474. [CrossRef]
8. Albert, C.H.; Thuiller, W.; Yoccoz, N.G.; Douzet, R.; Aubert, S.; Lavorel, S. A multi-trait approach reveals the structure and the relative importance of intra- vs. interspecific variability in plant traits. *Funct. Ecol.* **2010**, *24*, 1192–1201. [CrossRef]
9. Grime, J.P. *Plant Strategies, Vegetation Processes, and Ecosystem Properties*; Wiley: Hoboken, NJ, USA, 2001.
10. Westoby, M.; Falster, D.S.; Moles, A.T.; Vesk, P.A.; Wright, I.J. Plant Ecological Strategies: Some Leading Dimensions of Variation between Species. *Annu. Rev. Entomol.* **2002**, *33*, 125–159. [CrossRef]
11. Dayrell, R.L.C.; Arruda, A.J.; Pierce, S.; Negreiros, D.; Meyer, P.B.; Lambers, H.; Silveira, F.A.O.; Godoy, O. Ontogenetic shifts in plant ecological strategies. *Funct. Ecol.* **2018**, *32*, 2730–2741. [CrossRef]
12. Funk, J.L.; Larson, J.E.; Ames, G.M.; Butterfield, B.J.; Cavender-Bares, J.; Firn, J.; Laughlin, D.C.; Sutton-Grier, A.E.; Williams, L.; Wright, J. Revisiting the Holy Grail: Using plant functional traits to understand ecological processes. *Biol. Rev. Camb Philos. Soc.* **2017**, *92*, 1156–1173. [CrossRef] [PubMed]
13. McGill, B.J.; Enquist, B.J.; Weiher, E.; Westoby, M. Rebuilding community ecology from functional traits. *Trends Ecol. Evol.* **2006**, *21*, 178–185. [CrossRef] [PubMed]
14. Díaz, S.; Kattge, J.; Cornelissen, J.H.C.; Wright, I.J.; Lavorel, S.; Dray, S.; Reu, B.; Kleyer, M.; Wirth, C.; Colin Prentice, I.; et al. The global spectrum of plant form and function. *Nature* **2016**, *529*, 167–171. [CrossRef] [PubMed]
15. Grime, J.P. Vegetation classification by reference to strategies. *Nature* **1974**, *250*, 26–31. [CrossRef]
16. Grime, J.P. Evidence for the Existence of Three Primary Strategies in Plants and Its Relevance to Ecological and Evolutionary Theory. *Am. Nat.* **1977**, *111*, 1169–1194. [CrossRef]
17. Caccianiga, M.; Luzzaro, A.; Pierce, S.; Ceriani, R.M.; Cerabolini, B. The functional basis of a primary succession resolved by CSR classification. *Oikos* **2006**, *112*, 10–20. [CrossRef]
18. Pierce, S.; Brusa, G.; Vagge, I.; Cerabolini, B.E.L. Allocating CSR plant functional types: The use of leaf economics and size traits to classify woody and herbaceous vascular plants. *Funct. Ecol.* **2013**, *27*, 1002–1010. [CrossRef]
19. Hodgson, J.G.; Wilson, P.J.; Hunt, R.; Grime, J.P.; Thompson, K. Allocating C-S-R Plant Functional Types: A Soft Approach to a Hard Problem. *Oikos* **1999**, *85*, 282–294. [CrossRef]
20. Kunstler, G.; Falster, D.; Coomes, D.A.; Hui, F.; Kooyman, R.M.; Laughlin, D.C.; Poorter, L.; Vanderwel, M.; Vieilledent, G.; Wright, S.J.; et al. Plant functional traits have globally consistent effects on competition. *Nature* **2016**, *529*, 204–207. [CrossRef]
21. Rosado, B.H.P.; Mattos, E.A.; Baltzer, J. On the relative importance of CSR ecological strategies and integrative traits to explain species dominance at local scales. *Funct. Ecol.* **2017**, *31*, 1969–1974. [CrossRef]
22. Negreiros, D.; Le Stradic, S.; Fernandes, G.W.; Rennó, H.C. CSR analysis of plant functional types in highly diverse tropical grasslands of harsh environments. *Plant Ecol.* **2014**, *215*, 379–388. [CrossRef]
23. Cerabolini, B.E.L.; Brusa, G.; Ceriani, R.M.; De Andreis, R.; Luzzaro, A.; Pierce, S. Can CSR classification be generally applied outside Britain? *Plant Ecol.* **2010**, *210*, 253–261. [CrossRef]
24. Kattenborn, T.; Fassnacht, F.E.; Pierce, S.; Lopatin, J.; Grime, J.P.; Schmidtlein, S.; Paruelo, J. Linking plant strategies and plant traits derived by radiative transfer modelling. *J. Veg. Sci.* **2017**, *28*, 717–727. [CrossRef]
25. de Paula, L.F.A.; Negreiros, D.; Azevedo, L.O.; Fernandes, R.L.; Stehmann, J.R.; Silveira, F.A.O. Functional ecology as a missing link for conservation of a resource-limited flora in the Atlantic forest. *Biodivers. Conserv.* **2015**, *24*, 2239–2253. [CrossRef]
26. Hu, G.; Jin, Y.; Liu, J.; Yu, M. Functional diversity versus species diversity: Relationships with habitat heterogeneity at multiple scales in a subtropical evergreen broad-leaved forest. *Ecol. Res.* **2014**, *29*, 897–903. [CrossRef]
27. Huston, M.A.; Smith, T. Plant Succession: Life History and Competition. *Am. Nat.* **1987**, *130*, 168–198. [CrossRef]
28. Markesteijn, L.; Poorter, L. Seedling root morphology and biomass allocation of 62 tropical tree species in relation to drought- and shade-tolerance. *J. Ecol.* **2009**, *97*, 311–325. [CrossRef]
29. Barba-Escoto, L.; Ponce-Mendoza, A.; García-Romero, A.; Calvillo-Medina, R.P. Plant community strategies responses to recent eruptions of Popocatépetl volcano, Mexico. *J. Veg. Sci.* **2019**, *30*, 375–385. [CrossRef]
30. Lohbeck, M.; Poorter, L.; Lebrija-Trejos, E.; Martínez-Ramos, M.; Meave, J.A.; Paz, H.; Pérez-García, E.A.; Romero-Pérez, I.; Tauro, A.; Bongers, F. Successional changes in functional composition contrast for dry and wet tropical forest. *Ecology* **2013**, *94*, 1211–1216. [CrossRef]
31. Malizia, A.; Easdale, T.A.; Grau, H.R. Rapid structural and compositional change in an old-growth subtropical forest: Using plant traits to identify probable drivers. *PLoS ONE* **2013**, *8*, e73546. [CrossRef]
32. Heim, A.; Lundholm, J. Changes in plant community composition and functional plant traits over a four-year period on an extensive green roof. *J. Environ. Manag.* **2022**, *304*, 114154. [CrossRef] [PubMed]
33. Kraft, N.J.B.; Valencia, R.; Ackerly, D.D. Functional traits and niche-based tree community assembly in an amazonian forest. *Science* **2008**, *322*, 580–582. [CrossRef] [PubMed]
34. Hodáňová, D. Plant strategies and vegetation processes. *Biol. Plant* **1981**, *23*, 254. [CrossRef]
35. Mouillot, D.; Villeger, S.; Scherer-Lorenzen, M.; Mason, N.W. Functional structure of biological communities predicts ecosystem multifunctionality. *PLoS ONE* **2011**, *6*, e17476. [CrossRef]
36. Ding, Y.; Zang, R.G. Effects of Logging on the Diversity of Lianas in a Lowland Tropical Rain Forest in Hainan Island, South China. *Biotropica* **2009**, *41*, 618–624. [CrossRef]

37. Lu, X.; Zang, R.; Ding, Y.; Huang, J. Changes in biotic and abiotic drivers of seedling species composition during forest recovery following shifting cultivation on Hainan Island, China. *Biotropica* **2016**, *48*, 758–769. [CrossRef]
38. Condit, R. The CTFS and the Standardization of Methodology. In *Tropical Forest Census Plots: Methods and Results from Barro Colorado Island, Panama and a Comparison with Other Plots*; Springer: Berlin/Heidelberg, Germany, 1998; pp. 3–7.
39. Pérez-Harguindeguy, N.; Díaz, S.; Garnier, E.; Lavorel, S.; Poorter, H.; Jaureguiberry, P.; Bret-Harte, M.S.; Cornwell, W.K.; Craine, J.M.; Gurvich, D.E.; et al. New handbook for standardised measurement of plant functional traits worldwide. *Aust. J. Bot.* **2013**, *61*, 167–234. [CrossRef]
40. Magneville, C.; Loiseau, N.; Albouy, C.; Casajus, N.; Claverie, T.; Escalas, A.; Leprieur, F.; Maire, E.; Mouillot, D.; Villéger, S. mFD: An R package to compute and illustrate the multiple facets of functional diversity. *Ecography* **2022**, *2022*. [CrossRef]
41. Han, X.; Huang, J.; Zang, R. Soil nutrients and climate seasonality drive differentiation of ecological strategies of species in forests across four climatic zones. *Plant Soil* **2022**, *473*, 517–531. [CrossRef]
42. Garnier, E.; Cortez, J.; Billès, G.; Navas, M.-L.; Roumet, C.; Debussche, M.; Laurent, G.; Blanchard, A.; Aubry, D.; Bellmann, A.; et al. Plant functional markers capture ecosystem properties during secondary succession. *Ecology* **2004**, *85*, 2630–2637. [CrossRef]
43. Han, X.; Huang, J.; Yao, J.; Xu, Y.; Ding, Y.; Zang, R. Effects of logging on the ecological strategy spectrum of a tropical montane rain forest. *Ecol. Indic.* **2021**, *128*, 107812. [CrossRef]
44. R Core Team. *R: A Language and Environment for Statistical Computing*; R Foundation for Statistical Computing: Vienna, Austria, 2015; Volume 1, pp. 12–21.
45. Rosenfield, M.F.; Müller, S.C.; Overbeck, G.E. Short gradient, but distinct plant strategies: The CSR scheme applied to subtropical forests. *J. Veg. Sci.* **2019**, *30*, 984–993. [CrossRef]
46. Liu, X.; Swenson, N.G.; Zhang, J.; Ma, K.; Thompson, K. The environment and space, not phylogeny, determine trait dispersion in a subtropical forest. *Funct. Ecol.* **2013**, *27*, 264–272. [CrossRef]
47. Cerabolini, B.E.L.; Pierce, S.; Verginella, A.; Brusa, G.; Ceriani, R.M.; Armiraglio, S. Why are many anthropogenic agroecosystems particularly species-rich? *Plant Biosyst.* **2014**, *150*, 550–557. [CrossRef]
48. Gutiérrez-Girón, A.; Gavilán, R. Plant functional strategies and environmental constraints in Mediterranean high mountain grasslands in central Spain. *Plant Ecol. Divers.* **2013**, *6*, 435–446. [CrossRef]
49. Grime, J.P. Trait convergence and trait divergence in herbaceous plant communities: Mechanisms and consequences. *J. Veg. Sci.* **2006**, *17*, 255–260. [CrossRef]
50. Read, Q.D.; Moorhead, L.C.; Swenson, N.G.; Bailey, J.K.; Sanders, N.J.; Fox, C. Convergent effects of elevation on functional leaf traits within and among species. *Funct. Ecol.* **2014**, *28*, 37–45. [CrossRef]
51. Yu, R.; Huang, J.; Xu, Y.; Ding, Y.; Zang, R. Plant Functional Niches in Forests Across Four Climatic Zones: Exploring the Periodic Table of Niches Based on Plant Functional Traits. *Front. Plant Sci.* **2020**, *11*, 841. [CrossRef]
52. Han, X.; Huang, J.; Zang, R. Shifts in ecological strategy spectra of typical forest vegetation types across four climatic zones. *Sci. Rep.* **2021**, *11*, 14127. [CrossRef]
53. Zhang, J.; Zhao, N.; Liu, C.; Yang, H.; Li, M.; Yu, G.; Wilcox, K.; Yu, Q.; He, N. C:N:P stoichiometry in China's forests: From organs to ecosystems. *Funct. Ecol.* **2018**, *32*, 50–60. [CrossRef]
54. Lebrija-Trejos, E.; Perez-Garcia, E.A.; Meave, J.A.; Poorter, L.; Bongers, F. Environmental changes during secondary succession in a tropical dry forest in Mexico. *J. Trop. Ecol.* **2011**, *27*, 477–489. [CrossRef]
55. Lasky, J.R.; Sun, I.F.; Su, S.H.; Chen, Z.S.; Keitt, T.H.; Canham, C. Trait-mediated effects of environmental filtering on tree community dynamics. *J. Ecol.* **2013**, *101*, 722–733. [CrossRef]
56. Dudova, K.V.; Dzhatdoeva, T.M.; Dudov, S.V.; Akhmetzhanova, A.A.; Tekeev, D.K.; Onipchenko, V.G. Competitive Strategy of Subalpine Tall-Grass Species of the Northwestern Caucasus. *Mosc. Univ. Biol. Sci. Bull.* **2019**, *74*, 140–146. [CrossRef]
57. Hou, D.; He, W.; Liu, C.; Qiao, X.; Guo, K. Litter accumulation alters the abiotic environment and drives community successional changes in two fenced grasslands in Inner Mongolia. *Econ. Evol.* **2019**, *9*, 9214–9224. [CrossRef] [PubMed]

Article

Relationships between Bird Assemblages and Habitat Variables in a Boreal Forest of the Khentii Mountain, Northern Mongolia

Zoljargal Purevdorj [1,2], Munkhbaatar Munkhbayar [2], Woon Kee Paek [3], Onolragchaa Ganbold [2], Ariunbold Jargalsaikhan [2], Erdenetushig Purevee [2], Tuvshinlkhagva Amartuvshin [2], Uranchimeg Genenjamba [2], Batbayar Nyam [2] and Joon Woo Lee [1,*]

[1] Department of Forest & Environmental Resources, Chungnam National University, Daejeon 34134, Korea; zoljargal@msue.edu.mn

[2] Department of Biology, Mongolian National University of Education, Ulaanbaatar 210685, Mongolia; mmunkhbaatar@msue.edu.mn (M.M.); f.naumanni13@gmail.com (O.G.); ariunbold@msue.edu.mn (A.J.); e_tushig@msue.edu.mn (E.P.); tuvshinlkhagva@msue.edu.mn (T.A.); forestbirdurnaa@gmail.com (U.G.); batbayrnka@gmail.com (B.N.)

[3] Daegu National Science Museum of Korea, Daegu 43014, Korea; paekwk@naver.com

* Correspondence: jwlee@cnu.ac.kr; Tel.: +82-010-5405-5749

Abstract: In order to determine the relationships between bird assemblages and forest habitat, we conducted surveys for bird assemblages in different forest habitats in the Khentii Mountain region, Northern Mongolia. A total of 1730 individuals belonging to 71 species from 23 families of 11 orders were recorded. Our findings revealed that passeriformes are the most species-rich order, accounting for 86.2% of the total species. The dominant species were *Anthus hodgsoni*, *Parus major*, *Poecile palustris*, and *Sitta europaea* in study area. Non-metric multidimensional scaling (NMDS) and permutation multivariate analysis of variance (PERMANOVA) showed that bird assemblages were affected by forest habitat types. Our findings also showed significant relationships between bird assemblages and canopy height and ground cover vegetation structure, whereas there were no relationships between altitude and other habitat variables. Thus, maintaining diverse forest habitats or restoring forest would play a key role in bird conservation and sustainable management of forest areas.

Keywords: bird assemblages; environmental variables; habitat; boreal forest

Citation: Purevdorj, Z.; Munkhbayar, M.; Paek, W.K.; Ganbold, O.; Jargalsaikhan, A.; Purevee, E.; Amartuvshin, T.; Genenjamba, U.; Nyam, B.; Lee, J.W. Relationships between Bird Assemblages and Habitat Variables in a Boreal Forest of the Khentii Mountain, Northern Mongolia. *Forests* **2022**, *13*, 1037. https://doi.org/10.3390/f13071037

Academic Editors: Konstantinos Poirazidis and Panteleimon Xofis

Received: 3 June 2022
Accepted: 24 June 2022
Published: 1 July 2022

Publisher's Note: MDPI stays neutral with regard to jurisdictional claims in published maps and institutional affiliations.

1. Introduction

The boreal biome is located in the Northern Hemispheres, between 50° and 60° N latitudes [1]. The boreal forest covers northern Europe and Asia, and stretches from Far East Siberia to Scandinavia in the west [2]. Temperate conifers with varying proportions of deciduous trees dominate the boreal forest landscapes [2]. Mongolia is one of the world's least forested regions [3]. Forest covers approximately 12% of Mongolia; 84% of the forest is mostly coniferous and deciduous, but also 16% saxaul forest [4–6]. In Mongolia, boreal forests mostly exist in the northern parts, which form a transitional zone between the Siberian Boreal Taiga and the Central Asian steppe [6]. These forests have relatively few tree species and are composed mainly of Siberian larch (*Larix sibirica* Ledeb.), Scots pine (*Pinus sylvestris* L.), Siberian pine (*Pinus sibirica* Du Tour.), Siberian fir (*Abies sibirica* Ledeb.), and Birch (*Betula* spp.), along with some deciduous tree species [5]. Mongolian boreal forest is characterized by a low human population density and a relatively low level of anthropogenic impact compared to other countries with Boreal forest regions [5]. Nevertheless, human population increases have led to an ever-increasing demand for forest products, timber harvesting, forest fires, increases in livestock numbers, degradation, and pests progressively depleting the forest cover [7,8]. Forest depletion totaled four million hectares in the last three decades, and the rate of deforestation increased to approximately 60,000 ha per year [5]. Deforestation by legal and illegal logging for especially conifer trees—sawmilling targets the largest trees—increased from the beginning of the 1970s [9,10]. Studies in the Boreal biome and tropical deforestation show that global bird diversity is tending to go down [11–15], and reforestation is a key action for bird conservation [16]. The relationships between bird assemblages and forest habitat have been the focus of many studies worldwide, and it is evident that forest habitat is an important determinant of the condition of bird assemblages in the boreal forest biome [17]. Bird species diversity has been shown to increases as forest tree diversity increases [18,19]. In addition, environmental variables such as habitat heterogeneity, canopy cover, tree size, and seasonal and climate changes are important effects on the growth and reproduction of the forest bird assemblages [20].

Most of the research is bird composition on boreal forest has concentrated on countries with boreal biomes, such as those in Europe, North America, and several Russian regions [2]. In Mongolia, the study of forest habitat and its biological communities (especially birds) has just started, and a limited number of studies on forest birds and their habitats are available. Most studies on bird diversity and biological communities have focused on waterfowl and threatened bird species, but there are only a few studies on the relationship between forest habitat and bird assemblages [10,21]. Unfortunately, ecological studies of the forest bird in the area are rare or have been mostly published in the Russian or Mongolian journals in their native languages and thus are hardly accessible to other scientists. Moreover, as noted by Bold's (2003) [22], the early ecological finds from the forest bird assemblages of different habitats were reported based on bird species distribution as a result of a Russian–Mongolian joint expedition. His research focuses on bird species distribution and avifauna (Bold, 1973) [23]. In 1969, Bold [24] described some additional ecological and behavioral characteristics of some forest birds in different habitats of Mongolia. [25,26]. However, there are research gaps in Mongolia, especially regarding biological and ecological characteristics, such as forest bird species density, population dynamics, and the heterogeneity of micro habitat variables [10,21]. This his study aimed to quantify the forest structure, bird species composition, community assemblages, and occurrences of birds in different habitats of boreal forests, and to investigate the relationships between the avian community and forest conditions.

2. Materials and Methods

2.1. Study Sites

The research mainly focused on the surrounding regions of the Khentii Mountains (Figure 1). The Khentii Mountains are known as a forest region that is in northern Mongolia [27]. The annual average precipitation is 250–320 mm [3,28], and approximately 50–60% of precipitation is recorded in the summer [29]. The average annual temperature varies between −1.9 and −3.8 °C [30]. The forest of Khentii Mountains accounts for 33.5% (3755.2 ha, thousand) of the total forest area of Mongolia [24]. In the Khentii region, the western Siberian dark taiga forests with Spruce-*Picea obovata*, Fir-*Abies sibirica*, *Pinus sibirica*, and Siberian larch-*Larix sibirica* meet the eastern Siberian light taiga forests composed of species such as Birch-*Betula platyphylla* and related species, Larch-*Larix* sp. and Scotch pine-*Pinus sylvestris* [3,31–33]. Tree species of Khentii Mountain, especially mixed conifer and deciduous trees, and the plants, are relatively different in ecological regions from the other parts of Mongolia [34].

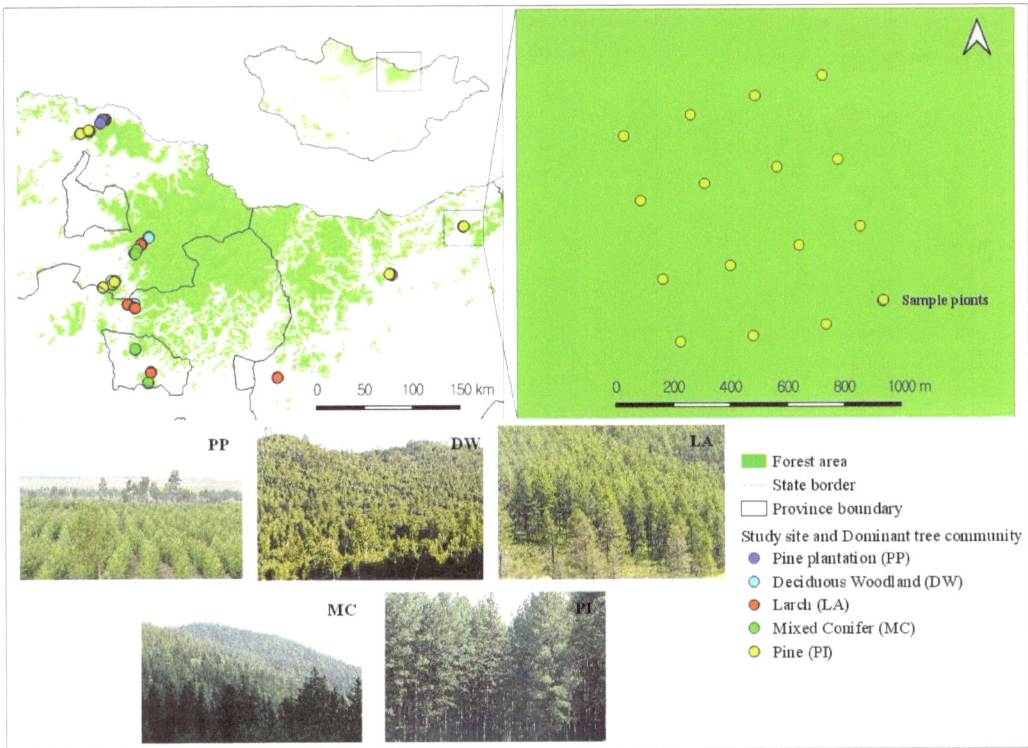

Figure 1. Study area location and sampling sites in a forested area of the Khentii Mountains.

2.2. Sample Collection and Analyses

The survey was carried out at twenty plot sites on Kentii Mountain for native forested areas and five plots within a pine plantation area (Tujiin Nars). Five plots at plantation areas were selected based on the restorative time period since reforestation [35]. Bird surveys were conducted from June to August in 2019, a period recognized as breeding season for most species in the region [24]. All birds were counted and recorded via visual observation [36,37]. Each species was assigned to several functional groups based on migrations and diet preferences (Table 1) [21]. Field surveys were carried out from 6:00 to 11:00 AM during suitable weather conditions (without rain and with wind less than 5 m/s) and birds were recorded in point-counts of 10 min with a 50 or 80 m radius [36,37]. Survey plots were 1 km^2 area with a 4 × 4 grid of 16 points, and a total of 400 point-counts were completed (Figure 1) [38,39]. The distance between points within plots was 250 m [37,38], and when we selected points at the forest's edge, the distance between a point and the interior of the forest was 50 m. The occurrences of bird species and study sites were recorded using a global positioning system (GPS; German map62), and these positions were used to map in QGIS version 3.22 [40].

Table 1. List of bird species and their abundances and occurrences in different habitat types in the forest of the Khentii Mountains.

Scientific Name	English Name	Diet	Status	DW	LA	PI	PP	MC	Frequency of Occurrence (%)	Relative Abundance (%)
Accipitriformes-Accipitridae										
Accipiter gentilis	Northern Goshawk	Carn	M		1				0.25	0.06
Aquila nepalensis	Steppe Eagle	Carn	M			1			0.25	0.06
Buteo buteo	Eurasian Buzzard	Carn	M	3	1		2		1.25	0.35
Hieraaetus pennatus	Booted Eagle	Carn	M	1	1				0.5	0.12
Milvus migrans	Black Kite	Carn	M	12	20	7	11	3	9	3.06
Aegypius monachus	Cinereous Vulture	Carn	R	1					0.25	0.06
Anseriformes-Anatidae										
Tadorna ferruginea	Ruddy Shelduck	Omni	M				9		1.5	0.52
Bucerotiformes-Upupidae										
Upupo epops	Eurasian Hoopoe	Inse	M		1	3	1		1	0.29
Caprimulgiformes-Apodidae										
Apus apus	Common Swift	Inse	M			5	21		1.25	1.50
Ciconiiformes-Ciconiidae										
Ciconia nigra	Black Stork	Carn	M			1		1	0.5	0.12
Columbiformes-Columbidae										
Streptopelia orientalis	Oriental Turtle Dove	Gran	M				15		2.5	0.87
Cuculiformes-Cuculidae										
Cuculus canorus	Common Cuckoo	Inse	M		5		26		6.75	1.79
Cuculus saturatus	Oriental Cuckoo	Inse	M		1				0.25	0.06
Falconiformes-Falconidae										
Falco cherrug	Saker Falcon	Carn	M			1	1		0.5	0.12
Falco tinnunculus	Common Kestrel	Carn	M	1	4				0.75	0.29
Falco amurensis	Amur Falcon	Carn	M		4				0.75	0.23
Falco columbarius	Merlin	Carn	M		3				0.5	0.17
Galliformes-Phasianidae										
Lyrurus tetrix	Black Grouse	Inse	R	12					1.5	0.69
Gruiformes-Gruidae										
Anthropoides virgo	Demoiselle Crane	Omni	M			2	5		1.25	0.40
Antigone vipio	White-naped Crane	Omni	M				3		0.25	0.17
Passeriformes-Aegithalidae										
Aegithalos caudatus	Long tailed Tit	Inse	R	14	1			2	1.75	0.98
Alaudidae										

Table 1. *Cont.*

Scientific Name	English Name	Diet	Status	Bird Individuals in Different Habitat Types					Frequency of Occurrence (%)	Relative Abundance (%)
				DW	LA	PI	PP	MC		
Alauda arvensis	Eurasian Skylark	Inse	M				28		1.75	1.62
Corvidae										
Corvus dauuricus	Daurian Jackdaw	Omni	M		6				1	0.35
Corvus corax	Northern Raven	Omni	R	4	5	5	14	2	4	1.73
Corvus corone	Carrion Crow	Omni	R	11	12	9	18	9	10	3.41
Cyanopica cyanus	Azure-winged Magpie	Omni	R				14		0.75	0.81
Garrulus glandarius	Eurasian Jay	Omni	R	9	3	5		10	3	1.56
Pica pica	Eurasian Magpie	Omni	R	3	9	15			2.5	1.56
Pyrrhocorax pyrrhocorax	Red-billed Chough	Omni	R			1			0.25	0.06
Emberizidae										
Emberiza cioides	Meadow Bunting	Inse	M		18	26	25		8.75	3.99
Emberiza leucocephalos	Pine Bunting	Inse	M		14	24	67		10.25	6.07
Emberiza pusilla	Little Bunting	Inse	M	7					0.75	0.40
Emberiza pallasi	Pallas's Bunting	Inse	M		2				0.5	0.12
Fringillidae										
Fringilla montifringilla	Brambling	Omni	M			1			0.25	0.06
Carpodacus erythrinus	Common Rosefinch	Gran	M		10				1	0.58
Carpodacus roseus	Pallas's Rosefinch	Gran	R	1					0.25	0.06
Coccothrraustes coccothrraustes	Hawfinch	Gran	R			6			1.25	0.35
Laniidae										
Lanius cristatus	Brown Shrike	Inse	M		12	2			2.5	0.81
Motacillidae										
Anthus hodgsoni	Olive Backed Pipit	Inse	M	15	62	41	94	8	26.25	12.72
Anthus richardi	Richard's Pipit	Inse	M				1		0.25	0.06
Anthus trivialis	Tree Pipit	Inse	M	10	1	27	28		8.5	3.82
Motacilla alba	White Wagtail	Inse	M	1	3				1	0.23
Muscicapidae										
Ficedula albicilla	Taiga Flycatcher	Inse	M	8	16	4			3.5	1.62
Muscicapa sibirica	Dark-sided Flycatcher	Inse	M			13			2	0.75
Oenanthe oenanthe	Northern Wheatear	Inse	M		7				1	0.40
Phoenicurus phoenicurus	Common Redstart	Inse	M	15	10	3		2	5	1.73
Saxicola torquatus	Common Stonechat	Inse	M	2	10	4	2		2	1.04
Muscicapa dauurica	Asian Brown Flycatcher	Inse	M		1	8			1	0.52

Table 1. *Cont.*

Scientific Name	English Name	Diet	Bird Individuals in Different Habitat Types						Frequency of Occurrence (%)	Relative Abundance (%)
			Status	DW	LA	PI	PP	MC		
Oenanthe pleschanka		Inse	M				2		0.25	0.12
Oenanthe isabellina	Isabelline Wheatear	Inse	M	2	4		23		2.5	1.68
Phoenicurus auroreus	Daurian Redstart	Inse	M		9	1	6		2.5	0.92
Phoenicurus erythrogastrus	White winged Radstart	Inse	R		1				0.25	0.06
Paridae										
Cyanistes cyanus	Azure Tit	Inse	R		14				1.25	0.81
Parus major	Great Tit	Inse	R	50	27	32	4	31	16.75	8.32
Periparus ater	Coal Tit	Inse	R	26	9	1		7	6.75	2.49
Poecile montanus	Willow Tit	Gran	R	9	32	42	1	14	9.75	5.66
Poecile palustris	Marsh Tit	Gran	R	72	31	20		4	15.25	7.34
Passeridae										
Passer domesticus	House Sparrow	Inse	R		5				0.5	0.29
Passer montanus	Eurasian Tree Sparrow	Omni	R			9	8		1	0.98
Phylloscopidae										
Phylloscopus borealis	Arctic Warbler	Inse	M		1				0.25	0.06
Phylloscopus fuscatus	Dusky Warbler	Inse	M		5				0.75	0.29
Phylloscopus proregulus	Palla's leaf Warbler	Inse	M	26	5	5		1	4.5	2.14
Sittidae										
Sitta europaea	Wood Nuthach	Inse	R	11	31	66		11	13.25	6.88
Turdidae										
Turdus naumanni	Naumanns Thrush	Omni	M		5				0.5	0.29
Turdus ruficollis	Red Throated Trush	Omni	M	5					0.75	0.29
Piciformes-Picidae										
Dendrocopos leucotos	White backed Woodpecker	Inse	R	5	2				1.5	0.40
Dendrocopos major	Great spotted Woodpecker	Inse	R	5	3	2	11	3	4.75	1.39
Dryobates minor	Lesser spotted Woodpecker	Inse	R		7	7	2		3.25	0.92
Dryocopus martius	Black Woodpecker	Inse	R	1		1			0.5	0.12
Strigiformes-Strigidae										
Aegolius funereus	Boreal Owl	Carn	R	2					0.25	0.12
Bubo bubo	Eurasian Eagle Owl	Carn	R			1		1	0.5	0.12
	Abundance			344	434	401	442	109	-	-
	Species richness			31	46	36	28	16	-	-

Notes: Diet: Carn—carnivores, Omni—omnivores, Inse—insectivores, Gran—granivores. Migratory status: R—resident, M—migrant. Forest types are deciduous woodland (DW), larch (LA), pine (PI), pine plantation (PP), and mixed conifer (MC).

We estimated forest tree (height, diameter at breast height—dbh) cover, and ground-cover variables (percentage of vegetation cover, bare ground, and down wood) of each point in a 50 m radius [39], and snag samples were 20 m in radius. Tree height was measured at ≥ 10 dominant trees at each point (around the center points). These data were grouped as mean values at the point-count level. The forest vegetation was sampled immediately after the end of bird counts. Forest characteristics were described as forest pattern, average tree height, cover, and tree component in the study area, the habitat was classified into five groups: the forest patterns and structure variables of each habitat type are shown in Table 2.

The relative abundance (RA) of a bird species was determined using the following expression: (number of individuals for species n/N total number of individuals) $\times$ 100%. To find differences in environmental factors and bird community attributes (i.e., species richness, abundance, and forest structure variables) among different habitat types, we performed analysis of variance (ANOVA) and Tukey's post hoc test using the function aov in R software [41].

Differences in bird species richness among forest habitats were assessed through the rarefaction and extrapolation method based on sample coverage [42]. The species richness was calculated for each forest habitat based on the lowest sample coverage among the five habitats obtained within the 95% confidence intervals. This analysis was performed with the iNEXT and devtools package, using R [43,44].

To visualize the differences in bird assemblage composition between habitat types, we used permutational multivariate analysis of variance (PERMANOVA). Before performing PERMANOVA, the multivariate homogeneity of group dispersions was tested by the function betadisper, which indicated that there was a difference in dispersion between groups (F = 4.95, $p < 0.01$). The habitat type was used as an explanatory factor for PERMANOVA, which was tested using the function Adonis in the R package vegan [45]. The five forest types were considered a fixed effect in the analysis. Differences in species composition among samples collected in each site and forest type are presented in non-metric multidimensional scaling (NMDS) using the function metaMDS in the R package vegan 2.6-2 [46]. Moreover, similarity percentage (SIMPER) was also used to determine species that contributed most to the dissimilarities observed. All recorded species were included in the analysis. For SIMPER, we reported species that contributed to 83% of the bird community assemblages [45].

Redundancy analysis (RDA) was used as a direct gradient approach in order to determine how much variation in bird assemblages could be explained by environmental variables. Then, bird abundance data with total species were Hellinger transformed [47] using the function decosdtand in the R package vegan, in order to reduce the weights of abundant species while preserving the BrayCurtis index between samples in multidimensional space [45]. We performed RDA using the function RDA and tested the significance using the function ANOVA. In order to reduce the number of environmental variables entering the RDA, we used forward selection to get a parsimonious model. The forward selection was performed using the function ordiR2step with a permutation test (999 permutations) via the R package vegan [45]. The level of significance for all results was set to $p < 0.05$. All statistical analyses were performed using R version 3.5.1 [41].

Table 2. Structural variables were used to characterize the forest habitats of sampling sites where bird assemblages were described.

Structural Variables	Code	Description
Altitude (m)	Alt	Point elevation (m.a.s.l)
Number of snags	Sna	Counted the numbers of snags ($\geq$15 cm dbh, trees that are completely dead) and stems those are $\geq$3 m high and within a 20 m radius of
Number of stem	Ste	the center survey point.
Dbh (cm)	Dbh	Measured the dominant tree's average diameter at breast height (dbh) of overstore trees within a number of $\geq$10 stem.
Over story cover (%)	Osc	Estimated the total percent coverage and dominant tree's average height of all overstore trees within a 50 m radius.
Over story height (m)	Osh	
Understory cover (%)	Usc	Estimated the percent cover and species makeup of any woody vegetation (including seedling trees) that is $\geq$0.5 m high and <3.0 m high
Understory height (m)	Ush	of the understory layer.
Bare/Litter cover (%)	Bal	
Dead and down (%)	Dad	
Dead standing grass cover (%)	Dsg	
Herbaceous cover (%)	Her	The percentage of the ground surface covered by shrubs 0–0.5 m high, litter, down wood, forbs, grasses, and moss was estimated visually
Live grass cover (%)	Lig	within 50 m radius plots (total 100%).
Moss cover (%)	Mos	
Dead standing grass height (cm)	Dsh	
Live grass&herb. height (cm)	Lgh	
Deciduous Woodland	DW	
Larch	LA	
Pine	PI	Forested habitat type of dominant tree species with $\geq$50% present in the overstore. The overstore cover should be $\geq$10% trees within
Plantation Pine	PP	a 50 m radius.
Mixed Conifer	MC	

3. Results

3.1. Bird Assemblage Composition

A total of 1730 birds were recorded during this study, belonging to 71 species from 23 families and 11 orders (Table 2). Of these, Passeriformes was the most species-rich order, accounting for 89% of the total species. The dominant species were *Anthus hodgsoni*—olive backed pipit (12.72%), *Parus major*—great tit (8.32%), *Poecile palustris*—marsh tit (7.34%), *Sitta europaea*—wood nuthatch (6.88%), *Emberiza leucocephalos*—pine bunting (6.07%), and *Poecile montanus*—willow tit (5.66%) which comprised 46.99% of the total bird count in this study. The dominant species varied among habitat types. For instance, *A. hodgsoni*, *P. major*, and *P. montanus* were generalists, which were dominant species in the most habitats. Moreover, *E. leucocephalos* was dominant in larix, pine, and pine plantation habitats, whereas *P. palusris* and *S. europaea* were dominant species in most habitats, but these species were not recorded in habitats with pine plantations. The high values of estimated sample coverage (ranges from 0.97 in LA and MC to 0.99 in PP) indicate that the sampling was sufficient to detect most species (Figure 2 and Table 3). In the forest habitats, species richness was higher in the larch (46 species) and pine-dominated habitats (31 species), compared to the pine plantation (28 species) and mixed conifer (16 species). We grouped the 71 species into four guilds: insectivores (38 species), omnivores (14), carnivores (13), and granivores (6) (Table 3). Insectivores were the most abundant group (68.09%), followed by granivores (14.85%), omnivores (12.2%), and carnivores (4.86%).

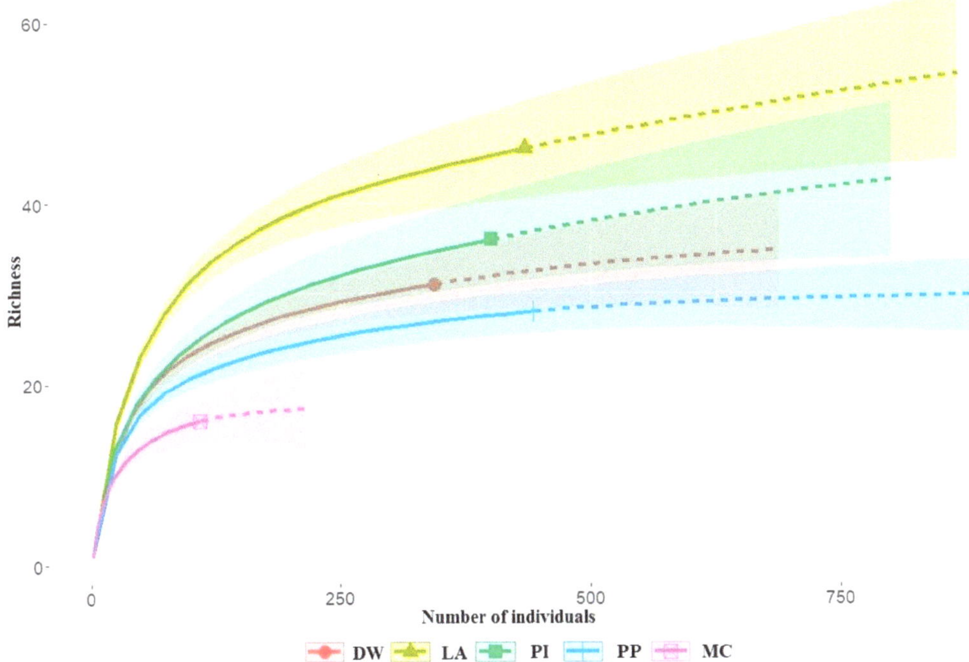

Figure 2. Species richness of birds was recorded during this survey in the Khentii Mountains. Habitat variables are deciduous woodland (DW), larch (LA), pine (PI), pine plantation (PP), and mixed conifer (MC). Solid and dashed lines are interpolated and extrapolated data, respectively, based on rarefaction and extrapolation methods, with their associated 95% confidence intervals.

Table 3. Bird richness in areas with different forest habitats in Khentii Mountains.

Habitat Type	Observed Richness	Sample Coverage	S.LCL	CI
Table 1 Deciduous Woodland	31	0.983	27.4	32–64.5
Larch	46	0.977	41.0	51–168.9
Pine	36	0.978	31.7	38.8–98.7
Pine Plantation	28	0.991	25.2	28.2–42.3
Mixed Conifer	16	0.973	13.4	16.1–28.7

Note: S.LCL = richness based on the lowest sample coverage for that forest habitat type; CI = 95% confidence interval. Differences between letters next to the CI indicate significant differences between forest types.

3.2. Correlations between Bird Assemblage and Habitat Types

Results of one-way ANOVA showed that the bird abundance per point for pine plantations was significantly higher than that for mixed conifer forest habitats ($F (4, 395) = 3.14$, $p < 0.05$). Bird species richness per point in the pine plantation habitat was significantly higher than for other habitat types ($F (4, 395) = 3.97$, $p < 0.05$) (Figure 3A,B). Permutation analysis of variance and NMDS (stress = 0.22) revealed bird assemblages were significantly affected by habitat type (PERMANOVA, pseudo-F = 2.59, $p < 0.001$) (Figure 4). In addition, SIMPER analysis confirmed these dissimilarities and revealed that *A. hodgsoni*, *P. major*, *P. palustris*, *E. leucocephalos*, *S. europaea*, and *P. montanus* are major contributors to dissimilarities between habitat types (Table 4).

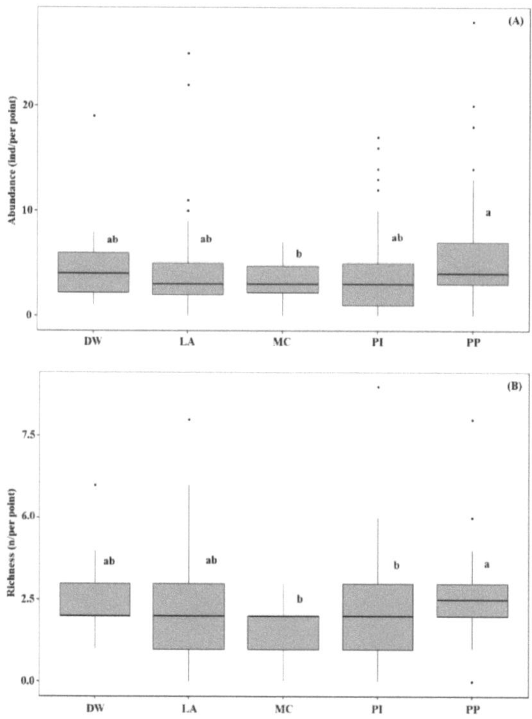

Figure 3. Analysis of variance (ANOVA) for abundance (**A**) and species richness (**B**) of birds in the number of birds per point among selected habitats. Forest types are deciduous woodland (DW), larch (LA), pine (PI), pine plantation (PP), and mixed conifer (MC).

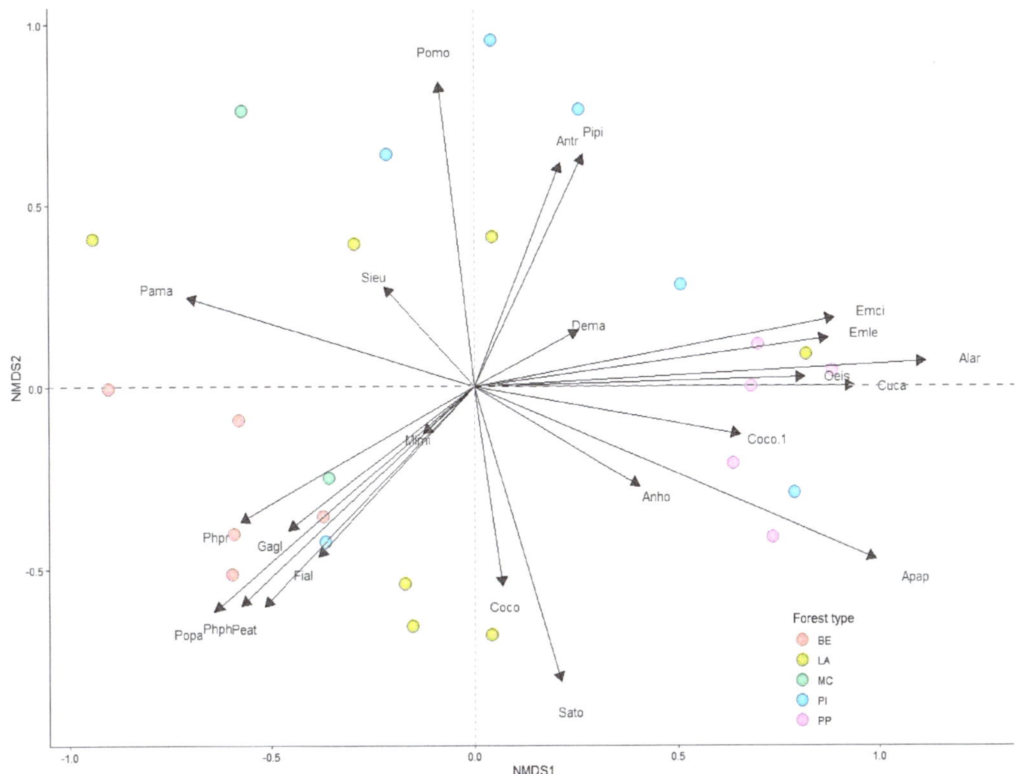

Figure 4. Non-metric multidimensional scaling (NMDS) ordination of selected bird assemblages in different habitat types in the forest of the Khentii Mountains. Forest types are deciduous woodland (DW), larch (LA), pine (PI), pine plantation (PP), and mixed conifer (MC). Note: Species names are generic acronyms or the four letters of the scientific names in Table 1.

Table 4. SIMPER analysis of dissimilarity among forest habitat types by most abundant species.

Species	Av. Dissim	Contrib. %	Cumulative %	Mean DW	Mean LA	Mean MC	Mean PI	Mean PP
Anthus hodgsoni	10.95	11.70	11.70	0.18	0.63	0.28	0.47	1.25
Parus major	8.28	8.85	20.55	0.61	0.28	1.07	0.36	0.05
Poecile palustris	7.32	7.82	28.37	0.88	0.32	0.14	0.23	-
Sitta europaea	5.65	6.04	34.41	0.13	0.32	0.38	0.75	-
Emberiza leucocephalos	5.23	5.59	40.00	-	0.14	-	0.27	0.89
Poecile montanus	5.08	5.43	45.42	0.11	0.33	0.48	0.48	0.01
Emberiza cioides	3.96	4.23	49.66	-	0.18	-	0.30	0.33
Corvus corone	3.88	4.15	53.80	0.13	0.12	0.31	0.10	0.24
Anthus trivialis	3.45	3.69	57.49	0.12	0.01	-	0.31	0.37
Milvus migrans	2.80	2.99	60.49	0.15	0.20	0.10	0.08	0.15
Periparus ater	2.57	2.75	63.24	0.32	0.09	0.24	0.01	-
Phylloscopus proregulus	1.98	2.12	65.35	0.32	0.05	0.03	0.06	-
Phoenicurus phoenicurus	1.97	2.10	67.46	0.18	0.10	0.07	0.03	-
Cuculus canorus	1.91	2.04	69.49	-	0.05	-	-	0.35
Ficedula albicilla	1.71	1.82	71.32	0.10	0.16	-	0.05	-
Dendrocopos major	1.66	1.77	73.09	0.06	0.03	0.10	0.02	0.15
Garrulus glandarius	1.60	1.71	74.80	0.11	0.03	0.35	0.06	-
Corvus corax	1.42	1.51	76.32	0.05	0.05	0.07	0.06	0.19

Table 4. *Cont.*

Species	Av. Dissim	Contrib. %	Cumulative %	Mean DW	Mean LA	Mean MC	Mean PI	Mean PP
Pica pica	1.16	1.24	77.56	0.04	0.09	-	0.17	-
Apus apus	1.12	1.19	78.75	-	-	-	0.06	0.28
Alauda arvensis	1.04	1.11	79.87	-	-	-	-	0.37
Oenanthe isabellina	1.04	1.11	80.97	0.02	0.04	-	-	0.31
Phoenicurus auroreus	1.03	1.11	82.08	-	0.09	-	0.01	0.08
Saxicola torquatus	1.01	1.08	83.15	0.02	0.10	-	0.05	0.03

3.3. Relationships between Bird Assemblages and Environmental Variables

One-way ANOVA revealed that mixed conifer had the highest average altitude (1688.6 ± 204.6 m), followed by larch-dominant habitats (1374.7 ± 267.9 m), and the pine plantation habitat altitude was lower still (701.7 ± 18.3 m). The forest habitat types were significantly different at different altitudes ($F_{4,\ 395}$ = 240.1, $p < 0.01$). The average numbers of snags of deciduous woodland and larch habitats were significantly higher than other habitat types ($F_{4,\ 395}$ = 10.95, $p < 0.05$). The dbh sizes in deciduous woodland, larch, and mixed conifer habitats were significantly larger than in other habitat types ($F_{4,\ 395}$ = 4.71, $p < 0.05$). The overstory, understory, and ground-level were significantly different in habitat structural variables (Table 5).

Table 5. Results of one-way analysis of variance (ANOVA) with Tukey's HSD test for environmental variables (mean ± SD) among different forest habitat types in the Khentii Mountains.

Habitat Types	Alt (m)	Sna (n)	Dbh (cm)	Ste (n)	Osh (m)	Osc (%)	Usc (%)	Ush (m)
			Overstory Level				**Understory Level**	
DW	1139.4 ± 119.9 [c]	4.6 ± 4.3 [a]	15.4 ± 6.9 [a]	46.8 ± 46.2 [b]	7.7 ± 3.4 [b]	25.8 ± 20.7 [c]	32.5 ± 15.5 [a]	2.3 ± 1 [a]
LA	1374.7 ± 267.9 [b]	3.8 ± 3.8 [a]	20.5 ± 8.7 [a]	66.5 ± 69.4 [ab]	10.8 ± 4 [a]	32.5 ± 25.5 [bc]	12.4 ± 10.5 [b]	1.4 ± 0.8 [bc]
MC	1688.6 ± 204.6 [a]	3.3 ± 3.1 [ab]	20.5 ± 8 [a]	86.2 ± 52.3 [a]	12.0 ± 4.8 [a]	53.1 ± 28.1 [a]	9.8 ± 7 [b]	1.8 ± 0.6 [ab]
PI	956.7 ± 205.29 [d]	2.4 ± 2.8 [b]	18.7 ± 8.7 [b]	63.5 ± 62.6 [ab]	10.5 ± 4.3 [a]	39.7 ± 29 [ab]	13.4 ± 12.7 [b]	1.8 ± 1.1 [c]
PP	701.7 ± 18.3 [e]	1.6 ± 1.6 [b]	13.0 ± 4.6 [b]	88.1 ± 85.9 [a]	7.2 ± 2.6 [b]	36.9 ± 31.1 [bc]	16.7 ± 16.9 [b]	2.2 ± 0.6 [a]

Habitat Types	DaD (%)	Her (%)	Bal (%)	Lig (%)	Mos (%)	Lgh (cm)
			Ground level			
DW	7.8 ± 6 [a]	31.7 ± 11.3 [a]	19.9 ± 13 [a]	32.1 ± 14.2 [bc]	1.3 ± 2.3 [c]	27.8 ± 8.9 [a]
LA	5.5 ± 6 [b]	31.3 ± 16.8 [a]	14.0 ± 10.6 [a]	38.220.5 [b]	3.8 ± 9.8 [bc]	21.0 ± 8.3 [b]
MC	8.6 ± 7.3 [a]	23.3 ± 17.7 [b]	12.9 ± 8.5 [b]	26.1 ± 16.2 [c]	26.6 ± 28.2 [a]	16.0 ± 7.5 [c]
PI	5.5 ± 6.5 [b]	16.8 ± 12.7 [b]	19.3 ± 13.5 [b]	39.0 ± 21.1 [b]	7.0 ± 11.3 [b]	16.3 ± 9.8 [c]
PP	0.5 ± 1.1 [c]	6.9 ± 6 [c]	5.4 ± 2.9 [c]	63.5 ± 5.8 [a]	0.0 [c]	28.2 ± 10 [a]

Notes: Different habitat types with different letters (a, b, c, d, e) in the same column indicate significant differences ($p < 0.05$). Forest types are deciduous woodland (DW), larch (LA), pine (PI), pine plantation (PP), and mixed conifer (MC). Habitat structural variable names were used by generic acronyms letters of the names in Table 2.

A total of 23 bird species were selected for redundancy analysis (RDA) with frequency of occurrence, and seven environmental variables were selected after a forward stepwise selection, including overstory, live grass, and dead grass height; and herbaceous, dead grass, live grass, and dead down wood cover. The first two axes (RDA1 and RDA2) accounted for 22.75% and 15.14% of the variation of 23 bird species, respectively (Figure 5). Different bird species preferred different environmental variables, which supports hypothesis ii. For instance, *Sitta europaea*, *Parus major*, and *Pica pica* were positively correlated with overstory height and dead wood, but negatively related to ground cover grass and height. *Ficedula albicilla* was positively related to herbaceous cover, but negatively related to dead standing grass cover. Moreover, *Emberiza cioides*, *Emberiza leucocephalos*, and *Anthus hodgsoni* also were positively correlated with grass cover dominant and pine plantation habitats, and negatively correlated with overstory height and dead down wood habitats.

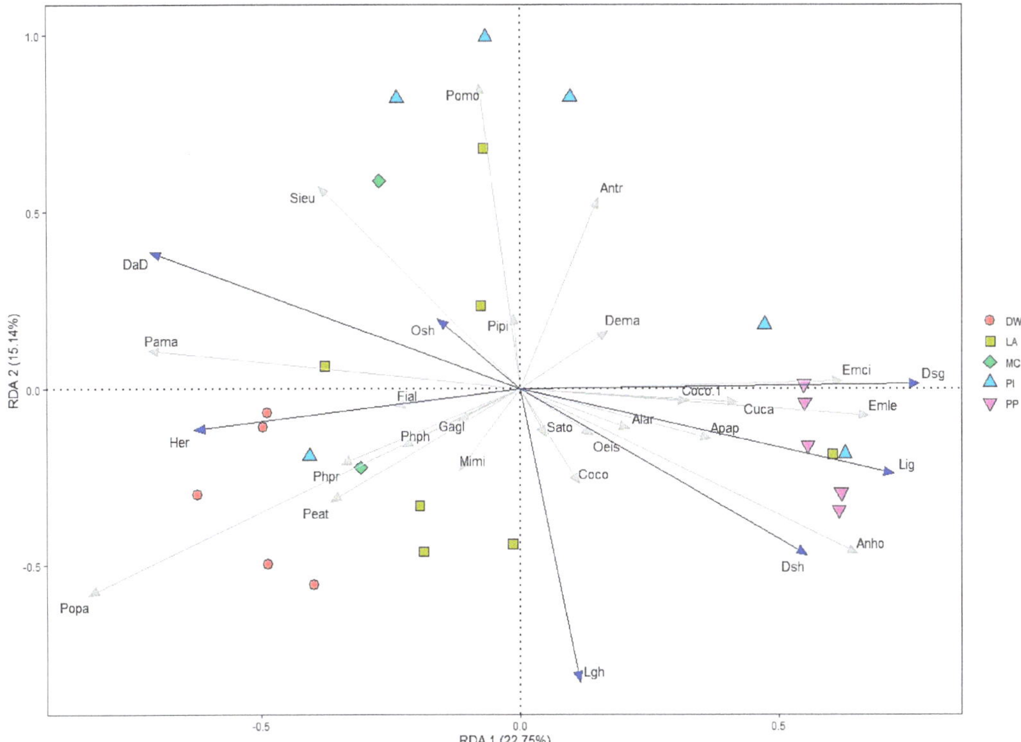

Figure 5. Redundancy analysis (RDA) ordination among bird assemblage and habitat types and environmental variables in the forest of the Khentii Mountains. Forest types are deciduous woodland (DW), larch (LA), pine (PI), pine plantation (PP), and mixed conifer (MC). Note: Species and habitat structural variable names are generic acronyms or the four or three letters of the names in Tables 1 and 2.

4. Discussion and Conclusions

Our study demonstrates the species richness and composition of bird assemblages, and the relations of forest habitat patterns of the Khentii Mountain boreal forest. Other studies have provided overviews of the taxonomic richness and composition of the avifauna across the Northern Mongolia [24–26,48]. Nevertheless, these only focused on avifauna, which resulted in forest regions remaining understudied. In total, 71 species were seen by the end of the survey, which were recorded in 400 point-counts, and on average, 3.14 species were seen per point. Birds in the study area account for 36.4% of the total species richness in the Khentii Mountains region and contribute to bird diversity [24], suggesting that forest habitat variables play an important role in bird diversity in forest areas [10,25]. Of this total, fourteen bird species occurred in all habitat types. Bird assemblages in the Khentii forested area were dominated mainly by *Anthus hodgsoni, Parus major, Poecile palustris, Sitta europaea, Emberiza leucocephalos,* and *Poecile montanus.* These species had relative abundance values of 5 to 12%, thereby contributing a lot to bird diversity in this area, accounting for 45% of total species richness. The bird species richness increased from the taiga zone to the forest-steppe zone. More specifically, our study found that bird richness was higher in larch dominated forests. Light mixed deciduous and larch forest communities have highly diverse bird assemblages, more so than taiga forest communities [24]. Larch forests, especially those burned by wildfires, change their landscapes and increase the forest layers of young birch, shrubs, and grasses [3]. The presence of deciduous trees in conifer-dominated forests generally allows for higher bird diversity compared to pure conifer stands [49,50].

In this study, forest habitat variables (altitude, overstory, understory, and ground composition) changed markedly by forest type. Different habitats are essential for bird assemblages in this forested area. The total species richness in the Khentii forested area was found to play an important role in the breeding and growth of the bird community, and a refuge for bird assemblages in the breeding season, for 57.3% of Mongolian bird species [25,26]. Among all types of habitats, the avifauna abundances were similar: the resident and migrant species accounted for 47.2% and 52.8%, respectively. The guild types also have a certain impact on the habitat selection of birds [25]; thus, insectivores (DW, LA, PI, PP, MC) were the dominant guild within each habitat type, and in our study area, there were 38 (53.5%) insectivores out of the 71 species sampled. The researchers have mentioned that insectivores are commonly observed to dominate forest habitats in terms of species richness and abundance; they are very important to forest regeneration [25,26,51].

Moreover, the present study showed bird assemblages in forest habitats of the Khentii Mountains were structured by a combination of overstory height and ground-layer features (e.g., dead down wood, herbaceous, and grass). Among these, ground layer-factors acted as the most important contributors to forest habitat types. The importance of forest habitat heterogeneity to the compositions of bird assemblages has been demonstrated by numerous studies [51–55]. The habitat variability and the diversity of the bird community are strongly influenced by the structure of the vegetation heterogeneity [56,57]. Our study found a significant difference in the tree and vegetation cover between forest type and canopy layer-based forested areas. Thus, in our study areas, ground-nesting bird species associated with grass (e.g., *Anthus hodgsoni*, *Emberiza leucocephalos*, *Anthus trivialis*, *Alauda arvensis*, and *Emberiza cioides*) tended to occupy areas with tall grass, such as pine plantations, and pines were often high in these habitats. The species that prefer an open forest canopy structure often tend to dominate the assemblages of dry Scots pine stands with sparse crown cover [20,58]. The high canopy-dominated conifer tree sites (LA, PI, MC) provide habitats for secondary cavity-nester bird species associated with low grasses and herbaceous cover (e.g., *Sitta europaea*, *Parus major*, and *Poecile montanus*). Those habitats are the most abundant in dead and down trees. Moreover, these species did not occur in pine plantation sites, and some secondary cavity-nester species were less abundant. Natural mature forests are highly suitable for cavity nesters or bark feeding species [59,60] and the abundance of the insectivores responds positively to an increase in basal area or dead wood volume [11]. On the other hand, the *Phylloscopus proregulus*, *Poecile palustris*, and *Periparus ater* have been recorded in high-herbaceous-cover areas dominated by deciduous woodland and some conifer habitats. As well, four species (*Poecile palustris*, *Periparus ater*, *Phoenicurus phoenicurus*, and *Ficedula albicilla*) are hole-nesters, and two species are low canopy nesters (*Phylloscopus proregulus*, and *Garrulus glandarius*), suggesting that the presence of deciduous trees is suitable for nesting for these species. Especially, species that generally require the presence of deciduous trees within stands are woodpeckers, tits, some nuthatches, and songbirds [54,61,62]. In deciduous woodland and deciduous mixed conifer forests, tits and warblers represented by far the most abundant portions of the community, and in pine and pine plantations, Emberiza species were more abundant than tits. However, woodpeckers were often the smallest fraction of the community. Our results showed that some species are more associated with sites with taller vegetation, whereas other species occupy sites with shorter vegetation and diverse herbaceous cover.

This study was the first to describe the forest-type variations of bird assemblages in the Khentii Mountains forest in Northern Mongolia. Identifying the assemblage pattern is useful for the conservation of not only birds but also biodiversity in general. In the course of the investigation, 71 bird species from 23 families and 11 orders were registered in the forest habitats of the Khentii partially of the forest area, and the Passeriformes order dominated; this result is the most up-to-date and systematically collected baseline data for future forest bird research. Information obtained from this study will enhance our understanding of the variation in bird assemblages, and then help to develop strategies for future forest bird conservation in such areas. Among these, insectivore birds dominate in all forest habitats. Many studies have also shown that insectivorous birds are more sensitive to habitat disturbance and loss than other feeding guilds [51,55,62–64]. From 2004 to 2014, the burned forest in the area increased by 1.4 million hectares, and the logging and insect-affected forest area increased by 340 thousand hectares [6]. Unfortunately, at the same time, the reforestation area increased by 2 thousand hectares [6], which shows a lack of reforestation. The bird community structure was affected by many factors, such as vegetation, the size and structure of the forest, and forest type.

In conclusion, according to our findings, Khentii forest areas are outstanding sites for migratory and resident birds. The results from this study show the importance of forest habitat structure for the abundance and diversity of birds in mixed tree and conifer forests. The bird species diversity and distribution of Khentii region can be currently regarded as moderately well studied. However, considering the bird assemblages of forested regions in Mongolia, further study is needed to fine-tune the species population estimates. In the future, we aim to study the co-effects of habitat disturbance and temporal variation on bird communities and bird density in different forest habitats. Thus, further studies on the relationship between temporal variation and bird density are important not only for ecological theory, but also for the scientific fundamentals of forest management and environmental protection in Mongolia.

Author Contributions: This paper received individual contributions from each author as specified: Conceptualization, Z.P. and J.W.L.; Methodology, Z.P.; Software, Z.P.; Validation, Z.P., J.W.L., W.K.P. and O.G.; Formal Analysis, Z.P.; Investigation, Z.P., J.W.L., M.M., O.G., U.G., E.P., B.N., T.A. and A.J.; Resources, Z.P. and J.W.L.; Data Curation, Z.P. and U.G.; Writing—Original Draft Preparation, Z.P.; Writing—Review and Editing, Z.P., O.G., M.M. and J.W.L.; Visualization, Z.P. and J.W.L.; Supervision, J.W.L. All authors have read and agreed to the published version of the manuscript.

Funding: This research was funded by the research fund of Chungnam National University: CNU; 2020-0532-01.

Institutional Review Board Statement: Not applicable.

Informed Consent Statement: Not applicable.

Data Availability Statement: The data may be available upon request to the corresponding author, subject to the approval.

Acknowledgments: The authors appreciate the researchers and students of the Mongolian National University of Education, who participated in the field work. We would also like to thank our research colleagues from the Mongolian National University of Education (MNUE). J. Ariunbold partially supported.

Conflicts of Interest: The authors declare no conflict of interest.

References

1. Olson, D.M.; Dinerstein, E.; Wikramanayake, E.D.; Burgess, N.D.; Powell, G.V.; Underwood, E.C.; D'amico, J.A.; Itoua, I.; Strand, H.E.; Morrison, J.C.; et al. Terrestrial Ecoregions of the World: A New Map of Life on Earth A new global map of terrestrial ecoregions provides an innovative tool for conserving biodiversity. *BioScience* **2001**, *51*, 933–938. [CrossRef]
2. Roberge, J.M.; Virkkala, R.; Monkkonen, M. Boreal 6 r Forest Bird Assemblages and Their Conservation. *Ecol. Conserv. For. Birds* **2018**, *29*, 183. [CrossRef]
3. Dugarjav, C.H. *Larch Forests of Mongolia*; Bembi San: Ulan Bator, Mongolia, 2006; p. 317. (In Mongolian)
4. FAO. *Mongolia—Global Forest Resources Assessment 2015—Country Report*; Food and Agriculture Organization (FAO) of the United Nations: Rome, Italy, 2014; p. 97.
5. Batkhuu, N.O.; Lee, D.K.; Tsogtbaatar, J. Forest and forestry research and education in Mongolia. *J. Sustain. For.* **2011**, *30*, 600–617. [CrossRef]
6. Government of Mongolia. *Mongolia's Forest Reference Level Submission to the United Nations Framework Convention on Climate Change*; UN-REDD Mongolia National Programme, Ministry of Environment and Tourism: Ulaanbaatar, Mongolia, 2018; p. 62.
7. Tsogtbaatar, J. Deforestation and reforestation of degraded forestland in Mongolia. In *the Mongolian Ecosystem Network*; Springer: Tokyo, Japan, 2013; pp. 83–98. [CrossRef]
8. Sainnemekh, S.; Isabel, C.B.; Bulgamaa, D.; Brandon, B.; Ása, L.A. Rangeland degradation in Mongolia: A systematic review of the evidence. *J. Arid. Environ.* **2022**, *196*, 104654. [CrossRef]

9. Erdenechuluun, T. Wood Supply in Mongolia: The Legal and Illegal Economies. In *Mongolia Discussion Papers, East Asia and Pacific Environment and Social Development Department*; World Bank: Washington, DC, USA, 2006.
10. Wildlife Science and Conservation Center, Institute of Biology and BirdLife International. *Directory of Important Bird Areas in Mongolia: Key Sites for Conservation*; Nyambayar, B., Tseveenmyadag, N., Eds.; Wildlife Science and Conservation Center, Institute of Biology and BirdLife International: Ulaanbaatar, Mongolia, 2009.
11. Collar, N.J.; Crosby, M.J.; Stattersfield, A.J. *Birdlife International. Birds to Watch 2: The World List of Threatened Birds: The Official Source for Birds on the Iucn Red List*; BirdLife International: Cambridge, UK, 1994.
12. Balestrieri, R.; Basile, M.; Posillico, M.; Altea, T.; De Cinti, B.; Matteucci, G. A guild-based approach to assessing the influence of beech forest structure on bird communities. *For. Ecol. Manag.* **2015**, *356*, 216–223. [CrossRef]
13. Kamp, J.; Oppel, S.; Heldbjerg, H.; Nyegaard, T.; Donald, P.F. Unstructured citizen science data fail to detect long-term population declines of common birds in Denmark. *Divers. Distrib.* **2016**, *22*, 1024–1035. [CrossRef]
14. Donald, P.F.; Green, R.E.; Heath, M.F. Agricultiral intensification and the collapse of Europe's farmland bird populations. *Proc. R. Soc. Lond.* **2001**, *268*, 25–29. [CrossRef]
15. Brinkert, A.; Hölzel, N.; Sidorova, T.V.; Kamp, J. Spontaneous steppe restoration on abandoned cropland in Kazakhstan: Grazing affects successional pathways. *Biodivers. Conserv.* **2016**, *25*, 2543–2561. [CrossRef]
16. Kwok, H.K.; Corlett, R.T. The bird communities of a natural secondary forest and a *Lophostemon confertus* plantation in Hong Kong, South China. *For. Ecol. Manag.* **2000**, *130*, 227–234. [CrossRef]
17. Lott, C.A.; Akresh, M.E.; Costanzo, B.E.; D'Amato, A.W.; Duan, S.; Fiss, C.J.; Fraser, J.S.; He, H.S.; King, D.I.; McNeil, D.J.; et al. Do Review Papers on Bird–Vegetation Relationships Provide Actionable Information to Forest Managers in the Eastern United States. *Forests* **2021**, *12*, 990. [CrossRef]
18. Thompson, P.S.; Greenwood, J.D.; Greenaway, K. Birds in European gardens in the winter and spring of 1988–89. *Bird Stud.* **1993**, *40*, 120–134. [CrossRef]
19. Tu, H.M.; Fan, M.W.; Ko, J.C. Different habitat types affect bird richness and evenness. *Sci. Rep.* **2020**, *10*, 1–10. [CrossRef] [PubMed]
20. Jang, W.; Seol, A.; Chung, O.-S.; Sagong, J.; Lee, J.K. Avian Reporting Rates in Chugcheongnam Province, South Korea Depend on Distance from Forest Edge, Size of Trees, and Size of Forest Fragments. *Forests* **2019**, *10*, 364. [CrossRef]
21. Gombobaatar, S.; Monks, E.M.; Seidler, R.; Sumiya, D.; Tseveenmyadag, N.; Bayarkhuu, S.; Baillie, J.E.; Boldbaatar, S.; Uuganbayar, C. *Regional Red List Series Vol. 7. Birds*; Zoological Society of London, National University of Mongolia, and Mongolian Ornithological Society: Ulaanbaatar, Mongolia, 2011; (In English and Mongolian).
22. Bold, A. Mongolian Birds. In *Mongolia Today Science, Culture, Environment and Development*; Badarch, D., Zilinskas, R.A., Balint, P., Eds.; RoutledgeCurzon: London, UK, 2003; pp. 143–171.
23. Bold, A. Mongolian Birds study of past 5 years. *Sci. Proceed. Inst. Biol. Mong. Acad. Sci.* **1973**, *7*, 143–171. (In Mongolian)
24. Bold, A. Birds of Hentii mountain region. *Sci. Proc. Inst. Biol. Mong. Acad. Sci.* **1969**, *3*, 4–26. (In Mongolian)
25. Bold, A. Birds of Hentii Mountain Region and Their Practical Importance. Ph.D. Dissertation, Department of Biology, National University of Mongolia, Ulaanbaatar, Mongolia, 1977. (In Russian).
26. Bold, A. Census result of game birds in Hentii. *Sci. Proc. Inst. Biol. Mong. Acad. Sci.* **1970**, *4*, 19–27. (In Mongolian)
27. Jambaajamts, A. *Climate Brief Overview of the Republic of Mongolia*; National Press Office: Ulaanbaatar, Mongolia, 1989. (In Mongolian)
28. Tsegmid, S. *Physical Geography of Mongolia*; Mongolian Academy of Sciences, Institute of Geography and Permafrost, National publishing: Ulaanbaatar, Mongolia, 1969; p. 405. (In Mongolian)
29. Sato, T.; Kimura, F.; Kitoh, A. Projection of global warming onto regional precipitation over Mongolia using a regional climate model. *J. Hydrol.* **2007**, *333*, 144–154. [CrossRef]
30. Tsegmid, C. Some results of studies on microclimate and soil humidity of microassociations in mossy Larix forest of the eastern Khentey. *Tesisi Docl Nauchnoi Konf. Posveshennie Vopr. Vozobnov. Resur. Lesa MNR* **1989**, 170–176.
31. Ermakov, N.; Cherosov, M.; Gogoleva, P. Classification of ultracontinental boreal forests in central Yakutia. *Folia Geobot.* **2002**, *37*, 419–440. [CrossRef]
32. Dulamsuren, C. Floristische Diversität, Vegetation und Standortbedingungen in der Gebirgstaiga des Westkhentej, Nordmongolei. *Ber Forsch. Wald. A* **2004**, *191*, 1–290.
33. Mühlenberg, M.; Appelfelder, J.; Hoffmann, H.; Ayush, E.; Wilson, K.J. Structure of the montane taiga forests of West Khentii, Northern Mongolia. *J. For. Sci.* **2012**, *58*, 45–56. [CrossRef]
34. Dulamsuren, C.; Hauck, M.; Mühlenberg, M. Vegetation at the taiga forest-steppe borderline in the western Khentej Mountains, northern Mongolia. *Ann. Bot. Fenn.* **2005**, *42*, 411–426.
35. Bowman, J. Tujiin Nars: A Story of the Forest. Independent Study Project (ISP) Collection. 2012. Available online: https://digitalcollections.sit.edu/isp_collection/1453 (accessed on 20 June 2021).
36. Bibby, C.J.; Burgess, N.D.; Hill, D.A. *Bird Census Techniques*; Academic Press: London, UK, 1992.
37. Bibby, C.J.; Jones, M.; Marsden, S. *Bird Surveys*; Expedition Advisory Centre: London, UK, 1998.
38. Hanni, D.J.; White, C.M.; van Lanen, N.J.; Birek, J.J.; Berven, J.M.; McLaren, M.F. *Integrated Monitoring in Bird Conservation Regions (Imbcr): Field Protocol for Spatially-Balanced Sampling of Landbird Populations, Unpublished Report*; Bird Conservancy of the Rockies: Brighton, CO, USA, 2016.
39. Blakesley, J.A.; Hanni, D.J. *Monitoring Colorado's Birds, 2008. Technical Report M-MCB08-01*; Rocky Mountain Bird Observatory: Brighton, CO, USA, 2009.

40. QGIS. QGIS Geographic Information System. *Open Source Geospatial Foundation Project*. Available online: https://qgis.org/en/site/ (accessed on 20 June 2021).

41. R Core Team. *R: A Language and Environment for Statistical Computing*; R Foundation for Statistical Computing: Vienna, Austria, 2015; Available online: https://www.R-project.org/ (accessed on 15 September 2021).

42. Chao, A.; Jost, L. Coverage-based rarefaction and extrapolation: Standardizing samples by completeness rather than size. *Ecology* **2012**, *93*, 2533–2547. [CrossRef] [PubMed]

43. Hsieh, T.C.; Ma, K.H.; Chao, A. iNEXT: An R package for rarefaction and extrapolation of species diversity (Hill numbers). *Methods Ecol. Evol.* **2016**, *7*, 1451–1456. [CrossRef]

44. Chao, A.; Gotelli, N.J.; Hsieh, T.C.; Sander, E.L.; Ma, K.H.; Colwell, R.K.; Ellison, A.M. Rarefaction and extrapolation with Hill numbers: A framework for sampling and estimation in species diversity studies. *Ecol Monogr.* **2014**, *84*, 45–67. [CrossRef]

45. Oksanen, J.; Simpson, G.L.; Blanchet, F.G.; Kindt, R.; Legendre, P.; Minchin, P.R.; O'Hara, R.B.; Solymos, P.; Stevens, M.H.H.; Szoecs, E.; et al. Vegan: Community Ecology Package. 2019. Available online: https://cran.r-project.org/package=vegan (accessed on 5 January 2022).

46. Anderson, M.J. A new method for non-parametric multivariate analysis of variance. *Austral Ecol.* **2001**, *26*, 32–46. [CrossRef]

47. Legendre, P.; Gallagher, E.D. Ecologically meaningful transformations for ordination of species data. *Oecologia* **2001**, *129*, 271–280. [CrossRef]

48. Stenzel, T.; Stubbe, M.; Samjaa, R.; Gombobaatar, S. *Quantitative Investigations on Bird Communities in Different Habitats in the Orkhon-Selenge-Valley in Northern Mongolia*; Erforschung biologischer Ressourcen der Mongolei/Exploration into the Biological Resources of Mongolia: Ulaanbaatar, Mongolia, 2005; p. 127. ISSN 0440-1298.

49. Stokland, J.N. Representativeness and Efficiency of Bird and Insect Conservation in Norwegian Boreal Forest Reserves: Representatividad y Eficiencia en la Conservación de Aves e Insectos en las Reservas de Bosque Boreal de Noruega. *Conserv. Biol.* **1997**, *11*, 101–111. [CrossRef]

50. Felton, A.; Andersson, E.; Ventorp, D.; Lindbladh, M. A comparison of avian diversity in spruce monocultures and spruce-birch polycultures in Southern Sweden. *Silva Fenn.* **2011**, *45*, 1143–1150. [CrossRef]

51. Duco, R.A.; Fidelino, J.S.; Duya, M.V.; Ledesma, M.M.; Ong, P.S.; Duya, M.R. Bird Assemblage and Diversity along Different Habitat Types in a Karst Forest Area in Bulacan, Luzon Island, Philippines. *Philipp. J. Sci.* **2020**, *15*, 399–414.

52. Berg, Å. Diversity and abundance of birds in relation to forest fragmentation, habitat quality and heterogeneity. *Bird Study* **1997**, *44*, 355–366. [CrossRef]

53. Bersier, L.F.; Meyer, D.R. Bird assemblages in mosaic forests: The relative importance of vegetation structure and floristic composition along the successional gradient. *Acta Oecologica* **1994**, *15*, 561–576.

54. Estades, C.F.; Temple, S.A. Deciduous-forest bird communities in a fragmented landscape dominated by exotic pine plantations. *Ecol. Appl.* **1999**, *9*, 573–585. [CrossRef]

55. Mansor, M.S.; Sah, S.A. The influence of habitat structure on bird species composition in lowland malaysian rain forests. *Trop. Life Sci. Res.* **2012**, *23*, 1–14. [PubMed]

56. Deconchat, M.; Balent, G. Vegetation and bird community dynamics in fragmented coppice forests. *Forestry* **2001**, *74*, 105–118. [CrossRef]

57. Haapanen, A. Bird fauna of the Finnish forests in relation to forest succession. In *I. InAnnales Zoologici Fennici*; Finnish Zoological and Botanical Publishing Board: Helsinki, Finland, 1965; Volume 2, pp. 153–196.

58. Bergner, A.; Avcı, M.; Eryiğit, H.; Jansson, N.; Niklasson, M.; Westerberg, L.; Milberg, P. Influences of forest type and habitat structure on bird assemblages of oak (*Quercus* spp.) and pine (*Pinus* spp.) stands in southwestern Turkey. *For. Ecol. Manag.* **2015**, *336*, 137–147. [CrossRef]

59. Nikolov, S.C. Effect of stand age on bird communities in late successional Macedonian pine forests in Bulgaria. *For. Ecol. Manag.* **2009**, *257*, 580–587. [CrossRef]

60. Morozov, N.S. Breeding forest birds in the Valdai Uplands, north-west Russia: Assemblage composition, interspecific associations and habitat amplitudes. In *InAnnales Zoologici Fennici*; Finnish Zoological Publishing Board, formed by the Finnish Academy of Sciences, Societas Biologica Fennica Vanamo, Societas pro Fauna et Flora Fennica, and Societas Scientiarum Fennica: Helsinki, Finland, 1992; pp. 7–28.

61. Enoksson, B.; Angelstam, P.; Larsson, K. Deciduous forest and resident birds: The problem of fragmentation within a coniferous forest landscape. *Landsc. Ecol.* **1995**, *10*, 267–275. [CrossRef]

62. Laurance, S.G.; Stouffer, P.C.; Laurance, W.F. Effects of road clearings on movement patterns of understory rainforest birds in central Amazonia. *Conserv. Biol.* **2004**, *18*, 1099–1109. [CrossRef]

63. Tvardíková, K. Bird abundances in primary and secondary growths in Papua New Guinea: A preliminary assessment. *Trop. Conserv. Sci.* **2010**, *3*, 373–388. [CrossRef]

64. Şekercioğlu, Ç.H.; Ehrlich, P.R.; Daily, G.C.; Aygen, D.; Goehring, D.; Sandí, R.F. Disappearance of insectivorous birds from tropical forest fragments. *Proc. Natl. Acad. Sci. USA* **2002**, *99*, 263–267. [CrossRef] [PubMed]

Article

Rocky Area Inhabiting Daddy Long-Legs Spiders, *Pholcus* Walckenaer, 1805 (Araneae: Pholcidae) in Mountainous Mixed Forests from South Korea

Chang-Moon Jang [1,2,3], Jung-Sun Yoo [4], Seung-Tae Kim [5,*] and Yang-Seop Bae [1,2,3,*]

[1] Division of Life Sciences, College of Life Sciences and Bioengineering, Incheon National University, Incheon 22012, Republic of Korea
[2] Convergence Research Center for Insect Vectors, Incheon 22012, Republic of Korea
[3] Bio-Resource and Environmental Center, Division of Life Sciences, College of Life Sciences and Bioengineering, Incheon National University, Songdo-dong, Incheon 22012, Republic of Korea
[4] Species Diversity Research Division, National Institute of Biological Resources, Incheon 22689, Republic of Korea
[5] Life and Environment Research Institute, Konkuk University, Seoul 05029, Republic of Korea
[*] Correspondence: stkim2000@hanmail.net (S.-T.K.); baeys@inu.ac.kr (Y.-S.B.)

Abstract: Two new spider species of the genus *Pholcus* Walckenaer, 1805, *Pholcus deokjeok* **sp. nov.** and *Pholcus gangneung* **sp. nov.** in the family Pholcidae C. L. Koch, 1850 are newly described from South Korea. The present new species belong to the *phungiformes* group in the genus. They are found on rock walls in mountainous mixed forests. This work provides diagnoses, detailed descriptions, and taxonomic photographs for these new species. The unusual shaped and strongly sclerotized embolus of *P. gangneung* **sp. nov.** in the *Pholcus phungiformes* group is the first to be reported.

Keywords: diagnosis; habitat; new species; *phungiformes* group; rock wall; taxonomy

Citation: Jang, C.-M.; Yoo, J.-S.; Kim, S.-T.; Bae, Y.-S. Rocky Area Inhabiting Daddy Long-Legs Spiders, *Pholcus* Walckenaer, 1805 (Araneae: Pholcidae) in Mountainous Mixed Forests from South Korea. *Forests* **2023**, *14*, 538. https://doi.org/10.3390/f14030538

Academic Editors: Konstantinos Poirazidis, Panteleimon Xofis and Timothy A. Martin

Received: 19 January 2023
Revised: 24 February 2023
Accepted: 6 March 2023
Published: 9 March 2023

1. Introduction

Pholcidae C. L. Koch, 1850 is one of the most diversified and largest families comprising 1836 species in 97 genera within the order Araneae Clerck, 1757 [1]. To date, 43 species of pholcid spiders have been described in 3 genera from various ecosystems in Korea [1–9]. Pholcidae is one of the taxa that has not been explored much yet in Korea. The genus *Pholcus* Walckenaer, 1805 among the largest genera in the family is known to mainly thrive on dusky, humid spaces such as rock walls and road drains in mountainous regions [8,10]. As of 2021, Korea's forest area was 6298 ha, accounting for 62.7% of the total national area; most of them are mountainous areas composed of conifer and deciduous mixed forests [11], and most mountainous forests have many rock walls suitable for the habitat of *Pholcus* spiders. The genus *Pholcus* can be distinguished from other genera by the combination of the male chelicerae with a pair of proximal frontal apophyses and epigynum sclerotized with a knob [12]. The *Pholcus phungiformes* group can be distinguished from other species group by the male chelicera with frontal apophysis, male palpal tibia with prolatero-ventral modification, male genital bulb without an appendix or having a pseudoappendix, dorsal spine on the procursus, and epigynum sclerotized with a knob [9,10,12]. In Korea, 37 *Pholcus* species belonging to this species group have been described [9]. Two new spiders belonging to the *P. phungiformes* group were collected during a seasonal survey on the spider fauna of mountainous mixed forests in 2022 (Figure 1) and are described with measurements, illustrations, and a diagnosis.

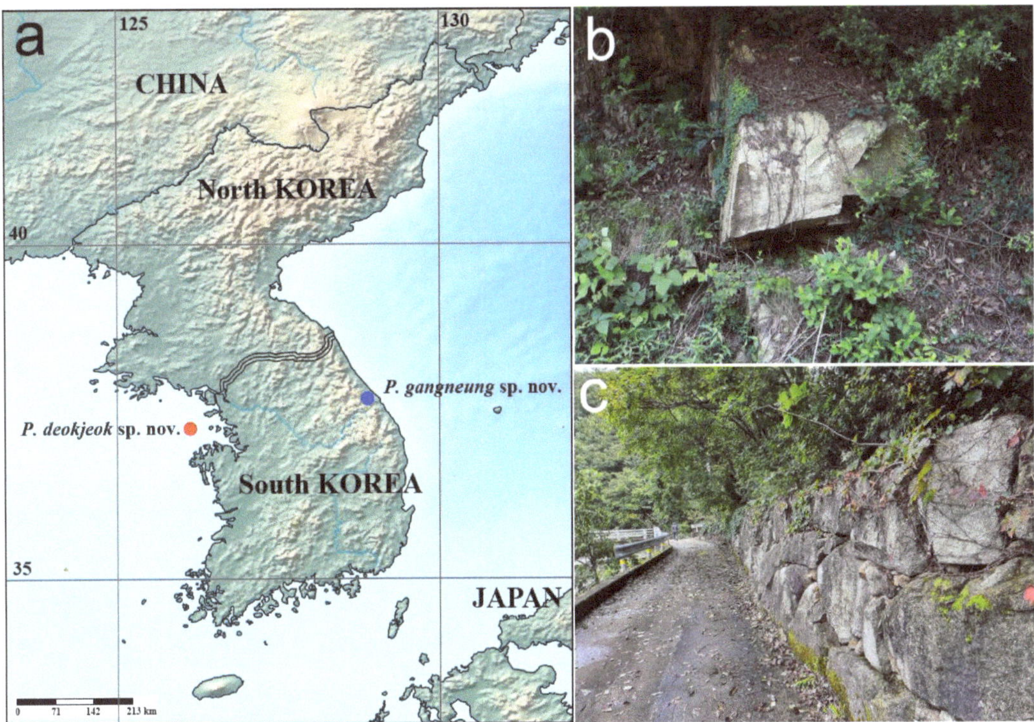

Figure 1. Distribution map and habitats of new *Pholcus* species. (**a**) Distribution of the new *Pholcus* in South Korea. (**b**) Habitat of *Pholcus deokjeok* **sp. nov.** (**c**) Habitat of *Pholcus deokjeok* **sp. nov.**

2. Materials and Methods

All specimens were preserved in 98% ethyl alcohol, and external morphology was examined under a Leica S8APO (Leica, Singapore) stereomicroscope. Images were captured with a Tucsen Dhyana 400DC digital camera (Fuzhou Tucsen Photonics Co., Ltd., Fuzhou, China) mounted on a Leica S8APO and assembled using Helicon Focus 8.2.0 image stacking software [13]. The female epigynum was dissected and cleared in 10% KOH for 2 h to examine the internal genitalia before illustration. Leg measurements are shown as follows: total length (femur, patella, tibia, metatarsus, tarsus). The morphological terminology follows Huber [12]. The holotype specimens studied are deposited in the Animal Resources Division of the National Institute of Biological Resources, Incheon (NIBR), and the paratype specimens are deposited in the Life and Environment Research Institute of Konkuk University, Seoul (KKU), South Korea. The distribution map was produced by modifying SimpleMappr [14]. The following abbreviations are used in the descriptions: ALE = anterior lateral eye; AME = anterior median eye; PLE = posterior lateral eye; PME = posterior median eye; ALE–AME = distance between ALE–AME; ALE–PME = distance between ALE–PME; AME–AME = distance between AMEs; AME–PME = distance between AME–PME; PLE–PME = distance between PLE–PME; PME–PME = distance between PMEs in the eye region; L/d = length/diameter in the leg measurement.

3. Results

Taxonomic accounts (Figures 2 and 3)
Family Pholcidae C. L. Koch, 1850
Genus *Pholcus* Walckenaer, 1805
***Pholcus phungiformes* species group**

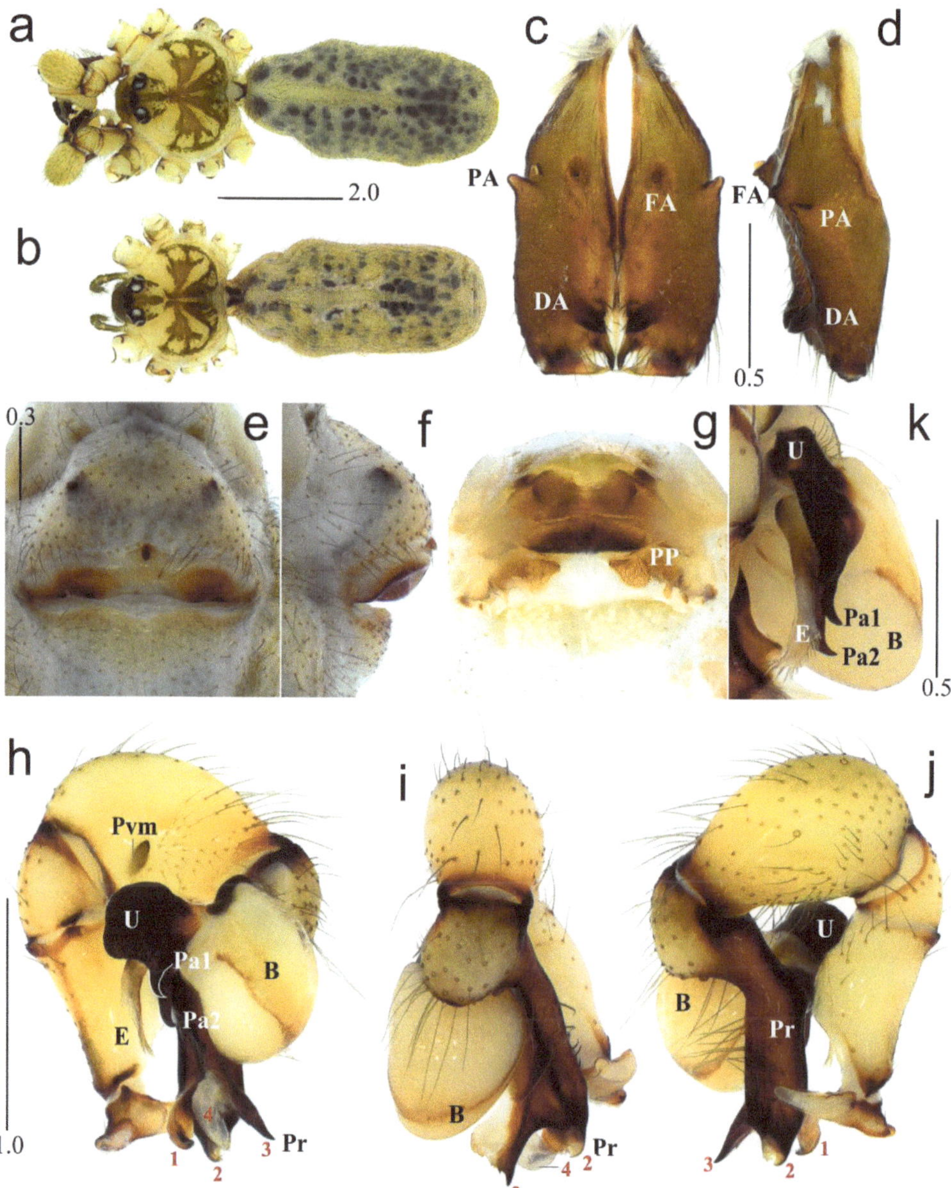

Figure 2. *Pholcus deokjeok* **sp. nov.** (**a**) Holotype male (Habitus). (**b**) Paratype female (Habitus). (**c**) Male chelicerae, frontal view. (**d**) *Ditto*, lateral view. (**e**) Female epigynum, ventral view. (**f**) *Ditto*, lateral view. (**g**) Female internal genitalia, dorsal view. (**h**) Male palp, prolateral view. (**i**) *Ditto*, frontal view. (**j**) *Ditto*, retrolateral view. (**k**) Embolic division (1 = ventral apophysis, 2 = retrolateral apophysis, 3 = dorsal apophysis, 4 = prolateral apophysis, B = bulb, DA = distal apophysis, E = embolus, FA = frontal apophysis, PA = proximo-lateral apophysis, Pa = pseudoappendix, PP = pore plate, Pr = procursus, Pvm = prolatero-ventral modification, U = uncus). Scale bars in mm.

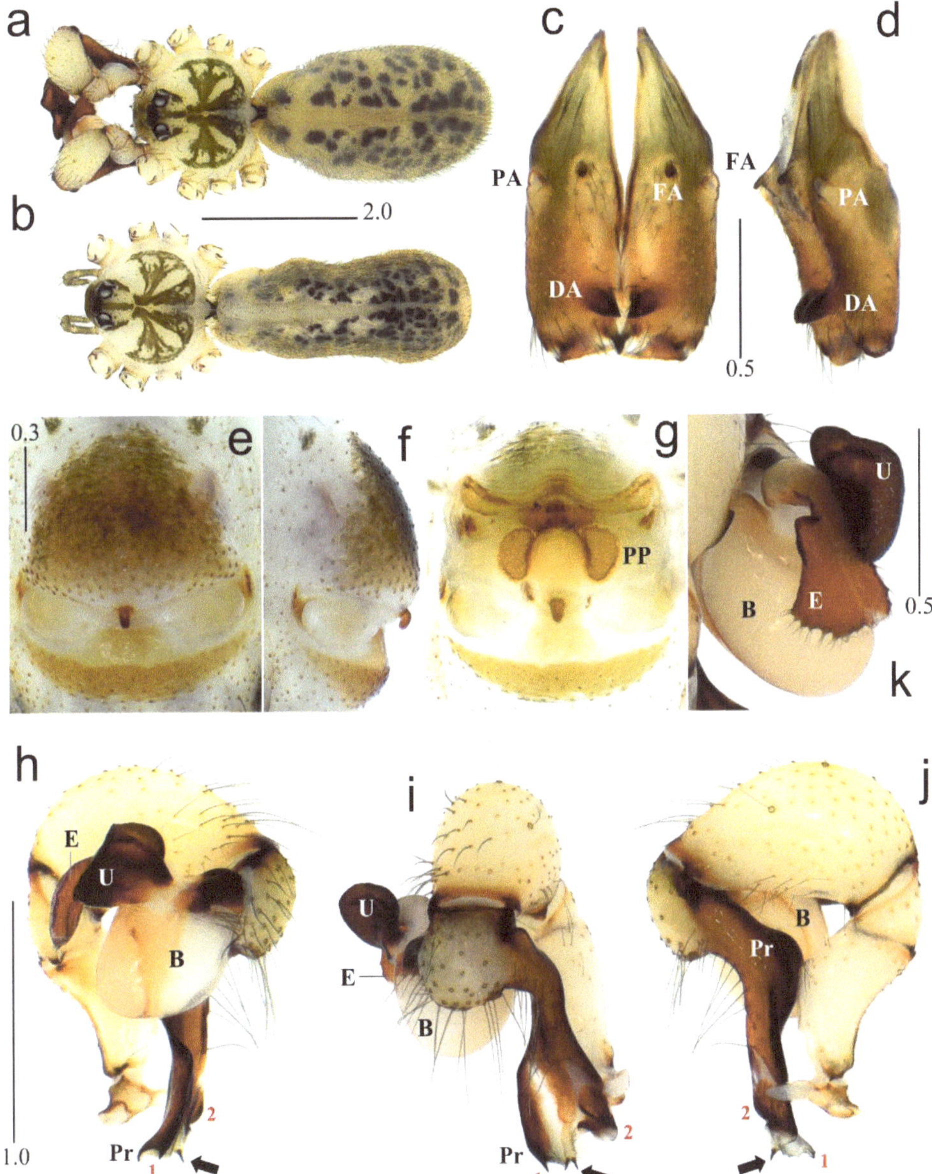

Figure 3. *Pholcus gangneung* **sp. nov.** (**a**) Holotype male (Habitus). (**b**) Paratype female (Habitus). (**c**) Male chelicerae, frontal view. (**d**) *Ditto*, lateral view. (**e**) Female epigynum, ventral view. (**f**) *Ditto*, lateral view. (**g**) Female internal genitalia, dorsal view. (**h**) Male palp, prolateral view. (**i**) *Ditto*, frontal view. (**j**) *Ditto*, retrolateral view. (**k**) Embolic division (1 = prolatero-ventral apophysis, 2 = retrolateral apophysis, B = bulb, DA = distal apophysis, E = embolus, FA = frontal apophysis, PA = proximo-lateral apophysis, PP = pore plate, Pr = procursus, U = uncus). Scale bars in mm.

3.1. Pholcus deokjeok sp. nov.

(Figure 1a,b and Figure 2)

Type material. Holotype: ♂, Bijobong Peak, Deokjeokdo Island, Jin-ri, Deokjeok-myeon, Ongjin-gun, Incheon-si, South Korea (37.214236N, 126.124386E, alt. 65 m), 26

July 2022, C.M. Jang leg. (NIBR, #WGJTIV0000000568). **Paratypes:** 6♀♀, same data as the holotype (KKU, #Ara_Phol_Deokjeok_202201~06_PT).

Etymology. The specific name is a noun in apposition and refers to the type locality, Deokjeokdo Island.

Diagnosis. The new species is similar to *Pholcus chiakensis* Seo, 2014 in the shape of the palpal organ and body appearance but can be easily distinguished from the latter by the combination of the following characters: Male—uncus with two short and long hook-shaped pseudoappendices (Figure 2h,k); procursus with four distal apophyses (two thick ventral and retrolateral apophyses with hooked tips numbered 1 and 2 in Figure 2h–j; one spear-shaped and broad dorsal apophysis with appointed numbered 3 in Figure 2h–j; one thick, finger-shaped, and membranous prolateral apophysis with a pointed tip numbered 4 in Figure 2h,i) versus uncus with two slender and knife-shaped, and pointed short pseudoappendices; procursus with four distal apophyses (one small pointed ventral apophysis; one retrolateral apophysis with a hooked tip; one spear-shaped and broad dorsal apophysis with a twisted tip; and one thick, irregular-shaped, and membranous prolateral apophysis with a blunt tip) in *P. chiakensis* (Figure 2A–C) [15]. Female—a pair of semicircle-shaped pore plates and projected inward in the internal genitalia (Figure 2e–g) versus a pair of European pear-shaped pore plates in the internal genitalia in *P. chiakensis* (Figure 2G,H) [15].

Description. Male (holotype). Habitus as in Figure 2a. Total length 5.56. Carapace: 1.84 long/1.95 wide. Eyes: AER 0.70, PER 0.74, ALE 0.18, AME 0.11, PLE 0.18, PME 0.16, ALE–PLE contiguous, AME–ALE 0.06, AME–AME 0.08, AME–PME 0.05, PME–PLE 0.04, PME–PME 0.30. Chelicera: 1.18 long/0.34 wide. Endite: 0.58 long/0.35 wide. Labium: 0.30 long/0.39 wide. Sternum: 0.90 long/1.19 wide. Legs: I 48.15 (12.19, 0.71, 12.62, 20.27, 2.36)/II 31.78 (8.96, 0.68, 8.02, 12.64, 1.48)/III 22.51 (6.46, 0.68, 5.55, 8.54, 1.28)/IV 28.19 (8.16, 0.64, 7.16, 11.04, 1.19), tibia I L/d 83. Palp: 3.71 (0.78, 0.41, 1.15, -, 1.37). Abdomen: 3.72 long/1.68 wide.

Carapace pale yellowish brown, cephalic region with pale blackish brown median and marginal bands, thoracic region with pale blackish brown radial and marginal bands (Figure 2a). Chelicera with three apophyses; blunt proximo-lateral apophysis diagonally upward and protrudent out of chelicera, small and pointed frontal apophysis protrudent forward, and thick and pointed distal apophysis diagonally downward (Figure 2c,d). Legs yellowish brown, retrolateral trichobothrium on tibia I at 4% proximally, tarsus I with >40 pseudosegments (only distally about 10 visible), 3/4 of femora brown or pale blackish brown proximally with one brown distal annulus, tibia I with one brown annulus at proximal end and two distal annuli, tibiae II–IV with two brown proximal annuli and two distal annuli, metatarsi with one brown annulus at proximal end, leg formula I–II–IV–III. Abdomen elliptical, pale greenish yellow with a long cardiac pattern and many black irregular spots (Figure 2a). Palp (Figure 2h–k): trochanter with blunt and bulged retrolatero-ventral apophysis, shorter than femur; palpal tibia with finger-shaped prolatero-ventral modification (Figure 2h); bulb pale yellowish brown, cordiform, appendix absent (Figure 2h); uncus dark blackish brown and fist-shaped with fine scales, edge finely serrated, two hook-shaped pseudoappendices present, pseudoappendix 1 shorter than pseudoappendix 2 (Figure 2h,k); embolus unmodified, weakly sclerotized with some semi-transparent distal fringed processes (Figure 2h,k); procursus large and brown with dark blackish brown margin, distinct ventral knee present, four distal apophyses present, two thick ventral and retrolateral apophyses with hooked tips (numbered 1, 2 in Figure 2h–j); one spear-shaped (in lateral view) and broad (in frontal view) dorsal apophysis partly sclerotized and membranous with many distal processes (numbered 3 in Figure 2h–j); one thick, finger-shaped, and membranous prolateral apophysis with a pointed tip (numbered 4 in Figure 2h,i); dorsal spine absent.

Female. General appearance similar to male, habitus as in Figure 2b. Total length 4.92. Carapace: 1.63 long/1.88 wide. Eyes: AER 0.64, PER 0.70, ALE 0.17, AME 0.12, PLE 0.18, PME 0.15, ALE–PLE contiguous, AME–ALE 0.07, AME–AME 0.05, AME–PME 0.06, PME–PLE 0.06, PME–PME 0.23. Chelicera: 1.01 long/0.32 wide. Endite: 0.55 long/0.29

wide. Labium: 0.24 long/0.37 wide. Sternum: 0.87 long/1.14 wide. Legs: I 35.25 (8.88, 0.69, 8.98, 14.54, 2.16)/II 28.35 (8.04, 0.63, 7.15, 11.24, 1.29)/III 17.31 (5.02, 0.57, 4.24, 6.54, 0.94)/IV 24.08 (7.01, 0.58, 6.05, 9.25, 1.19), tibia I L/d 60. Palp: 1.17 (0.38, 0.17, 0.21, -, 0.41). Abdomen: 3.29 long/1.58 wide. Epigynum: 1.14 wide.

Legs yellowish brown, femora with two pale brown proximal annuli and two brown distal annuli, tibiae with two brown proximal annuli and two brown distal annuli, metatarsi with one brown annulus at proximal end, leg formula I–II–IV–III. Epigynum (Figure 2e,f): sclerotized, anterior epigynal plate strongly protrudent, anterior arch with median portion almost straight; small and short knob with a blunt tip. Internal genitalia (Figure 2g): anterior arch straight with a pair of semicircle-shaped pore plates and projected inward.

Habitat. The species was collected by hand on rock walls and under rocks in a hilly mixed forest (Figure 1b).

Distribution. South Korea (Deokjeokdo Island, Incheon-si) (Figure 1a).

3.2. Pholcus gangneung sp. nov.

(Figure 1a,c and Figure 3)

Type material. Holotype: ♂, Mt. Jabyeongsan, Sangye-ri, Okgye-myeon, Gangneung-si, Gangwon-do, South Korea (37.573733N, 128.961348E, alt. 119 m), 19 September 2022, C.M. Jang & S.T. Kim leg. (NIBR, #NIBRIV0000901598). **Paratypes:** 11♀♀4♂♂, same data as the holotype (KKU, #Ara_Phol_Gangneung_202201~15_PT).

Etymology. The specific name is a noun in apposition and refers to the type locality, Gangneung-si.

Diagnosis. This species can be distinguished from the other *Pholcus phungiformes* group members by the combination of the following characters: Male—embolus highly modified, sclerotized with 16–17 distal spiny process (Figure 3h,k); procursus with two simple distal apophyses present (one bifurcate and broad prolatero-ventral apophysis with pointed corners numbered 1 in Figure 3h–j, one round retrolateral apophysis numbered 2 in Figure 3h–j); two ventral spines present at the distal end (arrowed in Figure 3h–j). Female—epigynal anterior arch with median portion procurved, a small knob with a blunt tip curved downward; anterior arch chevron-shaped with a pair of kidney-shaped pore plates in the internal genitalia (Figure 3e–g). The unusual shaped and strongly sclerotized embolus of *P. gangneung* **sp. nov.** in the *Pholcus phungiformes* group is the first to be reported.

Description. Male (holotype). Habitus as in Figure 3a. Total length 4.49. Carapace: 1.54 long/1.62 wide. Eyes: AER 0.71, PER 0.77, ALE 0.18, AME 0.15, PLE 0.19, PME 0.18, ALE–PLE contiguous, AME–ALE 0.05, AME–AME 0.07, AME–PME 0.05, PME–PLE 0.05, PME–PME 0.29. Chelicera: 1.05 long/0.30 wide. Endite: 0.47 long/0.29 wide. Labium: 0.28 long/0.37 wide. Sternum: 0.84 long/1.11 wide. Legs: I 37.78 (9.71, 0.61, 9.75, 15.57, 2.14)/II 28.45 (7.32, 0.67, 6.63, 12.50, 1.33)/III 19.72 (5.36, 0.54, 4.54, 8.22, 1.06)/IV 25.55 (6.66, 0.60, 6.01, 11.02, 1.26), tibia I L/d 66. Palp: 3.39 (0.66, 0.38, 1.01, -, 1.34). Abdomen: 2.95 long/1.71 wide.

Carapace pale yellowish brown, cephalic region with pale blackish brown median and marginal bands, thoracic region with pale blackish brown radial and marginal bands (Figure 3a). Chelicera with three apophyses; blunt proximo-lateral apophysis diagonally upward and unprotrudent out of chelicera, small and pointed frontal apophysis protrudent forward, and thick and pointed distal apophysis slightly diagonally downward (Figure 3c,d). Legs pale yellowish brown, retrolateral trichobothrium on tibia I at 4% proximally, tarsus I with >35 pseudosegments (only distally about 10 visible), femora with two faint proximal annuli and two blackish brown distal annuli, tibia I with one blackish brown proximal annulus and one distal annulus, tibiae II–IV with two blackish brown proximal annuli and two distal annuli, metatarsi with one blackish brown annulus at proximal end, leg formula I–II–IV–III. Abdomen elliptical, pale yellowish red with a long cardiac pattern and many black irregular spots (Figure 3a). Palp (Figure 3h–k): trochanter with blunt and finger-like retrolatero-ventral apophysis, shorter than femur; palpal tibia with finger-shaped prolatero-ventral modification hidden by uncus; bulb pale yellowish brown, pocket-shaped, appendix absent (Figure 3h); uncus dark blackish brown and fist-shaped with a truncated edge and fine scales, edge finely serrated, pseudoappendix absent; embolus strongly modified, brown, thick, sclerotized, and elongated elliptical in prolateral view, broad spatula-shaped with 16–17 spiny processes distally such as a cactus in frontal view (Figure 3h,k); procursus large, simple, and brown with dark blackish brown margin, distinct ventral knee present, two distal apophyses present, one bifurcate (in lateral view) and broad (in frontal view) prolatero-ventral apophysis partly sclerotized and membranous with pointed corners (numbered 1 in Figure 3h–j), one round retrolateral apophysis (numbered 2 in Figure 3h–j); dorsal spine absent; two ventral spines present at distal end (arrowed in Figure 3h–j).

Female. General appearance similar to male, habitus as in Figure 3b. Total length 4.89. Carapace: 1.57 long/1.72 wide. Eyes: AER 0.59, PER 0.66, ALE 0.17, AME 0.11, PLE 0.15, PME 0.16, ALE–PLE contiguous, AME–ALE 0.03, AME–AME 0.06, AME–PME 0.04, PME–PLE 0.06, PME–PME 0.22. Chelicera: 0.94 long/0.31 wide. Endite: 0.47 long/0.28 wide. Labium: 0.26 long/0.34 wide. Sternum: 0.78 long/1.04 wide. Legs: I 33.62 (8.47, 0.66, 8.48, 13.84, 2.17)/II 23.06 (6.24, 0.65, 5.75, 9.04, 1.38)/III 16.56 (4.71, 0.60, 3.98, 6.20, 1.07)/IV 22.3 (6.35, 0.60, 5.40, 8.58, 1.37), tibia I L/d 57. Palp: 1.39 (0.39, 0.17, 0.26, -, 0.45). Abdomen: 3.32 long/1.44 wide. Epigynum: 1.12 wide.

Legs yellowish brown, femora with two pale brown proximal annuli and two blackish brown distal annuli, tibiae with two brown proximal annuli and two brown distal annuli, metatarsi with a blackish brown annulus at the proximal end, leg formula I–II–IV–III. Epigynum (Figure 3e,f): anterior epigynal plate slightly protrudent, anterior arch with median portion procurved; a small knob with a blunt tip curved downward. Internal genitalia (Figure 3g): sclerotized, anterior arch chevron-shaped with a pair of kidney-shaped pore plates.

Habitat. The species was collected by hand on artificially constructed rock walls in a mountainous mixed forest (Figure 1c).

Distribution. South Korea (Gangneung-si, Gangwon-do) (Figure 1a).

4. Discussion

This study taxonomically describes two new species belonging to the *Pholcus phungiformes* group in the genus *Pholcus* and is considered to be an important contribution to understand the Korean spider fauna. Huber described that the *P. phungiformes* group is distinguished from other species groups by the combination of the following diagnostic characters: male chelicerae with proximo-frontal apophyses (absent in *P. beijingensis*), male palpal tibia with prolatero-ventral modification, procursus with dorsal spines (absent in *P. alloctospilus*, *P. beijingensis*, *P. chiakensis*, *P. palgongensis*, *P. piagolensis*, *P. uiseongensis*, and *P. yeongwol*), appendix absent, sometimes with a pseudoappendix, and epigynum sclerotized with a knob [12,15–18]. Of the newly described spiders in the present study, *P. gangneung* **sp. nov.** has proximo-frontal apophyses on male chelicerae (Figure 3c,d), prolatero-ventral modification on male palpal tibia, and sclerotized epigynum with a knob (Figure 3e,f). In view of these diagnostic characters, *P. gangneung* **sp. nov.** is considered to belong to the *P. phungiformes* group. However, Huber described that the embolus in this species group is weakly sclerotized [12], while that of *P. gangneung* **sp. nov.** is strongly sclerotized (Figure 3h,k). Nevertheless, we tentatively placed this new species in the *P. phungiformes* group instead of erecting a new species group only with a degree of sclerotization because the diagnostic characters of *P. gangneung* **sp. nov.** are still close to this species group in the current taxonomic status. The *P. phungiformes* group is largely restricted to northeastern China, the Korean Peninsula, and Russia (Far East) [12,18]. Recently, taxonomic studies belonging to this species group have been actively conducted in Korea and China, and many species have been newly described in the past decade [8–10,16–18]. Despite many recent studies on this species group, Pholcidae is still one of the taxa that has not been explored much yet, and many species that have not yet been taxonomically studied appear to be waiting for new names in Korea. Further studies will be needed to provide more definitive answers to the difference of the embolus of *P. gangneung* **sp. nov.** within the *P. phungiformes* group.

The *phungiformes* species in the genus *Pholcus* are mostly found around rocky areas in mountainous mixed forests and is therefore an important natural enemy of forest insect pests flying around such environments. In addition to exploring the species that have not yet been discovered, it would also be important to accumulate further biological information of the *Pholcus* spiders for the understanding of the various ecological roles they play within their ecosystems.

Author Contributions: C.-M.J.: conceptualization, methodology, investigation, collection, identification, and original draft preparation. S.-T.K.: conceptualization, methodology, investigation, collection, identification, original draft preparation, project administration, and funding acquisition. J.-S.Y.: conceptualization, review and editing, and funding acquisition. Y.-S.B.: conceptualization, methodology, and review and editing. All authors have read and agreed to the published version of the manuscript.

Funding: This work was supported by grants from the National Institute of Biological Resources (NIBR) funded by the Ministry of Environment (MOE) (NIBR202203105 and 202203202), the National Research Foundation of Korea (NRF) funded by the Ministry of Education (2020R1A6A1A03041954), and the R&D Program for Forest Science Technology (2017042B10-2223-CA01) funded by Korea Forest Service (Korea Forestry Promotion Institute) of the Republic of Korea.

Data Availability Statement: Not applicable.

Acknowledgments: We are grateful to Sue Yeon Lee (Life and Environment Research Institute, Konkuk University) and Ulziijargal Bayarsaikhan (Division of Life Sciences, College of Life Sciences and Bioengineering, Incheon National University) for contributing to the fieldwork and valuable suggestions with critical comments on the original draft.

Conflicts of Interest: The authors declare no conflict of interest.

References

1. World Spider Catalog. World Spider Catalog. Version 23.5. 2022. Natural History Museum Bern. Available online: http://wsc.nmbe.ch (accessed on 11 January 2023).
2. Kim, J.P.; Yoo, S.H. One new species of genus *Pholcus* (Araneae: Pholcidae) from Korea. *Kor. Arachnol.* **2008**, *24*, 1–6.
3. Kim, J.P.; Park, Y.C. One new species of genus *Pholcus* (Araneae: Pholcidae) from Korea. *Kor. Arachnol.* **2009**, *25*, 99–103.
4. Kim, J.P.; Lee, J.H.; Lee, J.G. Two new species of the genus *Pholcus* Walckenaer, 1805 (Araneae: Pholcidae), with redescription of *P. zichyi* Kulczyński, 1901 from Korea. *Kor. Arachnol.* **2015**, *31*, 53–71.
5. Kim, J.P.; Ye, S.H. A new species of the genus *Pholcus* Walckenaer, 1805 (Araneae: Pholcidae) from Korea. *Kor. Arachnol.* **2015**, *31*, 73–80.
6. Kim, J.P.; Kim, T.W. Redescription of *Pholcus kwangkyosanensis* Kim & Park, 2009 from Korea. *Kor. Arachnol.* **2016**, *32*, 23–28.
7. Kim, S.T. Araneae. In *National Institute of Biological Resources (NIBR) (Ed.) National Species List of Korea II, Vertebrates, Invertebrates, Protozoans*; National Institute of Biological Resources: Incheon, Republic of Korea, 2019; pp. 412–443.
8. Lee, J.G.; Lee, J.H.; Choi, D.Y.; Park, S.J.; Kim, A.Y.; Kim, S.K. Four new species of the genus *Pholcus* Walckenaer (Araneae, Pholcidae) from Korea. *J. Spec. Res.* **2021**, *10*, 86–98.
9. Lee, J.G.; Lee, J.H.; Choi, D.Y.; Park, S.J.; Kim, A.Y.; Kim, S.K. Five new species of the genus *Pholcus* Walckenaer (Araneae: Pholcidae) from South Korea. *Zootaxa* **2021**, *5052*, 61–77. [CrossRef] [PubMed]
10. Yao, Z.Y.; Wang, X.; Li, S.Q. Tip of the iceberg: Species diversity of *Pholcus* spiders (Araneae, Pholcidae) in the Changbai Mountains, northeast China. *Zool. Res.* **2021**, *42*, 267–271. [CrossRef] [PubMed]
11. Korea Forest Service. *Basic Forest Statistics 2020*; Korea Forest Service: Daejeon, Republic of Korea, 2021; 371p.
12. Huber, B.A. Revision and cladistic analysis of *Pholcus* and closely related taxa (Araneae, Pholcidae). *Bonn. Zool. Monogr.* **2011**, *58*, 1–509.
13. Khmelik, V.V.; Kozub, D.; Glazunov, A. Helicon Focus. Version 8.2.0. 2006. Available online: http://www.heliconsoft.com/heliconfocus.html (accessed on 8 November 2022).
14. Shorthouse, D.P. SimpleMappr, an Online Tool to Produce Publication-Quality Point Maps. 2010. Available online: https://www.simplemappr.net (accessed on 5 January 2023).
15. Seo, B.K. Four new species of the genus *Pholcus* (Araneae: Pholcidae) from Korea. *Kor. J. Appl. Entomol.* **2014**, *53*, 399–408. [CrossRef]
16. Zhang, B.S.; Zhang, F.; Liu, J.Z. Three new spider species of the genus *Pholcus* from the Taihang Mountains of China (Araneae, Pholcidae). *ZooKeys* **2016**, *600*, 53–74. [CrossRef]

17. Seo, B.K. New species and records of the spider families Pholcidae, Uloboridae, Linyphiidae, Theridiidae, Phrurolithidae, and Thomisidae (Araneae) from Korea. *J. Spec. Res.* **2018**, *7*, 251–290. [CrossRef]
18. Lu, Y.; Chu, C.; Zhang, X.Q.; Li, S.Q.; Yao, Z.Y. Europe vs. China: *Pholcus* (Araneae, Pholcidae) from Yanshan-Taihang Mountains confirms uneven distribution of spiders in Eurasia. *Zool. Res.* **2022**, *43*, 532–534. [CrossRef] [PubMed]

Article

Allometric Equation for Aboveground Biomass Estimation of Mixed Mature Mangrove Forest

Hazandy Abdul-Hamid [1,2,*], **Fatin-Norliyana Mohamad-Ismail** [2,*], **Johar Mohamed** [1], **Zaiton Samdin** [2,3], **Rambod Abiri** [1,2], **Tuan-Marina Tuan-Ibrahim** [4], **Lydia-Suzieana Mohammad** [4], **Abdul-Majid Jalil** [2,5] **and Hamid-Reza Naji** [6]

[1] Department of Forestry Science and Biodiversity, Faculty of Forestry and Environment, Universiti Putra Malaysia, UPM Serdang 43400, Selangor, Malaysia; johar.mohamed@upm.edu.my (J.M.); rambod.abiri@gmail.com (R.A.)

[2] Laboratory of Bioresource Management, Institute of Tropical Forestry and Forest Products, Universiti Putra Malaysia, UPM Serdang 43400, Selangor, Malaysia; zaisa@upm.edu.my (Z.S.); majid@frim.gov.my (A.-M.J.)

[3] Faculty of Economy and Management, Universiti Putra Malaysia, UPM Serdang 43400, Selangor, Malaysia

[4] Forest Economics Section, Forestry Department Peninsular Malaysia, Headquarters, Jalan Sultan Salahuddin, Kuala Lumpur 50660, Federal Territory of Kuala Lumpur, Malaysia; marina@forestry.gov.my (T.-M.T.-I.); lydia@forestry.gov.my (L.-S.M.)

[5] Natural Products Division, Forest Research Institute Malaysia, Kuala Lumpur 52109, Selangor, Malaysia

[6] Department of Forest Science, Faculty of Agriculture, Ilam University, Ilam 69315-516, Iran; h.naji@ilam.ac.ir

* Correspondence: hazandy@upm.edu.my (H.A.-H.); fatinnorliyana22@gmail.com (F.-N.M.-I.)

Abstract: The disturbance of mangrove forests could affect climate regulation, hydrological cycles, biodiversity, and many other unique ecological functions and services. Proper biomass estimation and carbon storage potential are needed to improve forest reference on biomass accumulation. The establishment of a site-specific allometric equation is crucial to avert destructive sampling in future biomass estimation. This study aimed to develop a site-specific allometric equation for biomass estimation of a mix-mature mangrove forest at Sungai Pulai Forest Reserve, Johor. A stratified line transect was set up and a total of 1000 standing trees encompassing seven mangrove tree species were inventoried. Destructive sampling was conducted using the selective random sampling method on 15 standing trees. Five allometric equations were derived by using diameter at breast height (D), stem height (H), and wood density (ρ) which were then compared to the common equation. Simulations of each allometric equation regarding species were performed on 1000 standing trees. Results showed that the single variable (D) equation provided an accurate estimation, which was slightly improved when incorporated with the H variable. Both D and H variables, however, gave inconsistent results for large-scale data and imbalance of sampled species. Meanwhile, the best fit either for small-scale or large-scale data, as well as for imbalanced sample species was achieved following the inclusion of the ρ variable when developing the equation. Hence, excluding the H variable while including the ρ variable should be considered as an important determinant in mixed mangrove species and uneven-aged stand for aboveground biomass estimation. This valuation can both improve and influence decision-making in forest development and conservation.

Keywords: mangrove; aboveground biomass; tree component; allometric equation; power function

Citation: Abdul-Hamid, H.; Mohamad-Ismail, F.-N.; Mohamed, J.; Samdin, Z.; Abiri, R.; Tuan-Ibrahim, T.-M.; Mohammad, L.-S.; Jalil, A.-M.; Naji, H.-R. Allometric Equation for Aboveground Biomass Estimation of Mixed Mature Mangrove Forest. *Forests* **2022**, *13*, 325. https://doi.org/10.3390/f13020325

Academic Editors: Konstantinos Poirazidis and Panteleimon Xofis

Received: 13 December 2021
Accepted: 10 February 2022
Published: 16 February 2022

Publisher's Note: MDPI stays neutral with regard to jurisdictional claims in published maps and institutional affiliations.

1. Introduction

Mangrove forests play unique ecological functions in subtropical coastal regions [1–4]. It takes years for the ecosystem to reach the maturity phase to facilitate the provision of providing essential services, such as fisheries, timber and fuelwood production [5,6], habitat protection [7], coastal defense [8], and carbon sink production in the tropics [9–11]. In compliance with the Reduced Emission from Deforestation and Land Degradation (REDD+) [12,13], the Malaysian Government must provide the national Forest Reference

Level of mangrove forest biomass productivity and carbon stock to the United Nations Framework Convention on Climate Change (UNFCC). The proper and accurate estimation of forest biomass is one of the 22 elements to determine the Total Economic Valuation (TEV) for forest ecosystems and services based on the framework by The Economics of Ecosystem and Biodiversity (TEEB) [14] as explained by De Groot et al. [15].

The mangrove forest geomorphological condition is limited in terms of accessibility, time consumption, and in posing a threat to worker's safety. In relation to the tidal water, mangrove root systems such as the numerous massive stilt roots, knee roots and pneumatophores systems that outgrow the trunk and grow vertically above the soil, these may threaten worker's safety. Nevertheless, the aboveground biomass of mangrove forests still needs to be estimated, considering their exceptional roles and services to the environment.

There is a growing interest in estimating forest composition by using Unmanned Aerial Vehicles (UAVs), such as drones and remote sensing in capturing forest imagery based on forest canopy that can be further described based on color features. The use of UAVs is advantageous, especially in forest areas that are difficult to access. Additionally, the use of drones and remote sensing is vital in managing and monitoring forests from undesirable practices, such as illegal logging. However, UAVs are unreliable is estimating certain crucial data for determining forest carbon stock. Examples of these data include wood density and biomass by tree component (trunk, branch, leaves, flowers, and fruits) that support the aboveground biomass accumulation. On the other hand, different definitions of forest canopy height (such as the mean height of all trees, basal-area weighted height, or height of the tallest tree within a certain area) that are generated from airborne lasers/Lidar (ALS), lead to contracting results from the same datasets, [16–18] as regular inundation might result in an error in the height retrievals [19].

Mathematically, various methods have been developed by forest ecologists in estimating forest biomass, involving the relationship between the biomass of whole trees and, their components, as well as some readily measured parameters such as the diameter of the stem at breast height (DBH), stem height (H), and wood density (ρ). One of the methods is by developing allometric equations with destructive biomass, which requires a small number of tree samples to be harvested and the estimation may be performed by either the whole or partial tree weight from the measurable tree dimension. This method was preferred instead of the destructive and mean-tree methods that require the harvest of all the trees in developing the equation [11,20–25].

Allometric relationships between the aboveground biomass and the DBH parameter have been reported for specific mangrove species, such as *Rhizophora apiculata* (L.) Blume [26,27] and *Bruguiera parviflora* (R.) Wight [26]. The present study focused on determining the biomass accumulation on a natural mature mangrove forest occupied by mixed species. This study will improve the current knowledge on forest conditions facing other land areas (Singapore) or mangrove islands, where possibly seed sources are received from the nearest forest, rather than the areas directly facing the vast ocean of rough tides, huge waves, strong winds, and tropical storms, such as typhoons and hurricanes. By using primary data, the verified single developed equation (with several relationships) that fits the available mangrove species could help in efficiently managing forest. It will also provide the relevant figures of forest canopy layers and forest profiles regarding biomass accumulation and carbon stock at this forest area, instead of applying single species models that may require higher cost and are time consuming in data inventory, gathering, and presenting.

Nonetheless, in ensuring the validity of the equation that will be developed, there is a need to have an existing equation to rely on by considering the geographical origin and species that make up the data set of the derived equation. A previous study conducted by Hazandy et al. [28] developed allometric equations for estimating aboveground biomass in the Matang Forest Reserve (northern part of Peninsular Malaysia), however, the researchers found the equation to be less suitable for application in the present study. Moreover, the equation developed by Hazandy et al. [28] focused on the even-age planted mangrove. In this study, the equation developed by Komiyama et al. [11] known as 'the common equation' was applied for both practical and comparison application due to the segregation of the species and the similarity of the study site conditions to that of the Asian region.

Apart from the D and H, the wood density (ρ) is one of the important enablers in estimating aboveground biomass as it differs significantly among various mangroves species [21]. A lower difference of ρ is only found for various individuals within a species [29]. This study aimed to develop a site-specific allometric equation by considering tree wood density ρ in relationship to DBH and H for a mixed mature mangrove forest in the Sungai Pulai Forest Reserve in the southern part of the Peninsular Malaysia.

2. Materials and Methods

2.1. Sampling

The site for the study is located at Sungai Pulai Forest Reserve; in the southeast of Pontian and Johor Bahru district (01°27′ N, 103°33′ E). It is the largest mangrove forest in Johor state and the second largest in Peninsular Malaysia. Ground-truthing (visual assessment) was conducted to identify high, medium, and low standing tree distribution to ensure that the range of biomass is sampled [30,31]. The transect line was set up across Compartments 16, 259A, 412B, and 453B Sungai Pulai Forest Reserve, Johor (Figure 1). Plot establishments were performed near the river or the estuary. For each compartment, two plot designs of 50 × 50 m each (total plot area = 2 hectares) were randomly established from the marine to the center of the compartment [31]. Parameters such as D and H were measured for trees of 5 cm diameter and above, and the species were identified. In the total of 2 hectares (ha), a total of 1386 standing trees were inventoried in the plotted area of 2 hectares (ha). Thereafter, 1000 of the standing trees were randomly selected to derive the perform equation.

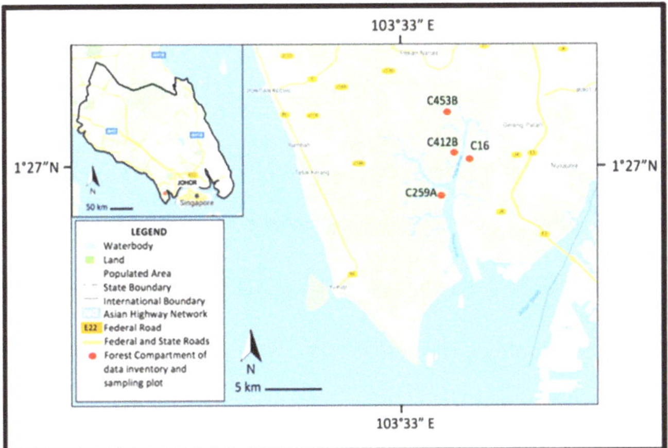

Figure 1. Study area in the Sungai Pulai Forest Reserve showing the forest compartment of the data inventory sampling plot.

The D was measured using the Forestry Suppliers Metric Fabric Diameter Tape (Model 283D/10M, Jackson, MS, USA), whereas the H was measured using the Suunto Compass Clinometer (Model PM-5, Tammiston Kauppatie, Vaanta, Findland). Main stem tree diameter was measured at diameter breast height, approximately 1.37 m above the ground. Stem diameter is often measured above the highest stilt root for stilt rooted trees such as *Rhizophora* spp. [29,32]. Two-compartments (Compartment 16 and 259A) were then selected to perform the destructive sampling by considering the suitability of the accuracy of biomass distribution, sampling time, and cost efficiency [31]. A selective random sampling method was used in selecting 15 standing trees that represented seven mangrove species (Table 1) based on the normal distribution of diameter at breast height, stem height (Figure A1 in Appendix B), and species composition.

Table 1. The number of sampled tree species and range of the diameter and height.

Species	Local Name	Sample Trees	D (cm)	H (m)	Tree Density
Bruguiera cylindrica (L.) Blume	Bakau putih	3	17.13 ± 3.85	17.17 ± 1.23	715.62 ± 10.2
Bruguiera parviflora (R.) Wight	Lenggadai	1	18.7 ± 0	18.0 ± 0	736.81 ± 0
Bruguiera sexangula (L.) Poir	Tumu putih	2	24.9 ± 1.10	20.50 ± 0.50	713.72 ± 7.87
Ceriops tagal (P.) Rob	Tengar	1	15.5 ± 0	14.5 ± 0	745.45 ± 0
Rhizophora apiculata (L.) Blume	Bakau minyak	5	20.08 ± 4.21	19.92 ± 1.69	842.22 ± 28.5
Rhizophora mucronata (L.) Lam	Bakau kurap	1	16.8 ± 0	18.1 ± 0	801.41 ± 0
Xylocarpus granatum (J.) Koenig	Nyireh bunga	2	14.70 ± 4.10	13.88 ± 2.63	708.65 ± 17.9

2.2. Sample Preparation

The fresh weight of each component (trunk, branches, and leaves) from the 15 destructive trees was measured in situ using the Brecknell 235 10X 300 kg mechanical hanging scale. For all trunk sample trees, the trunk diameters were measured at the lower, middle, and upper part of the log that was cut from ground level in the following lengths: (0–0.3 m, 0.3–1.3 m, 3.3–5.3 m and followed then for every 2 m lengths) [29]. The log was cut into several cutting logs—modified from Doruska et al. [33] to avoid exceeding the load of the weight balance as the trunk was assumed to be conical in shape [29] and to standardize the cutting log length for each tree. The lower part from each log sample was cut into a disc form (2 cm to 3 cm thickness) to obtain the range of ρ because the variation of ρ in individual species correlates with carbon allocation [34] and effective vertical stem expansion [35]. This could reduce the error with the different scaling factors among vertical log positions such as butt log, middle log, and upper log [33]. Thereafter, 500 g from the fresh weight of branch, twig, and leaf components were taken to the laboratory [29]. The samples were then oven-dried for 15 days at 70 °C until a constant weight was attained to calculate the dry weight conversions for each component [36]. The ρ of the wooden disc (mass contained in a unit volume) [37] was determined from the stem and the largest branches [28].

2.3. Approach in Biomass Estimation

The biomass of each component was calculated by $B = M_{Fre} \times (M_{sam,Dry}/M_{sam,Fre})$, where M_{Fre} is the fresh mass of each component and $M_{sam,Dry}$ and $M_{sam,Fre}$ are the dry mass and fresh mass of the samples of each component, respectively. The total aboveground biomass was obtained by summing the biomasses of each component.

The data were fitted to a non-linear regression model in the form of an intrinsically linear model of the power function, whereas the accuracy of the calculation was based on the coefficient of determination (R). The equation was derived as a single parameter by combining the diameters at breast height, stem height, and wood density to determine the variability explained by the model [38]. The equations were simplified as in the model shown below:

$$M = aD^b e \tag{1}$$

$$M = aD^2H^b e \tag{2}$$

where,

M = Aboveground biomass
D = Diameter at breast height
H = Stem height
a and b = constant
e = error term

The allometric relationship of Equation (3) in a linear form was derived by taking the natural logarithms of both sides of the equation [38]:

$$\ln M = \ln a + b \ln \rho + b \ln D + \ln e \tag{3}$$

where,

ρ = wood density

In this form, linear regression can be used to build the regression model that fits the biomass data. The calculated result is aimed at presenting the known values (common equation), whereas the percent error formula is used to determine the precision of the calculations. The experimental value is the calculated value while, the theoretical values are the common equation value. All the parameters used to develop the equation are influenced by measurement error, notably having different effects on the model parameters [39]. Meanwhile, the absolute error is the magnitude of the difference between the actual value and the estimated value. According to Bellasen and Stephan [40], an error value is acceptable if it is lower than 10% at a 90% confidence interval with an uncertainty factor of 1.5%.

The formula of the percentage of error is presented below:

$$\left[\frac{(\text{Theoretical} - \text{Experimental})}{\text{Theoretical}} \right] \times 100 \tag{4}$$

where,

Theoretical = Known value
Experimental = Calculated value

2.4. Statistical Analysis

Further, the allometric equations of 15 destructive samples were derived using the General Linear Model, IBM SPSS statistic software version 25.0, IBM Corp, Armonk, NY, USA using the following: diameter at D, H, and ρ variables, the aboveground biomass (kg) comparison between the observed value and predicted value for the equation using D, D^2H and ρ variables, homoscedasticity of residuals, and the developed equations of aboveground biomass (kg) estimation regarding species of 1000 trees in a 1 ha plot inventory of the Sungai Pulai Forest Reserve. The developed equations of aboveground biomass (kg) estimation regarding species of 1000 trees in the 1 ha plot inventory of Sungai Pulai Forest Reserve were derived using SigmaPlot Version 12.5, Systat Software, Inc., San Jose, CA, USA.

3. Results and Discussion

3.1. Allometric Equations Derived from 15 Destructive Trees

The allometric equation derived for the relationship between D, H, and ρ variables from 15 destructive sampling trees of seven species was in the range of DBH from 9.0 cm to 33.0 cm. Equations (1) and (2) were derived for each part of the tree samples (mass of stem, leaves, branch, and twigs) in a single D variable and a combination with the H variable (Table 2). The combination of the variables in the second equation was derived to study the level or degree of variability explained in the biomass accumulation on the stem, leaves, branch, and twigs. The equation derived is accepted to be normal distributed as the standardized and unstandardized residual normality were of the same value and greater than 0.05 (Table 2). Regardless of species, the results in Table 2 indicate that the R^2 values were in the range 0.5834 to 0.9543 for Equation (1) (single D variable) and from 0.553 to 0.9556 for Equation (2) (incorporating the D and H variables).

Table 2. Summary of the allometric equation derived from 15 destructive samples using diameter at breast height, stem height, and wood density variables.

Model	No. of Equation	Component		Equation	R^2	Standardized and Unstandardized Residual Normality
$M = aD^b$	1	(a)	M of stem	$0.1761D^{2.3769}$	0.9223	0.107
		(b)	M of branches and twigs	$0.0553D^{2.3055}$	0.7792	0.499
		(c)	M of leaves	$0.0347D^{1.9762}$	0.5834	0.206
$M = aD^2H^b$	2	(a)	M of stem	$0.0355D^2H^{0.9778}$	0.9579	0.833
		(b)	M of branches and twigs	$0.0221D^2H^{0.8745}$	0.6880	0.853
		(c)	M of leaves	$0.0125D^2H^{0.7767}$	0.5530	0.246
$M = aD^b$	1		Total biomass	$0.2999D^{2.3001}$	0.9543	0.117
$M = aD^2H^b$	2		Total biomass	$0.0739D^2H^{0.9291}$	0.9556	0.813
$M = a\rho^{b1}D^{b2}$	3		Total biomass	$0.00475\rho^{0.6309}D^{2.28787}$	0.9697	0.724

Note: M = biomass; D = diameter at breast height; H = stem height; ρ = wood density; a and b = constant.

The results revealed that the biomass of different tree components could be estimated using power equations based on a single D variable (stem R^2 = 0.9223; branch R^2 = 0.7792; and leaves R^2 = 0.5834) and the combination of D and H variables (stem R^2 = 0.9579; branch R^2 = 0.688; and leaves R^2 = 0.553). Equations (1) and (2) explained the aboveground biomass accumulation in tree components, in which the stem biomass was allocated the biggest biomass, followed by branches and twigs. This finding depicts that by either ignoring or excluding branch and twig samples during destructive sampling, the resulting equation becomes invalid. Meanwhile, the lowest R^2 value for both Equations (1) and (2) (Table 2) was from the leaf component, showing that crown distribution plays a minimum role in biomass allocation as highlighted by Komiyama et al. [11]. The exclusion of leaf biomass might be considered, however, the development of the equation remains inaccurate.

The equations that incorporated the H variable (Equation (2)) were found to be slightly higher compared to the equations using a single variable, D (Equation (1)). Meanwhile, the equation that incorporated both D and ρ variables (Equation (3)) resulted in the highest R^2 value (0.9697) compared to the equation that incorporated both the D and H variables, and the single D variable. Independent *t*-test (Table 3) also indicates that the allometric equations derived were statistically significantly different. Thus, the inclusion of the ρ variable to develop an allometric equation is not significant for aboveground biomass estimation that involves low species variation (planted forest) [36], nonetheless, the ρ variable must be considered in the biomass estimation of a variety of species, especially for uneven age mangrove forest [32,41,42].

Table 3. Summary of the independent samples t-test of the allometric equation derived from 15 destructive samples using diameter at breast height, stem height, and wood density variables.

No. of Equation		N	Mean	SD	SE	t-Value	p-Value	95% Lower Bound	95% Upper Bound
1	(a)	15	5.0923	0.8731	0.1914	12.4186	1.3858×10^{-8}	1.9634	2.7904
	(b)	15	3.7284	0.8470	0.3404	6.7726	1.3161×10^{-5}	1.5701	3.0410
	(c)	15	2.3163	0.7260	0.4632	4.2667	0.0009	0.9756	2.9768
2	(a)	15	5.0923	0.8898	0.0569	17.1905	2.5414×10^{-10}	0.8549	1.1007
	(b)	15	3.7293	0.7942	0.1636	5.3369	0.0001	0.5216	1.2274
	(c)	15	2.3163	0.7068	0.1937	4.0105	0.0015	0.3583	1.1951
1		15	5.4042	0.8449	0.1395	16.4836	4.2895×10^{-10}	1.9986	2.6015
2		15	5.4042	0.8455	0.0555	16.7277	3.572×10^{-10}	0.8091	1.0491
3		15	5.4042	0.8517	0.1184; 0.2555	19.3310; 2.4701	2.0743×10^{-10}; 0.0295	1.9838; −1.2527	1.6592; 2.6144

The scatter plot (Figure 2) of predicted value and residual value shows how much of an error the regression equation made with respect to predicting individual values in the dataset. The result indicates that the distribution of the residual values (dependent variable) is distributed uniformly and does not have any clusters forming together. The average proportions of biomass allocation of 15 destructive samples were 74% in stems, 20% in branches, and 6% in leaves. This result is consistent with the study conducted by Gong and Ong [43] on other mangrove forests in Peninsular Malaysia. Biomass accumulation of standing trees is relatively higher for the structural tissue and lower for the leaves [26,44–46]. This is due to a gradual increase in the absolute mass of stem and branches while the absolute mass of leaves tends to stabilize or be shed as litter upon attaining a certain tree size [47–49].

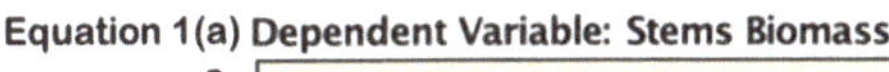

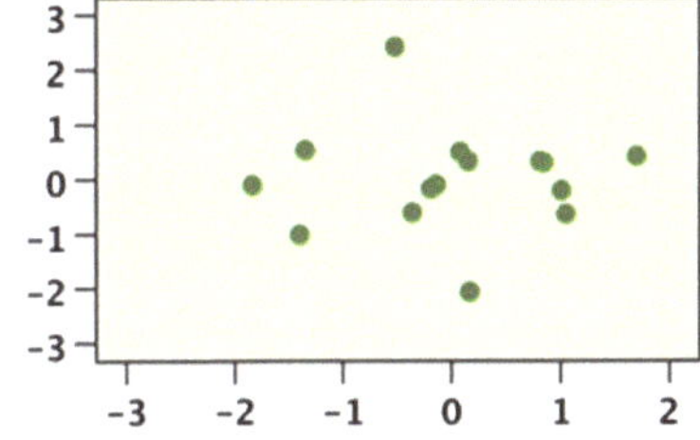

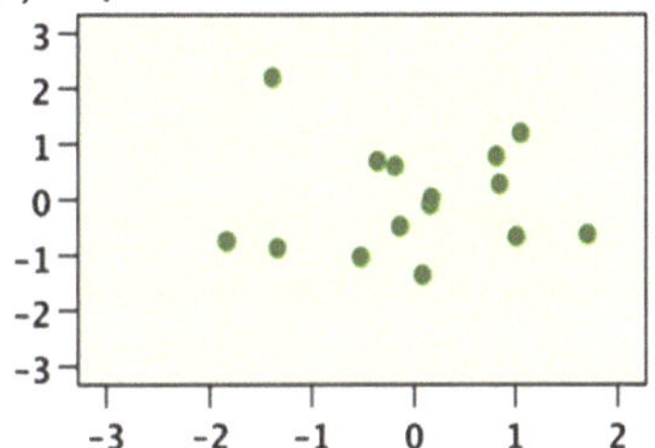

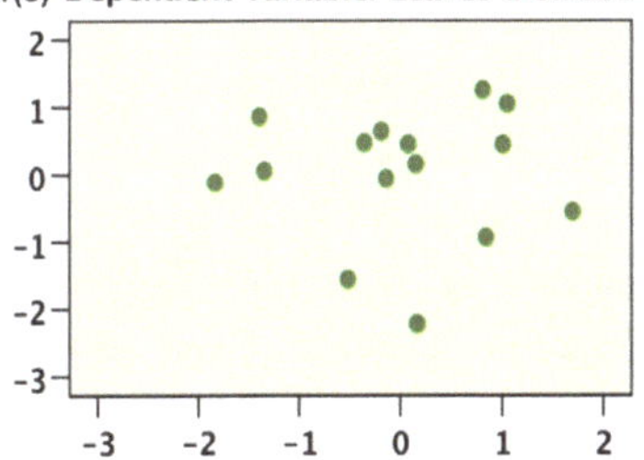

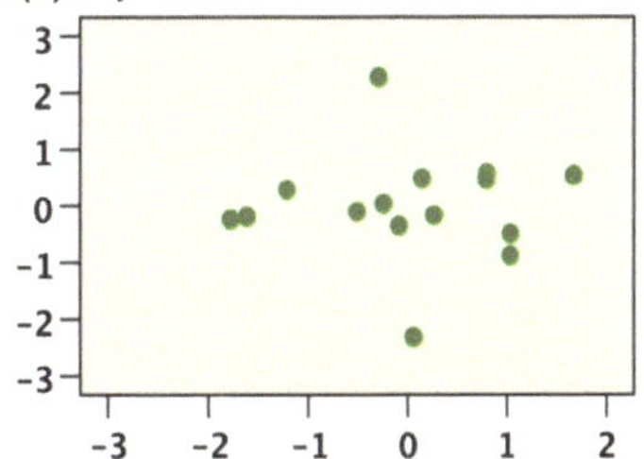

Figure 2. *Cont.*

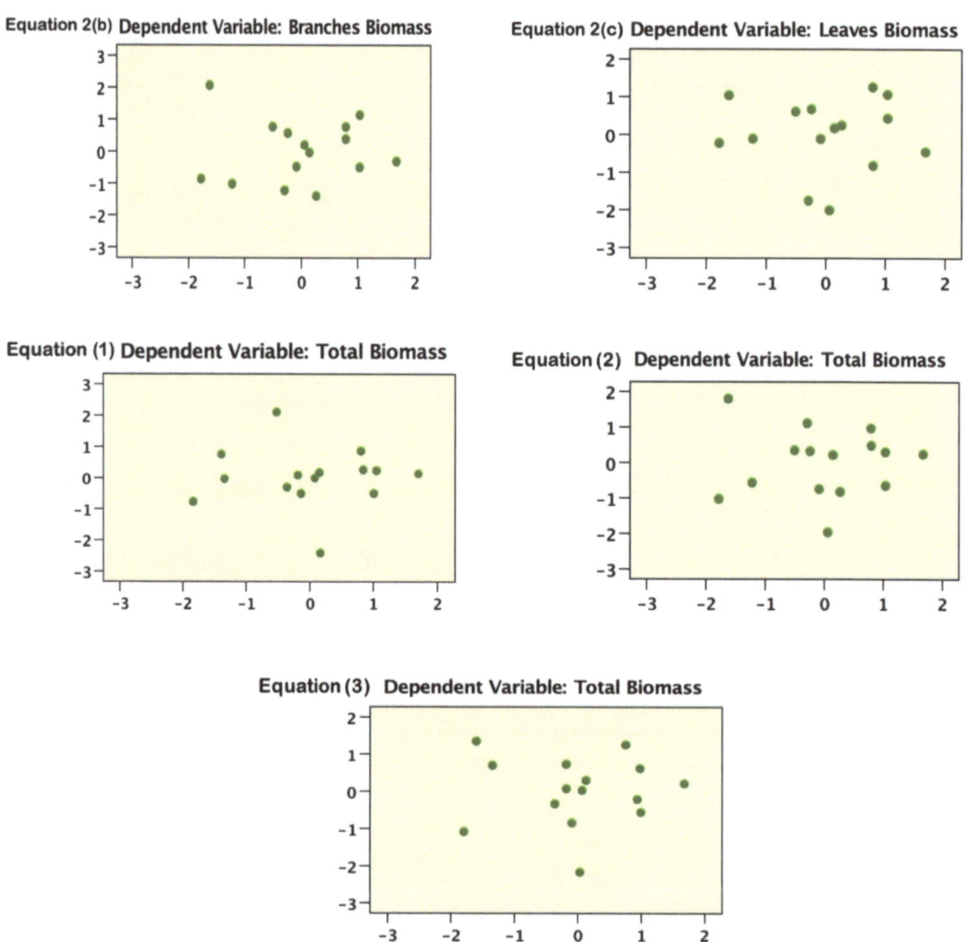

Figure 2. The test of homoscedasticity of residuals.

Five developed equations of 15 destructive trees were then regressed between the observed value (total biomass collected from the field) and the predicted value (D, H, and ρ value that were applied to the developed equation) (Figure 3). Regardless of species, the R^2 single variables (Equation (1)) were in the range of 0.9717 to 0.9718 and were found to be slightly increased (from 0.9717 to 0.972) for incorporated variables (Equation (2)). Meanwhile, the estimation of total aboveground biomass was similar upon applying the single D variable (Equation (1)) and incorporated variables (Equation (2)).

These two estimations (observed versus predicted) indicate the presence of a strong relationship between the tree variables (D and D^2H) for mixed species and uneven age of mangrove stand. Smith and Whelan [50] also found a good relationship between stem height and tree biomass in Florida mangroves and similar results were obtain when the diameter variable was used. Henry and Aarssen [51] reviewed the regressions between stem diameter and height and reported a lack of uniformity that might be influenced by biomechanical constraints and near neighbor effects. Comparatively, Equation (3) (0.9760) revealed that small destructive sampling (15 trees) with the inclusion of ρ variable provided a reliable and validated equation instead of the common equation ($R^2 = 0.9730$) (Figure 3) developed by Komiyama et al. [29] that used 104 destructive sampling trees (Appendix A).

To ascertain if the developed equation could fit large-scale data [51], the equation was applied to the primary data of 1000 standing trees.

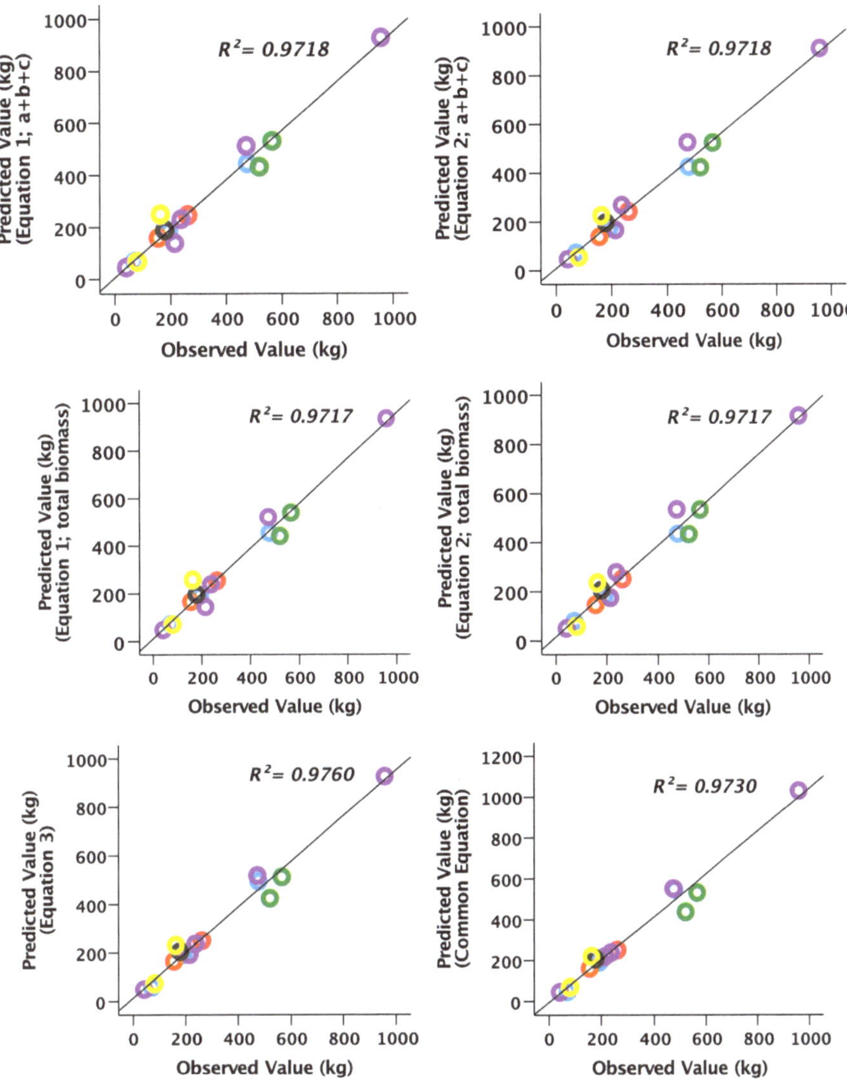

Figure 3. Comparison of the observed value and predicted value of aboveground biomass (kg) for a single variable (D) and incorporated variables (D^{2}H, and D and ρ variables) of 15 destructive trees. Blue: *Bruguiera cylindrica* (L.) Blume, red: *Bruguiera parviflora* (R.) Wight; green: *Bruguiera sexangula* (L.) Poir; orange: *Ceriops tagal* (P.) Rob; purple: *Rhizophora apiculata* (L.) Blume; dark grey: *Rhizophora mucronata* (L.) Lam; and yellow: *Xylocarpus granatum* (J.) Koenig.

3.2. Allometric Equations Applied to 1000 Standing Trees

Primary data inventory of seven species with a total of 1000 standing trees in Sungai Pulai Forest Reserve are listed in Table 4 with normal distributions of DBH and H as in Appendix B. Regardless of species, the range of D of 1000 standing trees in the study site was smaller but was still in the range of those used by Komiyama et al. [29] to develop a common equation (Appendix A). The reason for the limited size in the study site could be due to the regeneration process of an uneven-aged stand from the previous harvest of charcoal production [52]. The *R. apiculata* species (553 individual trees) dominated a total of 1 ha area, followed by *Bruguiera cylindrica* (L.) Blume (185 trees), *Bruguiera sexangula* (L.) Poir (131 trees), *Rhizophora mucronata* (L.) Lam (44 trees), *B. parviflora* (40 trees), *Xylocarpus granatum* (J.) Koenig (39 trees), and *Ceriops tagal* (P.) Rob (8 trees). Species distribution in the study site revealed that *R. apiculata* accounted for 65.5% of the 1000 individual trees (Table 3).

Table 4. Summary of descriptive statistics for aboveground biomass 1000 standing trees in Sg. Pulai Forest Reserve.

Species	No. of Tree	Diameter (cm)			Height (m)		
		Min	Max	Mean	Min	Max	Mean
B. cylindrica	185	5	31.5	[1] 15.09 ± 10.38	4.2	30	15.99 ± 0.33
B. parviflora	40	6.9	34	18.05 ± 1.13	6.5	28	18.75 ± 0.77
B. sexangula	131	6.1	34	16.35 ± 0.39	4.1	28	17.25 ± 0.30
C. tagal	8	13	25.2	19.45 ± 1.40	15	26	19.32 ± 1.26
R. apiculata	553	9.9	40.5	20.03 ± 0.24	10	35	19.51 ± 0.16
R. mucronata	44	11	27.8	17.38 ± 0.65	11	24	18.16 ± 0.42
X. granatum	39	[2] 2.4	43.2	19.74 ± 1.21	5.2	24	14.83 ± 0.71

Note: [1] ± represent standard error, [2] The minimum value for *X. granatum* is below 5 cm diameter because the data were obtained from multiple leader trees.

A bigger dimensional size (mean D = 20 cm; mean H = 19.5 m) was also found from *R. apiculata* species. Likewise, previous studies showed that *R. apiculata* was the dominant species in other mangrove forests in Peninsular Malaysia and East Malaysia [26,53–56]. *B. cylindrica*, on the other hand, was documented to have the smallest dimensional size (mean D = 15 cm; mean H = 15.9 m) despite being the second-highest in the total number of individual trees. Plot establishments were performed near to the river, and the estuary because species distribution may be dominant for the seaward zone and mid-zone of mangrove species. However, it yielded an unbiased effect to the development of the equation due to the un-even age of the natural regeneration mix species. Moreover, the number of destructive samples on each species was based on the number of trees of each species in the primary inventory.

To perform in-depth study, to ascertain the internal consistency of the developed equation, the equation was applied to the primary data of 1000 standing trees regardless of species and in comparison, to the common equation (Figure 4). The best fit was found for Equation (3) with an R^2 value of 0.9979 and the lowest percentage error of 5.93% (Figure 4). The fitting (R^2 value) for both Equation (1) (R^2 = 0.9892) and Equation (3) (R^2 = 0.9888) were slightly lower than the Equation (3) with corresponding percentages error of 10.06% and 9.94%, respectively. The inclusion of the H variable for Equation (2) yielded the lowest data fit (R^2 = 0.9205 and 0.9196 respectively) and the highest percentage error (15.42% and 15.62% respectively). Haase and Haase [57], Rayachhetry et al. [58], and Chave et al. [41] reported that it is not advisable to include the H variable for equation development. Furthermore, Novitzky [59] found that the H increases with an increase in temperature and precipitation as the latitude decreases. On the other hand, Kodikara et al. [60] stressed that the effect varied with the soil salinity.

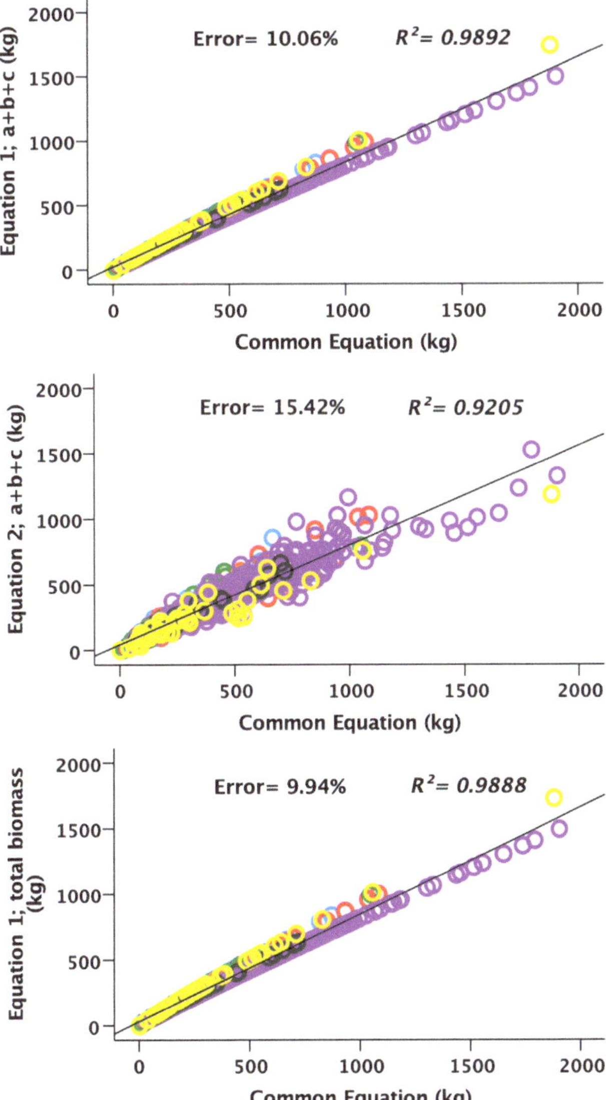

Figure 4. *Cont.*

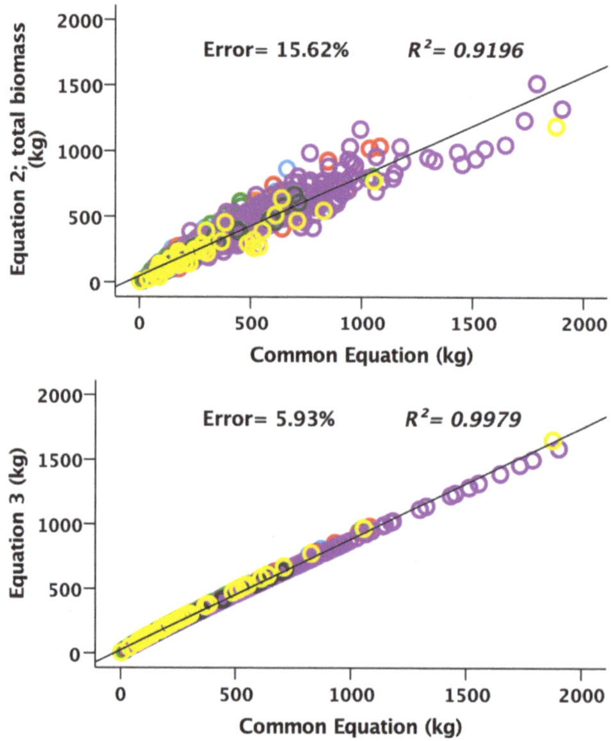

Figure 4. Five developed equations of aboveground biomass (kg) estimation of 1000 standing trees regarding species in the 1 ha plot inventory of Sungai Pulai Forest Reserve. Blue: *B. cylindrica*, red: *B. parviflora*; green: *B. sexangula*; orange: *C. tagal*; purple: *R. apiculata*; dark grey: *R. mucronata*; and yellow: *X. granatum*.

The H variable is not a good estimator for biomass estimation, and it produced the highest percentage error (>15%) when implemented in a large data set (Figure 4). Conversely, the combination of D and ρ variables resulted in the largest improvement (R^2 values close to one) for aboveground biomass estimation and showed the lowest percentage error (<6%) (Figure 4). These findings align with the earlier observations by Putz and Chan [27] and Ong et al. [26] that D and ρ variables for mangrove species provide a reliable means of estimating aboveground biomass (Equation (3)).

Further regression analysis was performed for the developed equation regarding the species to depict the values of which species were overestimated or underestimated for large-scale data estimation. Figure 5 indicates that the Equation (3) recorded the highest R^2 value (0.9943), followed by both equations of the single D variable (Equation (1) R^2 = 0.9337; Equation (3) R^2 = 0.9355). The inclusion of the H variable for Equation (2) yielded the lowest R^2 value (0.8388 and 0.8367, respectively). Equation (2) in Figure 5 showed inconsistent results when the H variable was incorporated. This may be due to the difficulty of stem height measurement in situ, the mangrove soil condition (muddy soil), and the tidal water that always introduce a bias and greater inaccuracy of measurements. Meanwhile, the regression line indicates the equation over-estimated and under-estimated biomass at low observed value and high observed value, respectively (Figure 5).

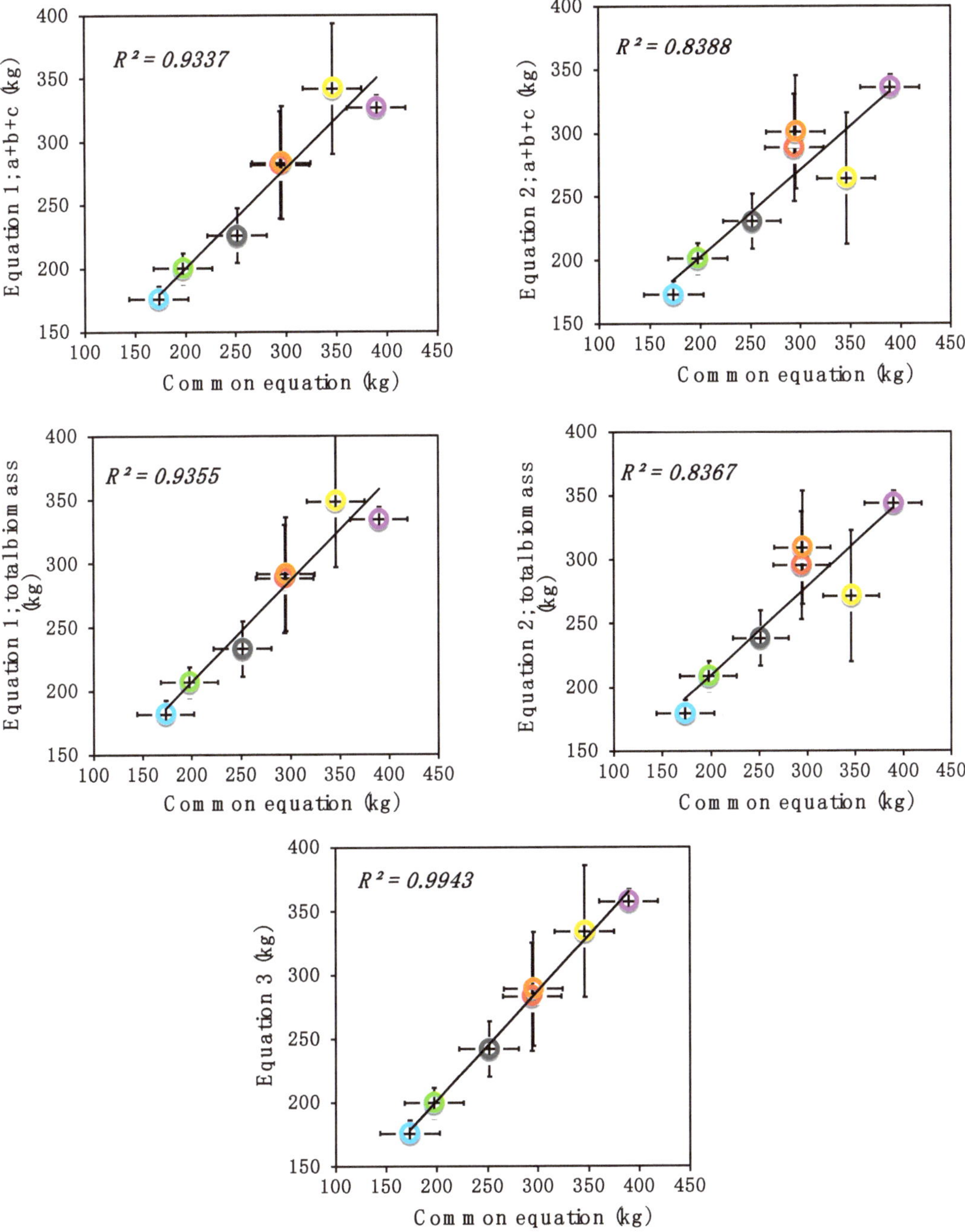

Figure 5. Five developed equations of aboveground biomass (kg) estimation of 1000 standing trees regarding species in the 1 ha plot inventory of Sungai Pulai Forest Reserve. Blue: *B. cylindrica*, red: *B. parviflora*; green: *B. sexangula*; orange: *C. tagal*; purple: *R. apiculata*; dark grey: *R. mucronata*; and yellow: *X. granatum*.

The estimation point (species) located under the regression line yielded an underestimate of the biomass, meanwhile the point (species) located above the line yielded an over-estimate of the biomass. All equations showed both underestimation and overestimation of the total biomass of the seven species except Equation (3), which yielded well-fitted data for the seven species (Figure 5). Meanwhile, Figure 4 indicates that the biomass estimation was accurately estimated in the large data set regardless of the species by considering the ρ variable. Other equations fitted the data only when the ρ variable was included. In contrast, excluding the ρ variable in the developed equation yields an inaccurate model both for small-scale (15 trees) and large-scale data (1000 trees), as the species sample becomes imbalanced. Thus, the ρ variable is important in reducing the error in the biomass estimation. Therefore, the best-fit equation for all seven species was found for Equation (3). The developed equation of mixed species and uneven age in the study site provides a different successful establishment of the equation (Table 2), as compared to the single species equation [61,62].

4. Conclusions

The single variable (D) equation provides an accurate estimation, which is slightly improved when incorporated with the H variable. The exclusion of the variable H might be considered on time consuming grounds and for difficult events, however, both D and H variables show inconsistent results for large-scale data and imbalanced sample species. Meanwhile, the best fit either for small-scale or large scale-data, as well as for imbalanced sample species was achieved following the inclusion of the ρ variable. We suggest that the ρ variable should be considered as an important determinant variable in mixed mangrove species and uneven-aged stand for aboveground biomass estimation. This valuation can both improve and influence decision-making in forest development and conservation.

Author Contributions: Conceptualization, H.A.-H., F.-N.M.-I., J.M., Z.S., T.-M.T.-I. and L.-S.M.; methodology, H.A.-H., F.-N.M.-I., R.A., Z.S. and H.-R.N.; data acquisition, H.A.-H., F.-N.M.-I., J.M. and A.-M.J.; formal analysis, H.A.-H. and F.-N.M.-I.; writing-original draft preparation, H.A.-H., F.-N.M.-I. and J.M.; Writing-review and editing, H.A.-H., F.-N.M.-I., J.M., Z.S., R.A., T.-M.T.-I., L.-S.M., A.-M.J. and H.-R.N.; supervision, H.A.-H. All authors have read and agreed to the published version of the manuscript.

Funding: This research received no external funding.

Acknowledgments: We would like to express our gratitude to the Forestry Department Peninsular Malaysia for the assistance and permission to conduct this study at Sungai Pulai Forest Reserve, Johor, Malaysia.

Conflicts of Interest: The authors declare no conflict of interest.

Appendix A

Table A1. List of allometric equations for estimating aboveground biomass (ABG) by using DBH (D), height (H) and wood density (ρ).

Species Group	Equation	R^2	N	Data Origin	D Max (cm)	Source
General equation	$B = \rho \times \exp\,[-1.349 + 1.980 \times \ln(D) + 0.207 \times (\ln(D))^2 - 0.0281 \times (\ln(D))^3]$	unknown	84	Americas	42.0	Chave et al. [41]
General equation	$B = 0.168 \times \rho \times (D)^{2.471}$	0.99	84	Americas	42.0	Chave et al. [41]; Komiyama et al. [11]
General equation	$B = 0.251\,\rho\,(D)^{2.46}$	0.98	104	Asia	49.0	Komiyama et al. [29]

Table A1. *Cont.*

Species Group	Equation	R^2	N	Data Origin	D Max (cm)	Source
	Specific tree equations—Asia-Pacific region					
Rhizophora apiculata	$B = 0.1709D^{2.516}$	0.98	20	Malaysia	30.0	Putz & Chan [27]
Rhizophora apiculata	$B = 0.043D^{2.63}$	0.97	34	Indonesia	40.0	Amira [63]
Rhizophora apiculata (wood mass)	$B_{wood} = 0.0695D^{2.644} \times \rho$	0.89	191	Micronesia	60.0	Modified from Cole et al. [64]; Kauffman & Cole [65]
Xylocapus granatum	$B = 0.1832D^{2.21}$	0.95	30	Indonesia	41.0	Tarlan [66]

Appendix B

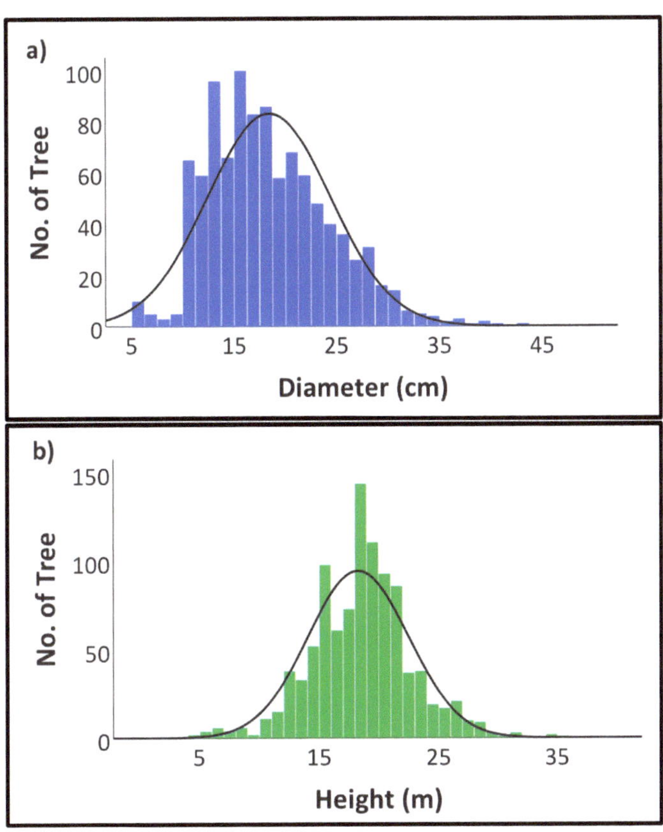

Figure A1. Histogram of (**a**) the diameter and (**b**) the height distribution of 1000 standing trees in the 2-ha plot of Sungai Pulai Forest Reserve.

References

1. Ball, M.C. Ecophysiology of mangroves. *Trees* **1988**, *2*, 129–142. [CrossRef]
2. Twilley, R.R. Coupling of mangroves to the productivity of estuarine and coastal waters. In *Coastal-Offshore Ecosystem Interactions: Lecture Notes on Coastal and Estuarine Studies*; Jansson, B.O., Ed.; American Geophysical Union Publisher: Washington, DC, USA, 1988; pp. 165–202. [CrossRef]
3. Baran, E.; Hambrey, J. Mangrove conservation and coastal management in Southeast Asia: What impact on fishery resources? *Mar. Pollut. Bull.* **1999**, *37*, 431–440. [CrossRef]
4. Primavera, J.H. Tropical shrimp farming and its sustainability. *Trop. Maric.* **1998**, *8*, 257–289. [CrossRef]
5. Aburto-Oropeza, O.; Ezcurra, E.; Danemann, G.; Valdez, V.; Murray, J.; Sala, E. Mangroves in the Gulf of California increase fishery yields. *Proc. Natl. Acad. Sci. USA* **2008**, *105*, 10456–10459. [CrossRef]
6. Dahdouh-Guebas, F.; Mathenge, C.; Kairo, J.G.; Koedam, N. Utilization of mangrove wood products around Mida Creek (Kenya) amongst subsistence and commercial users. *Econ. Bot.* **2000**, *54*, 513–527. [CrossRef]
7. Kathiresan, K.; Bingham, B.L. Biology of mangroves and mangrove ecosystems. *Adv. Mar. Biol.* **2001**, *40*, 81–251. [CrossRef]
8. Zhang, K.; Liu, H.; Li, Y.; Xu, H.; Shen, J.; Rhome, J.; Smith, T.J. The role of mangroves in attenuating storm surges. *Estuar. Coast. Shelf Sci.* **2012**, *102–103*, 11–23. [CrossRef]
9. Cahoon, D.R.; Hensel, P.; Rybczyk, J.; McKee, K.L.; Proffitt, C.E.; Perez, B.C. Mass tree mortality leads to mangrove peat collapse at Bay Islands, Honduras after Hurricane Mitch. *J. Ecol.* **2003**, *91*, 1093–1105. [CrossRef]
10. Bouillon, S.; Borges, A.V.; Castañeda-Moya, E.; Diele, K.; Dittmar, T.; Duke, N.C.; Kristensen, E.; Lee, S.Y.; Marchand, C.; Middleburg, J.J.; et al. Mangrove production and carbon sinks: A revision of global budget estimates. *Glob. Biogeochem. Cycles* **2008**, *22*, 1–12. [CrossRef]
11. Komiyama, A.; Ong, J.E.; Poungparn, S. Allometry, biomass, and productivity of mangrove forests: A review. *Aquat. Bot.* **2008**, *89*, 128–137. [CrossRef]
12. Ministry of Natural Resources and Environment (MNRE), Malaysia. Malaysia's Submission on Reference Levels for REDD+ Results Based Payments under UNFCCC, 2015. Available online: https://redd.unfccc.int/files/2018_frel_submission_malaysia.pdf (accessed on 7 January 2022).
13. Nomura, K.; Mitchard, E.T.A.; Bowers, S.J.; Patenaude, G. Missed carbon emissions from forests: Comparing countries' estimates submitted to UNFCCC to biophysical estimates. *Environmental Res. Lett.* **2019**, *14*, 4015. [CrossRef]
14. The Economics of Ecosystem and Biodiversity, TEEB. *An Interim Report*; European Communities: Brussel, Belgium, 2008.
15. De Groot, R.; Brander, L.; Van Der Ploeg, S.; Costanza, R.; Bernard, F.; Braat, L.; Christie, M.; Crossman, N.; Ghermandi, A.; Hein, L.; et al. Global estimates of the value of ecosystems and their services in monetary units. *Ecosyst. Serv.* **2012**, *1*, 50–61. [CrossRef]
16. Zolkos, S.; Goetz, S.; Dubayah, R. A meta-analysis of terrestrial aboveground biomass estimation using Lidar remote sensing. *Remote Sens. Environ.* **2013**, *128*, 289–298. [CrossRef]
17. Duncanson, L.; Dubayar, R.; Cook, B.; Rosette, J.; Parker, G. The importance of spatial detail: Assessing the utility of individual crown information and scaling approaches for Lidar-based biomass density estimation. *Remote Sens. Environ.* **2015**, *168*, 102–112. [CrossRef]
18. Taylor, P.; Asner, G.; Dahlin, K.; Anderson, K.; Knapp, D.; Martin, R.; Mascaro, J.; Chazdon, R.; Cole, R.; Wanek, W. Landscape-scale controls on aboveground forest carbon stocks on the Osa Peninsula, Costa Rica. *PLoS ONE* **2015**, *10*, e0126748. [CrossRef]
19. Fatoyinbo, T.E.; Simard, M. Height and biomass of mangroves in Africa from ICESat/ GLAS and SRTM. *Int. J. Remote Sens.* **2013**, *34*, 668–681. [CrossRef]
20. Brown, S.; Gillespie, A.J.R.; Lugo, A.E. Biomass estimation methods for tropical forests with applications to forest inventory data. *For. Sci.* **1989**, *35*, 881–902. [CrossRef]
21. Nelson, B.W.; Mesquita, R.; Pereira, J.L.G.; De-Souza, S.G.A.; Batista, G.T.; Couto, L.B. Allometric regressions for improved estimate of secondary forest biomass in the central Amazon. *For. Ecol. Manag.* **1999**, *117*, 149–167. [CrossRef]
22. Basuki, T.M.; Van-Laake, P.E.; Skidmore, A.K.; Hussin, Y.A. Allometric equations for estimating the aboveground biomass in tropical lowland Dipterocarp forests. *For. Ecol. Manag.* **2009**, *257*, 1684–1694. [CrossRef]
23. Djomo, A.N.; Ibrahima, A.; Saborowski, J.; Gravenhorst, G. Allometric equations for biomass estimations in Cameroon and pan moist tropical equation including biomass data from Africa. *For. Ecol. Manag.* **2010**, *260*, 1873–1885. [CrossRef]
24. Fayolle, A.; Doucet, J.L.; Gillet, J.F.; Bourland, N.; Lejeune, P. Tree allometry in Central Africa: Testing the validity of pantropical multi-species allometric equations for estimating biomass and carbon stocks. *For. Ecol. Manag.* **2013**, *305*, 29–37. [CrossRef]
25. Chave, J.; Réjou-Méchain, M.; Búrquez, A.; Chidumayo, E.; Colgan, M.S.; Delitti, W.B.C.; Duque, A.; Eid, T.; Fearnside, P.M.; Goodman, R.C.; et al. Improved allometric models to estimate the aboveground biomass of tropical trees. *Glob. Change Biol.* **2014**, *20*, 3177–3190. [CrossRef] [PubMed]
26. Ong, J.E.; Gong, W.K.; Wong, C.H. Allometry and partitioning of the mangrove, *Rhizophora apiculata*. *For. Ecol. Manag.* **2004**, *188*, 395–408. [CrossRef]
27. Putz, F.E.; Chan, H.T. Tree Growth, dynamics, and productivity in a mature mangrove forest in Malaysia. *For. Ecol. Manag.* **1986**, *17*, 211–230. [CrossRef]

28. Hazandy, A.H.; Ahmad-Ainuddin, N.; Zaiton, S.; Arifin, A.; Tuan-Marina, T.I.; Lydia-Suzieana, M. Quantifying aboveground carbon stock of 30-year-old mangrove forest and its economic value in Matang Mangrove Forest. In Proceedings of the National Conference on "Forest Resource Economics Assessment", Putrajaya, Malaysia, 2–5 September 2014; pp. 188–213.
29. Komiyama, A.; Poungparn, S.; Kato, S. Common allometric equations for estimating the tree weight of mangroves. *J. Trop. Ecol.* **2005**, *21*, 471–477. [CrossRef]
30. Jachowski, N.R.A.; Quak, M.S.Y.; Friess, D.A.; Duangnamon, D.; Webb, E.L.; Ziegler, A.D. Mangrove biomass estimation in Southwest Thailand using machine learning. *Appl. Geogr.* **2013**, *45*, 311–321. [CrossRef]
31. Kauffman, J.B.; Donato, D.C. *Protocols for the Measurement, Monitoring and Reporting of Structure, Biomass and Carbon Stocks in Mangrove Forests*; Working Paper 86; Center for International Forestry Research (CIFOR): Bogor, Indonesia, 2012; 37p.
32. Clough, B.F.; Scott, K. Allometric relationships for estimating above-ground biomass in six mangrove species. *For. Ecol. Manag.* **1989**, *27*, 117–127. [CrossRef]
33. Doruska, P.F.; Patterson, D.W.; Posey, T.E. Variation in wood density by stand origin and log position for Loblolly Pine sawtimber in the Coastal Plain of Arkansas. In *Proceedings of the 13th Biennial Southern Silvicultural Research Conference on General Technical Report, SRS-92*; Kristina, C.F., Ed.; Department of Agriculture, Forest Service, Southern Research Station: Asheville, NC, USA, 2006; pp. 341–343.
34. Preston, K.A.; Cornwell, W.K.; De-Noyer, J.L. Wood density and vessel traits as distinct correlates of ecological strategy in 51 California coast range angiosperms. *New Phytol. Trust.* **2006**, *170*, 807–818. [CrossRef]
35. Iida, Y.; Lourens, P.; Sterk, F.J.; Kassim, A.R.; Kubo, T.; Potts, M.D.; Kohyama, T.S. Wood density explains architectural differentiation across 145 co-occuring tropical tree species. *Funct. Ecol.* **2012**, *26*, 274–282. [CrossRef]
36. Medeiros, T.C.C.; Sampaio, E.V.S.B. Allometry of aboveground biomasses in mangrove species in Itamaracá, Pernambuco, Brazil. *Wetl. Ecol. Manag.* **2008**, *16*, 323–330. [CrossRef]
37. Fearnside, P.M. Wood density for estimating forest biomass in Brazilian Amazonia. *For. Ecol. Manag.* **1997**, *90*, 59–87. [CrossRef]
38. Banaticla, M.R.N.; Sales, R.F.; Lasco, R.D. Biomass equation for tropical tree plantation species in young stands using secondary data from the Philippines. *Ann. Trop. Res.* **2007**, *29*, 73–90. [CrossRef]
39. Qin, L.; Liu, Q.; Zhang, M.; Saeed, S. Effect of measurement errors on the estimation of tree biomass. *Can. J. For. Res.* **2019**, *49*, 1371–1378. [CrossRef]
40. Bellasen, V.; Stephan, N. *Accounting for Carbon*; Cambridge University Press: Cambridgr, UK, 2015; 540p.
41. Chave, J.; Andalo, C.; Brown, S.; Cairns, M.A.; Chambers, J.Q.; Eamus, D.; Folster, H.; Fromard, F.; Giguchi, N.; Kira, T.; et al. Tree allometry and improved estimation of carbon stocks and balance in Tropical Forests. *Oecologia* **2005**, *145*, 87–99. [CrossRef]
42. Komiyama, A.; Jintana, V.; Sangtiean, T.; Kato, S. A common allometric equation for predicting stem weight of mangroves growing in secondary forests. *Ecol. Res.* **2002**, *17*, 415–418. [CrossRef]
43. Gong, W.K.; Ong, J.E. Plant biomass and nutrient flux in a managed mangrove forest in Malaysia. *Estuar. Coast. Shelf Sci.* **1990**, *31*, 519–530. [CrossRef]
44. Kahn, M.N.I.; Suwa, R.; Hagihara, A. Allometric relationships for estimating the aboveground phytomass and leaf area of mangrove *Kandelia candel* (L.) Druce trees in the Manko Wetland, Okinawa Island, Japan. *Trees* **2005**, *19*, 266–272. [CrossRef]
45. Soares, M.L.G.; Schaeffer-Novelli, Y. Above-ground biomass of mangrove species. I. Analysis of models. *Estuar. Coast. Shelf Sci.* **2005**, *65*, 1–18. [CrossRef]
46. Jalil, A.-M.; Abdul-Hamid, H.; Mohamad-Ali, N.-A.; Mohamed, J. Allometric models for estimating aboveground biomass and carbon stock in planted *Aquilaria malaccensis* stand. *J. Trop. For. Sci.* **2021**, *33*, 240–246. [CrossRef]
47. Clough, B.F. Primary Productivity and Growth of Mangrove Forests. In *Tropical Mangrove Ecosystem*; Robertson, A.I., Alongi, D.M., Eds.; American Geophysical Union: Washington, DC, USA, 1992; pp. 225–250.
48. Fromard, F.; Puig, H.; Mougin, E.; Marty, G.; Betoulle, J.L.; Cadamuro, L. Structure, above-ground biomass and dynamics of mangrove ecosystems: New data from French Guiana. *Oecologia* **1998**, *115*, 39–53. [CrossRef]
49. King, D.A. Linking tree form, allocation and growth with an allometrically explicit model. *Ecol. Model.* **2005**, *185*, 77–91. [CrossRef]
50. Smith, T.J., III; Whelan, K.R.T. Development of allometric relations for three mangrove species in South Florida for use in the Greater Everglades Ecosystem restoration. *Wetl. Ecol. Manag.* **2006**, *14*, 409–419. [CrossRef]
51. Henry, H.A.L.; Aarssen, L.W. The interpretation of stem diameter–height allometry in trees: Biomechanical constraints, neighbour effects, or biased regressions? *Ecol. Lett.* **1999**, *2*, 89–97. [CrossRef]
52. Latiff, A. Conservation strategies for endangered mangrove swamp forests in Malaysia. *Pak. J. Bot.* **2012**, *44*, 27–36.
53. Norhayati, A.; Latiff, A. Biomass and species composition of a mangrove forest in Pulau Langkawi, Malaysia. *Malays. Appl. Biol.* **2001**, *30*, 75–80.
54. Chandra, I.A.; Seca, G.; Hena, M.K.A. Aboveground biomass production of *Rhizophora apiculata* Blume in Sarawak mangrove forest. *Am. J. Agric. Biol. Sci.* **2011**, *6*, 469–474.
55. Norilani, W.W.I.; Juliana, W.W.A.; Salam, M.R.; Latiff, A. Community structure at two compartments of a disturbed mangrove forests at Pulau Langkawi. *Am. Inst. Phys. AIP Conf. Proc.* **2014**, *1614*, 790–794. [CrossRef]
56. Rozainah, M.Z.; Nazri, M.N.; Sofawi, A.B.; Hemati, Z.; Juliana, W.A. Estimation of carbon pool in soil, above and below ground vegetation at different types of mangrove forests in Peninsular Malaysia. *Mar. Pollut. Bull.* **2018**, *137*, 237–245. [CrossRef]
57. Haase, R.; Haase, P. Above-ground biomass estimates for invasive trees and shrubs in the Pantanal of Mato Grosso, Brazil. *For. Ecol. Manag.* **1995**, *73*, 29–35. [CrossRef]

58. Rayachhetry, M.B.; Van, T.K.; Center, T.D.; Laroche, F. Dry weight estimation of the aboveground components of *Melaleuca quinquenervia* trees in southern Florida. *For. Ecol. Manag.* **2001**, *142*, 281–290. [CrossRef]
59. Novitzky, P. Analysis of Mangrove Structure and Latitudinal Relationships on the Gulf Coast of Peninsular Florida. Master's Thesis, University of South Florida, Tampa, FL, USA, 2010.
60. Kodikara, K.A.S.; Jayatissa, L.P.; Huxham, M.; Dahdouh-Guebas, F.; Koedam, N. The effects of salinity on growth and survival of mangrove seedlings changes with age. *Acta Bot. Bras.* **2018**, *32*, 37–46. [CrossRef]
61. Phan-Khanh, L. Wave Attenuation in Coastal Mangroves: Mangrove Squeeze in the Mekong Delta. Doctoral's Thesis, Delft University of Technology, Delf, Netherlands, 2019.
62. Vinh, T.V.; Marchand, C.; Linh, T.V.K.; Vinh, D.D.; Allenbach, M. Allometric models to estimate above-ground biomass and carbon stocks in *Rhizophora apiculata* tropical managed mangrove forests (Southern Viet Nam). *For. Ecol. Manag.* **2019**, *434*, 131–141. [CrossRef]
63. Amira, S. Estimation of *Rhizophora apiculata B1* Biomass in Mangrove Forest in Batu Ampar Kubu Raya Regency, West Kalimantan. Honor's Thesis, Bogor Agricultural University, Bogor, Indonesia, 2008.
64. Cole, T.G.; Ewel, K.C.; Devoe, N.N. Structure of mangrove trees and forests in Micronesia. *For. Ecol. Manag.* **1999**, *117*, 95–109. [CrossRef]
65. Kauffman, J.B.; Cole, T.G. Micronesian mangrove forest structure and tree responses to a severe typhoon. *Wetlands* **2010**, *30*, 1077–1084. [CrossRef]
66. Tarlan, M.A. Biomass Estimation of Nyirih (*Xylocarpus granatum* Koenig. 1784) in Primary Mangrove Forest in Batu Ampar, West Kalimantan. Undergraduate Thesis, Bogor Agricultural University, Bogor, Indonesia, 2008.

Article

Landscape Characteristics in Relation to Ecosystem Services Supply: The Case of a Mediterranean Forest on the Island of Cyprus

George Kefalas [1,†], Roxanne Suzette Lorilla [2], Panteleimon Xofis [1], Konstantinos Poirazidis [3] and Nicolas-George Homer Eliades [4,5,*,†]

1   Department of Forestry and Natural Environment, International Hellenic University, GR-66100 Drama, Greece; gkefalas@emt.ihu.gr (G.K.); pxofis@for.ihu.gr (P.X.)
2   Department of Geography, Harokopio University of Athens, El. Venizelou 70, GR-17676 Kallithea, Greece; rslorilla@hua.gr
3   Department of Environmental Sciences, Ionian University, GR-29100 Zakynthos, Greece; kpoiraz@ionio.gr
4   Frederick Research Center, Filokyprou 7-9, Palouriotissa, Nicosia CY-1036, Cyprus
5   Nature Conservation Unit, Frederick University, P.O. Box 24729, Nicosia CY-1303, Cyprus
*   Correspondence: res.en@frederick.ac.cy or niceliades@gmail.com
†   These authors contributed equally to this work.

Abstract: The Mediterranean area is one of the most significantly altered biodiversity hotspots on the Earth's surface; it has been intensively affected by anthropogenic activity for millennia, forming complex socioecological systems. In parallel, the long history of natural ecological processes and the deep interlinking with human populations led to landscape patterns, such as spatial heterogeneity, that facilitate the provision of essential ecosystem services (ESs). As such, a comprehensive understanding of the underlying factors that influence the supply of ESs is of paramount importance for effective forest management policies that ensure both ecological integrity and human welfare. This study aimed at identifying local specific interactions across three different spatial scales between landscape metrics and ESs using global and geographical random forest models. The findings showed that dense forest cover may have a positive effect on the supply of ESs, such as climate regulation and timber provision. Although landscape heterogeneity is considered among the main facilitators of ecosystem multifunctionality, this did not fully apply for the Marathasa region, as forest homogeneity seems to be linked with provision of multiple services. By assessing under which landscape conditions and characteristics forest ESs thrive, local stakeholders and managers can support effective forest management to ensure the co-occurrence of ESs and societal wellbeing.

Keywords: ecosystem services; landscape structure; random forest (RF); geographical random forest (GRF); Mediterranean forest; Marathasa; Cyprus

Citation: Kefalas, G.; Lorilla, R.S.; Xofis, P.; Poirazidis, K.; Eliades, N.-G.H. Landscape Characteristics in Relation to Ecosystem Services Supply: The Case of a Mediterranean Forest on the Island of Cyprus. *Forests* **2023**, *14*, 1286. https://doi.org/10.3390/f14071286

Academic Editor: Pablo Vergara

Received: 15 May 2023
Revised: 17 June 2023
Accepted: 19 June 2023
Published: 21 June 2023

1. Introduction

The composition and structure of landscapes have been shown to have a significant impact on ecological processes and the multifunctionality of ecosystems [1]. The specific characteristics of a given landscape are mainly formed by anthropogenic activities (such as deforestation, urbanisation, agriculture, tourism, and mining) along with climatic conditions, geomorphological characteristics, and natural and human disasters. The Mediterranean region, where human–nature interactions have been ongoing for millennia, is characterized by highly heterogeneous vegetation, habitat types, and landscapes, resulting in rich biodiversity and a vast supply of ecosystem services (ESs) [2–5]. Mediterranean forests are an example of dynamic ecosystems with a variety of ESs to human society [6,7]. However, increasing environmental changes and alterations in the intensity of human activities have led to two main patterns. The first pattern has resulted from increased human pressures, leading to landscape diversification and the creation of isolated patches that function as biodiversity islets and support important ESs [8]. The second pattern has arisen from the abandonment of agricultural and pastoral activities, resulting in the gradual homogenization of forests and often in a negative impact on biodiversity and several ESs [9–11].

Landscape structure has extensively been used to assess the integrity of ecosystems (especially regarding their capacity to maintain biodiversity) and to study the influence of humans in its transition processes [12]. In parallel, several studies have investigated the relationships between the configuration of a landscape and the supply of essential ESs [5,13–16]. Considering the dependency between Ess and a landscape's composition and structure, Termorshuizen and Opdam (2009) [17] introduced the concept of "landscape services (LSs)", describing them as "the range of functions that are or can be retrieved by a landscape and be valued by humans for economic, sociocultural, and ecological reasons". Although the concepts of LSs and Ess overlap, the difference between these is that for the former, the supply of ESs is linked to the landscape's patterns and its social dimension; while for the latter, the supply of ESs is based solely on ecosystems and their spatial configuration [18–20]. Given that the spatial characteristics and patterns of a landscape are considered among the main principles for the provision of ESs, the identification and understanding of how specific landscape features act in favour of ES supply is of primary importance [21,22]. Furthermore, the quantification of a landscape's structure delivers crucial knowledge about the impacts of landscape feature on ecological processes, and for this reason, it can be used as a tool to systematically monitor the effects of landscape changes on ecosystems [23].

Landscape metrics are the means used for quantifying the spatial configuration of a given landscape, and they are usually employed to assess the impact of landscape structure on ecological functionality and biodiversity [12,24]. The ease of obtaining landscape metrics over large geographical regions and their rapid calculation feasibility compared to data- and time-intensive models, such as species distribution models, field data assessments, and/or other modeling procedures, has led to their widespread use [25,26]. By using a land use/cover map, landscapes can be analyzed at three levels (namely, patch, class, and landscape level) allowing for the characterization of patch size, shape, and/or the connectivity land use/cover classes, and the diversity of an entire landscape [27,28]. It is well documented that in order to completely comprehend a landscape's structure it is necessary to estimate and evaluate a set of multiple landscape metrics, as each metric studies a unique characteristic of the landscape. However, in most cases, landscape metrics are highly correlated, leading to the misinterpretation of findings, hindering their use in multivariate statistical analyses [26,29]. To address these issues and select a representative set of metrics, many statistical and theoretical frameworks have been developed. Such frameworks are based on expert knowledge and literature review [4,30–32], principal component analysis and factor analysis [29,33,34], regression models [35], and, more recently, on machine learning (ML) algorithms that are used in ecological applications [5,36].

Machine learning algorithms have become one of the main tools used in Earth and life sciences, such as remote sensing [37], GIS and environmental spatial analysis [38,39], ESs [40,41], and ecology [42,43]. Compared to traditional statistical techniques, ML has several advantages, such as accurate predictive and classification capabilities, increased ability to manage complex relationships (non-linear), and the capability to automate tasks [44]. There are two types of ML approaches, namely, supervised and the unsupervised algorithms; and multiple methods/algorithms, such as tree-based methods, neural networks, support vector machines, genetic algorithm, fuzzy inference systems, and Bayesian methods [41,45]. The most frequently used ML method in ecological applications is random forest (RF) due to its versatility and high accuracy in responding to different research questions [36]. Among the advantages of RF are included the internal self-testing procedures, the high predictive accuracy, and the ability to estimate the importance of each input variable [46]. However, despite the important advantages of all ML algorithms, in most cases, the spatial aspect of the data cannot be used in the model calibration procedure. Considering the spatial heterogeneity of ecological datasets, the use of "global" algorithms/techniques limits the identification of spatially varying relationships within the study area. To address this issue, geographical weighted regression (GWR) was developed by Fotheringham et al. [47], which is based on the traditional regression technique. Georganos et al. [48] and Georganos and Kalogirou [49] have introduced the geographical random forest (GRF) approach, which is based on GWR principals and is increasingly used in various scientific fields, including epidemiology [50], socioeconomic studies [48], natural disasters [51], forestry [52], and agriculture [53], where in all cases GRF outperformed RF.

The recent advances in remote sensing data and methods allow the thematically and spatially accurate mapping of land uses and habitats across various spatial scales. This, in turn, allows the analysis of landscape structure and composition in a quantitative manner using landscape metrics. However, the relationship between landscape structure and composition and the supply of ecosystem services in forest dominated landscapes is poorly understood. This study implements a spatially explicit approach using random forest and geographical random forest algorithms to assess ES supply in relation to landscape metrics across three spatial scales. The aim of the study is to investigate the validity of using landscape spatial characteristics as indicators of ES supply. The specific objectives of the study are (a) to generate a thematically and spatially accurate land use/cover map of the area using readily available remote sensing data and well documented methods; (b) to assess the performance of two modeling approaches for identifying relationships between landscape spatial characteristics and supply of ES; and (c) to identify the spatial characteristics that appear to be most highly related to the supply of ES. The results of the study are expected to provide a useful tool for the monitoring of ES supply and valuable information regarding the specific landscape characteristics that can support ecosystem multifunctionality by enabling co-occurrence of multiple ES.

2. Materials and Methods

2.1. Study Area

Marathasa is a geographical area of the mountainous mass of the Troodos Mountain range and occupies an area of ~208 Km^2; that is, 2.2% of the total area of Cyprus (data source: https://www.data.gov.cy/, accessed on 1 January 2023). The largest part of the Marathasa area is covered by state forests, while the geological structure of the area is that of the Troodos ophiolite (igneous rocks which make up the oceanic crust). The landscape's morphology is characterized by a mountainous topography with high and variable mountain peaks, often over 1000 m in altitude (altitude range: 163 m to 1932 m). The Marathasa region is particularly characterized by gorges, which have resulted from the changes of the sea surface and those of the sea and the renewal of the rivers and the marine terraces. The landscape contributed to the shape of a dense network of river systems which are often of a dendritic pattern and which maintain their flow over a long period [54]. The natural ecosystems have evolved through both environmental conditions and human impact. The presence of human communities in Marathasa is observed as early as the 6th–9th century (647–965 AD), where several small settlements (5–10 houses) were established on the mountains of Marathasa when the inhabitants of the coastal areas sought safer places in the mainland [55]. During the past two centuries, the region of Marathasa has been defined by the administrative boundaries of 14 communities (villages). The northern and western communities showed a population increase over time in relation to the communities of southern Marathasa [55]. The region's professional activity has been directly linked to the primary sector of economy, namely, agricultural activities and timber sale [54]. However, since 1980, urbanization contributed to the population reduction in the area (inhabitants in 1982: 4341; inhabitants in 2011: 1523). Data from CORINE Land Cover (CLC) inventory in 2018 show that most of the land is occupied by forests (Figure 1).

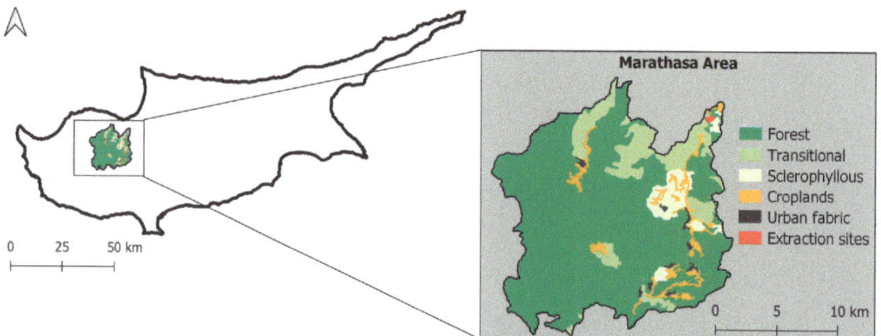

Figure 1. Map of the study area and the main LULC classes according to CORINE.

2.2. Mapping Land Cover and Ecosystem Services Supply

The land use/cover (LULC) dataset was produced using a pre-processed Level 1 Landsat 8 OLI image acquired in 2021 and an object-oriented classification scheme (OBIA) developed and evaluated by Kefalas et al. [56]. The applied OBIA scheme is stepwise, and it is based solely on crisp or fuzzy rules that use vegetation indices (Table A1). Through this procedure, the study area was divided into seven LULC classes (three density-based vegetation classes, open/rocky areas, croplands, settlements, and inland water). The accuracy assessment was based on the use of ground-truth data, verified in situ, and the estimation of statistical measures such as Kappa index (K) and overall accuracy (OA).

Based on data availability and the merit for estimating ESs to achieve effective forest management on the study area, four ESs were selected and mapped covering all ES categories (following the Common International Classification of Ecosystem Services (CICES)) [57]. Specifically, one ES refers to the Provisioning ES section, two in the Regulating and Maintenance section, and one in the Cultural section (Table 1).

Table 1. The estimated ecosystem services and their indicators/proxies.

ES Section	Ecosystem Service	Indicator/Proxy	Data Source
Provisioning	Materials from timber (MT)	Presence of forest and agroforest land	1
Regulating and Maintenance	Climate regulation (CR)	Below and above ground carbon storage	1 & 2
	Erosion protection (EP)	Soil erosion prevention	1, 3, 4, & 5
Cultural	Recreation (RC)	Recreation potential	1, 3, 6, 7

Data sources code: (1) Land Cover data; (2) Carbon Dioxide Information Analysis Centre (15 December 2022) [58]; (3) ASTER Digital Elevation Model (DEM) (15 December 2022) [59]; (4) Soil erodibility index (K-factor) (15 December 2022) [60]; (5) Rainfall erosivity index (R-factor) (15 December 2022) [60]; (6) Hiking trails (available in: https://www.prettymap.gr/troodos/geotourism/el.html (10 December 2022); https://www.data.gov.cy/ (accessed on 12 January 2023); mapping by the project WaterWays); (7) Sites with touristic and/or cultural merit (mapping by the WaterWays project).

Provisioning ESs are all nutritional, material, and energetic outputs that are derived from a living ecosystem [57]. Materials from timber (MT) represent the products from trees harvested from natural forests and plantations [61], and to map this service, the presence of forest and agroforest land was considered.

Regulating and maintenance ESs are defined as the way in which local ecosystems control the biotic and abiotic features of the environment in order to enhance human well-being [57,61]. The Intergovernmental Panel on Climate Change (IPCC) reported that transitions and processes in a given landscape, such as LULC changes, soil degradation, and deforestation, play a crucial role in the emissions of greenhouse gases [62]. Natural ecosystems, and especially forests, regulate climatic conditions through several processes including carbon sequestration, moisture production, and temperature control [63]. The Climate regulation (CR) service was mapped considering the carbon pool table derived from the INVEST Carbon Storage and Sequestration model. This model combines the amount of carbon stored in four carbon pools (aboveground biomass, belowground biomass, soil organic matter, and dead organic matter) based on LULC. Erosion prevention (EP) represents the capacity of ecosystems to prevent erosion, and was calculated using the soil erosion prevention framework based on the RUSLE equation [64]:

$$Es = Y - \beta e, \{Y = R \times LS \times K, \beta e = Y \times a\},$$

where Es represents the actual ES provision (tons of soil not eroded), Y represents the structural impact, βe represents the mitigated impact (where $a = C$ and $e_s = 1 - a$), R represents the rainfall erosivity factor, LS represents the topographic factor, K represents the soil erodibility factors, and C represents the vegetation cover factor.

The non-material services offered from ecosystems that affect the physical and mental state of people are defined as cultural services. To estimate and map the potential of the Marathasa forest to offer recreational activities (RC), a multicriterial model was developed considering two main factors: (a) the biophysical factor; and (b) the cultural factor. The biophysical factor was a combination of four indicators that characterize ecosystems in terms of natural attractiveness; these indicators were the Normalized Difference Vegetation Index (estimated from Landsat 8 OLI image), the Shannon's Landscape Diversity Index (estimated using the LULC thematic map), and the Geodiversity index [65]. The cultural factor was defined by the presence of hiking trails and sites with touristic and/or cultural merit (such as camping sites, churches, old bridges, and water mills) as those indicators were either point or line features, while a kernel density tool was used to create a continuous raster layer.

To estimate and map the total ES supply, each ES map was standardized to a scale between 0 and 1, based on the minimum and maximum values (higher values corresponded to a greater magnitude of ES) and combined.

2.3. Landscape Characteristics That Contribute to ES Supply

This study identified the landscape characteristics that are related to ES supply across three scales. The unit of analysis was based on hexagonal grids of three sizes, with apothem of 50, 250, and 500 m. For each hexagonal grid, a set of 17 landscape metrics were estimated at class and landscape level (56 metrices in total). The selected metrics refer to the area, the core area, and the edges of a given landscape and class (Total Area (TA), Total Core Area (TCA), Total Edge (TE), Edge Density (ED), Largest Patch Index (LPI)), the shape of a landscape (Landscape Shape Index (LSI)), the aggregation of landscape and classes (Patch Density (PD), Cohesion, Division, Effective Mesh Size (MESH), and Contagion), and the landscape diversity (Patch Richness (PR), Patch Richness Density (PRD), Relative Patch Richness (RPR), Shannon's Diversity Index (SHDI), and Simpson's Diversity Index (SIDI)).

The identification of landscape metrics related to ESs was based on global and local random forest (RF) algorithms. Landscape metrics were used as explanatory variables. while the total ES supply formed the response variables (Figure A1). Random forest is a non-parametric machine learning approach which is used for classification and regression purposes. It is suitable when the relationship among variables is non-linear and when multicollinearity is evident [66–68]. The first step when employing an RF model is the creation of a training set, on which the model development is based, and a test set, which is used to validate the initial model. In the case of Marathasa forest, we created the training dataset from 70% of randomly selected samples of the initial dataset. The remaining 30%, namely, the out-of-bag (OOB) set, is excluded from the model training and is used to estimate the RF's model efficiency [69]. The total number of randomly selected samples used for training and testing vary according to the spatial scale, and they were 36,817 in the apothem of 50 m, 1586 in the apothem of 250 m, and 428 in the apothem of 500 m. The second step is to parameterize the RF model by setting the appropriate (1) number of randomly selected predictors at each tree (mtry); (2) the minimum number of records contained in a leaf (nodesize); and (3) the number of trees (ntrees). The optimal number for mtry was estimated using the function "rf.mtry.optim" given from the spatialML

R package [49], while the node size was set to 1, as suggested by Breiman (2001) and Lorilla et al. (2020) [5,66]. Regarding the ntrees parameter, different values were tested, and the ones with the highest accuracy were selected as more appropriate for their use in the RF model (Table A3). The third step was the implementation of the RF model. In the case of a regression, the main outputs are the variables' importance and error rate that are estimated by the OOB method [50].

Despite the advantages of RF models, the spatial heterogeneity of the data cannot be assessed and validated. To address this issue, Georganos et al. [48] extended the "traditional" RF model by developing the geographical random forest (GRF), which is a disaggregation containing several local sub-models. The fundamental principles of GRF are same to the ones of geographical weighed regression (GWR) [47], where the model is calibrated locally rather than globally. Thus, using a GRF model, for each spatial unit i, an RF local model is estimated, taking into consideration an n number of neighbour observations. Each local model has its own performance, predictive power, and variables importance. A simplistic GRF equation is [48] as follows:

$$Y_i = a(u_i, v_i)x_i + e, i = 1 : n,$$

where Y_i is the value of the dependent variable for the ith observation, $a(u_i, v_i)x$ is the prediction of an RF model calibrated on location i, (u_i, v_i) are the coordinates, and e is an error term.

Similar to GWR, important factors in a GRF are the "kernel" or "neighbourhood" (the area that the local model considers in calibration) and the "bandwidth" (the maximum distance away from the RF location) [48,49]. There are two types of kernels (a) the "fixed kernel" and (b) the "adaptive kernel", where in the former, the neighbourhood is defined by a circle; while in the latter, as the number of nearest neighbours that are to be included in the modeling procedure [70]. The size of the bandwidth also plays a significant role in the modeling procedure as it determines the distance limit at which observations are considered to fall in the sub-model. If the bandwidth is large, then local models would be turned into a global model as all observations are used. In contrast, if the bandwidth is small, the independent variables are less biased [71]. In this study, we used adaptive kernel with the appropriate bandwidth size estimated using an automated function offered in spatialML R package [72], which re-testes various bandwidth sizes to obtain the highest R^2 value for the local model. The performance of both global and local RF models was assessed by estimating the coefficient of determination R^2 and the mean square error. The modeling procedures were implemented using the open-source R version 4.2.2, R Studio version 2022.12.0, and the R packages: rgdal [73], spatialML [72], cli [74], and caret [75].

It is worth noting that GRF produces a map for each independent variable, presenting the local importance to the dependent variable. To present the direction of the relationship between ESs supply and LM, the local bivariate relationship mapping procedure was used, considering the total ES supply and the two most important LMs for each scale of analysis. The output of this procedure is a map presenting six classes: (a) not significant; (b) positive linear; (c) negative linear; (d) concave; (e) convex; and (f) undefined complex.

3. Results

3.1. Distribution of LULC and ES Supply

After applying the OBIA classification scheme in Marathasa forest, the overall accuracy assessment and the Kappa statistics index were 91% and 0.89, respectively (Table A2). The LULC class with the highest thematic and spatial accuracy was the open and rocky areas (producer accuracy: 0.97) followed by high-density natural vegetation (producer accuracy 0.91) (Table A2). The accuracy of LULC classes varied between 0.80 in the agricultural areas and 0.97 for open/rocky areas (Table A2). The dominant LULC class was the medium-density natural vegetation covering more than 40% of the study area (Table 2). Areas covered by high- or low-density vegetation were less intense, as in both cases they extended approximately at the 25% of Marathasa (Table 2). High- and medium-density natural vegetation mainly occupied the central part of the area, while small stands of dense vegetation were located in the northwest part of the Marathasa (Figure 2). Open and rocky areas are distributed mainly in northern and eastern parts, covering 1200 ha (~5% of the area), while the agricultural areas, shaping four agricultural zones (three on the east and one on the west), covered almost 1640 ha (little more than 5% of the area) (Table 2 and Figure 2).

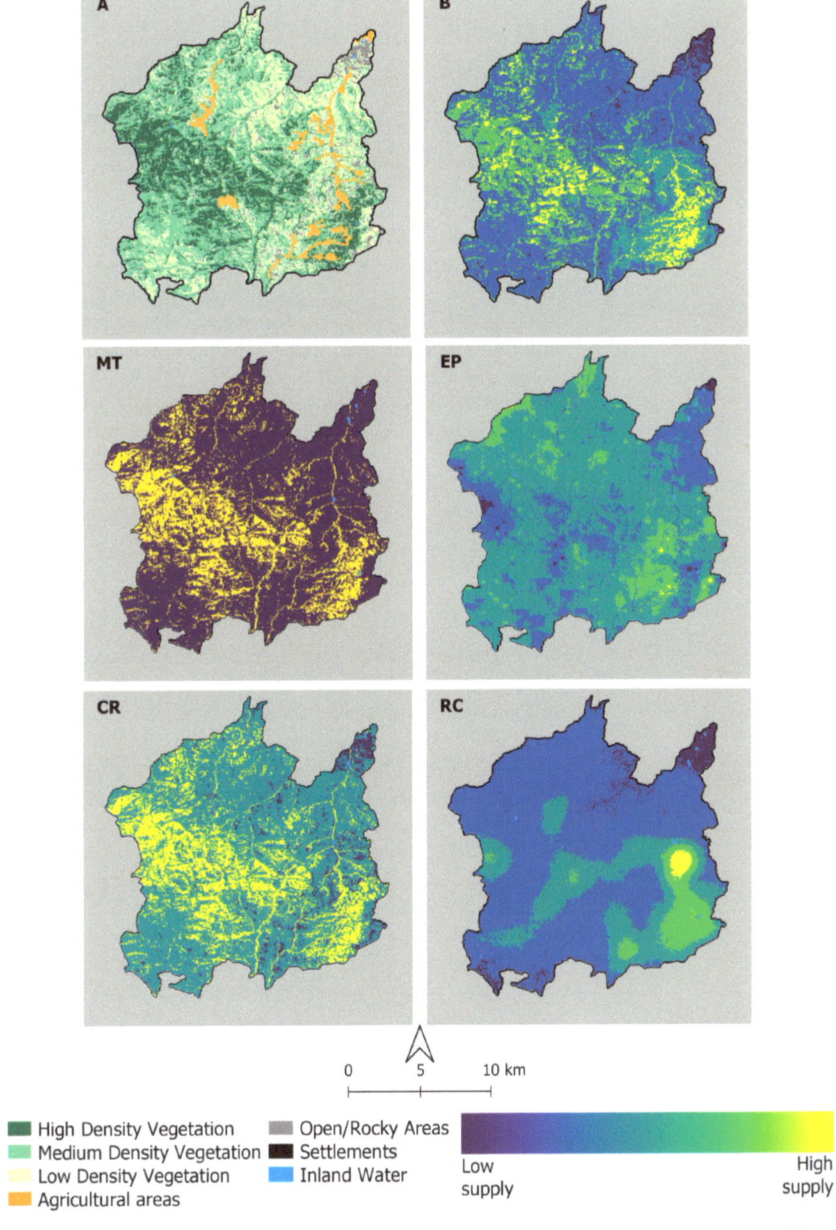

Figure 2. Distribution of (**A**) land use/cover and (**B**) total ecosystem services supply and the individual ESs (MT: Material from Timber; CR: Climate Regulation; EP: Erosion Prevention; RC: Recreation Potential).

Table 2. Part A: extend of LULC; Part B: the mean value of ESs supply.

Part A			Part B	
LULC	**Area (ha)**	**%**	**ES**	**Mean Value**
HDNV	7571.91	24.22	MT	0.48
MDNV	12,973.78	41.50	EP	0.55
LDNV	7824.86	25.03	CR	0.63
OR	1252.06	4.01	RC	0.28
AA	1639.63	5.24		
TOTAL	31,262.23			

HDNV: High Density Natural Vegetation; MDNV: Medium Density Natural Vegetation; LDNV: Low Density Natural Vegetation; OR: Open and Rocky areas; AA: Agricultural Area; MT: Material from Timber; EP: Erosion Prevention; CR: Climate Regulation; RC: Recreation Potential.

The total ES supply exhibited lower values in the northern part of the study area, which is occupied mainly by low-density vegetation, while at the central forested part, the mean total value showed higher ES supply (Figure 2). Due to the forest's nature to support ecological functions, ESs presented high mean value (Table 2). Depending on the relation between individual ESs and forest cover, ESs presented two different spatial distribution patterns. The first relates to the "materials from timber" and "climate regulation" services, where the distribution of their higher values was aligned with the distribution of the high-density natural vegetation (Figure 2); the second spatial pattern is linked to "erosion prevention" and "recreation potential" services. In the case of "erosion prevention", higher supply was located in the eastern and in the northwest parts of the study area (Figure 2); those areas are not only characterized by forest cover but also by a smooth surface relief. Regarding the service of "recreation potential", higher values were observed in the eastern part of the Marathasa forest, where sites with touristic and/or cultural merit were located (Figure 2).

3.2. Contributing Landscape Characteristics to ES Supply

Global and geographical random forest was used to identify the specific landscape characteristics that are correlated to the total supply of ES. Both global and local RF models for the total ES supply had excellent performance, as indicated by the R^2 measures, which were close to 90% across all scales (Table 3). Specifically, for the global and local RF models for the 250 m and 500 m apothem grids, R^2 exceeded 87%, while in the case of the 50 m apothem, R^2 values were slightly lower, reaching 86.64% and 87.63% for the global and local models, respectively (Table 3). The pseudo-local coefficient of determination showed different patterns among the spatial scales (Figure 3). Specifically, at the 50 m scale, the performance of GRF was excellent in the majority of the area (pseudo-R^2 > 80%), while in areas occupied by sparse vegetation or characterized as open/rocks, the performance was poor (pseudo-R^2 < 40%). This pattern was also evident at the 500 m scale, where better performance was found in areas occupied by dense forests. Finally, at the 250 m scale, higher performance was observed in areas primarily covered by medium and low-density vegetation, while in areas occupied by forest, the performance was moderate (40% < pseudo-R^2 < 60%) (Figure 3).

Figure 4 shows the ranking of the ten most important landscape metrics (LMs). Across all three scales of analysis, the majority of the most important LM variables were related to the high-density vegetation class. These metrics mainly express aggregation (estimated through the Effective Mesh Size, Largest Patch Index, Cohesion, and Division) and configuration (estimated through the Total Core Area and Edge Density). Exceptions to this pattern were the metrics Contagion, Proximity, and Simpson's Diversity Index calculated at the landscape level, and the metrics of Edge Density and Cohesion calculated for the medium-density natural vegetation class. The former metrics express the landscape's aggregation and were found to be important in the case of 250 m, while the latter metrics express the class aggregation and configuration and were found to be important in the case of 500 m.

Table 3. Performance of global and local random forest models on the different scales of analysis.

	RF Model		GRF Model	
	Mean Total ES Supply		**Mean Total ES Supply**	
	R^2 (OOB) MSE	R^2 (Not OOB) MSE	R^2 (OOB) MSE	R^2 (Not OOB) MSE
50 m	86.64% 0.01	93.45 0.01	87.63% 0.01	98.71 0.00
250 m	87.45% 0.02	95.78% 0.0.	87.84% 0.01	98.85% 0.00
500 m	87.64% 0.02	97.56% 0.01	88.98% 0.01	99.58% 0.00

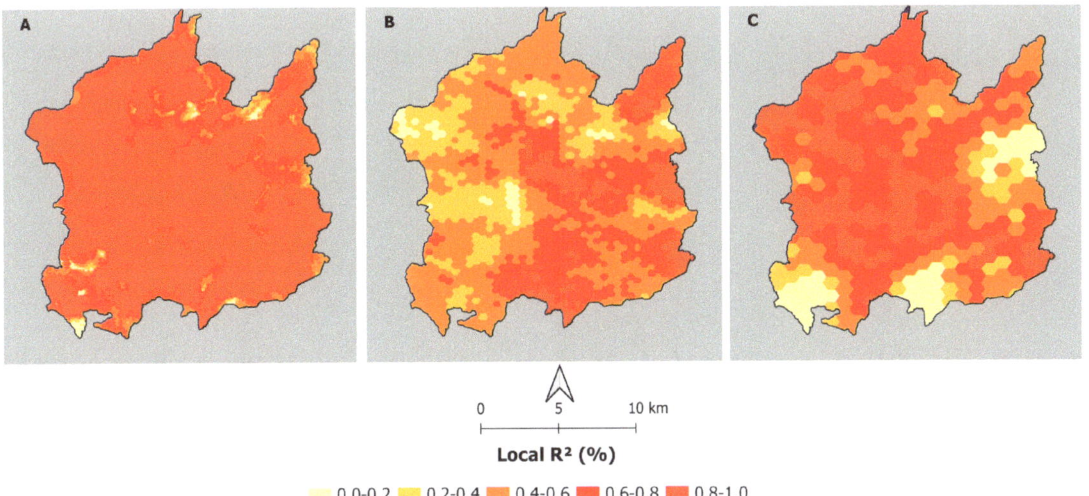

Figure 3. Pseudo-local coefficient of determination of GRF: (**A**): 50 m; (**B**): 250 m; and (**C**): 500 m apothem.

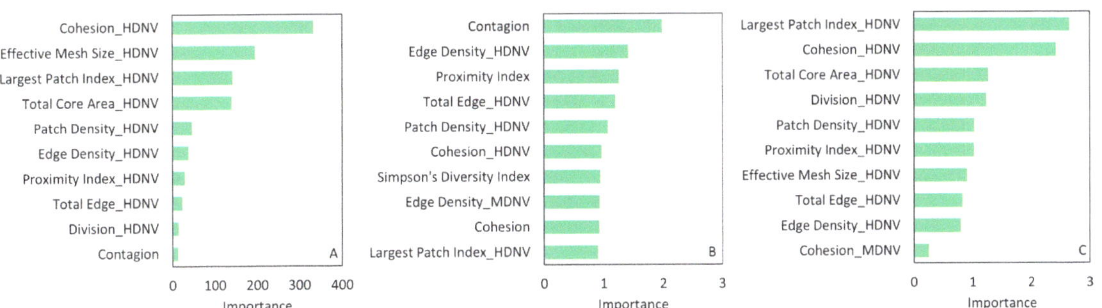

Figure 4. Ten most important variables for mean total ES supply per each scale of analysis: (**A**): 50 m; (**B**): 250 m; and (**C**): 500 m apothem.

To further analyze the spatial distribution of local variable importance, the magnitude (Figure 5 and Table A4) and direction (Figure 6) were mapped for the two most important landscape metrics (LMs) at each scale of analysis. At the 500 m scale, the Largest Patch

Index and Cohesion LMs for the HDNV class were found to be highly important and positively related in two main parts of the Marathasa forest, which were characterized by high-density vegetation. At the 250 m scale, the LM contagion, which expresses landscape aggregation, was found to be highly important mainly in the central part of the study area, characterized by high landscape heterogeneity. However, in the central part of the forest, the relationship was negative, while in areas around the main forest body, the relationship was significant but complex (neither positive nor negative). Finally, at the 50 m scale, one can identify specific places where the importance of LMs was high. The importance of individual and smaller forest stands can be recognized, considering the relatively higher values of importance of LMs in areas where high-density natural vegetation was not in excess (Figures 2 and 5). In parallel, the direction of the relationships was positive (linear positive or convex) in the MESH index, indicating that extended forest stands favourably affect ES supply. The Cohesion Index for the 50 m scale was positively related in areas mainly characterized by high-density natural vegetation, while in areas where medium- or low-density vegetation was the primary land cover type, the relationship was negative (linear negative or concave) (Figures 5 and 6).

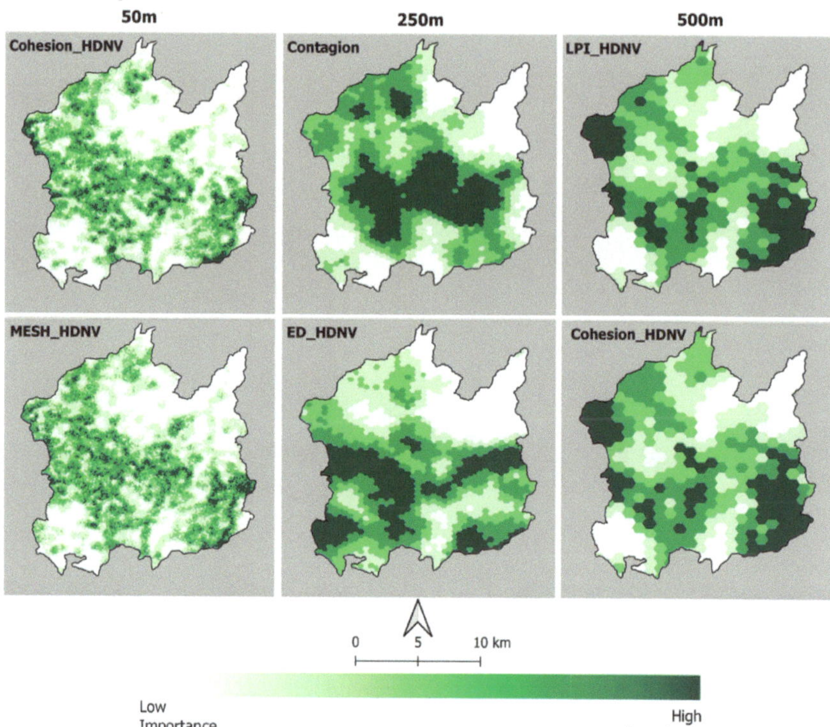

Figure 5. Maps of the two most important landscape metrics for mean total ES supply per each scale of analysis.

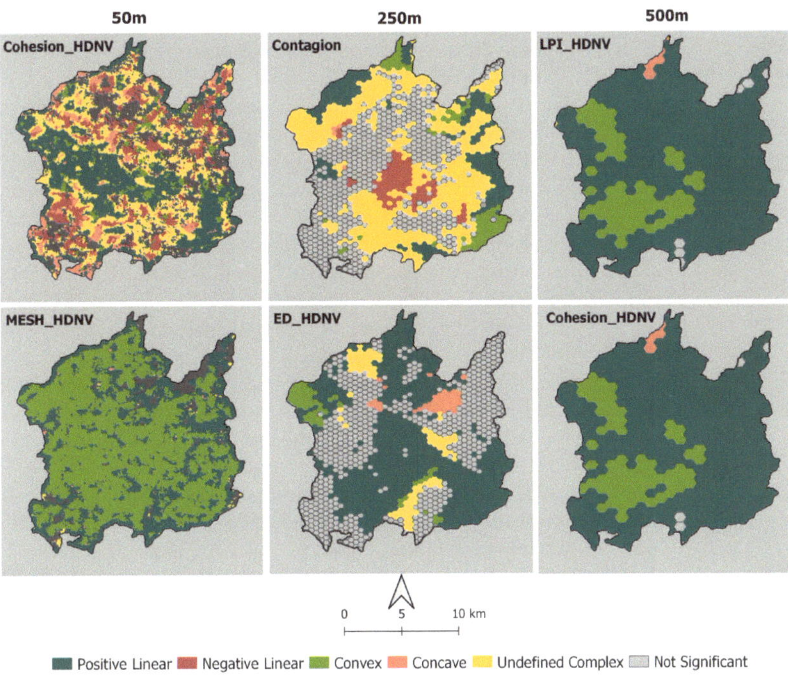

Figure 6. Local bivariate relationships among the two most important landscape metrics and the mean total ES supply per each scale of analysis.

4. Discussion

Forests in the Mediterranean basin have been characterized as a biodiversity and ecosystem services (ESs) hotspot primarily due to the complex and heterogeneous landscapes shaped by long-term interactions among human activities, natural environment, and climatic conditions [2,76]. As the material and non-material products derived from forests are strongly linked to human well-being, mapping the provision of essential services, as well as examining the conditions under which these ESs thrive, provides essential knowledge towards effective forest management policies that secure both ecological integrity and human welfare [77]. From this perspective, several studies on Mediterranean forests have shown that previous or existing forest management policies have played a significant role in the provision of ES, especially supporting various ecological processes [6,8].

The analysis of the landscape's composition and structure provides information required for developing conservation measures and sustainable management strategies for natural ecosystems [78], as well as data that can be used to further develop products indicating the status, the processes, and the functioning of an area. For this reason, a state-of-the-art object-oriented image (OBIA) classification procedure was applied providing land use/cover data with high thematic and spatial accuracy, similar to products produced via OBIA for other Mediterranean landscapes [57,79]. It is worth noting that in areas where landscape is characterized by high complexity, OBIA outperforms traditional pixel-based classification procedures [77], while in parallel, it can be used as a tool for systematic monitoring [11]. The composition of Marathasa's landscape is characterized mainly by medium density vegetation, i.e., shrubs and maquis; followed by low density vegetation, i.e., phryganic vegetation and natural grasses; and finally, by high density vegetation i.e., dense stands of maquis and pines. Considering the dynamics of Mediterranean forest [4,11,79], and especially the dynamics and status of Cypriot forests [80–84], this composition seems to be the result of the progressive evaluation of the more degraded ecosystems transitioning gradually from sparse vegetation to more dense forest stands through secondary ecological succession. In parallel, in the western part of the study area, where dense forest stands mainly occur, several management actions have been applied aiming to conserve the narrowly-distributed Cedar forest (*Cedrus brevifolia*). The implemented measures consisted of silvicultural interventions (i.e., forest thinning and natural and artificial plantations) shaping a dense and heterogenous forest landscape [81].

The Marathasa region was found to provide essential forest-related ESs that are strongly linked to both physical characteristics and the presence of anthropogenic activities and cultural elements. Overall, areas occupied by high-density vegetation provide multiple ES, while areas characterized by low-density vegetation or open rocks have lower supply values. The individual ESs presented two different patterns regarding their spatial distribution. The services of "materials from timber" and "climate regulation" tend to co-occur across the study area and reach their highest supply in areas where dense forest is dominant. This pattern is rather expected, given the crucial role of forests as carbon sinks with significant impact in mitigating the effects of climate change [76,81,84] and the importance of forested areas in providing materials to local communities. The second spatial pattern referred to the services of "soil erosion prevention" and "recreation potential"; these did not strictly follow the distribution of high-density forest, as both services were mapped following multicriteria modeling procedures which take into consideration various geospatial data [64,65]. In the case of EP, as expected, higher values were not only observed in forested areas but also in areas where the surface relief is smooth, which is in line with other studies that examined erosion risk and control at various scales [9,65,82]. Similarly, the RC supply had greater values in areas covered by high-density vegetation as a result of the higher weight value given to the degree of the biophysical factor. At the same time, the RC value is higher in areas in which sites with tourist and cultural merit are present, and, as indicated by De Valck et al. [85], mixed landscapes that include forest, farmlands, infrastructure, and cultural elements are assessed positively by visitors [85].

As previously mentioned, examining the conditions under which ESs are maximized offers fundamental knowledge to support effective forest management and planning. One way to estimate and assess the status, dynamics, and shaping factors of a given landscape is through landscape metrics [12], which were employed in this study as explanatory variables of ES supply. The implementation of the cutting-edge machine learning algorithms of random forests and geographical random forests at three different spatial scales proved effective in identifying the spatial and thematic parameters that lead to increased provision of significant ES. Overall, both RF and GRF showed excellent performance in all scales, with GRF outperforming the aspatial RF, as has been observed in other cases where both spatial and aspatial RF models have been used [48–50]. However, the main advantage of GRF models is not their performance (predictive or explanatory) but their ability to produce maps presenting a possible spatial interaction among explanatory and response variables [48]. Regardless of the scale of analysis, the outputs from the modeling procedure showed that the configuration and aggregation of dense vegetation positively influence the supply of ESs, which aligns with the findings of a previous study demonstrating that regulating and recreational services are significantly and positively affected by homogenous forests [5]. Although the aforementioned results are in contrast with the main ESs concept which states that areas with diverse landscapes supply a high number of ESs [86], in the current study, the mapped ESs are strongly linked to forest cover. This result indicates that landscape diversity alone does not necessarily lead to the supply of multiple ESs, and that the dominant vegetation that characterizes a given landscape plays a fundamental role in the provision of ESs [5,87].

The holistic framework used here offers a comprehensive understanding of how the composition and structure of a given landscape contribute to the supply of ES. Since landscapes are mainly influenced by the presence or absence of humans, considering landscape structure in the ES context offers useful insights towards understanding the complex relationship between humans and nature. Additionally, contemporary mapping and modeling procedures can assist spatial planners, conservationists, and decision makers in developing and applying strategies, actions, and measures that consider both the global and local characteristics of an area, thereby ensuring effective management.

Limitations of the Study

Due to the complexity of Mediterranean forest landscapes, revealing the interlinkages between ecosystem services and landscape characteristics is not a straightforward task. Here, we attempted to assess these relationships by combining the conceptual framework of ecosystem and landscape services and a cutting-edge methodological approach. However, there are some limitations and improvements that should be considered when interpreting the findings of such approaches.

In terms of the use of random forest, such models can be prone to overfitting if the number of trees is too large or if the model is overly complex. This can result in an underestimation or overestimation of the relationship between landscape metrics and ecosystem services, leading to the generalization of the results. While random forest models can provide variable importance measures, the interpretation of these measures for landscape metrics may be challenging as their complex interactions with ESs may not be clearly evident through variable importance rankings. Finally, the accuracy and availability of data used to derive landscape metrics and measure ESs can significantly impact the results. Incomplete or biased data may introduce errors and uncertainties in the model, affecting the reliability of the relationship assessment.

In regard to the ESs studied in Marathasa forest, future research should consider multiple ESs to provide more insights into landscape multifunctionality and the relevant underlying processes. Our study, due to the limited available data, focused on four forest-related ESs and did not consider other important ESs, such as food provision, which could be related to the agricultural regions distributed across the studied region. Furthermore, the relationships between ESs and landscape metrices were assessed for a single year, which restricted the extraction of knowledge on the dynamic nature of Mediterranean forest landscapes.

5. Conclusions

The spatial arrangement and configuration of different land-cover types within a landscape can influence the provision and distribution of ES. By using global and geographically weighted random forest models, this study aimed to assess the provision of ESs in relation to the composition and structure of a forest-dominated landscape.

Prior to the mapping of ES, an object-based image classification was applied to identify the LULC evident in the study area. More than 90% of the landscape was occupied by natural vegetation with medium density vegetation (shrubs and maquis) covering approximately 41% of the total area, followed by low (phryganic vegetation) and high-density vegetation (dense stands of maquis and pines) reaching ~25% and ~24%, respectively. This composition of land cover classes resulted from the succession of vegetation where ecological processes and management interventions facilitated the transition from sparse vegetation to more dense forest stands.

The holistic approach applied in this study offers improves our understanding of how the composition and structure of a given landscape contribute to the supply of ESs. The existence of high-density forests, although it may reduce landscape heterogeneity, leads at the same time to the maximization of ESs related to climate regulation and provision to local communities. Future research should focus on providing an overview of Ess not studied here for a comprehensive understanding of how a landscape's composition can influence the multifunctionality of forests and the human population that depends on healthy ecosystems.

Author Contributions: Conceptualization: G.K., R.S.L. and N.-G.H.E.; methodology: G.K. and R.S.L.; software: G.K.; validation: R.S.L., N.-G.H.E., K.P. and P.X.; formal analysiss: G.K.; writing—original draft preparation: G.K., R.S.L. and P.X.; writing—review and editing: P.X., K.P. and N.-G.H.E.; funding acquisition: N.-G.H.E. All authors have read and agreed to the published version of the manuscript.

Funding: The study was carried out by the WaterWays project. The project WaterWays is co-funded by the European Union (ERDF) and national funds from Greece and Cyprus within the Cooperation Programme Interreg V-A Greece—Cyprus 2014–2020.

Data Availability Statement: No new data were created or analyzed in this study. Data sharing is not applicable to this article.

Conflicts of Interest: The authors declare no conflict of interests.

Appendix A

Table A1. Vegetation indices used in LULC mapping.

Vegetation Index	Formula
Normalized Differencing Vegetation Index (NDVI)	$NDMI = \frac{NIR-RED}{NIR+RED}$
Modified Soil-Adjusted Vegetation Index (EVI)	$MSAVI = \frac{2\rho_{NIR}+1-\sqrt{(2\rho_{NIR}+1)^2-8(\rho_{NIR}-\rho_{RED})}}{2}$
Normalized Differencing Moisture Index (NDMI)	$NDMI = \frac{NIR-SWIR1}{NIR+SWIR1}$

Where: NIR: Near-infrared band; RED: Red band; and $SWIR1$: Short-wave infrared band.

Table A2. Accuracy assessment.

	HDNV	MDNV	LDNV	OR	AA	Total	UA
HDNV	32	3	0	0	0	35	0.91
MDNV	3	31	1	0	1	36	0.86
LDNV	0	1	31	1	4	33	0.94
OR	0	0	2	34	2	36	0.94
AA	0	1	2	0	28	31	0.81
Total	35	35	35	35	35	156	
PA	0.91	0.89	0.89	0.97	0.80		
OA	0.91						
Kappa	0.89						

HDNV: High-Density Natural Vegetation; MDNV: Medium-Density Natural Vegetation; LDNV: Low-Density Natural Vegetation; OR: Open and Rocky Areas; AA: Agricultural Area, OA: Overall Accuracy, PA: Producer's Accuracy, UA: User's Accuracy.

Table A3. RF models accuracy setting different values on "ntrees" parameter.

ntrees	50 m		250 m		500 m	
	R^2 (OOB)	R^2 (NOT OOB)	R^2 (OOB)	R^2 (NOT OOB)	R^2 (OOB)	R^2 (NOT OOB)
100	83.43%	92.86%	86.97%	95.42%	86.99%	96.84%
200	83.88%	92.88%	87.24%	95.53%	87.09%	97.11%
300	84.04%	92.92%	87.38%	95.69%	87.28%	97.26%
400	84.18%	93.16%	87.45%	95.78%	87.52%	97.49%
500	86.64%	93.45%	87.42%	95.74%	87.64%	97.56%
600	86.47%	93.34%	87.37%	95.67%	87.63%	97.51%
700	86.42%	93.12%	87.26%	95.53%	87.56%	97.42%
800	85.64%	92.83%	87.14%	95.41%	87.37%	97.36%
900	85.51%	92.74%	87.01%	95.39%	87.09%	97.17%
1000	84.98%	92.68%	86.92%	95.22%	87.00%	97.05%

Table A4. Descriptive statistics of the selected two explanatory variables outputting from GRF models.

			Min	Max	Mean	SD
50		Cohesion_HDNV	0.00	1.57	0.27	0.24
		MESH_HDNV	0.00	1.46	0.24	0.20
250		Contagion	0.02	0.49	0.15	0.07
		ED_HDNV	0.00	0.23	0.07	0.04
500		LPI_HDNV	0.00	0.10	0.02	0.02
		Cohesion_HDNV	0.00	0.09	0.02	0.02

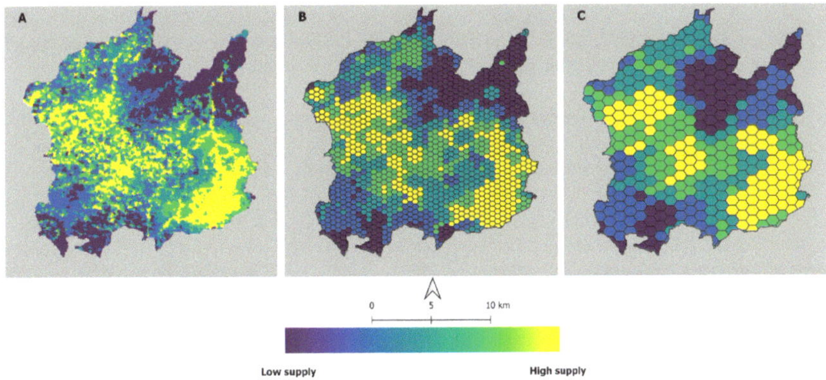

Figure A1. Graphs (**A–C**) are the ES supply maps at 50 m, 250 m, and 500 m scale analysis, respectively.

References

1. Turner, M.G. Landscape Ecology: What Is the State of the Science? *Annu. Rev. Ecol. Evol. Syst.* **2005**, *36*, 319–344. [CrossRef]
2. Bajocco, S.; De Angelis, A.; Perini, L.; Ferrara, A.; Salvati, L. The Impact of Land Use/Land Cover Changes on Land Degradation Dynamics: A Mediterranean Case Study. *Environ. Manag.* **2012**, *49*, 980–989. [CrossRef] [PubMed]
3. Geri, F.; Amici, V.; Rocchini, D. Human Activity Impact on the Heterogeneity of a Mediterranean Landscape. *Appl. Geogr.* **2010**, *30*, 370–379. [CrossRef]
4. Kefalas, G.; Kalogirou, S.; Poirazidis, K.; Lorilla, R.S. Landscape Transition in Mediterranean Islands: The Case of Ionian Islands, Greece 1985–2015. *Landsc. Urban Plan.* **2019**, *191*, 103641. [CrossRef]
5. Lorilla, R.S.; Poirazidis, K.; Detsis, V.; Kalogirou, S.; Chalkias, C. Socio-Ecological Determinants of Multiple Ecosystem Services on the Mediterranean Landscapes of the Ionian Islands (Greece). *Ecol. Model.* **2020**, *422*, 108994. [CrossRef]
6. Aznar-Sánchez, J.A.; Belmonte-Ureña, L.J.; López-Serrano, M.J.; Velasco-Muñoz, J.F. Forest Ecosystem Services: An Analysis of Worldwide Research. *Forests* **2018**, *9*, 453. [CrossRef]
7. Jenkins, M.; Schaap, B. Forest Ecosystem Services: Background Analytical Study. Global Forest Goals. United Nations Forum on Forests. 2018. Available online: https://www.un.org/esa/forests/wp-content/uploads/2018/05/UNFF13_BkgdStudy_ForestsEcoServices.pdf (accessed on 18 June 2023).
8. Gounaridis, D.; Newell, J.P.; Goodspeed, R. The Impact of Urban Sprawl on Forest Landscapes in Southeast Michigan, 1985–2015. *Landsc. Ecol.* **2020**, *35*, 1975–1993. [CrossRef]
9. Lorilla, R.S.; Poirazidis, K.; Kalogirou, S.; Detsis, V.; Martinis, A. Assessment of the Spatial Dynamics and Interactions among Multiple Ecosystem Services to Promote Effective Policy Making across Mediterranean Island Landscapes. *Sustainability* **2018**, *10*, 3285. [CrossRef]
10. Vacchiano, G.; Garbarino, M.; Lingua, E.; Motta, R. Forest Dynamics and Disturbance Regimes in the Italian Apennines. *For. Ecol. Manag.* **2017**, *388*, 57–66. [CrossRef]
11. Xofis, P.; Poirazidis, K. Combining Different Spatio-Temporal Resolution Images to Depict Landscape Dynamics and Guide Wildlife Management. *Biol. Conserv.* **2018**, *218*, 10–17. [CrossRef]
12. Uuemaa, E.; Mander, Ü.; Marja, R. Trends in the Use of Landscape Spatial Metrics as Landscape Indicators: A Review. *Ecol. Indic.* **2013**, *28*, 100–106. [CrossRef]
13. Almeida, D.; Rocha, J.; Neto, C.; Arsénio, P. Landscape Metrics Applied to Formerly Reclaimed Saltmarshes: A Tool to Evaluate Ecosystem Services? *Estuar. Coast. Shelf Sci.* **2016**, *181*, 100–113. [CrossRef]
14. Duarte, G.T.; Santos, P.M.; Cornelissen, T.G.; Ribeiro, M.C.; Paglia, A.P. The Effects of Landscape Patterns on Ecosystem Services: Meta-Analyses of Landscape Services. *Landsc. Ecol.* **2018**, *33*, 1247–1257. [CrossRef]
15. Grafius, D.R.; Corstanje, R.; Harris, J.A. Linking Ecosystem Services, Urban Form and Green Space Configuration Using Multivariate Landscape Metric Analysis. *Landsc. Ecol.* **2018**, *33*, 557–573. [CrossRef] [PubMed]
16. Syrbe, R.-U.; Walz, U. Spatial Indicators for the Assessment of Ecosystem Services: Providing, Benefiting and Connecting Areas and Landscape Metrics. *Ecol. Indic.* **2012**, *21*, 80–88. [CrossRef]
17. Termorshuizen, J.W.; Opdam, P. Landscape Services as a Bridge between Landscape Ecology and Sustainable Development. *Landsc. Ecol.* **2009**, *24*, 1037–1052. [CrossRef]
18. Bastian, O.; Grunewald, K.; Syrbe, R.-U.; Walz, U.; Wende, W. Landscape Services: The Concept and Its Practical Relevance. *Landsc. Ecol.* **2014**, *29*, 1463–1479. [CrossRef]
19. Frank, S.; Fürst, C.; Koschke, L.; Makeschin, F. A Contribution towards a Transfer of the Ecosystem Service Concept to Landscape Planning Using Landscape Metrics. *Ecol. Indic.* **2012**, *21*, 30–38. [CrossRef]
20. Hodder, K.H.; Newton, A.C.; Cantarello, E.; Perrella, L. Does Landscape-Scale Conservation Management Enhance the Provision of Ecosystem Services? *Int. J. Biodivers. Sci. Ecosyst. Serv. Manag.* **2014**, *10*, 71–83. [CrossRef]

21. Mitchell, M.G.E.; Suarez-Castro, A.F.; Martinez-Harms, M.; Maron, M.; McAlpine, C.; Gaston, K.J.; Johansen, K.; Rhodes, J.R. Reframing Landscape Fragmentation's Effects on Ecosystem Services. *Trends Ecol. Evol.* **2015**, *30*, 190–198. [CrossRef] [PubMed]

22. Turner, M.G.; Donato, D.C.; Romme, W.H. Consequences of Spatial Heterogeneity for Ecosystem Services in Changing Forest Landscapes: Priorities for Future Research. *Landsc. Ecol.* **2013**, *28*, 1081–1097. [CrossRef]

23. Braimoh, A.K. Random and Systematic Land-Cover Transitions in Northern Ghana. *Agric. Ecosyst. Environ.* **2006**, *113*, 254–263. [CrossRef]

24. Adler, K.; Jedicke, E. Landscape Metrics as Indicators of Avian Community Structures—A State of the Art Review. *Ecol. Indic.* **2022**, *145*, 109575. [CrossRef]

25. Curd, A.; Chevalier, M.; Vasquez, M.; Boyé, A.; Firth, L.B.; Marzloff, M.P.; Bricheno, L.M.; Burrows, M.T.; Bush, L.E.; Cordier, C.; et al. Applying Landscape Metrics to Species Distribution Model Predictions to Characterize Internal Range Structure and Associated Changes. *Glob. Chang. Biol.* **2023**, *29*, 631–647. [CrossRef] [PubMed]

26. Schindler, S.; Poirazidis, K.; Wrbka, T. Towards a Core Set of Landscape Metrics for Biodiversity Assessments: A Case Study from Dadia National Park, Greece. *Ecol. Indic.* **2008**, *8*, 502–514. [CrossRef]

27. Li, Z.; Han, H.; You, H.; Cheng, X.; Wang, T. Effects of Local Characteristics and Landscape Patterns on Plant Richness: A Multi-Scale Investigation of Multiple Dispersal Traits. *Ecol. Indic.* **2020**, *117*, 106584. [CrossRef]

28. McGarigal, K.; Marks, B.J. FRAGSTATS: Spatial Pattern Analysis Program for Quantifying Landscape Structure. In *General Technical Report. PNW-GTR-351*; US Department of Agriculture, Forest Service: Portland, OR, USA, 1995.

29. Riitters, K.H.; O'Neill, R.V.; Hunsaker, C.T.; Wickham, J.D.; Yankee, D.H.; Timmins, S.P.; Jones, K.B.; Jackson, B.L. A Factor Analysis of Landscape Pattern and Structure Metrics. *Landsc. Ecol.* **1995**, *10*, 23–39. [CrossRef]

30. Botequilha Leitão, A.; Ahern, J. Applying Landscape Ecological Concepts and Metrics in Sustainable Landscape Planning. *Landsc. Urban Plan.* **2002**, *59*, 65–93. [CrossRef]

31. Lechner, A.M.; Reinke, K.J.; Wang, Y.; Bastin, L. Interactions between Landcover Pattern and Geospatial Processing Methods: Effects on Landscape Metrics and Classification Accuracy. *Ecol. Complex.* **2013**, *15*, 71–82. [CrossRef]

32. Roces-Díaz, J.V.; Díaz-Varela, E.R.; Álvarez-Álvarez, P. Analysis of Spatial Scales for Ecosystem Services: Application of the Lacunarity Concept at Landscape Level in Galicia (NW Spain). *Ecol. Indic.* **2014**, *36*, 495–507. [CrossRef]

33. Plexida, S.G.; Sfougaris, A.I.; Ispikoudis, I.P.; Papanastasis, V.P. Selecting Landscape Metrics as Indicators of Spatial Heterogeneity—A Comparison among Greek Landscapes. *Int. J. Appl. Earth Obs. Geoinf.* **2014**, *26*, 26–35. [CrossRef]

34. Schindler, S.; von Wehrden, H.; Poirazidis, K.; Hochachka, W.M.; Wrbka, T.; Kati, V. Performance of Methods to Select Landscape Metrics for Modelling Species Richness. *Ecol. Model.* **2015**, *295*, 107–112. [CrossRef]

35. Chen, A.; Yao, L.; Sun, R.; Chen, L. How Many Metrics Are Required to Identify the Effects of the Landscape Pattern on Land Surface Temperature? *Ecol. Indic.* **2014**, *45*, 424–433. [CrossRef]

36. Stupariu, M.-S.; Cushman, S.A.; Pleşoianu, A.-I.; Pătru-Stupariu, I.; Fürst, C. Machine Learning in Landscape Ecological Analysis: A Review of Recent Approaches. *Landsc. Ecol.* **2022**, *37*, 1227–1250. [CrossRef]

37. Maxwell, A.E.; Warner, T.A.; Fang, F. Implementation of Machine-Learning Classification in Remote Sensing: An Applied Review. *Int. J. Remote Sens.* **2018**, *39*, 2784–2817. [CrossRef]

38. Aburas, M.M.; Ahamad, M.S.S.; Omar, N.Q. Spatio-Temporal Simulation and Prediction of Land-Use Change Using Conventional and Machine Learning Models: A Review. *Environ. Monit. Assess.* **2019**, *191*, 205. [CrossRef]

39. Nikparvar, B.; Thill, J.-C. Machine Learning of Spatial Data. *ISPRS Int. J. Geo-Inf.* **2021**, *10*, 600. [CrossRef]

40. Scowen, M.; Athanasiadis, I.N.; Bullock, J.M.; Eigenbrod, F.; Willcock, S. The Current and Future Uses of Machine Learning in Ecosystem Service Research. *Sci. Total Environ.* **2021**, *799*, 149263. [CrossRef]

41. Willcock, S.; Martínez-López, J.; Hooftman, D.A.P.; Bagstad, K.J.; Balbi, S.; Marzo, A.; Prato, C.; Sciandrello, S.; Signorello, G.; Voigt, B.; et al. Machine Learning for Ecosystem Services. *Ecosyst. Serv.* **2018**, *33*, 165–174. [CrossRef]

42. Humphries, G.; Magness, D.R.; Huettmann, F. (Eds.) *Machine Learning for Ecology and Sustainable Natural Resource Management*; Springer International Publishing: Cham, Switzerland, 2018; ISBN 978-3-319-96976-3.

43. Lucas, T.C.D. A Translucent Box: Interpretable Machine Learning in Ecology. *Ecol. Monogr.* **2020**, *90*, e01422. [CrossRef]

44. Zhong, S.; Zhang, K.; Bagheri, M.; Burken, J.G.; Gu, A.; Li, B.; Ma, X.; Marrone, B.L.; Ren, Z.J.; Schrier, J.; et al. Machine Learning: New Ideas and Tools in Environmental Science and Engineering. *Environ. Sci. Technol.* **2021**, *55*, 12741–12754. [CrossRef] [PubMed]

45. Thessen, A. Adoption of Machine Learning Techniques in Ecology and Earth Science. *One Ecosyst.* **2016**, *1*, e8621. [CrossRef]

46. Sarica, A.; Cerasa, A.; Quattrone, A. Random Forest Algorithm for the Classification of Neuroimaging Data in Alzheimer's Disease: A Systematic Review. *Front. Aging Neurosci.* **2017**, *9*, 329. [CrossRef] [PubMed]

47. Fotheringham, A.S.; Brunsdon, C.; Charlton, M. Charlton Geographically Weighted Regression: The Analysis of Spatially Varying Relationships | Wiley. Available online: https://www.wiley.com/en-us/Geographically+Weighted+Regression%3A+The+Analysis+of+Spatially+Varying+Relationships+-p-9780471496168 (accessed on 9 February 2023).

48. Georganos, S.; Grippa, T.; Niang Gadiaga, A.; Linard, C.; Lennert, M.; Vanhuysse, S.; Mboga, N.; Wolff, E.; Kalogirou, S. Geographical Random Forests: A Spatial Extension of the Random Forest Algorithm to Address Spatial Heterogeneity in Remote Sensing and Population Modelling. *Geocarto Int.* **2021**, *36*, 121–136. [CrossRef]

49. Georganos, S.; Kalogirou, S. A Forest of Forests: A Spatially Weighted and Computationally Efficient Formulation of Geographical Random Forests. *ISPRS Int. J. Geo-Inf.* **2022**, *11*, 471. [CrossRef]

50. Grekousis, G.; Feng, Z.; Marakakis, I.; Lu, Y.; Wang, R. Ranking the Importance of Demographic, Socioeconomic, and Underlying Health Factors on US COVID-19 Deaths: A Geographical Random Forest Approach. *Health Place* **2022**, *74*, 102744. [CrossRef]
51. Quevedo, R.P.; Maciel, D.A.; Uehara, T.D.T.; Vojtek, M.; Rennó, C.D.; Pradhan, B.; Vojteková, J.; Pham, Q.B. Consideration of Spatial Heterogeneity in Landslide Susceptibility Mapping Using Geographical Random Forest Model. *Geocarto Int.* **2021**, *37*, 8190–8213. [CrossRef]
52. Santos, F.; Graw, V.; Bonilla, S. A Geographically Weighted Random Forest Approach for Evaluate Forest Change Drivers in the Northern Ecuadorian Amazon. *PLoS ONE* **2019**, *14*, e0226224. [CrossRef]
53. Khan, S.N.; Li, D.; Maimaitijiang, M. A Geographically Weighted Random Forest Approach to Predict Corn Yield in the US Corn Belt. *Remote Sens.* **2022**, *14*, 2843. [CrossRef]
54. Kountouris, P.; Kousiappa, I.; Papasavva, T.; Christopoulos, G.; Pavlou, E.; Petrou, M.; Feleki, X.; Karitzie, E.; Phylactides, M.; Fanis, P.; et al. The Molecular Spectrum and Distribution of Haemoglobinopathies in Cyprus: A 20-Year Retrospective Study. *Sci. Rep.* **2016**, *6*, 26371. [CrossRef]
55. The Geology of Cyprus—Cultural Foundation. Available online: https://www.boccf.org/about/projects/The-Geology-of-Cyprus/ (accessed on 21 February 2023).
56. Kefalas, G.; Poirazidis, K.; Xofis, P.; Kalogirou, S. Mapping and Understanding the Dynamics of Landscape Changes on Heterogeneous Mediterranean Islands with the Use of OBIA: The Case of Ionian Region, Greece. *Sustainability* **2018**, *10*, 2986. [CrossRef]
57. Haines-Young, R.; Potschin, M. Guidance on the Application of the Revised Structure. 2018. Available online: https://cices.eu/content/uploads/sites/8/2018/01/Guidance-V51-01012018.pdf (accessed on 18 June 2023).
58. Olson's Major World Ecosystem Complexes Ranked by Carbon in Live Vegetation: An Updated Database Using the GLC2000 Land Cover Product (Tables 1–4). Available online: https://cdiac.ess-dive.lbl.gov/epubs/ndp/ndp017/table_b.html#table4 (accessed on 8 February 2023).
59. ASTER Global Digital Elevation Map. Available online: https://asterweb.jpl.nasa.gov/gdem.asp (accessed on 8 February 2023).
60. ESDAC—European Commission. Available online: https://esdac.jrc.ec.europa.eu/ (accessed on 8 February 2023).
61. Maes, J.; Paracchini, M.L.; Zulian, G.; Dunbar, M.B.; Alkemade, R. Synergies and Trade-Offs between Ecosystem Service Supply, Biodiversity, and Habitat Conservation Status in Europe. *Biol. Conserv.* **2012**, *155*, 1–12. [CrossRef]
62. Pörtner, H.-O.; Scholes, R.J.; Agard, J.; Archer, E.; Arneth, A.; Bai, X.; Barnes, D.; Burrows, M.; Chan, L.; Cheung, W.L.; et al. Scientific Outcome of the IPBES-IPCC Co-Sponsored Workshop on Biodiversity and Climate Change. Available online: https://zenodo.org/record/5101125 (accessed on 15 December 2022).
63. Ciais, P.; Schelhaas, M.J.; Zaehle, S.; Piao, S.L.; Cescatti, A.; Liski, J.; Luyssaert, S.; Le-Maire, G.; Schulze, E.-D.; Bouriaud, O.; et al. Carbon Accumulation in European Forests. *Nat. Geosci.* **2008**, *1*, 425–429. [CrossRef]
64. Guerra, C.A.; Maes, J.; Geijzendorffer, I.; Metzger, M.J. An Assessment of Soil Erosion Prevention by Vegetation in Mediterranean Europe: Current Trends of Ecosystem Service Provision. *Ecol. Indic.* **2016**, *60*, 213–222. [CrossRef]
65. Benito-Calvo, A.; Pérez-González, A.; Magri, O.; Meza, P. Assessing Regional Geodiversity: The Iberian Peninsula. *Earth Surf. Process. Landf.* **2009**, *34*, 1433–1445. [CrossRef]
66. Breiman, L. Random Forests. *Mach. Learn.* **2001**, *45*, 5–32. [CrossRef]
67. Catani, F.; Lagomarsino, D.; Segoni, S.; Tofani, V. Landslide Susceptibility Estimation by Random Forests Technique: Sensitivity and Scaling Issues. *Nat. Hazards Earth Syst. Sci.* **2013**, *13*, 2815–2831. [CrossRef]
68. Luo, Y.; Yan, J.; McClure, S. Distribution of the Environmental and Socioeconomic Risk Factors on COVID-19 Death Rate across Continental USA: A Spatial Nonlinear Analysis. *Environ. Sci. Pollut. Res.* **2021**, *28*, 6587–6599. [CrossRef]
69. Cutler, D.R.; Edwards, T.C., Jr.; Beard, K.H.; Cutler, A.; Hess, K.T.; Gibson, J.; Lawler, J.J. Random Forests for Classification in Ecology. *Ecology* **2007**, *88*, 2783–2792. [CrossRef]
70. Kalogirou, S. *SPATIAL ANALYSIS*; Hellenic Academic Libraries Link: Athens, Greece, 2015; ISBN 978-960-603-285-1.
71. Propastin, P.A. Spatial Non-Stationarity and Scale-Dependency of Prediction Accuracy in the Remote Estimation of LAI over a Tropical Rainforest in Sulawesi, Indonesia. *Remote Sens. Environ.* **2009**, *113*, 2234–2242. [CrossRef]
72. Kalogirou, S.; Georganos, S. R Package: SpatialML. R Cran. 2022. Available online: https://cran.r-project.org/web/packages/SpatialML/index.html (accessed on 20 January 2023).
73. Roger, B.; Tim, K.; Barry, R.; Edzer, P.; Michael, S.; Robert, H.; Daniel, B.; Even, R.; Frank, W.; Jeroen, O.; et al. R Package: Rgdal. R Cran 2023. Available online: https://cran.r-project.org/web/packages/rgdal/index.html (accessed on 20 January 2023).
74. Gábor, C.; Hadley, W.; Kirill, M. R Package: Cli. R Cran 2023. Available online: https://cran.r-project.org/web/packages/cli/index.html (accessed on 1 April 2023).
75. Max, K.; Jed, W.; Steve, W.; Andre, W.; Chris, K.; Allan, E.; Tony, C.; Zachary, M.; Brenton, K.; Michael, B.; et al. R Package: Carret. R Cran 2023. Available online: https://cran.r-project.org/web/packages/caret/index.html (accessed on 1 April 2023).
76. Morán-Ordóñez, A.; Ramsauer, J.; Coll, L.; Brotons, L.; Ameztegui, A. Ecosystem Services Provision by Mediterranean Forests Will Be Compromised above 2 °C Warming. *Glob. Chang. Biol.* **2021**, *27*, 4210–4222. [CrossRef] [PubMed]
77. Lorilla, R.S.; Kalogirou, S.; Poirazidis, K.; Kefalas, G. Identifying Spatial Mismatches between the Supply and Demand of Ecosystem Services to Achieve a Sustainable Management Regime in the Ionian Islands (Western Greece). *Land Use Policy* **2019**, *88*, 104171. [CrossRef]

78. Galidaki, G.; Gitas, I. Mediterranean Forest Species Mapping Using Classification of Hyperion Imagery. *Geocarto Int.* **2015**, *30*, 48–61. [CrossRef]

79. Xofis, P.; Buckley, P.G.; Takos, I.; Mitchley, J. Long Term Post-Fire Vegetation Dynamics in North-East Mediterranean Ecosystems. The Case of Mount Athos Greece. *Fire* **2021**, *4*, 92. [CrossRef]

80. Médail, F. The Specific Vulnerability of Plant Biodiversity and Vegetation on Mediterranean Islands in the Face of Global Change. *Reg. Environ. Chang.* **2017**, *17*, 1775–1790. [CrossRef]

81. Lorilla, R.S.; Kefalas, G.; Christou, A.K.; Poirazidis, K.; Homer Eliades, N.-G. Enhancing the Conservation Status and Resilience of a Narrowly Distributed Forest: A Challenge to Effectively Support Ecosystem Services in Practice. *J. Nat. Conserv.* **2023**, *73*, 126414. [CrossRef]

82. Petrou, P.; Milios, E. Establishment and Survival of Pinus Brutia Ten. Seedlings over the First Growing Season in Abandoned Fields in Central Cyprus. *Plant Biosyst.-Int. J. Deal. All Asp. Plant Biol.* **2012**, *146*, 522–533. [CrossRef]

83. Petrou, P.; Milios, E. Investigation of the Factors Affecting Artificial Seed Sowing Success and Seedling Survival in Pinus Brutia Natural Stands in Middle Elevations of Central Cyprus. *Forests* **2020**, *11*, 1349. [CrossRef]

84. Reyers, B.; O'Farrell, P.J.; Cowling, R.M.; Egoh, B.N.; Le Maitre, D.C.; Vlok, J.H.J. Ecosystem Services, Land-Cover Change, and Stakeholders: Finding a Sustainable Foothold for a Semiarid Biodiversity Hotspot. *Ecol. Soc.* **2009**, *14*, 38. [CrossRef]

85. De Valck, J.; Landuyt, D.; Broekx, S.; Liekens, I.; De Nocker, L.; Vranken, L. Outdoor Recreation in Various Landscapes: Which Site Characteristics Really Matter? *Land Use Policy* **2017**, *65*, 186–197. [CrossRef]

86. Queiroz, C.; Meacham, M.; Richter, K.; Norström, A.V.; Andersson, E.; Norberg, J.; Peterson, G. Mapping Bundles of Ecosystem Services Reveals Distinct Types of Multifunctionality within a Swedish Landscape. *Ambio* **2015**, *44* (Suppl. S1), S89–S101. [CrossRef] [PubMed]

87. Yapp, G.; Walker, J.; Thackway, R. Linking Vegetation Type and Condition to Ecosystem Goods and Services. *Ecol. Complex.* **2010**, *7*, 292–301. [CrossRef]

MDPI AG

Grosspeteranlage 5

4052 Basel

Switzerland

Tel.: +41 61 683 77 34

Forests Editorial Office

E-mail: forests@mdpi.com

www.mdpi.com/journal/forests